Lipopolysaccharides (LPSs)

Lipopolysaccharides (LPSs)

Selected Articles Published by MDPI

MDPI • Basel • Beijing • Wuhan • Barcelona • Belgrade

This is a reprint of articles published online by the open access publisher MDPI from 2016 to 2019 (available at: http://www.mdpi.com). The responsibility for the book's title and preface lies with Juan M. Tomás, who compiled this selection.

For citation purposes, cite each article independently as indicated on the article page online and as indicated below:

LastName, A.A.; LastName, B.B.; LastName, C.C. Article Title. *Journal Name* **Year**, *Article Number*, Page Range.

ISBN 978-3-03928-256-2 (Pbk)
ISBN 978-3-03928-257-9 (PDF)

Cover image courtesy of Juan M. Tomás.

Contents

Preface to "Lipopolysaccharides (LPSs)"

The outer membrane of Gram-negative bacteria reacts to changes in the environment, inhibits the entrance of toxic compounds (such as antibiotics), plays an important role in nutrient transport, and mediates the physiological and pathophysiological interaction of bacteria with host organisms. Lipopolysaccharides (LPSs) play a fundamental role in the organization and integrity of the outer membrane structure, and are critical for the maintenance of barrier function, preventing the entrance of small hydrophobic compounds and reducing the effectiveness of antibiotics and polycationic peptides. Because of their exposed position, LPSs also represent an ideal target for the recognition by both antibodies and other immunological or pharmacological agents. Lipopolysaccharide (LPS) has been considered an essential component of outer membrane biogenesis and cell viability based on pioneering studies in the model Gram-negative organisms Escherichia coli and Salmonella.

LPSs are endotoxins, one of the most potent class of activators of the mammalian immune system; they can be released from cell surfaces of bacteria during multiplication, lysis, and death. Low doses of endotoxins are thought to be beneficial for the host, causing immunostimulation and enhanced resistance to infection. On the other hand, the presence of a large amount of endotoxin in the bloodstream, as observed during severe systemic infections, leads to strong pathophysiological reactions such as multiorgan failure, shock, and potentially death. The medical significance of this endotoxin and its wide range of biological properties are grounds for the ever-expanding scientific interest in this molecule, from the standpoint of their structure, biosynthesis, genetic organization, and biological action.

Juan M. Tomás

International Journal of
Molecular Sciences

Review

Coxiella burnetii Lipopolysaccharide: What Do We Know?

Prasad Abnave [1], Xavier Muracciole [2] and Eric Ghigo [3,*

[1]   Department of Zoology, University of Oxford, Tinbergen Building, South Parks Road, Oxford OX1 3PS, UK; prasadabnave@gmail.com
[2]   Department of Radiotherapy Oncology, CHU de la Timone, Assistance Publique-Hopitaux Marseille, 13385 Marseille, France; xavier.muracciole@gmail.com
[3]   Unité de Recherche sur les Maladies Infectieuses et Tropicales Emergentes (URMITE), Institut Hospitalier Universitaire Méditerranée-Infection, 19-21 Bd Jean Moulin, CEDEX 05, 13385 Marseille, France
*   Correspondence: eric.ghigo@gmail.com; Tel.: +33-(0)4-137-32405

Received: 29 September 2017; Accepted: 21 November 2017; Published: 23 November 2017

Abstract: A small gram-negative bacterium, *Coxiella burnetii* (*C. burnetii*), is responsible for a zoonosis called Q fever. *C. burnetii* is an intracellular bacterium that can survive inside microbicidal cells like monocytes and macrophages by hijacking several functions of the immune system. Among several virulence factors, the lipopolysaccharide (LPS) of *C. burnetii* is one of the major factors involved in this immune hijacking because of its atypical composition and structure. Thus, the aim of this mini-review is to summarize the repressive effects of *C. burnetii* LPS on the antibacterial immunity of cells.

Keywords: lipopolysaccharide; *Coxiella burnetii*; Q fever; phagosome; virenose

---

1. Introduction

Coxiella burnetii is an intracellular bacterium responsible for a worldwide zoonosis known as Q fever [1,2]. After primary infection, approximately 60% of humans remain asymptomatic, while 40% manifest clinical signs consisting of isolated fever, hepatitis, and pneumonia [3]. The principal clinical manifestation of Q fever is endocarditis with a lethal prognosis without treatment. The treatment involves a combination of doxycycline and hydroxychloroquine [1,3]. However, this long-term treatment carries the persistent risk of relapse [4].

C. burnetii is a small bacterium measuring approximately 0.2 to 0.4 μm wide and 0.4 to 1 μm long, and it has been classified in the *Proteobacteria* subdivision based on its 16S ribosomal RNA sequence. As *C. burnetii* harbours lipopolysaccharide (LPS) in its membrane, it is defined as a gram-negative bacterium. Though *C. burnetii* is not stained by Gram stain, it can be stained by Gimenez stain [5]. *C. burnetii* primarily infects domestic ruminants and pets, but arthropods have also been found to be infected. In animals, the infection is asymptomatic but induces abortions in livestock. Both abortion and parturition contribute to the bacteria spreading into the environment, since the placenta of infected animals contains large amounts of *C. burnetii*. Contamination via aerosols also remains the major route of infection in both animals and humans [3,6]. *C. burnetii* has been categorized as a biological weapon due to its high infectivity, the possibility of producing large quantities of bacteria, its environmental stability through a sporulation-like mechanism, and its dispersion via aerosolization [7,8]. *C. burnetii* was likely used as a bio-weapon during World War II, as a Q fever outbreak was observed during this time among army troops [9].

C. burnetii resides primarily within myeloid cells (monocytes and macrophages) [10,11] but has also been shown to infect trophoblasts [12] and adipocytes [13]. The adaptation of *C. burnetii* to its environment is probably critical for its survival. To survive within its host, *C. burnetii* interferes with the host's antimicrobial response (immunity and phagolysosome biogenesis). For this purpose, *C. burnetii* has an arsenal of virulence factors [14–17], including LPS [18]. The molecular variations observed in *C. burnetii* LPS, a major component of its outer membrane, contribute to its pathogenic properties [19–21]. Moreover, the intracellular fate of virulent *C. burnetii* in myeloid cells is also determined by its LPS composition [18].

2. *Coxiella burnetii* LPS: Structure and Composition

As observed in enterobacteria, *C. burnetii* displays antigenic variations, from a smooth-rough form called Phase I to a rough form known as Phase II. The Phase I form is isolated from natural sources and defined as a virulent form of *C. burnetii*. It is characterized by full-length LPS and survives inside monocytes and macrophages [10,11]. After several passages of the virulent *C. burnetii* in embryonated eggs or tissue culture, an irreversible modification is observed in the molecular weight of *C. burnetii* LPS. *C. burnetii* harbouring a truncated LPS is defined as an avirulent microorganism and eliminated by monocytes and macrophages [22,23]. This avirulent form does not exist in the natural environment. It was shown that this LPS modification occurs due to a genomic deletion [3]. The difference between the virulent and avirulent forms of *C. burnetii* lies in the *O*-antigen; specifically, LPS from virulent *C. burnetii* has an *O*-antigen that contains unusual sugars, L-virenose, dihydrohydroxystreptose, and galactosamine uronyl-α-(1,6)-glucosamine residues, whereas LPS from the avirulent form does not have any *O*-antigen [19–21,24–32]. Virenose and dihydrohydroxystreptose have not been found in any other enterobacterial LPSs and are thus unique biomarkers of virulent *C. burnetii*. Interestingly, the lipid A molecules of both virulent and avirulent *C. burnetii* display the same ionic species and fragmentation profiles in mass spectrometry, suggesting that they have very similar and likely identical structures. The *C. burnetii* lipid A structure differs considerably from the published standard form of enterobacterial lipid A. An analysis of lipid A from *C. burnetii* identified two major tetra-acylated molecular species sharing the classical backbone of a dephosphorylated GlcN (acylated D-glucosamine residues) disaccharide in which both GlcN I and GlcN II carry an amide-linked iso or normal (n) C16:0(3-OH) [24,33]. The core polysaccharide is conserved between virulent and avirulent *C. burnetii* LPSs and contains a heptasaccharide localized in the proximal region of lipid A. The heptasaccharide is formed by two terminal D-mannoses (Man), 2- and 3,4-linked D-*glycero*-D-*manno*-heptoses, and terminal 4- and 4,5-linked 3-deoxy-D-*manno*-oct-2-ulosonic acid residues [20,29]. It is important to note that a third *C. burnetii* LPS has been identified as an intermediate-length LPS at the surface of the Nine Mile Crazy strain [34]. Large chromosomal deletions have been found in these avirulent *C. burnetii* Nine Mile and Nine Mile Crazy strains [35]. These deletions eliminate open reading frames involved in the biosynthesis of *O*-antigen sugars, including the rare sugar virenose [35]. The description of the virenose biosynthesis pathway suggests the formation of GDP-β-D-virenose via the modification of GDP-L-fucose by the addition of a methyl group at position C3″, and perhaps the open reading frame CBU0691, and the inversion of the stereochemistry at position C2″ [36].

3. *C. burnetii* LPS Interferes with Phagocytosis

It is known that phagocytosis efficiency depends on the activation of phagocytic receptor CR3 (complement receptor-3) through αvβ3 integrin and CD47 (integrin-associated protein). *C. burnetii*, via its LPS, subverts receptor-mediated phagocytosis [22] by inhibiting the interplay between integrins, including CR3, remodelling the actin cytoskeleton organization, and activating protein tyrosine kinases. This strategy possibly determines the evolution of Q fever. *C. burnetii*, via its LPS, interacts with macrophages through αvβ3 integrins, and avoids internalization by inhibiting the interaction between αvβ3 integrins and CR3, which is essential for bacterial uptake [22,37]. Inhibition of the interplay between αvβ3 integrins and CR3 leads to poor internalization of virulent *C. burnetii* compared with its avirulent form, which harbours a truncated LPS and is largely internalized by monocytes and macrophages. Interestingly, the inhibitory mechanism mediated by virulent *C. burnetii* through its LPS does not target CD47 [22]. Note that CR3, not αvβ3 integrin, is excluded from the cytoskeleton protrusions formed during the cytoskeleton reorganization induced by virulent *C. burnetii* LPS, thus decreasing the efficiency of phagocytosis [22,37,38]. An in-depth analysis has demonstrated that the uptake of avirulent *C. burnetii* requires both CD11b/CD18 and CR3, whereas virulent organism internalization does not involve CR3. It has been shown that the LPS from virulent *C. burnetii* prevents the activation of CR3 by interfering with its lectin sites [22]. This leads to conformational changes in the I domain and in the exposure of activation epitopes and cytoskeleton reorganization [39].

Finally, virulent *C. burnetii* induces early protein tyrosine kinase activation as well as the tyrosine phosphorylation of two Src-related kinases: Hck and Lyn [40]. By contrast, the avirulent form does not stimulate protein tyrosine kinases. Tyrosine-phosphorylated proteins co-localize with F-actin inside protrusions. Cell membrane protrusions are induced via the activation of protein tyrosine kinases by *C. burnetii* LPS, which in turn down-modulates *C. burnetii* uptake [40,41]. The use of protein tyrosine kinase inhibitors rescues *C. burnetii* phagocytosis. It has been hypothesized that the membrane ruffling induced by protein tyrosine kinase activation may interfere with the co-localization of CR3 with $\alpha v \beta 3$ integrin and *C. burnetii* [42,43]. It has also been shown that *C. burnetii* LPS interferes with Toll Like Receptor (TLR)-2 and TLR-4 signalling through cytoskeleton reorganization [38,41,42]. Indeed, cytoskeleton reorganization induces a redistribution of TLR-2 and TLR-4 on the membrane of macrophages. This redistribution disrupts the colocalization between TLR-2 and TLR-4, in contrast to what is observed in macrophages challenged with LPS from the avirulent strain of *C. burnetii*. Co-immunoprecipitation experiments have revealed that a possible physical link between TLR-2 and TLR-4 is broken in cells challenged with virulent *C. burnetii* LPS. As a consequence, p38α Mitogen-Activated Protein Kinase (MAPK) is not activated in macrophages challenged with virulent *C. burnetii* and LPS extracted from virulent *C. burnetii* [18,41,44]. However, the existence of a *TLR2/TLR4/p38α* MAPK axis in *C. burnetii* infection remains to be demonstrated.

4. *C. burnetii* LPS Interferes with the Antibacterial Immune Response

Macrophage immune polarization is reoriented by *C. burnetii* to deactivate the macrophage microbicidal response [45,46]. Indeed, *C. burnetii* is responsible for atypical M2 macrophage activation, and it has been shown to induce expression of M2 polarization-related genes (transforming growth factor-$\beta 1$, interleukin (IL)-1 receptor antagonist, Chemokine (C-C motif) ligand (CCL)18, mannose receptor, arginase-1). By contrast, the expression of genes related to M1 polarization (tumor necrosis factor, CD80, C-C chemokine receptor type (CCR)7) is inhibited. It is interesting to note that the expression of arginase-1 is associated with the absence of nitric oxide production, while the expression of the *Interleukin (IL)-6* and *Chemokine (C-X-C motif) ligand (CXCL)8* genes (M1-related genes) is increased, although their proteins are weakly secreted [45]. In addition, monocytes produce high levels of IL-10 in response to *C. burnetii* or its LPS. IL-10 favours the persistence of *C. burnetii* by down-regulating the expression of tumor necrosis factor [47–49]. It is also responsible for the expression of *Programmed cell Death protein (PD)-1* by monocytes in vitro, and most likely, in patients with Q fever endocarditis. The LPS of *C. burnetii* does not induce the expression of PD-1 by monocytes. PD-1 delivers an inhibitory signal to T cells [50,51], and its expression in Q fever contributes to the immune suppression observed in Q fever endocarditis [52].

5. *C. burnetii* LPS as a Determinant Factor in Phagolysosome Biogenesis

In human macrophages, it has been observed that, in contrast to virulent *C. burnetii*, the avirulent form is quickly eliminated in degradative phagolysosome-like compartments [11,47]. Their replication is partially controlled in resident mouse peritoneal macrophages [53]. Immediately after phagocytosis, both virulent and avirulent forms of *C. burnetii* are localized within an early phagosome, transiently harbouring EEA1 (early endosome auto-antigen-1). This early phagosome undergoes a maturation process and is transformed into a late phagosome, presenting the markers Lamp-1, CD63, mannose-6-phosphate receptor, and V-H+ATPase and possessing an acidic pH. The major difference between the compartments containing virulent and avirulent forms of *C. burnetii* is the absence of the small GTPase Rab7 at the surface of the phagosome containing the virulent *C. burnetii* [11,17,18,23]. In contrast to the vacuole with avirulent bacteria, the phagosome containing the virulent strain of *C. burnetii* does not mature in phagolysosomes [23]. Surprisingly, the intracellular trafficking of *C. burnetii* LPS is similar to the trafficking of intact bacteria. Indeed, the LPSs from virulent and avirulent *C. burnetii* traffic through early phagosomes characterized by the presence of the small GTPase Rab5 and EEA1 [18,54]. Nevertheless, endosomes containing LPS purified from avirulent bacteria develop into late endosomes (Rab7, Lamp1) and then into lysosomes containing the lysosomal enzyme cathepsin D. The endosomes transporting LPS isolated from virulent bacteria mature from early to late endosomes but do not become lysosomes. Interestingly, in terms of intact *C. burnetii*, late endosomes containing LPS do not express the Rab7 protein on their surface [18,23,55]. This result suggests that LPS is responsible for blocking phagolysosome maturation induced by *C. burnetii*. Investigations of *C. burnetii* LPS have demonstrated that the LPS from pathogenic *C. burnetii* does not induce the phosphorylation of p38α MAPK by Mitogen-Activated Protein Kinase Kinase (MKK)6. This defect in the activation of p38α MAPK prevents the serine phosphorylation (S796E) of Vps41. In the absence of phosphorylation, Vps41 does not promote the targeting of the HOPS (homotypic fusion and protein sorting) complex to endosome–vacuole fusion sites, and thus it fails to recruit the GTP-bound Rab7 required for phagosome–lysosome fusion [56–61]. The absence of p38α MAPK activation is most likely due to the engagement of TLR4 by two unusual sugars, virenose and dihydrohydroxystreptose, present in the LPS of pathogenic *C. burnetii*. Thus, LPS from virulent *C. burnetii* acts as an antagonist of TLR-4.

6. Concluding Remarks

Collectively, this evidence highlights the importance of LPS and its composition in the strategies used by *C. burnetii* to infect cells and develop an efficient infection that leads to Q fever. It is interesting to observe that the particular composition of *C. burnetii* LPS allows several axes of the immune response to be modulated, ranging from phagocytosis to vesicular trafficking. Certainly, the virulence of *C. burnetii* does not only depend on LPS, as other virulence factors have been identified in *C. burnetii* [11,14]. The recent successful culturing of *C. burnetii* in axenic conditions might significantly develop our understanding of *C. burnetii* infection by facilitating the identification of new virulence factors [62,63]. Further work is required to understand the mechanisms implied in anti-microbicidal response hijacking. It might be interesting to generate transgenic *Escherichia coli* expressing the LPSs from both the virulent and avirulent *C. burnetii* to better understand LPS action. Similarly, as several new *C. burnetii* strains that cause severe Q fever have been isolated [2], it will be interesting to analyse their LPS composition to determine if the virulence and clinical issues observed are linked to any particular structure or composition of LPS.

Acknowledgments: This work was supported by IHU Méditerranée Infection.

Author Contributions: Prasad Abnave, Xavier Muracciole and Eric Ghigo wrote the manuscript.

Conflicts of Interest: The authors declare no conflict of interest.

References

1. Maurin, M.; Benoliel, A.M.; Bongrand, P.; Raoult, D. Phagolysosomal alkalinization and the bactericidal effect of antibiotics: The *Coxiella burnetii* paradigm. *J. Infect. Dis.* **1992**, *166*, 1097–1102. [CrossRef] [PubMed]
2. Eldin, C.; Melenotte, C.; Mediannikov, O.; Ghigo, E.; Million, M.; Edouard, S.; Mege, J.L.; Maurin, M.; Raoult, D. From Q fever to *Coxiella burnetii* infection: A paradigm change. *Clin. Microbiol. Rev.* **2017**, *30*, 115–190. [CrossRef] [PubMed]
3. Maurin, M.; Raoult, D. Q fever. *Clin. Microbiol. Rev.* **1999**, *12*, 518–553. [PubMed]
4. Houpikian, P.; Habib, G.; Mesana, T.; Raoult, D. Changing clinical presentation of Q fever endocarditis. *Clin. Infect. Dis.* **2002**, *34*, E28–E31. [CrossRef] [PubMed]
5. Gimenez, D.F. Staining rickettsiae in yolk-sac cultures. *Stain Technol.* **1964**, *39*, 135–140. [CrossRef] [PubMed]

6. Melenotte, C.; Lepidi, H.; Nappez, C.; Bechah, Y.; Audoly, G.; Terras, J.; Raoult, D.; Bregeon, F. Mouse model of *Coxiella burnetii* aerosolization. *Infect. Immun.* **2016**, *84*, 2116–2123. [CrossRef] [PubMed]

7. McCaul, T.F.; Williams, J.C. Developmental cycle of *Coxiella burnetii*: Structure and morphogenesis of vegetative and sporogenic differentiations. *J. Bacteriol.* **1981**, *147*, 1063–1076. [PubMed]

8. Oswald, W.; Thiele, D. A sporulation gene in *Coxiella burnetii*? *Zent. Vet. Reihe B J. Vet. Med.* **1993**, *40*, 366–370. [CrossRef]

9. Madariaga, M.G.; Rezai, K.; Trenholme, G.M.; Weinstein, R.A. Q fever: A biological weapon in your backyard. *Lancet* **2003**, *3*, 709–721. [CrossRef]

10. Baca, O.G.; Li, Y.P.; Kumar, H. Survival of the Q fever agent *Coxiella burnetii* in the phagolysosome. *Trends Microbiol.* **1994**, *2*, 476–480. [CrossRef]

11. Ghigo, E.; Colombo, M.I.; Heinzen, R.A. The *Coxiella burnetii* parasitophorous vacuole. *Adv. Exp. Med. Biol.* **2012**, *984*, 141–169. [PubMed]

12. Ben Amara, A.; Ghigo, E.; Le Priol, Y.; Lepolard, C.; Salcedo, S.P.; Lemichez, E.; Bretelle, F.; Capo, C.; Mege, J.L. *Coxiella burnetii*, the agent of Q fever, replicates within trophoblasts and induces a unique transcriptional response. *PLoS ONE* **2010**, *5*, e15315. [CrossRef] [PubMed]

13. Bechah, Y.; Verneau, J.; Ben Amara, A.; Barry, A.O.; Lepolard, C.; Achard, V.; Panicot-Dubois, L.; Textoris, J.; Capo, C.; Ghigo, E.; et al. Persistence of *Coxiella burnetii*, the agent of Q fever, in murine adipose tissue. *PLoS ONE* **2014**, *9*, e97503. [CrossRef] [PubMed]

14. Qiu, J.; Luo, Z.Q. Legionella and coxiella effectors: Strength in diversity and activity. *Nat. Rev.* **2017**, *15*, 591–605. [CrossRef] [PubMed]

15. Weber, M.M.; Chen, C.; Rowin, K.; Mertens, K.; Galvan, G.; Zhi, H.; Dealing, C.M.; Roman, V.A.; Banga, S.; Tan, Y.; et al. Identification of *Coxiella burnetii* type IV secretion substrates required for intracellular replication and coxiella-containing vacuole formation. *J. Bacteriol.* **2013**, *195*, 3914–3924. [CrossRef] [PubMed]

16. Newton, H.J.; Roy, C.R. The *Coxiella burnetii* dot/icm system creates a comfortable home through lysosomal renovation. *mBio* **2011**, *2*. [CrossRef] [PubMed]

17. Ghigo, E.; Pretat, L.; Desnues, B.; Capo, C.; Raoult, D.; Mege, J.L. Intracellular life of *Coxiella burnetii* in macrophages. *Ann. N. Y. Acad. Sci.* **2009**, *1166*, 55–66. [CrossRef] [PubMed]

18. Barry, A.O.; Boucherit, N.; Mottola, G.; Vadovic, P.; Trouplin, V.; Soubeyran, P.; Capo, C.; Bonatti, S.; Nebreda, A.; Toman, R.; et al. Impaired stimulation of p38α-MAPK/VPS41-hops by LPS from pathogenic *Coxiella burnetii* prevents trafficking to microbicidal phagolysosomes. *Cell Host Microbe* **2012**, *12*, 751–763. [CrossRef] [PubMed]

19. Lukacova, M.; Barak, I.; Kazar, J. Role of structural variations of polysaccharide antigens in the pathogenicity of gram-negative bacteria. *Clin. Microbiol. Infect.* **2008**, *14*, 200–206. [CrossRef] [PubMed]

20. Toman, R.; Skultety, L.; Ihnatko, R. *Coxiella burnetii* glycomics and proteomics—Tools for linking structure to function. *Ann. N. Y. Acad. Sci.* **2009**, *1166*, 67–78. [CrossRef] [PubMed]

21. Stulik, J.; Toman, R.; Butaye, P.; Ulrich, R.G. Lipopolysaccharides of *Coxiella burnetii*: Chemical composition and structure, and their role in diagnosis of Q fever. In *BSL3 and BSL4 Agents: Proteomics, Glycomics, and Antigenicity*; Hoboken, N., Ed.; Wiley-Blackwell: Hoboken, NJ, USA, 2011; Volume 1166, pp. 115–123.

22. Capo, C.; Lindberg, F.P.; Meconi, S.; Zaffran, Y.; Tardei, G.; Brown, E.J.; Raoult, D.; Mege, J.L. Subversion of monocyte functions by *Coxiella burnetii*: Impairment of the cross-talk between αvβ3 integrin and CR3. *J. Immunol.* **1999**, *163*, 6078–6085. [PubMed]

23. Ghigo, E.; Capo, C.; Tung, C.H.; Raoult, D.; Gorvel, J.P.; Mege, J.L. *Coxiella burnetii* survival in THP-1 monocytes involves the impairment of phagosome maturation: IFN-γ mediates its restoration and bacterial killing. *J. Immunol.* **2002**, *169*, 4488–4495. [CrossRef] [PubMed]

24. Alexander, C.; Rietschel, E.T. Bacterial lipopolysaccharides and innate immunity. *J. Endotoxin Res.* **2001**, *7*, 167–202. [CrossRef] [PubMed]

25. Schramek, S.; Radziejewska-Lebrecht, J.; Mayer, H. 3-C-branched aldoses in lipopolysaccharide of phase I *Coxiella burnetii* and their role as immunodominant factors. *FEBS J.* **1985**, *148*, 455–461.

26. Skultety, L.; Toman, R.; Patoprsty, V. A comparative study of lipopolysaccharides from two *Coxiella burnetii* strains considered to be associated with acute and chronic Q fever. *Polymers* **1996**, *35*, 189–194.

27. Toman, R.; Hussein, A.; Palkovic, P.; Ftacek, P. Structural properties of lipopolysaccharides from *Coxiella burnetii* strains henzerling and S. *Ann. N. Y. Acad. Sci.* **2003**, *990*, 563–567. [CrossRef] [PubMed]

28. Toman, R.; Hussein, A.; Slaba, K.; Skultety, E. Further structural characteristics of the lipopolysaccharide from *Coxiella burnetii* strain nine mile in low virulent phase II. *Acta Virol.* **2003**, *47*, 129–130. [PubMed]

29. Toman, R.; Skultety, L. Structural study on a lipopolysaccharide from *Coxiella burnetii* strain nine mile in avirulent phase II. *Carbohydr. Res.* **1996**, *283*, 175–185. [CrossRef]

30. Toman, R.; Skultety, L.; Ftacek, P.; Hricovini, M. NMR study of virenose and dihydrohydroxystreptose isolated from *Coxiella burnetii* phase I lipopolysaccharide. *Carbohydr. Res.* **1998**, *306*, 291–296. [CrossRef]

31. Vadovic, P.; Ihnatko, R.; Toman, R. Composition and structure of lipid A of the intracellular bacteria piscirickettsia salmonis and *Coxiella burnetii*. In *BSL3 and BSL4 Agents*; Stulik, J., Toman, R., Butaye, P., Ulrich, R.G., Eds.; Wiley-Blackwell: Hoboken, NJ, USA, 2011; pp. 139–144.

32. Vadovic, P.; Slaba, K.; Fodorova, M.; Skultety, L.; Toman, R. Structural and functional characterization of the glycan antigens involved in immunobiology of Q fever. *Ann. N. Y. Acad. Sci.* **2005**, *1063*, 149–153. [CrossRef] [PubMed]

33. Toman, R.; Garidel, P.; Andra, J.; Slaba, K.; Hussein, A.; Koch, M.H.; Brandenburg, K. Physicochemical characterization of the endotoxins from *Coxiella burnetii* strain priscilla in relation to their bioactivities. *BMC Biochem.* **2004**, *5*, 1. [CrossRef] [PubMed]

34. Hackstadt, T.; Peacock, M.G.; Hitchcock, P.J.; Cole, R.L. Lipopolysaccharide variation in *Coxiella burnetii*: Intrastrain heterogeneity in structure and antigenicity. *Infect. Immun.* **1985**, *48*, 359–365. [PubMed]

35. Thompson, H.A.; Hoover, T.A.; Vodkin, M.H.; Shaw, E.I. Do chromosomal deletions in the lipopolysaccharide biosynthetic regions explain all cases of phase variation in *Coxiella burnetii* strains? An update. *Ann. N. Y. Acad. Sci.* **2003**, *990*, 664–670. [CrossRef] [PubMed]

36. Narasaki, C.T.; Mertens, K.; Samuel, J.E. Characterization of the GDP-D-mannose biosynthesis pathway in *Coxiella burnetii*: The initial steps for GDP-β-D-virenose biosynthesis. *PLoS ONE* **2011**, *6*, e25514. [CrossRef] [PubMed]

37. Capo, C.; Moynault, A.; Collette, Y.; Olive, D.; Brown, E.J.; Raoult, D.; Mege, J.L. *Coxiella burnetii* avoids macrophage phagocytosis by interfering with spatial distribution of complement receptor 3. *J. Immunol.* **2003**, *170*, 4217–4225. [CrossRef] [PubMed]

38. Honstettre, A.; Ghigo, E.; Moynault, A.; Capo, C.; Toman, R.; Akira, S.; Takeuchi, O.; Lepidi, H.; Raoult, D.; Mege, J.L. Lipopolysaccharide from *Coxiella burnetii* is involved in bacterial phagocytosis, filamentous actin reorganization, and inflammatory responses through toll-like receptor 4. *J. Immunol.* **2004**, *172*, 3695–3703. [CrossRef] [PubMed]

39. Vetvicka, V.; Thornton, B.P.; Ross, G.D. Soluble beta-glucan polysaccharide binding to the lectin site of neutrophil or natural killer cell complement receptor type 3 (CD11b/CD18) generates a primed state of the receptor capable of mediating cytotoxicity of IC3B-opsonized target cells. *J. Clin. Investig.* **1996**, *98*, 50–61. [CrossRef] [PubMed]

40. Meconi, S.; Capo, C.; Remacle-Bonnet, M.; Pommier, G.; Raoult, D.; Mege, J.L. Activation of protein tyrosine kinases by *Coxiella burnetii*: Role in actin cytoskeleton reorganization and bacterial phagocytosis. *Infect. Immun.* **2001**, *69*, 2520–2526. [CrossRef] [PubMed]

41. Conti, F.; Boucherit, N.; Baldassarre, V.; Trouplin, V.; Toman, R.; Mottola, G.; Mege, J.L.; Ghigo, E. *Coxiella burnetii* lipopolysaccharide blocks p38α-MAPK activation through the disruption of TLR-2 and TLR-4 association. *Front. Cell. Infect. Microbiol.* **2015**, *4*, 182. [CrossRef] [PubMed]

42. Meconi, S.; Jacomo, V.; Boquet, P.; Raoult, D.; Mege, J.L.; Capo, C. *Coxiella burnetii* induces reorganization of the actin cytoskeleton in human monocytes. *Infect. Immun.* **1998**, *66*, 5527–5533. [PubMed]

43. Patil, S.; Jedsadayanmata, A.; Wencel-Drake, J.D.; Wang, W.; Knezevic, I.; Lam, S.C. Identification of a talin-binding site in the integrin β3 subunit distinct from the nply regulatory motif of post-ligand binding functions. The talin N-terminal head domain interacts with the membrane-proximal region of the β3 cytoplasmic tail. *J. Biol. Chem.* **1999**, *274*, 28575–28583. [CrossRef] [PubMed]

44. Boucherit, N.; Barry, A.O.; Mottola, G.; Trouplin, V.; Capo, C.; Mege, J.L.; Ghigo, E. Effects of *Coxiella burnetii* on mapkinases phosphorylation. *FEMS Immunol. Med. Microbiol.* **2012**, *64*, 101–103. [CrossRef] [PubMed]

45. Benoit, M.; Barbarat, B.; Bernard, A.; Olive, D.; Mege, J.L. *Coxiella burnetii*, the agent of Q fever, stimulates an atypical M2 activation program in human macrophages. *Eur. J. Immunol.* **2008**, *38*, 1065–1070. [CrossRef] [PubMed]

46. Benoit, M.; Desnues, B.; Mege, J.L. Macrophage polarization in bacterial infections. *J. Immunol.* **2008**, *181*, 3733–3739. [CrossRef] [PubMed]

47. Ghigo, E.; Honstettre, A.; Capo, C.; Gorvel, J.P.; Raoult, D.; Mege, J.L. Link between impaired maturation of phagosomes and defective *Coxiella burnetii* killing in patients with chronic Q fever. *J. Infect. Dis.* **2004**, *190*, 1767–1772. [CrossRef] [PubMed]

48. Honstettre, A.; Imbert, G.; Ghigo, E.; Gouriet, F.; Capo, C.; Raoult, D.; Mege, J.L. Dysregulation of cytokines in acute Q fever: Role of interleukin-10 and tumor necrosis factor in chronic evolution of Q fever. *J. Infect. Dis.* **2003**, *187*, 956–962. [CrossRef] [PubMed]

49. Ghigo, E.; Capo, C.; Raoult, D.; Mege, J.L. Interleukin-10 stimulates *Coxiella burnetii* replication in human monocytes through tumor necrosis factor down-modulation: Role in microbicidal defect of Q fever. *Infect. Immun.* **2001**, *69*, 2345–2352. [CrossRef] [PubMed]

50. Parry, R.V.; Chemnitz, J.M.; Frauwirth, K.A.; Lanfranco, A.R.; Braunstein, I.; Kobayashi, S.V.; Linsley, P.S.; Thompson, C.B.; Riley, J.L. CTLA-4 and PD-1 receptors inhibit t-cell activation by distinct mechanisms. *Mol. Cell. Biol.* **2005**, *25*, 9543–9553. [CrossRef] [PubMed]

51. Freeman, G.J.; Wherry, E.J.; Ahmed, R.; Sharpe, A.H. Reinvigorating exhausted HIV-specific T cells via PD-1-PD-1 ligand blockade. *J. Exp. Med.* **2006**, *203*, 2223–2227. [CrossRef] [PubMed]

52. Ka, M.B.; Gondois-Rey, F.; Capo, C.; Textoris, J.; Million, M.; Raoult, D.; Olive, D.; Mege, J.L. Imbalance of circulating monocyte subsets and PD-1 dysregulation in Q fever endocarditis: The role of IL-10 in PD-1 modulation. *PLoS ONE* **2014**, *9*, e107533. [CrossRef] [PubMed]

53. Zamboni, D.S.; Mortara, R.A.; Freymuller, E.; Rabinovitch, M. Mouse resident peritoneal macrophages partially control in vitro infection with *Coxiella burnetii* phase II. *Microbes Infect.* **2002**, *4*, 591–598. [CrossRef]

54. Brumell, J.H.; Scidmore, M.A. Manipulation of Rab GTPase function by intracellular bacterial pathogens. *Microbiol. Mol. Biol. Rev.* **2007**, *71*, 636–652. [CrossRef] [PubMed]

55. Bucci, C.; Thomsen, P.; Nicoziani, P.; McCarthy, J.; van Deurs, B. Rab7: A key to lysosome biogenesis. *Mol. Biol. Cell* **2000**, *11*, 467–480. [CrossRef] [PubMed]

56. Barry, A.O.; Mege, J.L.; Ghigo, E. Hijacked phagosomes and leukocyte activation: An intimate relationship. *J. Leukoc. Biol.* **2011**, *89*, 373–382. [CrossRef] [PubMed]

57. Cabrera, M.; Ostrowicz, C.W.; Mari, M.; LaGrassa, T.J.; Reggiori, F.; Ungermann, C. VPS41 phosphorylation and the Rab YPT7 control the targeting of the hops complex to endosome-vacuole fusion sites. *Mol. Biol. Cell* **2009**, *20*, 1937–1948. [CrossRef] [PubMed]

58. Nickerson, D.P.; Brett, C.L.; Merz, A.J. VPS-C complexes: Gatekeepers of endolysosomal traffic. *Curr. Opin. Cell Biol.* **2009**, *21*, 543–551. [CrossRef] [PubMed]

59. Peralta, E.R.; Martin, B.C.; Edinger, A.L. Differential effects of tbc1d15 and mammalian VPS39 on Rab7 activation state, lysosomal morphology, and growth factor dependence. *J. Biol. Chem.* **2010**, *285*, 16814–16821. [CrossRef] [PubMed]

60. Rink, J.; Ghigo, E.; Kalaidzidis, Y.; Zerial, M. Rab conversion as a mechanism of progression from early to late endosomes. *Cell* **2005**, *122*, 735–749. [CrossRef] [PubMed]

61. Zick, M.; Wickner, W. Phosphorylation of the effector complex hops by the vacuolar kinase YCK3P confers rab nucleotide specificity for vacuole docking and fusion. *Mol. Biol. Cell* **2012**, *23*, 3429–3437. [CrossRef] [PubMed]

62. Sandoz, K.M.; Sturdevant, D.E.; Hansen, B.; Heinzen, R.A. Developmental transitions of *Coxiella burnetii* grown in axenic media. *J. Microbiol. Methods* **2013**, *96*, 104–110. [CrossRef] [PubMed]

63. Kuley, R.; Smith, H.E.; Frangoulidis, D.; Smits, M.A.; Jan Roest, H.I.; Bossers, A. Cell-free propagation of *Coxiella burnetii* does not affect its relative virulence. *PLoS ONE* **2015**, *10*, e0121661. [CrossRef] [PubMed]

International Journal of
Molecular Sciences

Article

Structural Masquerade of *Plesiomonas shigelloides* Strain CNCTC 78/89 *O*-Antigen—High-Resolution Magic Angle Spinning NMR Reveals the Modified D-galactan I of *Klebsiella pneumoniae*

Karolina Ucieklak, Sabina Koj, Damian Pawelczyk and Tomasz Niedziela *

Hirszfeld Institute of Immunology and Experimental Therapy, Polish Academy of Sciences, 53-114 Wroclaw, Poland; karolina.ucieklak@iitd.pan.wroc.pl (K.U.); sabina.koj@iitd.pan.wroc.pl (S.K.); damianpawelczyk@yahoo.pl (D.P.)

* Correspondence: tomasz.niedziela@iitd.pan.wroc.pl; Tel.: +48-71-337-1172

Received: 8 November 2017; Accepted: 25 November 2017; Published: 29 November 2017

Abstract: The high-resolution magic angle spinning nuclear magnetic resonance spectroscopy (HR-MAS NMR) analysis of *Plesiomonas shigelloides* 78/89 lipopolysaccharide directly on bacteria revealed the characteristic structural features of the *O*-acetylated polysaccharide in the NMR spectra. The *O*-antigen profiles were unique, yet the pattern of signals in the, spectra along with their ^{1}H,^{13}C chemical shift values, resembled these of D-galactan I of *Klebsiella pneumoniae*. The isolated *O*-specific polysaccharide (O-PS) of *P. shigelloides* strain CNCTC 78/89 was investigated by ^{1}H and ^{13}C NMR spectroscopy, mass *spectrometry* and chemical methods. The analyses demonstrated that the *P. shigelloides* 78/89 *O*-PS is composed of →3)-α-D-Gal*p*-(1→3)-β-D-Gal*f*2OAc-(1→ disaccharide repeating units. The *O*-acetylation was incomplete and resulted in a microheterogeneity of the *O*-antigen. This *O*-acetylation generates additional antigenic determinants within the *O*-antigen, forms a new chemotype, and contributes to the epitopes recognized by the *O*-serotype specific antibodies. The serological cross-reactivities further confirmed the inter-specific structural similarity of these *O*-antigens.

Keywords: *Plesiomonas shigelloides*; *O*-antigen; lipopolysaccharide; *O*-acetylation; D-galactan I; HR-MAS; NMR spectroscopy

1. Introduction

Plesiomonas shigelloides is a facultative anaerobic Gram-negative flagellated, rod-shaped bacterium belonging to the *Enterobacteriaceae* family [1]. It is widely distributed in nature, but predominantly isolated from aquatic environments and animals [2]. These bacteria are not part of the natural human microflora. Human infections with *P. shigelloides* are generally related to visiting countries with low sanitary standards [3,4], drinking unpurified water, and eating uncooked shellfish [5,6]. These bacteria are potent inducers of an invasive shigellosis-like disease [7], gastroenteritis [8], and diarrheal disease [9]. Although the pathogenicity of *P. shigelloides* is not entirely understood, lipopolysaccharide (LPS) is considered the main virulence factor. LPS is a major component of the outer leaflet of the external membrane of Gram-negative bacteria. These are amphiphilic molecules isolated from smooth bacterial strains (S-LPS). Structurally, they can be divided into three distinct regions: lipid A, core oligosaccharide, and *O*-specific polysaccharide. These segments are important for the biological activity [10] and take part in host–bacterium interactions [11]. The hydrophobic lipid A constitutes the most conserved part of LPS, yet lipid A structures and the endotoxic activities that they imply vary substantially between different species of Gram-negative bacteria [12]. The core oligosaccharides (OS) have structures that are generally conserved within bacterial species. The *O*-specific polysaccharide (*O*-antigen, *O*-specific chain) determines bacterial *O*-serotype and constitutes a fingerprint of bacteria [13]. The high structural diversity of the *O*-specific polysaccharides provides the serological distinction between bacterial strains. The variability of *O*-antigen structures is one of the strategies used by bacteria to avoid recognition by host organisms and to hamper the host's defenses. Despite this variability, serological cross-reactivities between various species occur, indicating the presence of common structural epitopes within *O*-antigens. To date only several LPS structures out of 102 *O*-serotypes of *P. shigelloides* [14] have been analyzed and reported [15–25]. All these studies exposed several characteristic features of *P. shigelloides* LPSs, i.e., the lack of phosphate groups, the presence of uronic acid residues in the core oligosaccharides, and the unusual hydrophobicity of the *O*-specific polysaccharides [22]. Most of the polysaccharides are unique to the *Plesiomonas* species and distinguish them from other members of the *Enterobacteriaceae* family. However, some *O*-antigens of *P. shigelloides* have shown cross-reactivity with antisera directed against LPS of *Shigella* spp. The structure of *P. shigelloides* serotype O17 was found to be identical to the *Shigella sonnei* phase I *O*-chain [19,26,27]. Two other strains of *P. shigelloides* share a type-specific antigen with *S. flexneri* and *S. dysenteriae* [15,28]. The structural element α-L-Rha*p*(1→2)-α-L-Rha*p* described in *S. flexneri* serotype 6 is shared with *P. shigelloides* and *Klebsiella pneumoniae* *O*-antigens [29]. In some aspects, this cross-reactivity of antibodies against the *O*-antigens is desirable as it can provide broad protection against heterologous bacteria.

We have now identified a new *O*-antigen of *P. shigelloides*, structurally similar to this of *K. pneumoniae* strain Kp20. Herein, we present the chemotyping of the *O*-antigen of *P. shigelloides* strain CNCTC 78/89 by high-resolution magic angle spinning (HR-MAS) nuclear magnetic resonance (NMR) spectroscopy in situ alongside the structural analysis of the isolated *O*-specific polysaccharide, which supports this preliminary observations. The shared epitopes of *Klebsiella* and *Plesiomonas* *O*-antigens, responsible for their serological cross-reactivities, have also been determined.

2. Results

2.1. HR-MAS NMR Analysis of P. shigelloides 78/89 Bacteria and LPS

The *O*-antigens of *P. shigelloides* 78/89 were initially investigated by HR-MAS NMR spectroscopy as the technique allows for the direct identification of the flexible *O*-antigen molecules on the bacterial cells in situ. The screening of the whole bacteria of *P. shigelloides* 78/89 using HR-MAS NMR technique provided data on the *O*-PS spectral pattern of this strain. The HR-MAS NMR spectra of *P. shigelloides* 78/89 bacteria were complex and contained signals for anomeric and ring protons, as well as resonances of other surface molecules and metabolites (Figure S1). The observed ^{1}H resonances and ^{1}H,^{13}C-correlations in the HR-MAS NMR spectra of bacteria were further complemented by the HR-MAS NMR analysis of the isolated LPS. Both sets of data revealed distinct structural features of the *O*-acetylated polysaccharide in the *O*-antigen of *P. shigelloides* 78/89. The spectra of the isolated *P. shigelloides* 78/89 LPS contained main signals for three anomeric protons, resonances of the ring protons and a distinct signal in the region of acetyl groups (δ_H 2.11 ppm). The observed pattern of resonances was compared with the *O*-antigen structural data available in our laboratory, including LPSs of various *Klebsiella* strains, and the data published previously. This *O*-antigen profile was unique, however, in that a subset of signals in the 1D and HSQC-DEPT spectra as well as the ^{1}H, ^{13}C chemical shift values were similar to these of *K. pneumoniae* strain Kp20 LPS (Figure 1). To unscramble this similarity, the LPS of *P. shigelloides* 78/89 was subjected to further structural analyses.

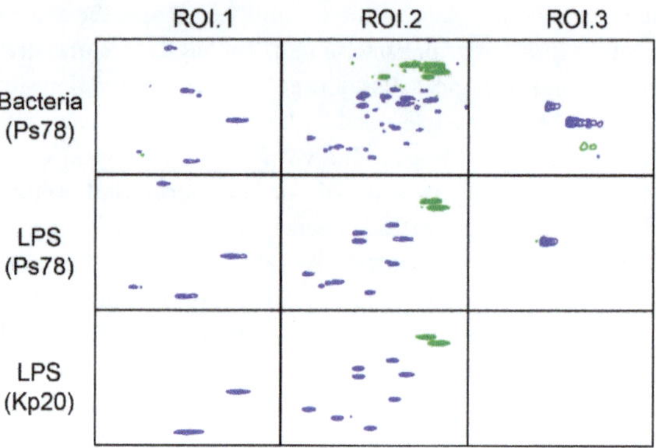

Figure 1. Regions of interest (ROI) extracted from the high-resolution magic angle spinning (HR-MAS) HSQC-DEPT NMR spectra of the *O*-antigens of *P. shigelloides* strain CNCTC 78/89 (Ps78) bacteria and LPS compared to these of *K. pneumoniae* strain Kp20 LPS. The regions were selected directly from the SPARKY processed spectra using the rNMR software. The compared regions and their chemical shift ranges include: anomeric signals (ROI.1, δ_H 5.45–4.96 ppm, δ_C 112.6–79.5 ppm), the ring resonances (ROI.2, δ_H 4.57–3.51 ppm, δ_C 88.9–54.4 ppm) and acetyl-group resonances (ROI.3, δ_H 2.36–1.81 ppm, δ_C 27.6–13.5 ppm). The HSQC-DEPT NMR spectra of bacteria (~4 mg dry mass) and LPS (~3 mg) suspensions in ^2H_2O (total volume of ~30 µL in the Bruker Kel-F inserts) were obtained using a Bruker 4 mm HR-MAS probe on an Avance III 600 MHz spectrometer. The experiments were carried out using a ZrO_2 rotor at a spin rate of 4 kHz at 27 °C (the actual temperature of the bearing gas).

2.2. Isolation of LPS and O-Antigen Fractions

The LPS of *P. shigelloides* CNCTC 78/89 was extracted from bacterial mass by the hot phenol/water method and purified by ultracentrifugation. The heteropolysaccharide components were released by mild acid hydrolysis of the LPS and isolated by gel filtration on Bio-Gel P-10, yielding four main fractions. The fractions were analyzed by matrix-assisted laser desorption ionization time-of-flight (MALDI-TOF) mass spectrometry and NMR spectroscopy, and identified as the *O*-specific polysaccharide fraction (PSI), fraction composed of short *O*-specific chains substituted by core oligosaccharide (OSII) and the core oligosaccharide (OSIII and OSIV) (Figure S2).

As the attempts to obtain the MALDI-TOF spectra of the intact PSI failed, the mass of the repeating unit of *P. shigelloides* strain 78/89 has been deduced from the analysis of the partially hydrolyzed PSI fraction. The PSI fraction was subjected to a partial acid hydrolysis with 0.5 M TFA. The MALDI-TOF mass spectrum (Figure 2) showed the clusters of ions corresponding to oligosaccharide fragments consisting of 3 up to 6 repeating units. The main signal at *m/z* 1013.63 Da corresponded to an oligosaccharide fragment comprising three repeating units and it was accompanied by a minor signal (*m/z* 1055.59) of the *O*-acetyled variant of the oligosaccharide. The observed mass differences indicated the disaccharide (Δ 324 Da) and the *O*-acetylated disaccharide (Δ 366 Da) *O*-repeats.

The combined NMR and MS data of the more abundant core oligosaccharide fraction (OSIV) indicated the presence of 11 sugar residues, together having a monoisotopic mass of 2015.65 Da. The chemical shift values of the spin systems of the OSIV were compared with these for other core types of *P. shigelloides* described to date. The acquired NMR data of the OSIV appeared very much like the previously published core oligosaccharide structure of *P. shigelloides* O33:H3 (strain CNCTC 34/89) [30], with two noticeable differences. The -4)-α-Gal*p*NAc-(1-residue in the core of *P. shigelloides* 78/89 is not *O*-acetylated. The disaccharide element in the outer core is built of -4)-α-Gal*p*NAc-(1→6)-α-Glc*p*N-(1-, and there is no heterogeneity related to the presence of Glc*p* instead of Glc*p*N, as was observed previously in the core oligosaccharide of *P. shigelloides* O33:H3 (Table S1 and Figure S3). The OSIII structure is also identical to the core oligosaccharide identified in the *P. shigelloides* O51 (strain CNCTC

110/92) [31]. The structural identity of the core oligosaccharide was further confirmed serologically. In the immunoblotting analysis the antiserum specific for the OS of *P. shigelloides* O51 reacted vividly with the fast migrating bands of the SDS-PAGE-separated LPS fractions composed of the core oligosaccharide linked to lipid A. Reactions were also observed for the LPS fractions composed of the lipid A-core oligosaccharides substituted by the increasing number of the *O*-repeats (Figure S4).

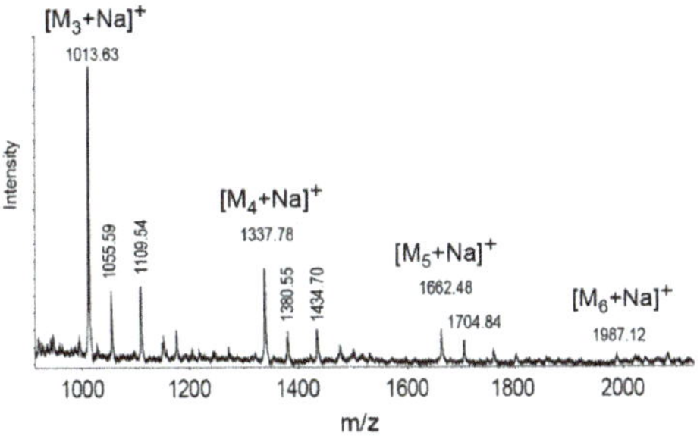

Figure 2. Matrix-assisted laser desorption ionization time-of-flight (MALDI-TOF) mass spectrum of the partially hydrolyzed polysaccharide fraction of *Plesiomonas shigelloides* 78/89 LPS. The MALDI-TOF mass spectrum was obtained in a positive linear mode. 2,5-Dihydroxybenzoic was used as matrix. M_3, M_4, M_5, and M_6 represent the mass of the respective number of repeating units.

2.3. Structure Analysis of the O-Specific Polysaccharide

Initial analysis of the *P. shigelloides* 78/89 bacteria and LPS using HR-MAS NMR spectroscopy provided a structural fingerprint of the *O*-antigen chemotype, including the ^{1}H and ^{13}C chemical shift values for the *O*-specific polysaccharide in situ. Composition analysis of the PSI polysaccharide fraction together with determination of the absolute configuration confirmed the presence of D-galactose residues in galacto-furanose and galacto-pyranose ring forms. The NMR spectra of the isolated PSI contained main signals for three anomeric protons and a distinct signal (δ_H 2.11 ppm) in the region of acetyl groups. The spin systems for each residue (denoted as uppercase letters through the entire text, tables, and figures) were assigned using COSY, TOCSY, HSQC-DEPT, HSQC-TOCSY, HMBC, and NOESY NMR experiments (Table 1 and Figure 3).

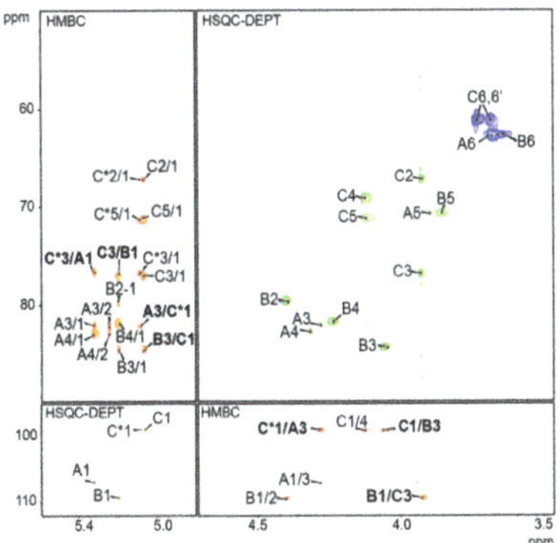

Figure 3. Selected $^1J_{H,C}$- and $^3J_{H,C}$-connectivities in HSQC-DEPT and HMBC spectra of the isolated PSI of *Plesiomonas shigelloides* 78/89 LPS. The inter-residue connectivities from the anomeric atoms in HMBC spectra are marked in bold. The spectra were obtained for ^2H_2O solutions at 600 MHz and 30 °C. The uppercase letters refer to sugar residues in the *O*-specific polysaccharide, as described in Table 1.

The coupling patterns of the identified spin systems in the COSY and TOCSY spectra indicated the *galacto*-configuration of all residues.

Table 1. ^{1}H and ^{13}C NMR chemical shifts of the *O*-specific polysaccharide of *P. shigelloides* strain CNCTC 78/89 [a].

Residue	Chemical Shifts (ppm)					
	H-1	H-2	H-3	H-4	H-5	H-6, H-6'
	C-1	C-2	C-3	C-4	C-5	C-6
A →3)-β-D-Gal*f*2OAc-(1→	5.32	5.25 [b]	4.28	4.32	3.89	3.68, 3.65
	107.1	81.8	82.1	82.9	70.8	62.7
B →3)-β-D-Gal*f*-(1→	5.20	4.40	4.06	4.24	3.86	3.66, 3.68
	109.4	79.8	84.5	81.8	70.8	62.7
C →3)-α-D-Gal*p*-(1→	5.07	3.93	3.93	4.12	4.12	3.73, 3.68
	99.5	67.2	76.9	69.2	71.3	61.1
C' α-D-Gal*p*-(1→ [c]	5.11	3.90	4.01	4.12	4.27	3.8, 3.75
	95.4	68.5	69.6	70.3	71.2	60.7
C* →3)-α-D-Gal*p*-(1→	5.08	3.94	3.95	4.11	4.11	3.69, 3.62
	99.4	67.2	76.7	69.1	71.3	60.9

[a] Spectra were obtained for ^{2}H$_2$O solutions at 30 °C. Acetone was used as internal reference (δ_H/δ_C 2.225/31.05 ppm); [b] The *O*-2 of this residue is *O*-acetylated (δ_{CH3CO} 2.11/20.5 ppm, 173.1 ppm); [c] The non-reducing terminal variant of residue C.

Residue **A** (δ_{H1}/δ_{C1} 5.32/107.1 ppm) was identified as a 3-substituted 2-*O*-acetyl-β-D-galactofuranose (β-D-Gal*f*2OAc) on the basis of the high chemical shift values of anomeric carbon (δ 107.1), C-2 (δ 81.8), C-3 (δ 82.1) and C-4 (δ 82.9) as well as the similarities of the chemical shift values of H-2, H-3, and H-4 to the published data [32]. This was further supported by the observed correlation of H-1 with both C-3 and C-4, and a lack of correlation with C-5 in the HMBC spectrum (Figure 3). The characteristic large downfield shift of the H-2 signal (δ_H 5.25) and its multiple bond correlation with a carbonyl carbon resonance at δ_C 173.1 ppm in the HMBC spectrum indicated an ester substitution at this position by an *O*-acetyl group (δ_H/δ_C 2.11/20.5). Similarly, residue **B** (δ_H/δ_C 5.2/109.4 ppm, $^1J_{C1,H1}$ ~171 Hz) was recognized as the 3-substituted-β-D-galactofuranose devoid of the *O*-acetyl. The chemical shift values except the H-2 were similar to these of residue A. The relatively high chemical shift of the C-3 signal (δ 84.5) indicated a substitution by another residue. Residue **C** with the H1/C1 signal at δ_H/δ_C 5.07/99.5 ppm, $^1J_{C1,H1}$ ~175 Hz was assigned as a 3-substituted α-D-galactopyranose (α-D-Gal*p*) based on the large vicinal couplings between H-2 and H-3, the small vicinal coupling constants between H-3, H-4 and H-5, as well as relatively high chemical shift value of the C-3 signal (δ 76.9). The heteronuclear multiple bond correlations between H-1 and both C-3 and C-5 observed in the HMBC spectrum confirmed the pyranose ring size. The presence of the **C*** variant (δ_H/δ_C 5.08/99.4 ppm) of residue **C** reflected the changes of chemical shifts induced by the *O*-acetylation of the disaccharide repeating unit and indicated that two forms of the disaccharide repeating units exist in the *O*-specific polysaccharide: the *O*-acetylated A-C* and non-*O*-acetylated B-C.

The minor anomeric signal (δ_H/δ_C 5.11/95.4 ppm, residue **C'**) and the corresponding spin system observed at a lower contour level in the 2D spectra of the PSI showed chemical shift values similar to the unsubstitited α-D-galactopyranose [33] and was attributed to a non-reducing terminal variant of residue **C**, defining the biological repeating unit of the *O*-specific polysaccharide.

The NOESY and HMBC (Table 2, Figure 3) spectra showed inter-residue cross-peaks between the transglycosidic protons, and between the anomeric protons and carbons at the linkage position as well as carbons and the protons at the linkage position, respectively. The inter-residue NOEs were identified between H-1 of residue **A** and H-3 of residue **C*** and between H-1 of residue **C*** and H-3 of residue **A** for the *O*-acetylated RU as well as between H-1 of residue **B** and H-3 of residue **C**, and between H-1 of residue **C** and H-3 of residue **B** for the non-acetylated form of the repeating unit.

Table 2. Selected inter-residue NOE and $^3J_{H,C}$-connectivities from the anomeric atoms of the *O*-antigen unit of PS *P. shigelloides* strain CNCTC 78/89.

Residue	Atom H-1/C-1 (ppm)	Connectivities to		Inter-Residue Atom/Residue
		δ_C	δ_H	
A →3)-β-D-Gal*f*2OAc-(1→	5.32/107.1	76.7	3.95	C-3, H-3 of **C***
B →3)-β-D-Gal*f*-(1→	5.20/109.4	76.9	3.93	C-3, H-3 of **C**
C →3)-α-D-Gal*p*-(1→	5.07/99.5	84.5	4.06	C-3, H-3 of **B**
C* →3)-α-D-Gal*p*-(1→	5.08/99.4	82.1	4.28	C-3, H-3 of **A**

The combined data indicated the disaccharide repeating unit of *P. shigelloides* 78/89 *O*-PS with the following structure:

$$
\left[\begin{array}{cc}
 & (OAc) \\
 & \downarrow \\
 & 2 \\
\rightarrow\text{3)-}\alpha\text{-D-Gal}p\text{-(1}\rightarrow\text{3)-}\beta\text{-D-Gal}f\text{-(1}\rightarrow \\
\textbf{(C*)C} & \textbf{(A)B}
\end{array}\right]_n
$$

O-deacetylation of the *O*-polysaccharide rendered only signals for two anomeric protons of residues **B** and **C** in the spectrum. Thus, the PSI NMR profile reverted to that of the D-galactan I, and the NMR spectrum became identical to the HR-MAS NMR spectrum of *K. pneumoniae* Kp20 LPS (Figure S5).

The *O*-acetylation of the *O*-specific polysaccharide in the PSI fraction was not complete. The degree of the *O*-acetylation was determined by integration of the anomeric signal of the 2-*O*-Ac-Gal*f* residue (δ_H 5.32 ppm) in the ^{1}H NMR spectrum of PSI relative to the resolved resonance of an anomeric proton at 5.20 ppm (integral value of 1, corresponding to a single proton). The ratios of the integral values have shown that ~32% of the repeating units were *O*-acetylated in the PSI, generating the *O*-acetylation-related heterogeneity. This level of substitution was sufficient to disguise the chemotype of pure D-galactan I. Interestingly, the NMR spectra also indicated that there was no *O*-acetylation of the repeating units in the OSII fraction (Figure S2). As the isolation procedures can cleave off or induce migration of the *O*-acetyl groups to neighboring positions we have also compared volumetric integration values for the H-1/C-1 resonances in the HR-MAS HSQC-DEPT NMR spectra in situ. The relative content of the *O*-acetyl groups was calculated as 29.9% in the isolated *O*-specific polysaccharides, 29.4% in the isolated LPS and 25.7% in the LPS directly on bacterial cells.

2.4. Serological Analysis

Reactivities of the serotype O12-specific serum with the homologous *P. shigelloides* 78/89 and *K. pneumoniae* Kp20 LPSs were investigated by immunoblotting, ELISA experiments and ELISA inhibition assays. The silver stained SDS-PAGE of *P. shigelloides* strain 78/89 LPS showed both low and high molecular mass LPS bands (Figure 4A). In the immunoblotting analysis (Figure 4B) the antiserum against *P. shigelloides* O12 reacted vividly with the high molecular mass bands of the homologous *O*-antigen of *P. shigelloides* 78/89.

We also observed a cross-reaction of the anti-*P. shigelloides* O12 serum with the high molecular mass bands of *K. pneumoniae* Kp20 LPS. However, this reaction covered a broader range of the *O*-specific chain length. In ELISA assay both lipopolysaccharides reacted with the serum specific for the serotype O12, although the profiles differed (Figure 5).

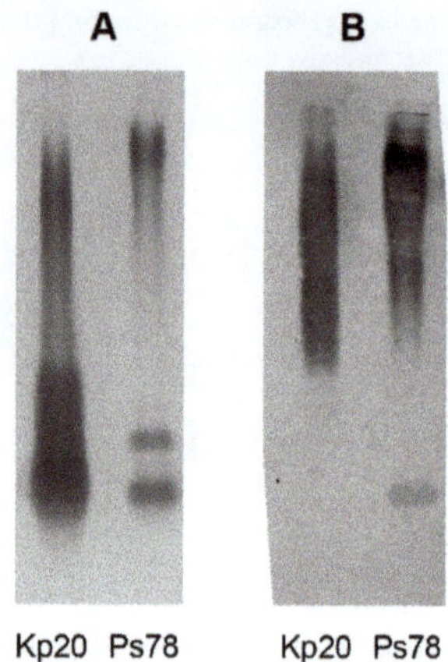

Figure 4. SDS-PAGE analysis and reactivities of *P. shigelloides* serotype O12-specific polyclonal antibodies with LPSs of *P. shigelloides* 78/89 and *K. pneumoniae* Kp20 in immunoblotting. LPSs of *P. shigelloides* 78/89 (Ps78) and *K. pneumoniae* Kp20 (Kp20) were analyzed by SDS-PAGE (1.25 µg/lane) using 15% separating gel and visualized by silver staining method (**A**) or transblotted onto nitrocellulose (**B**). Polyclonal antibodies specific for the O12 serotype were 200-fold diluted.

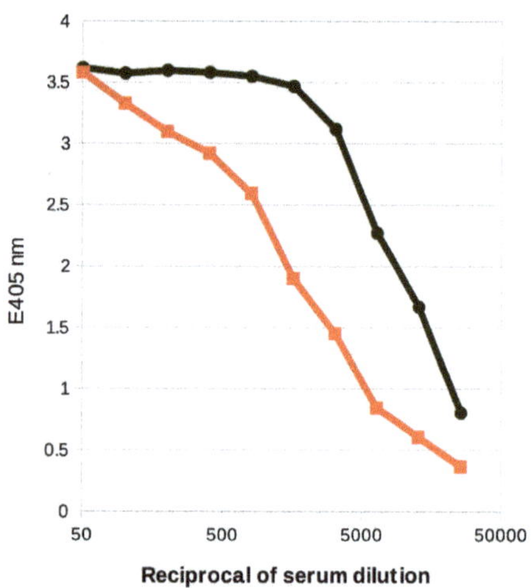

Figure 5. ELISA profiles of polyclonal antibodies specific for the *P. shigelloides* serotype O12 with LPSs of *P. shigelloides* 78/89 (black line) and *K. pneumoniae* Kp20 (red line) as solid-phase antigens (10 µg/mL). The depicted ELISA values are the means of four replicates.

This cross-reaction indicates the presence of antibodies that recognize the common structural element—the non-*O*-acetylated disaccharide repeating unit segments in the *O*-antigens of *P. shigelloides* 78/89 and *K. pneumoniae* Kp20. To further assess the role of the *O*-acetylation in the recognition by antibodies we used ELISA inhibition assay. The inhibitory effects of the *O*-deacetylated *O*-specific polysaccharide of *P. shigelloides* strain 78/89 on the reactivity of the anti-*P. shigelloides* O12 serum with LPS of *P. shigelloides* strain 78/89 and *K. pneumoniae* Kp20 were compared. The *O*-specific polysaccharide devoid of the *O*-acetyl groups inhibited the reactions of the anti-*P. shigelloides* O12 antibodies with LPS of *K. pneumoniae* Kp20 to a maximal value of ~84% at the inhibitor concentration of 62.5 µg/mL, indicating the presence of antibodies specific for the epitopes different from the pure D-galactan I. The reaction of anti-*P. shigelloides* O12 antibodies with LPS of *P. shigelloides* strain 78/89 was inhibited by the *O*-deacetylated *O*-PS to a maximal value of ~94% at the inhibitor concentration of 250 µg/mL (Figure S6). Thus the combined data from ELISA and ELISA inhibition tests indicate the presence of a pool of antibodies specific for the D-galactan I in the anti-*P. shigelloides* O12 serum, but more importantly, the analysis suggests that the *O*-acetylation creates additional *O*-serotype-related antigenic determinants within this *O*-antigen that are recognized by the O12-specific serum. However, the exact contribution of the epitopes comprising the *O*-acetyl groups could not be determined as neither the linear nor spatial distribution of the *O*-acetyls in the PSI was known.

3. Discussion

The *O*-acetylation of the *O*-antigens is typically investigated using the isolated *O*-specific polysaccharides, obtained by mild acidic hydrolysis of the ketosidic linkage between the Kdo residue of the heteropolysaccharide and lipid A. This isolation procedure may result in partial cleavage of the *O*-acetyl groups and their migration to neighboring positions. Non-stoichiometric *O*-acetylation of the *O*-specific polysaccharides, both native and introduced by the isolation methods, makes the NMR spectra more complex, alters the signals of structure reporter groups for known chemotypes and introduces new ones. Here we have presented HR-MAS NMR as a technique that allows for a comparison of the degree of *O*-acetylation of the *O*-antigens at different stages of the isolation procedure, as the changes in the HR-MAS NMR spectra provide immediate distinction of varying structures [18]. This approach to the structural analysis of the *O*-antigens revealed that the mere presence of the *O*-acetyl groups can mask the dominant chemotype. The structure of *P. shigelloides* strain 78/89 LPS is unique among *Plesiomonas* strains. However, the pattern in the HR-MAS spectrum of this LPS appeared similar to the HR-MAS profile of *K. pneumoniae* Kp20 LPS. The latter is built of a disaccharide repeating unit →3)-β-D-Gal*f*-(1→3)-α-D-Gal*p*-(1→ which is identical to the non-*O*-acetylated repeating unit variant of *P. shigelloides* strain 78/89 LPS.

The use of NMR spectroscopy, MALDI-TOF mass spectrometry, and immunochemical analysis provided detailed structural data on the *O*-antigen of *P. shigelloides* strain 78/89 LPS. The *O*-specific polysaccharide isolated from the LPS by conventional methods showed a pattern typical for smooth-type enterobacterial lipopolysaccharide. The identified disaccharide →3)-α-D-Gal*p*-(1→3)-β-D-Gal*f*2OAc-(1→ unit exhibits *O*-acetylation-related heterogeneity and constitutes a new chemotype among *O*-antigens of the *P. shigelloides* strains. As indicated by the HR-MAS NMR analysis in situ and on the isolated LPS, the relative degree of the *O*-acetylation and the position of the *O*-acetyl group was consistent throughout the different stages of the preparations and analysis. The serological analysis confirmed that the strain represents serotype O12 [14]. Thus the structure of *P. shigelloides* serotype O12 LPS has now been established.

We have also observed that the *O*-specific polysaccharide repeating units were nearly identical to these of *K. pneumoniae* Kp20 LPS [34]. However, it is worth noting that their *O*-specific chain length differed and that the *O*-antigen of *P. shigelloides* 78/89 showed higher-degree of polymerization, as indicated by SDS-PAGE and immunoblotting analysis.

The shared structural element in the *O*-specific polysaccharide that occurs in those two strains also explains the observed cross-reactivity with the antiserum specific for the *O*-antigen of *P. shigelloides* O12. Since only 32% of β-D-Gal*f* molecules in *P. shigelloides* 78/89 *O*-PS are *O*-acetylated, it is no surprise that a pool of antibodies in the anti-*P. shigelloides* O12 serum recognizes the D-galactan I of *K.pneumoniae* and the *O*-acetylation effects the cross-reactivity of the strains by creating an additional epitope recognized by the O12-specific antibodies. The role of nonstoichiometric substitutions on O antigenicity is commonly acknowledged. It is a well-known phenomenon among *Salmonella* and *Shigella* as the introduction of acetyl or glucose can form new *O*-types [10].

The β-D-galactofuranosyl-(1→3)-D-galactopyranose known as D-galactan I [35] is present in the LPS of the genus *Klebsiella*. This disaccharide is considered a valuable model for studies of biochemical pathways for formation of the Gal*f*-contaning glycans present in the bacterial cell wall complex (extracellular glycocalyx) [36]. These glycans may play a critical role in the survival and pathogenicity of microorganisms; therefore, their biosynthetic pathways could be attractive targets for drugs that act by inhibiting cell wall biosynthesis [36]. *P. shigelloides* 78/89 expresses LPS with the modified D-galactan I structure of the *O*-antigen.

Klebsiella and *Plesiomonas* are Gram-negative organisms that are frequently co-isolated from intestinal infections. They are also isolated from different clinical specimens including respiratory tract. Their common residence and a niche-specific selection may have contributed to the fact that the strains share epitopes. The cross-reactivities of *P. shigelloides* *O*-antigens with antisera raised against some strains of the genus *Shigella* have been previously reported [15,28]. These strains of *Plesiomonas* and *Shigella* either had some common epitopes or structures of their *O*-antigens were identical. The structural determinant →3)-α-D-Gal*p*-(1→3)-β-D-Gal*f*-(1→ described here for *P. shigelloides* 78/89 LPS is a part of the O1 antigen [35], which is the most clinically prevalent serotype of *K. pneumoniae*. The type-specific cross-reactivities among non-closely related bacteria are a subject of special interest, particularly in the case of antibiotic-resistant pathogens such as *Klebsiella*. Furthermore, it was reported that *K. pneumoniae* strains that belong to O1 and O2 (termed D-galactan I) serogroups cause over 50% of all *Klebsiella* infections worldwide [37].

Typically, *O*-serotypes of *Klebsiella* and *Plesiomonas* were classified using cross-absorbed sera, indicating the differences among the epitopes recognized by polyclonal antibodies with no immediate correlation to the chemotypes of their *O*-antigens.

The *O*-antigens are appealing targets for active and passive immunizations with a potential for therapeutic applications. However, it is important to uncover subtle structural attributes of LPS molecules that may have impact on the desired specificity of antibodies [38]. It is worth noting that all such structural features could be identified using HR-MAS NMR data alone. The results of our study demonstrate a potential application of the HR-MAS NMR technique for screening of the bacterial *O*-antigen structures in situ, providing robust information on the chemotype of the *O*-antigens.

4. Materials and Methods

4.1. Bacteria

P. shigelloides strain CNCTC 78/89, classified as serovar O12:H12 according to Aldova's antigenic scheme [14], was obtained from the Institute of Hygiene and Epidemiology, Prague, Czech Republic. The bacteria were grown on a Davies medium enriched with glucose for 24 h in 37 °C, killed with 0.5% phenol, and harvested as described previously, yielding 6.48 g of freeze-dried bacteria. *K. pneumoniae* strain 20 (Kp20, serogroup O2) is a bloodstream clinical isolate and was kindly provided by Szilvia Melegh (Department of Medical Microbiology and Immunology, Medical School, University of Pecs, Pecs, Hungary) [34]. Samples of freeze-dried bacteria and the isolated LPS used for the HR-MAS analysis were generously supplied by Jolanta Lukasiewicz (Hirszfeld Institute of Immunology and Experimental Therapy, Wroclaw, Poland).

4.2. Lipopolysaccharides and O-Specific Polysaccharide Fractions

LPS was isolated from bacterial cells by the hot phenol/water extraction [39] and purified by ultracentrifugation [40]. Subsequently, *O*-specific polysaccharide was isolated by mild acid hydrolysis (1.5% acetic acid containing 2% SDS at 100 °C for 15 min) of LPS (50 mg). SDS was removed by extraction with 96% ethanol and the pellet was suspended in water and centrifuged. The supernatant was fractionated by gel permeation chromatography, on Bio-Gel P-10 (1.6 × 100 cm) equilibrated with 0.05 M pyridine/acetic acid buffer, pH 5.6. The chromatography yielded a main fraction containing *O*-specific polysaccharide (PSI, 7 mg), separated from shorter *O*-specific polysaccharide chains linked to the core (OSII, <<1 mg) and two fractions containing unsubstituted core oligosaccharides (OSIII,

1.2 mg and OSIV, 5.4 mg). The eluates were monitored with a Knauer differential refractometer and all fractions were checked by NMR spectroscopy MALDI-TOF mass spectrometry (MS).

4.3. Analytical Procedures

Monosaccharides were analyzed as their alditol acetates by GC-MS. Methylation was performed on the isolated PS according to the method described by Hakomori [41]. Alditol acetates and partially methylated alditol acetates were analyzed by GC-MS using an ITQ 700 (Thermo Scientific, Waltham, MA, USA) system, equipped with a HP-1 fused-silica capillary column (0.2 mm $\times$ 12.5 m) and a temperature gradient from 150 to 270 °C at 12 °C·min^{-1}. The absolute configurations of the sugars were determined as described by Gerwig et al. [42] using ($-$)-2-butanol for the formation of 2-butyl glycosides. The trimethylsilylated butyl glycosides were then identified by comparison with authentic samples on GC-MS.

4.4. O-Deacetylation of Polysaccharide

Polysaccharide (5 mg) was treated with aqueous 12.5% NH_3 (1 mL) at room temperature for 16 h, after which the solution was diluted with water and freeze-dried. The product was analyzed by NMR spectroscopy.

4.5. Partial Acid Hydrolysis

The polysaccharide (0.5 mg) was used for hydrolysis with 0.5 M trifluoroacetic acid (1 mL) at 80 °C. A sample (20 µL) was taken every 30 min for 2 h and the progress of hydrolysis was checked by MALDI-TOF MS.

4.6. SDS-PAGE and Serological Analysis

The LPS was analyzed by SDS-PAGE according to the method of Laemmli with modifications [43]. The LPS bands were visualized by the silver staining method [44] and by immunoblot using polyclonal rabbit antisera specific for the *O*-antigen of *P. shigelloides* O12 and for the OS of *P. shigelloides* O33 in separate experiments. Immunoblotting was done as previously described [40]. An ELISA, using LPS as the solid-phase antigen, was performed by a modification [45] of the method described by Voller et al. [46]. The detection systems consisted of a goat anti-rabbit IgG conjugated with alkaline phosphatase (Bio-Rad, Richmond, CA, USA) as a second antibody and 5-bromo-4-chloro-3-indolylphosphate-nitroblue tetrazolium and p-nitrophenylphosphate for immunoblotting and the ELISA test, respectively. ELISA inhibition test was performed as described in [47]. Briefly, the serum (100 µL) at a concentration twice as high as the one giving E_{405nm} value in the range (0.8–2.2) was mixed with 100 µL of inhibitor solution (inhibitor concentration range from 250 µg/mL down to 0.49 µg/mL) and incubated for 1 h at 37 °C. The mixture (100 µL) was then transferred to a microtiter plate with LPS, incubated for 15 min with shaking and followed by the subsequent steps of standard ELISA protocol.

4.7. Mass Spectrometry

MALDI-TOF MS spectra of the PSI, the partially hydrolyzed PSI and the OS fractions were acquired using Bruker Autoflex III (Bruker Daltonics, Bremen, Germany) time-of-flight instrument. Spectra were recorded in positive and negative modes. 2,5-Dihydroxybenzoic acid was used as matrix.

4.8. NMR Spectroscopy

NMR spectra of bacteria (~4 mg dry mass) and LPS (~3 mg) suspensions in ^2H_2O (total volume of ~30 µL in the Bruker Kel-F inserts) were obtained using the HR-MAS technique on a Bruker Avance III 600 MHz spectrometer (Bruker BioSpin, Rheinstetten, Germany). HR-MAS NMR experiments were carried out at a spin rate of 4 kHz at 27 °C (the actual temperature of the bearing gas) using a Bruker 4 mm HR-MAS probe and a ZrO_2 rotor, as previously described [48]. NMR spectra of the isolated polysaccharide and oligosaccharides were recorded for ^2H_2O solutions at 30 °C on Bruker Avance III 600 MHz spectrometer using a 5 mm inverse detection QCI cryoprobe and 1.7 mm TXI microprobe. The polysaccharide and oligosaccharide fractions were repeatedly exchanged with ^2H_2O, with intermediate lyophilization. The data were acquired and processed using Topspin 3.1 (Bruker BioSpin, Rheinstetten, Germany). The processed spectra were assigned with the help of the SPARKY program [49]. The signals were assigned by one- and two-dimensional experiments (COSY, TOCSY, NOESY, HMBC, HSQC-DEPT, and HSQC-TOCSY). In the TOCSY experiments the mixing times used were 30, 60, and 100 ms. The delay time in the HMBC experiment was 60 ms and the mixing time in the NOESY experiment was 200 ms. Regions of interest (ROI) in the HR-MAS HSQC-DEPT spectra were visualized using the rNMR program [50].

The relative content of the *O*-acetyl groups in the *O*-specific polysaccharides was determined by integration of the relevant signals from their 1D ^1H NMR spectra acquired with a 30° pulse and a relaxation delay of 1 s. Volumetric integration of the selected resonances in the HSQC-DEPT spectra was performed by the Gaussian fit method using built-in functions of the SPARKY program.

Supplementary Materials: Supplementary materials can be found at www.mdpi.com/1422-0067/18/12/2572/s1.

Acknowledgments: The authors wish to thank Szilvia Melegh for providing the *K. pneumoniae* clinical isolate and Jolanta Lukasiewicz for supplying the samples of freeze-dried bacteria and isolated LPS. Preliminary structural data included in this work were presented at BIO 2016—2nd Congress of Polish Biochemistry, Cell biology, Biotechnology and Bioinformatics, 13–16 September 2016, Wroclaw, Poland. This work was supported by the statutory funds of the Laboratory of Microbial Immunochemistry and Vaccines of the Hirszfeld Institute of Immunology & Experimental Therapy, Polish Academy of Sciences. The costs to publish in open access were covered by Wroclaw Centre of Biotechnology, programme the Leading National Research Centre (KNOW) for years 2014–2018.

Author Contributions: Sabina Koj, Damian Pawelczyk, and Tomasz Niedziela performed the HR-MAS NMR experiments. SabinaKoj performed NMR experiments on the isolated fractions, analyzed the data, and wrote the relevant segments of the paper. Damian Pawelczyk cultured bacteria and isolated LPS and LPS-derived fractions; Karolina Ucieklak performed and analyzed all the serological experiments; Sabina Koj and Karolina Ucieklak performed and analyzed the MALDI-TOF experiments; Tomasz Niedziela designed the experiments and wrote the paper.

Conflicts of Interest: The authors declare no conflict of interest.

Abbreviations

LPS	Lipopolysaccharide
PS	O-specific polysaccharide
OS	Oligosaccharide
HR-MAS	High-Resolution Magic Angle Spinning
NMR	Nuclear Magnetic Resonance
COSY	Correlation spectroscopy
TOCSY	Total correlation spectroscopy
NOESY	Nuclear Overhouser Effect spectroscopy
HSQC	Heteronuclear Single Quantum Coherence
DEPT	Distortionless Enhancement by polarization transfer
HMBC	Heteronuclear Multiple Bond Correlation
GC	Gas chromatography
MS	Mass spectrometry
MALDI-TOF	Matrix-assisted laser desorption ionization time-of-flight
SDS	Sodium dodecyl sulfate

References

1. Garrity, G.M.; Bell, J.A.; Lilburn, T.G. Taxonomic Outline of the Prokaryotes. In *Bergey's Manual of Systematic Bacteriology*, 2nd ed.; Springer: New York, NY, USA, 2004; pp. 1–399.
2. Farmer, J.; Arduino, M.J.; Hickman-Brenner, F.W. The genera Aeromonas and Plesiomonas. In *The Prokaryotes*; Springer: New York, NY, USA, 1992; Volume 3, pp. 3012–3043.
3. Yamada, S.; Matsushita, S.; Dejsirilert, S.; Kudoh, Y. Incidence and clinical symptoms of Aeromonas-associated travellers' diarrhoea in Tokyo. *Epidemiol. Infect.* **1997**, *119*, 121–126. [CrossRef] [PubMed]
4. Rautelin, H.; Sivonen, A.; Kuikka, A.; Renkonen, O.V.; Valtonen, V.; Kosunen, T.U. Enteric *Plesiomonas shigelloides* infections in Finnish patients. *Scand. J. Infect. Dis.* **1995**, *27*, 495–498. [CrossRef] [PubMed]
5. Centers for Disease Control and Prevention (CDC). *Plesiomonas shigelloides* and Salmonella serotype Hartford infections associated with a contaminated water supply—Livingston County, New York, 1996. *Morb. Mortal. Wkly. Rep.* **1998**, *47*, 394–396.
6. Levy, D.; Bens, M.; Craun, G.; Calderon, R.; Herwaldt, B. Surveillance for waterborne-disease outbreaks—United States, 1995–1996. *MMWR CDC Surveill. Summ.* **1998**, *47*, 1–34. [PubMed]
7. McNeeley, D.; Ivy, P.; Craft, J.C.; Cohen, I. Plesiomonas: Biology of the organism and diseases in children. *Pediatr. Infect. Dis.* **1984**, *3*, 176–181. [CrossRef] [PubMed]
8. Mandal, B.K.; Whale, K.; Morson, B.C. Acute colitis due to *Plesiomonas shigelloides*. *Br. Med. J. (Clin. Res. Ed.)* **1982**, *285*, 1539–1540. [CrossRef]
9. Brenden, R.A.; Miller, M.A.; Janda, J.M. Clinical disease spectrum and pathogenic factors associated with *Plesiomonas shigelloides* infections in humans. *Rev. Infect. Dis.* **1988**, *10*, 303–316. [CrossRef] [PubMed]
10. Raetz, C.R.; Whitfield, C. Lipopolysaccharide endotoxins. *Annu. Rev. Biochem.* **2002**, *71*, 635–700. [CrossRef] [PubMed]
11. Munford, R.S. Sensing Gram-Negative Bacterial Lipopolysaccharides: A Human Disease Determinant? *Infect. Immun.* **2008**, *76*, 454–465. [CrossRef] [PubMed]
12. Molinaro, A.; Holst, O.; di Lorenzo, F.; Callaghan, M.; Nurisso, A.; D'Errico, G.; Zamyatina, A.; Peri, F.; Berisio, R.; Jerala, R.; et al. Chemistry of lipid A: At the heart of innate immunity. *Chem. Eur. J.* **2015**, *21*, 500–519. [CrossRef] [PubMed]
13. Caroff, M.; Karibian, D. Structure of bacterial lipopolysaccharides. *Carbohydr. Res.* **2003**, *338*, 2431–2447. [CrossRef] [PubMed]
14. Aldová, E.; Shimada, T. New O and H antigens of the international antigenic scheme for *Plesiomonas shigelloides*. *Folia Microbiol. (Praha)* **2000**, *45*, 301–304. [CrossRef] [PubMed]
15. Linnerborg, M.; Widmalm, G.; Weintraub, A.; Albert, M.J. Structural elucidation of the O-antigen lipopolysaccharide from two strains of *Plesiomonas shigelloides* that share a type-specific antigen with *Shigella flexneri* 6, and the common group 1 antigen with *Shigella flexneri* spp and *Shigella dysenteriae* 1. *Eur. J. Biochem.* **1995**, *231*, 839–844. [CrossRef] [PubMed]
16. Czaja, J.; Jachymek, W.; Niedziela, T.; Lugowski, C.; Aldova, E.; Kenne, L. Structural studies of the O-specific polysaccharide from *Plesiomonas shigelloides* strain CNCTC 113/92. *Eur. J. Biochem.* **2000**, *267*, 1672–1679. [CrossRef] [PubMed]
17. Niedziela, T.; Lukasiewicz, J.; Jachymek, W.; Dzieciatkowska, M.; Lugowski, C.; Kenne, L. Core oligosaccharides of *Plesiomonas shigelloides* O54:H2 (strain CNCTC 113/92): Structural and serological analysis of the lipopolysaccharide core region, the O-antigen biological repeating unit, and the linkage between them. *J. Biol. Chem.* **2002**, *277*, 11653–11663. [CrossRef] [PubMed]
18. Niedziela, T.; Dag, S.; Lukasiewicz, J.; Dzieciatkowska, M.; Jachymek, W.; Lugowski, C.; Kenne, L. Complete lipopolysaccharide of *Plesiomonas shigelloides* O74:H5 (strain CNCTC 144/92). 1. Structural analysis of the highly hydrophobic lipopolysaccharide, including the O-antigen, its biological repeating unit, the core oligosaccharide, and the linkage between them. *Biochemistry* **2006**, *45*, 10422–10433. [CrossRef] [PubMed]
19. Kubler-Kielb, J.; Schneerson, R.; Mocca, C.; Vinogradov, E. The elucidation of the structure of the core part of the LPS from *Plesiomonas shigelloides* serotype O17 expressing O-polysaccharide chain identical to the Shigella sonnei O-chain. *Carbohydr. Res.* **2008**, *343*, 3123–3127. [CrossRef] [PubMed]

20. Pieretti, G.; Corsaro, M.M.; Lanzetta, R.; Parrilli, M.; Canals, R.; Merino, S.; Tomás, J.M. Structural Studies of the *O*-Chain Polysaccharide from *Plesiomonas shigelloides* Strain 302–73 (Serotype O1). *Eur. J. Org. Chem.* **2008**, *2008*, 3149–3155. [CrossRef]

21. Pieretti, G.; Carillo, S.; Lindner, B.; Lanzetta, R.; Parrilli, M.; Jimenez, N.; Regué, M.; Tomás, J.M.; Corsaro, M.M. The complete structure of the core of the LPS from *Plesiomonas shigelloides* 302–73 and the identification of its *O*-antigen biological repeating unit. *Carbohydr. Res.* **2010**, *345*, 2523–2528. [CrossRef] [PubMed]

22. Maciejewska, A.; Lukasiewicz, J.; Kaszowska, M.; Man-Kupisinska, A.; Jachymek, W.; Lugowski, C. Core oligosaccharide of *Plesiomonas shigelloides* PCM 2231 (Serotype O17) lipopolysaccharide—Structural and serological analysis. *Mar. Drugs* **2013**, *11*, 440–454. [CrossRef] [PubMed]

23. Kaszowska, M.; Jachymek, W.; Lukasiewicz, J.; Niedziela, T.; Kenne, L.; Lugowski, C. The unique structure of complete lipopolysaccharide isolated from semi-rough *Plesiomonas shigelloides* O37 (strain CNCTC 39/89) containing (2S)-*O*-(4-oxopentanoic acid)-α-D-Glc*p* (α-D-Lenose). *Carbohydr. Res.* **2013**, *378*, 98–107. [CrossRef] [PubMed]

24. Kaszowska, M.; Jachymek, W.; Niedziela, T.; Koj, S.; Kenne, L.; Lugowski, C. The novel structure of the core oligosaccharide backbone of the lipopolysaccharide from the *Plesiomonas shigelloides* strain CNCTC 80/89 (serotype O13). *Carbohydr. Res.* **2013**, *380*, 45–50. [CrossRef] [PubMed]

25. Lundqvist, L.C.E.; Kaszowska, M.; Sandström, C. NMR study of the *O*-specific polysaccharide and the core oligosaccharide from the lipopolysaccharide produced by *Plesiomonas shigelloides* O24:H8 (strain CNCTC 92/89). *Molecules* **2015**, *20*, 5729–5739. [CrossRef] [PubMed]

26. Batta, G.; Lipták, A.; Schneerson, R.; Pozsgay, V. Conformational stabilization of the altruronic acid residue in the *O*-specific polysaccharide of Shigella sonnei/*Plesiomonas shigelloides*. *Carbohydr. Res.* **1997**, *305*, 93–99. [CrossRef]

27. Kenne, L.; Lindberg, B.; Petersson, K.; Katzenellenbogen, E.; Romanowska, E. Structural studies of the *O*-specific side-chains of the *Shigella sonnei* phase I lipopolysaccharide. *Carbohydr. Res.* **1980**, *78*, 119–126. [CrossRef]

28. Albert, M.J.; Ansaruzzaman, M.; Qadri, F.; Hossain, A.; Kibriya, A.K.; Haider, K.; Nahar, S.; Faruque, S.M.; Alam, A.N. Characterisation of *Plesiomonas shigelloides* strains that share type-specific antigen with Shigella flexneri 6 and common group 1 antigen with *Shigella flexneri* spp. and *Shigella dysenteriae* 1. *J. Med. Microbiol.* **1993**, *39*, 211–217. [CrossRef] [PubMed]

29. Ansaruzzaman, M.; Albert, M.J.; Holme, T.; Jansson, P.-E.; Rahman, M.M.; Widmalm, G. A *Klebsiella pneumoniae* Strain that Shares a Type-Specific Antigen with Shigella flexneri Serotype 6. *Eur. J. Biochem.* **1996**, *237*, 786–791. [CrossRef] [PubMed]

30. Nestor, G.; Lukasiewicz, J.; Sandström, C. Structural Analysis of the Core Oligosaccharide and the *O*-Specific Polysaccharide from the *Plesiomonas shigelloides* O33:H3 (Strain CNCTC 34/89) Lipopolysaccharide. *Eur. J. Org. Chem.* **2013**, *2014*, 1241–1252. [CrossRef]

31. Maciejewska, A. Structural and Serological Studies of Bacterial Endotoxins of *Plesiomonas Shigelloides* O17 and O51. Ph.D. Thesis, Hirszfeld Institute of Immunology & Experimental Therapy, Wroclaw, Poland, 2008.

32. Kelly, R.F.; Severn, W.B.; Richards, J.C.; Perry, M.B.; MacLean, L.L.; Tomás, J.M.; Merino, S.; Whitfield, C. Structural variation in the *O*-specific polysaccharides of *Klebsiella pneumoniae* serotype O1 and O8 lipopolysaccharide: Evidence for clonal diversity in rfb genes. *Mol. Microbiol.* **1993**, *10*, 615–625. [CrossRef] [PubMed]

33. Gorin, P.A.J.; Mazurek, M. Further Studies on the Assignment of Signals in 13C Magnetic Resonance Spectra of aldoses and derived Methyl Glycosides. *Can. J. Chem.* **1975**, 1212–1223. [CrossRef]

34. Szijártó, V.; Guachalla, L.M.; Hartl, K.; Varga, C.; Banerjee, P.; Stojkovic, K.; Kaszowska, M.; Nagy, E.; Lukasiewicz, J.; Nagy, G. Both clades of the epidemic KPC-producing *Klebsiella pneumoniae* clone ST258 share a modified galactan *O*-antigen type. *Int. J. Med. Microbiol.* **2016**, *306*, 89–98. [CrossRef] [PubMed]

35. Whitfield, C.; Richards, J.C.; Perry, M.B.; Clarke, B.R.; MacLean, L.L. Expression of two structurally distinct D-galactan O antigens in the lipopolysaccharide of *Klebsiella pneumoniae* serotype O1. *J. Bacteriol.* **1991**, *173*, 1420–1431. [CrossRef] [PubMed]

36. Wang, H.; Zhang, G.; Ning, J. First synthesis of β-D-Gal*f*-(1→3)-D-Gal*p*—The repeating unit of the backbone structure of the *O*-antigenic polysaccharide present in the lipopolysaccharide (LPS) of the genus Klebsiella. *Carbohydr. Res.* **2003**, *338*, 1033–1037. [CrossRef]

37. Hansen, D.S.; Mestre, F.; Albertí, S.; Hernández-Allés, S.; Álvarez, D.; Doménech-Sánchez, A.; Gil, J.; Merino, S.; Tomás, J.M.; Benedí, V.J. *Klebsiella pneumoniae* Lipopolysaccharide O Typing: Revision of Prototype Strains and *O*-Group Distribution among Clinical Isolates from Different Sources and Countries. *J. Clin. Microbiol.* **1999**, *37*, 56–62. [PubMed]

38. Stojkovic, K.; Szijártó, V.; Kaszowska, M.; Niedziela, T.; Hartl, K.; Nagy, G.; Lukasiewicz, J. Identification of D-Galactan-III as Part of the Lipopolysaccharide of *Klebsiella pneumoniae* Serotype O1. *Front. Microbiol.* **2017**, *8*, 684. [CrossRef] [PubMed]

39. Westphal, O.; Jann, K. Extraction with phenol-water and further applications of the procedure. *Methods Carbohydr. Chem.* **1965**, *5*, 83–89.

40. Petersson, C.; Niedziela, T.; Jachymek, W.; Kenne, L.; Zarzecki, P.; Lugowski, C. Structural Studies of the *O*-Specific Polysaccharide of Hafnia alvei Strain PCM 1206 Lipopolysaccharide Containing D-Allothreonine. *Eur. J. Biochem.* **1997**, *244*, 580–586. [CrossRef] [PubMed]

41. Hakomori, S. A Rapid Permethylation of Glycolipid, and Polysaccharide Catalyzed by Methylsulfinyl Carbanion in Dimethyl Sulfoxide. *J. Biochem.* **1964**, *55*, 205–208. [PubMed]

42. Gerwig, G.J.; Kamerling, J.P.; Vliegenthart, J.F. Determination of the absolute configuration of mono-saccharides in complex carbohydrates by capillary G.L.C. *Carbohydr. Res.* **1979**, *77*, 10–17. [CrossRef]

43. Laemmli, U.K. Cleavage of Structural Proteins during the Assembly of the Head of Bacteriophage T4. *Nature* **1970**, *227*, 680–685. [CrossRef] [PubMed]

44. Tsai, C.M.; Frasch, C.E. A sensitive silver stain for detecting lipopolysaccharides in polyacrylamide gels. *Anal. Biochem.* **1982**, *119*, 115–119. [CrossRef]

45. Jennings, H.J.; Lugowski, C. Immunochemistry of groups A, B, and C meningococcal polysaccharide-tetanus toxoid conjugates. *J. Immunol.* **1981**, *127*, 1011–1018. [PubMed]

46. Voller, A.; Draper, C.; Bidwell, D.E.; Bartlett, A. Microplate enzyme-linked immunosorbent assay for Chagas' disease. *Lancet* **1975**, *305*, 426–428. [CrossRef]

47. Jachymek, W.; Czaja, J.; Niedziela, T.; Lugowski, C.; Kenne, L. Structural studies of the *O*-specific polysaccharide of Hafnia alvei strain PCM 1207 lipopolysaccharide. *Eur. J. Biochem.* **1999**, *266*, 53–61. [CrossRef] [PubMed]

48. Jachymek, W.; Niedziela, T.; Petersson, C.; Lugowski, C.; Czaja, J.; Kenne, L. Structures of the *O*-specific polysaccharides from *Yokenella regensburgei* (*Koserella trabulsii*) strains PCM 2476, 2477, 2478, and 2494: High-resolution magic-angle spinning NMR investigation of the *O*-specific polysaccharides in native lipopolysaccharides and directly on the surface of living bacteria. *Biochemistry* **1999**, *38*, 11788–11795. [CrossRef] [PubMed]

49. Goddard, T.D.; Kneller, D.G. *Sparky*, 3rd ed.; University of California: San Francisco, CA, USA, 2001.

50. Lewis, I.A.; Schommer, S.C.; Markley, J.L. rNMR: Open source software for identifying and quantifying metabolites in NMR spectra. *Magn. Reson. Chem.* **2009**, *47*, S123–S126. [CrossRef] [PubMed]

International Journal of
Molecular Sciences

Article

Salmonella O48 Serum Resistance is Connected with the Elongation of the Lipopolysaccharide O-Antigen Containing Sialic Acid

Aleksandra Pawlak [1,*], Jacek Rybka [2], Bartłomiej Dudek [1], Eva Krzyżewska [2], Wojciech Rybka [2], Anna Kędziora [1], Elżbieta Klausa [3] and Gabriela Bugla-Płoskońska [1,*]

[1] Department of Microbiology, Institute of Genetics and Microbiology, University of Wrocław, 51-148 Wrocław, Poland; bartlomiej.dudek@uwr.edu.pl (B.D.); anna.kedziora@uwr.edu.pl (A.K.)

[2] Department of Immunology of Infectious Diseases, Hirszfeld Institute of Immunology and Experimental Therapy, Polish Academy of Sciences, 53-114 Wrocław, Poland; rybka@iitd.pan.wroc.pl (J.R.); eva.krzyzewska@iitd.pan.wroc.pl (E.K.); wrybka@iitd.pan.wroc.pl (W.R.)

[3] Regional Centre of Transfusion Medicine and Blood Bank, 50-345 Wrocław, Poland; e.klausa@rckik.wroclaw.pl

* Correspondence: aleksandra.pawlak@uwr.edu.pl (A.P.); gabriela.bugla-ploskonska@uwr.edu.pl (G.B.-P.); Tel.: +48-71-375-62-28 (A.P.); +48-71-375-63-23 (G.B.-P.)

Received: 7 August 2017; Accepted: 12 September 2017; Published: 21 September 2017

Abstract: Complement is one of the most important parts of the innate immune system. Some bacteria can gain resistance against the bactericidal action of complement by decorating their outer cell surface with lipopolysaccharides (LPSs) containing a very long O-antigen or with specific outer membrane proteins. Additionally, the presence of sialic acid in the LPS molecules can provide a level of protection for bacteria, likening them to human cells, a phenomenon known as molecular mimicry. *Salmonella* O48, which contains sialic acid in the O-antigen, is the major cause of reptile-associated salmonellosis, a worldwide public health problem. In this study, we tested the effect of prolonged exposure to human serum on strains from *Salmonella* serogroup O48, specifically on the O-antigen length. After multiple passages in serum, three out of four tested strains became resistant to serum action. The gas-liquid chromatography/tandem mass spectrometry analysis showed that, for most of the strains, the average length of the LPS O-antigen increased. Thus, we have discovered a link between the resistance of bacterial cells to serum and the elongation of the LPS O-antigen.

Keywords: serum resistance; complement; *Salmonella*; lipopolysaccharide; sialic acid; reptile-associated salmonellosis

1. Introduction

Complement, a component of blood serum of vertebrates, is one of the most important parts of the immune system, playing a decisive role in the defense of the host against infections. Its activation during infections can lead to sepsis [1,2]. Complement can be activated via the classical, alternative, or lectin pathways. Because of the presence of complement and other parts of the immune system, e.g., lysozymes, human serum is an extremely unfavorable environment for bacterial survival and growth. However, several bacteria have established a number of strategies protecting them from these conditions, one being the molecular mimicry phenomenon [3–5]. For instance, the LPSs of some bacteria contain sialic acid (*N*-acetylneuraminic acid, NeuAc), which enables them to mimic human cells and avoid the bactericidal action of serum [3–5]. Non-typhoidal *Salmonella* (NTS) organisms pose a significant epidemiological problem all over the world [6–12]. Each year *Salmonella* rods are the causative agents of 93.8 million cases of gastroenteritis worldwide, of which 155,000 are fatal [6]. According to the CDC (Center for Disease Control and Prevention) only 1–5% of *Salmonella* infections are laboratory confirmed and reported, so the real number of salmonellosis cases is much higher [13]. Salmonellosis is the most common illness among food-borne diseases, in some instances leading to hospitalization and death [9–11,13,14]. NTS infections usually cause diarrhea, although they can also lead to extra-intestinal infections, including bacteremia, sepsis, and in rare cases miscarriage [15–18]. Salmonellosis is a zoonotic infection; the animal sources of *Salmonella* rods are inter alia: birds, cattle, pigs, horses, rodents, dogs, cats, reptiles, or amphibians [13,16,19], with reptiles being reported as the most common source of *Salmonella* causing RAS (reptile-associated salmonellosis). Most reptiles are asymptomatic carriers of *Salmonella* [17,20–22]. However, transmission of these bacteria from reptiles to humans occurs frequently, sometimes leading to sepsis, affecting mainly children under 5 years of age, immunocompromised patients, and AIDS (acquired immune deficiency syndrome) patients [13,16,17,20–22]. Recently, the term REPAS (reptile exotic pet-associated salmonellosis) was proposed for these infections in the literature [20], to point out that the main source of salmonellosis in humans are non-native reptiles, as the data show that the main carriers of *Salmonella* are reptiles kept in captivity as pets, not wild ones [20,22]. RAS/REPAS cases lead more frequently to hospitalizations than salmonellosis not connected to reptiles. There is an increasing trend in RAS/REPAS incidence, especially among children under 3 years of age [20], which is probably connected with the current trend for breeding reptiles in households [23–25]. A common causative agent of RAS/REPAS is *Salmonella* from the O48 serogroup, containing sialic acid in the structure of lipopolysaccharides (LPSs) [26–28]. As mentioned before, the presence of NeuAc, a common constituent in outer structures of both higher organisms and bacteria, can represent the molecular mimicry phenomenon, as it can protect bacterial cells from the complement lytic action [3–5]. Sialic acid can also bind the complement factor H, resulting in the inhibition of the alternative pathway activation [29]. LPS is a component of the outer cell membrane, characteristic of most Gram-negative bacteria. LPS is built up of three main parts: lipid A, the inner and outer core, and the O-specific polysaccharide chain. Lipid A, anchoring the structure in the hydrophobic membrane, is linked to the non-repeating oligosaccharide core, while the O-specific chain linked to the core oligosaccharide can contain up to 70–100 repeating oligosaccharide units among different *Salmonella* strains. Each repeating unit is built up of two to eight monosaccharide residues, e.g., mannose, rhamnose, galactose, or sialic acid [19,28,30]. The presence of terminal NeuAc in the O-antigen units [28,31] renders the LPS structure similar to human glycosphingolipids and enables it to take part in the molecular mimicry phenomenon. Due to its high structural variability, the O-polysaccharide chain is used in serological classification as an O-antigenic determinant. The basic function of the O-chain is the protection of bacteria from host immune response (especially the alternative complement cascade and phagocytosis) [30,32]. A shortened O-specific chain of LPS is a possible reason for its bacterial sensitivity to human serum [33–35]. In our previous study [19], we demonstrated that various serovars of *Salmonella* O48 with the same structure of the O-specific antigen differed in the number of repeating units (measured as the NeuAc/Kdo ratio). However, the average length of the O-antigen did not correlate with bacterial cells' susceptibility to human blood serum action. In the present study, we concentrated on the role of sialic acid in bacterial resistance to serum. Serum, as a challenging environment for bacterial growth, can enforce various modifications in bacterial outer structures, which can build up the protection against complement activity. In this study, we investigated whether prolonged contact of *Salmonella* O48 cells with human serum (multiple passages) can lead to any changes in the average length of the LPS O-specific antigen, measured as the NeuAc/Kdo ratio, and changes in bacteria susceptibility to human blood serum.

2. Results

2.1. The C3 Concentration in Serum

The level of C3 in the human serum used in this study was 1240 mg/L. This result is within the normal range for males (970–1576 mg/L) and females (1032–1495 mg/L) (Human Complement C3 & C4: "Nl" BindaridTM Radial Immunodiffusion Kit; The Binding site Group Ltd., Birmingham, UK).

2.2. Passages of Bacteria in NHS (Normal Human Serum)

The strains used in the following experiments were previously tested for their initial length of the LPS O-specific antigen [19]. All chosen strains were initially sensitive to 50% human serum action (which was confirmed three times) and differed substantially in the O-antigen average length: *S*. Hammonia with the highest NeuAc/Kdo ratio of 256%, *S*. Erlangen and *S*. Bongori with an intermediate ratio (87% and 41%, respectively), and *S*. Isaszeg with the lowest NeuAc/Kdo ratio (at the detection limit). These results were precisely described in our previous study [19]. Here, we found that the susceptibility of the tested strains (*S*. Erlangen, *S*. Isaszeg, *S*. Hammonia, and *S*. Bongori) to normal human serum (NHS) action changed during the passages in NHS (Figure 1, Table 1). At the start of the experiments, all tested strains were sensitive to bactericidal action of 50% NHS. The survival after 3 h (T3) of incubation in serum was 0.001% for *S*. Erlangen, 0.04% for *S*. Isaszeg, $\leq$0.00009% for *S*. Bongori, and 0.06% for *S*. Hammonia. During the passages, three tested strains (*S*. Erlangen, *S*. Isaszeg, and *S*. Hammonia) showed a tendency to a biphasic process, shown in Figure 1. They significantly increased the survival percentage, twice during nine passages. After prolonged contact of bacterial cells with serum (nine passages in serum), three out of four tested strains (*S*. Erlangen, *S*. Isaszeg, *S*. Hammonia) became resistant to serum. After nine passages in serum, the percentage survival of *S*. Erlangen, *S*. Isaszeg, and *S*. Hammonia cells was 200%, 183.78%, and 491.80%, respectively. *S*. Bongori was still not considered as resistant to serum after nine passages, although its survival after prolonged contact with NHS increased substantially up to 20,000 times (from $\leq$0.00009 in the first passage to 1.80% in the ninth passage). All tested strains incubated for 3 h in NHS or heated at 56 °C for 30 min (control) proliferated very intensively (Table 2), as the heating removed the bactericidal activity of the serum. Bacterial cells of all tested strains obtained after the ninth passage in NHS were transferred into fresh LB (lysogeny broth) medium with glycerol and frozen in -70 °C. After three months, all frozen strains were re-examined and their susceptibility to bactericidal activity of 50% NHS (survival percent) was compared to results obtained after passage 9. The results indicated that all tested strains (*S*. Erlangen, *S*. Isaszeg, *S*. Hammonia, and *S*. Bongori) generally maintained the resistance achieved by prolonged contact with serum. The survival of bacterial cells after 3 h of incubation with serum was higher than 100% for *S*. Erlangen (643.68%), for *S*. Isaszeg (112.77%), and *S*. Hammonia (733.33%), while for *S*. Bongori the resistance remained weak (0.05%). Serum heated at 56 °C for 30 min was used as a control. In these conditions, all tested strains proliferated intensively (Table 2).

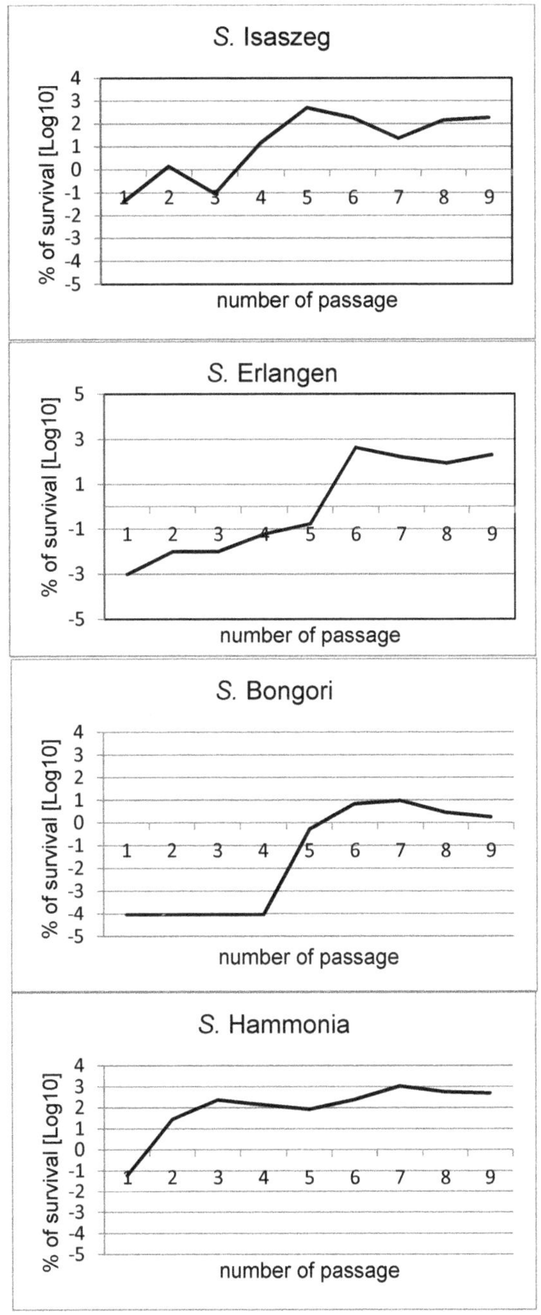

Figure 1. Survival percent of *Salmonella* O48 strains during nine passages in 50% NHS (normal human serum).

Table 1. Bactericidal activity against *Salmonella* O48 strains during passages in 50% NHS.

Number of Passages	S. Hammonia CFU/mL *		Survival of Cells at T3 (%)	S. Isaszeg CFU/mL		Survival of Cells at T3 (%)	S. Bongori CFU/mL		Survival of Cells at T3 (%)	S. Erlangen CFU/mL		Survival of Cells at T3 (%)
	T0	T3		T0	T3		T0	T3		T0	T3	
1	1.7×10^6	9.6×10^2	0.06	5.7×10^6	2.0×10^3	0.04	6.9×10^6	$\leq 10^0$	≤ 0.00009	5.1×10^6	5.5×10^1	0.001
2	1.0×10^6	2.7×10^5	27.00	6.5×10^6	9.0×10^4	1.38	3.5×10^6	$\leq 10^0$	≤ 0.00009	7.1×10^6	7.1×10^2	0.01
3	2.3×10^6	5.4×10^6	234.78	5.4×10^6	5.1×10^3	0.09	3.4×10^6	$\leq 10^0$	≤ 0.00009	8.4×10^6	9.2×10^2	0.01
4	4.7×10^6	6.4×10^5	136.17	4.8×10^6	7.2×10^5	15.00	2.8×10^6	$\leq 10^0$	≤ 0.00009	4.2×10^6	2.6×10^3	0.06
5	1.7×10^6	1.4×10^6	82.35	3.2×10^6	1.6×10^7	500.00	1.5×10^6	8.0×10^3	0.53	5.0×10^6	8.6×10^3	0.17
6	1.0×10^6	2.4×10^6	240.00	3.7×10^6	6.8×10^6	183.78	3.2×10^6	2.2×10^5	6.88	2.0×10^6	8.5×10^6	425.00
7	1.7×10^6	1.8×10^7	1058.82	3.4×10^6	7.8×10^5	22.94	4.5×10^6	4.2×10^5	9.33	3.2×10^6	5.1×10^6	159.38
8	1.3×10^6	7.3×10^6	561.64	4.1×10^6	5.8×10^6	141.46	5.0×10^5	1.4×10^4	2.80	4.5×10^6	3.8×10^6	84.44
9	6.1×10^6	3.0×10^7	491.80	3.7×10^6	6.8×10^6	183.78	5.0×10^6	8.1×10^4	1.80	3.0×10^6	6.0×10^6	200.00

* CFU/mL—colony-forming units in milliliter; NHS: normal human serum.

Table 2. Bactericidal activity of 50% NHS decomplemented by heating at 56 °C for 30 min against *Salmonella* O48 strains.

S. Hammonia CFU/mL *		Survival of Cells at T3 (%)	S. Isaszeg CFU/mL		Survival of Cells at T3 (%)	S. Bongori CFU/mL		Survival of Cells at T3 (%)	S. Erlangen CFU/mL		Survival of Cells at T3 (%)
T0	T3		T0	T3		T0	T3		T0	T3	
4.6×10^6	5.1×10^7	1108.69	3.7×10^6	2.7×10^7	729.73	4.2×10^6	3.6×10^7	857.15	3.6×10^6	3.1×10^7	861.11

* CFU/mL—colony-forming units in milliliter.

2.3. SDS-PAGE (SDS-Polyacrylamide Gel Electrophoresis) of LPS

As the next step, the SDS-PAGE analysis of LPS isolated from bacterial cells before the passages and after ninth passage in 50% NHS (normal human serum) was performed. The results showed that LPS isolated from all tested strains produced very long O-antigen (VL-OAg: more than 100 O-specific units in the O-chain; Figure 2). Such behavior is described in the literature as a factor influencing bacterial resistance to human serum [35,36]. Moreover, the comparison of the LPS profile of *Salmonella* O48 before passages in 50% NHS (BP) and after the ninth passage (AP) shows distinct differences in the quantitative proportions between the short and long O-antigen regions. In the case of *S.* Hammonia, there was a distinct increase in the LPS of medium length (L-OAg LPS).

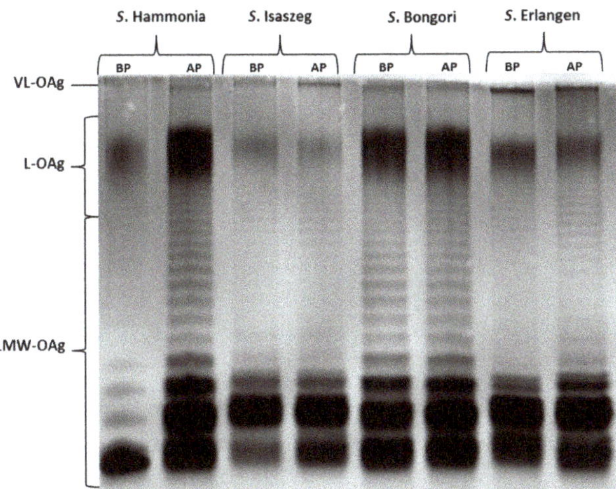

Figure 2. Lipopolysaccharide (LPS) profiles of four *Salmonella* O48 strains before (BP) and after (AP) nine passages in 50% normal human serum. VL-OAg: very long O-antigen; L-OAg: long O-antigen; LMW-OAg: low molecular weight O-antigen.

2.4. GLC-MS/MS (Gas Liquid Chromatography-Mass Spectrometry) Analysis

In the lipopolysaccharide molecule of O48 serotype, the number of Kdo (3-Deoxy-D-manno-octulosonic acid) residues is constant, while the number of repeating units in the O-antigen changes. Since each repeating unit of the O-antigen possesses one NeuAc (sialic acid) residue, the elongation of the O-antigen is directly linked to the increase in NeuAc content in the molecule. Therefore, we assume that NeuAc/Kdo proportions provide important information about the average length of the O-specific chain in the tested samples of bacteria. In our previous study, we analyzed the NeuAc/Kdo ratio in bacterial cells in minimal Falcov' medium using GLC-MS [19]. In the present study, we used a more sensitive GLC-MS/MS method; therefore, we could measure the NeuAc/Kdo ratio before and after passages in NHS in single bacterial colonies on agar plates (Figures 3 and 4). Each time, the amount of NeuAc was compared with the amount of Kdo in bacterial cells. The results showed that for all tested strains, the average length of the LPS O-antigen (measured as the NeuAc/Kdo ratio) increased after the passages of bacterial cells in 50% NHS, with relatively high variability of the LPS O-antigen length among colonies from the same strain (Figure 4). This indicates that results usually obtained for samples of bacterial cells mass are only an average of single cells values, which differ substantially.

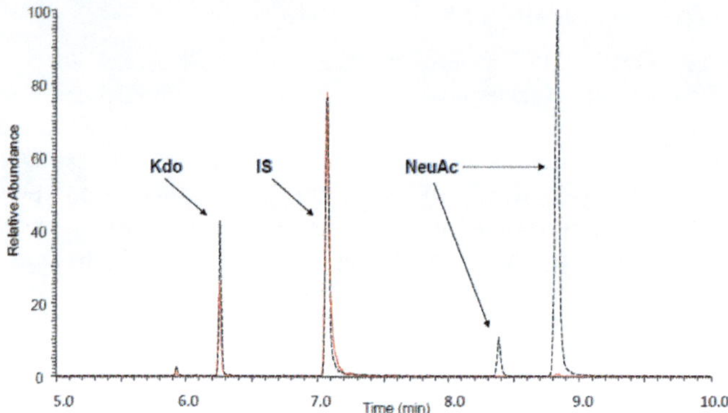

Figure 3. *S.* Erlangen results of analysis of two single colonies as an example showing high and low NeuAc (sialic acid)/Kdo (3-Deoxy-D-manno-octulosonic acid) ratio: colony 3 (black, dashed line) and colony 4 (red, dotted line) differing in NeuAc/Kdo ratio, superimposed for better view. MS/MS simultaneous analysis of Kdo (marker ion of m/z = 195), NeuAc (ion of m/z = 386), and perseitol (ion of m/z = 128) as an internal standard (IS).

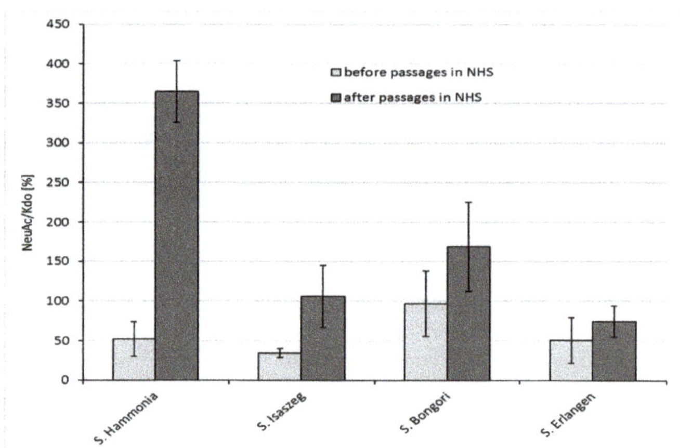

Figure 4. Results of GLC-MS/MS (Gas Liquid Chromatography-Mass Spectrometry) analysis of NeuAc/Kdo content (%) in *Salmonella* O48 strains measured in single randomly selected colonies on agar plates (averaged results; n = 5).

3. Discussion

Salmonella O48 are rare but very dangerous bacteria, especially for children under 5 years of age and immunocompromised patients. Gastroenteritis or sepsis caused by these strains is mainly related to contact with reptiles [13,24–27]. Complement plays a very important role in protecting the host from sepsis caused by pathogens, especially Gram-negative bacteria [2]. Passaging bacteria in NHS is a model of interaction between the host and pathogenic organism. Bacterial resistance to complement proteins is not yet fully explained. As some authors described a potential role of LPS in the serum resistance phenomenon [32,34,35], we tested whether changes in the LPS O-chain length can be a way of adaptation and resistance generation by *Salmonella* O48.

In our previous study [19], we did not find a clear correlation between the LPS O-antigen length (NeuAc/Kdo ratio) and the resistance of *Salmonella* O48 to bactericidal action of serum.

Among the strains that were sensitive to serum action, we found both strains with long and short O-antigen. From that group of sensitive strains, we selected four strains of *Salmonella* O48 which varied substantially in the LPS O-antigen length for the experiments described in the present work. Here, we show that prolonged contact of bacteria with NHS results in the extension of the average LPS O-chain length, and that a very long O-antigen (VL-OAg) might protect bacteria from NHS action. The research results of Murray et al. [37,38] and Bravo et al. [35] showed that VL-OAg provides serum resistance of *S.* Typhimurium. Crawford et al. [39] proved the key role of VL-OAg in *Salmonella* Typhimurium resistance to bile. The O-antigen length is under the control of the *wzz* gene. *wzz* is under the control of OMPs (outer membrane proteins): OMP FepE controls VL-OAg production, and OMP Wzz controls L-OAg production [35]. However, it is still unclear as to what extent the genes and/or the environment are involved in this process. It is possible that both changes in the LPS length and the OMPs arrangement in OM (outer membrane) are responsible for bacterial adaptation and survival in the presence of serum proteins. Our experiments also include an analysis of *Salmonella* Enteritidis OMP changes in the serum resistance phenomenon. The results are described in our previous works [40–42]. Considering these results as well as those of our present findings, we suggest that it is possible that both long LPSs and certain OMPs (e.g., PgtE in *S.* Enteritidis) together provide resistance to the bactericidal action of serum. Apart from the serum, there are other environmental factors influencing the length of the LPS O-chain, e.g., growth temperature, lower amount of Mg^{2+} and Fe^{3+} in the environment, or the growth phase of bacterial cells [35]. In our study, single colonies of the same strain from one agar plate differed in the NeuAc/Kdo ratio. As Bravo et al. [35] showed that the longest O-chain is produced at the late exponential and stationary phase of growth, we suppose that this difference may be due to the growth phase or, alternatively, it can reflect the innate variability of the strains. Our results of the NeuAc/Kdo ratio in bacterial colonies show the average length of LPS, however, the variation of the O-antigen length between particular colonies of one strain is relatively high. One of the tested strains (*S.* Bongori) did not become resistant to serum after nine passages (survival rate of bacterial cells in T3 < 100%), although its susceptibility substantially changed. Before the passages, the strain was much more sensitive to serum than other tested *Salmonella* O48 serovars. During the passages, *S.* Bongori became more than four-fold more resistant to serum compared to before the passages. Also, the average length of the O-chain of this strain increased (Figure 4), suggesting a role of LPSs in bacterial adaptation to harsh environments.

4. Materials and Methods

4.1. Bacterial Strains and Growth Conditions

The study was carried out on four strains of *Salmonella* O48 serogroup: *Salmonella bongori* from serovar Bongori and *Salmonella enterica* from serovars: Erlangen, Isaszeg, and Hammonia. The strains were obtained from Polish Collection of Microorganisms (PCM), Hirszfeld Institute of Immunology and Experimental Therapy, Polish Academy of Sciences, Wroclaw, Poland. The list of the tested strains, showing their origin and antigenic characteristics, is presented in Table 3.

Table 3. The origin and antigenic characteristics of the *Salmonella* O48 strains used in this study.

Species	Subspecies	Serovar	Somatic (O) Antigen	Source
Salmonella bongori	-	Bongori	48	PCM * 2547
Salmonella enterica	*enterica*	Isaszeg	48	PCM * 2550
Salmonella enterica	*salamae*	Erlangen	48	PCM * 2533
Salmonella enterica	*salamae*	Hammonia	48	PCM * 2535

* PCM—Polish Collection of Microorganisms, Hirszfeld Institute of Immunology and Experimental Therapy, Polish Academy of Sciences, Wroclaw, Poland.

4.2. Media

For the bactericidal assay of human serum and LPS extraction, bacterial cells were grown in liquid medium (YP yeast extract-peptone broth: bactopeptone (Difco), yeast extract (Difco) and NaCl (POCh, Gliwice, Poland), pH 7.0) at 37 °C for 18 h in a water bath with shaking. For GLC-MS/MS analysis, the bacteria were grown in minimal Falcov' medium (K_2HPO_4, KH_2PO_4, $MgSO_4$, $(NH_4)_2SO_4$, glucose and NaCl, (all POCh, Gliwice, Poland)) and on nutrient agar plates (BIOCORP, Warszawa, Poland). For bacteria storage in −70 °C, LB medium with 50% glycerol (POCh, Gliwice, Poland) was used.

4.3. Sera

Normal human serum (NHS) was obtained from the Regional Center of Transfusion Medicine and Blood Bank, Wroclaw, Poland (120 donors) and was approved by the authors' institutional review board (Elżbieta Klausa, DG-G/2739/11, 18.05.2011). This was conducted according to the principles expressed in the Law on the public service of blood of 20 May 2016 and in the Directive 2002/98/EC of the European Parliament and of the Council of 27 January 2003, establishing standards of quality and safety for the collection, testing, processing, storage, and distribution of human blood and blood components. Blood samples were collected into sterile tubes with clot activator and gel for serum separation. The samples were then stored at room temperature (RT) for 30 min. After that, the samples were centrifuged for 5 min at 3000 rpm. Only the serum samples without hemolysis and lipemia were used for experiments. The C3 concentration in the mixed serum was quantified by radial immunodiffusion (Human Complement C3 & C4 "Nl" BindaridTM Radial Immunodiffusion Kit; The Binding site Group Ltd., Birmingham, UK). The collected serum was frozen in 0.5-mL aliquots at −70 °C for no longer than six months. The required volume of serum was thawed immediately before each passage and each portion was used only once.

4.4. Bactericidal Assay of NHS

The bactericidal action of NHS against the tested strains was determined as described previously [19,29] with slight modification. Bacterial cells were grown for 18 h in 5.0 mL of YP medium with shaking. Then 0.05 mL of the overnight bacterial culture was transferred to 3.0 mL of fresh YP medium and incubated at 37 °C for 1 h in water bath with shaking. Next, bacterial cells were centrifuged (4000 rpm for 20 min at 4 °C) and the pellet was suspended in physiological saline (0.9% NaCl). Then, 1.0 mL of the suspension was transferred into 5.0 mL of physiological saline and, after shaking, 0.5 mL of the resulting suspension was mixed with 0.5 mL of freshly thawed NHS. The purpose of such preparation was to obtain an early log-phase culture of bacteria mixed with serum and the initial colony forming units in 1.0 mL of medium was 10^6 CFU/mL (colony-forming units in milliliter), which was verified each time. Bacterial cells mixed with serum were incubated in water bath with shaking at 37 °C for 3 h. The cells were collected after 0 (T0) and 3 h (T3). The collected samples were diluted and cultured on nutrient agar plates at 37 °C for 18 h. After 18 h of incubation, the average number of colonies was estimated from agar plates, then the CFU/mL was calculated and the value of CFU/mL at T0 was taken as 100% of bacterial cells' survival. According to this value, the survival percent of bacterial cells in T3 was estimated. When the survival percent of bacterial cells in T3 was >100%, the cells were considered resistant, and those with survival rates <100% were considered susceptible to 50% NHS bactericidal action. NHS decomplemented by heating at 56 °C for 30 min was used as a control (Table 2).

4.5. Passages of Bacterial Cells in NHS

Passages of bacterial cells in 50% NHS were based on the method of bactericidal action of NHS described above. Every single passage was considered as an independent experiment on the bactericidal action of serum. Each tested strain was passaged in serum nine times. The number of passages was estimated in our previous study [43] as sufficient for changing bacterial susceptibility to NHS. The first passage was the first contact of the cells with complement. After conducting the first passage, 10 randomly chosen bacterial colonies obtained on agar plates from T3 were used to prepare the bacterial culture for passage 2. Those colonies were transferred into another 5 mL of fresh YP medium, and then the experiment was carried out as described for the method of bactericidal assay of NHS. The usage of fresh medium was necessary, as the NHS selective pressure was so high that after 3 h of incubation of *Salmonella* with NHS the survival ratio of the cells was too low to continue the experiment using NHS. That is why the cells from T3 were grown in medium without selective pressure, and they were again prepared to obtain 10^6 CFU/mL and transferred back to NHS. Serial passages of bacterial cells in 50% NHS were performed nine times, each time using colonies obtained on agar plates from T3 in the previous passage. The control was a parallel test using medium containing the same composition, however NHS was decomplemented by heating, as described above, to remove the selective pressure. Bacterial cells from passage 9 were transferred into fresh LB medium with glycerol and frozen in $-70\ ^\circ$C for further analysis.

4.6. Isolation of Lipopolysaccharides and Analysis by SDS-PAGE (SDS-Polyacrylamide Gel Electrophoresis)

Lipopolysaccharides were extracted from all of the tested *Salmonella* O48 strains before the passages and after ninth passage in 50% NHS. The extraction was done using a commercial RNA isolating reagent according to Yi and Hackett [44]. Briefly, 10 mg of lyophilized bacterial cells were suspended in 200 µL of Tri-Reagent (Sigma-Aldrich, St. Louis, MO, USA). The cell suspension was then incubated at room temperature for 10 min for complete cell homogenization. After incubation, 200 µL of chloroform was added to create a phase separation. The mixture was than vigorously vortexed and incubated at room temperature for an additional 10 min. The resulting mixture was centrifuged at 14,000 rpm (Minispin Plus, Eppendorf, Hamburg, Germany) for 10 min to separate the aqueous and organic phase. The aqueous phase was transferred to a new 1.5-mL centrifuge tube. Distilled water (100 µL) was added to the organic phase. The mixture was vortexed, incubated at room temperature for 10 min, and centrifuged at 14,000 rpm for 10 min. The upper aqueous phases from both steps were combined. The water extraction steps were repeated twice. The combined aqueous phase was lyophilized. After lyophilization, we used the cold magnesium precipitation procedure according to Darveau and Hancock for the purification of LPS [45]. LPS was dissolved in 500 µL of 0.375 M magnesium chloride (POCh) in 95% ethanol, stored at $-20\ ^\circ$C, followed by centrifugation at 14,000 rpm for 15 min. The pellet was suspended in 200 µL of distilled water and lyophilized.

LPS extracts were analyzed by discontinuous SDS-PAGE using a Laemmli buffer system [46]. Samples were applied to the slabs after mixing with Laemmli buffer (composed of 10 mM Tris-HCL, glycerol, SDS, bromophenol blue) and heating at 98 $^\circ$C for 7 min. Gel electrophoresis was performed using 6% polyacrylamide stacking gel and 15% separating gel. The SDS-PAGE separation of LPS was performed at a constant voltage (120 V), for 90 min using a Mini-Protean Tetra Cell apparatus (Bio-Rad, Hercules, CA, USA). The separated LPS was visualized using silver staining according to Tsai and Frasch [47] with Fomsgaard [48] and our own slight modifications [40]. The gels containing separated LPS were photographed using a GelDoc XR imaging system (Bio-Rad, Hercules, CA, USA) under white light.

4.7. Preparation of Samples for GLC-MS/MS Analysis

For the analysis of NeuAc content, a sample of bacteria was placed in a screw-capped tube and an internal standard (10 µg of perseitol (Koch-Light Laboratories Ltd., Suffolk, UK)) was added. The lyophilized sample was methanolized with 2 M HCl in CH_3OH (Sigma-Aldrich, St. Louis, MO, USA) for 1 h at 80 $^\circ$C, evaporated with a stream of N_2 at 40 $^\circ$C, and acetylated with 100 µL of acetic anhydride (Sigma-Aldrich) and 20 µL of pyridine (Sigma-Aldrich, St. Louis, MO, USA) at 80 $^\circ$C for 30 min. After acetylation, the sample was dried with N_2, dissolved in 100 µL of ethyl acetate (POCh, Gliwice, Poland), and 1 µL was taken for GLC-MS/MS analysis.

4.8. GLC-MS/MS Analysis

Samples were analyzed by the GLC-MS/MS system: Thermo FOCUS GC with ITQ 700 ion trap detector with external ionization, equipped with Rxi-5 ms column: 30 m, 0.25 mm ID (Restek, Bellefonte, PA, USA). In the GC method, the ion source temperature was set at 250 °C with automatic ionization energy. Then, 1 µL of the sample was injected with split injection (split = 10). The MS/MS analysis of peracetylated methyl ester of NeuAc methyl glycoside was performed with an ion of m/z 446 as a primary ion, which was isolated and fragmented. The secondary fragment of m/z 386 was used for the quantitation of NeuAc derivative in the sample. Kdo analysis was performed as described previously [49]. Soon after that, a primary ion of m/z 375 was isolated from the mass spectrum of peracetylated methyl ester ethyl glycoside of Kdo and fragmented; a secondary ion of m/z 195 was used for the quantitation of Kdo.

4.9. Standard Curve for NeuAc Determination by GLC-MS/MS Method

NeuAc standard (1, 3, 10, 30, 100, and 300 µg/mL) with perseitol as an internal standard (1000 ng) were methanolized (2 M HCl/CH$_3$OH, 80 °C, 1 h), dried with a stream of N$_2$, and acetylated with 200 µL acetic anhydride and 20 µL pyridine at 80 °C for 30 min. Samples were dissolved in 100 µL of ethyl acetate for GLC-MSMS analysis.

In our previous study [19], we performed GLC-MS/MS analysis of the NeuAc/Kdo ratio in bacterial *Salmonella* O48 cells in minimal Falcov' medium. In this study, we decided to increase the spectrum of research. We measured the NeuAc/Kdo ratio before and after passages in NHS in bacterial cells grown in minimal Falcov' medium, and single bacterial colonies taken from agar plates.

5. Conclusions

Our study shows that prolonged contact of *Salmonella* cells with serum results in the adaptation of the bacteria to adverse environmental conditions. The resistance of *Salmonella* O48 to the bactericidal action of the serum has a multifactorial basis. However, our results show that LPS O-chain elongation plays a significant role in this phenomenon. Present findings constitute a significant basis for the huge field of exploration, that will involve more detailed studies on the possible changes in the length and structure of LPS upon challenging bacteria with complement proteins (especially the acetylation pattern of the resulting lipopolysaccharide molecule).

In our opinion, the results clearly show that multiple passages of bacteria in human serum lead to the resistance of the cells to the serum. There is a very significant change in the LPS O-chain length between cells before and after multiple passages in serum. That is why we deduced that the elongation of LPS protects bacteria from the bactericidal action of serum. We did not show the exact mechanism of how this elongation protects bacteria from serum action; however, we can presume that either the increased amount of sialic acid causes the molecular mimicry phenomenon or the VL-OAg creates a barrier around the cell, so the complement proteins cannot reach the cell surface. An explanation of this process is our team's goal for the next investigation.

Acknowledgments: The authors thank: Prof. Dr hab. A. Gamian (Polish Academy of Sciences, Wroclaw, Poland) for the *Salmonella* O48 strains from the Polish Collection of Microorganisms, and Dr Kamila Myka (Department of Microbiology and Immunology, Columbia University, New York, NY, USA) for English changes. This research was supported in part by a grant funded by University of Wroclaw—project nr 2015/M/IGM/12. Publication was also supported by Wroclaw Centre of Biotechnology, programme The Leading National Research Centre (KNOW) for years 2014–2018. The funding agency (KNOW) had no direct role in the conduct of the study, the collection, management, and interpretation of the data, preparation, nor approval of the manuscript.

Author Contributions: Jacek Rybka, Gabriela Bugla-Płoskońska, Aleksandra Pawlak conceived and designed the experiments; Jacek Rybka, Gabriela Bugla-Płoskońska, Aleksandra Pawlak obtained funding, Aleksandra Pawlak, Bartłomiej Dudek, Eva Krzyżewska, Wojciech Rybka performed the experiments; Aleksandra Pawlak, Bartłomiej Dudek, Eva Krzyżewska, Jacek Rybka, Gabriela Bugla-Płoskońska analyzed the data; Aleksandra Pawlak, Gabriela Bugla-Płoskońska, Jacek Rybka, Eva Krzyżewska, Elżbieta Klausa contributed reagents/materials/analysis tools; Aleksandra Pawlak, Bartłomiej Dudek, Eva Krzyżewska, Anna Kędziora,

Jacek Rybka, Gabriela Bugla-Płoskońska wrote the paper. Gabriela Bugla-Płoskońska and Jacek Rybka provided study supervision. All co-authors revised and approved the final manuscript.

Conflicts of Interest: The authors declare no conflicts of interest.

Abbreviations

LPS	lipopolysaccharide
VL-OAg	very long O-antigen
Kdo	3-Deoxy-D-manno-octulosonic acid
L-OAg	long O-antigen
CDC	Center for Disease Control and Prevention
NeuAc	sialic acid
NTS	non-typhoidal salmonellosis
RAS	reptile-associated salmonellosis
REPAS	reptile exotic pet associated salmonellosis
NHS	normal human serum
CFU	colony forming units
OMP	outer membrane protein
GLC-MS	gas liquid chromatography-mass spectrometry
PCM	Polish Collection of Microorganisms
SDS-PAGE	SDS-polyacrylamide gel electrophoresis

References

1. Markiewski, M.M.; DeAngelis, R.A.; Lambris, J.D. Complexity of complement activation in sepsis. *J. Cell. Mol. Med.* **2008**, *12*, 2245–2254. [CrossRef] [PubMed]
2. Haeney, M.R. The role of the complement cascade in sepsis. *J. Antimicrob. Chemother.* **1998**, *41* (Suppl. A), 41–46. [CrossRef] [PubMed]
3. Heikema, A.P.; Koning, R.I.; Duarte dos Santos Rico, S.; Rempel, H.; Jacobs, B.C.; Endtz, H.P.; van Wamel, W.J.B.; Samsom, J.N. Enhanced, Sialoadhesin-dependent uptake of Guillain-Barre syndrome-associated *Campylobacter jejuni* strains by human macrophages. *Infect. Immun.* **2013**, *81*, 2095–2103. [CrossRef] [PubMed]
4. Spinola, S.M.; Li, W.; Fortney, K.R.; Janowicz, D.M.; Zwickl, B.; Katz, B.P.; Munson, R.S. Sialylation of lipooligosaccharides is dispensable for the virulence of *Haemophilus ducreyi* in humans. *Infect. Immun.* **2012**, *80*, 679–687. [CrossRef] [PubMed]
5. Bax, M.; Kuijf, M.L.; Heikema, A.P.; van Rijs, W.; Bruijns, S.C.; García-Vallejo, J.J.; Crocker, P.R.; Jacobs, B.C.; van Vliet, S.J.; van Kooyk, Y. *Campylobacter jejuni* lipooligosaccharides modulate dendritic cell-mediated T cell polarization in a sialic acid linkage-dependent manner. *Infect. Immun.* **2011**, *79*, 2681–2689. [CrossRef] [PubMed]
6. Majowicz, S.E.; Musto, J.; Scallan, E.; Angulo, F.J.; Kirk, M.; O'Brien, S.J.; Jones, T.F.; Fazil, A.; Hoekstra, R.M. International Collaboration on Enteric Disease "Burden of Illness" studies The global burden of nontyphoidal *Salmonella* gastroenteritis. *Clin. Infect. Dis Off. Publ. Infect. Dis. Soc. Am.* **2010**, *50*, 882–889. [CrossRef] [PubMed]
7. Rondini, S.; Micoli, F.; Lanzilao, L.; Gavini, M.; Alfini, R.; Brandt, C.; Clare, S.; Mastroeni, P.; Saul, A.; MacLennan, C.A. Design of glycoconjugate vaccines against invasive African *Salmonella enterica* serovar Typhimurium. *Infect. Immun.* **2015**, *83*, 996–1007. [CrossRef] [PubMed]
8. LaRock, D.L.; Chaudhary, A.; Miller, S.I. *Salmonellae* interactions with host processes. *Nat. Rev. Microbiol.* **2015**, *13*, 191–205. [CrossRef] [PubMed]
9. Onsare, R.S.; Micoli, F.; Lanzilao, L.; Alfini, R.; Okoro, C.K.; Muigai, A.W.; Revathi, G.; Saul, A.; Kariuki, S.; MacLennan, C.A.; et al. Relationship between antibody susceptibility and lipopolysaccharide O-antigen characteristics of invasive and gastrointestinal nontyphoidal *Salmonellae* isolates from Kenya. *PLos Negl. Trop. Dis.* **2015**, *9*, e0003573. [CrossRef] [PubMed]
10. Marshall, J.M.; Gunn, J.S. The O-Antigen Capsule of *Salmonella enterica* serovar Typhimurium Facilitates serum Resistance and surface Expression of FliC. *Infect. Immun.* **2015**, *83*, 3946–3959. [CrossRef] [PubMed]

11. Tennant, S.M.; Wang, J.Y.; Galen, J.E.; Simon, R.; Pasetti, M.F.; Gat, O.; Levine, M.M. Engineering and preclinical evaluation of attenuated nontyphoidal *Salmonella* strains serving as live oral vaccines and as reagent strains. *Infect. Immun.* **2011**, *79*, 4175–4185. [CrossRef] [PubMed]

12. Keestra-Gounder, A.M.; Tsolis, R.M.; Bäumler, A.J. Now you see me, now you don't: The interaction of *Salmonella* with innate immune receptors. *Nat. Rev. Microbiol.* **2015**, *13*, 206–216. [CrossRef] [PubMed]

13. Hoelzer, K.; Moreno Switt, A.I.; Wiedmann, M. Animal contact as a source of human non-typhoidal salmonellosis. *Vet. Res.* **2011**, *42*, 34. [CrossRef] [PubMed]

14. Liu, B.; Knirel, Y.A.; Feng, L.; Perepelov, A.V.; Senchenkova, S.N.; Reeves, P.R.; Wang, L. Structural diversity in *Salmonella* O antigens and its genetic basis. *FEMS Microbiol. Rev.* **2014**, *38*, 56–89. [CrossRef] [PubMed]

15. Parry, C.M.; Thomas, S.; Aspinall, E.J.; Cooke, R.P.D.; Rogerson, S.J.; Harries, A.D.; Beeching, N.J. A retrospective study of secondary bacteraemia in hospitalised adults with community acquired non-typhoidal *Salmonella* gastroenteritis. *BMC Infect. Dis.* **2013**, *13*, 107. [CrossRef] [PubMed]

16. Van Meervenne, E.; Botteldoorn, N.; Lokietek, S.; Vatlet, M.; Cupa, A.; Naranjo, M.; Dierick, K.; Bertrand, S. Turtle-associated *Salmonella* septicaemia and meningitis in a 2-month-old baby. *J. Med. Microbiol.* **2009**, *58*, 1379–1381. [CrossRef] [PubMed]

17. Schneider, L.; Ehlinger, M.; Stanchina, C.; Giacomelli, M.-C.; Gicquel, P.; Karger, C.; Clavert, J.-M. *Salmonella enterica* subsp. *arizonae* bone and joints sepsis. A case report and literature review. *Orthop. Traumatol. Surg. Res.* **2009**, *95*, 237–242. [CrossRef] [PubMed]

18. Gyang, A.; Saunders, M. *Salmonella* Mississippi: A rare cause of second trimester miscarriage. *Arch. Gynecol. Obstet.* **2008**, *277*, 437–438. [CrossRef] [PubMed]

19. Bugla-Płoskońska, G.; Rybka, J.; Futoma-Kołoch, B.; Cisowska, A.; Gamian, A.; Doroszkiewicz, W. Sialic acid-containing lipopolysaccharides of *Salmonella* O48 strains—potential role in camouflage and susceptibility to the bactericidal effect of normal human serum. *Microb. Ecol.* **2010**, *59*, 601–613. [CrossRef] [PubMed]

20. Pees, M.; Rabsch, W.; Plenz, B.; Fruth, A.; Prager, R.; Simon, S.; Schmidt, V.; Munch, S.; Braun, P. Evidence for the transmission of *Salmonella* from reptiles to children in Germany, July 2010 to October 2011. *Euro Surveill.* **2013**, *18*. Available online: http://www.eurosurveillance.org/ViewArticle.aspx?ArticleId=20634 (accessed on 1 August 2017). [CrossRef]

21. Friedman, C.R.; Torigian, C.; Shillam, P.J.; Hoffman, R.E.; Heltzel, D.; Beebe, J.L.; Malcolm, G.; DeWitt, W.E.; Hutwagner, L.; Griffin, P.M. An outbreak of salmonellosis among children attending a reptile exhibit at a zoo. *J. Pediatr.* **1998**, *132*, 802–807. [CrossRef]

22. Geue, L.; Löschner, U. *Salmonella enterica* in reptiles of German and Austrian origin. *Vet. Microbiol.* **2002**, *84*, 79–91. [CrossRef]

23. Mermin, J.; Hoar, B.; Angulo, F.J. Iguanas and *Salmonella* marina infection in children: A reflection of the increasing incidence of reptile-associated salmonellosis in the United States Pediatrics. *Pediatrics* **1997**, *99*, 399–402. [CrossRef] [PubMed]

24. Cain, C.; Tyre, D.; Ferraro, D. Incidence of *Salmonella* on Reptiles in the Pet Trade. *Rev. Undergrad. Res Agric. Life Sci.* **2009**, *4*, 1.

25. O'Byrne, A.M.; Mahon, M. Reptile-associated salmonellosis in residents in the south East of Ireland 2005–2007. *Eurosurveillance* **2008**, *13*, 1854–1861.

26. Schröter, M.; Roggentin, P.; Hofmann, J.; Speicher, A.; Laufs, R.; Mack, D. Pet snakes as a reservoir for *Salmonella* enterica subsp. diarizonae (serogroup IIIb): A prospective study. *Appl. Environ. Microbiol.* **2004**, *70*, 613–615.

27. Warwick, C.; Lambiris, A.J.; Westwood, D.; Steedman, C. Reptile-related salmonellosis. *J. R. Soc. Med.* **2001**, *94*, 124–126. [CrossRef] [PubMed]

28. Gamian, A.; Jones, C.; Lipiński, T.; Korzeniowska-Kowal, A.; Ravenscroft, N. Structure of the sialic acid-containing O-specific polysaccharide from *Salmonella enterica* serovar Toucra O48 lipopolysaccharide. *Eur. J. Biochem. FEBs* **2000**, *267*, 3160–3167. [CrossRef]

29. Ram, S.; Sharma, A.K.; Simpson, S.D.; Gulati, S.; McQuillen, D.P.; Pangburn, M.K.; Rice, P.A. A Novel sialic Acid Binding site on Factor H Mediates serum Resistance of sialylated *Neisseria gonorrhoeae*. *J. Exp. Med.* **1998**, *187*, 743–752. [CrossRef] [PubMed]

30. Ilg, K.; Zandomeneghi, G.; Rugarabamu, G.; Meier, B.H.; Aebi, M. HR-MAs NMR reveals a pH-dependent LPS alteration by de-O-acetylation at abequose in the O-antigen of *Salmonella enterica* serovar *Typhimurium*. *Carbohydr. Res.* **2013**, *382*, 58–64. [CrossRef] [PubMed]

31. Basu, S.; Schlecht, S.; Wagner, M.; Mayer, H.L. The sialic acid-containing lipopolysaccharides of *Salmonella djakarta* and *Salmonella isaszeg* (serogroup O: 48): Chemical characterization and reactivity with a sialic acid-binding lectin from *Cepaea hortensis. FEMS Immunol. Med. Microbiol.* **1994**, *9*, 189–197. [CrossRef] [PubMed]

32. Matsuura, M. Structural Modifications of Bacterial Lipopolysaccharide that Facilitate Gram-Negative Bacteria Evasion of Host Innate Immunity. *Front. Immunol.* **2013**, *4*, 109. [CrossRef] [PubMed]

33. Munn, C.B.; Ishiguro, E.E.; Kay, W.W.; Trust, T.J. Role of surface components in serum resistance of virulent *Aeromonas salmonicida. Infect. Immun.* **1982**, *36*, 1069–1075. [PubMed]

34. Taylor, P.W. Bactericidal and bacteriolytic activity of serum against gram-negative bacteria. *Microbiol. Rev.* **1983**, *47*, 46–83. [PubMed]

35. Bravo, D.; Silva, C.; Carter, J.A.; Hoare, A.; Alvarez, S.A.; Blondel, C.J.; Zaldívar, M.; Valvano, M.A.; Contreras, I. Growth-phase regulation of lipopolysaccharide O-antigen chain length influences serum resistance in serovars of *Salmonella. J. Med. Microbiol.* **2008**, *57*, 938–946. [CrossRef] [PubMed]

36. Vimr, E.R. Unified theory of bacterial sialometabolism: How and why bacteria metabolize host sialic acids. *ISRN Microbiol.* **2013**, *2013*, 816713. [CrossRef] [PubMed]

37. Murray, G.L.; Attridge, S.R.; Morona, R. Regulation of *Salmonella typhimurium* lipopolysaccharide O antigen chain length is required for virulence; identification of FepE as a second Wzz. *Mol. Microbiol.* **2003**, *47*, 1395–1406. [CrossRef] [PubMed]

38. Murray, G.L.; Attridge, S.R.; Morona, R. Inducible serum resistance in *Salmonella typhimurium* is dependent on wzz (fepE)-regulated very long O antigen chains. *Microbes Infect.* **2005**, *7*, 1296–1304. [CrossRef] [PubMed]

39. Crawford, R.W.; Keestra, A.M.; Winter, S.E.; Xavier, M.N.; Tsolis, R.M.; Tolstikov, V.; Bäumler, A.J. Very long O-antigen chains enhance fitness during *Salmonella*-induced colitis by increasing bile resistance. *PLoS Pathog.* **2012**, *8*, e1002918. [CrossRef] [PubMed]

40. Dudek, B.; Krzyżewska, E.; Kapczyńska, K.; Rybka, J.; Pawlak, A.; Korzekwa, K.; Klausa, E.; Bugla-Płoskońska, G. Proteomic Analysis of Outer Membrane Proteins from *Salmonella* Enteritidis strains with Different sensitivity to Human serum. *PLoS ONE* **2016**, *11*, e0164069. [CrossRef] [PubMed]

41. Bugla-Płoskońska, G.; Korzeniowska-Kowal, A.; Guz-Regner, K. Reptiles as a source of *Salmonella* O48—Clinically important bacteria for children: The relationship between resistance to normal cord serum and outer membrane protein patterns. *Microb. Ecol.* **2011**, *61*, 41–51. [CrossRef] [PubMed]

42. Futoma-Koloch, B.; Bugla-Ploskonska, G.; Sarowska, J. Searching for Outer Membrane Proteins Typical of serum-sensitive and serum-Resistant Phenotypes of *Salmonella*. In *Salmonella-Distribution, Adaptation, Control Measures and Molecular Technologies*; InTech: Rijeka, Croatia, 2012; pp. 265–290.

43. Skwara, A.; Dudek, B.; Rybka, J.; Rybka, W.; Klausa, E.; Doroszkiewicz, W.; Gamian, A.; Bugla-Płoskońska, G. Zmiany w ilości kwasu sjalowego w lipopolisacharydach pałeczek *Salmonella* O48 po pasażach w surowicy ludzkiej (in polish). In Proceedings of the Zjazd PTM, Lublin, Poland, 5–8 September 2012.

44. Yi, E.C.; Hackett, M. Rapid isolation method for lipopolysaccharide and lipid A from gram-negative bacteria. *Analyst* **2000**, *125*, 651–656. [CrossRef] [PubMed]

45. Darveau, R.P.; Hancock, R.E. Procedure for isolation of bacterial lipopolysaccharides from both smooth and rough Pseudomonas aeruginosa and *Salmonella* typhimurium strains. *J. Bacteriol.* **1983**, *155*, 831–838. [PubMed]

46. Laemmli, U.K. Cleavage of structural proteins during the assembly of the head of bacteriophage T4. *Nature* **1970**, *227*, 680–685. [CrossRef] [PubMed]

47. Tsai, C.M.; Frasch, C.E. A sensitive silver stain for detecting lipopolysaccharides in polyacrylamide gels. *Anal. Biochem.* **1982**, *119*, 115–119.

48. Fomsgaard, A.; Freudenberg, M.A.; Galanos, C. Modification of the silver staining technique to detect lipopolysaccharide in polyacrylamide gels. *J. Clin. Microbiol.* **1990**, *28*, 2627–2631. [PubMed]

49. Rybka, J.; Gamian, A. Determination of endotoxin by the measurement of the acetylated methyl glycoside derivative of Kdo with gas–liquid chromatography-mass spectrometry. *J. Microbiol. Methods* **2006**, *64*, 171–184. [CrossRef] [PubMed]

 International Journal of
Molecular Sciences

Article

Bordetella holmesii: Lipid A Structures and Corresponding Genomic Sequences Comparison in Three Clinical Isolates and the Reference Strain ATCC 51541

Valérie Bouchez [1], Sami AlBitar-Nehmé [2,†], Alexey Novikov [3], Nicole Guiso [1] and Martine Caroff [2,3,*]

[1] Institut Pasteur, Unité de Prévention et Thérapies Moléculaires des Maladies Humaines, 25 rue du Dr Roux, 75724 Paris, France; valerie.bouchez@pasteur.fr (V.B.); nicole.guiso@pasteur.fr (N.G.)

[2] Institute for integrative Biology of the Cell (I2BC), Commissariat à l'Energie Atomique (CEA), Centre National de la Recherche Scientifique (CNRS), Université Paris-Sud, Université Paris-Saclay, 91405 Orsay, France; sami.nehme@uniroma1.it

[3] LPS-BioSciences, I2BC, Bâtiment 409, Université de Paris-Sud, 91405 Orsay, France; alexey.novikov@lpsbiosciences.com

* Correspondence: martine.caroff@u-psud.fr; Tel.: +33-1-6915-7191

† Present address: Dipartimento di Biologia e Biotecnologie "C. Darwin", Sapienza-Università di Roma, Piazzale Aldo Moro 5, 00185 Roma, Italy.

Academic Editor: Juan M. Tomás
Received: 10 March 2017; Accepted: 11 May 2017; Published: 18 May 2017

Abstract: *Bordetella holmesii* can cause invasive infections but can also be isolated from the respiratory tract of patients with whooping-cough like symptoms. For the first time, we describe the lipid A structure of *B. holmesii* reference strain ATCC 51541 (alias NCTC12912 or CIP104394) and those of three French *B. holmesii* clinical isolates originating from blood (Bho1) or from respiratory samples (FR4020 and FR4101). They were investigated using chemical analyses, gas chromatography–mass spectrometry (GC–MS), and matrix-assisted laser desorption ionization–mass spectrometry (MALDI–MS). The analyses revealed a common bisphosphorylated β-(1→6)-linked D-glucosamine disaccharide with hydroxytetradecanoic acid in amide linkages. Similar to *B. avium*, *B. hinzii* and *B. trematum* lipids A, the hydroxytetradecanoic acid at the C-2′ position are carrying in secondary linkage a 2-hydroxytetradecanoic acid residue resulting of post-traductional biosynthesis modifications. The three clinical isolates displayed characteristic structural traits compared to the ATCC 51541 reference strain: the lipid A phosphate groups are more or less modified with glucosamine in the isolates and reference strain, but the presence of 10:0(3-OH) is only observed in the isolates. This trait was only described in *B. pertussis* and *B. parapertussis* strains, as well as in *B. petrii* isolates by the past. The genetic bases for most of the key structural elements of lipid A were analyzed and supported the structural data.

Keywords: Bordetellae; *Bordetella holmesii*; endotoxin; lipid A; structure; mass spectrometry; genomic

---

1. Introduction

The *Bordetella* genus contains at the moment a dozen species, of which at least five are responsible for respiratory diseases in humans and/or animals. Classical Bordetellae consist of *Bordetella pertussis*, a strict human pathogen responsible for whooping cough; *B. parapertussis* responsible for mild whooping-cough symptoms in humans, also described as a sheep pathogen; and *B. bronchiseptica*, able to infect a broad range of hosts. The Bordetellae virulence factors include toxins—such as pertussis toxin for *B. pertussis* only, adenylate cyclase-hemolysin, and lipopolysaccharide (LPS)—and adhesins—such as filamentous hemagglutinin, fimbriae and pertactin—, all involved in the binding to ciliated epithelial cells in the host upper respiratory tract. *B. bronchiseptica* and *B. pertussis* endotoxins LPS have been shown to be implicated in virulence [1–4]; therefore, it is important to compare the structures of *Bordetella* LPSs purified from other *Bordetella* pathogenic species and, particularly, from the recent *B. pertussis* relative *B. holmesii* [5].

B. holmesii was first described in 1995 following its isolation from the blood of a patient with septicemia [6]. At that time, this bacterium was only originating from invasive infections in immunocompromised patients. In the past years, increasing reports of the presence of *B. holmesii* in the respiratory tract of patients with pertussis-like symptoms have been published [7–13]. However, it is not known whether this bacterium is an opportunistic or a pathogenic one, able to induce pertussis-like symptoms in humans [14–16].

For the moment, it is not possible to differentiate *B. holmesii* isolates recovered from blood from isolates recovered from respiratory samples [17–19]. About 21 genomes of *B. holmesii* are available on The National Center for Biotechnology Information (NCBI) [20]. First considered as close to *B. pertussis* on the basis of 16S DNA analysis, *B. holmesii* is now described in the same clade as *B. hinzii*, *B. avium* and *B. trematum* on the basis of whole genome single nucleotide polymorphism (SNP)-based analysis [21]. Most virulence factors usually produced by the classical *Bordetella* seem to be missing in *B. holmesii* except a *Bordetella* master virulence regulatory system (*bvg*), a filamentous hemagglutinin (FHA)-like protein, a bvg-intermediate phaseA protein (*bipA*) ortholog, and an alcaligin operon [19,22–25]. *B. holmesii* LPS have only been roughly studied by Van den Akker in 1998 who found them phenotypically and immunologically distinct from those of *B. pertussis* [26].

We report here the detailed lipid A structures of three *B. holmesii* isolates, as compared to those of the *B. holmesii* reference strain ATCC 51541.

2. Results

2.1. Fatty Acids Composition

Total fatty acid analyses performed by gas chromatography—mass spectrometry (GC–MS) revealed the presence of 3-hydroxytetradecanoic acid [14:0(3-OH)], 2-hydroxytetradecanoic acid 14:0(2-OH), 2-hydroxydodecanoic acid 12:0(2-OH), and 3-hydroxydecanoic acid 10:0(3-OH) as well as traces of tetradecanoic acid 14:0 and dodecanoic acid 12:0 in lipids A extracted from all tested strains and isolates. They were found to be present in the relative corresponding proportions: 2.8:1:1:0.5 for ATCC51541, Bh01, and FR 4020 differing from the FR 4101 isolate having the following proportions of 2:1:1:1.2.

2.2. Matrix-Assisted Laser Desorption Ionization–Mass Spectrometry Structural Analyses

2.2.1. Interpretation of the Main Molecular Species in the Different Lipid a Spectra

The negative-ion spectrum of the di-phosphoryl ATCC 51541 reference strain lipid A was heterogeneous, containing two main molecular ion signals at m/z 1376.9 and 1603.7 as illustrated in Figure 1A. Composition of the corresponding molecular species were attributed on the basis of the overall chemical composition: m/z 1603.7 would correspond to two glucosamine (GlcN), two phosphates, three 14:0(3-OH), one 14:0(2-OH), and one 12:0(2-OH); and m/z 1376.9 corresponds to m/z 1603.7 minus one 14:0(3-OH). A molecular species corresponding to m/z 1404.9 can be explained by some microheterogeneity at the level of the 12:0(2-OH) fatty acid versus the 14:0(2-OH). The same difference was observed between molecular species at m/z 1575.4 and 1603.7. Each of the molecular species presented a twin species at -16, expressing the previously described heterogeneity and peculiarity of the *Bordetella* genus, the reduced enzymes specificity at different positions carrying, in some species, 2-hydroxylated fatty acids in secondary linkage [1,27–29]. The latter being known as a late structural modification of the structure occurring in the membrane and leading to increased robustness of the bacterial outer membrane barrier [30].

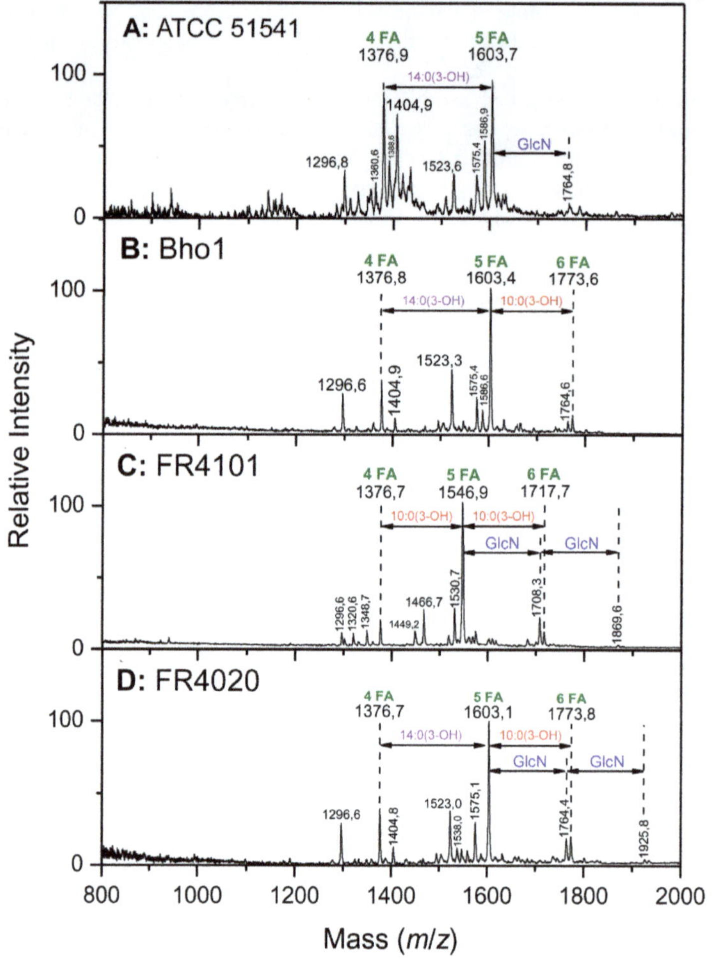

Figure 1. Negative-ion spectrum of lipid A from *Bordetella holmesii* strains and isolates: (**A**) ATCC 51541 reference strain; (**B**) Bho1 isolate; (**C**) FR4101 isolate; and (**D**) FR4020 isolate. FA: Fatty acid; GlcN: Glucosamine.

2.2.2. Distribution of the Fatty Acids on the D-Glucosamine Residues of the Clinical Isolates

The negative-ion mode matrix-assisted laser desorption ionization (MALDI) mass spectra of lipids A isolated from the three *B. holmesii* isolates displayed three major peaks at *m/z* 1603, 1546 and 1376 (Figure 1B–D). The first and third peaks were identical to those found in ATCC 51541 lipid A and corresponded, as explained for the reference strain, to penta- and tetra-acylated lipid A molecular species, respectively. For isolate FR4101, the ion at *m/z* 1546, absent in the ATCC strain spectrum, corresponded to the molecular species at *m/z* 1376 carrying an additional 10:0(3-OH) (170 units).

The peaks mentioned above were surrounded by smaller ones at −16 u, such as those at *m/z* 1587, 1360 and 1531. These peaks in which the 16-u difference represented hydroxyl modifications, were attributed to the same structures whose secondary fatty acids linked at C-2 and C-2′, 12:0(2-OH) or 14:0(2-OH) were non-hydroxylated (14:0 or 12:0, respectively). This modification was not observed in the FR4020 lipid A mass spectrum and might be a specific trait of this isolate, with biosynthetic relevance. In addition, another type of micro-heterogeneity, regarding the length of fatty acids, was observed in the first major peaks (at *m/z* 1603 in Figure 1A,B,D, at *m/z* 1376 in Figure 1A–D and at *m/z* 1546 in Figure 1C) that were also doubled by minor ones at −28 u or +28 u, i.e., at *m/z* 1575, 1349 and 1405. The mass difference of 28 u was attributed to 2 × CH2, and related peaks were not found systematically in the three *B. holmesii* lipid A mass spectra. Of note, in the Bho1 lipid A mass spectrum, this type of heterogeneity was only observed at plus 28 u from the major peak, at *m/z* 1376. In the high-mass molecular ions region, two minor peaks were observed at *m/z* 1773 and 1764 in Bho1 (with a really minor contribution for this isolate) and FR4020 lipid A mass spectra and, at *m/z* 1717 and 1708 in FR4101 mass spectrum. The peak at *m/z* 1773, as well as the one at *m/z* 1717, corresponded to 10:0 (3-OH) (170 u), resulting in a hexa-acyl lipid A molecular species in Bho1 and FR4020, and FR4101, respectively. Interestingly, the mass difference between peaks at *m/z* 1764 and 1708 and penta-acyl lipid A molecular species at *m/z* 1603 and 1546.9, corresponded to 161 units for GlcN. Moreover, other two small peaks (*m/z* 1925 and 1869) in FR4020 and FR4101 lipid A mass spectra gave the same mass difference with *m/z* 1764 and 1708, suggesting that these lipid A molecular species were carrying 2 GlcN, one at each phosphate groups. This modification is described here for the fifth time in lipid A species of the *Bordetella* genus [27,31–34]. Analogous to the FR4020 mass spectrum with the exception of peaks related to GlcN addition at *m/z* 1538, 1869 and 1925 (Figure 2) the mass spectrum for lipid A from Bho1 displayed the same peaks as FR4020 mass spectrum. This indicates that both isolates might have similar lipid A structures with the exception that Bho1 lipid A phosphate groups displayed a lower amount of GlcN. The reference ATCC strain also showed a peak corresponding to a very small amount of GlcN, but its presence was confirmed by statistical experiments allowing to obtain persistent and slightly increased signals at this level.

2.2.3. Distribution of the Fatty Acids on the D-Glucosamine Residues of the Isolates and the Reference Strain

The reference strain ATCC 51541 lipid A positive-ion MALDI mass spectrum presented in Figure 3 gave peaks in the lower field at *m/z* 920.6 and 904.6. The peak at *m/z* 920.6 was interpreted to correspond to an ion containing one GlcN, one phosphate, three hydroxytetradecanoic acid residues—two 14:0(3-OH) and one 14:0(2-OH)—, the second at *m/z* 904.6. From this fragmentation pattern, we concluded that these signals corresponded to the GlcN II (non-reducing) part of the lipids A and that the 14:0(2-OH) acid was linked at C-2′ in secondary acylation. The presence of the two peaks differing by 16 u illustrates the incomplete hydroxylation at position 2 of the 14:0 branched fatty acid.

The positive-ion MALDI mass spectra of Bho1 and FR4020 isolates also displayed a prominent fragment peak at *m/z* 920.6 whereas the major peak was at *m/z* 864.4 in the FR4101 mass spectrum. The first peak is attributed to a lipid A fragment that contained, according to the fragmentation pattern [35], the distal glucosamine (GlcN II) + one phosphate + three 14:0(OH): (two 14:0(3-OH) + one 14:0(2-OH) in secondary position). This confirms that 14:0(2-OH) was the acyloxyacyl linked at C-2′ and consequently 12:0(2-OH) is the acyloxyacyl linked at C-2 in the three isolates. Furthermore, it showed that 14:0(3-OH) is linked at C-3′ in Bho1 and FR4020. Concerning FR4101, the second peak is representing a lipid A fragment composed of GlcN II + one phosphate + one 10:0(3-OH) and two 14:0(OH) (one 14:0(3-OH) plus one 14:0(2-OH) in secondary position). This indicated that 10:0(3-OH) was ester-linked at C-3′ in GlcN II of FR4101 lipid A and confirmed that 14:0(2-OH) and 12:0(2-OH) acyl-oxy-acyls were linked at C-2′ and C-2, respectively. A small peak at *m/z* 694 was observed in all isolates spectra and was attributed to the lipid A fragment at 920 minus 14:0(3-OH) in Bho1, and FR4020, and minus 10:0(3-OH) in Bho1 (not shown).

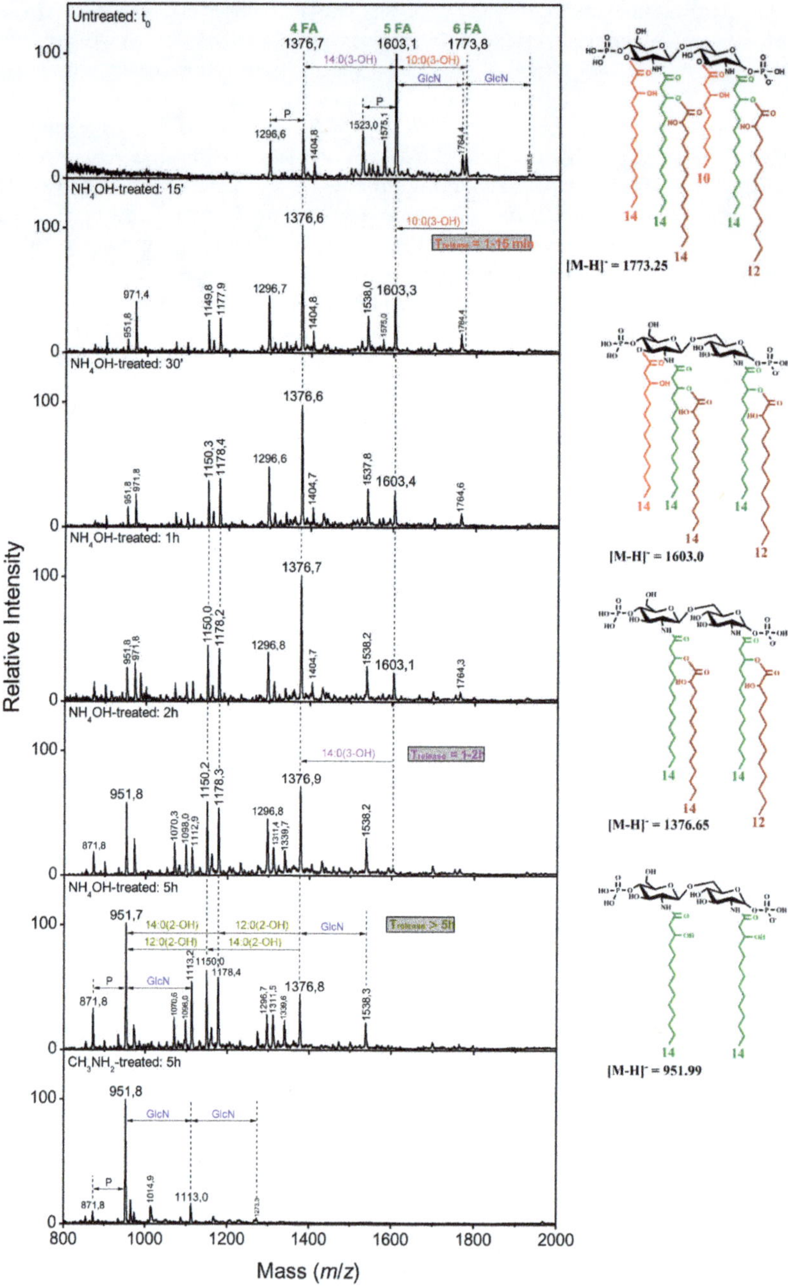

Figure 2. Kinetics of ester-linked fatty acids release in FR4020 in alkaline conditions and corresponding structures. Bho1 isolate and ATCC51541 behaved similarly and independently of the presence of GlcN, accordingly their kinetics are not presented here. Structures are displayed here without their GlcN substitution in order to better focus on the fatty acid pattern. However, the complete structures are presented in Figure 5.

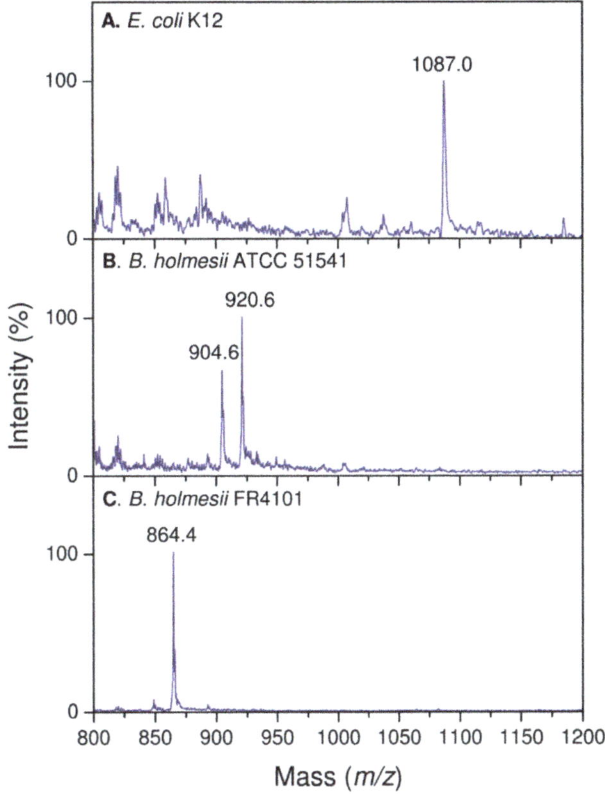

Figure 3. Comparison of positive-ion lipid A fragments obtained for: (**A**) *Escherichia coli* K12 taken as a reference, (**B**) *B. holmesii* ATCC 51541, and (**C**) *B. holmesii* FR4101.

2.2.4. Linkage of Fatty Acids in *B. holmesii* Lipids A

It was previously shown that sequential liberation of FAs was efficient to determine their position in the lipid A structure [32]. As shown in Table 1, the FA in ester linkages at the C-3 position are the first ones to be liberated, followed by those at C-3′, and the acyloxy-acyl secondary ester-linked FAs come as the last ones, this being due to steric hindrance conditions. Mass spectra at different NH_4OH treatment with length times of 15 min, 30 min, 1 h, 2 h and 5 h were recorded and compared to the initial lipid A mass spectra at the starting point, t_0 (Figures 2 and 4). The partial *O*-deacylation pattern, obtained by 28% NH_4OH treatment, was different in *B. holmesii* isolates; it was related to their linkage position. On the one hand, the time release of 10:0(3-OH) at C-3 ranged from 1 to 15 min in FR4020 and FR4101 lipids A and was illustrated by the disappearance of the hexa-acyl lipid A related peaks (at *m/z* 1773.8 and 1717.7 respectively) which were transformed to penta-acyl lipid A molecular species at *m/z* 1603.1 and 1546.9, respectively. On the other hand, the 10:0(3-OH) liberation in Bho1 took a longer time, 1–30 min (Figure 2). Then, we observed by following kinetics of sequential liberation: at the C3 position, 14:0(3-OH) was completely liberated after 1–2 h in all three clinical isolates (Figures 2 and 4). This led to convert penta-acyl lipids A at *m/z* 1603.1 and 1546.9 to tetra-acyl ones at *m/z* 1376.6. In fact, two peaks at *m/z* 1150.2 and 1177.9 appeared after 15 min in all three clinical isolates, corresponding to *m/z* 1376.6 −226 u and 198 u, respectively. These mass differences were attributed to 14:0(2-OH) and 12:0(2-OH) secondary fatty acids at C-2′and C-2, respectively. Another important peak appeared at *m/z* 952 after 1 h in Bho1, 30 min in FR4101 and 15 min in FR4020. The structure related to this peak corresponded to a diacyl lipid A with two amide-linked acyl chains [14:0(3-OH)] at C-2 and C-2′on the lipid A core made of two GlcN and two phosphate residues.

Table 1. Kinetics of ester fatty acids release from all isolates and ATCC strain. (**A**) Kinetics of ester fatty acids release from lipids A from *B. holmesii* strains ATCC 51541, Bho1 and FR4020 isolates. Round numbers are given for reflecting that three experiments were summarized in this table. (**B**) Kinetics of ester fatty acids release from *B. holmesii* FR4101 lipid A.

Treatment and Time Point	[M-H]⁻ Ions *m/z*			Fatty Acid Released	
				Type	Position
A					
t_0	1376	1603	1773		
NH₄OH, 15 min	1376	1603	1603	10:0(3-OH)	C3
NH₄OH, 2 h	1376	1376	1376	14:0(3-OH)	C3′
NH₄OH, 5 h	1178	1178	1178	12:0(2-OH)	secondary C2 partial release
	1150	1150	1150	14:0(2-OH)	secondary C2′ partial release
CH₃NH₂, 5 h	952	952	952	14:0(2-OH)	secondary C2′
				12:0(2-OH)	secondary C2
B					
t_0	1376.6	1547	1717		
NH₄OH, 15 min	1376.6	1547	1547	10:0(3-OH)	C3
NH₄OH, 2 h	1376.6	1376.6	1376.6	10:0(3-OH)	C3′
NH₄OH, 5 h	1178	1178	1178	12:0(2-OH)	secondary C2 partial release
	1150	1150	1150	14:0(2-OH)	secondary C2′ partial release
CH₃NH₂, 5 h	952	952	952	14:0(2-OH)	secondary C2′
				12:0(2-OH)	secondary C2

t_0: Starting time point.

Of particular note, the ion peak at *m/z* 1538 arose from addition of GlcN (161 u) to the ion at *m/z* 1376.6, lasting from 15 min to 5 h in FR4020 and from 15 min to 2 h in the FR4101 lipid A mass spectrum. The peak associated to the glucosamine modification was weak in the Bho1 mass spectrum, but confirmed with repeated experiments. The complete *O*-deacylation was carried out by releasing secondary ester-linked fatty acids with methylamine treatment (41% CH₃NH₂ at 37 °C). The negative-ion MALDI mass spectrum from Bho1 [36] displayed two peaks, after 5 h, a major one at *m/z* 952 (diacyl lipid A), accompanied by a minor one at *m/z* 872 (a mono-phosphorylated lipid A). No peak at plus GlcN was observed because of the very small amount of this molecular species present in the native molecule. In addition to these two peaks, further ones were observed in FR4101 and FR4020 mass spectra (Figures 2 and 4), such as *m/z* 1113, which corresponded to 953 plus 161 units (GlcN) in both isolates, and a very minor not indicated peak at *m/z* 1274 (1113 plus 161 units), thereby demonstrating that both lipid A phosphate groups were carrying GlcN.

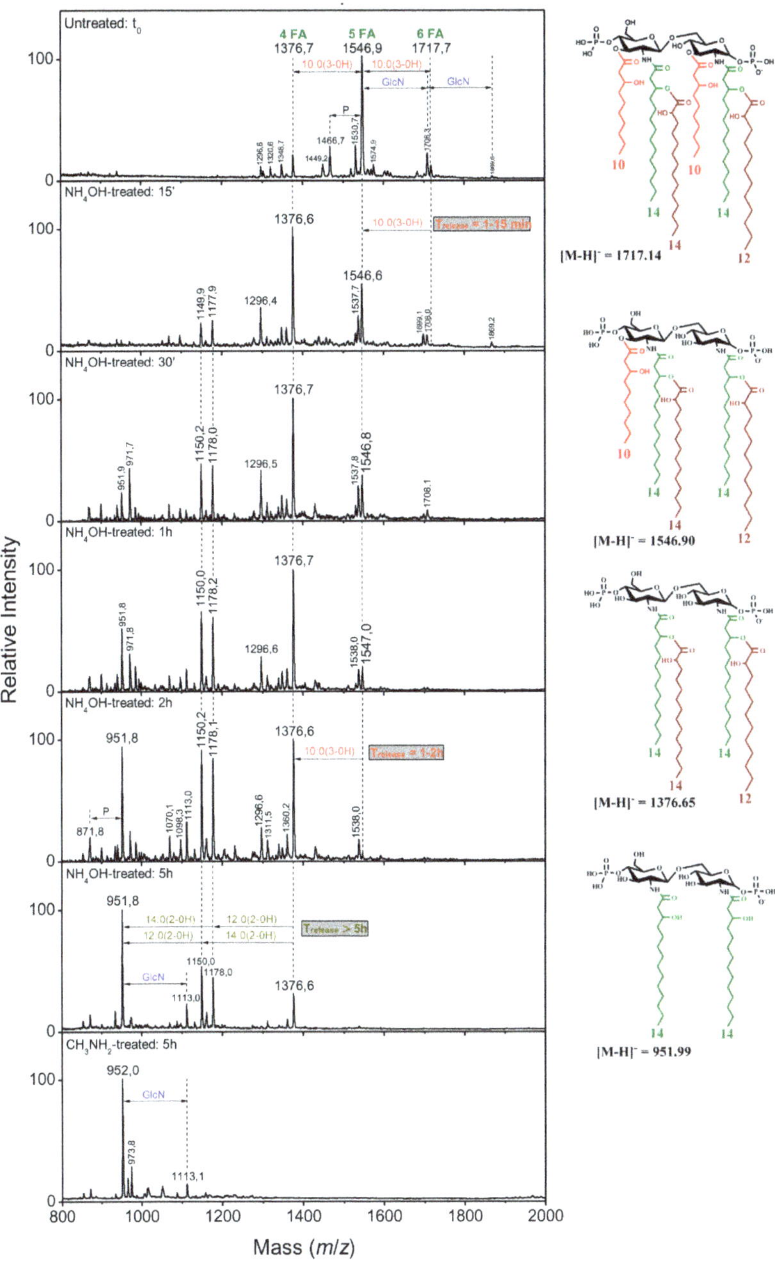

Figure 4. Kinetics of ester-linked fatty acid sequential release in FR4101 lipid A.

2.3. Genomic Analyses of Genes Involved in Lipid A Biosynthesis

We analyzed the sequence of genes involved in lipid A biosynthesis, more particularly focusing on the identification of the genomic basis for the two main differences observed in the lipid A structures of the three *B. holmesii* isolates in the present study, i.e., the presence/absence of glucosamines and the length of the carbon chain FA at C-3′ [14:0(3-OH) or 10:0(3-OH)]. We thus explored the Lipid A GlcN modification locus (*lgm* locus), the Acyl-[acyl-carrier-protein]–UDP-N-acetylglucosamine O-acyltransferase (*lpxA*), the Lipid A deacylase (*pagL*), the LipidA palmitoyltransferase (*pagP*), two oxygenases *lpxO1* and *lpxO2* and acyltransferase *lpxL1* genes (Table 2):

(i)   An orthologous sequence corresponding to the *lgm* locus was identified in the genome of the three clinical isolates as in the ATCC 51541 reference strain (position 1966909..1970398). We observe no sequence difference for this locus comprising *ArnT* (position 1967952..1969541) between ATCC type strain and the three clinical isolates.

(ii) Bho1 and FR4020 *lpxA* sequence is identical to that of ATCC 51541 reference strain (position 3231726..3232520) but FR4101 *lpxA* sequence displays a non-synonymous SNP in position 508 leading to an amino acid modification (S instead of G in position 170 of LpxA protein). We also sequenced *lpxA* gene from 16 additional *B. holmesii* isolates collected from blood or from respiratory samples and found that this SNP was not linked to the origin of the isolates (unpublished data).

(iii) *pagL* sequence is identical for the three clinical isolates and the ATCC 51541 reference strain (position 2144078..2144623). We did not observe any sequence difference between the three tested isolates for *pagP* but a difference was observed in ATCC 51541 reference strain *pagL* sequence (position 630144..630686) corresponding to an additional G leading to a frameshift.

(iv) We found two homologs of *lpxO* in the genome of the three isolates of *B. holmesii* as in the ATCC 51541 reference strain (position 2899145..2900041 for the first one and 2402394..2403293 for the second). They respectively displayed 84% nucleotidic identity with the KF214918 *lpxO1* of *B. avium* ATCC 35086 and 84% nucleotidic identity with the KF214919 *LpxO2* of *B. avium* ATCC 35086. We then identified a *lpxL1* homolog in the three isolates as in ATCC 51541 reference strain (position 1194726..1195580) with no sequence difference between them.

Table 2. Genomic basis for the structural differences observed within lipid A of Bho1, FR4020, FR4101 and the ATCC51541 reference strain. Genes of interest/position within ATCC51541 reference strain (Accession Number: CP007494.1)/sequence differences observed between isolates.

Genes of Interest	Position within ATCC 51541 Reference Strain (Accession Number: CP007494.1)	Sequences Differences Observed Between Isolates
Lgm locus	1966909..1970398	None
ArnT	1967952..1969541	None
lpxA	3231726..3232520	No sequence differences between Bho1, FR4020 and ATCC 51541
		Non-synonymous SNP in position 508 for FR4101 (*)
pagL	2144078..2144623	None
pagP	630144..630686	No sequence differences between Bho1, FR4020 and FR4101
		Additional G compared to ATCC51541 leading to a frameshift
lpxO1 (**)	2899145..2900041	None
lpxO2 (***)	2402394..2403293	None
lpxL1	1194726..1195580	None

(*) We also sequenced *lpxA* gene from 16 additional *B. holmesii* isolates collected from blood or from respiratory samples and found that this SNP was not linked to the origin of the isolates (unpublished data); (**) 84% Nucleotidic identity with the KF214918 *lpxO1* of *B. avium* ATCC 35086; (***) 84% Nucleotidic identity with the KF214919 *LpxO2* of *B. avium* ATCC 35086.

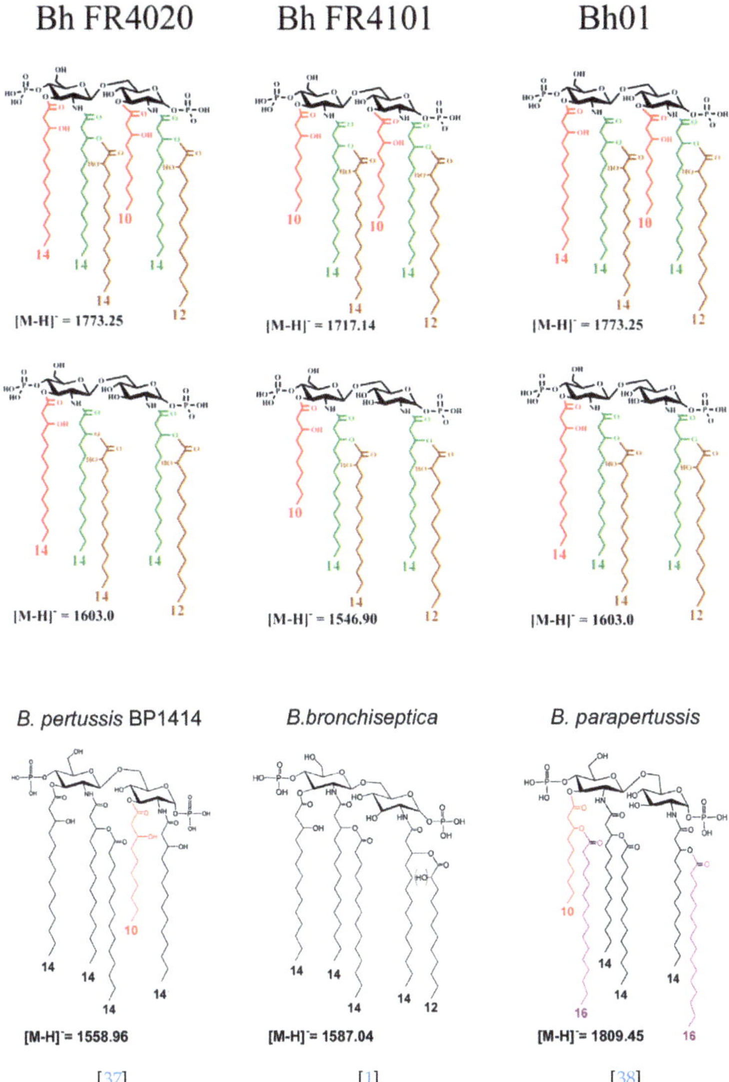

Figure 5. Proposed structures of the three main molecular species of higher masses present in *B. holmesii* lipid A isolates compared to those present in *B. pertussis* 1414, *B. bronchiseptica* and *B. parapertussis*.

3. Discussion

In the present study, we focused our analyses on the lipid A of different *B. holmesii* isolates originating either from blood or respiratory samples and on the ATCC 51541 reference strain.

Mass spectrometry analyses first led to the observation of a difference in the amount of modification with GlcN on the lipid A phosphate groups. The two clinical isolates FR4020 and FR4101 displayed a higher degree of modification compared to Bho1 and the ATCC strain. Such an hexosamine modification has been described in other Bordetellae species such as *B. pertussis*, *B. bronchiseptica* and *B. avium* and was attributed to a *lgm* locus encoding a glycosyl transferase *ArnT* [31,35,39]. An orthologous sequence corresponding to *lgm* locus was identified in the genome of the three clinical isolates as in the ATCC 51541 reference-strain genome with no sequence difference between them. In *B. pertussis* and *B. bronchiseptica*, the *lgm* locus has been reported to be *bvg* regulated [33]. Previous studies showed that Bho1 isolate and ATCC 51541 reference strain both have a particular non-functional *bvg* system because of a non-functional BvgA protein (due to an A insertion within *bvgA* sequence, leading to a frameshift [19,22]) that could explain why less GlcN is substituting their lipid A phosphate groups on contrary to FR4020 and FR4101. Recent sequencing and annotation of ATCC 51541 strain genome do not report this additional A within BvgA (position 3314780..3315403) suggesting that the mutation can occur frequently at random, according to subcultures or different ways of storage. Moreover, this difference is not a specific trait of respiratory samples as a functional *bvg* system is found in other *B. holmesii* isolates collected from blood [19]. All these data show that *B. holmesii* lipid A can be decorated with GlcN what has already been shown to have consequences on the modulation of host immunity via a different Toll-like receptor 4 (TLR4) activation for *B. pertussis*.

In addition, other specificities and differences deduced from the structural analysis seem to be supported by the genomic analysis:

(i) The differences in length of the carbon FA at C3', 14:0(3-OH) for Bho1, FR4020 and ATCC 51541 or 10:0(3-OH) for FR4101. LpxA is the first enzyme of the lipid A biosynthesis pathway catalyzing the addition of an acyl chain onto the C3' carbon [40]. We identified *lpxA B. holmesii* gene and observed a non-synonymous SNP in FR4101, leading to an amino acid change, as compared to ATCC 51541, FR4020 and Bho1 *lpxA* sequences. Thus, we concluded that the difference observed in the length of the carbon FA at C3' between FR4101 and the other isolates could be due to this difference within *lpxA* gene. Analysis of additional French *B. holmesii* isolates also led to the conclusion that this was not related to the blood or respiratory origin of the sample which was also confirmed by the *lpxA* sequence analysis of isolates with available genomes on NCBI.

(ii) The presence of 10:0(3-OH) at C-3, like in *B. pertussis* [1] and *B. petrii* [41] is also a common trait between the three clinical isolates. In *B. pertussis*, this is the consequence of the lack of activity of the C-3 de-*O*-acylase PagL, which has been shown to result in a lower cytokine induction capacity of the *Bordetella* human pathogens [29,42]. The presence of this fatty acid was interpreted as resulting in the facility of such bacteria, with short-chain fatty acid structures, and low acylation pattern, to escape the human host defense through the MD2–TLR4 complex [42,43]. We found that the *pagL* gene sequence is the same for the three isolates and the ATCC 51541 strain. Both H and S residues in the C-terminal part of PagL protein, described as essential for catalytic activity are present in *B. holmesii* [44]. In *B. pertussis*, *pagL* is a pseudogene because of a frameshift due to a deletion of "CA" bases. Such a deletion is not found in the *B. holmesii* sequence. Further investigations are necessary to understand why PagL is inactive in these isolates.

(iii) The presence of a secondary palmitoyl chain at two positions of lipid A (3' and 2). We previously described the presence of such palmitoyl chains at C-3' in *B. avium*, and in both C3' and C2 in *B. parapertussis* [27,45] as a result of PagP action. In the present study, we observed no *pagP* sequence difference between the three tested isolates. They nevertheless display an additional G insertion as compared to ATCC 51541 reference strain, leading to a frameshift that probably explains why no palmitoyl acid is observed in these isolates.

(iv) The hydroxylation of 12:0 and 14:0 at position 2 of the two secondary acylated residues was observed in the lipid A of the three isolates (Figure 5). In Novikov et al. [27], such hydroxylations have been observed in *B. avium*, *B. hinzii* and *B. trematum* and were attributed to two homologs of the LpxO enzyme. We found two homologs of *lpxO* in the genome of the three isolates of *B. holmesii* as in the ATCC 51541 type strain that support the degree of hydroxylation observed.

(v) The presence of a 12:0(2-OH) residue in a secondary position at C-2 on the amide-linked 14:0(3-OH) is a common trait between all the *B. holmesii* structures. LpxL is a lipid A lauroyl acyltransferase. Geurtsen et al. [46] showed that LpxL1, a homolog of the classical LpxL2 lauroyl acytransferase, is present but usually poorly expressed in *B. pertussis*, compared to LpxL2 [40,46]. However, through overexpressing it in *B. pertussis* B213 strain, they showed that this enzyme led to the presence of an extra secondary 12:0(2-OH) chain at the C-2 position, like we describe here in the *B. holmesii* isolates, and they reported that this type of acylation was required for efficient infection in human macrophages and could help host infection. We found a *lpxL1* homolog in the three isolates as in ATCC 51541 type strain supporting this substitution.

All these results allowed identification of differences within lipid A structures in the different *B. holmesii* strain and isolates tested supported by both structural and genomic analysis. As illustrated in Figure 5, lipid A structures differ in their fatty acid substitution. The difference in the amount of free amino GlcN derivatives on the phosphate groups is described herein but not shown in the figure as only the peak intensity can be illustrating such differences.

4. Materials and Methods

4.1. Strains and Isolates

Strain ATCC 51541 was originating from the National Research Council (NRC) collection and was grown as previously described [47]. The three human isolates were selected from the French National Reference Center for Whooping Cough and other Bordetelloses [19]. Bho1 was isolated in 1996 from the blood of a 20 years old man with sickle-cell anemia [48]. FR4020 and FR4101 were isolated, respectively; from nasopharyngeal swabs of an adult in 2008, and of an adolescent in 2009, both presenting Pertussis-like symptoms [19,49]. Bacteria were grown on Bordet–Gengou agar (Difco by Becton Dickinson, NJ, USA) supplemented with 15% sheep defibrinated blood (Biomerieux, Marcy l'Etoile, France) at 36 °C for 72 h, plated again for 18 h and then grown in enriched Stainer Scholte medium [50].

4.2. LPS Preparation

LPS were extracted by the isobutyric-M ammonium hydroxide method in a 5:3 (*v/v*) ratio [51]. LPS preparations were further extracted with solvents $CHCl_3$:MeOH (1:2 *v/v*) and $CHCl_3$: MeOH: H_2O (3:2:0.25, *v/v/v*) in order to remove phospholipids and lipopeptides, then treated with enzymes to remove DNA, RNA and protein contaminants, as described [32].

4.3. Lipid A Isolation from Whole Cells

Briefly, cells (10 mg) were washed twice with 400 µL of a fresh, single phase mixture of $CHCl_3$:MeOH (1:2 *v/v*) (Sigma, St. Louis, MO, USA) and once with 400 µL of $CHCl_3$:MeOH: H_2O (3:2:0.25 *v/v/v*). The insoluble material, corresponding to washed cells, was recovered by centrifugation in the pellet, and the supernatants were discarded. The washed cells were suspended in 400 µL of isobutyric acid/ammonium hydroxide (M) (5:3 *v/v*), and kept for 1.5 h at 100 °C in a screw cap test tube under magnetic stirring. The mixture was cooled in iced water, and centrifuged (2000× *g* for 15 min). The supernatant was diluted with water (1:3 *v/v*), and lyophilized. The sample was then washed twice with 400 µL of methanol, and centrifuged (2000× *g* for 15 min). Finally, the insoluble lipid A was solubilized and extracted once with 200 µL of a mixture of chloroform: methanol:water (3:1.5:0.25 *v/v/v*) [52].

4.4. Lipid A Isolation from LPS as Performed by the Triethylamine Citrate (TEA) Method

A concentration of 0.01M TEA-citrate (1:1 molar ratio, pH 3.6) was used. Samples were suspended in Eppendorf® tubes (Eppendorf AG, Hamburg, Germany) in the above-mentioned reagents at concentrations of 5 µg/µL, or 10 µg/µL depending to their solubility. After agitation and homogenization using an ultrasonic bath, the tubes were incubated for 1 h in a Thermomixer system (Eppendorf AG) under stirring at 1000 rpm and 100 °C [53].

4.5. Thin-Layer Chromatography

Thin-layer chromatography (TLC) was done on aluminum-backed silica-gel plates (Merck, Darmstadt KGaA, 64271, Germany) in the solvent [54] chloroform:methanol:water:triethylamine (12:6:1:0.04). Spots were visualized by charring after spraying with 10% sulfuric acid in ethanol for checking samples purity.

4.6. Fatty Acid

Fatty acids were analyzed after hydrolysis of the LPS or lipids A with 4 M HCl for 2 h at 100 °C, and neutralization, followed by treatment with 2 M NaOH 2 h at 100 °C [55], extraction with ethyl acetate, methylation of the extract with a mixture of anhydrous methanol and acetyl chloride (10:1.5 *v/v*) [56]. They were identified by GC–MS for the fine characterization and confirmation of FA composition as visualized by MALDI–MS (mass spectrometry).

4.7. O-Deacylation

Ester-linked fatty acids were likewise characterized after treating the lipids A with NH_4OH during 5 h at 50 °C [32] for liberation of primary ester-linked fatty acids and with methylamine for 5 h at 50 °C for the secondary ester-linked fatty acids. GC–MS analysis of the methylated fatty acids was performed as previously described. Arachidonic acid (C_{20}, Sigma), ,was used as a standard [56].

4.8. Mass Spectrometry

MALDI negative-ion mass spectrometry analyses were performed on a PerSeptive Voyager-DE STR model time-of-flight mass spectrometer of Applied Biosystem, (I2BC, Université de Paris Sud XI). The apparatus is equipped with a 337 nm nitrogen laser and spectra were recorded in the linear negative-ion mode with delayed extraction.

The ion-accelerating voltage was set at 20 kV. Dihydroxybenzoic acid (DHB) (Sigma) suspended in 0.1 M citric acid was used as a matrix. A few microliters of lipid A were dissolved in a mixture of chloroform:methanol:water (3:1.5:0.25 *v/v/v*) at 1 mg/mL and desalted with a few grains of ion-exchange resin (Dowex 50W-X8, Sigma, St. Louis, MO, USA) (H^+), in an Eppendorf tube. A 1 µL aliquot of the solution (50 µL) was deposited on the target and covered with the same amount of the matrix suspended at 10 mg/mL in the same mixture of solvents [39]. Different ratios between the samples and DHB were tested when necessary. *B. pertussis* lipid A was used as an external standard.

4.9. Genomic Analysis

Partial genome sequences of Bho1, FR4020 and FR4101 isolates were available from Roche 454 sequencing [19]. Sequences of genes involved in lipid A biosynthesis were identified using the NCBI nucleotide and protein blast programs (blastn and blastp) (https://blast.ncbi.nlm.nih.gov). For *lgm* locus and *lpxA* gene, sequence differences were checked by polymerase chain reaction (PCR) and classical Sanger sequencing using primers Lgm-bho-F: 5′-CACATGAGCGACGAGCTCTA-3′, Lgm-bho-R: 5′-GGCTGCACTTGAACCTGTCT-3′, LpxA-bho-F: 5′-CGCACCATCTGCAAGTATCA-3′, LpxA-bho-R: 5′-CCACCATACCAATGGACAGA-3′. PCR amplifications were done from genomic DNA extracted with DNeasy Blood and Tissue kit (Qiagen, Hilden, Germany) according to the manufacturer's instructions.

5. Conclusions

The *B. holmesii* lipid A structures established in the present study differ from one to the other:

(i)   Presence of free amino GlcN on the phosphate groups of lipid A structure: in Bho1 as in ATCC 51541, the failure of BvgA to function leads to a weaker substitution than in the two other isolates.

(ii)   Presence of 10:0(3-OH) at C-3, which we already observed in other *Bordetella* species such as
       *B. pertussi*s [31], is thought to be the consequence of the lack of activity of the 3 *O*-deacylase PagL,
       even if genomic explanation could not be found to support this hypothesis.

(iii) The 14:0(2-OH) and 12:0(2-OH) fatty acids found in all the presented structures were previously
       shown to be important structural features for the infection of human macrophages [42].

The combined structural virulent traits, characteristic of the *B. holmesii* lipid A structures,
demonstrate the potential strength of this bacterium in its capacity to adapt for escaping to the
host immune defenses. The efficiency of these traits, and especially the presence of the short-chain
10:0(3-OH) is already attested by the virulence of the whooping-cough pathogens *B. pertussis* and
B. parapertussis [31,52] and were also recently characterized in *B. petrii* lipid A isolates [41]. However,
no real difference of lipid A structure was observed between isolates from blood or respiratory origins
which correlates with data from other studies [18,19]. This also correlates with the fact that, in cases of
bacteremia, nasal carriage should be assessed as shown in a previous study [13].

The lipid A structures we characterized along the years on numerous strains and species always
gave important information confirmed by genomics; lipid A is a powerful taxonomic tool and this
work is another example of the kind. We demonstrated here the correlation between *B. holmesii* and
B. hinzii, B. avium and *B. trematum* also recently shown to belong to the same clade as based on whole
genome SNP-based analysis.

As already mentioned, lipid A are good bacterial markers, and their characterization in small
biological samples can now help to differentiate the human *Bordetella* pathogens.

Acknowledgments: This paper is dedicated to Martine Jégu-Belliveau recently deceased, much too young.
She spent lots of time and energy together with her husband Vincent in the "Clara Belliveau Association" raising
funds for research on whooping-cough, and explaining the necessity to use vaccine boosters in adults for their
protection and that of newborns. Funds from Institut Pasteur Fondation, Clara Belliveau association and CNRS
are acknowledged.

Author Contributions: Martine Caroff and Nicole Guiso. conceived and designed the experiments; Valérie
Bouchez contributed to genomic data, writings, corrections and shaping of the paper; Sami Al Bitar-Nehme,
Alexey Novikov, and Martine Caroff contributed equally to mass spectrometry data and structural analyses;
Martine Caroff, Valérie Bouchez, Nicole Guiso and Alexey Novikov wrote the paper.

Conflicts of Interest: The authors declare no conflict of interest.

Abbreviations

LPS	Lipopolysaccharide
MALDI	Matrix-ssisted laser desorption
MS	Mass spectrometry
FA	Fatty acid

1. Basheer, S.M.; Guiso, N.; Tirsoaga, A.; Caroff, M.; Novikov, A. Structural modifications occurring in lipid A
    of *Bordetella bronchiseptica* clinical isolates as demonstrated by matrix-assisted laser desorption/ionization
    time-of-flight mass spectrometry. *Rapid Commun. Mass Spectrom.* **2011**, *25*, 1075–1081. [CrossRef] [PubMed]
2. Flak, T.A.; Goldman, W.E. Signalling and cellular specificity of airway nitric oxide production in pertussis.
    Cell microbiol. **1999**, *1*, 51–60. [CrossRef] [PubMed]
3. Preston, A.; Maxim, E.; Toland, E.; Pishko, E.J.; Harvill, E.T.; Caroff, M.; Maskell, D.J. Bordetella bronchiseptica
    Pagp is a Bvg-regulated lipid A palmitoyl transferase that is required for persistent colonization of the mouse
    respiratory tract. *Mol. Microbiol.* **2003**, *48*, 725–736. [CrossRef] [PubMed]
4. Schaeffer, L.M.; McCormack, F.X.; Wu, H.; Weiss, A.A. Interactions of pulmonary collectins with *Bordetella
    bronchiseptica* and *Bordetella pertussis* lipopolysaccharide elucidate the structural basis of their antimicrobial
    activities. *Infect. Immun.* **2004**, *72*, 7124–7130. [CrossRef] [PubMed]

5. Rietschel, E.T.; Schade, U.; Jensen, M.; Wollenweber, H.W.; Luderitz, O.; Greisman, S.G. Bacterial endotoxins: Chemical structure, biological activity and role in septicaemia. *Scand. J. Infect. Dis. Suppl.* **1982**, *31*, 8–21. [CrossRef] [PubMed]

6. Weyant, R.S.; Hollis, D.G.; Weaver, R.E.; Amin, M.F.; Steigerwalt, A.G.; O'Connor, S.P.; Whitney, A.M.; Daneshvar, M.I.; Moss, C.W.; Brenner, D.J. *Bordetella holmesii* sp. Nov., a new gram-negative species associated with septicemia. *J. Clin. Microbiol.* **1995**, *33*, 1–7. [PubMed]

7. Bottero, D.; Griffith, M.M.; Lara, C.; Flores, D.; Pianciola, L.; Gaillard, M.E.; Mazzeo, M.; Zamboni, M.I.; Spoleti, M.J.; Anchart, E.; et al. *Bordetella holmesii* in children suspected of pertussis in Argentina. *Epidemiol. Infect.* **2013**, *141*, 714–717. [CrossRef] [PubMed]

8. Rodgers, L.; Martin, S.W.; Cohn, A.; Budd, J.; Marcon, M.; Terranella, A.; Mandal, S.; Salamon, D.; Leber, A.; Tondella, M.L.; et al. Epidemiologic and laboratory features of a large outbreak of pertussis-like illnesses associated with cocirculating *Bordetella holmesii* and *Bordetella pertussis*–Ohio, 2010–2011. *Clin. Infect. Dis.* **2013**, *56*, 322–331. [CrossRef] [PubMed]

9. Kamiya, H.; Otsuka, N.; Ando, Y.; Odaira, F.; Yoshino, S.; Kawano, K.; Takahashi, H.; Nishida, T.; Hidaka, Y.; Toyoizumi-Ajisaka, H.; et al. Transmission of *Bordetella holmesii* during pertussis outbreak, Japan. *Emerg. Infect. Dis.* **2012**, *18*, 1166–1169. [CrossRef] [PubMed]

10. Miranda, C.; Porte, L.; Garcia, P. *Bordetella holmesii* in nasopharyngeal samples from chilean patients with suspected *Bordetella pertussis* infection. *J. Clin. Microbiol.* **2012**, *50*, 1505; Author reply 1506. [CrossRef] [PubMed]

11. Mooi, F.R.; Bruisten, S.; Linde, I.; Reubsaet, F.; Heuvelman, K.; van der Lee, S.; King, A.J. Characterization of *Bordetella holmesii* isolates from patients with pertussis-like illness in the netherlands. *FEMS Immunol. Med. Microbiol.* **2012**, *64*, 289–291. [CrossRef] [PubMed]

12. Njamkepo, E.; Bonacorsi, S.; Debruyne, M.; Gibaud, S.A.; Guillot, S.; Guiso, N. Significant finding of *Bordetella holmesii* DNA in nasopharyngeal samples from french patients with suspected pertussis. *J. Clin. Microbiol.* **2011**, *49*, 4347–4348. [CrossRef] [PubMed]

13. Nguyen, L.B.; Epelboin, L.; Gabarre, J.; Lecso, M.; Guillot, S.; Bricaire, F.; Caumes, E.; Guiso, N. Recurrent *Bordetella holmesii* bacteremia and nasal carriage in a patient receiving rituximab. *Emerg. Infect. Dis.* **2013**, *19*, 1703–1705. [CrossRef] [PubMed]

14. Guiso, N.; Hegerle, N. Other bordetellas, lessons for and from pertussis vaccines. *Expert Rev. Vaccines* **2014**, *13*, 1125–1133. [CrossRef] [PubMed]

15. Pittet, L.F.; Posfay-Barbe, K.M. *Bordetella holmesii* infection: Current knowledge and a vision for future research. *Expert. Rev. Anti. Infect. Ther.* **2015**, *13*, 965–971. [CrossRef] [PubMed]

16. Pittet, L.F.; Posfay-Barbe, K.M. *Bordetella holmesii*: Still emerging and elusive 20 years on. *Microbiol. spectr.* **2016**, *4*. [CrossRef]

17. Planet, P.J.; Narechania, A.; Hymes, S.R.; Gagliardo, C.; Huard, R.C.; Whittier, S.; Della-Latta, P.; Ratner, A.J. *Bordetella holmesii*: Initial genomic analysis of an emerging opportunist. *Pathog. Dis.* **2013**, *67*, 132–135. [CrossRef] [PubMed]

18. Tatti, K.M.; Loparev, V.N.; Ranganathanganakammal, S.; Changayil, S.; Frace, M.; Weil, M.R.; Sammons, S.; Maccannell, D.; Mayer, L.W.; Tondella, M.L. Draft genome sequences of *Bordetella holmesii* strains from blood (F627) and nasopharynx (H558). *Genome Announc.* **2013**, *1*, e0005613. [CrossRef] [PubMed]

19. Bouchez, V.; Guiso, N. *Bordetella holmesii*: Comparison of two isolates from blood and a respiratory sample. *Adv. Infect. Dis.* **2013**, *3*, 123–133. [CrossRef]

20. NCBI, *Bordetella pertussis* genome and annotation report.

21. Linz, B.; Ivanov, Y.V.; Preston, A.; Brinkac, L.; Parkhill, J.; Kim, M.; Harris, S.R.; Goodfield, L.L.; Fry, N.K.; Gorringe, A.R.; et al. Acquisition and loss of virulence-associated factors during genome evolution and speciation in three clades of *Bordetella* species. *BMC Genom.* **2016**, *17*, 767. [CrossRef] [PubMed]

22. Gerlach, G.; Janzen, S.; Beier, D.; Gross, R. Functional characterization of the BvgAS two-component system of *Bordetella holmesii*. *Microbiology* **2004**, *150*, 3715–3729. [CrossRef] [PubMed]

23. Diavatopoulos, D.A.; Cummings, C.A.; van der Heide, H.G.; van Gent, M.; Liew, S.; Relman, D.A.; Mooi, F.R. Characterization of a highly conserved island in the otherwise divergent *Bordetella holmesii* and *Bordetella pertussis* genomes. *J. Bacteriol.* **2006**, *188*, 8385–8394. [CrossRef] [PubMed]

24. Hiramatsu, Y.; Saito, M.; Otsuka, N.; Suzuki, E.; Watanabe, M.; Shibayama, K.; Kamachi, K. BipA is associated with preventing autoagglutination and promoting biofilm formation in *Bordetella holmesii*. *PLoS ONE* **2016**, *11*, e0159999. [CrossRef] [PubMed]

25. Link, S.; Schmitt, K.; Beier, D.; Gross, R. Identification and regulation of expression of a gene encoding a filamentous hemagglutinin-related protein in *Bordetella holmesii*. *BMC Microbiol.* **2007**, *7*, 100. [CrossRef] [PubMed]

26. Van den Akker, W.M. Lipopolysaccharide expression within the genus *Bordetella*: Influence of temperature and phase variation. *Microbiology* **1998**, *144 Pt 6*, 1527–1535. [CrossRef] [PubMed]

27. Novikov, A.; Shah, N.R.; Albitar-Nehme, S.; Basheer, S.M.; Trento, I.; Tirsoaga, A.; Moksa, M.; Hirst, M.; Perry, M.B.; Hamidi, A.E.; et al. Complete *Bordetella avium*, *Bordetella hinzii* and *Bordetella trematum* lipid A structures and genomic sequence analyses of the loci involved in their modifications. *Innate Immun.* **2013**. [CrossRef] [PubMed]

28. Caroff, M.; Aussel, L.; Zarrouk, H.; Martin, A.; Richards, J.C.; Therisod, H.; Perry, M.B.; Karibian, D. Structural variability and originality of the *Bordetella* endotoxins. *J. Endotoxin Res.* **2001**, *7*, 63–68. [CrossRef] [PubMed]

29. MacArthur, I.; Jones, J.W.; Goodlett, D.R.; Ernst, R.K.; Preston, A. Role of *pagl* and *lpxo* in *Bordetella bronchiseptica* lipid A biosynthesis. *J. Bact.* **2011**, *193*, 4726–4735. [CrossRef] [PubMed]

30. Kawasaki, K. Complexity of lipopolysaccharide modifications in *Salmonella enterica*: Its effects on endotoxin activity, membrane permeability, and resistance to antimicrobial peptides. *Food Res. Int.* **2012**, *45*, 493–501. [CrossRef]

31. Albitar-Nehme, S.; Basheer, S.M.; Njamkepo, E.; Brisson, J.R.; Guiso, N.; Caroff, M. Comparison of lipopolysaccharide structures of *Bordetella pertussis* clinical isolates from pre- and post-vaccine era. *Carbohyd. Res.* **2013**. [CrossRef] [PubMed]

32. Tirsoaga, A.; El Hamidi, A.; Perry, M.B.; Caroff, M.; Novikov, A. A rapid, small-scale procedure for the structural characterization of lipid A applied to *Citrobacter* and *Bordetella* strains: Discovery of a new structural element. *J. Lipid Res.* **2007**, *48*, 2419–2427. [CrossRef] [PubMed]

33. Marr, N.; Tirsoaga, A.; Blanot, D.; Fernandez, R.; Caroff, M. Glucosamine found as a substituent of both phosphate groups in *Bordetella* lipid A backbones: Role of a BvgAS-activated Arnt ortholog. *J. Bact.* **2008**, *190*, 4281–4290. [CrossRef] [PubMed]

34. Geurtsen, J.; Dzieciatkowska, M.; Steeghs, L.; Hamstra, H.J.; Boleij, J.; Broen, K.; Akkerman, G.; El Hassan, H.; Li, J.; Richards, J.C.; et al. Identification of a novel lipopolysaccharide core biosynthesis gene cluster in *Bordetella pertussis*, and influence of core structure and lipid A glucosamine substitution on endotoxic activity. *Infect. Immun.* **2009**, *77*, 2602–2611. [CrossRef] [PubMed]

35. Karibian, D.; Brunelle, A.; Aussel, L.; Caroff, M. 252Cf-plasma desorption mass spectrometry of unmodified lipid A: Fragmentation patterns and localization of fatty acids. *Rapid Commun. Mass Spectrom.* **1999**, *13*, 2252–2259. [CrossRef]

36. Albitar-Nehme, S. Endotoxins of the *Bordetella* Genus: Structure, Evolution and Impact on Bacterial Virulence. Ph.D. Thesis, Université de Paris-Sud, Orsay, France, 13 June 2014.

37. Caroff, M.; Deprun, C.; Richards, J.C.; Karibian, D. Structural characterization of the lipid A of *Bordetella pertussis* 1414 endotoxin. *J. Bacteriol.* **1994**, *176*, 5156–5159. [CrossRef] [PubMed]

38. El Hamidi, A.; Novikov, A.; Karibian, D.; Perry, M.B.; Caroff, M. Structural characterization of *Bordetella parapertussis* lipid A. *J. Lipid Res.* **2009**, *50*, 854–859. [CrossRef] [PubMed]

39. Therisod, H.; Labas, V.; Caroff, M. Direct microextraction and analysis of rough-type lipopolysaccharides by combined thin-layer chromatography and MALDI mass spectrometry. *Anal. Chem.* **2001**, *73*, 3804–3807. [CrossRef] [PubMed]

40. Shah, N.R.; Albitar-Nehme, S.; Kim, E.; Marr, N.; Novikov, A.; Caroff, M.; Fernandez, R.C. Minor modifications to the phosphate groups and the C3′ acyl chain length of lipid a in two *Bordetella pertussis* strains, BP338 and 18–323, independently affect Toll-like receptor 4 protein activation. *J. Biol. Chem.* **2013**, *288*, 11751–11760. [CrossRef] [PubMed]

41. Basheer, S.M.; Bouchez, V.; Novikov, A.; Augusto, L.A.; Guiso, N.; Caroff, M. Structure activity characterization of *Bordetella petrii* lipid A, from environment to human isolates. *Biochimie* **2016**, *120*, 87–95. [CrossRef] [PubMed]

42. Marr, N.; Hajjar, A.M.; Shah, N.R.; Novikov, A.; Yam, C.S.; Caroff, M.; Fernandez, R.C. Substitution of the *Bordetella pertussis* lipid A phosphate groups with glucosamine is required for robust NF-kB activation and release of proinflammatory cytokines in cells expressing human but not murine Toll-like receptor 4-MD-2-CD14. *Infect. Immun.* **2010**, *78*, 2060–2069. [CrossRef] [PubMed]

43. Marr, N.; Novikov, A.; Hajjar, A.M.; Caroff, M.; Fernandez, R.C. Variability in the lipooligosaccharide structure and endotoxicity among *Bordetella pertussis* strains. *J. Infect. Dis.* **2010**, *202*, 1897–1906. [CrossRef] [PubMed]

44. Geurtsen, J.; Steeghs, L.; Hove, J.T.; van der Ley, P.; Tommassen, J. Dissemination of lipid A deacylases (Pagl) among gram-negative bacteria: Identification of active-site histidine and serine residues. *J. Biol. Chem.* **2005**, *280*, 8248–8259. [CrossRef] [PubMed]

45. Hittle, L.E.; Jones, J.W.; Hajjar, A.M.; Ernst, R.K.; Preston, A. *Bordetella parapertussis* Pagp mediates the addition of two palmitates to the lipopolysaccharide lipid A. *J. Bacteriol.* **2015**, *197*, 572–580. [CrossRef] [PubMed]

46. Geurtsen, J.; Angevaare, E.; Janssen, M.; Hamstra, H.J.; ten Hove, J.; de Haan, A.; Kuipers, B.; Tommassen, J.; van der Ley, P. A novel secondary acyl chain in the lipopolysaccharide of *Bordetella pertussis* required for efficient infection of human macrophages. *J. Biol. Chem.* **2007**, *282*, 37875–37884. [CrossRef] [PubMed]

47. Di Fabio, J.L.; Caroff, M.; Karibian, D.; Richards, J.C.; Perry, M.B. Characterization of the common antigenic lipopolysaccharide O-chains produced by *Bordetella bronchiseptica* and *Bordetella parapertussis*. *FEMS Microbiol. Lett.* **1992**, *76*, 275–281. [CrossRef] [PubMed]

48. Njamkepo, E.; Delisle, F.; Hagege, I.; Gerbaud, G.; Guiso, N. *Bordetella holmesii* isolated from a patient with sickle cell anemia: Analysis and comparison with other *Bordetella holmesii* isolates. *Clin. Microbiol. Infect.* **2000**, *6*, 131–136. [CrossRef] [PubMed]

49. Tizolova, A.; Guiso, N.; Guillot, S. Insertion sequences shared by *Bordetella* species and implications for the biological diagnosis of pertussis syndrome. *Eur. J. Clin. Microbiol. Infect. Dis.* **2013**, *32*, 89–96. [CrossRef] [PubMed]

50. Stainer, D.W.; Scholte, M.J. A simple chemically defined medium for the production of phase I *Bordetella pertussis*. *J. Gen. Microbiol.* **1970**, *63*, 211–220. [CrossRef] [PubMed]

51. Caroff, M. Brevet Français. Brevet International Patent wo2004/062690a1, 9 December 2002.

52. El Hamidi, A.; Tirsoaga, A.; Novikov, A.; Hussein, A.; Caroff, M. Microextraction of bacterial lipid A: Easy and rapid method for mass spectrometric characterization. *J. Lipid Res.* **2005**, *46*, 1773–1778. [CrossRef] [PubMed]

53. Chafchaouni-Moussaoui, I.; Novikov, A.; Bhrada, F.; Perry, M.B.; Filali-Maltouf, A.; Caroff, M. A new rapid and micro-scale hydrolysis, using triethylamine citrate, for lipopolysaccharide characterization by mass spectrometry. *Rapid Commun. Mass Spectrom.* **2011**, *25*, 2043–2048. [CrossRef] [PubMed]

54. Caroff, M.; Tacken, A.; Szabo, L. Detergent-accelerated hydrolysis of bacterial endotoxins and determination of the anomeric configuration of the glycosyl phosphate present in the "isolated lipid A" fragment of the bordetella pertussis endotoxin. *Carbohydr. Res.* **1988**, *175*, 273–282. [CrossRef]

55. Haeffner, N.; Chaby, R.; Szabo, L. Identification of 2-methyl-3-hydroxydecanoic and 2-methyl-3-hydroxytetradecanoic acids in the 'Lipid X' fraction of the *Bordetella pertussis* endotoxin. *Eur. J. Biochem.* **1977**, *77*, 535–544. [CrossRef] [PubMed]

56. Wollenweber, H.W.; Rietschel, E.T. Analysis of lipopolysaccharide (lipid A) fatty acids. *J. Microbiol. Methods* **1990**, *11*, 195–211. [CrossRef]

International Journal of
Molecular Sciences

Article

The New Structure of Core Oligosaccharide Presented by *Proteus penneri* 40A and 41 Lipopolysaccharides

Agata Palusiak [1,†], Anna Maciejewska [2,†], Czeslaw Lugowski [2,3], Antoni Rozalski [4] and Marta Kaszowska [2,*]

[1]   Laboratory of General Microbiology, Institute of Microbiology, Biotechnology and Immunology, University of Lodz, PL-90-237 Lodz, Poland; agata.palusiak@biol.uni.lodz.pl
[2]   Hirszfeld Institute of Immunology and Experimental Therapy, Polish Academy of Sciences, R. Weigla 12, PL-53-114 Wroclaw, Poland; anna.maciejewska@iitd.pan.wroc.pl (A.M.); lugowski@iitd.pan.wroc.pl (C.L.)
[3]   Department of Biotechnology and Molecular Biology, University of Opole, PL-45-035 Opole, Poland
[4]   Department of Biology of Bacteria, Institute of Microbiology, Biotechnology and Immunology, University of Lodz, 90-237 Lodz, Poland; antoni.rozalski@biol.uni.lodz.pl
*   Correspondence: marta.kaszowska@iitd.pan.wroc.pl; Tel.: +48-71-370-99-27
†   These authors contributed equally to this work.

Received: 29 January 2018; Accepted: 22 February 2018; Published: 28 February 2018

Abstract: The new type of core oligosaccharide in *Proteus penneri* 40A and 41 lipopolysaccharides has been investigated by ^{1}H and ^{13}C NMR spectroscopy, electrospray ionization mass spectrometry and chemical methods. Core oligosaccharides of both strains were chosen for structural analysis based on the reactivity of LPSs with serum against *P. penneri* 40A core oligosaccharide–diphtheria toxoid conjugate. Structural analyses revealed that *P. penneri* 40A and 41 LPSs possess an identical core oligosaccharide.

Keywords: anti-conjugate serum; core oligosaccharide; lipopolysaccharide; NMR spectroscopy; ESI MS; *Proteus penneri*

1. Introduction

P. penneri are human-opportunistic pathogens causing, in preferred conditions, several types of infections among which urinary tract and wound infections are predominant. These Gram-negative bacteria produce many virulence factors including LPS (endotoxin), which seems to be the most dangerous due to its contribution to septic shock [1]. LPS of smooth bacterial strains consists of three regions: lipid A, a core oligosaccharide (OS) and an *O*-polysaccharide (*O*-PS, *O*-antigen). Only the last two have been described for the *P. penneri* species [2–4]. Although the *P. penneri* core oligosaccharide is characterized by lower structural diversity than the *O*-PS part (over 26 chemotypes), it is still structurally heterogeneous (12 different structures of the outer core region and a few variants of its inner part) [2–6]. The LPS core region may be masked by an *O*-polysaccharide, but its exposition on bacterial cells is still accessible for specific immunoglobulins. This fact was confirmed by the detection, in polyclonal rabbit antisera against *Proteus* strains, of anti-core-specific antibodies recognizing low-molecular-mass LPS species not only of homologous but also heterologous antigens [7–9]. This observation encouraged the examination of different *P. penneri* LPSs, in search of antigen groups with identical or similar serological activities of their core oligosaccharides, which would complete the *Proteus* classification scheme with the data on the core region serotypes. To date, 11 groups of LPS presenting one core serotype have been classified [10]. In this paper, the results of serological studies and structural analysis are presented to show another type among *P. penneri* LPSs with a common sero- and chemotype of their core oligosaccharides.

2. Results

2.1. Serological Studies

The rabbit polyclonal serum against the *P. penneri* 40A core oligosaccharide–diphtheria toxoid conjugate (anti-conjugate serum) was obtained and tested by ELISA assay and immunoblotting (Western blot) with the homologous and 40 other *Proteus* spp. LPSs. The heterologous LPSs (rough *P. penneri* strains: (serotypes 4, 5, 11), R mutant of *P. mirabilis* (serotype 6) and smooth *P. penneri* strains (O8, O17, O19a,b, O31a, O31a,b, O52, O58, O59, O61, O62, O63, O64a,b,c, O64a,c,e, O65, O67–O71, O72a, O73a,b, O73a,c) representing different O serogroups and subgroups of the genus were selected as described previously [7]. In ELISA, two LPSs, *P. penneri* 40A and 41, reacted to the titer the most strongly with the tested serum (1:16,000); two LSPs, *P. penneri* 1 and 4, cross-reacted to the titer (1:8000); two other LPSs, *P. penneri* 27 and 71 showed the lowest serum reactivity titers (1:2000). Residual *Proteus* spp. LPSs were not cross-reacted with the tested serum. The cross-reactivity of the tested antiserum with *P. penneri* 1, 4, 27, 71 LPSs indicates the presence in their core oligosaccharides of similar epitope(s) common with core oligosaccharide of the homologous LPS. The strongly reacting LPSs, *P. penneri* 40A and 41, were chosen for further study. In the Western blot technique, all reactions concerned the low-molecular-mass LPS fragments consisting of the core-lipid A moieties of tested antigens (Figure 1). The tested antiserum was adsorbed a few times with an alkali-treated cross-reacting or homologous antigen and checked once more in ELISA with the same LPS preparations. The adsorption of *P. penneri* 40A anti-conjugate serum with each of the reacting LPSs completely abolished the reactions with tested LPSs. *P. penneri* 40A and 41 LPSs, which reacted strongly and similarly in all assays, have been selected for structural studies by ESI mass spectrometry and NMR spectroscopy to check the similarity of these core oligosaccharides.

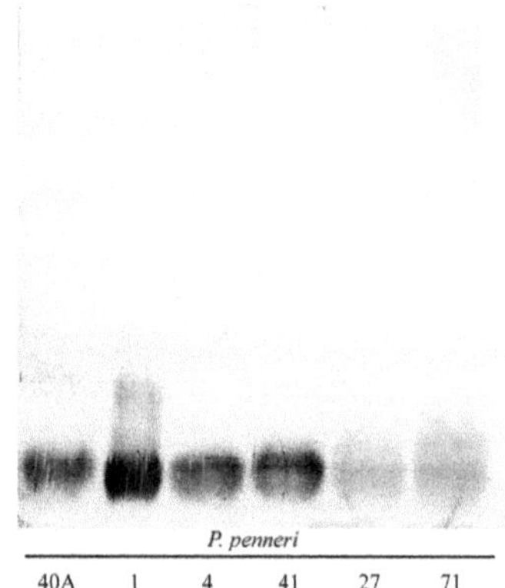

Figure 1. Western blot of *P. penneri* LPSs with the *P. penneri* 40A anti-conjugate serum.

2.2. Structural Studies

The mass spectra obtained for the *P. penneri* 40A and 41 core oligosaccharides showed a high degree of similarity (Figure 2A,B). To avoid unnecessary duplication, only data concerning the *P. penneri* 40A core oligosaccharide have been presented in the text. Table 1 presents an interpretation of all ions in core oligosaccharide fractions which have been identified by ESI MS. The major fraction represented by the ions at m/z 1180.42 [M+Ac+2H]$^{2+}$ and m/z 1171.40 [M+Ac-H$_2$O+2H]$^{2+}$ corresponded to the core oligosaccharide containing two hexoses (Glc and Gal); five heptoses (Hep); hexuronic acid (GalA); hexosamine (GalN); *N*-acetylated hexosamine (GlcNAc); 4-amino-4-deoxyarabinose (Ara4N); 3-deoxy-D-*manno*-oct-2-ulosonic acid (Kdo); phosphoethanolamine (PEtn) and a one *O*-acetyl group (OAc). Additionally, both core oligosaccharides were de-*O*-acetylated and checked by ESI MS.

The differences between core oligosaccharides and their de-*O*-acetylated fractions were related to the removal of the *O*-acetyl group from the structure. The major fraction represented by the ion at *m/z* 1159.30 [M+2H]$^{2+}$ (Figure 2D) corresponded to the structure without an *O*-acetyl group in contrast with the ion at *m/z* 1180.42 [M+Ac+2H]$^{2+}$ (Figure 2C). These two ions were selected for further analysis by use of positive ion mode ESI MS/MS. The main daughter ions detected in the ESI MS/MS spectra were explained. The ion at *m/z* 366.43 corresponds to the GalGlcNAc fragment, while the ion at *m/z* 407.48 was explained by the GlcNAcGalN-OAc (Figure 2C). The daughter ion with the highest *m/z* 569.59 was subsequently attributed to the GalGlcNAcGalN-OAc fragment. These observations, in comparison with NMR data, indicate that an *O*-acetyl group substitutes at Gal*p*N (residue **K′**).

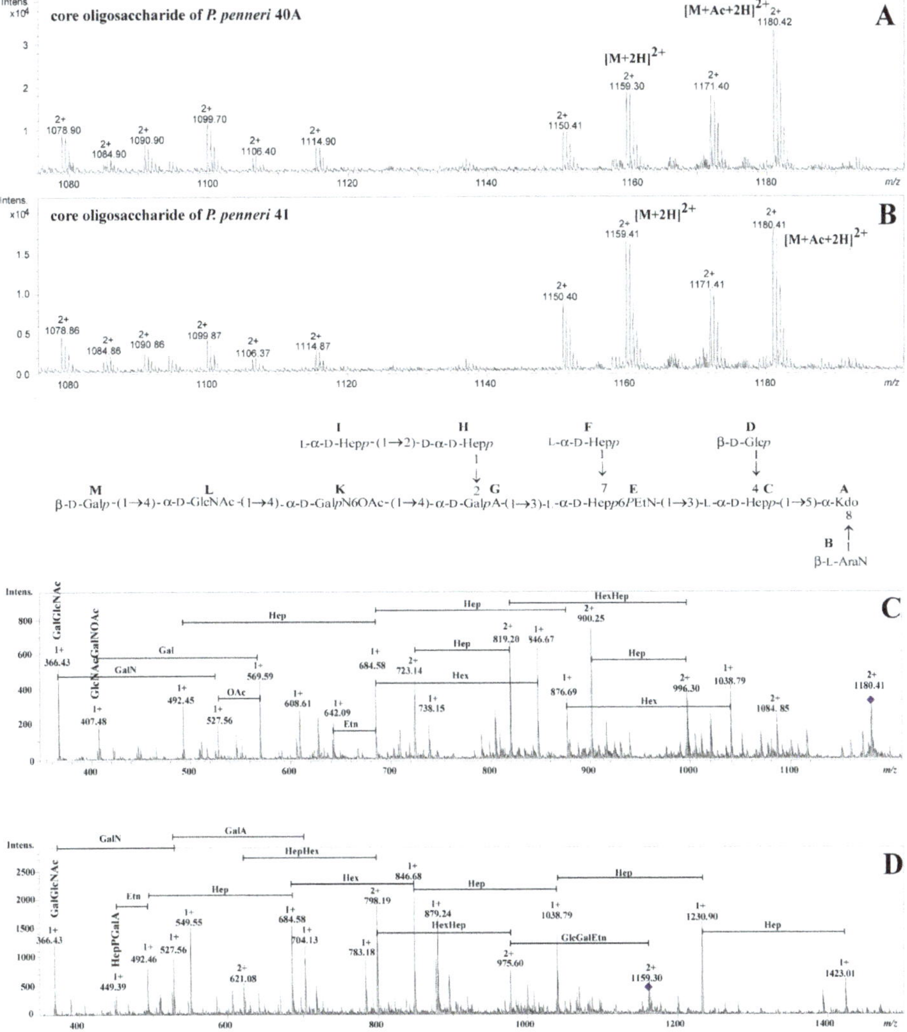

Figure 2. Positive ion mode ESI mass spectra of the core oligosaccharides from *P. penneri* 40A (**A**) and 41 (**B**); (**C**) Positive ion mode ESI MS/MS of the core oligosaccharide from *P. penneri* 40A represented by ions at *m/z* 1180.42 and (**D**) at *m/z* 1159.30.

Table 1. ESI MS data obtained for the core oligosaccharides of *P. penneri* 40A and 41.

The Observed Ion (*m/z*)/The Calculated Mass		The Ion Interpretation
P. penneri 40A	*P. penneri* 41	
1180.42^{2+}/2358.84	1180.41^{2+}/2358.82	$[M+Ac+2H]^{2+}$
1171.40^{2+}/2340.80	1171.41^{2+}/2340.82	$[M+Ac-H_2O+2H]^{2+}$
1159.30^{2+}/2316.60	1159.41^{2+}/2316.82	$[M+2H]^{2+}$
1150.41^{2+}/2298.82	1150.40^{2+}/2298.80	$[M-H_2O+2H]^{2+}$
1114.90^{2+}/2227.80	1114.87^{2+}/2227.74	$[M-Ara4N+Ac+2H]^{2+}$
1106.40^{2+}/2210.80	1106.37^{2+}/2210.74	$[M-Ara4N+Ac-H_2O+2H]^{2+}$
1099.70^{2+}/2197.40	1099.87^{2+}/2197.74	$[M-Hex+Ac+2H]^{2+}$
1090.90^{2+}/2179.80	1090.86^{2+}/2179.72	$[M-Hex+Ac-H_2O+2H]^{2+}$
1084.90^{2+}/2167.80	1084.86^{2+}/2167.72	$[M-Hep+Ac+2H]^{2+}$
1078.90^{2+}/2155.80	1078.86^{2+}/2155.72	$[M-Hex+2H]^{2+}$

The initial NMR investigation indicated the presence of hexosamine residue among the constituents of the core oligosaccharide; therefore, methylation analysis was performed on *N*-acetylated oligosaccharide. Methylation indicated the presence of 3,7-disubstituted Hep*p*, 3,4-disubstituted Hep*p*, 2-substituted Hep*p*, terminal Hep*p*, 4-substituted Glc*p*N, 4-substituted Gal*p*N, terminal Glc*p* and terminal Gal*p*.

The ^{1}H NMR spectrum of the *P. penneri* 40A core oligosaccharide (Figure 3A) contained the main signals for eleven anomeric protons, as well as signals characteristic for the deoxy protons from the Kdo residue. The ^{1}H-^{1}H COSY, TOCSY with different mixing times, ^{1}H-^{13}C HSQC-DEPT and HSQC-TOCSY spectra allowed for the assignments of the H-1 to H-6 (H-7,7′, H-8,8′) signals for each residue (marked as uppercase letters) of the core oligosaccharide (Table 2, Figure 3).

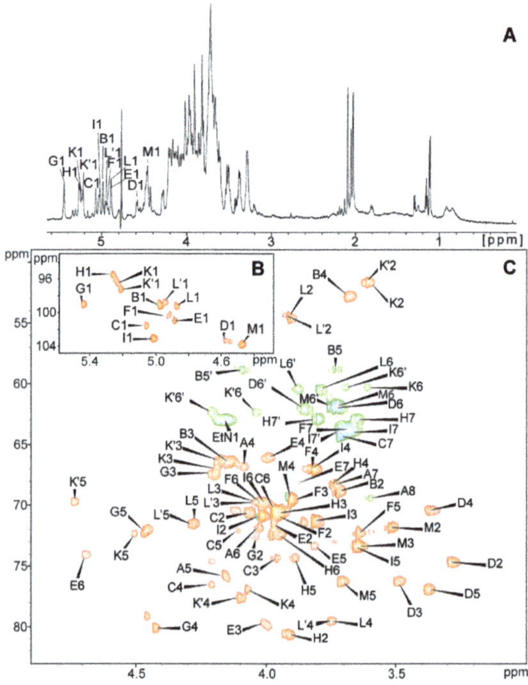

Figure 3. (**A**) The ^{1}H NMR spectrum and (**B,C**) selected regions of the ^{1}H-^{13}C HSQC-DEPT spectrum of the core oligosaccharide *P. penneri* 40A LPS. The cross-peaks are labeled as shown in the text.

Table 2. ^{1}H and ^{13}C NMR chemical shifts of the core oligosaccharide of *P. penneri* 40A LPS.

Residues	Chemical Shifts (ppm)							
	H1/C1	H2/C2	H3(H3ax,eq)/C3	H4/C4	H5,5′/C5	H6,6′/C6	H7,7′/C7	H8,8′/C8 (NAc, OAc)
A→5,8)-Kdo			(1.80, 2.19) 35.4	4.09 66.8	4.16 75.7	4.03 71.7	3.74 68.2	3.60, 3.92 69.4
B β-L-Ara4N-(1→	4.98 99.1	3.72 68.8	4.14 66.5	3.68 52.8	3.73, 4.10 58.8			
C→3,4)-L-*glycero*-α-D-*manno*-Hep*p*-(1→	5.07 101.6	4.06 70.6	3.96 74.4	4.21 76.5	4.11 72.1	3.97 70.5	3.72 64.3	
D β-D-Glc*p*-(1→	4.58 103.3	3.30 74.7	3.49 76.3	3.37 70.4	3.37 76.9	3.74, 3.86 62.0		
E→3,7)-L-*glycero*-α-D-*manno*-Hep*p*-(1→	4.89 101.0	3.98 71.6	4.01 79.7	3.99 66.1	3.81 73.3	4.69 74.1	3.88 68.2	
F L-*glycero*-α-D-*manno*-Hep*p*-(1→	4.93 100.4	3.96 70.9	3.89 69.5	3.82 67.1	3.65 72.4	4.02 69.8	3.69 63.8	
G→2,4)-α-D-Gal*p*A-(1→	5.45 99.0	4.04 72.7	4.21 67.3	4.43 80.1	4.46 72.2	176.0		
H→2)-D-*glycero*-α-D-*manno*-Hep*p*-(1→	5.27 95.5	3.91 80.6	4.00 70.6	3.75 68.3	3.89 74.3	3.95 72.4	3.65, 3.81 62.9	
I L-*glycero*-α-D-*manno*-Hep*p*-(1→	5.02 103.1	4.02 70.8	3.82 71.3	3.83 67.0	3.66 73.4	4.01 69.8	3.67, 3.72 64.0	
K→4)-α-D-Gal*p*N-(1→	5.23 96.5	3.60 51.7	4.19 66.9	4.08 76.9	4.50 72.4	3.61, 3.70 60.3		
K′→4)-α-D-Gal*p*N6OAc-(1→	5.22 97.3	3.62 51.7	4.18 66.3	4.11 77.6	4.73 69.7	4.04, 4.21 62.3		(2.10) (21.0, 174.5)
L→4)-α-D-Glc*p*NAc-(1→	4.88 99.2	3.92 54.6	4.04 69.7	3.75 79.5	4.28 71.5	3.79,3.88 60.5		(2.04) (22.7, 175.3)
L′→4)-α-D-Glc*p*NAc-(1→	4.95 98.8	3.93 54.4	4.05 69.8	3.78 79.6	71.5 4.29			(2.06) (22.7, 175.4)
M β-D-Gal*p*-(1→	4.48 103.7	3.53 71.8	3.64 73.4	3.91 69.5	3.71 76.3	3.72, 3.75 61.9		
PEtN	4.16 63.0	3.28 41.0						

Residue **A** was recognized as a 5,8-disubstituted Kdo based on the characteristic deoxy proton signals of H-3ax (δ_H 1.80 ppm), H-3eq (δ_H 2.19 ppm) and a high chemical shift of the C-5 (δ_C 75.7 ppm) signal.

Residue **B** at δ_H/δ_C 4.98/99.1 ppm, $^1J_{C\text{-}1,H\text{-}1}$ ~170 Hz, was assigned as the terminal β-L-Ara4N residues based on the chemical shift of the C-4 (δ_C 52.8 ppm) and characteristic H5, H5′/C5 (δ_H 3.73, δ_H/δ_C 4.10/58.8 ppm) signals.

Residue **C** at δ_H/δ_C 5.07/101.6 ppm, $^1J_{C\text{-}1,H\text{-}1}$ ~170 Hz, was recognized as a 3,4-disubstituted L-*glycero*-α-D-*manno*-Hep*p* based on the relatively high chemical shift of the C-3 (δ_C 74.4 ppm) and C-4 (δ_C 76.5 ppm) signals.

Residue **D** at δ_H/δ_C 4.58/103.3 ppm, $^1J_{C\text{-}1,H\text{-}1}$ ~162 Hz, was assigned as the terminal β-D-Glc*p* based on the large vicinal couplings between all protons in the sugar ring.

Residue **E** at δ_H/δ_C 4.89/101.0 ppm, $^1J_{C\text{-}1,H\text{-}1}$ ~172 Hz, was recognized as the 3,7-disubstituted L-*glycero*-α-D-*manno*-Hep*p* from the relatively high chemical shifts of the C-3 (δ_C 79.7 ppm) and C-7 (δ_C 68.2 ppm) signals.

Residue **F** at δ_H/δ_C 4.93/100.4 ppm, $^1J_{C\text{-}1,H\text{-}1}$ ~168 Hz, as well as residue **I** at δ_H/δ_C 5.02/103.1 ppm, $^1J_{C\text{-}1,H\text{-}1}$ ~170 Hz, were assigned as terminal L-*glycero*-α-D-*manno*-Hep*p*.

Residue **G** at δ_H/δ_C 5.45/99.0 ppm, $^1J_{C\text{-}1,H\text{-}1}$ ~168 Hz, was assigned as the 2,4-disubstituted α-D-Gal*p*A residues based on the characteristic five proton spin systems, and the high ^{13}C chemical shift of the C-2 (δ_C 72.7 ppm), C-4 (δ_C 80.1 ppm) and C-6 (δ_C 176.0 ppm) signals.

Residue **H** at δ_H/δ_C 5.27/95.5 ppm, $^1J_{C\text{-}1,H\text{-}1}$ ~173 Hz, was recognized as a 2-substituted D-*glycero*-α-D-*manno*-Hep*p* based on the relatively high chemical shift of the C-2 (δ_C 80.6 ppm) signal.

Additionally, the characteristic chemical shift of the C-6 (δ_C 72.4 ppm) signal indicates the D-*glycero*-D-*manno* configuration [11].

Residue **K′** at δ_H/δ_C 5.22/97.3 ppm, $^1J_{C\text{-}1,H\text{-}1}$ ~175 Hz, was recognized as the 4-substituted α-D-Gal*p*N6OAc based on the chemical shift of the C-2 (δ_C 51.7 ppm) signal, the relatively high chemical shift C-4 (δ_C 77.6 ppm) signal, and the downfield shift of the C-6 (δ_C 62.3 ppm) signal, consistent with an *O*-acetyl group at position 6. The presence of an *O*-acetyl group was supported by de-*O*-acetylation. The spectrum of the de-*O*-acetylated core oligosaccharide contained only one signal from the *N*-acetyl group of residue **L** whereas OS also possesses an additional signal from the *O*-acetyl group at Gal*p*N (residue **K′**). Residue **K** at δ_H/δ_C 5.23/96.5 ppm, $^1J_{C\text{-}1,H\text{-}1}$ ~174 Hz, represented a variant of residue **K′** caused by the lack of the *O*-acetyl group at *O*-6 of residue **K′**. Residue **K** was thus identified as 4-substituted α-D-Gal*p*N.

Residue **L** at δ_H/δ_C 4.88/99.2 ppm, $^1J_{C\text{-}1,H\text{-}1}$ ~177 Hz, was recognized as the 4-substituted α-D-Glc*p*NAc based on the chemical shift of the C-2 (δ_C 54.6 ppm) signal, the relatively high chemical shift of the C-4 (δ_C 79.5 ppm) signal. Residues **L′** at δ_H/δ_C 5.40/102.4 ppm, $^1J_{C\text{-}1,H\text{-}1}$ ~175 Hz, were recognized as variants of residue **L** (4-substituted α-D-Glc*p*NAc) due to the absence of the *O*-acetyl group located at position 6 of residue **K′**.

Residue **M** at δ_H/δ_C 4.48/103.7 ppm, $^1J_{C\text{-}1,H\text{-}1}$ ~162 Hz, was recognized as the terminal β-D-Gal*p* based on chemical shifts in good agreement with those of previously reported β-D-Gal*p* [12].

Each disaccharide element of the core oligosaccharide was identified by ^{1}H-^{1}H NOESY (Figure 4) and ^{1}H-^{13}C HMBC experiments. The NOESY spectrum showed strong inter-residue cross-peaks between the transglycosidic protons: H-1 of **B**/H-8 of **A**, H-1 of **C**/H-5 of **A**, H-1 of **D**/H-4 of **C**, H-1 of **E**/H-3 of **C**, H-1 of **G**/H-3 of **E**, H-1 of **H**/H-2 of **G**, H-1 of **I**/H-2 of **H**, H-1 of **K**/H-4 of **G**, H-1 of **L**/H-4 of **K′**, H-1 of **L′**/H-4 of **K** and H-1 of **M**/H-4 of **L** (Figure 4). These data confirmed the substitution positions of the monosaccharide residues and demonstrated their sequence in the core oligosaccharide *P. penneri* 40A (structure inserted into Figure 2).

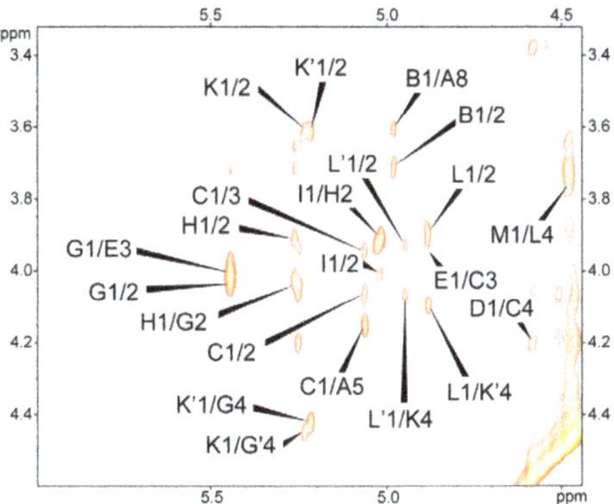

Figure 4. Selected part of the NOESY spectrum of the core oligosaccharide of *P. penneri* 40A LPS. The cross-peaks are labeled as shown in the text.

3. Discussion

This work provides the serological and chemical characterization of a new type of the core region presented by *P. penneri* 40A and 41 LPSs. In ELISA, anti-conjugate serum *P. penneri* 40A reacted differently with three groups: (I) *P. penneri* 40A and 41—showing the strongest serological activity; (II) *P. penneri* 4 and 1—weaker reactions; and (III) *P. penneri* 27 and 71—the weakest serological activity. The weakest activity of the last two LPSs was also confirmed by the results of the Western blot technique (Figure 1). The LPS whose binding-pattern of low-molecular-mass LPS species distinguished itself from the patterns of the tested residual LPSs was *P. penneri* 1. The LPS, *P. penneri* 4, reacted in Western blotting similarly to *P. penneri* 40A and 41 but its reactivity titer in ELISA was twice as low as the titers of *P. penneri* 40A and 41 LPSs. These differences in serological activity of *P. penneri* 1, 4, 27 and 71 LPS core oligosaccharides compared to that observed for *P. penneri* 40A and 41 LPSs suggest that these two groups of LPSs share common epitopes but do not necessarily present the same sero- and chemotype of the core region. In many cases, LPSs, presenting one chemotype of the core region, showed similar binding-patterns of low-molecular-mass LPS species in the Western blot technique and reacted in ELISA up to the same value of the antiserum reactivity titers [7,8]. Due to the fact that *P. penneri* 40A and 41 LPSs reacted similarly in all serological assays, they were chosen for the structural analyses.

The results of mass spectra analyses of the core oligosaccharides from *P. penneri* 40A and *P. penneri* 41 LPS were able to reveal the structure, which had not been previously identified for *Proteus* LPS core regions. The new type of structure is typical for *P. penneri* core regions in its inner part containing five Hep*p* residues, Glc*p*, Gal*p*A, Kdo and Ara4N residues and presenting III glycoform [2,4]. Only in six *P. penneri* LPSs 12, 13, 14, 37, 42 and 44, can structural variations of the inner core region be observed [2,4]. The uniqueness of *P. penneri* 40A and 41 core regions is found in its outer part defined in the literature as R substituent [4]. To date, 20 different structures of R substituent have been determined for *Proteus* spp. strains, among which 12 are presented by *P. penneri* strains [4,13]. These structures contain from one (e.g., *P. penneri* 12, 13) to four sugar residues or their *N*-acetylated forms (e.g., *P. penneri* 7, 14, 15). One structure of R substituent can be represented by one or a few *P. penneri* strains [4]. The R substituent of *P. penneri* 40A (O64a,b,d) and the 41 (O62) LPS core region is built of three residues (β-D-Gal*p*-(1→4)-α-D-Glc*p*NAc-(1→4)-α-D-Gal*p*N6OAc) and it is similar to the outer core region of only one strain, *P. penneri* 103 (O73a,b) (β-D-Glc*p*-(1→4)-α-D-Glc*p*NAc-(1→4)-α-D-Gal*p*N6OAc) [4]. These two fragments differ from each other in the terminal residue. The importance of the terminal residue in the serospecificity of the outer core oligosaccharide region can be confirmed by the fact that *P. penneri* 40A anti-conjugate serum did not react with *P. penneri* 103 LPS. *P. penneri* 103 was classified into serotype group no.

10 together with *P. penneri* 75 LPS, recognized by anti-core-specific antibodies present in *P. penneri* 103 antiserum [10]. A similar situation also occurs in the case of two core oligosaccharide serotypes: R1 (α-D-Glc*p*-(1→4)-α-D-Gal*p*NAc-(1→2)-α-DD-Hep*p*-(1→6)-α-D-Glc*p*NGly) presented by *P. penneri* 7, 14 LPSs and R2 presented by *P. penneri* 8 LPS (α-D-Gal*p*NAc-(1→2)-α-DD-Hep*p*-(1→6)-α-D-Glc*p*NGly) differing in the lack of terminal residue [4,14]. The serological studies performed by use of *P. penneri* 7 core-specific antiserum and *P. penneri* 7, 8, 14 and 15 LPSs proved a crucial role of the terminal residue from the outer core region in its serospecificity [8]. As a result, *P. penneri* 8 LPS and *P. penneri* 7, 14 and 15 LPSs were classified into two different core oligosaccharide serotypes [10]. The 6-*O*-acetylation of the Gal*p*N residue is also unique for the *Proteus* spp. LPS core region. In *P. penneri* 16 and 18, Gal*p*N is substituted by the phosphoethanolamine group at position 6 [4].

The results presented in this work will allow *P. penneri* 40A and 41 LPSs to be classified into a new core oligosaccharide serotype group extending the core oligosaccharide serotypes scheme [10]. It is another example of two *P. penneri* LPSs of one core serotype but presenting two O serotypes: *P. penneri* 40A (O64a,b,d) and *P. penneri* 41 (O62)—the first representative of this O serogroup in the core types classification scheme.

Finding a new structure and serotype of the *P. penneri* LPS core region confirmed the huge structural heterogeneity of *P. penneri* LPSs, a unique phenomenon among other *Enterobacteriaceae*. Extension of the core serotype scheme with other representatives may be helpful in the identification of the most common R and O serotypes needed for the selection of vaccine antigens to obtain cross-reactive and cross-protective antibodies [10].

4. Materials and Methods

4.1. Bacterial Strains

P. penneri 40A (O64a,b,d) and 41 (O62) are clinical isolates from patients in Toronto (Canada) but their isolation sources remain unknown. The other strains, whose LPSs were checked with the tested serum, have been presented in another article [7]. All tested strains belong to the collection of the Laboratory of General Microbiology, University of Lodz (Poland), where they are stored in a glycerol mixture at −80 °C.

4.2. Lipopolysaccharide

The LPSs were extracted from dried bacterial cells, as previously described [7], by the phenol–water procedure according to the method of Westphal [15] and purified with aqueous 50% trichloroacetic acid. Alkali-treated LPSs used for the sera adsorption were prepared as described in detail elsewhere [16].

The LPSs of *P. penneri* 40A and 41 were degraded by treating with 1.5% acetic acid at 100 °C for 1 h and the carbohydrate portions were fractionated and monitored as described previously [7]. The fractions (*O*-PS, OS, and the mixture of low molecular mass) were eluted, freeze-dried and checked by ESI mass spectrometry and NMR spectroscopy.

4.3. De-O-Acetylation of the Core Oligosaccharide

The *P. penneri* 40A, 41 core oligosaccharides (5 mg) were treated with aqueous 12.5% NH_3 (1 mL) at 23 °C for 16 h and then the solution was freeze-dried. The products were analyzed by ESI mass spectrometry and NMR spectroscopy.

4.4. P. penneri 40A Core Oligosaccharide Conjugate

Conjugation of the *P. penneri* 40A core oligosaccharide with diphtheria toxoid was performed by the method of H. J. Jennings and C. Lugowski based on the reaction of reductive amination, which was described in detail elsewhere [17]. The *P. penneri* 40A anti-conjugate serum was obtained by the immunization of New Zealand white rabbits as described previously [7].

4.5. Serological Assays

Purified LPS samples were tested with rabbit antisera in an enzyme-linked immunosorbent assay (ELISA), and Western blot procedure after sodium dodecyl sulfate polyacrylamide gel electrophoresis (SDS-PAGE) with non-adsorbed antisera and/or antisera adsorbed with selected alkali-treated LPSs. All assays were performed as previously described [16] with some modifications [7].

4.6. Chemical Method

Methylation analysis was performed according to the method of Ciucanu and Kerek [18]. Partially methylated alditol acetates were analyzed by gas chromatography-mass spectrometry using a Thermo Scientific ITQ system using a Zebron™ ZB-5HT (Thermo Fisher Scientific, Waltham, USA), GC Capillary Column (30 m × 0.25 mm × 0.25 μm) and with temperature rising from 150 to 270 °C at 8 °C/min.

4.7. Instrumental Methods

ESI MS analyses were performed using a Bruker microTOF-QII mass spectrometer (Bruker Brema, Germany) in a positive ion mode. The samples were dissolved in an acetonitrile–water–formic acid solution (50:50:0.1, $v/v/v$). The spectra were scanned in the m/z 200–2200 range. The mass isolation window for the precursor ion selection was set to 4 Da in the MS^2 analyses.

All NMR spectra were recorded using a Bruker Avance III 600 Hz spectrometer equipped with a 2.5 mm microprobe, incorporating gradients along the z-axis. The measurements were performed at 298 K. The signals were assigned by one- and two-dimensional experiments: ^{1}H-^{1}H COSY, TOCSY (with mixing time: 30, 60, 100 ms), NOESY and ^{1}H-^{13}C HSQC-DEPT, HSQC-TOCSY, and HMBC. The data were acquired and processed using standard Bruker software. The processed spectra were assigned with the help of the SPARKY program [19].

Acknowledgments: The authors wish to thank the Laboratory of Biomedical Chemistry "Neolek" for the possibility to use their microTOF-QII spectrometer and Jaroslaw Ciekot for technical assistance. The research was supported by Wroclaw Research Centre EIT+ under the project "Biotechnologies and advanced medical technologies"—BioMed (POIG.01.01.02-02-003/08) financed from the European Regional Development Fund (Operational Programme Innovative Economy, 1.1.2). Publication supported by Wroclaw Centre of Biotechnology, programme The Leading National Research Centre (KNOW) for years 2014–2018.

Author Contributions: Agata Palusiak, Anna Maciejewska and Marta Kaszowska designed research, performed experiments, analyzed data and wrote the manuscript. Antoni Rozalski, Czeslaw Lugowski analyzed data and wrote the manuscript.

Conflicts of Interest: The authors declare no conflict of interest.

References

1. Drzewiecka, D.; Sidorczyk, Z. Characterization of *Proteus penneri* species–Human opportunistic pathogens. *Post. Mikrobiol.* **2005**, *44*, 113–126.
2. Palusiak, A. Immunochemical properties of *Proteus penneri* lipopolysaccharides—One of the major *Proteus* sp. virulence factors. *Carbohydr. Res.* **2013**, *380*, 16–22. [CrossRef] [PubMed]
3. Knirel, Y.A.; Perepelov, A.V.; Kondakova, A.N.; Senchenkova, S.N.; Sidorczyk, Z.; Rozalski, A.; Kaca, W. Structure and serology of O-antigens as the basis for classification of *Proteus* strains. *Innate Immun.* **2011**, *17*, 70–96. [CrossRef] [PubMed]
4. Vinogradov, E.; Sidorczyk, Z.; Knirel, Y.A. Structure of the lipopolysaccharide core region of the bacteria of the genus *Proteus*. *Aust. J. Chem.* **2002**, *55*, 61–67. [CrossRef]
5. Holst, O. The structures of core regions from enterobacterial lipopolysaccharides—An update. *FEMS Microbiol. Lett.* **2007**, *271*, 3–11. [CrossRef] [PubMed]
6. Siwińska, M.; Levina, E.A.; Ovchinnikova, O.G.; Drzewiecka, D.; Shashkov, A.S.; Różalski, A.; Knirel, Y.A. Classification of a *Proteus penneri* clinical isolate with a unique O-antigen structure to a new *Proteus* serogroup, O80. *Carbohydr. Res.* **2015**, *407*, 131–136. [CrossRef] [PubMed]

7. Palusiak, A.; Dzieciątkowska, M.; Sidorczyk, Z. Application of two different kinds of sera against the *Proteus penneri* lipopolysaccharide core region in search of epitopes determining cross-reactions with antibodies. *Arch. Immunol. Ther. Exp.* **2008**, *56*, 135–140. [CrossRef] [PubMed]

8. Palusiak, A.; Sidorczyk, Z. Characterization of epitope specificity of *Proteus penneri* 7 lipopolysaccharide core region. *Acta Biochim. Pol.* **2010**, *57*, 529–532. [PubMed]

9. Palusiak, A.; Sidorczyk, Z. Serological characterization of the core region of lipopolysaccharides of rough *Proteus* sp. strains. *Arch. Immunol. Ther. Exp.* **2009**, *57*, 303–310. [CrossRef] [PubMed]

10. Palusiak, A. Classification of *Proteus penneri* lipopolysaccharides into core region serotypes. *Med. Microbiol. Immunol.* **2016**, *205*, 615–624. [CrossRef] [PubMed]

11. Vinogradov, E.; Bock, K. The structure of the core part of *Proteus vulgaris* OX2 lipopolysaccharide. *Carbohydr. Res.* **1999**, *320*, 239–243. [CrossRef]

12. Jansson, P.E.; Kenne, L.; Widmalm, G. Structure of the O-antigen polysaccharide from *Escherichia coli* O18ac: A revision using computer-assisted structural analysis with the program CASPER. *Carbohydr. Res.* **1989**, *193*, 322–325. [CrossRef]

13. Vinogradov, E. Structure of the core part of the lipopolysaccharide from *Proteus mirabilis* genomic strain HI4320. *Biochemistry* **2011**, *76*, 803–807. [CrossRef] [PubMed]

14. Vinogradov, E.; Sidorczyk, Z.; Knirel, Y.A. Structure of the core part of the lipopolysaccharides from *Proteus penneri* strains 7, 8, 14, 15, and 21. *Carbohydr. Res.* **2002**, *337*, 643–649. [CrossRef]

15. Westphal, O.; Jann, K. Bacterial lipopolysaccharides extraction with phenol-water and further applications of the procedure. *Methods Carbohydr. Chem.* **1965**, *5*, 83–91.

16. Sidorczyk, Z.; Zych, K.; Toukach, F.V.; Arbatsky, N.P.; Zabłotni, A.; Shashkov, A.S.; Knirel, Y.A. Structure of the *O*-polysaccharide and classification of *Proteus mirabilis* strain G1 in *Proteus* serogroup O3. *Eur. J. Biochem.* **2002**, *269*, 1406–1412. [CrossRef] [PubMed]

17. Jennings, H.J.; Lugowski, C. Immunochemistry of groups A, B, and C meningococcal polysaccharide-tetanus toxoid conjugates. *J. Immunol.* **1981**, *127*, 1011–1018. [PubMed]

18. Ciucanu, I.; Kerek, F. A simple and rapid method for the permethylation of carbohydrates. *Carbohydr. Res.* **1984**, *131*, 209–217. [CrossRef]

19. Goddard, T.D.; Kneller, D.G. *SPARKY 3*; University of California: San Francisco, CA, USA, 2001.

International Journal of
Molecular Sciences

Article

The Complete Structure of the Core Oligosaccharide from *Edwardsiella tarda* EIB 202 Lipopolysaccharide

Marta Kaszowska [1,*,†], Elena de Mendoza-Barberá [2,†], Anna Maciejewska [1], Susana Merino [2], Czeslaw Lugowski [1,3] and Juan M. Tomás [2]

1   Department of Immunochemistry, Hirszfeld Institute of Immunology and Experimental Therapy, Polish Academy of Sciences, R. Weigla 12, PL-53-114 Wroclaw, Poland; aniaaugustyniuk@iitd.pan.wroc.pl (A.M.); lugowski@iitd.pan.wroc.pl (C.L.)
2   Department of Microbiology, University of Barcelona, Diagonal 643, 08071 Barcelona, Spain; edemendoza@ub.edu (E.d.M.-B.); smerino@ub.edu (S.M.); jtomas@ub.edu (J.M.T.)
3   Department of Biotechnology and Molecular Biology, University of Opole, PL-45-035 Opole, Poland
*   Correspondence: marta.kaszowska@iitd.pan.wroc.pl; Tel.: +48-71-370-99-27
†   These authors contributed equally to this work.

Academic Editor: William Chi-shing Cho
Received: 18 February 2017; Accepted: 24 May 2017; Published: 31 May 2017

Abstract: The chemical structure and genomics of the lipopolysaccharide (LPS) core oligosaccharide of pathogenic *Edwardsiella tarda* strain EIB 202 were studied for the first time. The complete gene assignment for all LPS core biosynthesis gene functions was acquired. The complete structure of core oligosaccharide was investigated by ^{1}H and ^{13}C nuclear magnetic resonance (NMR) spectroscopy, electrospray ionization mass spectrometry MS^{n}, and matrix-assisted laser-desorption/ionization time-of-flight mass spectrometry. The following structure of the undecasaccharide was established:

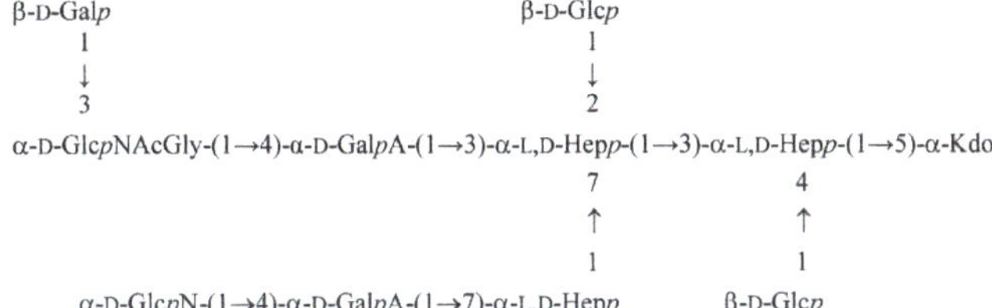

The heterogeneous appearance of the core oligosaccharide structure was due to the partial lack of β-D-Gal*p* and the replacement of α-D-Glc*p*NAcGly by α-D-Glc*p*NGly. The glycine location was identified by mass spectrometry.

Keywords: *Edwardsiella tarda*; core oligosaccharide; MALDI-TOF MS; ESI MS^{n}; NMR; genomic

---

1. Introduction

Edwardsiella tarda is a Gram-negative bacterium and a pathogen of farmed fish. It is the etiological agent of a systematic disease called edwardsiellosis, which has been reported to affect a wide range of freshwater and marine fish [1,2]. In addition to fish, *E. tarda* is also an occasional human pathogen and known to cause both gastroenteritis and extraintestinal infections in humans [3,4]. A number of virulence-associated systems and factors, such as the type III and type VI secretion systems, the LuxS/AI-2 quorum sensing system, and hemolysin systems, have been identified in *E. tarda* [5]. Additionally, a sialidase shows a potential pathogenicity and immunogenicity [6].

In Gram-negative bacteria, the lipopolysaccharide (LPS) is one of the major structural and immunodominant molecules of the outer membrane. It consists of three moieties: lipid A, core oligosaccharide, and *O*-specific polysaccharide (*O*-antigen). The *O*-antigen is the external component of LPS, and its structure consists of different number of repeating units. The *O*-specific polysaccharide chains are transferred to lipid A-core to form LPS, in a step involving WaaL, the putative bifunctional enzyme named *O*-antigen ligase. Another interesting feature is the high chemical variability shown by the *O*-antigen LPS, leading to a similar genetic variation in the genes involved in their biosynthesis, the so-called *wb* cluster (for a review, see [7]). Despite the emerging importance of this pathogenic microorganism, until now only four LPS structures of *E. tarda* strains were investigated [8–11].

In studies of several *Enterobacteriaceae* such as *Escherichia coli*, *Salmonella enterica*, and *Klebsiella pneumoniae*, genes involved in LPS core biosynthesis are usually found clustered in a region of the chromosome, the *waa* gene cluster [12,13]. On the other hand, a careful analysis of several full sequenced genomes suggested that genes for the LPS core biosynthesis may not be clustered and may be distributed between several regions, e.g., as in *Yersinia pestis* [14] or *Proteus mirabilis* [15]. In other cases, only a single gene involved in LPS core biosynthesis is out of the *waa* gene cluster, for instance, *Plesiomonas shigelloides* [16]. Nothing is known about the genomics or the LPS core structure from any *E. tarda* strain, besides the role played by the *waaL* (*O*-antigen ligase) characterized from strain EIB 202 [17]. *E. tarda* strain EIB 202 was isolated from moribund fish *Scophthalmus maximus* in a marine culture farm in China [18], and its full genome sequenced [19].

Here, the chemical structure of the core oligosaccharide in a pathogenic strain of *E. tarda* EIB 202 to proceed with the genomics of the core biosynthesis is reported for the first time.

2. Results

2.1. Isolation of the Core Oligosaccharide

LPS of *E. tarda* EIB 202 was isolated from bacterial mass with a yield of 0.5%. The mild acid hydrolysis of the LPS yielded eight polysaccharide (PS) and oligosaccharide (OS) fractions: PSI-VI consisting of a core oligosaccharide substituted by several repeating units, and OSVII and OSVIII—the unsubstituted core oligosaccharide fractions. The high yield of PSI-VI suggested the smooth (S-LPS) type of *E. tarda* EIB 202 LPS. The data presented herein concern the OSVIII fraction. The differences between OSVII and OSVIII fractions are presented herein based on MALDI-TOF MS (matrix assisted laser desorption/ionization-time of flight mass spectrometry) and ESI MS^n (electrospray ionization mass spectrometry) analysis.

2.2. Structure Analysis of the Core Oligosaccharide Fractions

The chemical analyses of OSVIII showed the presence of 2,3,7-trisubstituted L,D-Hep*p*, 3,4-disubstituted L,D-Hep*p*, 7-substituted L,D-Hep*p*, two terminal D-Glc*p*N, two terminal D-Glc*p*, two 4-substituted D-Gal*p*A, and 5-substituted Kdo*p*. The analyses of OSVII showed the presence of monosaccharides identified for OSVIII and additionally two sugar residues: the terminal D-Gal*p*, and 3-substituted D-Glc*p*NAc was identified instead of the terminal D-Glc*p*N in OSVIII.

The ^{1}H NMR (nuclear magnetic resonance) spectrum of the OSVIII contained main signals for nine anomeric protons, and signals characteristic for the deoxy protons of Kdo*p* residue belongs to part of the core oligosaccharide (residues **A–J**). The HSQC-DEPT (heteronuclear single-quantum correlation-distortionless enhancement by polarization transfer) spectra obtained for the OSVIII fraction contained signals for nine major anomeric protons and carbons, and Kdo spin systems, respectively (Figure 1 and Table 1).

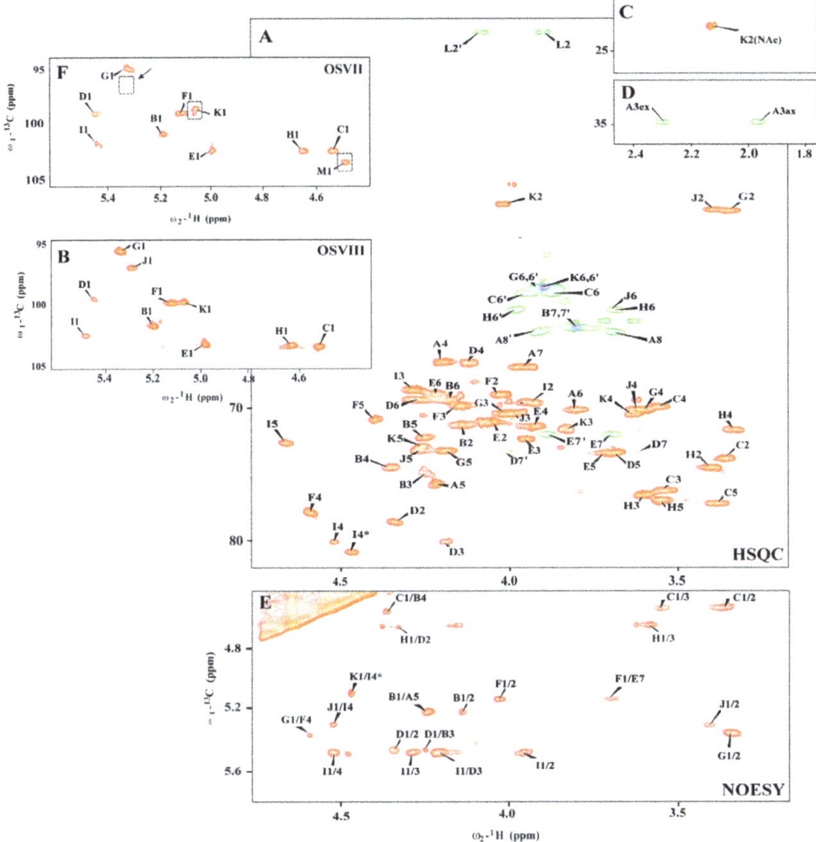

Figure 1. (**A–D**) Selected regions of the ^{1}H–^{13}C HSQC-DEPT (heteronuclear single-quantum correlation–distortionless enhancement by polarization transfer) and (**E**) ^{1}H–^{1}H NOESY (nuclear overhauser spectroscopy) spectra of the fraction OSVIII of *Edwardsiella tarda* EIB 202 lipopolysaccharide (LPS); (**F**) anomeric region of the ^{1}H–^{13}C HSQC-DEPT spectrum of the fraction OSVII with marked difference in comparison with OSVIII. The cross-peaks are labeled as shown in the text.

Residue **A** was identified as the 5-substituted Kdo on the basis of characteristic deoxy proton signals at δ_H 1.96 ppm (H-3*ax*) and δ_H 2.29 ppm (H-3*eq*), as well as a high chemical shift of the C-5 signal (δ_C 75.7 ppm). Residue **B** (δ_H/δ_C 5.19/101.7 ppm, $^1J_{C-1,H-1}$ ~175 Hz) was recognized as the 3,4-disubstituted L-*glycero*-α-D-*manno*-Hep*p* residue on the basis of the small vicinal couplings between H-1, H-2, and H-3 and relatively high chemical shifts of the C-3 (δ_C 74.7 ppm) and the C-4 (δ_C 74.4 ppm) signals. Residue **C** (δ_H/δ_C 4.51/103.3 ppm, $^1J_{C-1,H-1}$ ~163 Hz) and residue **H** (δ_H/δ_C 4.63/103.2 ppm, $^1J_{C-1,H-1}$ ~166 Hz) were recognized as the β-D-Glc*p* on the basis of the large vicinal couplings between all protons in the sugar ring. Residue **D** (δ_H/δ_C 5.44/99.6 ppm, $^1J_{C-1,H-1}$ ~175 Hz) was recognized as the 2,3,7-trisubstituted L-*glycero*-α-D-*manno*-Hep*p* residue from the ^{1}H and ^{13}C chemical shift values, small vicinal couplings between H-1, H-2, and H-3 and relatively high chemical shift values of the C-2 (δ_C 78.6 ppm), C-3 (δ_C 80.0 ppm), and C-7 (δ_C 73.3 ppm) signals. Residue **E** (δ_H/δ_C 4.98/103.2 ppm, $^1J_{C-1,H-1}$ ~173 Hz) was recognized as the 7-substituted L-*glycero*-α-D-*manno*-Hep*p* from the ^{1}H and ^{13}C chemical shifts, the small vicinal couplings between H-1, H-2, and H-3 and the relatively high chemical shift value of the C-7 (δ_C 72.0 ppm) signal. Residues **F** (δ_H/δ_C 5.12/99.8 ppm, $^1J_{C-1,H-1}$ ~175 Hz) was recognized as the 4-substituted α-D-Gal$p$A based on the characteristic five proton spin system, the high chemical shifts of the H-3 (δ_H 4.16 ppm), H-4 (δ_H 4.62), H-5 (δ_H 4.41), and C-4 (δ_C 77.9 ppm) signals, the large vicinal couplings between H-2 and H-3 and small vicinal coupling between H-3, H-4, and H-5. Residue **I** (δ_H/δ_C 5.47/102.5 ppm, $^1J_{C\text{-}1,H\text{-}1}$ ~174 Hz) was also recognized as the 4-substituted α-D-Gal$p$A residue based on the similar characteristic five proton spin system. Residues **G** (δ_H/δ_C 5.33/95.6 ppm, $^1J_{C\text{-}1,H\text{-}1}$ ~176 Hz) and **J** (δ_H/δ_C 5.29/97.0 ppm, $^1J_{C\text{-}1,H\text{-}1}$ ~176 Hz) were recognized as the terminal α-D-Glc$p$N due to the large coupling between H-1, H-2, and H-3 and the small vicinal coupling between H-3, H-4, and H-5, as well as the chemical shift value of the C-2 (δ_C 55.1 and δ_C 55.1 for **G** and **J**, respectively). The 1D ^{31}P NMR spectrum showed no indication of phosphate groups in the OSVIII.

Additionally, the residue **K** (δ_H/δ_C 5.06/99.8 ppm, $^1J_{C\text{-}1,H\text{-}1}$ ~165 Hz) was recognized as the terminal α-D-Glc$p$NAc from a low ^{13}C chemical shift of the C-2 signal (δ_C 54.6 ppm), and the large vicinal couplings between all ring protons. The *N*-acetyl group at δ_H/δ_C 2.13/23.2 ppm (δ_C 175.9 ppm) was identified. The presence of heterogeneity in OSVIII was due to partial replacement of α-D-Glc$p$N (residue **J**) by α-D-Glc$p$NAc (residue **K**). The last sugar residue **M** (δ_H/δ_C 4.45/103.6 ppm, $^1J_{C\text{-}1,H\text{-}1}$ ~163 Hz), identified only in OSVII, was recognized as the terminal β-D-Gal$p$ due to the large vicinal couplings between H-1, H-2, and H-3 and the small vicinal couplings between H-3, H-4, and H-5. The terminal residue **M** in OSVII is linked to C-3 of $\rightarrow$ 3)-α-D-Glc$p$NAc (residue **K**) corresponding to the terminal form of this residue in the OSVIII.

In the HSQC-DEPT spectra of OSVIII (at δ_H/δ_C 3.90/41.8 ppm), additional negative CH$_2$ signals were detected. These resonances showed correlation with a carbonyl carbon signals at δ_C 168.0 ppm in the HMBC (heteronuclear multiple bond correlation) spectra, suggesting the presence of glycine (residue **L**). This residue was also confirmed by mass spectrometry.

The monosaccharide sequence in OSVIII was established using a NOESY (nuclear overhauser spectroscopy) and HMBC experiments. NOESY spectra showed strong inter-residue cross-peaks between the following transglycosidic protons: H-1 of **B**/H-5 of **A**, H-1 of **D**/H-3 of **B**, H-1 of **C**/H-4 of **B**, H-1 of **E**/H-7,7' of **D**, H-1 of **F**/H-7,7' of **E**, H-1 of **G**/H-4 of **F**, H-1 of **H**/H-2 of **D**, H-1 of **I**/H-3 of **D**, and H-1 of **J**/H-4 of **I**. (Figure 1E). The HMBC spectrum of OSVIII confirmed substitution positions of all of the monosaccharide residues (data not shown). Additionally, NOESY showed the cross-peak between H-1 of **K** and H-4 of **I***, and it also provided evidence for the heterogeneity of OSVIII with the presence of α-D-Glc$p$NAc and as a substitution of 4-substituted α-D-Gal$p$A (residue **I**) in OSVIII.

The OSVIII and OSVII fractions were analyzed by ES MS^n and MALDI-TOF MS/MS. Ten sugar residues: two Glc, three Hep, two GalA, two GlcN, and Kdo, give together a monoisotopic mass of 1812.567 Da (M_{OSVIII}). Eleven sugar residues from OSVII give together a monosotopic mass of 2016.630 Da. The negative ESI MS^n (Figure 2A) mass spectrum of OSVIII showed the main ion at m/z 905.2 $[M_{OSVIII}\text{-}2H]^{2-}$ correspond to core oligosaccharide (OSVIII), and ions correspond to core substituted with the glycine (Gly) residue at m/z 933.7 $[M_{OSVIII}\text{+}Gly\text{-}2H]^{2-}$ and m/z 924.7 $[M_{OSVIII}\text{+}Gly\text{-}H_2O\text{-}2H]^{2-}$. The negative ESI MS^n mass spectrum of OSVII showed the main ion at m/z 1035.8 $[M_{OSVII}\text{+}Gly\text{-}2H]^{2-}$, which represented the complete core oligosaccharide substituted by Gly (Figure 2B).

Table 1. ^{1}H and ^{13}C NMR (nuclear magnetic resonance) chemical shifts of the core oligosaccharide of *E. tarda* EIB 212 LPS.

Residues	Oligosaccharide		Chemical Shifts (ppm)							
	OSVIII	OSVII	H1/C1	H2/C2	H3(ax,eq)/C3	H4/C4	H5/C5	H6,6'/C6	H7,7'/C7	H8,8'/C8 (NAc)
A →5)-Kdo	*	*	- nd	- 97.7	1.96, 2.29 34.7	4.21 66.5	4.25 75.7	3.81 70.1	3.97 66.9	3.69, 3.92 64.3
B →3,4)-L-*glycero*-α-D-*manno*-Hep*p*-(1→	*	*	5.19 101.7	4.15 71.3	4.25 74.7	4.35 74.4	4.26 72.2	4.18 69.4	3.80 63.9	
C →β-D-Glc*p*-(1→	*	*	4.51 103.3	3.36 73.9	3.55 76.2	3.56 69.9	3.55 76.2	3.87, 3.95 61.5		
D →2,3,7)-L-*glycero*-α-D-*manno*-Hep*p*-(1→	*	*	5.44 99.6	4.34 78.6	4.19 80.0	4.12 66.6	3.67 73.4	4.28 69.3	3.62, 4.01 73.3	
E →7)-L-*glycero*-α-D-*manno*-Hep*p*-(1→	*	*	4.98 103.2	4.05 71.1	3.95 72.3	3.93 71.4	3.72 73.4	4.23 69.1	3.69, 4.88 72.0	
F →4)-α-D-Gal*p*A-(1→	*	*	5.12 99.8	4.05 69.0	4.16 69.9	4.62 77.9	4.41 70.9	 176.7		
G α-D-Glc*p*N-(1→	*	*	5.33 95.6	3.34 55.1	4.02 70.4	3.60 76.6	4.19 73.2	3.91^a 60.0		
H α-D-Glc*p*-(1→	*	*	4.63 103.2	3.40 74.5	3.56 76.3	3.33 71.6	3.60 76.5	3.69, 3.98 62.7		
I →4)-α-D-Gal*p*A-(1→	*	*	5.47 102.5	3.93 69.6	4.26 72.2	4.48 80.9	4.67 72.5	- 175.6		
J α-D-Glc*p*N-(1→	*		5.29 97.0	3.39 55.1	3.99 70.5	3.63 70.5	4.27 72.8	3.69, 3.97 62.5		
K α-D-Glc*p*NAc-(1→		*	5.06 99.8	4.02 54.6	3.83 71.6	3.63 70.2	4.28 73.2	3.90^a 61.0		2.13 23.2
L Gly	*	*	 168.0	3.90, 4.09 41.8						
M →β-D-Gal*p*-(1→		*	4.45 103.6	3.61 71.0	3.73 72.7	3.93 71.4	3.59 75.1	3.74, 3.80 63.2		

ax: Axial position; eq: Equatorial position; nd: Not resolved.

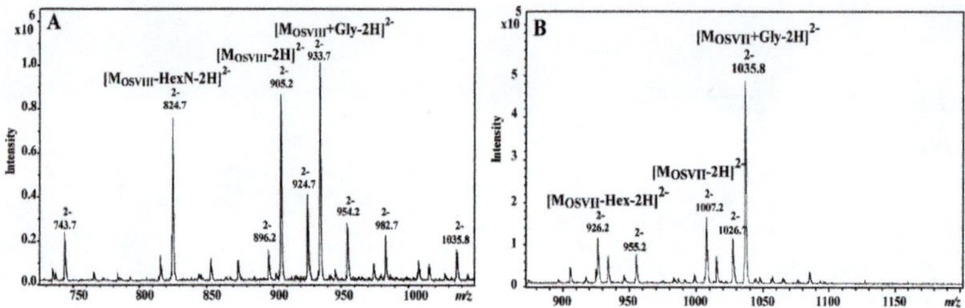

Figure 2. ESI (electrospray ionization) mass spectra of the core oligosaccharide fractions (**A**) OSVIII and (**B**) OSVII of *E. tarda* EIB 202.

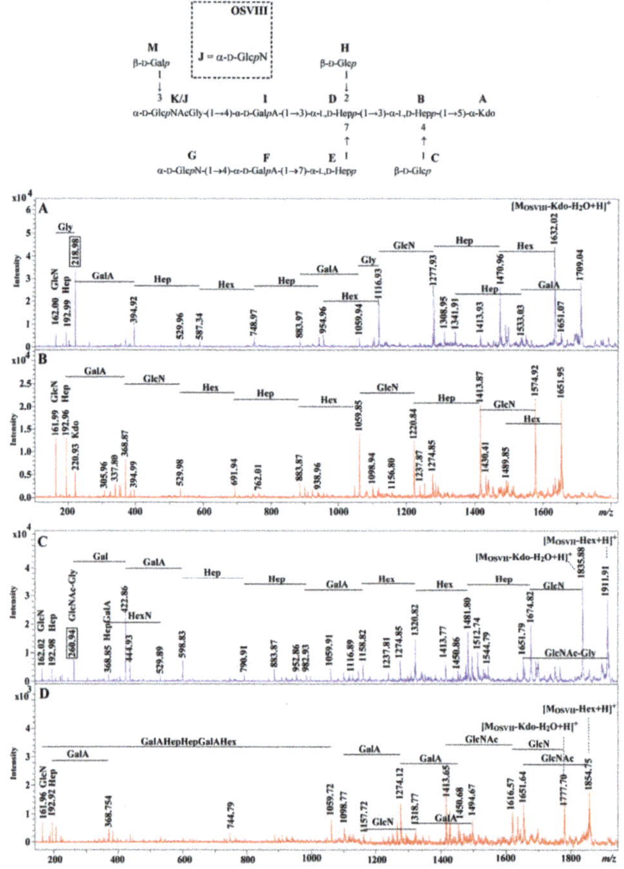

Figure 3. MALDI-TOF (matrix assisted laser desorption/ionization-time of flight) MS/MS fragmentation mass spectra of the ions (**A**) at m/z 1870.16 [M$_{OSVIII}$+Gly+H]$^+$, (**B**) at m/z 1813.15 [M$_{OSVIII}$+H]$^+$, (**C**) at m/z 2074.19 [M$_{OSVII}$+Gly+H]$^+$, and (**D**) at m/z 2017.19 [M$_{OSVII}$+H]$^+$ of *E. tarda* EIB 202, differing in the presence of glycine.

The location of glycine was determined by the positive ion mode MALDI-TOF MS/MS. The ions at m/z 1813.15 [M$_{OSVIII}$+H]$^+$, at m/z 1870.16 [M$_{OSVIII}$+Gly+H]$^+$, at m/z 2017.19 [M$_{OSVII}$+H]$^+$, and at

m/z 2074.19 [M$_{OSVIII}$+Gly+H]$^+$ were selected for further fragmentations by the use of positive ion mode MALDI-TOF MS/MS. The main daughter ions detected in the MALDI-TOF MS/MS spectra were explained. In Figure 3A, the ion at m/z 162.00 corresponds to GlcN, while the ion at m/z 218.98 was explained by the GlcN–Gly. The similar pair of fragment ions at m/z 1116.93 and at m/z 1059.94 with the mass difference corresponding to the glycine residue was also identified. These ions were not identified on the spectrum of the ion m/z 1813.15 fragmentation (Figure 3B). In Figure 3C, the daughter ion at m/z 260.94 was subsequently attributed to the GlcNAc-Gly fragment. The similar pair of fragment ions at m/z 1911.91 and at m/z 1651.79 with the mass difference corresponding to the glycine residue was also identified. These ions were not identified on the spectrum of the ion at m/z 2017.19 fragmentation (Figure 3D). These observations indicate that glycine substitutes GlcN (residue J) in OSVIII and GlcNAc (residue K) in OSVII. The positions of glycine in OSVIII and OSVII were not determined.

2.3. Organization of the E. tarda Strain EIB 202 waa Gene Cluster

In most *Enterobacteriaceae* studied so far, the genes involved in core LPS biosynthesis were found clustered (*waa* gene cluster). When we inspected the currently available *E. tarda* strain EIB 202 genome, we found a clear region with the *waa* gene cluster (proteins encoded ETAE_0083 to ETAE_0072). This *waa* region, like in the majority of *Enterobacteriaceae*, is started by the *hldE* (encoded protein ETAE_0083), which codifies for the ADP-L-*glycero*-D-mannoheptose-6-epimerase, and the end flanked by the *coaD* (encoded protein ETAE_0071) codifying for phosphopantetheine adenylyltransferase [20].

Despite the genome annotation, it seems that more of the genes are shared by different *Enterobacteriaceae* mainly *K. pneumoniae* or *P. shigelloides*, which were previously characterized by us [7,14]. Table 2 shows proteins encoded from *E. tarda* EIB 202 *waa*.

Table 2. Characteristics of the proteins encoded from *E. tarda* EIB 202 *waa*.

Protein	Homologus Protein	Predicted Function	% Identity/Similarity
ETAE_0083	HldE *Enterobacteriaceae*	ADP-L-*glycero*-D-*manno* Heptose-6-epimerase	85/91
ETAE_0082	WaaF *Enterobacteriaceae*	ADP-heptose:LPS heptosyl transferase II	76/86
ETAE_0081	WaaC *Enterobacteriaceae*	ADP-heptose:LPS heptosyl transferase I	78/84
ETAE_0080	WaaL *Klebsiella pneumoniae*	*O*-antigen ligase	29/46
ETAE_0079	Similar only to WabK *Klebsiella pneumoniae*	unknown	34/53
ETAE_0078	WapC *Plesiomonas shigelloides*	UDP-galacturonic a transferase $\alpha(1{\rightarrow}7)$ to HepIII acid	76/89
ETAE_0077	WapB *Plesiomonas shigelloides*	UDP-*N*-acetyl glucosamine $\alpha(1{\rightarrow}4)$ to GalAII	61/82
ETAE_0076	WabN *Klebsiella pneumoniae*	Protein deacetilase	77/89
ETAE_0075	WaaQ *Klebsiella pneumoniae*	ADP-heptose: LPS heptosyl	72/83
ETAE_0074	WabG *Klebsiella pneumoniae*	UDP-galacturonic acid transferase $\alpha(1{\rightarrow}3)$ to HepIII	78/87
ETAE_0073	WabH *Klebsiella pneumoniae*	UDP-*N*-acetyl glucosamine transferase $\alpha(1{\rightarrow}4)$ to GalAI	70/83
ETAE_0072	WaaA *Enterobacteriaceae*	3-deoxy-D-*manno*-octulosonic acid transferase	88/96
ETAE_0071	WapE *Plesiomonas shigelloides*	UDP-galactose transferase $\beta(1{\rightarrow}4)$ to l-HepI	82/97
ETAE_0070	CoaD *Enterobacteriaceae*	Phosphopantetehine adenylyltransferase	81/89

3. Discussion

Here, the chemical structure and genomics of the complete undecasaccharide core structure of *E. tarda* EIB 202 LPS are presented for the first time. This core oligosaccharide is heterogeneous. The heterogeneity corresponded to the partial lack of β-D-Gal*p* and the replacement of α-D-Glc*p*NAcGly by α-D-Glc*p*NGly. The functions of the genes found in the *waa* gene cluster from the *E. tarda* strain EIB 202 seems to be in agreement with the chemical structure of the LPS-core. The *E. tarda* core LPS structure is highly similar to that of *K. pneumoniae* at least up to the outer-core residue GlcNI and *P. shigelloides* 302-73 in practically up to the last monosaccharide residue that links the *O*-antigen LPS [7,14]. WabH and WapB are enzymes that transfer GlcNAc to a GalA in different acceptor substrates of LPS-core in an α(1→4) linkage. WabG and WapC are enzymes that transfer GalA to a Hep also in different LPS-core substrates and with different linkage, α(1→3) and α(1→7), respectively. It is important to note that besides performing the same enzymatic functions the acceptor substrate differences determine that the enzymes showed very little homology; furthermore, WabG and WapC are more similar among them (26 identity and 47% similarity) than WabH and WapB are (24% identity and 46% similarity), except that the latter ones showed identical linkage. This point indicates the importance of the substrates in the enzymatic reactions to build-up the LPS-core molecules. *K. pneumoniae* WabK is a glycosyltransferase that incorporates a Glc residue in a β(1→4) to GlcN. *E. tarda waa* ORF4, according to the *E. tarda* strain EIB 202 LPS-core established, as well as their low homology but unique to WabK, could be the galatosyltransferase that incorporates a Gal residue in a β(1→4) to GlcNAc *E. tarda waa* Orf5 encoding for ETAE_0079 (Table 2). All genes from the *E. tarda waa* cluster were found in the *E. ictaluri* genomes, except for *wapB* and *wapC*, which seems to be unique for the species *E. tarda*. All of the other genes from the *E. tarda waa* cluster (*hldE*, *waaF*, *waaC*, *ETAE_0079*, *waaN*, *waaQ*, *wabG*, *wabH*, *waaA*, and *wapE*) show 98% or more identity to the related *E. ictaluri* LPS-core biosynthetic genes according to their genomes. Of course, *E. ictaluri waaL*, which is the *O*-antigen ligase, is a bifunctional enzyme recognizing the *O*-antigen LPS and the LPS-core, as it usual shows a reduced identity (56%) compared to *E. tarda waaL*. Nevertheless, the *E. ictaluri waaL*-encoded protein shows the typical transmembrane domains (data not shown).

The *E. tarda* LPS motif β-Glc-(1→2)-α-L-HepII seems not to be encoded by any of the glycosyltransferases found in the *waa* cluster. This LPS motif is identical to a previously studied by us in the *P. shigelloides* strain 302-73 encoded by WapG. For this reason, we decided to blastx the *P. shigelloides* 302-73 WapG [16] against the *E. tarda* strain EIB 202 genome. We found a clear unique candidate, the gene encoding ETAE_1955, which showed 58% identity and 74% similarity to WapG. We suggest that it could be responsible for β-Glc-residue linked to HepII-[(1→2)-α-L-HepII].

4. Materials and Methods

4.1. Growth Conditions and Isolation of the Lipopolysaccharide and the Polysaccharide

Bacteria *E. tarda* EIB 202 was obtained from the Y. Zhang laboratory [18]. The bacteria were grown and harvested as described previously [21]. The LPS was extracted from bacterial cells of *E. tarda* EIB 202 by the hot phenol/water method [22]. LPS (200 mg) was degraded by treatment with 1.5% acetic acid at 100 °C for 45 min. The supernatant was fractionated on a column (1.6 × 100 cm) of Bio-Gel P-10, equilibrated with 0.05 M pyridine/acetic acid buffer, pH 5.4. Eluates were monitored with a Knauer differential refractometer, and all fractions were checked by NMR spectroscopy and mass spectrometry (MALDI-TOF and ESI MS^n).

4.2. Chemical Methods

Methylation of oligosaccharide fractions was performed according to the method described by Ciucanu and Kerek [23]. The absolute configurations of the monosaccharides were determined as described by Gerwig et al. [24]. Alditol acetates and partially methylated alditol acetates were analyzed by gas chromatography GC-MS with the Thermo Scientific TSQ system using an RX5 fused-silica capillary column (0.2 mm by 30 m) and a temperature program of 150 → 270 °C at 12 °C/min.

4.3. Instrumental Methods

All NMR spectra were recorded on a Bruker Avance III 600 MHz spectrometer equipped with a 5 mm QCI cryoprobe with z-gradient. The measurements were performed at 303 K without sample spinning and using the acetone signal (δ_H/δ_C 2.225/31.05 ppm) as an internal reference. The signals were assigned by one- and two-dimensional experiments: ^{1}H-^{1}H COSY (correlation spectroscopy), TOCSY (total correlated spectroscopy), NOESY, ^{1}H-^{13}C HSQC-DEPT, HSQC-TOCSY, and HMBC. In the TOCSY experiments, the mixing times were 30, 60, and 100 ms. The NOESY experiment was performed with the mixing time of 200 ms, and HMBC experiment with a delay of 80 ms. For observation of phosphorus atoms, one-dimensional ^{31}P NMR spectra were recorded. The data were acquired and processed using standard Bruker software. The processed spectra were assigned with the help of the SPARKY program [25].

Core oligosaccharide (1 mg/mL in mQ) fractions were analyzed using a MALDI-TOF Ultraflextreme (Bruker, Germany) instrument. The MALDI-TOF MS spectra were obtained in a positive ion mode. 2,5-Dihydroxybenzoic acid (10 mg/mL in 1:1 AcN/0.2 M citric acid [v/v]) was used as a matrix for analyses.

Negative-ion electrospray ionization mass spectra (ESI-MS^n) were recorded using an Amazon SL (Bruker Daltonics, Bremen, Germany) ion trap mass spectrometer. The samples were dissolved in a 50 μg/mL acetonitrile-water-formic acid solution (50:50:0.5 [$v/v/v$]).

4.4. Comparative Genomics

For each analyzed genome we gathered all coding sequence (CDS) and pseudo-CDS information by parsing NCBI GenBank records. When we obtained the UniProt Knowledge Base records for these loci using the cross-reference with Entrez GeneIDs and parsed them for gene names, functional annotations, and associated COG, PFAM, and TIGRFAM protein domains were studied. To annotate orthologs, we wrote custom scripts to analyze reference sequence alignments made to subject genomes with blastn and tblastn via NCBI's web application programming interface.

Acknowledgments: This work was partially funded by BIO2016-80329-P from the Spanish Ministerio de Economía y Competitividad and by the Generalitat de Catalunya (Centre de Referència en Biotecnologia). We thank Maite Polo for her technical assistance and the Servicios Científico-Técnicos from the University of Barcelona, Spain. Publication supported by Wroclaw Centre of Biotechnology, programme The Leading National Research Centre (KNOW) for years 2014-2018.

Author Contributions: Marta Kaszowska, Elena de Mendoza-Barberá, Susana Merino and Juan M. Tomás designed the research. Marta Kaszowska, Anna Maciejewska, and Elena de Mendoza-Barberá performed the experiments. Marta Kaszowska, Czeslaw Lugowski, Susana Merino and Juan M. Tomás wrote the manuscript. All authors analyzed the data.

Conflicts of Interest: The authors declare no conflict of interest.

1. Matsuyama, T.; Kamaishi, T.; Ooseko, N.; Kurohara, K.; Iida, T. Pathogenicity of motile and non-motile *Edwardsiella tarda* to some marine fish. *Fish Pathol.* **2005**, *40*, 133–135. [CrossRef]

2. Mohanty, B.R.; Sahoo, P.K. Edwardsiellosis in fish: A brief review. *J. Biosci.* **2007**, *32*, 1331–1344. [CrossRef] [PubMed]

3. Janda, J.M.; Abbott, S.L. Infections associated with the genus *Edwardsiella*: The role of *Edwardsiella tarda* in human disease. *Clin. Infect. Dis.* **1993**, *17*, 742–748. [PubMed]

4. Nelson, J.J.; Nelson, C.A.; Carter, J.E. Extraintestinal manifestations of *Edwardsiella tarda* infection: A 10-year retrospective review. *J. La. State Med. Soc.* **2009**, *161*, 103–106. [PubMed]

5. Leung, K.Y.; Siame, B.A.; Tenkink, B.J.; Noort, R.J.; Mok, Y.K. *Edwardsiella tarda* virulence mechanisms of an emerging gastroenteritis pathogen. *Microbes Infect.* **2012**, *14*, 26–34. [CrossRef] [PubMed]

6. Jin, R.P.; Hu, Y.H.; Sun, B.G.; Zhang, X.H.; Sun, L. *Edwardsiella tarda* sialidase: Pathogenicity involvement and vaccine potential. *Fish Shellfish Immunol.* **2012**, *33*, 514–521. [CrossRef] [PubMed]

7. Aquilini, E.; Tomás, J.M. (Eds.) *Lipopolysaccharides (Endotoxins)*; Reference Module in Biomedical Sciences; Elsevier: Amsterdam, The Netherlands, 2015. [CrossRef]

8. Katzenellenbogen, E.; Kocharova, N.A.; Shashkov, A.S.; Gorska-Fraczek, S.; Gamian, A.; Knirel, Y.A. Structure of the *O*-polysaccharide of *Edwardsiella tarda* PCM 1156. *Carbohydr. Res.* **2013**, *477*, 45–48. [CrossRef] [PubMed]

9. Katzenellenbogen, E.; Kocharova, N.A.; Shashkov, A.S.; Gorska-Fraczek, S.; Bogulska, M.; Gamian, A.; Knirel, Y.A. Structure of the *O*-polysaccharide of *Edwardsiella tarda* PCM 1150 containing an amide of D-glucuronic acid with L-alanine. *Carbohydr. Res.* **2013**, *347*, 84–88. [CrossRef] [PubMed]

10. Katzenellenbogen, E.; Kocharova, N.A.; Toukach, P.V.; Gorska, S.; Bogulska, M.; Gamian, A.; Knirel, Y.A. Structures of a unique *O*-polysaccharide of *Edwardsiella tarda* PCM 1153 containing an amide of galacturonic acid with 2-aminopropane-1,3-diol and an abequose-containing *O*-polysaccharide shared by *E. tarda* PCM 1145, PCM 1151 and PCM 1158. *Carbohydr. Res.* **2012**, *355*, 56–62. [CrossRef] [PubMed]

11. Vinogradov, E.; Nossova, L.; Perry, M.B.; Kay, W.W. Structural characterization of the *O*-polysaccharide antigen of *Edwardsiella tarda* MT 108. *Carbohydr. Res.* **2005**, *340*, 85–90. [CrossRef] [PubMed]
12. Heinrichs, D.E.; Yethon, J.A.; Whitfield, C. Molecular basis for structural diversity in the core regions of the lipopolysaccharides of *Escherichia coli* and *Salmonella enterica*. *Mol. Microbiol.* **1998**, *30*, 221–232. [CrossRef] [PubMed]
13. Regué, M.; Climent, N.; Abitiu, N.; Coderch, N.; Merino, S.; Izquierdo, L.; Altarriba, M.; Tomás, J.M. Genetic characterization of the *Klebsiella pneumoniae waa* gene cluster, involved in core lipopolysaccharide biosynthesis. *J. Bacteriol.* **2001**, *183*, 3564–3573. [CrossRef] [PubMed]
14. Knirel, Y.A.; Dentovskaya, S.V.; Bystrova, O.V.; Kocharova, N.A.; Senchenkova, S.N.; Shaikhutdinova, R.Z.; Titareva, G.M.; Bakhteeva, I.V.; Lindner, B.; Pier, G.B.; et al. Relationship of the lipopolysaccharide structure of *Yersinia pestis* to resistance to antimicrobial factors. *Adv. Exp. Med. Biol.* **2007**, *603*, 88–96. [PubMed]
15. Aquilini, E.; Azevedo, J.; Jimenez, N.; Bouamama, L.; Tomás, J.M.; Regué, M. Functional Identification of the *Proteus mirabilis* core lipopolysaccharide biosynthetic genes. *J. Bacteriol.* **2010**, *192*, 4413–4424. [CrossRef] [PubMed]
16. Aquilini, E.; Merino, S.; Regué, M.; Tomás, J.M. Genomic and proteomic studies of *Plesiomonas shigelloides* lipopolysaccharide core biosynthesis. *J. Bacteriol.* **2014**, *196*, 556–567. [CrossRef] [PubMed]
17. Xu, L.; Wang, Q.; Xiao, J.; Liu, Q.; Wang, X.; Chen, T.; Zhang, Y. Characterization of *Edwardsiella tarda waaL*: Roles in lipopolysaccharide biosynthesis, stress adaptation, and virulence toward fish. *Arch. Microbiol.* **2010**, *192*, 1039–1047. [CrossRef] [PubMed]
18. Xiao, J.F.; Wang, Q.Y.; Liu, Q.; Wang, X.; Liu, H.; Zhang, Y.X. Isolation and identification of fish pathogen *Edwardsiella tarda* from mariculture in China. *Aquac. Res.* **2008**, *40*, 13–17. [CrossRef]
19. Wang, Q.; Yang, M.; Xiao, J.; Wu, H.; Wang, X.; Lu, Y.; Xu, L.; Zheng, H.; Wang, S.; Zhao, G.; et al. Genome sequence of the versatile fish pathogen *Edwardsiella tarda* provides insights into its adaptation to broad host ranges and intracellular niches. *PLoS ONE* **2009**, *4*, e7646. [CrossRef] [PubMed]
20. Geerlof, A.; Lexendon, A.; Shaw, V. Purification and characterization of phosphopantheine adenyltransferase from *Escherichia coli*. *J. Biol. Chem.* **1999**, *274*, 27105–27111. [CrossRef] [PubMed]
21. Kaszowska, M.; Jachymek, W.; Jachymek, W.; Lukasiewicz, J.; Niedziala, T.; Kenne, L.; Lugowski, C. The unique structure of complete lipopolysaccharide isolated from semi-rough *Plesiomonas shigelloides* O37 (strain CNCTC 39/89) containing (2*S*)-*O*-(4-oxopentanoic acid)-α-D-Glc*p* (α-D-lenose). *Carbohydr. Res.* **2013**, *378*, 98–107. [CrossRef] [PubMed]
22. Westphal, O.; Jann, K. Bacterial lipopolysaccharide: Extraction with phenol-water and further application of the procedure. *Methods Carbohydr. Chem.* **1965**, *5*, 83–89.
23. Ciucanu, I.; Kerek, F. A simple and rapid method for the permethylation of carbohydrates. *Carbohydr. Res.* **1984**, *131*, 209–217. [CrossRef]
24. Gerwig, G.J.; Kamerling, J.P.; Vliegenthart, J.F.G. Determination of the D and L configuration of neutral monosaccharides by high-resolution capillary g.l.c. *Carbohydr. Res.* **1978**, *62*, 349–357. [CrossRef]
25. Goddard, T.D.; Kneller, D.G. *SPARKY 3*; University of California: San Francisco, CA, USA, 2001. Available online: https://www.cgl.ucsf.edu/home/sparky/ (accessed on 30 May 2017).

International Journal of
Molecular Sciences

Article

Structural Characterization of Core Region in *Erwinia amylovora* Lipopolysaccharide

Angela Casillo [1], Marcello Ziaco [1], Buko Lindner [2], Susana Merino [3], Elena Mendoza-Barberá [3], Juan M. Tomás [3] and Maria Michela Corsaro [1,*]

[1]   Department of Chemical Sciences, University of Naples "Federico II",
      Complesso Universitario Monte S. Angelo, Via Cintia 4, 80126 Naples, Italy;
      angela.casillo@unina.it (A.C.); marcello.ziaco@unina.it (M.Z.)
[2]   Division of Bioanalytical Chemistry, Research Center Borstel, Leibniz-Center for Medicine and Biosciences,
      Parkallee 10, D-23845 Borstel, Germany; blindner@fz-borstel.de
[3]   Department of Genética, Microbiología y Estadística, Universidad de Barcelona, Diagonal 643,
      08071 Barcelona, Spain; smerino@ub.edu (S.M.); elenademendoza@hotmail.com (E.M.-B.);
      jtomas@ub.edu (J.M.T.)
*   Correspondence: corsaro@unina.it; Tel.: +39-081-674-149

Academic Editor: David Arráez-Román

Received: 2 February 2017; Accepted: 28 February 2017; Published: 4 March 2017

Abstract: *Erwinia amylovora* (*E. amylovora*) is the first bacterial plant pathogen described and demonstrated to cause fire blight, a devastating plant disease affecting a wide range of species including a wide variety of *Rosaceae*. In this study, we reported the lipopolysaccharide (LPS) core structure from *E. amylovora* strain CFBP1430, the first one for an *E. amylovora* highly pathogenic strain. The chemical characterization was performed on the mutants *waaL* (lacking only the O-antigen LPS with a complete LPS-core), *wabH* and *wabG* (outer-LPS core mutants). The LPSs were isolated from dry cells and analyzed by means of chemical and spectroscopic methods. In particular, they were subjected to a mild acid hydrolysis and/or a hydrazinolysis and investigated in detail by one and two dimensional Nuclear Magnetic Resonance (NMR) spectroscopy and ElectroSpray Ionization Fourier Transform-Ion Cyclotron Resonance (ESI FT-ICR) mass spectrometry.

Keywords: lipopolysaccharide; *Erwinia amylovora*; NMR; ESI FT-ICR; structural determination

---

1. Introduction

Erwinia amylovora is the causal agent of fire blight, a disease of nutritionally important members of the family Rosaceae, such as apple and pear trees. The symptoms in apple plants are present on rootstocks, blossoms, shoots, and fruits [1]. On fruits, the disease provokes the development of ooze, which is composed of bacteria, polysaccharides, and plant sap [1]. Bacteria are transported by insects, rain, birds, wind-wipping, and hail from the ooze to flowers. From flower infection, it can also be transferred to lateral parts of the plant through an endophytic mechanism [2].

Two major virulence determinants are known for the pathogenesis of *Erwinia*: (i) one involves the *hrp/dsp* gene cluster, the role of which is to secrete and deliver proteins from bacteria to plant apoplasts and cytoplasm [1]; (ii) the second is the production of two types of exopolysaccharides (EPS), amylovoran and levan [3,4].

The outer membrane (OM) of almost all Gram-negative bacteria and of some cyanobacteria [5–8] contains lipopolysaccharides (LPSs), where they constitute approximately 75% of the outer surface. They are amphiphilic endotoxic molecules necessary for the viability and survival of Gram-negative bacteria, as they seriously contribute to the structural integrity of the OM and to the protection of the bacterial cell envelope [9].

The colony morphology of Gram-negative bacteria can appear as smooth or rough as a consequence of a different structure of the lipopolysaccharides, named smooth (S-LPS) or rough (R-LPS), respectively. The structure of a S-LPS molecule can be described as three covalently linked domains: the glycolipid portion, called lipid A; the intermediate core oligosaccharide region (core); and the O-specific polysaccharide (O-chain) [10]. Instead, the R-LPSs (named lipooligosaccharides, LOSs) completely lacks the O-specific polysaccharide chain, either due to genetic mutation or to the inherent nature of bacteria [11].

Bacterial lipopolysaccharides show multiple roles in plant–microbe interactions. LPS give a contribution to the low permeability of the outer membrane, which acts as a barrier to protect microorganisms from plant-derived antimicrobial substances. LPS-defective mutants display augmented in vitro sensitivity to antibiotics and antimicrobial peptides and, upon introduction into susceptible plants, the numbers of viable bacteria often decay very quickly [12–14].

In this study we characterized the lipopolysaccharide core structure from the strain CFBP1430, the first one completely sequenced for a highly pathogenic *E. amylovora*. In particular, the results about *waaL*, *wabH*, and *wabG* mutants are reported. We studied *wabH* and *wabG* mutants because we found these genes in the *Erwinia amylovora* strain CFBP1430 *wb*; they were fully characterized in *Klebsiella pneumoniae* and *Serratia marcescens*, and both correspond to the outer-LPS core [15]. The lipooligosaccharides were degraded both by mild hydrazinolysis (*O*-deacylation) and hot KOH (*N*-deacylation). Both products were investigated by chemical analysis, by ^{1}H and ^{13}C NMR spectroscopy, and by ESI FT-ICR spectrometry.

2. Results and Discussion

2.1. Preparation and Structural Characterization of Oligosaccharides from EaΔwaaL LPS

The LPSs from the *waaL*, *wabH*, and *wabG* mutants (EaΔwaaL, EawabH, and EaΔwabG, respectively) of the *Erwinia amylovora* strain CFBP1430 were extracted by the phenol-chloroform-light petroleum (PCP) method. WabG is responsible for the transfer of D-GalA to the O-3 position of L,D-Hep II, and WabH transfers a D-GlcNAc residue from UDP-GlcNAc to the D-GalA [15]. The monosaccharides composition was performed as already reported [16]. In particular, the Gas Chromatography-Mass Spectrometry (GC-MS) analysis of methyl and octyl glycosides for EaΔwaaL and EaΔwabH mutant LPSs revealed the following sugars; D-Glc, D-GalA, D-GlcN, L,D-Hep, D,D-Hep, and Kdo. When this analysis was performed on EaΔwabG LPS, the lack of D-GalA and D,D-Hep was observed.

The removal of fatty acids from EaΔwaaL LPS was performed both by strong alkaline and acetic acid hydrolyses. The obtained products were analyzed by ESI FT-ICR mass spectrometry. In addition, mono- and two-dimensional NMR spectroscopy (^{1}H,^{1}H DQF-COSY, ^{1}H,^{1}H TOCSY, ^{1}H,^{1}H ROESY, ^{1}H,^{13}C HSQC-DEPT, ^{1}H,^{13}C HSQC-TOCSY, and ^{1}H,^{13}C HMBC) was allowed to assign all the proton and carbon chemical shifts of each mutant. The anomeric configurations were identified on the basis of both ^{13}C chemical shift values and $^3J_{H1,H2}$ coupling constants. The sequence of the residues in all the oligosaccharides was obtained both by long-range scalar couplings and by nuclear Overhauser enhancement (NOE) data.

Starting from the totally deacylated product (L-OS$_{KOH}$, Scheme 1, structures **1a–c**), the negative ion mode ESI FT-ICR mass spectrum of the sample identified eight main species (Figure 1 and Table 1). Species **M1**, occurring at 2220.595 Da (calculated molecular mass 2220.586 Da), represented the higher molecular mass oligosaccharide chain containing the phosphorylated lipid A backbone and the core structure.

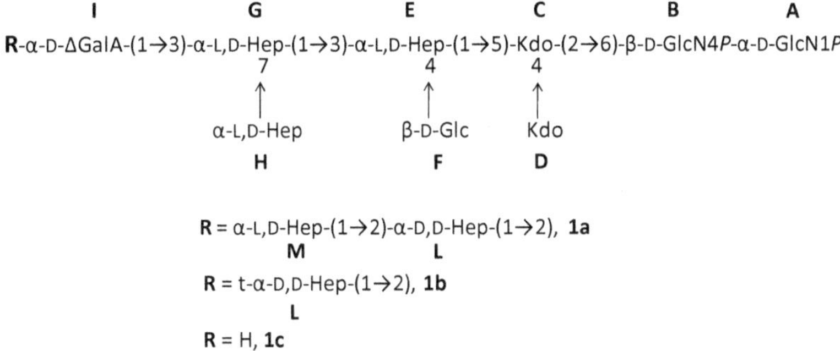

Scheme 1. Structures of the totally deacylated LPS fraction (L-OS$_{KOH}$) isolated from the *E. amylowora* EaΔwaaL mutant.

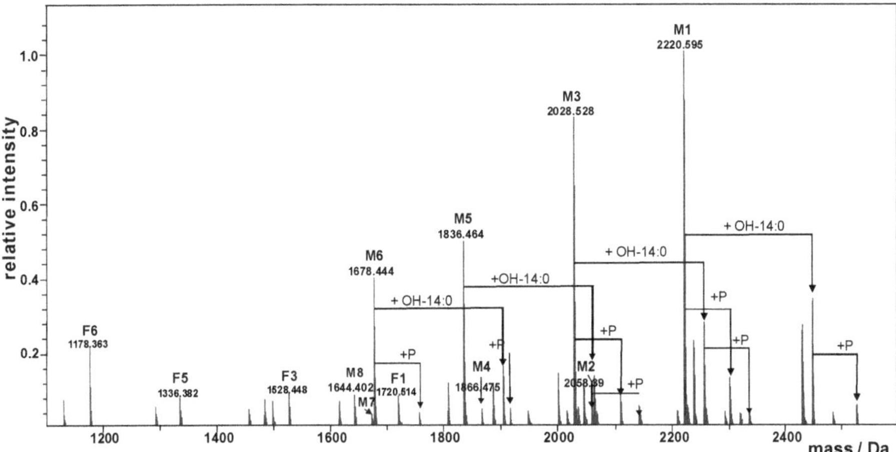

Figure 1. Charge deconvoluted ElectroSpray Ionization Fourier Transform-Ion Cyclotron Resonance (ESI FT-ICR) mass spectrum of the totally deacylated lipopolysaccharide (LPS) fraction (L-OS$_{KOH}$) isolated from the *E. amylowora* EaΔwaaL mutant. The spectrum was acquired in negative ion mode. The mass numbers given refer to the monoisotopic peak of the neutral molecular species.

Table 1. Composition of the main species observed in the charge deconvoluted ESI FT-ICR mass spectrum of the totally deacylated LPS (L-OS$_{KOH}$) from a *E. amylowora waaL* mutant (EaΔwaaL).

Species	Observed Mass (Da)	Calculated Mass (Da)	Composition
M1	2220.595	2220.589	$GlcN_2P_2Kdo_2Hep_5Glc_1\Delta HexA$
M2	2058.390	2058.536	$GlcN_2P_2Kdo_2Hep_5\Delta HexA$
M3	2028.528	2028.525	$GlcN_2P_2Kdo_2Hep_4Glc_1\Delta HexA$
M4	1866.475	1866.473	$GlcN_2P_2Kdo_2Hep_4\Delta HexA$
M5	1836.464	1836.462	$GlcN_2P_2Kdo_2Hep_3Glc_1\Delta HexA$
M6	1678.444	1678.440	$GlcN_2P_2Kdo_2Hep_3Glc_1$
M7	1674.410	1674.409	$GlcN_2P_2Kdo_2Hep_3\Delta HexA$
M8	1644.402	1644.399	$GlcN_2P_2Kdo_2Hep_2Glc_1\Delta HexA$

In particular, the following composition was attributed to the **M1** species: $GlcN_2P_2Kdo_2Hep_5Glc_1\Delta GalA$, where ΔGalA represents an unsaturated galacturonic acid. The presence of this acid unit in the **M1** species clearly indicated that the strong alkaline conditions determined the lack of one or more sugars from GalA, due to a β-elimination reaction [17]. Species **M3**, **M5**, and **M8** differed from **M1** in their heptoses composition, displaying four units in **M3** and three and two in **M5** and **M8**, respectively. The remaining species, **M2**, **M4**, and **M7**, lacked a glucose unit respect to **M1**, **M3**, and **M5**, respectively, thus confirming the terminal position for this residue, while the **M6** species lacked ΔGalA compared to **M5**. Further mass peaks originate from an additional phosphate group P (Δm = +79.966 u) and/or one linked C14:0(3OH) fatty acid (Δm = +226.193 u) due to incomplete deacylation. Finally, **F1**–**F6** are fragments induced during the ESI process leading to the cleavage of the lipid A backbone $GlcN_2P_2$ (Δm = −500.081 u).

Despite the complexity of the ^1H-^{13}C Heteronuclear Single Quantum Coherence-Distortionless Enhancement by Polarization Transfer (HSQC-DEPT) experiment (Figure 2), it was possible to assign the signals of the main species, **M1**, **M3**, and **M5**. All the NMR data (Table 2, structures **1a–c**) strongly suggested that the fractions L-OS$_{KOH}$ and the structure of the oligosaccharide OS1 obtained from the previously published *K. pneumoniae* 52145 *waaL* mutant [17] possessed a common structural element constituted by the residues **A**, **B**, **C**, **D**, **E**, **F**, **G**, **H**, and **I** (monosaccharide units are as shown in Table 2), in agreement with the strain genomics.

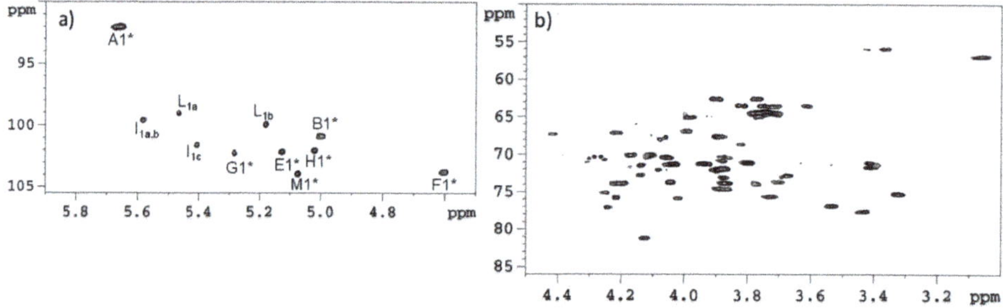

Figure 2. Anomeric (**a**) and carbinolic (**b**) region of ^1H-^{13}C HSQC-DEPT of the totally deacylated LPS (L-OS$_{KOH}$) from *E. amylowora waaL* mutant. All the correlation cross-peaks in the anomeric region indicate the signals of structures **1a–c**. The spectrum was recorded in D_2O at 298 K at 600 MHz.

Table 2. ^1H and ^{13}C chemical shifts (δ) of sugar residues of core oligosaccharides from *Erwinia amylovora* LPS. Nuclear Magnetic Resonance (NMR) spectra were recorded in D_2O at 600 MHz at 298 K.

Residue	Nucleus	H1 / C1	H2 / C2	H3 / C3	H4 / C4	H5 / C5	H6 / C6	H7 / C7	H8 / C8
α-GlcN1*P* A, 1a–c	^1H	5.65	3.37	3.94	3.41	4.21	3.88/4.27		
	^{13}C	92.0	55.9	71.2	71.1	73.8	70.2		
α-GlcN1*P* A, 3	^1H	5.53	3.28	3.80	3.51	4.05	3.65/4.20		
	^{13}C	92.0	55.9	71.0	70.9	73.4	71.0		
α-GlcN1*P* A,4	^1H	5.54	3.30	3.80	3.51	4.04	3.65/4.19		
	^{13}C	92.0	57.1	71.0	71.1	73.6	71.3		
β-GlcN4*P* B, 1a–c	^1H	4.99	3.06	3.88	3.88	3.72	3.62/3.75		
	^{13}C	100.9	56.9	73.1	74.6	75.6	63.5		
β-GlcN4*P* B, 3	^1H	4.72	2.93	3.73	3.72	3.60	3.35/3.57		
	^{13}C	100.9	56.9	70.9	74.0	72.3	63.7		
β-GlcN4*P* B, 4	^1H	4.73	2.94	3.72	3.73	3.63	3.35/3.60		
	^{13}C	100.9	57.1	70.9	73.8	75.3	64.0		
4,5-Kdo C, 1a-c	^1H	n.d.	n.d.	1.88/2.12	4.21	4.21	4.14	4.17	3.77/3.92
	^{13}C			36.0	73.9	75.7	72.9	70.2	64.9
4,5-Kdo C, 3	^1H	175.6	100.7	1.81/1.98	4.03	4.17	3.83	3.51	3.62-3.87
	^{13}C			36.0	71.6	70.1	74.0	73.4	64.3

Table 2. *Cont.*

| Residue | Nucleus | H1 | H2 | H3 | H4 | H5 | H6 | H7 | H8 |
		C1	C2	C3	C4	C5	C6	C7	C8
4,5-Kdo C, 4	[1]H	176.3	101.1	1.88/2.04	3.99	4.17	3.88	3.54	3.61-3.87
	[13]C			36.2	72.0	70.4	71.9	73.6	64.7
t-Kdo D, 1a-c	[1]H	n.d.	n.d.	1.82/2.18	4.09	4.06	3.71	4.09	3.77/3.99
	[13]C			35.9	68.1	67.7	73.7	72.1	65.0
t-Kdo D, 3	[1]H	176.1	102.6	1.67/2.06	3.95	3.95	3.60	3.56	3.56/3.87
	[13]C			35.7	67.4	67.4	73.9	73.1	64.5
t-Kdo D, 4	[1]H	176.1	102.8	1.66/2.00	3.96	3.94	3.62	3.56	3.61/3.78
	[13]C			36.3	67.7	67.9	73.3	73.3	65.1
3,4-α-L,D-Hep E, 1a–c	[1]H	5.12	4.13	4.24	4.25	4.10	4.17	3.76	
	[13]C	102.2	71.5	77.1	75.1	70.1	70.2	64.5	
3,4-α-L,D-Hep E, 2a,b	[1]H	5.12	4.07	4.23	4.23	4.10	4.15	3.75	
	[13]C	101.9	71.4	77.1	75.3	70.3	69.8	64.4	
3,4-α-L,D-Hep E, 3	[1]H	5.19	3.95	4.06	4.19	4.18	4.00	3.69	
	[13]C	100.1	71.8	75.1	74.4	72.2	70.2	65.1	
3,4-α-L,D-Hep E, 4	[1]H	5.16	4.00	3.99	4.15	4.12	3.98	3.68/3.89	
	[13]C	103.0	72.0	75.6	74.6	73.3	70.3	73.3	
t-β-Glc F, 1a-c	[1]H	4.60	3.33	3.53	3.41	3.42	3.77/3.90		
	[13]C	103.8	75.3	77.0	71.7	77.7	62.5		
t-β-Glc F, 2a,b	[1]H	4.62	3.31	3.54	3.41	3.42	3.77/3.90		
	[13]C	104.0	75.3	77.0	71.1	77.5	62.5		
t-β-Glc F, 3	[1]H	4.46	3.18	3.41	3.26	3.30	3.63/3.78		
	[13]C	103.2	75.0	76.5	71.1	77.3	62.7		
t-β-Glc F, 4	[1]H	4.43	3.19	3.39	3.23	3.36	3.61-3.85		
	[13]C	103.6	75.0	76.9	71.6	77.8	62.8		
3,7-α-L,D-Hep G, 1a–c	[1]H	5.28	4.26	4.13	3.99	3.78	4.21	3.80	
	[13]C	102.1	70.8	81.1	66.9	74.0	67.1	71.2	
3,7-α-L,D-Hep G, 2a,b	[1]H	5.26	4.12	4.03	4.01	3.90	4.17	3.77	
	[13]C	102.0	71.1	80.8	66.9	74.0	71.1	71.2	
3,7-α-L,D-Hep G, 3	[1]H	5.27	4.08	3.89	3.83	3.92	n.d.	3.69	
	[13]C	102.2	70.7	80.6	66.9	73.3		72.7	
t-α-L,D-Hep G, 4	[1]H	5.16	4.06	3.77	3.77	3.56	3.95	3.55	
	[13]C	100.1	71.0	72.0	67.6	73.2	70.3	64.7	
t-α-L,D-Hep H, 1a–c	[1]H	5.02	4.04	3.90	3.89	3.67	4.06	3.72	
	[13]C	102.0	73.6	70.9	67.6	72.9	70.5	63.6	
t-α-L,D-Hep H, 2a,b	[1]H	4.94	4.02	4.02	3.87	3.64	n.d	n.d	
	[13]C	101.8	73.1	71.3	67.6	72.7			
t-α-L,D-Hep H, 3	[1]H	4.84	3.88	3.91	3.88	3.63	3.93	3.64	
	[13]C	102.9	71.3	70.3	67.0	73.4	64.6	62.9	
2-Δ-GalA I, 1a,b	[1]H	5.58	4.02	4.53	5.82				
	[13]C	99.6	75.8	65.8	108.9				
t-Δ-GalA I, 1c	[1]H	5.40	3.87	4.42	n.d.				
	[13]C	101.6	71.9	67.3					
2,4-α-GalA I, 2a,b	[1]H	5.48	4.07	4.22	4.23	4.47			
	[13]C	99.5	80.4	68.5	77.2	72.6	176.2		
2-α-GalA I, 3	[1]H	5.39	3.91	3.96	4.21	4.34			
	[13]C	98.9	80.6	69.2	72.2	73.7	176.8		
2-α-D,D-Hep L, 1a	[1]H	5.35	4.06	3.97	3.89	n.d.	4.12	3.61	
	[13]C	98.9	80.0	71.4	68.5	n.d.	72.1	63.5	

Table 2. *Cont.*

Residue	Nucleus	H1	H2	H3	H4	H5	H6	H7	H8
		C1	C2	C3	C4	C5	C6	C7	C8
t-α-D,D-Hep L, 1b	^{1}H	5.18	4.03	3.90	3.83	3.87	4.04	3.82	
	^{13}C	99.9	71.3	72.1	68.7	73.9	71.4	63.5	
2-α-D,D-Hep L, 2a,b	^{1}H	5.28	3.99	3.99	3.81	3.99	4.05	3.84	
	^{13}C	96.9	80.8	71.4	68.9	75.0	71.4	63.6	
2-α-D,D-Hep L, 3	^{1}H	4.99	3.87	3.75	3.67	3.92	4.09	3.69	
	^{13}C	98.3	80.6	71.7	68.7	73.3	73.1	63.1	
t-α-L,D-Hep M, 1a	^{1}H	5.09	4.07	4.07	3.88	3.70	n.d.	n.d.	
	^{13}C	103.5	71.3	73.7	67.7	73.1			
t-α-L,D-Hep M, 2a,b	^{1}H	5.05	4.05	4.06	3.86	n.d.	n.d.	n.d.	
	^{13}C	103.5	71.4	73.8	67.7				
t-α-L,D-Hep M, 3	^{1}H	5.21	3.89	3.75	3.83	4.18	n.d.	3.56/3.78	
	^{13}C	101.9	71.5	70.1	66.8	72.1		64.8	
4-α-GlcN N, 2a,b	^{1}H	5.19	3.27	3.85	3.79	4.17	3.83		
	^{13}C	97.6	55.3	72.5	77.0	72.5	61.3		
6-α-Glc O, 2a,b	^{1}H	5.49	3.62	3.74	3.90	3.54	3.88/4.20		
	^{13}C	100.7	72.8	74.2	73.0	70.8	69.8		
t-β-Glc P, 2a	^{1}H	4.50	3.33	3.51	3.42	3.47	3.74/3.92		
	^{13}C	103.9	74.5	77.0	71.0	77.3	62.3		
2-β-Glc P, 2b	^{1}H	4.62	4.47	3.59	3.45	3.46	3.74/3.92		
	^{13}C	103.9	78.9	75.9	70.9	77.3	62.3		
t-β-Glc Q, 2b	^{1}H	4.36	3.61	3.75	3.47	4.17	n.d.		
	^{13}C	99.2	72.9	74.2	70.8	72.9			

All the species possessed α-GlcN1P residue at the reducing end (**A**, H1, δ 5.65 ppm; C1, δ 92.0 ppm), originating from the lipid A backbone. The second residue of glucosamine of the lipid A (**B**, H1, δ 4.99 ppm, C1, δ 100.9 ppm) was O-4 phosphorylated, as suggested by the downfield shift of both H4 and C4 chemical shifts [18]. In all the oligosaccharides corresponding to the **M1–M7** species, the core region was linked to the lipid A backbone through an α-Kdo residue **C**, which was substituted at O-4 position by a second residue of Kdo (**D**).

Residues **E**, **G**, **H**, **L**, and **M** were identified as five *manno*-configured α-heptopyranoses due to their small coupling constants $^3J_{\text{H1,H2}}$ and $^3J_{\text{H2,H3}}$ values. Downfield shifted carbon signals with respect to reference values [19] indicated substitutions at O-3 and O-4 of residue **E** (C3 at 77.1 and C4 at 75.1 ppm, respectively) and at O-3 and O-7 of residue **G** (C3 at 81.1 ppm and C7 at 71.2 ppm). Both residues **H** and **M** were identified as terminal units as none of their carbon signals were downfield shifted. All the heptose residues were found to be L,D-configured except for **L**; in fact, for this residue, the chemical shift of its C6 at 72.1 ppm suggested a D,D-configuration [20]. In addition, the downfield shift of its C2 resonance at 80.0 ppm clearly indicated that it was substituted at this position. Residue **I**, with H1/C1 at 5.58/99.6 ppm and H4/C4 at 5.82/108.9 ppm, was identified as a α-threo-hex-4-enuronopyranosyl unit (ΔGalA).

Finally, residue **F** was assigned to a β-glucose residue, on the basis of the proton multiplicities obtained by the DQF-COSY and TOCSY experiments. No downfield carbon chemical shifts were observed for this residue, thus indicating its terminal non reducing end position.

The sequence of the residues was deduced from the HMBC experiment, which showed correlations between C1 of **L** and H2 of **I**, H1 of **I** and C3 of **G**, H1 of **G** and C3 of **E**, C1 of **H** and H7 of **G**, and C1 of **F** and H4 of **E**.

In addition, residue **M** was found to be linked at the O-2 position of residue **L**, as shown by NOE contacts among H1 of **M** and both H1 and H2 of **L**, observed in the ROESY experiment. This last completed and confirmed the above sequence by inter-residual dipolar couplings. In the same spectrum, the intra-residue NOE contacts were in agreement with the assigned relative configuration of the monosaccharides.

All these data indicated for **M1** species of L-OS$_{KOH}$ the oligosaccharide structure **1a**. Structures **1b** and **1c**, corresponding to the species **M3** and **M5**, were identified on the basis of their terminal residue D,D-heptose and ΔGalA, respectively.

The LPS of EaΔwaaL mutant was then treated with 5% acetic acid at 100 °C, and the supernatant was analyzed (L-OS$_{AcOH}$, Scheme 2, structures **2a,b**). Gas Chromatography-Mass Spectrometry (GC-MS) analysis of the partially methylated alditol acetates of this sample indicated the presence of terminal L,D-heptose, terminal glucose, terminal D,D-heptose, 2-substituted D,D-heptose, 2-substituted glucose, 6-substituted glucose, 4-substituted glucosamine 3,4-substituted L,D-heptose, and 3,7-substituted L,D-heptose. Each sugar was recognized from both its Electronic Ionization (EI) mass spectrum and the GC column retention time, by comparison with standards. The Kdo was not revealed in this analysis, due to the presence of its anhydro form.

The negative mode ESI FT-ICR mass spectrum of **2** (Figure 3 and Table 3) and NMR data (Table 2) revealed a complex mixture of oligosaccharides. The main component of the mass spectrum **N2** species was revealed to possess a common part with the oligosaccharides of L-OS$_{KOH}$, constituted by the residues **E**, **F**, **G**, **H**, **L**, and **M**. In addition, the trisaccharide β-Glc-(1 → 6)-α-Glc-(1 → 4)-α-GlcN-(1 → was present at position O-4 of α-GalA **I**.

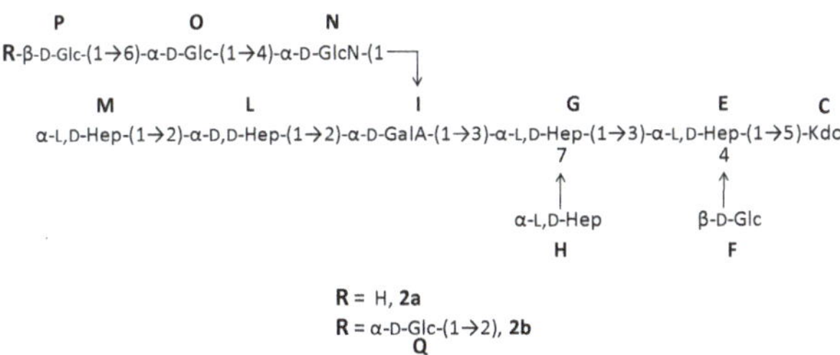

Scheme 2. Structures of acetic acid hydrolysed LPS fraction (L-OS$_{AcOH}$) isolated from the *E. amylowora* EaΔwaaL mutant.

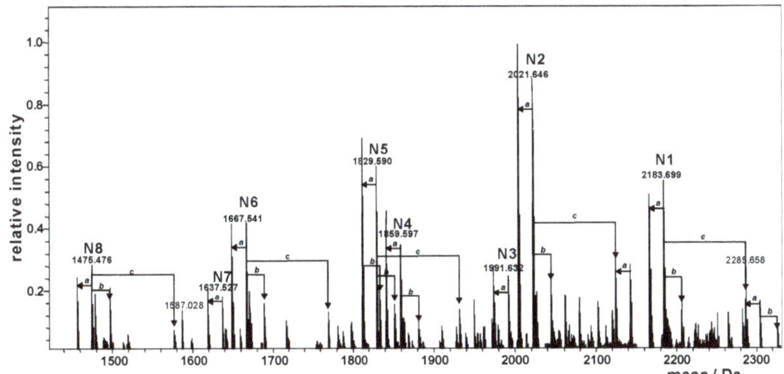

Figure 3. Charge deconvoluted ESI FT-ICR mass spectrum of supernatant of the acetic acid hydrolysed LPS fraction (L-OS$_{AcOH}$) isolated from *E. amylowora waaL* mutant. The spectrum was acquired in negative ion mode. The mass numbers given refer to the monoisotopic peak of the neutral molecular species. Further mass peaks originate from a: loss of H_2O (Δm = −18.010 u); b: sodium adducts [M − H + Na] (Δm = +21,979 u); c: an additional phosphate group (P − H + Na) (Δm = +101.948 u).

Table 3. Composition of the main species observed in the charge deconvoluted ESI FT-ICR mass spectrum of supernatant of the acetic acid hydrolysed LPS fraction (L-OS$_{AcOH}$) isolated from *E. amylowora waa* L. mutant.

Species	Observed Mass (Da)	Calculated Mass (Da)	Composition
N1	2183.699	2183.698	Kdo$_1$Hep$_5$Glc$_4$GalA$_1$GlcN$_1$
N2	2021.646	2021.645	Kdo$_1$Hep$_5$Glc$_3$GalA$_1$GlcN$_1$
N3	1991.632	1991.635	Kdo$_1$Hep$_4$Glc$_4$GalA$_1$GlcN$_1$
N4	1859.597	1859.592	Kdo$_1$Hep$_5$Glc$_2$GalA$_1$GlcN$_1$
N5	1829.590	1829.578	Kdo$_1$Hep$_4$Glc$_3$GalA$_1$GlcN$_1$
N6	1667.541	1667.529	Kdo$_1$Hep$_4$Glc$_2$GalA$_1$GlcN$_1$
N7	1637.527	1637.518	Kdo$_1$Hep$_3$Glc$_3$GalA$_1$GlcN$_1$
N8	1475.476	1475.466	Kdo$_1$Hep$_3$Glc$_2$GalA$_1$GlcN$_1$

We compared our NMR data with the data already published for deacylated oligosaccharides from *Serratia marcenscens* [19], and we found very similar chemical shifts, even if we recognized in *Erwinia* an additional minor component (**N1** species, Table 2 and Figure 3). In fact, the molecular mass of the **N1** species indicated an additional hexose with respect to that of **N2**. This residue (**Q**, Table 2) was identified as gluco-configurated with an α anomeric configuration, as suggested by its $^3J_{H1,H2}$ anomeric coupling constant of 3.5 Hz. A long-range heteronuclear scalar coupling between H1 of **Q** and C2 of **P** indicated that in the **N1** species the oligosaccharide substituting the O-4 of GalA had the structure: α-Glc-(1 $\rightarrow$ 2)-β-Glc-(1 $\rightarrow$ 6)-α-Glc-(1 $\rightarrow$ 4)-α-GlcN. All these data were in agreement with the methylation analysis, except for the 2,4-disubstituted galacturonic acid residue, which was absent in the GC-MS chromatogram. This fact could be due to the hindrance of both oligosaccharides substituting O-2 and O-4 positions of GalA, thus preventing its methylation.

2.2. Preparation and Structural Characterization of Oligosaccharides from EaΔwabH and EaΔwabG LPSs

The LPSs from EaΔwabH and EaΔwabG mutants were completely deacylated by hydrazine followed by the KOH reaction. After purification on a gel filtration Sephadex G10 column, the samples, named H-OS$_{KOH}$ and G-OS$_{KOH}$ respectively, were analysed by mono- and two-dimensional NMR experiments (Table 2). The acetic acid hydrolysis was not performed on these two mutants as no β-degradation was observed for these samples, indicating that no information was lost.

The ^{1}H NMR spectrum of the totally deacylated EaΔwabH LPS (H-OS$_{KOH}$, Figure S1, Supporting Information) showed nine main anomeric proton signals, assigned to residues **A–M**, in the range 5.6–4.4 ppm. Signals in the range 2.2–1.5 ppm, attributable to the presence of the Kdo units, were also present.

The galacturonic acid residue (residue **I**, Scheme 3, structure **3**) was recognized from the chemical shift of its C6 at δ 176.8 ppm (Table 2), which correlated with its H5 proton at δ 4.34 ppm in the Heteronuclear Multiple Bond Correlation (HMBC) experiment.

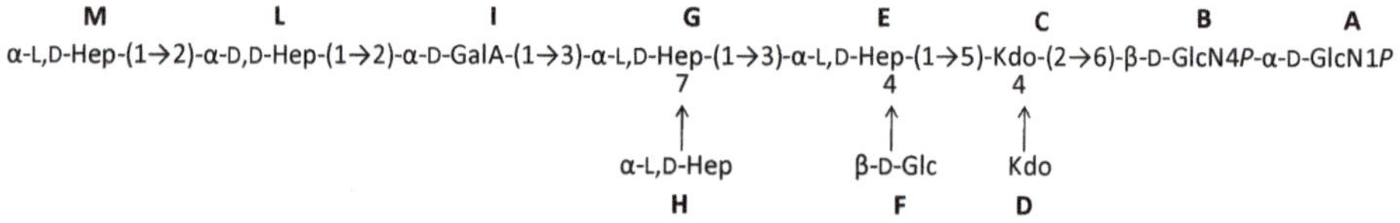

<pre>
 M   L   I   G   E   C   B   A
α-L,D-Hep-(1→2)-α-D,D-Hep-(1→2)-α-D-GalA-(1→3)-α-L,D-Hep-(1→3)-α-L,D-Hep-(1→5)-Kdo-(2→6)-β-D-GlcN4P-α-D-GlcN1P
                                               7   4   4
                                               ↑   ↑   ↑
                                          α-L,D-Hep   β-D-Glc   Kdo
                                               H   F   D
</pre>

Scheme 3. Structures of the totally deacylated LPS fraction (H-OS$_{KOH}$) isolated from the *E. amylowora* EaΔwabH mutant.

In addition, the O-substitution at position 2 of **I** was inferred by the long-range between H1 of **L** and C2 of **I**. The value of the coupling constant $^3J_{H1,H2}$ of the anomeric proton was found to be 3.5 Hz, thus indicating for residue **I** an α configuration. The chemical shifts of heptoses **E, G, H, L,** and **M** were in agreement with that already found for structure **1**, as well as that of terminal β-Glc **F**. Finally, both the HMBC (Figure S2) and ROESY experiments indicated the sequence shown in structure **3**, which is in agreement with strain genomics.

The ^{1}H-NMR spectrum of the totally deacylated EaΔwabG LPS, (G-OS$_{KOH}$, Figure S3, Supporting Information) showed only five anomeric proton signals. All the ^{1}H and ^{13}C chemical shifts of each residue were identified and assigned by two-dimensional NMR experiments (Table 2). NMR data, together with glycosyl analysis, indicated the lack of the trisaccharide α-Hep-(1 → 2)-α-Hep-(1 → 2)-α-GalA. This was in agreement with the hypothesis of assignment of the *wabG* gene to a galacturonic acid residue glycosyltransferase. In addition, the ^{1}H,^{13}C HSQC-DEPT NMR experiment (Figure S4) revealed the absence of the inner core heptose residue **H**, since only two L,D-heptose anomeric signals, **E** and **G**, were found (Scheme 4, structure **4**).

This fact suggested that the lack of GalA residue may preclude the addition of heptose **H** to the position O-7 of heptose **G**.

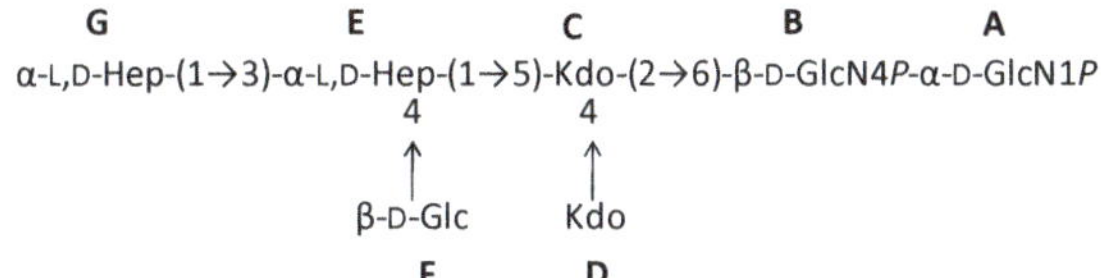

Scheme 4. Structures of the totally deacylated LPS fraction (G-OS$_{KOH}$) isolated from the *E. amylowora* EaΔwabG mutant.

3. Experimental Section

3.1. Bacteria Growth and LPS Isolation

Dried bacteria cells from EaΔwaaL (3.3 g), EaΔwabH (2.63 g), and EaΔwabG (3.58 g) mutants of *E. amylovora* strain CFBP1430 were all extracted by the PCP method [21] to give 208 mg of LPS (LPS$_{PCP}$ yield 6.3% *w/w* of dried cells) for the *waaL* mutant, 124 mg of LPS for the *wabH* mutant (LPS$_{PCP}$ yield 4.2% *w/w* of dried cells), and 170 mg of LPS (LPS$_{PCP}$ yield 4.7% *w/w* of dried cells) for the *wabG* mutant, respectively.

3.2. Sugar Analysis

The sugar analysis was performed as reported [16]. The absolute configurations of the sugars were determined by gas chromatography of the acetylated (*S*)-2-octyl glycosides [22]. The derivatized samples were injected into an Agilent Technologies gas chromatograph 6850A equipped with a mass selective detector 5973N and a Zebron ZB-5 capillary column (Phenomenex, Bologna, Italy, 30 m × 0.25 mm i.d., flow rate 1 mL/min, He as carrier gas). The following temperature programs were used; 140 °C for 3 min, 140 °C → 240 °C at 3 °C/min (acetylated methyl glycosides), 150 °C for 5 min, 150 °C → 240 °C at 6 °C/min, and 240 °C for 5 min (acetylated octyl glycosides).

3.3. Mild Acid Hydrolysis

The LPS of the EaΔwaaL mutant (20 mg) was treated with 5% aqueous CH$_3$COOH (2 mL, 100 °C for 4 h). After centrifugation (7500 rpm, 4 °C, 30 min), the pellet was washed twice with water. Then, the supernatant layers were pooled together and lyophilized. The sample was then fractionated on a Bio-Gel P-10 column (Biorad, 1.5 × 110 cm, flow rate 15 mL/h, fraction volume 2 mL) and eluted with water buffered with 0.05 M pyridine and 0.05 M AcOH. The fractions containing oligosaccharides were pooled and named L-OS$_{AcOH}$ (10.1 mg).

3.4. Deacylation of the LOSs

The LOS from each mutant (100 mg) was dried over phosphorus anhydride under a vacuum and treated with hydrazine (5 mL, at 37 °C for 2 h) [23]. Cold acetone was added, and the pellet was recovered after centrifugation at 4 °C and 7000 rpm for 30 min. After being washed three times with acetone, it was suspended in water and lyophilized, obtaining L_{LOS-OH} (68.6 mg), H_{LOS-OH} (72.4 mg), and G_{LOS-OH} (65.9 mg).

The partially deacylated LPS (LOS-OH) from each mutant was dissolved in KOH 4 M and incubated at 120 °C for 16 h. The KOH was neutralized with HCl, and the mixture was extracted three times with $CHCl_3$. The aqueous phases were recovered and desalted on a Sephadex G-10 column (Amersham Biosciences, Little Chalfont, UK, 2.5 × 43 cm, 31 mL·h^{-1}, fraction volume 2.5 mL, eluent NH_4HCO_3 10 mM). The eluted oligosaccharides mixture was then lyophilized (L-OS$_{KOH}$ 19 mg; H-OS$_{KOH}$ 13 mg; G-OS$_{KOH}$ 20 mg).

3.5. Methylation Analysis

The linkage positions of the monosaccharides were determined by GC-MS analysis of the partially methylated alditol acetates (PMAAs). The acetic acid oligosaccharides fraction of EaΔwaaL mutant (1 mg) was first reduced with $NaBD_4$ and then methylated with CH_3I (100 µL) and NaOH powder in dimethyl sulfoxide (DMSO) (300 µL) for 20 h [24].

The reduction of the carboxymethyl groups was obtained by treating the sample with sodium boro deuteride ($NaBD_4$). After the reaction working up, the sample was totally hydrolyzed with 2 M trifuoroacetic acid (TFA) at 120 °C for 2 h, reduced again with $NaBD_4$, and acetylated with Ac_2O and pyridine (50 µL each, 100 °C for 30 min) [25]. The PMAA mixture was analyzed by GC-MS with the following temperature program: 90 °C for 1 min, 90 → 140 °C at 25 °C/min, 140 °C → 200 °C at 5 °C/min, 200 → 280 °C at 10 °C/min, and 280 °C for 10 min.

3.6. Mass Spectrometry Analysis

Mass spectra were (ESI FT-ICR) were performed in negative ion mode using an APEX QE (Bruker Daltonics GmbH, Bremen, Germany) equipped with a 7 Tesla actively shielded magnet. The sample concentration was ~10 ng/µL. The solutions were sprayed at a flow rate of 2 µL/min and analyzed. The mass spectra obtained were charge-deconvoluted, and the mass numbers reported refer to the monoisotopic masses of the neutral molecules.

3.7. NMR Spectroscopy

^{1}H and ^{13}C NMR spectra were performed using a Bruker Avance 600 MHz spectrometer (Milano, Italy) equipped with a cryoprobe. All 2D homo- and heteronuclear experiments (double quantum-filtered correlation spectroscopy, DQF-COSY; total correlation spectroscopy TOCSY; rotating-frame nuclear Overhauser enhancement spectroscopy, ROESY; nuclear Overhauser effect spectroscopy, ^{1}H,^{13}C HSQC-DEPT; and heteronuclear multiple bond correlation, ^{1}H,^{13}C HMBC) were obtained using the standard pulse sequences available in the Bruker software (TopSpin 3.1 version). The TOCSY and ROESY experiments were obtained with mixing times of 100 ms. Chemical shifts were measured at 298 K in D_2O.

4. Conclusions

In this work, for the first time, we have characterised the complete structure of the core oligosaccharide from *E. amylowora* strain CFBP1430 LPS (Scheme 5, structure **5**). To this aim, we prepared three mutants, i.e. *waaL*, *wabH*, and *wabG*, the purified lipopolysaccharides of which were characterised. The reported data showed that the core oligosaccharides here reported share structural fragments with those of *Klebsiella pneumoniae* and *Serratia marcescens*. In particular, by comparison with the core of *S. marcescens* the present structure lacks the non-stoichiometric Ko residue, a feature considered useful to distinguish between the genera *Burkholderia* and *Pseudomonas*.

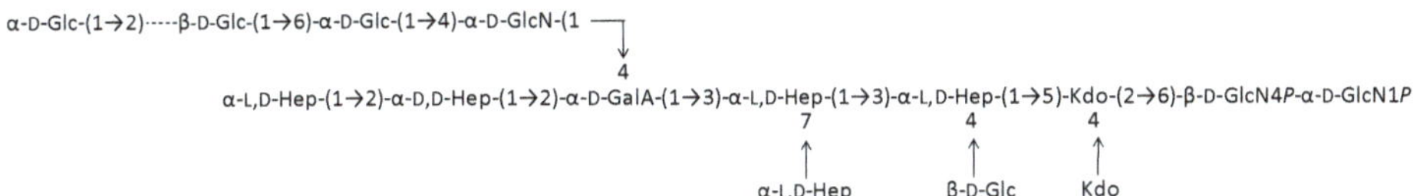

Scheme 5. Core oligosaccharide structures from *E. amylovora* strain CFBP1430 LPS.

Supplementary Materials: Supplementary materials can be found at www.mdpi.com/1422-0067/18/3/559/s1.

Acknowledgments: This work was partially funded by BIO2016-80329-P from the Spanish Ministerio de Economía y Competitividad, and from Generalitat de Catalunya (Centre de Referència en Biotecnologia). We thank Maite Polo for her technical assistance and the Servicios Científico-Técnicos from the University of Barcelona.

Author Contributions: Angela Casillo: Performed the experiments, suggested critical parameters in design of experiments and co-wrote the paper. Marcello Ziaco and Buko Lindner: Performed the experiments and co-wrote the paper. Susana Merino and Elena Mendoza-Barberá: Performed the experiments. Juan M. Tomás and Maria Michela Corsaro: Designed the experiments, provided advice in performance of experiments and wrote the paper.

Conflicts of Interest: The authors declare no conflict of interest.

1. Oh, C.; Beer, S.V. Molecular genetics of *Erwinia amylovora* involved in the development of fire blight. *FEMS Microbiol. Lett.* **2005**, *253*, 185–192. [CrossRef] [PubMed]

2. Thomoson, S.V. Epidemiology of Fire Blight. In *Fire Blight: The Disease and Its Causative Agent, Erwinia Amylovora*; Vanneste, J.L., Ed.; CABI: New York, NY, USA, 2000; pp. 9–36.

3. Bellemann, P.; Geider, K. Localization of transposon insertions in pathogenicity mutants of *Erwinia amylovora* and their biochemical characterization. *J. Gen. Microbiol.* **1992**, *138*, 931–940. [CrossRef] [PubMed]

4. Koczan, J.M.; McGrath, M.J.; Zhao, Y.; Sundin, G.W. Contribution of *Erwinia amylovora* exopolysaccharides amylovoran and levan to biofilm formation: Implications in pathogenicity. *Phytopathology* **2009**, *99*, 1237–1244. [CrossRef] [PubMed]

5. Wilkinson, S.C. Composition and Structure of Bacterial Lipopolysaccharides. In *Surface Carbohydrates of the Procaryotic Cell*; Sutherland, I.W., Ed.; Academic Press Inc.: New York, NY, USA, 1977; pp. 97–105.

6. Lüderitz, O.; Freudenberg, M.A.; Galanos, C.; Lehmann, V.; Rietschel, E.T.; Shaw, D.H. Lipopolysaccharides of Gram-Negative Bacteria. In *Current Topics in Membranes and Transport*; Razin, S., Rottem, S., Eds.; Academic Press Inc.: New York, NY, USA, 1982; pp. 79–151.

7. Westphal, O.; Lüderitz, O.; Galanos, C.; Mayer, H.; Rietschel, E.T. The Story of Bacterial Endotoxin. In *Advances in Immunopharmacology*; Chedid, L., Hadden, J.W., Spreafico, F., Eds.; Pergamon Press: Oxford, UK, 1986; pp. 13–34.

8. Carillo, S.; Pieretti, G.; Bedini, E.; Parrilli, M.; Lanzetta, R.; Corsaro, M.M. Structural investigation of the antagonist LPS from the cyanobacterium *Oscillatoria planktothrix* FP1. *Carbohydr. Res.* **2014**, *388*, 73–80. [CrossRef] [PubMed]

9. Alexander, C.; Rietschel, E.T. Invited review: Bacterial lipopolysaccharides and innate immunity. *J. Endotoxin Res.* **2001**, *7*, 167–202. [CrossRef] [PubMed]

10. Caroff, M.; Karibian, D. Structure of bacterial lipopolysaccharides. *Carbohydr. Res.* **2003**, *338*, 2431–2447. [CrossRef] [PubMed]

11. Lüderitz, O.; Galanos, C.; Risse, H.J.; Ruschmann, E.; Schlecht, S.; Schmidt, G.; Schulte-Holthausen, H.; Wheat, R.; Westphal, O.; Schlosshardt, J. Structural relationship of *Salmonella* O and R antigens. *Ann. N. Y. Acad. Sci.* **1966**, *133*, 349–374. [CrossRef] [PubMed]

12. Berry, M.C.; McGhee, G.C.; Zhao, Y.; Sundin, G.W. Effect of a *waaL* mutation on lipopolysaccharide composition, oxidative stress survival, and virulence in *Erwinia amylovora*. *FEMS Microbiol. Lett.* **2009**, *291*, 80–87. [CrossRef] [PubMed]

13. Kao, C.C.; Sequeira, L. A gene cluster required for coordinated biosynthesis of lipopolysaccharide and extracellular polysaccharide also affects virulence of *Pseudomonas solanacearum. J. Bacteriol.* **1991**, *173*, 7841–7847. [CrossRef] [PubMed]

14. Toth, I.K.; Thorpe, C.J.; Bentley, S.D.; Mulholland, V.; Hyman, L.J.; Perombelon, M.C.M.; Salmond, G.P.C. Mutation in a gene required for lipopolysaccharide and enterobacterial common antigen biosynthesis affects virulence in the plant pathogen *Erwinia carotovora* subsp. *atroseptica. Mol. Plant Microbe Interact.* **1999**, *12*, 499–507. [CrossRef] [PubMed]

15. Aquilini, E.; Tomás, J.M. Lipopolysaccharides (Endotoxins). *Ref. Modul. Biomed. Sci.* **2015**. [CrossRef]

16. Pieretti, G.; Corsaro, M.M.; Lanzetta, R.; Parrilli, M.; Canals, R.; Merino, S.; Tomás, J.M. Structural studies of the O-chain polysaccharide from *Plesiomonas shigelloides* strain 302–73 (serotype O1). *Eur. J. Org. Chem.* **2008**, *2008*, 3149–3155. [CrossRef]

17. Regué, M.; Izquierdo, L.; Fresno, S.; Piqué, N.; Corsaro, M.M.; Naldi, T.; de Castro, C.; Waidelich, D.; Merino, S.; Tomas, J. A second outer-core region in *Klebsiella pneumoniae* lipopolysaccharide. *J. Bacteriol.* **2005**, *187*, 4198–4206. [CrossRef] [PubMed]

18. Holst, O.; Muller-Leonnies, S.; Lindner, B.; Brade, H. Chemical structure of the lipid A of *Escherichia coli* J-5. *Eur. J. Biochem.* **1993**, *214*, 695–701. [CrossRef] [PubMed]

19. Vinogradov, E.; Lindner, B.; Seltmann, G.; Radziejewska-Lebrecht, J.; Holst, O. Lipopolysaccharides from *Serratia marcescens* possess one or two 4-Amino-4-deoxy-L-arabinopyranose 1-phosphate residues in the Lipid A and D-*glycero*-D-*talo*-Oct-2-ulopyranosonic Acid in the inner core region. *Chem. Eur. J.* **2006**, *12*, 6692–6700. [CrossRef] [PubMed]

20. Süsskind, M.; Brade, L.; Brade, H.; Holst, O. Identification of a novel heptoglycan of α1→2-linked D-*glycero*-D-*manno*-heptopyranose. *J. Biol. Chem.* **1998**, *273*, 7006–7017. [PubMed]

21. Galanos, C.; Lüderitz, O.; Westphal, O. A new method for the extraction of R lipopolysaccharides. *Eur. J. Biochem.* **1969**, *9*, 245–249. [CrossRef] [PubMed]

22. Leontein, K.; Lindberg, B.; Lönngren, J. Assignment of absolute configuration of sugars by GLC of their acetylated glycosides formed from chiral alcohols. *Carbohydr. Res.* **1978**, *62*, 359–362. [CrossRef]

23. Holst, O. Deacylation of Lipopolysaccharides and Isolation of Oligosaccharide Phosphates, In Bacterial Toxins: Methods and Protocols. *Methods Mol. Biol.* **2000**, *145*, 345–353. [PubMed]

24. Ciucanu, I.; Kerek, F. A simple and rapid method for the permethylation of carbohydrates. *Carbohydr. Res.* **1984**, *131*, 209–217.

25. Corsaro, M.M.; Pieretti, G.; Lindner, B.; Lanzetta, R.; Parrilli, E.; Tutino, M.L.; Parrilli, M. Highly phosphorylated core oligosaccharide structures from cold-adapted *Psychromonas arctica. Chem. Eur. J.* **2008**, *14*, 9368–9376. [CrossRef] [PubMed]

International Journal of
Molecular Sciences

Review

The Glycosyltransferases of LPS Core: A Review of Four Heptosyltransferase Enzymes in Context

Joy M. Cote and Erika A. Taylor *

Department of Chemistry, Wesleyan University, Middletown, CT 06459, USA; jcote@wesleyan.edu
* Correspondence: eataylor@wesleyan.edu; Tel.: +1-860-685-2739; Fax: +1-860-685-2211

Received: 29 September 2017; Accepted: 24 October 2017 ; Published: 27 October 2017

Abstract: Bacterial antibiotic resistance is a rapidly expanding problem in the world today. Functionalization of the outer membrane of Gram-negative bacteria provides protection from extracellular antimicrobials, and serves as an innate resistance mechanism. Lipopolysaccharides (LPS) are a major cell-surface component of Gram-negative bacteria that contribute to protecting the bacterium from extracellular threats. LPS is biosynthesized by the sequential addition of sugar moieties by a number of glycosyltransferases (GTs). Heptosyltransferases catalyze the addition of multiple heptose sugars to form the core region of LPS; there are at most four heptosyltransferases found in all Gram-negative bacteria. The most studied of the four is HepI. Cells deficient in HepI display a truncated LPS on their cell surface, causing them to be more susceptible to hydrophobic antibiotics. HepI–IV are all structurally similar members of the GT-B structural family, a class of enzymes that have been found to be highly dynamic. Understanding conformational changes of heptosyltransferases are important to efficiently inhibiting them, but also contributing to the understanding of all GT-B enzymes. Finding new and smarter methods to inhibit bacterial growth is crucial, and the Heptosyltransferases may provide an important model for how to inhibit many GT-B enzymes.

Keywords: LPS; lipopolysaccharide; heptosyltransferase; protein dynamics; glycosyltransferase; GT-B; inhibitor design

1. Introduction

Well before the discovery of penicillin, bacteria have been evolving to resist natural antibiotics and other extracellular threats [1]; however, advances in medical techniques and over use of antibiotics has lead to an exponential increase in resistance. The resulting bacteria that are resistant to multiple antimicrobial agents are regarded as one of the biggest threats to global health, food security and development by both the World Health Organization (WHO) and Centers for Disease Control and Prevention (CDC) [2,3]. Gram-negative bacteria are of particular concern because their peptidoglycan is protected behind the bacterium's outer membrane (OM). Furthermore, the physical properties of the OM enhance bacterial survival in diverse environments and while also limiting the uptake of many drugs [4].

The overall organization of the OM is largely conserved, despite some variability between different Gram-negative bacteria. Typically, the OM contains a phospholipid bilayer with the extracellular leaflet being composed of a mixture of lipopolysaccharides (LPS), lipoproteins, and oligosaccharides [5–7]. LPS are the primary component of the OM in most species of Gram-negative bacteria and have been shown to play an important role in cell motility, intestinal colonization, bacterial biofilm formation, and antibiotic resistance [8,9]. This makes understanding the role of LPS in host-pathogen interactions an area of great interest, especially in the development of therapeutic agents for the treatment of gram-negative bacterial infections [10].

The LPS is composed of three main sections: a hydrophobic lipid A anchored to the membrane, a core oligosaccharide containing octulose and heptose sugar moieties, and a repeating O-antigen region containing a diversity of sugars that are unique to bacterial cell surfaces (including pentoses, deoxy-hexoses, lactyl functionalized hexoses, heptoses and nonuloses) [11–13]; these components vary slightly between different bacteria [9,14]. The core oligosaccharide is further divided into the inner and outer core; the inner core is highly conserved and proximal to lipid A whereas the outer core is more variable. It is possible that the evolutionarily preserved structure of the inner core may be crucial for establishing the barrier function of the OM [15]. As can been seen in a schematic of *E. coli* OM biosynthesis, the complex synthesis and transportation of LPS involves many proteins (Figure 1) [7,16,17]. Sequential glycosyl transfer from nucleotide sugar precursors by membrane associated (or proximal) glycosyltransferases (GT) on the cytoplasmic face of the plasma membrane form the inner and outer core which is then transported to the periplasm where the fully formed O-antigen repeat is attached and the full LPS is exported to the outer leaflet [7,9,17]. Mutations in the biosynthesis of LPS are often lethal to bacteria, with the minimalistic structure required for secretion of LPS to the outer membrane being Kdo_2-lipid A (lipid A with two 3-deoxy-D-manno-octo-2-ulosonic acid (Kdo) sugar moieties attached) [9,18,19]. Truncation of the LPS by mutations to the inner core display a deep-rough phenotype and exhibit hypersensitivity to hydrophobic antibiotics and detergents [4,20,21].

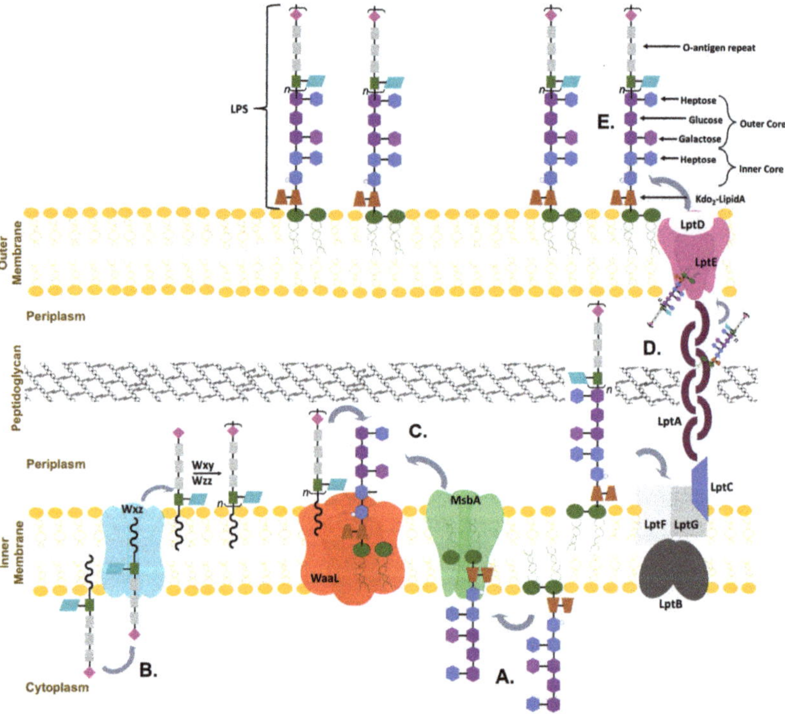

Figure 1. Representative organization of Gram-negative bacterium from *Escherichia coli* membrane. (**A**) demonstrates the sequential addition of inner core sugars to Kdo_2-lipid A anchored into the inner membrane; (**B**) Represents the formation of the O-antigen repeat also formed in the inner membrane; (**C**) Once both are complete, they are flipped into the periplasm and the O-antigen repeats are attached to the top of the core; (**D**) The whole lipopolysaccharides (LPS) is then transported across the periplasm and peptidoglycan layer; (**E**) finally embedding into the outer membrane.

2. Glycosyltransferases

Glycoslytransferases (GTs) are enzymes that catalyze the addition of various saccharides onto other biomolecules. GTs encompass a large group of enzymes that have similar structural scaffolds, but have evolved to utilize a vast diversity of substrates. Often, GTs act sequentially in order to build a complex polymer—the product of one GT will then be the acceptor substrate for the following GT. Many cellular functions such as: energy storage, cell wall structure, cell-cell interactions, signaling, host-pathogen, and protein glycosylation are dependent upon complex carbohydrates and polysaccharides. Due to this, biosynthesis of these chemically diverse oligosaccharides and polysaccharides require the use of multiple GTs [22–25].

2.1. Glycosyltransferase Structural Folds

Presently, there are over 300,000 known and putative GTs according to CAZY.org (Carbohydrate-Active enZYmes Database) and the number is ever growing [26]. Although GTs have diverse sequences, they can be characterized into three structural classes: GT-A, GT-B, and GT-C (Figure 2). Despite their differences, GTs catalyze the formation of a glycosidic bond, where a high-energy sugar nucleotide donates a monosaccharide to an acceptor molecule [27]. This acceptor can be a variety of molecules, such as oligosaccharides, monosaccharides, proteins, lipids, and others [23].

2.1.1. GT-A Structural Fold

SpsA from (*Bacillus subtilis*) was the first enzyme to be crystallized and characterized with a GT-A fold [28]. This structural family is characterized by two tightly packed domains, comprised of two βαβRossman-like folds, that are closely associated to form a continuous central β-sheet (Figure 2A). The close proximity of the folds lead many to describe the GT-A fold as a single domain, however there are distinct binding sites for the two substrates [25]. A short N-terminal domain binds the donor substrate and C-terminal domain is an open groove that binds an acceptor substrate.

GT-A enzymes typically contain two Asp residues separated by a non-conserved amino acid (DXD motif), that is located on a loop connecting the central β-sheet to an additional smaller β-sheet. A divalent cation interacts with one or both of the Asp residues and is essential for stabilization of the pyrophosphate group of the donor substrate. Typically absence of the cation renders the enzyme inactive, however there are a small number of GT-A enzymes where a DXD motif and cation are not required [29]. Additionally, in order for both substrates to bind and for catalysis to occur a conformational change is required. Specifically, the loops adjacent to the active site, often adopt a variety of conformations to assist in binding the substrate and performing chemistry [30].

2.1.2. GT-B Structural Fold

Similar to the GT-A structural class, the GT-B protein contains two βαβRossman-like domains. Unlike GT-A proteins, in the GT-B structural class the two domains are connected by a linker region with a deep cleft containing the active site separating the two domains (Figure 2B). Donor substrate binds to the C-terminal domain, while the N-terminal domain binds the acceptor substrates. A large domain movement is required for catalysis to occur in many GT-B enzymes [25,31–34]. There are no divalent metal ions or DXD motif in GT-B enzymes, and it is believed that the pyrophosphate is stabilized by charged and polar residues as well as the natural dipole of the α-helices located in the donor substrate binding site [24].

A DNA-modifying β-glucosyltransferase was the first GT-B to be structurally characterized, and was shown to have both an open and closed conformation. The closed conformation is at least in part caused by binding of donor substrate uridine diphosphate glucose (UDP-glucose) [32]. Nonetheless, donor substrate binding does not alway induce a conformational change in GT-B enzymes. For instance, heptosyltransferase I (HepI), which has been crystallized with and without an analog to its donor substrate, ADP-L-(glycero)-D-(manno)-heptose (ADPH), shows no donor substrate induced closure [31].

Some GT-B enzymes are closed with acceptor bound thus, the order or ligand state required for closure does not seem to be universal across the family.

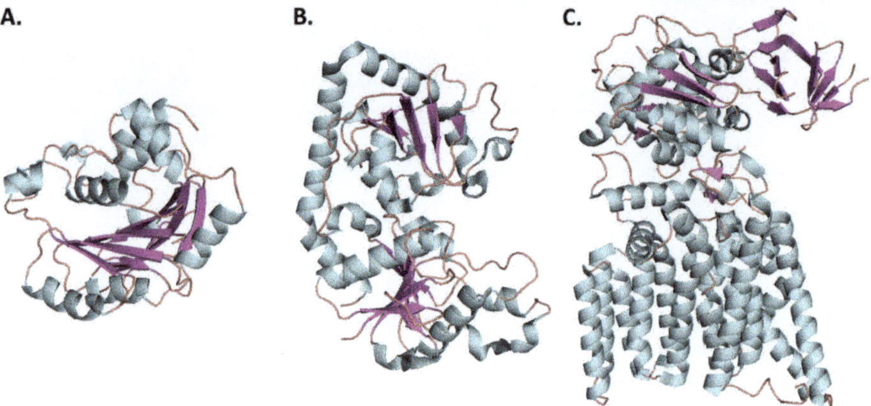

Figure 2. Representative folds of the first glycosyltransferase (GT) enzymes crystallized in each structural family: loops, α-helices, and β-sheets are colored salmon, cyan, and purple respectively. (**A**) GT-A fold represented by SpsA from *Bacillus subtilus*, PDB: 1QGQ; (**B**) GT-B fold represented by bacteriophage T4 β-glucosyltransferas, PDB: 1JG7; (**C**) GT-C fold represented by PglB from *Campylobacter lari*, PDB: 3RCE.

2.1.3. GT-C Structural Fold

Until recently, the third structural fold (GT-C) was only predicted on the bases of sequence analysis [35]. In 2011, the first GT-C structure was published for a bacterial oligosaccharyltransferase from *Campylobacter lari*, comprised of 2 domains: a 13 transmembrane domain and a periplasmic domain containing a mix of α/β folds [36]. Both GT-A and GT-C enzymes have a DXD motif, however the location of the DXD tripeptide in GT-C family is at the carboxy-terminal end of the first transmembrane helix. A small patch of hydrophobic amino acids following the helix is common. Although this arrangement is similar to that of the DXD signiture in GT-A structural fold, there is no conservation of sequence between these two regions [24,36].

2.1.4. Catalytic Mechanisms

Regardless of the structural fold, glycosyltransferases catalyze the transfer of a glycosyl group with either inversion or retention of the stereoconfiguration at the anomeric carbon. Both GT-A and GT-B families have been found to have inverting and retaining enzymes, however all GT-C enzymes are predicted to utilize an inverting mechanism. While literature often states that inverting GTs follow a S_N2-like mechanism, implying an uncharged transition state, it is generally accepted that the reaction has an oxocarbenium ion like transition state which is more correctly defined as a partially associated S_N1-like mechanism (Figure 3B) [37,38]. Unlike inverting GTs, there are multiple mechanism for retaining enzymes. Initially, it was thought that all retaining enzymes proceed via a double displacement mechanism with formation of a covalent glycosyl-enzyme intermediate (Figure 3C); nevertheless, only a small percentage of GTs contain a putative nucleophilic residue that is properly located in the active site to facilitate such a mechanism [25,39,40]. There is little direct evidence for a double displacement mechanism in the literature, however Soya et. al. was able to observe glycosyl-enzyme intermediates by mass spectrometry [41]. Further experimental and computational work has shown that a front face or S_Ni (substitution nucleophilic internal-like) mechanism is likely the primary pathway utilized [25,40,42,43]. While there is more support for an S_Ni mechanism, it is generally accepted that there are two classes of retaining GTs, that are classified based on the presence or absence of a nucleophile in the active site.

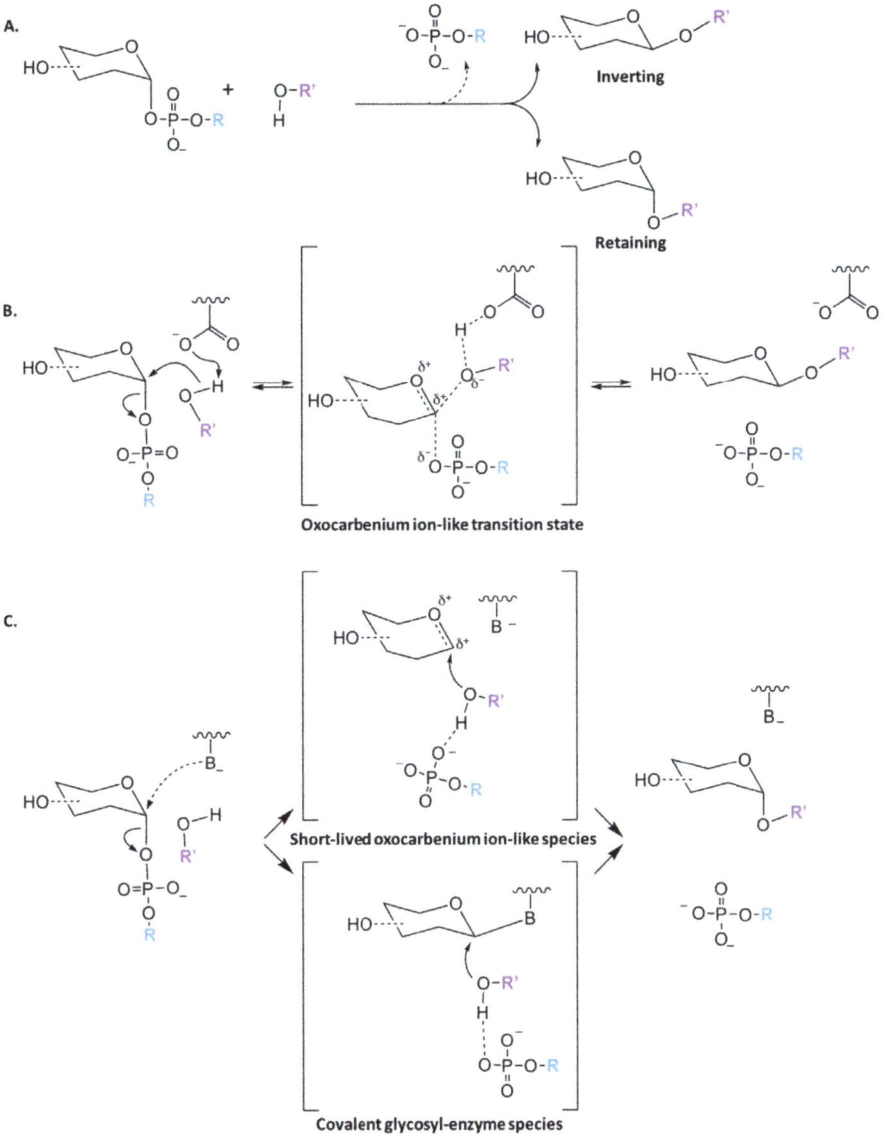

Figure 3. Proposed catalytic mechanism of GT enzymes. (**A**) The transfer of a sugar moiety is performed with either inversion or retention of the anomeric carbon in respect to the sugar donor substrate; (**B**) Schematic of S_N1-like mechanism for inverting GTs, where a single oxocarbenium ion-like transition state is formed; (**C**) There are currently two mechanisms for retaining GTs enzymes either through the formation of a short-lived oxocarbenium ion-like species or a covalent glycosyl-enzyme species.

3. Core Heptosyltransferase Enzymes

Many GTs have been extensively studied due to their biological and medical importance. Notable of these enzymes are the heptosyltransferases that are involved in the biosythesis of the LPS inner core (and in some bacteria outer core). Heptosytransferases catalyze the sequential addition of heptose moieties onto Kdo_2-lipid A (Figure 4) and are characterized as GT-B enzymes inverting reaction mechanism [44]. As mentioned earlier, in all Gram-negative bacteria, LPS is one of the major extracellular polymeric substances protecting the cell (a schematic of which is shown in Figure 1). For many bacteria the overall structure of LPS is highly conserved. However, as one moves away from the membrane the structure variability between bacterial species increases. Thus, the inner core of LPS has low variability where the outer core varies more between bacteria. Additionally, it has been shown that the less conserved regions are not required for bacterial viability [45]. In fact, the minimal structure required for bacterial survival is Kdo_2-lipid A - the acceptor substrate for HepI [9,18,19]. Truncation of the LPS increases the bacteria sensitivity to hydrophobic antibiotics and detergents, making the heptosyltransferases, especially HepI, novel drug targets [4,20,21].

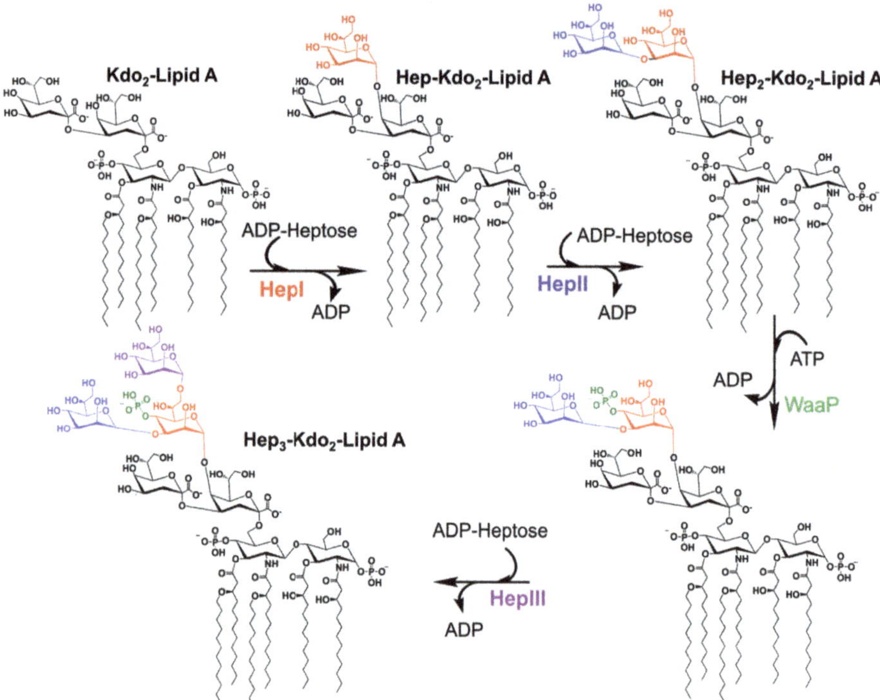

Figure 4. The sequential addition of three heptose moieties from ADPH catalyzed by Heptosytransferase I, II, and III (HepI, II, and III, respectively). Each moiety is color coded to match the enzyme that catalyzed the addition (red, blue, purple for HepI, II, and III, respectively). Prior to addition of the third heptose, WaaP phosphorylates the first heptose [46].

3.1. Multiple Sequence Alignment (MSA) of Heptosyltransferase Enzymes

Gram-negative bacteria have up to four heptosyltransferases; HepI and HepII are always present and catalyze the addition of the first two sugars of the inner core, whereas HepIII and HepIV are found only in some species (Figure 4). As a result HepI and HepII have been studied in many systems. To date, there has been little work on HepIII and even less on HepIV, despite both having been identified or suggested in *Vibrio cholerae, Escherichia coli, Yersinia pestis*, and *Klebsiella pneumoniae*. HepIII adds the third heptose to the inner core and HepIV adds a heptose moiety onto a glucose or galactose located within the outer core (Figure 1) [47,48]. A multiple seuqence alignment (MSA) of HepI–IV from *Vibrio cholerae, Escherichia coli, Yersinia pestis*, and *Klebsiella pneumoniae* shows the variability of sequence conservation among the heptosyltransferases (Figure 5). The average similarity for all 16 heptosyltransferases is about 30%, which is consistent to the percent similarity for the HepI–IV enzymes from the same organism. By comparing each homolog to *E. coli*, it can be concluded that HepI and HepII are highly conserved with percent similarities as high as 86%. HepIII and HepIV homologues have less then 46% similarity. It is perhaps unsurprising that each of the heptosyltransferases have the highest sequence similarity to their homologs rather than to the paralogs within an organism. HepIV are more divergent than HepI, most likely because the core region of LPS only varies only slightly, thus a HepI enzyme from *E. coli* and *V. cholerae* will bind more similar acceptor substrates than the corresponding HepIV enzymes [45].

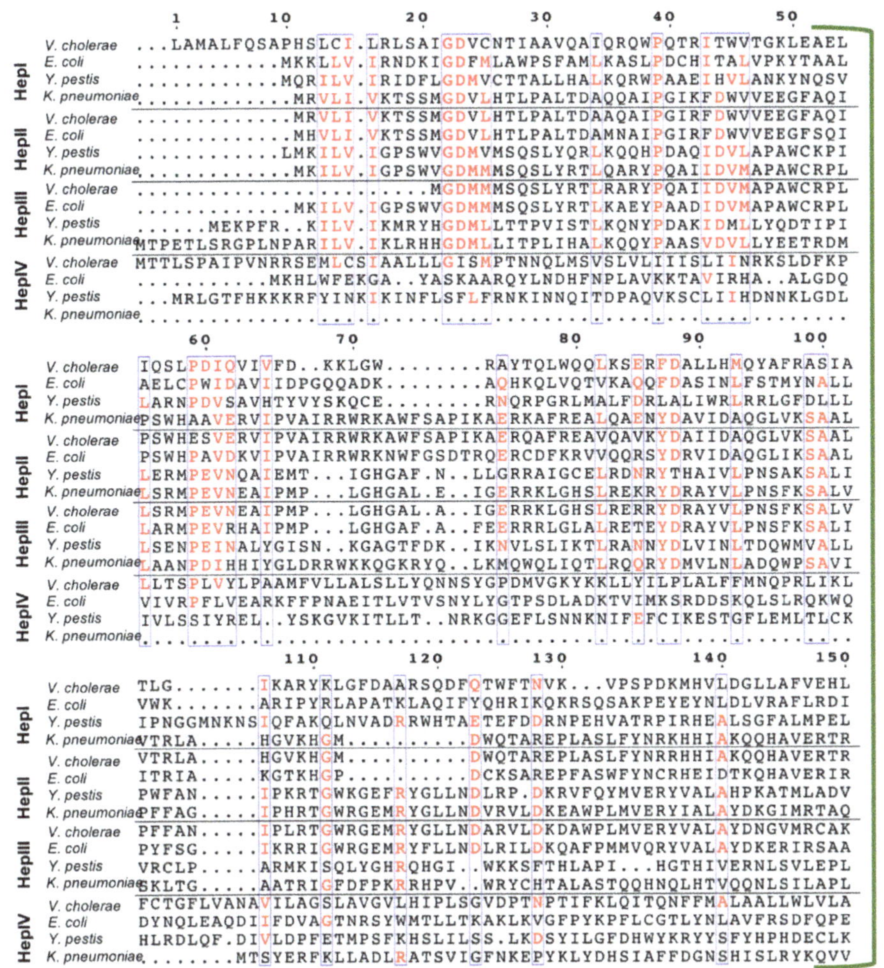

Figure 5. *Cont.*

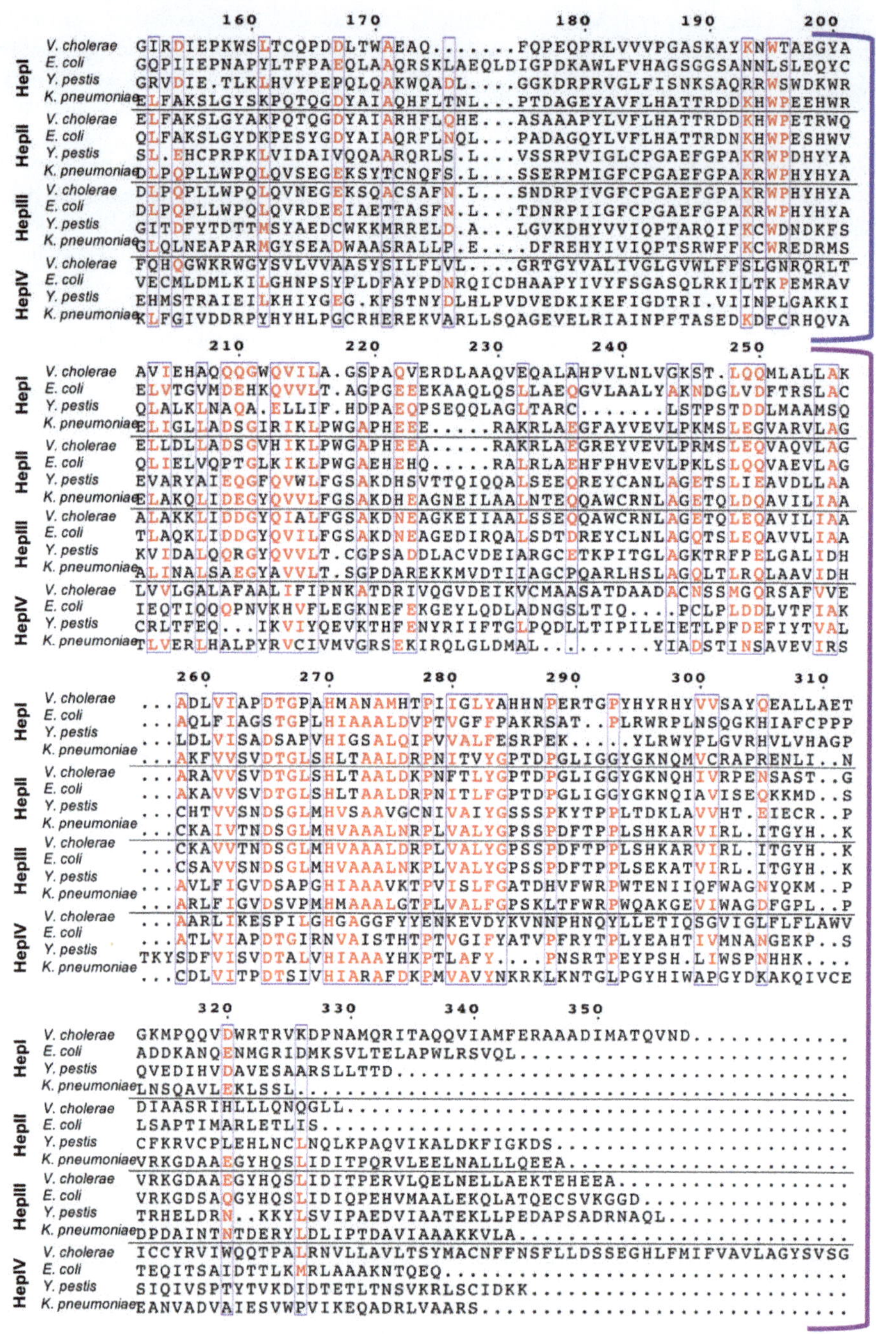

Figure 5. Muliple sequence alignment of HepI–HepIV from *Vibrio cholerae*, *Escherichia coli*, *Yersinia pestis*, and *Klebsiella pneumoniae* (domains for HepI *E. coli* HepI are annotated), using ClustalW 2.0 (https://www.ebi.ac.uk/Tools/msa/clustalw2/) alignment program and Esprit 3.0 (ESPript—http://espript.ibcp.fr) [49,50].

Although HepI–IV are variable in their sequence, the C-terminus has the most conservation followed by the N-terminus, while the linker is highly variable, and the overall structure of heptosyltrasferases are homologous (Figure 6). In *E. coli* HepI and HepII have been crystallized, and the structure of HepIII has been computationally predicted (Figure 6A–C) [51]. A computational model of *E. coli* HepIV was created using the I-Tasser protein structure prediction program (the resulting structure is shown in Figure 6D) and it appears similar to HepI and HepII crystal structure [52–54]. All are GT-B proteins with the the typical βαβRossman-like domains attached by a linker. The C-terminal domain for all, binds ADPH (the donor substrate) and look nearly identical, whereas the N-terminal domain varies slightly, likely due to their variation in acceptor substrates [31,51].

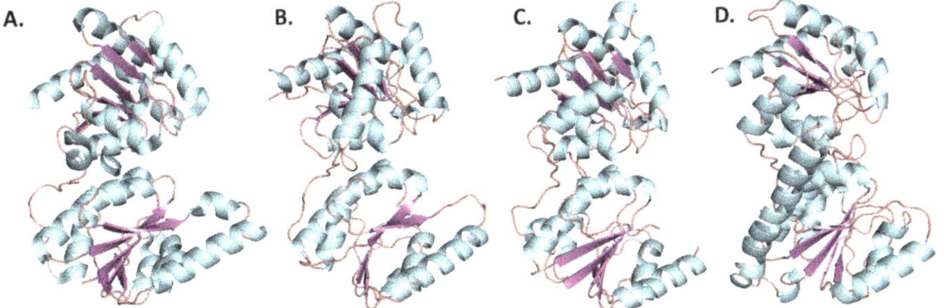

Figure 6. Structures of Heptosyltransferases (loops, α-helices, and β-sheets are colored salmon, cyan, and purple, respectively). (**A**) HepI from *E. coli*, PDB: 2H1H; (**B**) HepII from *E. coli*, PDB: 1PSW; (**C**) Computational model of HepIII from *E. coli* [51]; (**D**) Computational model of HepIV from *E. coli*.

To better compare the the structural variety of *E. coli* HepI–IV a sequence-based structural superposition was generated using HepI (PDB:2H1H) as the reference structure, with the VMD multiseq program. An overlay showing the conserved residues in the HepI–IV structures are displayed in Figure 7 (the blue areas indicate highly conserved regions and the red depicts non-conserved regions). By looking at the global conservation of heptosyltransferases, the interior is more conserved, while the surface residues are highly variable. Additionally, it is evident that the C-terminus (binding domain of ADPH for all heptosyltransferase enzymes) is more conserved than the N-terminal domain. The proposed catalytic base D13 is present in all the heptosyltransferases, suggesting that the mechanism of action for all heptosyltransferases are similiar [31]. Other specific residues, like K192 and D261 (which were shown by mutagenesis studies in HepI to be important for chemistry), are completely conserved for all *E. coli* heptosyltransferases as well as all *Vibrio cholerae*, *Escherichia coli*, *Yersinia pestis*, and *Klebsiella pneumoniae*. It is clear that all of the heptosyltransferase enzymes are structurally similar and many important residues are conserved not only between *E. coli* heptosyltransferase enzymes, but also in multiple bacterial species. Bacterial evolution to differentiate the LPS structure enhancing survival in different niches likely governs the sequence variability of heptosyltransferases.

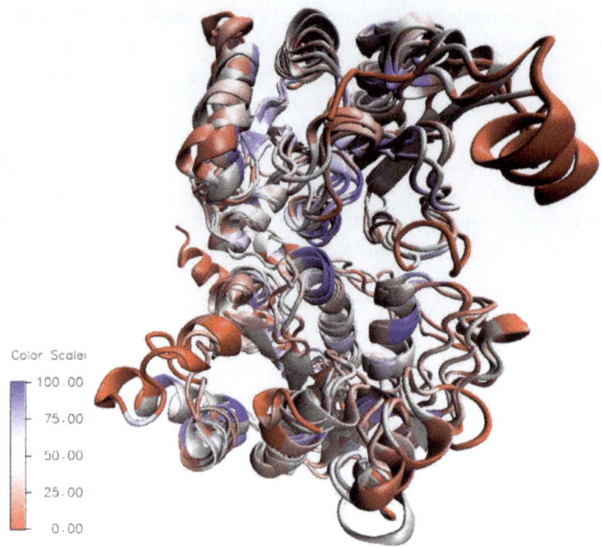

Figure 7. Overlay of Structurally aligned HepI–IV; residues are colored by sequence similarity [highly conserved (blue) non-conserved (red)].

3.2. Heptosyltransferase I

E. coli HepI, the most characterized heptosyltransferase, can reveal insights about the function of the other heptosyltransferase enzymes. The acceptor substrate of HepI, Kdo$_2$-lipid A, is the minimalistic structure required for LPS to be transported to the outer membrane. Mutations to *waaC* (*rfaC*), the gene that codes for HepI, leads to a rough phenotype LPS and an increase in sensitivity to hydrophobic antibiotics including: erthromycin, ampicillin, and novobiocin [10]. Early work on HepI sought to use alternative donor substrates as ADPH was not commercially available. ADP-mannose, GDP-manose, ADP-glucose, UDP-glucose, and UDP-galactose were tested for HepI transferase activity, only ADP-mannose was a viable alternative substrate. ADP-mannose was characterized by Kadrmas et al. to have an apparent V_{max} of 3 µmol/min/mg and a K_M of 1.47 mM. Kdo-lipid A (an analogue of Kdo$_2$-lipid A with only one Kdo) was a poor mannose acceptor substrate; this was unexpected since the second Kdo moiety was not expected to influence activity since the first Kdo is the one being modified by HepI [55]. Later work using the native substrate ADPH showed that Kdo-lipid A was in fact a competent acceptor substrate with a K_M of 46 µM [56]. Perhaps using two alternative substrates was the reason for the poor transferase activity, and in fact Kdo-lipid A may be sufficient for the continual formation of the inner core. Interestingly, the fatty acid chains were shown to not be important for catalysis, as has been demonstrated to be necessary in other LPS biosynthetic enzymes. The substrate analogue, Kdo$_2$-lipid IV$_A$, although missing three fatty acid chains normally present in Kdo$_2$-lipid A, gives a K_M of 4.7 µM demonstrating that the removal of fatty acid chains does not impair chemistry. Furthermore, HepI has activity with the fully deacylated and O-deacylated Kdo$_2$-lipid A (ODLA and FDLA respectively, Figure 8). Both were shown to be competent substrates; native substrate had a K_M of 29 µM where the analogues displayed a K_M of 1 µM (ODLA) and 0.3 µM (FDLA) [56]. Retrospectively, it is unsurprising that the fatty acid chains would be unimportant for catalytic efficiency since they are embedded into the inner membrane in vivo and therefore should not be accessible to influence binding. Taken together with the slightly better catalytic efficiency for deacylated Kdo$_2$-lipid A analogues, these observations suggest that the tetrasaccharide portion of the substrate provides HepI with the primary binding interactions required for acceptor substrate recognition.

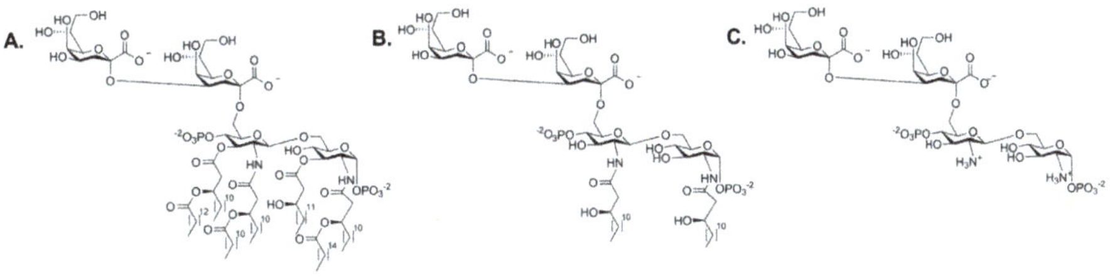

Figure 8. Strucutres of Kdo$_2$-lipid A and analogues: (**A**) *E. coli* Kdo$_2$-lipid A; (**B**) O-deacylated *E. coli* Kdo$_2$-lipid A (ODLA) and (**C**) fully deacylated *E. coli* Kdo$_2$-lipid A (FDLA).

3.2.1. Crystal Structures of HepI

As mentioned earlier, the structure of HepI has been previously determined. Three different structures are available in the PDB: 2GT1, 2H1H, and 2H1F corresponding to the Apo protein, HepI·ADP-2-deoxy-2-fluoro-heptose (ADPF) complex and HepI·ADP complex, respectively [31]. By comparing the three structures it seems that HepI does not undergo a domain rotation upon donor substrate binding like other GT-B's. However, as there is no crystal structure in complex with Kdo$_2$-lipid A or any of its analogues, and it is, therefore, possible that HepI closure is induced by the acceptor substrate or by formation of the ternary complex, as has been shown by other GT-B's. ADPF is a non-cleavable analogue to ADPH with a fluorine replacing a hydroxyl group in the 2′-position. ADPF has been shown to be an inhibitor of HepI with an IC$_{50}$ of 30 µM [31]. Attempts to crystallize HepI with ADPH lead to co-crystallization with ADP, suggesting that HepI is capable of hydrolysis in the absence of an acceptor, a phenomenon that was observed with other glycoslytransferases [34,57]. Upon crystallization of HepI, Grizot et al. performed site-directed mutagenesis to test the importance of numerous residues on catalysis and binding of ADPH. D13A and D261A exhibit a 4688-fold and 2027-fold drop in specific activity, respectively and were suggested to be catalytic residues. K192A had a 926-fold reduction in activity and due to its location proximal to the anomeric carbon of ADPH which is where the deprotonated hydroxyl of Kdo$_2$-lipid A attacks, and therefore may play an important role in binding or catalysis and binding [31].

3.2.2. Inhibition of HepI

One goal of understanding the heptosyltransferases is to learn how to effectively inhibit them, to date some work has been done to design inhibitors for HepI. Most inhibitors of GTs bind typically with low µM affinities, similar to K_M values of substrate. A structure-activity relationship (SAR) study was done by Moreau et. al. on a series of 2-aryl-5-methyl-4-(5-aryl-furan-2-yl-methylene)-2,4-dihydro-pyrazol-3-one analogues (Figure 9A) [58]. In this work, computational docking and biochemical assays were used to assess binding. All compounds bound with low µM IC$_{50}$'s and appeared to be preferentially bound close to acceptor site of Kdo$_2$-lipid A, specifically near where Kdo should bind. Residues R120, H139, A140, R143, and I287 were identified as important for inhibitor binding in this analysis, suggesting the potential importance of these residues for Kdo$_2$-lipid A binding. Further studies would need to be done however to test this hypothesis.

Additionally, Durka et al. published a library of synthesized multivalent glycosylated fullerene monomers and "balls" much larger than the proposed molecules previously discussed. Like the previous series, these compounds had inhibition constants in the low µM ranging from 7–47 µM, again on par K_M of substrates [59]. In 2016, the group published a second series of glycofullerenes slightly varying the fullerenes in an attempt to increase inhibition. The new fullerene compounds were competitive against Kdo$_2$-lipid A and uncompetitive towards ADPH, which was unexpected because the compounds were designed to mimic ADPH heptose moiety; similar to previously discussed work, the compounds showed low µM K_i. From these findings, glycoclusters mimicking Kdo with C$_{60}$ and other multivalent scaffolds were synthesized and IC$_{50}$ were calculated. Interestingly, high nanomolar inhibition was observed for Kdo fullerenes attached to C$_{60}$ scaffolds, a degree of inhibition never achieved for HepI and rarely for GTs [60].

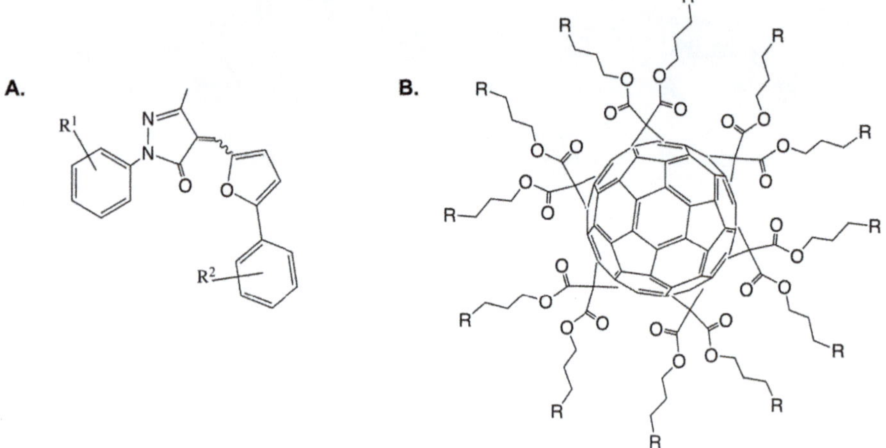

Figure 9. (**A**) Representative core structure of 2-aryl-5-methyl-4-(5-aryl-furan-2-yl-methylene)-2,4-dihydro-pyrazol-3-one analogues [58]; (**B**) Glycofullerene derivatives schematic [59].

3.2.3. Investigations of HepI Protein Dynamics

As mentioned earlier, a few GT-B glycosyltransferases have been shown to inter-convert between an "open" and "closed" structure [32–34]. Without a crystal structure of the ternary complex of HepI, it is unclear how/if HepI undergoes such an event. By looking at the crystal structure however, its is clear that the catalytic base (D13) is over 8 Å away from the anomeric carbon of ADPH which is too far away for efficient nucleophilic attack by deprotonated Kdo$_2$-lipid A [31]. Thus, it was hypothesized that HepI also undergoes a conformational change during the reaction. To assess protein dynamics, HepI steady state activity was tested in a variety of viscous buffers, specifically glycerol, ethylene glycol, and PEG 8000. Microviscogens; glycerol, and ethylene glycol, both had a strong impact on k_{cat} which could be explained by water reorganization being required for catalysis; this suggests that HepI conformational dynamics are partially rate-limiting [61].

Additionally, intrinsic tryptophan (Trp) fluorescence spectroscopy was employed to see if substrate binding induced changes in the protein fluorescence spectra [62–70]. HepI has 8 tryptophan residues, and by examination of a computational model of the closed structure of HepI, many of the Trp residues appear to become more buried in the protein, suggesting that upon substrate binding there may be a change in fluorescence spectrum (Figure 10A). Fluorescence spectra were obtained with and without substrates(Figure 10B). Consistent with crystal structures, no change was observed upon ADPH binding alone. ODLA binding resulted in a 6 nm blue shift, suggesting that ODLA binding induces conformational changes that lead to one or more the the Trp residues to becoming more buried in the protein [61].

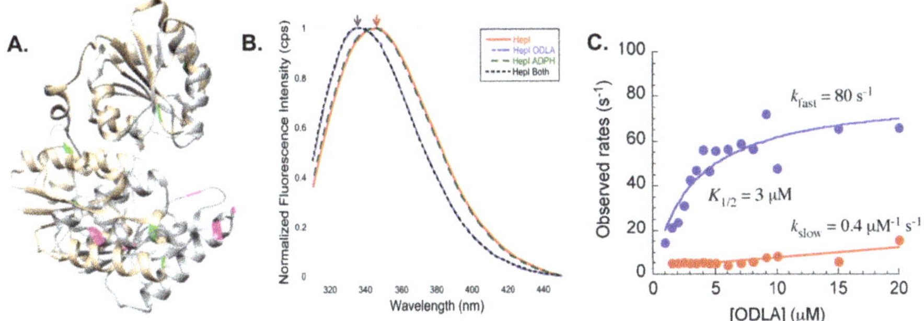

Figure 10. (**A**) HepI open structure (tan) and trypophan residues colored in green is superimposed with a structural model of closed HepI (gray) with tryptophans colored magenta; (**B**) Steady state intrinsic tryptophan emission spectra of HepI with and without substrates bound, blue shift is observed upon ODLA binding; (**C**) Pre-steady state kinetics of WT HepI titrated with ODLA (fast phase, blue; slow phase, red).

When stopped flow was used to monitor the kinetics of conformational changes in HepI, a concentration of ODLA dependent biphasic pre-steady state kinetics was observed. Two rates were observed, a fast rate which exhibits a hyperbolic dependence on ODLA concentration (saturates at $80\,s^{-1}$), and a concentration independent slow rate of ~5 s^{-1}. A two step binding mechanism of ODLA was suggested (initial collision complex between of HepI and ODLA followed one or more conformational change(s) to form the HepI·ODLA complex). Additionally, a catalytically impaired mutant (D13A) of HepI was tested and yielded the same pre-steady state kinetics. This suggests that the ODLA induced change in HepI must occur prior to chemistry [61]. Subsequently, work investigated which Trp residue(s) play(s) a role in the observed blue shift so as to better understand conformational change(s) that occur [71]. In this work, most of the eight Trp residues were mutated to phenylalanine (Phe). W62F and W116F both of which are located on the N-terminal ODLA binding domain, exhibited a reduced blue shift upon ODLA binding as compared to wild-type HepI. Additionally, these residues are located on dynamic loops (N-3 and N-7) suggesting that these loops may undergo conformational changes when ODLA binds, leading to a change in local environment of W62 and W116. Interestingly, the W217F mutant (Trp located on the C-terminal domain far from the ODLA binding site) resulted in a complete loss of the blue shift upon substrate binding. Although more experiments are need to fully understand the role of W217, upon ODLA binding, ADPH binding may be altered (ADPH is directly moved to impact W217 conformation), suggesting communication between the two domains.

In addition to fluorescence spectroscopy, circular dichroism (CD) experiments were used to investigate structural changes of HepI [72,73]. CD spectra for HepI demonstrate a characteristic spectra for a protein with primarily αcontent with a double minimum with peaks at 222 nm and 211 nm. Interestingly, upon binding of ODLA there is an increase in the intensity of the second minimum at around 211 nm consistent with a 12% increases in α-helicity (Figure 11A). The location of these conformational changes are unknown, but most likely this is the result of structural changes of disordered loops in HepI [71].

Protein stability was also explored by CD melts experiments (taking CD spectra at varying temperatures) and monitoring unfolding of protein. As can be seen in Figure 11B,C, apo HepI has a T_M of 40 °C, however upon ODLA binding there is a large increase in stability so that even at 95 °C the protein is still mostly folded [71]. It was concluded that the formation of HepI·ODLA complex must lead to formation of hydrogen bonds and/or salt bridges (ionic interactions) between ODLA and HepI. Without cyrstalographic evidence showing where ODLA binds, its hard to determine which interactions induce such a stabilization. Examination of the HepI structure reveals that there are many positively charged residues located on dynamic loops of the N-terminal domain (where ODLA binds) which could coordinate with negatively charged phosphate and carboxylate moieties of ODLA. High salt reverses some of the HepI·ODLA complex stabilization, suggesting that ionic interactions are essential for HepI·ODLA complex stabilization [71]. In sum, these data strongly support the hypothesis of heptosyltransferases undergoing open to closed transitions.

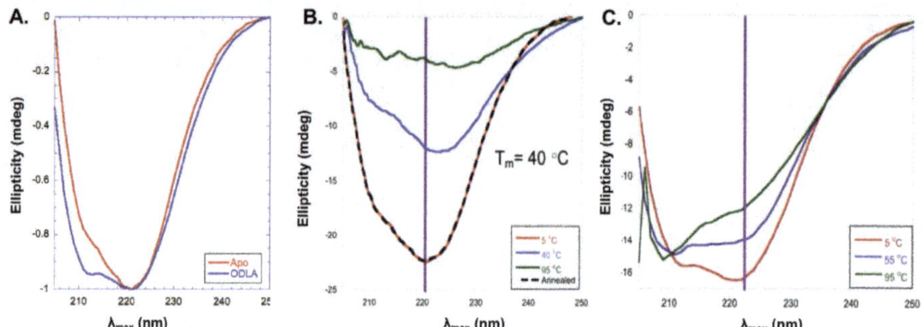

Figure 11. (**A**) Far-UV circular dichroism (CD) spectra of apo HepI (red), HepI with 100 μM ADPH (dark green), and HepI with 100 μM ODLA (blue) at 5 °C. Far-UV CD Melt spectra of (**B**) apo HepI at 5 °C (red), 40 °C (blue), 95 °C (green) and annealed at 5 °C (black) and (**C**) HepI with 100 μM ODLA 5 °C (red), 40 °C (blue) and 95 °C (green). Purple lines demonstate changes in ellipticity at 222 nm.

4. Conclusions

With the growing need for new antibiotics to treat antibiotic resistant (and multi-resistant) bacteria, it is essential for scientists to find new and smarter ways to inhibit bacterial growth. Heptosyltransferases are important for LPS biosynthesis and the resulting resistance. Structurally HepI–IV are very similar and all adopt a GT-B structural fold. HepI can be used as a model for other heptosyltransferases and GT-B enzymes to inform inhibition strategies. Perhaps disruption of GT-B dynamics with small molecules would be an effective new strategy for inhibitor development. Additionally, dynamics disruption could potentially allow for inhibition of multiple targets which undergo similar dynamical changes (with a single drug targeting multiple enzymes). Promising work has been done to understand the function of HepI and aid in designing such an inhibitor. Ultimately, heptosyltransferases provide useful information about the GT-B structural fold and provide a model for novel methods to inhibit many GT-B enzymes.

Author Contributions: J.M.C. and E.A.T. contributed equally to the conception and writing of this paper.

Conflicts of Interest: The authors declare no conflict of interest.

References

1. Barbier, F.; Luyt, C. Understanding resistance. *Intensiv. Care Med.* **2016**, *42*, 2080–2083.
2. Centers for Disease Control and Prevention (CDC). *Antibiotic Resistance Threats in the United States, 2013*; Technical Report; CDC; U.S. Department of Health and Human Services: Atlanta, GA, USA, 2013. Available online: http://www.cdc.gov/drugresistance/pdf/ar-threats-2013-508.pdf (accessed on 28 September 2017).
3. World Health Organization (WHO). *Antibiotic Resistance*; Fact Sheet; World Health Organization (WHO), 2016. Available online: http://www.who.int/mediacentre/factsheets/antibiotic-resistance/en/ (accessed on 28 September 2017).
4. Ruiz, N.; Kahne, D.; Silhavy, T.J. Advances in understanding bacterial outer-membrane biogenesis. *Nat. Rev. Microbiol.* **2006**, *4*, 57–66.
5. Nikaido, H. Molecular Basis of Bacterial Outer Membrane Permeability Revisited. *Microbiol. Mol. Biol. Rev.* **2003**, *67*, 593–656.

6. Piek, S.; Kahler, C.M. A comparison of the endotoxin biosynthesis and protein oxidation pathways in the biogenesis of the outer membrane of *Escerichia coli* and *Neisseria meningitidis*. *Front. Cell. Infect. Microbiol.* **2012**, *2*, 162.

7. Raetz, C.; Reynolds, M.C.; Trent, S.M.; Bishop, R.E. Lipid A Modification Systems in Gram-Negative Bacteria. *Annu. Rev. Biochem.* **2007**, *76*, 295–329.

8. Schnaitman, C.A.; Klena, J.D. Genetics of Lipopolysaccharide Biosynthesis in Enteric Bacteria. *Microbiol. Rev.* **1993**, *57*, 655–682.

9. Raetz, C.R.; Whitfield, C. Lipopolysaccharide endotoxins. *Annu. Rev. Biochem.* **2002**, *71*, 635–700.

10. Coleman, W.G.; Leive, L. Two mutations which affect the barrier function of the *Escherichia coli* K-12 outer membrane. *J. Bacteriol.* **1979**, *139*, 899–910.

11. Stenutz, R.; Weintraub, A.; Widmalm, G. The structures of *Escherichia coli* O-polysaccharide antigens. *FEMS Microbiol. Rev.* **2006**, *30*, 382–403.

12. Lerouge, I.; Vanderleyden, J. O-antigen structural variation: Mechanisms and possible roles in animal/ plant–microbe interactions. *FEMS Microbiol. Rev.* **2002**, *26*, 17–47.

13. Chatterjee, S.N.; Chaudhuri, K. Lipopolysaccharides of *Vibrio cholerae*: I. Physical and chemical characterization. *Biochim. Biophys. Acta* **2003**, *1639*, 65–79.

14. Caroff, M.; Karibian, D. Structure of bacterial lipopolysaccharides. *Carbohydr. Res.* **2003**, *338*, 2341–2347.

15. Heinrichs, D.E.; Yethon, J.A.; Whitfield, C. Molecular basis for structural diversity in the core regions of the lipopolysaccharides of *Escherichia coli* and *Salmonella enterica*. *Mol. Microbiol.* **1998**, *30*, 221–232.

16. Raetz, C.R.; Garrett, T.A.; Reynolds, C.M.; Shaw, W.A.; Moore, J.D.; Dale, C.; Smith, J.; Ribeiro, A.A.; Murphy, R.C.; Ulevitch, R.J.; et al. Kdo$_2$-Lipid A of *Escherichia coli*, a defined endotoxin that activates microphages via TLR-4. *J. Lipid Res.* **2006**, *47*, 1097–1111.

17. Whitfield, C.; Trent, M. Biosynthesis and export of bacterial lipopolysaccharides. *Annu. Rev. Biochem.* **2014**, *83*, 99–128.

18. Delcour, A.H. Outer membrane permeability and antibiotic resistance. *Biochim. Biophys. Acta* **2009**, *1794*, 808–816.

19. Klein, G.; Lindner, B.; Brabetz, W.; Brade, H.; Raina, S. *Escherichia coli* K-12 Suppressor-free Mutants Lacking Early Glycosyltransferase and Late Acyltransferase: Minimal Lipopolysaccharide Structure and Induction of Envelope Stress Response. *J. Biol. Chem.* **2009**, *284*, 15369–15389.

20. Kanipes, M.I.; Holder, L.C.; Corcoran, A.T.; Moran, A.P.; Guerry, P. A Deep-Rough Mutant of *Campylobacter jejuni* 81-176 Is Noninvasive for Intestinal Epithelial Cells. *Infect. Immun.* **2004**, *72*, 2452–2455.

21. Raetz, C.R. Bacterial lipopolysaccharides: A remarkable family of bioactive macroamphiphiles. In *Escherichia coli and Salmonella: Cellular and Molecular Biology*; Neidhardt, F.C., Curtiss, R., III, Ingraham, J.L., Lin, E.C.C., Low, K.B., Magasanik, B., Reznikoff, W.S., Riley, M., Schaechter, M., Umbarger, H.E., Eds.; American Society for Microbiology: Washington, DC, USA, 1996.

22. Coutinho, P.M.; Deleury, E.; Davies, G.J.; Henrissat, B. An Evolving Hierarchical Family Classification for Glycosyltransferases. *J. Mol. Biol.* **2003**, *328*, 307–317.

23. Varki, A.; Cummings, R.; Esko, J.; Freeze, H.; Hart, G.; Marth, J. (Eds) *Essentials of Glycobiology*; Cold Spring Harbor Laboratory Press: Cold Spring Harbor, NY, USA, 1999; Chapter 17 (Glycosyltransferases); pp. 253–262.

24. Albesa-Jové, D.; Giganti, D.; Jackson, M.; Alzari, P.; Guerin, M. Structure-function relationships of membrane-associated GT-B glycosyltransferases. *Glycobiology* **2014**, *24*, 108–124.

25. Lairson, L.; Henrissat, B.; Davies, G.; Withers, S. Glycosyltransferases: Structures, functions, and mechanisms. *Annu. Rev. Biochem.* **2008**, *77*, 521–555.

26. Lombard, V.; Golaconda Ramulu, H.; Drula, E.; Coutinho, P.M.; Henrissat, B. The carbohydrate-active enzymes database (CAZy) in 2013. *Nucleic Acids Res.* **2014**, *42*, D490–D495.

27. Sinnott, M.L. Catalytic mechanism of enzymic glycosyl transfer. *Chem. Rev.* **1990**, *90*, 1171–1202, doi:10.1021/cr00105a006.

28. Charnock, S.J.; Davies, G.J. Structure of the Nucleotide-Diphospho-Sugar Transferase, SpsA from *Bacillus subtilis*, in Native and Nucleotide-Complexed Forms. *Biochemistry* **1999**, *38*, 6380–6385, doi:10.1021/bi990270y.

29. Pak, J.E.; Arnoux, P.; Zhou, S.; Sivarajah, P.; Satkunarajah, M.; Xing, X.; Rini, J.M. X-ray Crystal Structure of Leukocyte Type Core 2 β1,6-*N*-Acetylglucosaminyltransferase: Evidencec for a Convergence of Metal Ion-independent Glycosyltransferase Mechanism. *J. Biol. Chem.* **2006**, *281*, 26693–26701.

30. Urresti, S.; Albesa-Jové, D.; Schaeffer, F.; Pham, H.T.; Kaur, D.; Gest, P.; van der Woerd, M.J.; Carreras-González, A.; López-Fernández, S.; Alzari, P.M.; et al. Mechanistic Insights into the Retaining Glucosyl-3-phosphoglycerate Synthase from Mycobacteria. *J. Biol. Chem.* **2012**, *287*, 24649–24661.
31. Grizot, S.; Salem, M.; Vongsouthi, V.; Durand, L.; Moreau, F.; Dohi, H.; Vincent, S.; Escaich, S.; Ducruix, A. Structure of the *Escherichia coli* Heptosyltransferase WaaC: Binary Complexes with ADP and ADP-2-deoxy-2-Fluoro Heptose. *J. Mol. Biol.* **2006**, *363*, 383–394.
32. Vrielink, A.; Rüger, W.; Driessen, H.P.; Freemont, P.S. Crystal structure of the DNA modifying enzyme beta-glucosyltransferase in the presence and absence of the substrate uridine diphosphoglucose. *EMBO J.* **1994**, *13*, 3413–3422.
33. Vetting, M.W.; Frantom, P.A.; Blanchard, J.S. Structural and Enzymatic Analysis of MshA from *Corynebacterium glutamicum*: Substrate-assisted Cataylsis. *J. Biol. Chem.* **2008**, *283*, 15834–15844.
34. Mulichak, A.M.; Losey, H.C.; Lu, W.; Wawrzak, Z.; Walsh, C.T.; Garavito, R.M. Structure of the TDP-epi-vancosaminyltransferase GtfA from the chloroeremomycin biosynthetic pathway. *Proc. Natl. Acad. Sci. USA* **2003**, *100*, 9238–9243.
35. Liu, J.; Mushegian, A. Three monophyletic superfamilies account for the majority of the known glycosyltransferases. *Protein Sci.* **2003**, *12*, 1418–1431.
36. Lizak, C.; Gerber, S.; Numao, S.; Aebi, M.; Locher, K.P. X-ray structure of a bacterial oligosaccharyltransferase. *Nature* **2011**, *474*, 350–355.
37. Kozmon, S.; Tvaroška, I. Catalytic Mechanism of Glycosyltranferases: Hybrid Quantum Mechanical/Molecular Mechanical Study of the Inverting N-Acetylglucosaminyltransferase I. *J. Am. Chem. Soc.* **2006**, *128*, 16921–16927.
38. Taylor Ringia, E.A.; Schramm, V.L. Transition states and inhibitors of the purine nucleoside phosphorylase family. *Curr. Top. Med. Chem.* **2005**, *5*, 1237–1258.
39. Rojas-Cervellera, V.; Ardèvol, A.; Boero, M.; Planas, A.; Rovira, C. Formation of a Covalent Glycosyl–Enzyme Species in a Retaining Glycosyltransferase. *Chem. Eur. J.* **2013**, *19*, 14018–14023.
40. Ardèvol, A.; Iglesias-Fernández, J.; Rojas-Cervellera, V.; Rovira, C. The reaction mechanism of retaining glycosyltransferases. *Biochem. Soc. Trans.* **2016**, *44*, 51–60.
41. Soya, N.; Fang, Y.; Palcic, M.M.; Klassen, J.S. Trapping and characterization of covalent intermediates of mutant retaining glycosyltransferases. *Glycobiology* **2011**, *21*, 547–552.
42. Lee, S.S.; Hong, S.Y.; Errey, J.C.; Izumi, A.; Davies, G.J.; Davis, B.G. Mechanistic evidence for a front-side, SNi-type reaction in a retaining glycosyltransferase. *Nat. Chem. Biol.* **2011**, *7*, 631–638.
43. Frantom, P.A.; Coward, J.K.; Blanchard, J.S. UDP-(5F)-GlcNAc Acts as a Slow-Binding Inhibitor of MshA, a Retaining Glycosyltransferase. *J. Am. Chem. Soc.* **2010**, *132*, 6626–6627.
44. Wang, X.; Quinn, P.J. Lipopolysaccharide: Biosynthetic pathway and structure modification. *Prog. Lipid Res.* **2010**, *49*, 97–107.
45. Schrinner, E. *Surface Structures of Microorganisms and Their Interactions with the Mammalian Host*. In Proceedings of the Eighteenth Workshop Conference, Hoechst, Schloss Ringberg, Germany, 20–23 October 1987; John Wiley & Sons: Hoboken, NJ, USA, 1988; Volume 18.
46. Zhao, X.; Wenzel, C.Q.; Lam, J.S. Nonradiolabeling Assay for WaaP, an Essential Sugar Kinase Involved in Biosynthesis of Core Lipopolysaccharide of *Pseudomonas aeruginosa*. *Antimicrob. Agents Chemother.* **2002**, *46*, 2035–2037.
47. Knirel, Y.; Anisimov, A. Lipopolysaccharide of *Yersinia pestis*, the Cause of Plague: Structure, Genetics, Biological Properties. *Acta Nat.* **2012**, *4*, 46–58.
48. Reeves, P.P.; Wang, L. Genomic organization of LPS-specific loci. *Curr. Top. Microbiol. Immunol.* **2002**, *264*, 109–135.
49. Robert, X.; Gouet, P. Deciphering key features in protein structures with the new ENDscript server. *Nucleic Acids Res.* **2014**, *42*, W320–W324.
50. Larkin, M.; Blackshields, G.; Brown, N.; Chenna, R.; McGettigan, P.; McWilliam, H.; Valentin, F.; Wallace, I.; Wilm, A.; Lopez, R.; et al. Clustal W and Clustal X version 2.0. *Bioinformatics* **2007**, *23*, 2947–2948.
51. Mudapaka, J.; Taylor, E.A. Cloning and Characterization of the *Escherichia coli* Heptosyltransferase III: Exploring Substrate Specificity in Lipopolysaccharide Core Biosynthesis. *FEBS Lett.* **2015**, *589*, 1423–1429.
52. Zhang, Y. I-TASSER server for protein 3D structure prediction. *BMC Bioinform.* **2008**, *9*, 40.
53. Roy, A.; Kucukural, A.; Zhang, Y. I-TASSER: A unified platform for automated protein structure and function prediction. *Nat. Protoc.* **2010**, *5*, 725–738.

54. Yang, J.; Yan, R.; Roy, A.; Xu, D.; Poisson, J.; Zhang, Y. The I-TASSER Suite: Protein structure and function prediction. *Nat. Methods* **2015**, *12*, 7–8.

55. Kadrmas, J.L.; Raetz, C.R.H. Enzymatic Synthesis of Lipopolysaccharide in *Escherichia coli*: Purification and Properties of Heptosyltransferase I. *J. Biol. Chem.* **1998**, *273*, 2799–2807.

56. Czyzyk, D.J.; Liu, C.; Taylor, E.A. Lipopolysaccharide Biosynthesis without the Lipids: Recognition Promiscuity of *Escherichia coli* Heptosyltransferase I. *Biochemistry* **2011**, *50*, 10570–10572.

57. Larivière, L.; Gueguen-Chaignon, V.; Moréra, S. Crystal Structures of the T4 Phage β-Glucosyltransferase and the D100A Mutant in Complex with UDP-glucose: Glucose Binding and Identification of the Catalytic Base for a Direct Displacement Mechanism. *J. Mol. Biol.* **2003**, *330*, 1077–1086.

58. Moreau, F.; Desroy, N.; Genevard, J.M.; Vongsouthi, V.; Gerusz, V.; Le Fralliec, G.; Oliveira, C.; Floquet, S.; Denis, A.; Escaich, S.; et al. Discovery of New Gram-negative Antivirulence Drugs: Structure and Properties of Novel *E. coli* WaaC Inhibitors. *Bioorg. Med. Chem. Lett.* **2008**, *18*, 4022–4026.

59. Durka, M.; Buffet, K.; Iehl, J.; Holler, M.; Nierengarten, J.F.; Vincent, S.P. The Inhibition of Liposaccharide Heptosyltransferase WaaC with Multivalent Glycosylated Fullerenes: A New Mode of Glycosyltransferase Inhibition. *Chem. Eur. J.* **2012**, *18*, 641–651.

60. Tikad, A.; Fu, H.; Sevrain, C.M.; Laurent, S.; Nierengarten, J.; Vincent, S.P. Mechanistic Insight into Heptosyltransferase Inhibition by using Kdo Multivalent Glycoclusters. *Chem. Eur. J.* **2016**, *22*, 13147–13155.

61. Czyzyk, D.J.; Sawant, S.; Ramirez-Mondragon, C.A.; Hingorani, M.M.; Taylor, E.A. *Escherichia coli* Heptosyltransferase I: Investigation of Protein Dynamics of a GT-B Structural Enzyme. *Biochemistry* **2013**, *52*, 5158–5160.

62. Garrity, J.D.; Pauff, J.M.; Crowder, M.W. Probing the Dynamics of a Mobile Loop above the Active Site of L1, a Metallo-β-lactamase from *Stenotrophomonas maltophilia*, via Site-directed Mutagenesis and Stopped-flow Fluorescence Spectroscopy. *J. Biol. Chem.* **2004**, *279*, 39663–39670.

63. Dumitraşcu, L.; Stănciuc, N.; Bahrim, G.E.; Ciumac, A.; Aprodu, I. pH and heat-dependent behaviour of glucose oxidase down to single molecule level by combined fluorescence spectroscopy and molecular modelling. *J. Sci. Food Agric.* **2016**, *96*, 1906–1914.

64. Hogue, C.W.V.; Doublié, S.; Xue, H.; Wong, J.T.; Carter, C.W., Jr.; Szabo, A.G. A Concerted Tryptophanyl-adenylate-dependent Conformational Change in *Bacillus subtilis* Tryptophanyl-tRNA Synthetase Revealed by the Fluorescence of Trp92. *J. Mol. Biol.* **1996**, *260*, 446–466.

65. Lakowicz, J.R. *Principles of Fluorescence Spectroscopy*; Springer: Berlin, Germany, 2007.

66. Meera, K.B.; Debra, A.K. Fluorescence spectroscopy of soluble *E. coli* SPase I Δ2-75 reveals conformational changes in response to ligand binding. *Proteins* **2014**, *82*, 596–606.

67. Haiyuan, D.; Ishita, M.; Donald, O. Lipid and Signal Peptide-Induced Conformational Changes within the C-Domain of *Escherichia coli* SecA Protein. *Biochemistry* **2001**, *40*, 1835–1843.

68. Imhof, N.; Kuhn, A.; Gerken, U. Substrate-Dependent Conformational Dynamics of the *Escherichia coli* Membrane Insertase YidC. *Biochemistry* **2011**, *50*, 3229–3239.

69. Wong, C.F.; McCammon, A.J. Dynamics and design of enzymes and inhibitors. *J. Am. Chem. Soc.* **1986**, *108*, 3830–3832.

70. Liu, X.; Cao, Y.F.; Ran, R.X.; Dong, P.P.; Gonzalez, F.J.; Wu, X.; Huang, T.; Chen, J.X.; Fu, Z.W.; Li, R.S.; et al. New insights into the risk of phthalates: Inhibition of UDP-glucuronosyltransferases. *Chemosphere* **2016**, *144*, 1966–1972.

71. Cote, J.M.; Ramirez-Mondragon, C.A.; Siegel, Z.S.; Czyzyk, D.J.; Gao, J.; Sham, Y.Y.; Mukerji, I.; Taylor, E.A. The Stories Tryptophans Tell: Exploring Protein Dynamics of Heptosyltransferase I from *Escherichia coli*. *Biochemistry* **2017**, *56*, 886–895, doi:10.1021/acs.biochem.6b00850.

72. Zhang, N.; Peng, K.C.; Chen, L.; Puett, D.; Pierce, M. Circular Dichroic Spectroscopy of *N*-Acetylglucosaminyltransferase V and Its Substrate Interactions. *J. Biol. Chem.* **1997**, *272*, 4225–4229.

73. Cavatorta, P.; Sartor, G.; Neyroz, P.; Farruggia, C.; Franzoni, L.; Szabo, A.G.; Spisni, A. Fluorescence and CD studies on the conformation of the gastrin releasing peptide in solution and in the presence of model membranes. *Biopolymers* **1991**, *31*, 653–661.

 International Journal of
Molecular Sciences

Review

Regulated Assembly of LPS, Its Structural Alterations and Cellular Response to LPS Defects

Gracjana Klein * and Satish Raina *

Unit of Bacterial Genetics, Gdansk University of Technology, Narutowicza 11/12, 80-233 Gdansk, Poland
* Correspondence: gracjana.klein@pg.edu.pl (G.K.); satish.raina@pg.edu.pl (S.R.);
Tel.: +48-58-347-2618 (G.K. & S.R.)

Received: 19 December 2018; Accepted: 13 January 2019; Published: 16 January 2019

Abstract: Distinguishing feature of the outer membrane (OM) of Gram-negative bacteria is its asymmetry due to the presence of lipopolysaccharide (LPS) in the outer leaflet of the OM and phospholipids in the inner leaflet. Recent studies have revealed the existence of regulatory controls that ensure a balanced biosynthesis of LPS and phospholipids, both of which are essential for bacterial viability. LPS provides the essential permeability barrier function and act as a major virulence determinant. In *Escherichia coli*, more than 100 genes are required for LPS synthesis, its assembly at inner leaflet of the inner membrane (IM), extraction from the IM, translocation to the OM, and in its structural alterations in response to various environmental and stress signals. Although LPS are highly heterogeneous, they share common structural elements defining their most conserved hydrophobic lipid A part to which a core polysaccharide is attached, which is further extended in smooth bacteria by *O*-antigen. Defects or any imbalance in LPS biosynthesis cause major cellular defects, which elicit envelope responsive signal transduction controlled by RpoE sigma factor and two-component systems (TCS). RpoE regulon members and specific TCSs, including their non-coding arm, regulate incorporation of non-stoichiometric modifications of LPS, contributing to LPS heterogeneity and impacting antibiotic resistance.

Keywords: LpxC; LapB; RpoE sigma factor; Rcs two-component system; lipid IV_A; lipid A modifications; Lpt transport system; noncoding small regulatory RNA

---

1. Introduction

The cell envelope of Gram-negative bacteria, including *Escherichia coli*, contains two distinct membranes, an inner (IM) and an outer (OM) membrane separated by the periplasm, a hydrophilic compartment that includes a layer of peptidoglycan. The OM is an asymmetric bilayer with phospholipids forming the inner leaflet and lipopolysaccharide (LPS) forming the outer leaflet. LPS is essential for bacterial viability. Because of the strong lateral chemical interactions, LPS provides the essential permeability barrier function. It is a complex glycolipid, a major component of the OM and highly heterogeneous in composition. Bacteria, like *E. coli* and *Salmonella*, contain approximately $2\text{-}3 \times 10^6$ molecules of LPS that cover more than 75% of the OM [1]. The biosynthesis, translocation and various modifications of LPS requires the function of more than 100 genes. Several of them are essential and unique to bacteria and hence they are excellent targets for the identification of their inhibitors for the development of new antibiotics.

The biosynthesis of LPS begins with the acylation of UDP-GlcNAc with *R*-3-hydroxymyristate derived from *R*-3-hydroxymyristoyl-ACP by LpxA [2]. *R*-3-Hydroxymyristoyl-ACP also serves as a precursor for the synthesis of phospholipids. The second reaction of the lipid A biosynthesis is catalysed by LpxC [UDP-3-*O*-(*R*-3-hydroxymyristoyl)-*N*-acetylglucosamine deacetylase] a Zn^{2+}-dependent deacetylase, constituting the first committed step in the LPS synthesis, as the equilibrium constant for the first reaction catalysed by LpxA is unfavourable. Following deacetylation, a second *R*-3-hydroxymyristate chain is added by LpxD leading to the synthesis of UDP-2,3-diacyl-GlcN. This serves as a substrate for LpxH to generate 2,3-diacyl-GlcN-1-phosphate, also called lipid X [3]. The next steps involve a condensation reaction generating the β,1′-6-linked disaccharide by LpxB, followed by its phosphorylation at 4′ position by LpxH. This generates the lipid IV_A precursor, which serves as an acceptor for the WaaA-mediated incorporation of two 3-deoxy-α-D-*manno*-oct-2-ulsonic acid (Kdo) residues. Up to the synthesis of Kdo_2-lipid IV_A, all the required seven enzymes are essential for the bacterial viability [4]. Kdo_2-lipid IV_A comprises a key intermediate in LPS biosynthesis that acts 2-fold as a specific substrate: (*i*) for acyltransferases that generate Kdo_2-lipid A moiety by the transfer of two additional fatty acids to the (*R*)-3-hydroxyl groups of both acyl chains, which are directly bound to position 2′ and 3′ of the non-reducing GlcN residue and (*ii*) for glycosyltransferases catalyzing further steps of the core oligosaccharide biosynthesis. The *E. coli* K-12 genome encodes three paralogous acyltransferases (LpxL, LpxM and LpxP), which catalyze acylation reactions using acyl carrier protein-activated fatty acids as co-substrates [5]. At ambient temperatures, a lauroyl residue is first transferred by LpxL to the OH group of the amide-bound (*R*)-3-hydroxymyristate residue at position 2′. This catalytic step is partially replaced at low temperature (12 °C) by LpxP, which transfers palmitoleate to the same position in approximately 80% of LPS molecules [6]. The free OH group of the ester-bound (*R*)-3-hydroxymyristate residue at position 3′ within both pentaacylated intermediates is then myristoylated by LpxM to give a hexaacylated lipid A moiety. However, it is noteworthy that under slow growth conditions and at low temperatures, the lipid IV_A precursor without the Kdo incorporation can be the substrate for a secondary acylation by LpxP, LpxL, and LpxM, as observed in Δ*waaA* grown at 21–23 °C [7].

In *Enterobacteriaceae* members, the core region of LPS is always attached to lipid A via a Kdo residue. The inner core usually contains residue(s) of Kdo and L-*glycero*-D-*manno*-heptose (L,D-Hep) [4,8]. Several of lipid A biosynthesis enzymes and core biosynthetic glycosyltransferases are either inner membrane-anchored or membrane-associated and hence the LPS synthesis occurs at the IM leaflet. After the completion of LPS synthesis on the inner leaflet of the IM, LPS is flipped by the essential IM-located MsbA transporter to the periplasmic side of the IM, where it is a substrate for the $LptB_2FG$ ABC transporter for translocation, using ATP as the energy, potentially in complex with LapA/B proteins.

2. Regulatory Steps in LPS Biosynthesis

Until recent discoveries, it was presumed that LPS biosynthesis occurs in a constitutive manner. However, recent studies have revealed regulatory controls exerted right from early steps in LPS biosynthesis till its final delivery in the OM as highlighted below: (a) Regulation of GlmS expression for the synthesis of UDP-GlcNAc, an essential metabolic precursor for LPS and peptidoglycan, by recruiting GlmZ/Y noncoding small regulatory RNAs (sRNAs). (b) Balanced synthesis of phospholipids and LPS by regulated turnover of LpxC by FtsH/LapB proteins, since they use *R*-3-hydroxymyristoyl-ACP as a common precursor [9]. (c) Bacteria also ensure that only completely synthesized LPS is delivered to the Lpt translocation system by preferential selectivity of hexaacylated LPS by the MsbA transporter and by recruiting LapA and LapB proteins as a scaffold for various LPS biosynthetic enzymes in the IM. (d) The transcriptional control by RfaH of the large *waaQ* operon encoding various LPS core biosynthetic enzymes and the *rfb* operon whose products are required for *O*-antigen biosynthesis [10–12]. (e) Regulation of the incorporation of non-stoichiometric modifications of LPS by the induction of genes, whose expression is controlled by BasS/R, PhoP/Q and PhoB/R two-component systems (TCSs) (Figure 1). The expression of some of these genes is regulated either at a transcriptional or post-transcriptional level or both [13]. (f) Some of the genes, whose products are involved in either lipid A biosynthesis, LPS translocation or phospholipid biosynthesis, are transcriptionally regulated by the RpoE sigma factor and the CpxA/R TCS [14]. (g) The LPS assembly in the OM requires a correctly folded outer membrane protein LptD. Its correct folding is SurA- and DsbA-dependent, which can be further fine-tuned by BepA for the removal of misfolded LptD.

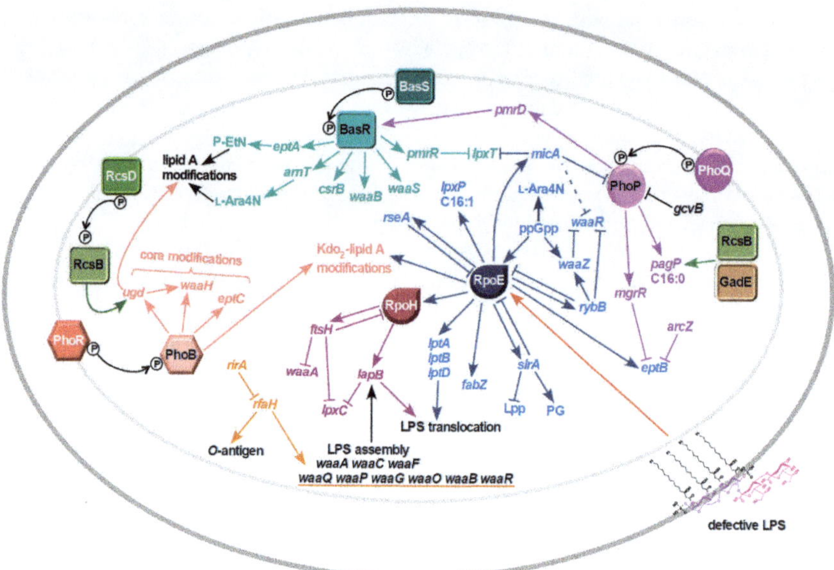

Figure 1. Networks of regulatory pathways that control the lipopolysaccharide (LPS) assembly and its non-stoichiometric modifications. The RpoE sigma factor responds to severe defects in LPS biosynthesis and is also required for transcription of genes involved in LPS biosynthesis/translocation and modifications via sRNAs like *rybB* and *micA*. RpoE also transcribes the *slrA* (*micL*) sRNA, which represses the Lpp synthesis and acts in a feedback manner to repress RpoE. Other regulatory controls involve two-component systems like BasS/R, PhoP/Q, PhoB/R and Rcs system, which are required for transcription of genes whose products are involved in lipid A and inner core modifications. The RpoH heat shock sigma factor transcribes *lapB* and *ftsH* genes, whose products control balanced biosynthesis of LPS and phospholipid by regulating LpxC levels. The unique transcriptional factor RfaH is required for overcoming antitermination, enhance transcriptional elongation and couple transcription/translation of *waaQ* and *rfb* LPS biosynthetic operons. The sRNA RirA binds to RfaH and abrogates its activity to maintain a balanced biosynthesis of LPS.

3. Essentiality of LPS and the Minimal LPS Structure

Generally, LPS is essential for the viability of vast majority of Gram-negative bacteria with few exceptions. Few limited exceptions include viable LPS-lacking mutants of *Acinetobacter baumannii* that have mutations in either *lpxA* or *lpxC* or *lpxD* genes [15]. However, in such mutant strains, the absence of LPS is compensated by increased expression of the Lol lipoprotein transport system to increase phospholipid export, enhanced expression of poly-β-1,6-*N*-acetylglucosamine and elevated expression Mla-retrograde phospholipid migration from the OM to the IM to maintain a balance in the essential constituents of the OM [15]. Another well-studied example of dispensability of LPS include construction of viable LPS deficient mutants of *Neisseria meningitides* [16].

It is well established that in bacteria, like *E. coli* and *Salmonella*, the minimal structure of LPS that can sustain the bacterial viability up to 42 °C is composed of hexaacylated lipid A-Kdo$_2$ (Re LPS "deep-rough mutants"). Thus, Δ*waaC* (lacking heptosyltransferase I) or Δ*gmhD* (absence of ADP-L-*glycero*-D-*manno*-heptose-6-epimerase) mutants are viable, although they exhibit temperature sensitive growth (unable to grow at temperatures above 43 °C), permeability defects, hypersensitivity to detergents, hyperactivated RpoE-regulated stress response, inability to colonize the host, sensitivity to antimicrobial peptides, defects in flagellar biosynthesis, compromised growth at low pH, and a constitutive induction of Rcs-dependent exopolysaccharide [7,17–21]. Consistent with the requirement of inner core heptose attachment to Kdo$_2$-lipid A for cell envelope integrity, a Δ(*waaC surA*) mutational combination confers synthetic lethality [7]. SurA is a major periplasmic folding factor required for the folding of outer membrane proteins (OMPs), including LptD [17,22]. Furthermore, the lack of other conserved core biosynthetic genes like *waaG*, *waaF* or genes involved in the pathway of synthesis of molecules of L-*glycero*-α-D-*manno*-heptose (heptose) such as *gmhA* and *gmhE*, also results in impairment of growth at high temperatures [23]. A deep-rough phenotype accompanied by hypersensitivity to detergents and antibiotics is also associated with the lack of WaaP kinase, which mediates phosphorylation of HepI and completion of core biosynthesis [24] (Figure 1). Underacylation of lipid A due to a lack of the LpxL lauroyl transferase is known to confer temperature sensitivity above 33 °C and strains lacking all three acyltransferases Δ(*lpxL lpxP lpxM*) cannot grow at even 30 °C on rich medium [5]. Moreover, Δ(*lpxL lpxP lpxM*) strains exhibit gross alterations in terms of accumulation of primarily glycoform IV/V with three Kdo residues and truncation of the terminal disaccharide [7]. Thus, complete synthesis of LPS is a requirement for bacterial fitness and survival.

During the detailed genetic construction of strains to define the minimal LPS structure in the absence of any suppressors, it was shown that strains with LPS composed of either Kdo_2-lipid IV_A Δ(*waaC lpxL lpxM lpxP*) or only lipid IV_A (absence of WaaA Kdo transferase) can be constructed under slow growth conditions on minimal medium at low temperatures (21–23 °C) [7]. Since lipid IV_A is a poor substrate for the MsbA IM LPS flippase, suppressor mutations that improved growth characteristics of Δ(*waaC lpxL lpxM lpxP*) or Δ*waaA* were found to map to the *msbA* gene [7]. It is likely that such MsbA variants exhibit altered binding properties of lipid A and might be more relaxed in substrate selectivity or changes in ATP binding/hydrolysis. Interestingly, Δ*waaA* suppressor-free strains synthesizing lipid IV_A were found to accumulate excess of phospholipids consistent with a balanced synthesis of LPS and phospholipids [7]. Moreover, such Δ*waaA* strains under slow growth conditions were also found to accumulate pentaacylated and hexaacylated species of lipid IV_A without any requirement for the Kdo presence. Thus, lipid IV_A derivatives with myristoyl, lauroyl, palmitolyl or palmitoleate chains could be identified from Δ*waaA* strains, indicating that under such conditions late acyltransferases can use lipid IV_A as a precursor without requirement for Kdo [7]. Consistent with these results, overexpression of the *lpxL* gene suppress the lethality of *waaA* deletions on nutrient broth up to 37 °C without the need for MsbA overproduction [25].

4. Regulation of Synthesis of UDP-GlcNAc-Precursor for LPS Biosynthesis

UDP-*N*-acetyl-D-glucosamine (UDP-GlcNAc) is a common metabolic precursor for LPS and peptidoglycan synthesis. Thus, regulation of biosynthesis of UDP-GlcNAc serves as an essential branch point in controlling and coupling the synthesis of major essential constituents of the cell envelope [13]. GlmS catalyzes the synthesis of glucosamine-6-phosphate (GlcN6P) from fructose-6-phosphate and glutamine, which constitutes the first committed step in the synthesis of UDP-GlcNAc. The amount of *glmS* transcript is regulated by a feedback mechanism in response to the GlcN6P level using homologous GlmZ and GlmY sRNAs [26,27]. These sRNAs act in a hierarchical manner to activate the *glmS* expression. Under GlcN6P limiting conditions, the GlmY sRNA accumulates and sequesters RNase adaptor protein RapZ, preventing GlmZ processing [26,27]. The GlmZ sRNA facilitates translation of the *glmS* mRNA through an anti-antisense mechanism and prevents the formation of an inhibitory structure that occludes the ribosome-binding site of *glmS*. Interestingly, transcription of the *glmY* gene is regulated by RpoN and RpoD sigma factors using the same transcription start site in an analogous manner to the transcriptional regulation of *rpoE*P2 and *rpoE*P3 promoters [11,28]. RpoN-regulated promoters of *glmY* and *rpoE* genes use QseF as an activator and thus this mode of regulation may be important to sense common signals and ensure cellular homeostasis in response to envelope stress.

5. Coupled Regulation of LPS and Phospholipids–Regulation of Amounts of Kdo$_2$-Lipid A Synthesis

Regulation of LpxC occurs by regulated proteolysis mediated by FtsH [29,30]. This proteolysis by FtsH requires the LPS assembly factor LapB [9,31]. Both FtsH and LapB are essential for bacterial growth and their depletion causes increased synthesis of LPS at the expense of phospholipids. This is due to stabilization of LpxC in either an *ftsH* or a *lapB* mutant, which causes diversion of a common precursor *R*-3-hydroxymyristoyl-ACP towards the LPS synthesis, since LpxC and FabZ compete for the same precursor, limiting the availability of phospholipids (Figure 1). Consistent with the notion of coupling of phospholipid and LPS synthesis, suppressors mapping to the *fabZ* gene, like *sfhC*21 that encodes a hyperactive variant of FabZ, can bypass the essentiality of either *ftsH* or *lapB* genes [9]. Additional evidence supporting this model is based on observations that inhibition of LpxC can be compensated by mutations that compromise the FabZ activity [32]. Similarly, overexpression of the *fabZ* gene is accompanied by an upregulation of the LpxC activity and vice versa [32]. Further supporting regulated LPS and phospholipid biosynthesis, an overexpression of noncoding sRNA *slrA* can bypass the lethality of the essential *lapB* gene [9]. The molecular basis of this suppression was attributed to translational repression of the gene encoding the most abundant protein Lpp, also called Braun's lipoprotein, with an abundance of 7×10^5 molecules per cell. Hence, SlrA is also called MicL [33]. Each Lpp molecule has three acyl chains {phosphatidylglycerol moieties (PG)} and therefore, when the Lpp amount is reduced due to overexpression of *slrA* sRNA, it causes an increase in the amount of PG that can restore a balance between phospholipids and LPS (Figure 1). SlrA can also act as a negative regulator of RpoE in a feedback manner, as its overproduction reduces the RpoE activity elevated due to LPS defects in Δ*lapB* mutants [9]. The gene encoding SlrA sRNA is transcribed from the RpoE-regulated promoter located within the *cutC* gene. This 80-nt sRNA is synthesized as a 307 nt precursor mRNA that is processed and is located within the 3′ end of the coding region of the *cutC* gene [9,33].

Although a role for LapB and FtsH for LpxC is now known, however how the proteolytic activity of FtsH is regulated remains to be elucidated. It is likely that proteolytic activity of FtsH might be regulated by concentration or forms of acyl-ACPs or lipid A disaccharide. Since LpxC stability is increased in a *fabI*(ts) mutant suggests that membrane fatty acids might influence proteolytic activity or add additional checkpoints in regulating lipid A and phospholipid amounts [29]. Although not fully elaborated, regulation at the level of LpxB and LpxK may serve as additional pathways of this co-regulation. Consistent with such a notion, LpxB has been shown earlier to co-purify with phospholipids [34]. A regulatory checkpoint has been postulated on the basis of a reduced LpxK activity upon reduced membrane fluidity when more saturated fatty acids are present [35]. LpxK mediates the last essential step that completes lipid IV$_A$ biosynthesis [4]. LpxK catalyzes the phosphorylation of the 4′ hydroxyl of the distal glucosoamine of lipid A disaccharide. Under conditions like compromised function of either LapB or FtsH, which stabilize LpxC leads to depletion of *R*-3-hydroxymyristoyl-ACP. This in turn decreases the synthesis of unsaturated fatty acids, which can decrease the LpxK activity, leading to accumulation of lipid A disaccharide [9,35]. This accumulation of lipid A disaccharide intermediate by the feedback mechanism causes increased proteolysis of LpxC, thereby decreasing flux of *R*-3-hydroxymyristoyl-ACP into a lipid A biosynthesis pathway, thus again to regain a balance between phospholipid and lipid A biosynthesis. WaaA is also one of the substrates of FtsH protease [36]. Regulation of WaaA turnover by FtsH may be yet another step in preventing an excessive synthesis of LPS over phospholipids. However, a direct impact of changes in WaaA concentration on either LPS synthesis or accumulation of different glycoforms remains to be understood.

6. Assembly of LPS Requires LapB

Besides a role for turnover of LpxC in concert with FtsH, LapB has been implicated in the assembly of LPS at the IM presumably along with LapA [9]. Genes encoding *lapA* and *lapB* are co-transcribed from three promoters, the distal promoter located upstream of the *pgsB* gene, the middle promoter recognized by the RpoH heat shock sigma factor and the last one resembling house-keeping promoters [9]. Overall, such a transcriptional organization suggests coupling of transcription with phospholipid metabolism using *pgpB* co-transcription. PgpB encodes phosphatidylglycerophosphatase, an enzyme that is part of the phosphatidylglycerol biosynthesis pathway, which is itself part of phospholipid metabolism. Transcription from the heat shock promoter ensures transcription of *lapA/B* genes at high temperatures, hence belonging to heat shock regulon, which comprises several chaperones and proteases.

The evidence that LapB plays a role for LPS assembly in the IM comes from experimental evidence that include: (a) The *lapB* gene is essential for bacterial growth and a deletion of the *lapB* gene can be constructed in the presence of suppressor that either restore phospholipid synthesis like *sfhC21*, or decrease the LPS synthesis like in the presence of mutations in an early lipid A biosynthesis pathway (*lpxA*, *lpxC*, *lpxD*) or when LPS is composed mostly of Kdo_2-lipid A derivatives like (*waaC lapB*) combination [9]. (b) The lack of LapB leads to accumulation of LPS precursor forms. These precursor forms represent pentaacylated lipid A species, the presence of Kdo_2-dilauroyl-lipid IV_A. (c) $\Delta lapB$ mutants have defects in the folding of LPS-specific enzymes like LpxM and conserved glycosyltransferases like WaaC, WaaO. A large proportion of these enzymes is present in aggregate form and hence LPS-specific enzymes like LpxM and core glycosyltransferase could be limiting explaining the accumulation of LPS precursor species [9]. (d) LapB also co-purifies with heptosyltransferase I. (e) LapA/B could function synergistically with classical chaperones like DnaK/DnaJ based on LapA/LapB co-purification with DnaK/J, multicopy suppression of growth defects of $\Delta(lap\ lapB)$ mutants, even more pronounced defects in the LPS composition when $\Delta(lapA\ lapB\ dnaK/J)$ were examined. (f) LapA/B co-purify with LPS, Lpt proteins and FtsH. (g) A $\Delta(lapA\text{-}lapB)$ mutation is synthetically lethal with a compromised LptD variant or when SurA is absent. (h) The absence of LapB induces a strong envelope stress response regulated by RpoE, Cpx and Rcs systems by inducing transcription of genes, whose products are required to maintain homeostasis in the cell envelope. LapB contains nine tetratricopeptide repeat (TPR) motifs and a C-terminal rubredoxin domain [9,37]. Mutations in the either rubredoxin domain or TPR/interface impair cell growth [9]. Thus, based on the essentiality of TPR repeats in LapB, which can mediate protein-protein interactions (interaction of LapA/B with LPS-specific enzymes) a model was presented wherein LapA/B could form a scaffold-like structure for delivery of various acyltransferases and glycosyltransferases to the site in IM where LPS is assembled and delivered to the Lpt complex. At such an assembly site, proteolysis of LpxC can occur given co-purification of LapA/B with FtsH to prevent excessive LPS buildup. This may serve as an essential purpose of preventing wasteful transfer of incompletely synthesized LPS. This model draws support from the IM association of LpxC in *Neisseria meningitidis*, which requires the presence of LapB counterpart called Ght [38]. Consistent with a broad essential function of LapB, a suppressor mutation *lapB*V43G in the *lapB* gene that confers protection against prolonged exposure to phosphate starvation and also suppresses the *mlaA*-dependent hyperproduction of LPS have been reported [39,40]. Thus, LapB plays an important role in maintaining cell envelope homeostasis and assembly of LPS.

7. Transport of LPS

7.1. MsbA-Mediated Transport of LPS Across the Inner Membrane

After the completion of core-lipid A synthesis, LPS is flipped across the IM by the essential ATP-binding cassette MsbA transporter. Several structures have been determined, which show that MsbA exists as a functional dimer, wherein core-lipid A binds the cytoplasmic open conformation [41,42]. Upon ATP binding, conformational change in transmembrane (TM) domains results into change from the cytoplasmic open to the periplasmic open state causing flipping of core-lipid A into the periplasmic leaflet of the IM. MsbA can return to the cytoplasmic open state after release of phosphate [42]. MsbA has a higher preference for hexaacylated lipid A derivatives, thereby providing an early checkpoint to prevent transport of early intermediates of LPS biosynthesis. Recent structural analysis of MsbA-core lipid A revealed that MsbA recognizes a bivalent phosphoglucosamine headgroup and correct acylation of lipid A to achieve substrate selectivity over competing bulk membrane phospholipids [43]. The tight packing of hexaacaylated LPS in the hydrophobic pocket suggests that MsbA packs acyl chains of correct length and number, providing the basis of this selectivity [43].

7.2. LPS Translocation and Assembly in the Outer Membrane

After the flipping of LPS by MsbA to the periplasmic side of IM, LPS is transported for its final localization in the OM which requires a complex of seven essential conserved proteins LptA-LptG, whose components reside in every cell compartment and form a single transenvelope complex spanning from the IM to the OM. Here, we briefly summarise key elements, as this topic has been aptly reviewed recently [44–47]. The Lpt system is organized into the IM complex comprised of LptB$_2$CFG and the OM component of 1:1 complex of LptDE, which are bridged by a periplasmic protein component LptA. This transenvelope complex acts as a single unit to transport LPS, since depletion of any component leads to LPS accumulation at the IM [44–47]. Mechanistically, the ABC transporter LptB$_2$FG extracts LPS from the IM in an ATP-dependent manner to deliver to LptC, and LPS transfer from LptC to LptA requires additional ATP hydrolysis [44–47], allowing LPS transit across the aqueous periplasm for its final delivery to LptD. The evidence from co-sedimentation, pull-down experiments, photo-cross-linking, mutational and structural analyses, assembly of liposomes from IM components and OM components in the presence or absence of LptA, shows that, while the N-terminus of LptA interacts with LptC at the IM, the C-terminus of LptA interacts with the periplasmic domain of the OM LptD, creating a continuous bridge of antiparallel β-strands between the IM and the OM [44–47]. The presence of β-jellyroll domains in LptG, LptF, LptC, LptA and LptD can sequester the acyl chains of LPS from the aqueous periplasm, using ATPase activity of LptB to transfer LPS over transenvelope bridge [48]. In support of the single stable bridge model of LPS transport, it was shown that LptA can promote association of liposomes containing the IM complex of LptBCFG with OM liposomes containing the OM LptDE [49]. Furthermore, ATP-dependent transfer of LPS from LptBFG to LptC and LptA has been demonstrated in liposomes and transfer to LptA was shown to be increased when LptC was present [49].

Structural studies of LptD and LptE revealed that LptD folds into two domains: a β-jellyroll and a β-barrel [50,51]. The β-jellyroll extends away from the OM and interacts with the β-jellyroll domain of LptA suitable for binding to lipid A, while leaving the LPS oligosaccharide exposed. LPS could be delivered directly into the cavity of the larger domain of LptD β-barrel, which is composed of 26 membrane-spanning β-strands. LptE adopts a roll-like structure located inside the barrel of LptD to form a unique two-protein 'barrel and plug' architecture [50,51]. The plugging of LptE inside of the LptD barrel causes a diminished lumen size to 45Å × 35Å on the periplasmic side. However, such a space is sufficient to accommodate LPS. In these structures, β-strands 1 and 2 are distorted and weak hydrogen bonding between β-strands 1 and 26 could support a lateral opening for LPS migration. In this process, positioning of the lipid A part could be assisted by LptE, as it is known to bind LPS and oligosaccharide could be placed in the hydrophilic cavity of LptD chamber.

Finally, LptD folding into the functional state can be a rate-limiting step. Several periplasmic folding catalysts like DsbA, DsbC, SurA, FkpA are known to play important roles in the folding of OMPs and periplasmic proteins [52]. Folding of LptD requires the SurA periplasmic folding catalyst and other periplasmic folding factors, like Skp and FkpA, could further modulate efficient folding of LptD [22]. LptD has two disulfide bridges between four non-consecutive cysteine residues, which require the periplasmic DsbA disulfide oxidoreductase and interaction of LptD with LptE to achieve correct folding with native disulfide bridges [53]. Additional factors like BepA may also be required for stimulating disulfide rearrangement of LptD and degradation of misfolded LptD [54]. From the transcriptional point of view, the *lptD* gene is transcribed by Eσ^E [14] and this discovery was the beginning of series of studies that led to subsequent elucidation of the Lpt system.

8. Regulated Structural Alterations in LPS

The LPS composition is highly heterogeneous and dynamically altered in response to various challenges like exposure to different stress conditions or changes in growth medium. Heterogeneity of LPS can be due to: modification of the lipid A part by the incorporation of phosphoethanolamine (P-EtN) and 4-amino-4-deoxy-L-arabinose (L-Ara4N) that mask negative charges, the addition or removal of acyl chains, changes in the inner core due to the incorporation of additional Kdo residue, rhamnose (Rha), uronic acid and P-EtN and changes in number of phosphate residues, and truncation of the outer core [55]. All these structural changes are highly regulated, important for resistance to cationic antimicrobial peptides, for bacterial virulence in pathogenic bacteria and may have adaptive significance in specific environmental niches. Certain modifications in the LPS structure can also contribute to biofilm tolerance of antimicrobial compounds [56]. Lipid A of *E. coli* under standard growth conditions is bisphosphorylated carbohydrate backbone disaccharide β-D-Glc$p$N4P-(1→6)-α-D-Glc$p$N1P, which is hexaacylated without any modifications. However, exposure to low pH, excess of Fe^{3+}, Zn^{2+}, Al^{3+}, change in divalent cations concentrations, challenge by antimicrobial peptides, treatment with the non-specific phosphatase inhibitor ammonium metavanadate (AMV), treatment with agents that disturb the OM symmetry like exposure to chelating agents like EDTA or genetic alterations that lead to the synthesis of tetraacylated derivatives can cause profound changes in the lipid A composition [7,55,57]. Some of these modifications are regulated at the transcriptional level, while some are subjected to a post-transcriptional control and certain modification occur at post-translational level [13].

8.1. Regulation of Lipid A Modifications

Most prevalent non-stoichiometric modifications that occur in the lipid A part involve either reducing the net negative charges of lipid A by modifying the 1 and/or 4′ ends of phosphate residues, or the addition or removal of acyl chains. Most commonly observed modifications of lipid A include the incorporation of P-EtN and L-Ara4N at 1 and/or 4′ ends, respectively [57]. Such substitutions are known to confer resistance to cationic antimicrobial peptides like polymyxin B. This non-stoichiometric incorporation of P-EtN and L-Ara4N residues requires IM-located EptA and ArnT transferases, respectively with active site facing the periplasm [57]. Genes encoding these transferases are part of operons, whose transcription is positively regulated by the BasS/R (PmrA/B) TCS with an overlap with the PhoP/Q system. These systems specifically respond to changes in Fe^{3+} and divalent cationic concentrations, respectively. The PhoP/Q system in *Salmonella* and in some pathogenic Gram-negative bacteria regulates the expression of virulence genes, including those encoded in pathogenic islands. Lipid A of strains synthesizing tetraacylated lipid A can have phosphate residues at 1 and/or 4′ ends modified by P-EtN, since they do not incorporate L-Ara4N due to defects in translocation and such lipid A species could be poor substrates for ArnT. Such modifications occur after the translocation at the periplasmic side and often serve as markers for LPS translocation with L-Ara4N incorporation as a more stringent signature [7]. In *E. coli* and *Salmonella*, the activation of PhoP/Q TCS upon depletion of Mg^{2+} and Ca^{2+} also leads to the BasS/R induction, which requires PmrD as the adaptor protein [58]. Such a cross talk between BasS/R and PhoP/Q TCSs allows integration of signals from different environmental cues and amplification of output response.

The majority of lipid A in *E. coli* K-12 contains monophosphate at positions 1 and 4′. Approximately one-third of lipid A molecules in the *E. coli* K-12 outer membrane contains a diphosphate unit at the 1 position. The enzyme LpxT is responsible for this phosphorylation at 1′ position, generating hexaacylated lipid A with two phosphate residues under ambient growth conditions in the absence of induction of lipid A modification systems increasing the negative charge [59]. However, upon induction of the BasS/R system, P-EtN is incorporated at 1′ position due to inhibition of the LpxT activity. This inhibition of LpxT activity upon BasS/R-inducing conditions is due to the expression of a short peptide PmrR, which directly bind to LpxT [60] and hence constitutes a post-translational control [13]. In *Salmonella*, PhoP activation can also positively regulate *lpxT* transcription. The PhoP-dependent *lpxT* expression induced in low Mg^{2+} results in 1-PP lipid A, favors further modification of lipid A phosphates with L-Ara4N and the exclusion of P-EtN. Thus, *Salmonella* favors lipid A modified with L-Ara4N under low Mg^{2+} and with both L-Ara4N and P-EtN when exposed to a mildly acidic pH [61].

As RpoE and LPS structural alterations are intricately linked, an interesting regulatory control of lipid A alteration by RpoE-regulated sRNAs has emerged. Structural analysis of LPS obtained from several different strains under simultaneous RpoE- and BasS/R-inducing conditions shows the absence of LpxT-dependent phosphorylation of lipid A, presumably due to the transcriptional induction of *micA* [62]. The RpoE-regulated MicA sRNA can also exert an influence on lipid A composition by connecting regulation of PhoP/Q TCS as well as LpxT to RpoE. Analysis of MicA targets revealed that besides known OMP-encoding genes, MicA represses PhoP synthesis by base pairing in the translation initiation region of *phoP* mRNA and inhibits its translation [63]. MicA sRNA could as well regulate the LpxT synthesis at a post-transcriptional level, since a base-paring region between the seed sequence in the *micA* sRNA and the *lpxT* mRNA has been predicted [64]. Another sRNA GcvB also represses *phoP* mRNA translation by base-pairing [65].

The lipid A part can also undergo non-stoichiometric modification by PagP-dependent palmitoylation due to the incorporation of a palmitate chain linked to the hydroxyl group of 3-hydroxymyristic acid at C-2 position of reducing-end [66]. Transcription of the *pagP* gene is positively regulated by the PhoP/Q TCS and via RcsB in biofilm environment or increase in osmolarity that requires the GadE auxiliary regulator independently of the Rcs phosphorelay cascade [56]. The PagP enzyme is usually inactive in the OM and is post-transcriptionally activated when the OM permeability is breached and this modification can dampen recognition by host immune responses. Another PhoP/Q-regulated modification is PagL-dependent deacylation. The *pagL* gene encodes a lipid A 3-*O*-deacylase and this deacylation of lipid A modifies its ability to induce immune response [67]. The PagL-dependent modification also occurs post-translationally after LPS is incorporated in the OM [13]. It is noteworthy that in *Pseudomonas aeruginosa* the expression of *pagL* is positively regulated by sRNA (Sr006) [68]. Regulation of another deacylase LpxR with a 3′-*O*-deacylase activity of relevance to pathogenicity is of interest and this modification also occurs in the OM [69,70]. The expression of LpxR is subjected to negative regulation by MicF *trans*-acting base-pairing RNA. The base-pairing of MicF within the coding sequence of the *lpxR* mRNA decreases its stability by promoting its degradation by RNase E [71]. Thus, variety of mechanisms exist to modify lipid A part of LPS which involve TCSs, the RpoE sigma factor and sRNAs and this regulation can occur at different steps of LPS biosynthesis transcriptionally or post-transcriptionally.

8.2. Regulation of Inner Core Modifications and Switches Between Different Glycoforms

The inner core of LPS in the majority of cases generally contains an α-(2-4)-linked Kdo disaccharide, to which L-*glycero*-D-*manno*-heptose residue (Hep) is attached at position 5 of KdoI. In *E. coli*, the inner core contains three Hep residues. Although the composition of the inner core is relatively conserved, Kdo, as well as Hep residues, can be non-stoichiometrically modified. These modifications are regulated by the RpoE sigma factor, PhoP/Q, PhoB/R TCSs, by alarmone ppGpp and recruitment of specific sRNAs, whose expression is regulated by these transcription factors (Figure 1). Like lipid A modifications, they are often important for antibiotic resistance OM permeability and contribute to diversity in the LPS composition [10,13,62]. Quite well studied modifications include the non-stoichiometric incorporation of a P-EtN residue on the second Kdo by EptB transferase, incorporation of a third Kdo by WaaZ transferase, the addition of Rha, which can be linked to either the second Kdo or the third Kdo, modification of phosphorylated HepI by P-EtN using the PhoB/R-regulated EptC and incorporation of glucuronic acid (GlcUA) by the PhoB/R-inducible WaaH glycosyltransferase with a concomitant loss of phosphate residue on HepII [55,62].

Transcription of the *eptB* gene encoding P-EtN transferase specific to the second Kdo is positively regulated by the RpoE sigma factor [7,72]. Hence, this modification of the second Kdo is quite pronounced when the RpoE activity is induced either in the absence of RseA anti-sigma factor or when LPS is defective as in $\Delta waaC$ or $\Delta waaF$ mutants [7,11,62]. Consistent with the induction of

RpoE and regulation of the *eptB* transcription, the lipid A part of Δ*waaC* or Δ*waaF* strains were found to lack P-EtN even under *eptA*-inducing conditions, but preferentially incorporating P-EtN on the second Kdo [7,11]. However, this incorporation requires the presence of Ca^{2+} in the growth medium. Furthermore, the synthesis of EptB is negatively regulated by two Hfq-dependent sRNAs MgrR and ArcZ, although in response to different environmental signals [72,73]. Under normal growth conditions, the PhoP-regulated MgrR sRNA via base-pairing with the *eptB* mRNA silences its expression due to translational repression of the *eptB* mRNA. Activation of PhoP/Q induces transcription of the *mgrR* sRNA. The EptB-dependent P-EtN modification of the second Kdo upon high Ca^{2+} concentration can be explained by the repression of transcription of PhoP/Q-regulated *mgrR* and promote the synthesis of active EptB under RpoE-inducing conditions. The Hfq-dependent ArcZ sRNA also inhibits translation and expression of *eptB* by base-pairing in an ArcA/B-dependent manner in response to oxygen concentration [72]. Quite like the P-EtN modification of lipid A, the incorporation of P-EtN on the second Kdo confers resistance to polymyxin B. Thus, RpoE regulon members like *eptB* and MicA sRNA, PhoP-regulated MgrR sRNA contribute to the incorporation of P-EtN on the second Kdo, lipid A and regulation of glycoform switches.

8.3. Modifications in the Heptose Region of the LPS Inner Core

The inner core of LPS exhibits the limited structural diversity, since it plays a crucial role in maintaining the OM stability. However, some non-stoichiometric substitutions have been demonstrated in this region of LPS. These include the incorporation or loss of phosphate residues and the addition of P-EtN, GlcN, or GlcUA. Among these, the phosphorylation of the HepI residue by WaaP is critical for the OM permeability and also provides the attachment sites for the other substituents and ensures the completion of core synthesis [11,24,55]. Accordingly, in *E. coli* K-12, phosphorylated HepI serves as an acceptor for P-EtN, which requires the EptC phosphoethanolamine transferase [55]. Furthermore, the HepIII residue, whose incorporation requires prior phosphorylation of HepI by WaaP, can be modified by GlcN (in *E. coli* R1 and R3 isolates) or by GlcUA (in *Salmonella* and *E. coli* B, K-12, R2 and R4 core types). Importantly, the modification of the HepIII residue is always accompanied by loss of phosphate residue at the HepII, thereby maintaining a net negative charge [55,74]. Structural studies revealed that GlcUA is attached to O-7 of the side-chain Hep, with the phosphate residue found at position O-4 of the HepII being absent [55] (Figure 2). WaaH shares 20% amino acid sequence similarity with WabO of *Klebsiella pneumoniae*. However, WabO in *K. pneumoniae* is responsible for the galacturonic acid (GalA) incorporation [75]. In *E. coli* K-12, *waaH* and *eptC* genes, whose products mediate transfer of GlcUA and P-EtN modifications of the Hep I and Hep III residues respectively, are positively regulated by the PhoB/R TCS [55]. However, the P-EtN-modification of HepI can also occur at the basal level without a requirement for induction of either PhoB/R or BasS/R systems. The EptC-dependent modification of HepI is important for the permeability function, as Δ*eptC* mutants exhibit sensitivity to exposure to sub-lethal Zn^{2+} concentration or the presence of SDS [55].

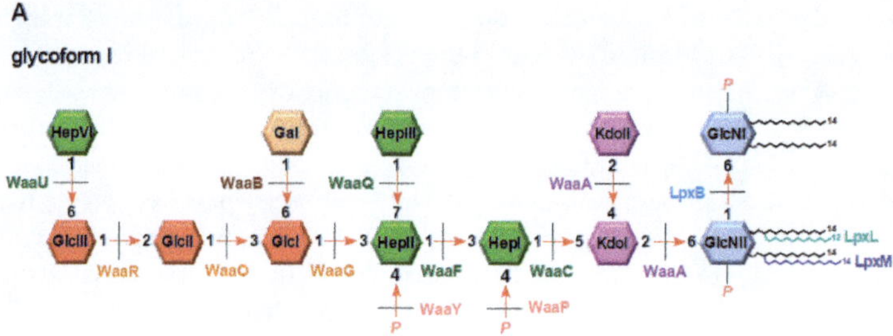

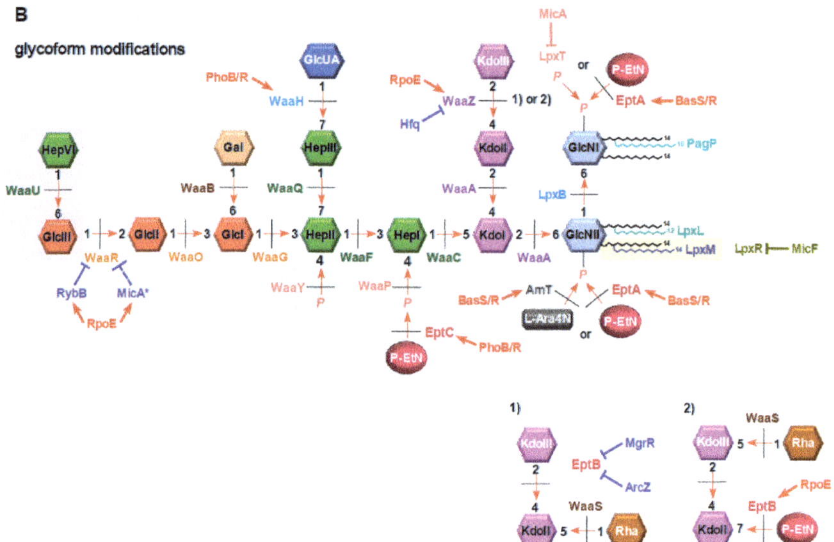

Figure 2. Schematic depiction of unmodified hexaacylated glycoform I and glycoform derivatives with various non-stoichiometric substitutions with a third Kdo-Rha disaccharide. Various genes whose products mediate different steps in LPS biosynthesis and incorporation of different modifications are indicated. Glycoform I constitutes the major LPS species under non-stress conditions (**A**). Upon the induction of RpoE sigma factor and conditions inducing BasS/R and PhoB/R TCSs, the lipid A part is modified by P-EtN and L-Ara4N at 1 and 4′ position by EptA and ArnT, respectively (**B**). The induction of RpoE causes increased synthesis of EptB overcoming MgrR silencing and incorporation of P-EtN on the second Kdo (panels B1 and B2). The induction of RpoE favors pathway of the third Kdo incorporation by increased synthesis of WaaZ and repression of WaaR by the RybB sRNA, leading to synthesis of glycoforms with a third Kdo with attachment of Rha on the third Kdo with a concomitant truncation of the terminal GlcIII-HepIV disaccharide shown in violet background. The PhoB/R induction leads to the incorporation of GlcUA at the expense of HepII phosphorylation (blue background). Lipid A can also be modified by PagP generating heptaacylated lipid A and by the removal of acyl chains by LpxR in the OM after translocation (light brown background). Various sRNA-mediated controls are shown in blue color.

8.4. Glycoform Switches

Up to now, at least seven structurally different glycoforms of *E. coli* K-12 have been characterized, whose relative abundance varies on growth conditions such as exposure to phosphate-limiting growth conditions, the induction of RpoE-regulated envelope stress response and activation of specific TCSs. These glycoforms differ due to the non-stoichiometric modifications such as P-EtN transfer to the second Kdo, incorporation of a third Kdo, addition of sugars like Rha, GlcN, uronic acids, alterations in the numbers of phosphate residues in the LPS core and truncation of outer core terminal disaccharide. In initial studies using optimal growth conditions, the wild-type *E. coli* K-12 was found to contain majority of LPS corresponding to glycoform I structure and only minor amounts of three additional glycoforms II, III and IV could be observed [76]. Glycoform I contains two Kdo residues in the inner core, and four heptoses and four hexoses attached in specific order in the inner core and the outer core (Figure 2). However, using strains or growth conditions that exhibit the induction of RpoE sigma factor or employing phosphate-limiting growth conditions supplemented with Fe^{3+} and Zn^{2+} (induction of PhoB/R and BasS/R TCSs) or treatment with AMV (induction of RpoE and non-specific induction of TCSs) revealed major shifts in the LPS composition and prevalence of different glycoforms not only in *E. coli* K-12 but also in *E. coli* strains with different core type and in *Salmonella* [55,62]. Using growth medium that induces BasS/R and PhoB/R TCSs, wild-type *E. coli* strains synthesize more glycoform IV/V derivatives as compared to glycoform I. However, when the RpoE induction is maximal, a near-exclusive synthesis of glycoform V and its derivatives are observed (Figure 2). This molecular switch to the synthesis of glycoform V derivatives requires ppGpp alarmone, induction of the RpoE-transcribed genes *eptB*, sRNAs *micA* and *rybB*, and the transcriptional upregulation of *waaZ* with a concomitant repression of WaaR synthesis [62] (Figure 2). *waaZ* and *waaS* genes encode the Kdo transferase required for the incorporation of the third Kdo and rhamnosyl transferase, respectively, while as the *eptB* gene encodes P-EtN transferase specific to the second Kdo [62,77] (Figure 2). Glycoforms IV and V have the same molecular masses, but are structurally different due to the incorporation of P-EtN on the second Kdo by the RpoE-regulated EptB and attachment of Rha to the terminal third Kdo defining glycoform V. Without RpoE induction, EptB synthesis is silenced by the PhoP/Q-dependent MgrR sRNA (no incorporation of P-EtN on the second Kdo, instead the attachment of Rha on the second Kdo) and hence the synthesis of glycofom IV derivatives.

One of the interesting structural features of glycoform V derivatives with a third Kdo is the concomitant truncation of the terminal disaccharide and the incorporation of P-EtN on the second Kdo with Rha on the third Kdo. The minimal LPS structure that can support incorporation of a third Kdo requires WaaO-mediated addition of glucose and hence serves as a branch point in determining switches between glycoform I and glycoform IV/V derivatives [62]. Primarily this switch is regulated by levels of WaaR and WaaZ, whose expression are regulated by induction of RpoE and PhoB/R TCS. Truncation of the terminal disaccharide suggested that the WaaR glycosyltransferase is limiting under RpoE-inducing conditions. This was indeed experimentally validated by observed repression of WaaR synthesis due to the RpoE-regulated RybB sRNA and to some extent by another RpoE-regulated sRNA MicA (Figure 2). At the same time, transcription of the *waaZ* gene is induced under such conditions. Physiologically, switch to the synthesis of glycoform derivatives with a truncation in the outer core, hence lacking terminal heptose, might be important for escaping detection by host, since such derivatives cannot incorporate *O*-antigen. Importance of the *O*-antigen incorporation can be critical, as it becomes essential for survival in *Salmonella* in the absence of the *rpoE* gene [78]. Thus, a coordinated molecular programming ensures increased synthesis of the WaaZ Kdo transferase under RpoE-inducing conditions, and the EptB P-EtN transferase, the induction of BasS/R-dependent WaaS synthesis with a simultaneous repression of WaaR and MgrR synthesis.

8.5. Transcriptional Regulation of Major LPS Biosynthetic Operons by RfaH

RfaH is a paralog of NusG family of universally conserved transcription factors. RfaH is unique in its specificity for recognition of only those operons that contain a short 8-nt conserved sequence (GGCGGTAG) in the 5' UTR called the *ops* (operon polarity suppressor) pause site. Thus, RfaH regulates the expression of genes that play important functions in virulence such as the synthesis of LPS core, *O*-antigen, haemolysin, capsule and for conjugation [11,12]. Recruitment of RfaH prevents transcriptional termination and enhances transcriptional elongation of long *waaQ* and *rfb* operons. Consistent with a role for regulation of expression of *waaQ* and *rfb* operons, they contain an 8-nt conserved *ops* site and the JUMPstart site in their 5′ UTR and also lack a ribosome-binding site. In the absence of RfaH, LPS is truncated, since the expression of the *waaQ* operon is severely compromised and Δ*rfaH* mutants exhibit permeability defects, the induction of RpoE-dependent envelope stress response and are avirulent in pathogenic bacteria [11]. RfaH without interaction with the *ops* site exists in an inactive form with a closed conformation with its C-terminal domain (CTD) folded in an α helical hairpin tightly packed against its N-terminal domain (NTD). In this conformation, the NTD of RfaH is not accessible for DNA-RNA polymerase interaction [79]. However, upon encountering the *ops* site in the presence of RNA polymerase, the CTD of RfaH folds into a β-barrel allowing recruitment of S10 ribosomal protein to couple transcription with translation and also enhance transcriptional elongation. Recently, a new RirA sRNA (RfaH interacting RNA) was identified while analyzing factors that induce transcription of the *rpoE* gene [11]. Overexpression of the 73-nt RirA sRNA, located in the 5′ UTR of the *waaQ* operon, increases transcription from the *rpoE*P3 promoter that specifically responds to LPS defects. RirA overexpression abrogates the synthesis of *O*-antigen, causes reduction of LPS amounts and truncation in the LPS core, thus mimicking the Δ*rfaH* phenotype. Indeed, RirA directly interacts with RfaH in the presence of RNA polymerase and this interaction is dependent on the presence of the *ops* site within the *rirA* RNA [11]. This RirA-RfaH interaction could lead to loss of the specificity for recognition of *ops*-containing operons, like *waaQ* and *rfb* operons, needed for LPS core and *O*-antigen biosynthesis.

9. Signal Transduction in Response to LPS Defects

Integrity of the OM and homeostasis of various overall cell envelope components is critical for growth and the viability of bacteria. This requires correct assembly of OMPs, a balance between synthesis of peptidoglycan, phospholipids and LPS. This in bacteria requires function of various regulon members of RpoE and Cpx TCS. Severe defects in LPS biogenesis in response to underacylation, defects in LPS when the LPS assembly factor LapB is missing or depletion of Lpt translocation system, truncation in the inner core or when LPS synthesis is compromised particularly when bacteria synthesize minimal LPS derivatives like Kdo_2-lipid A, Kdo_2 lipid IV_A or only lipid IV_A derivatives, cause major cell envelope perturbations leading to a signal response that causes activation of the RpoE sigma factor, induction of the Rcs phosphorelay and activation of the Cpx TCS [7,9–11]. The extent of induction of these stress responses varies depending upon the severity of defects. Early work addressing activation of RpoE, when the OMP assembly is compromised, revealed that strains lacking GmhD (HtrM), which synthesize Kdo_2-lipid A Re LPS, overexpress exopolysaccharide and have a constitutive elevation of RpoE response with a concomitant decrease in OMP content [17]. The exopolysaccharide synthesis is known to be positively regulated by the Rcs TCS by increasing transcription of *wca* genes [80]. Subsequently, during deciphering of the RpoE regulon, it was revealed that some of the genes involved in either LPS biosynthesis, LPS modifications, LPS transport and factors involved in OMP folding and insertion in the OM are part of this regulon [14]. In further studies, detailed mutational analysis of various genes, whose products are either involved in LPS biosynthesis or LPS modifications, coupled with mass spectrometric analysis of LPS of all such mutant derivatives and an impact on envelope stress response was undertaken in series of studies [7,9,11]. These studies reveal a full circuit wherein defects in LPS cause a major induction of RpoE and RpoE induction without any mutations in LPS biosynthetic/assembly pathways lead to several alterations leading to the accumulation of specific LPS glycoforms and various modifications (Figure 3). Some of the major conclusions and major pathways of stress response are discussed here. These studies showed that underacylation of LPS, when the core biosynthetic pathway is intact, does not cause any major induction of either RpoE or Cpx systems [7]. Only when all three acyltransferases are deleted, as in a strain Δ(*lpxL lpxM lpxP*), a modest induction of RpoE-dependent pathway is induced. In contrast, Δ*waaC* strains synthesizing Kdo_2-lipid A LPS exhibit a nearly 3-fold induction of RpoE pathway without impacting the Cpx system. However, tetraacylated derivatives synthesizing Kdo$_2$-lipid IV$_A$ like $\Delta(waaC\ lpxL)$ or $\Delta(waaC\ lpxL\ lpxM\ lpxP)$ exhibit a 3-4-fold induction of both RpoE and Cpx pathways even under permissive growth conditions. Consistent with these are the findings that $\Delta waaA$ mutants synthesizing lipid IV$_A$ precursor species exhibit maximal induction of RpoE and Cpx pathways. Given much higher induction of RpoE in $\Delta waaC$ mutants and lipid IV$_A$ synthesizing strains, as compared to its modest induction in $\Delta(lpxL\ lpxM\ lpxP)$ mutants, suggests an acute requirement for the incorporation of Kdo and initial steps in LPS core biosynthesis for cell envelope integrity. Interestingly, a $\Delta(lpxL\ lpxM\ lpxP)$ strain was found to synthesize LPS primarily of glycoform with a third Kdo residue and a truncation of outer core disaccharide, when grown in lipid A modifying conditions [7]. Furthermore, mutations in some of the conserved early glycosyltransferases encoding genes, like *waaF*, *waaG*, *waaO* or *waaP* whose product phosphorylates Heptose I, or in the absence of RfaH transcriptional factor whose product is required to ensure transcription/translation and prevent premature termination of transcription of the *waaQ* operon also induce the *rpoE* transcription [11].

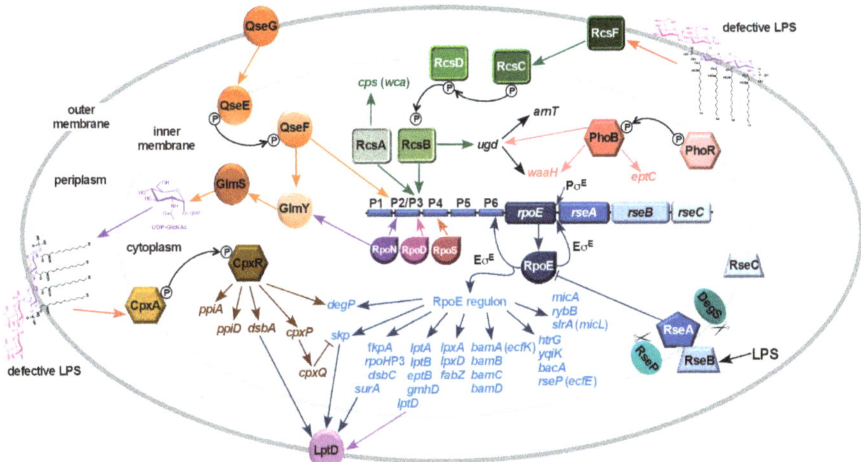

Figure 3. Pathways of sensing defects in LPS biosynthesis/assembly: sensing via the Rcs phosphorelay system, which leads to the induction of transcription from the sigma 70-regulated *rpoE*P3 promoter requiring RcsB as activator and the activation of Qse TCS induces the *rpoE*P2 promoter with QseF acting as an activator of RpoN. Severe defects in LPS induce the Rcs phosphorelay cascade leading to phosphorylation of RcsB. Phosphorylated RcsB acts as an activator for the transcription directed from the *rpoE*P3 promoter leading to the induction of RpoE-mediated stress response. The overlapping *rpoE*P2 promoter also responds to LPS defects albeit to a lower extent. Severe defects in LPS also induce the CpxR/A TCS. Eσ^E is required to maintain envelope integrity by regulating expression of genes whose products are involved in protein folding in the periplasm, assembly of OMPs (*bam* genes) and certain genes whose products are involved in LPS biosynthesis/modifications. Release of LPS from the Lpt system can lead to association of LPS with RseB, which relieves inhibition of RseA proteolysis and can lead to accumulation of free RpoE without RseA.

Consistent with sensing of correct assembly of LPS by RpoE, such a phenotype was used to identify LapA and LapB assembly factors, since the lack of LapB causes increased constitutive induction of RpoE and Cpx regulons quite like $\Delta waaA$ mutants [9]. As mentioned above, $\Delta(lapA\ lapB)$ mutants produce more LPS due to stabilization of LpxC, accumulate precursor species of LPS with incomplete core, defects in acylation of lipid A and reduced amounts of LptD in the OM. Thus, given the overwhelming evidence, it is fair to conclude that severe defects in either early steps in LPS core biogenesis or when LPS is composed of only Kdo$_2$-lipid IV$_A$ or when LPS assembly is defective lead to the induction of RpoE envelope stress response.

10. Mechanism of Signal Transduction

Here, we will discuss the activation of RpoE regulon in response to LPS defects with an overlap in the activation of the Rcs phosphorelay system, which jointly respond to LPS defects (Figure 3). We do not know how the Cpx system senses LPS defects as it can respond to several envelope stress signals in the IM, periplasm and OM. One of the main observations has been that only those structural LPS changes/defects that induce the Rcs system also induce RpoE. Thus, induction of PhoB/R-dependent GlcUA incorporation, BasS/R-dependent lipid A modifications, lack of either WaaY, or WaaZ or WaaS or truncation of outer core neither induce the Rcs system nor the RpoE-regulated stress response [11,55]. In a comprehensive study, a systematic approach of dissection of the transcriptional regulation of the *rpoE* gene and stimuli that induce or repress was taken based on previously observed RpoE induction in Δ*waaA* or strains synthesizing Kdo$_2$-lipid IV$_A$ derivatives [7,11]. When the *rpoE* gene was initially identified, two complementary studies showed that transcription of the *rpoE* gene is positively autoregulated from one of its promoters since it is recognized by Eσ^E polymerase and its activity is induced when OMP maturation is dysfunctional or due to an imbalance in OMP expression [81,82]. However, the regulation of distal promoter or promoter region (previously called the *rpoE*P1 promoter) remained elusive. Besides positive transcriptional regulation, RpoE is negatively regulated by RseA and RseB proteins, whose encoding genes are co-transcribed with the *rpoE* gene [83]. The RseA inner membrane protein associates via its cytoplasmic N-terminal domain with RpoE, thereby sequestering RpoE in the IM under non-stress conditions [83]. This inhibitory effect of RseA is relieved upon OMP misfolding or overexpression of some OMPs by sequential degradation of RseA initiated by DegS in the periplasm, followed by the second site inner membrane proteolysis by RseP (EcfE) and subsequent degradation of the N-terminal RseA in the cytoplasm [84]. The periplasmic C-terminal domain of RseA associates with RseB and prevents RseA degradation. Since the autoregulated *rpoE* promoter (P6 promoter), previously called *rpoE*P2 promoter, responds to effective concentration of available RpoE that complexes with core RNA polymerase, its activity reflects both transcriptional increase in the *rpoE* expression and its release from RseA and hence its activity or activity of reporter promoters of its regulon members (*rpoH*P3) does not directly reflect response to LPS defects.

During the studies aimed to understand the mechanism of *rpoE* induction in response to LPS defects and examination of transcriptional regulation of upstream element, it turned out that the *rpoE* gene is transcribed from six promoters and, among these, activity of *rpoE*P2 and *rpoE*P3 promoters is specifically induced when LPS is defective (Figure 3). Among these, transcription from the *rpoE*P3 promoter is significantly induced when either the LPS core biogenesis is defective or when LPS assembly is compromised or by challenge with LPS-binding antibiotic like polymyxin B or treatment with agents like AMV that induce non-specifically different TCS [11]. Global mutational screen identified insertion/deletion mutations mapping to genes involved in the early steps of LPS core biosynthesis such as *waaC*, *waaF*, *waaG*, *waaO* and *waaP*. Among these, maximal 7-fold induction of the *rpoE*P3 promoter was observed in Δ*waaC* mutants synthesizing Kdo$_2$-lipid IV$_A$ [11]. This activation was nearly abrogated in Δ(*waaC rcsB*) mutants suggesting that the signal of activation due to defective LPS requires the RcsB response regulator [11]. Interestingly, mutations in either *waaC* or *waaF* or *waaP* exhibit deep-rough phenotype accompanied by the induction of Rcs-regulated exopolysaccharide synthesis. Analysis of various other response regulators did not show any direct participation [11]. The activity of the same *rpoE*P3 promoter was found to be induced nearly 3–4 fold in Δ*rfaH* background and by the overexpression of *rirA* sRNA [11]. The 73 nt RirA sRNA contains *ops* element, located in the 5′ UTR of *waaQ* operon and acts by decreasing the RfaH activity, by directly binding to RfaH in the presence of RNA polymerase. The overexpression of *rirA* causes a reduction in the expression of operons that require the RfaH activity. When RirA is in excess, it causes a decrease in the expression of *waaQ* operon members leading to truncation of the LPS core and abolishes the incorporation of *O*-antigen, explaining the specific induction of LPS responsive *rpoE*P3 promoter [11]. Additional evidence that the *rpoE*P3 promoter responds to LPS-specific alterations is based on an increase in transcription from this promoter when bacteria are challenged in early growth phase by either polymyxin B or AMV.

The induction of RpoE regulon concomitant with Rcs pathway activation is also observed when any component of Lpt is depleted, since under such conditions LPS is modified by the Rcs-regulated colanic acid exopolysaccharide [85].

A direct proof that the *rpoEP3* promoter is positively regulated by the Rcs system comes from in vitro DNA binding assays with purified RcsB, establishing the presence of conserved RcsB motif in the *rpoEP3* promoter region [11]. This was supported by deletion and mutational analysis of RcsB-binding motif. Run-off assays established the authenticity of recognition by the house-keeping sigma factor and mutational analysis reinforced that RcsB acts as a positive regulator in response to LPS defects. Transcription from the *rpoEP3* promoter is also activated in Δ(*lapA-lapB*) which also requires RcsB. It should be noted that the RpoN-regulated *rpoEP2* promoter also responds to LPS defects albeit to a much-reduced extent (only a 70% to 2-fold increase as compared to the 7-fold induction by Δ*waaC*) and that too mainly upon entry into the stationary phase. Since the *rpoEP2* promoter requires QseF as an activator [11], it will be interesting to know how the Qse TCS senses LPS.

A role for sensing of LPS defects by the Rcs signalling pathway was proposed in earlier studies by examining a mutant strain carrying *rfa-1* mutation. Such a strain synthesizes LPS composed of only two heptose residues lacking any glucose in the core region and exhibits induction of *cps* genes leading to overexpression of exopolysaccharide [80]. Furthermore, it was shown that this LPS defect requires the outer membrane lipoprotein RcsF, since induction of cps transcription in *rfa-1* mutant was abolished in Δ*rcsF* mutants [80]. Interestingly, overexpression of *cps* genes due to LPS defects did not occur due to the increased *rcsF* transcription, rather could be the consequence of other mechanisms of the RcsF activation. More recent studies on RcsF revealed that it exists in complex with many OMPs [86] and such RcsF complex uses its positively charged, surface-exposed N-terminal domain to detect alterations in lateral interactions between LPS molecules (changes in phosphorylation status of the LPS core), which lead to the induction of Rcs system. This activation of Rcs pathway in turn can activate RcsB to induce the *rpoEP3* promoter when LPS defects are encountered (Figure 3).

Another model postulates that the periplasmic RseB binds to LPS that might be released in an off pathway during translocation via the Lpt system [84]. Usually, RseB association with RseA inhibits its proteolysis. However, LPS binding to RseB relives inhibition of proteolysis by disassociation of the RseA-RseB complex (Figure 3). In support of this model, authors show that RseA-RseB dissociation by LPS is inhibited in the presence of LptA, suggesting that the periplasmic accumulation of LPS, due to release from Lpt proteins in the periplasmic bridge, can promote the dissociation of RseB from RseA and cause RpoE induction. However, the validity of this model needs further studies as free LPS is not known to accumulate when the LPS core is truncated and severe defects in LPS, particularly upon accumulation in the IM when the Lpt system is depleted, also induce RpoE. Analysis of distribution of free RpoE vs. RpoE bound to the IM in Δ*waaA* has revealed that the elevated RpoE activity can be attributed to increased release of RpoE from the IM in such mutants [87]. However, a joint signaling when LPS is defective via the Rcs system and RseB can amplify RpoE induction and hence these models are not mutually exclusive.

In summary, comprehensive analysis of various LPS defects has revealed that RpoE activity is specifically induced when LPS biosynthesis or assembly is compromised by the activation of the *rpoEP3* promoter with phosphorylated RcsB acting as an activator and the sensing of LPS defects occurs via the Rcs phosphorelay system (Figure 3).

11. Conclusions

The LPS synthesis is subjected to a tight regulation at variety of steps. Regulation of LpxC amounts by FtsH and LapB constitutes an essential step to maintain a balanced synthesis of LPS and phospholipids. Further, MsbA by preferential selectivity of hexaacylated LPS provides an early checkpoint, preventing transport of precursor underacylated LPS species. LapB recruits LpxC, FtsH, glycosyltransferases, myristoyltrasferase LpxM and other proteins that are required for LPS synthesis at the IM, presumably acting as a scaffold to deliver completely synthesized LPS to its other interacting partners in the Lpt transport system. LPS is heterogeneous in composition and its composition is dynamically altered by a host of factors. These LPS alterations are regulated by the RpoE sigma factor, TCS like BasS/R, PhoP/Q, PhoB/R and Rcs system with an overwhelming role played by noncoding sRNAs. The Rcs system senses LPS defects resulting into phosphorylation of RcsB, which leads to activation of transcription from *rpoE*P3 promoter leading to RpoE induction. Finally, the transenvelope Lpt complex of seven essential proteins acts as a single unit and in an ATP-dependent manner ferries LPS across the periplasmic bridge comprised of LptA for delivery to the OM component of LptD/E in a continuous manner. Although tremendous progress has been achieved in the recent past, it is still unknown how the LptB cytoplasmic component of Lpt system is localized to the IM and how the proteolytic activity of FtsH is regulated by LapB protein and other factors. Further studies are also needed to demonstrate that LapB acts as the scaffold for LPS assembly and its coordination with the Lpt transport system, although LapA/LapB proteins have been demonstrated to co-purify with Lpt components. As LPS is essential, its essential steps in biosynthetic pathway and Lpt transport have been validated targets of discovery of new antibiotics. Already inhibitors of LpxC and a peptidomimetic POL7080 based on the cationic antimicrobial peptide that inhibits the LptD activity are in advanced clinical trails and many additional antimicrobial compounds using LPS synthesis/assembly are bound to be discovered in future.

Author Contributions: G.K. and S.R. participated in draft preparation, review and editing. All authors read and approved the final manuscript.

Funding: This research was funded by National Science Center (NCN) 2017/25/B/NZ6/02021 to SR.

Conflicts of Interest: The authors declare no conflict of interest.

Abbreviations

AMV	ammonium metavanadate
GlcUA	glucuronic acid
IM	inner membrane
Kdo	3-deoxy-α-D-*manno*-oct-2-ulsonic acid
L-Ara4N	4-amino-4-deoxy-L-arabinose
LPS	lipopolysaccharide
OM	outer membrane
OMP	outer membrane protein
ops	operon polarity suppressor
P-EtN	phosphoethanolamine
sRNA	noncoding small regulatory RNA
TCS	two-component system
TPR	tetratricopeptide repeats
UTR	untranslated region

References

1. Nikaido, H. Outer membrane. In *Escherichia coli and Salmonella: Cellular and Molecular Biology*, 2nd ed.; Neidhardt, F.C., Curtiss, R., Ingraham, J.L., Eds.; American Society for Microbiology Press: Washington, DC, USA, 1996; Volume 1, pp. 29–47.
2. Anderson, M.S.; Bulawa, C.E.; Raetz, C.R.H. The biosynthesis of Gram-negative endotoxin. Formation of lipid A precursors from UDP-GlcNAc in extracts of *Escherichia coli*. *J. Biol. Chem.* **1985**, *260*, 15536–15541.
3. Babinski, K.J.; Kanjilal, S.J.; Raetz, C.R.H. Accumulation of the lipid A precursor UDP-2,3-diacylglucosamine in an *Escherichia coli* mutant lacking the *lpxH* gene. *J. Biol. Chem.* **2002**, *277*, 25947–25956. [CrossRef] [PubMed]
4. Raetz, C.R.H.; Whitfield, C. Lipopolysaccharide endotoxins. *Annu. Rev. Biochem.* **2002**, *71*, 635–700. [CrossRef]

5. Vorachek-Warren, M.K.; Ramirez, S.; Cotter, R.J.; Raetz, C.R.H. A triple mutant of *Escherichia coli* lacking secondary acyl chains on lipid A. *J. Biol. Chem.* **2002**, *277*, 14194–14205. [CrossRef] [PubMed]

6. Carty, S.M.; Sreekumar, K.R.; Raetz, C.R.H. Effect of cold shock on lipid A biosynthesis in *Escherichia coli*. Induction at 12 °C of an acyltransferase specific for palmitoleoyl-acyl carrier protein. *J. Biol. Chem.* **1999**, *274*, 9677–9685. [CrossRef] [PubMed]

7. Klein, G.; Lindner, B.; Brabetz, W.; Brade, H.; Raina, S. *Escherichia coli* K-12 suppressor-free mutants lacking early glycosyltransferases and late acyltransferases. Minimal lipopolysaccharide structure and induction of envelope stress response. *J. Biol. Chem.* **2009**, *284*, 15369–15389. [CrossRef]

8. Holst, O. The structures of core regions from enterobacterial lipopolysaccharides—An update. *FEMS Microbiol. Lett.* **2007**, *271*, 3–11. [CrossRef]

9. Klein, G.; Kobylak, N.; Lindner, B.; Stupak, A.; Raina, S. Assembly of lipopolysaccharide in *Escherichia coli* requires the essential LapB heat shock protein. *J. Biol. Chem.* **2014**, *289*, 14829–14853. [CrossRef]

10. Klein, G.; Raina, S. Small regulatory bacterial RNAs regulating the envelope stress response. *Biochem. Soc. Trans.* **2017**, *45*, 417–425. [CrossRef] [PubMed]

11. Klein, G.; Stupak, A.; Biernacka, D.; Wojtkiewicz, P.; Lindner, B.; Raina, S. Multiple transcriptional factors regulate transcription of the *rpoE* gene in *Escherichia coli* under different growth conditions and when the lipopolysaccharide biosynthesis is defective. *J. Biol. Chem.* **2016**, *291*, 22999–23019. [CrossRef]

12. Bailey, M.J.; Hughes, C.; Koronakis, V. RfaH and the *ops* element, components of a novel system controlling bacterial transcription elongation. *Mol. Microbiol.* **1997**, *26*, 845–851. [CrossRef]

13. Klein, G.; Raina, S. Regulated control of the assembly and diversity of LPS by noncoding sRNAs. *Biomed. Res. Int.* **2015**, *2015*, 153561. [CrossRef]

14. Dartigalongue, C.; Missiakas, D.; Raina, S. Characterization of the *Escherichia coli* σ^E regulon. *J. Biol. Chem.* **2001**, *276*, 20866–20875. [CrossRef] [PubMed]

15. Henry, R.; Vithanage, N.; Harrison, P.; Seemann, T.; Coutts, S.; Moffatt, J.H.; Nation, R.L.; Li, J.; Harper, M.; Adler, B.; et al. Colistin-resistant, lipopolysaccharide-deficient *Acinetobacter baumannii* responds to lipopolysaccharide loss through increased expression of genes involved in the synthesis and transport of lipoproteins, phospholipids, and poly-β-1,6-*N*-acetylglucosamine. *Antimicrob. Agents Chemother.* **2012**, *56*, 59–69. [CrossRef] [PubMed]

16. Steeghs, L.; den Hartog, R.; den Boer, A.; Zomer, B.; Roholl, P.; van der Ley, P. Meningitis bacterium is viable without endotoxin. *Nature* **1998**, *392*, 449–450. [CrossRef] [PubMed]

17. Missiakas, D.; Betton, J.M.; Raina, S. New components of protein folding in extracytoplasmic compartments of *Escherichia coli* SurA, FkpA and Skp/OmpH. *Mol. Microbiol.* **1996**, *21*, 871–884. [CrossRef] [PubMed]

18. Crhanova, M.; Malcova, M.; Mazgajova, M.; Karasova, D.; Sebkova, A.; Fucikova, A.; Bortlicek, Z.; Pilousova, L.; Kyrova, K.; Dekanova, M.; et al. LPS structure influences protein secretion in *Salmonella enterica*. *Vet. Microbiol.* **2011**, *152*, 131–137. [CrossRef]

19. Ebbensgaard, A.; Mordhorst, H.; Aarestrup, F.M.; Hansen, E.B. The role of outer membrane proteins and lipopolysaccharides for the sensitivity of *Escherichia coli* to antimicrobial peptides. *Front. Microbiol.* **2018**. [CrossRef]

20. Vivijs, B.; Aertsen, A.; Michiels, C.W. Identification of genes required for growth of *Escherichia coli* MG1655 at moderately low pH. *Front. Microbiol.* **2016**, *7*, 1672. [CrossRef]

21. Wang, Z.; Wang, J.; Ren, G.; Li, Y.; Wang, X. Deletion of the genes *waaC*, *waaF*, or *waaG* in *Escherichia coli* W3110 disables the flagella biosynthesis. *J. Basic Microbiol.* **2016**, *56*, 1021–1035. [CrossRef]

22. Schwalm, J.; Mahoney, T.F.; Soltes, G.R.; Silhavy, T.J. Role for Skp in LptD assembly in *Escherichia coli*. *J. Bacteriol.* **2013**, *195*, 3734–3742. [CrossRef] [PubMed]

23. Murata, M.; Fujimoto, H.; Nishimura, K.; Charoensuk, K.; Nagamitsu, H.; Raina, S.; Kosaka, T.; Oshima, T.; Ogasawara, N.; Yamada, M. Molecular strategy for survival at a critical high temperature in *Escherichia coli*. *PLoS ONE* **2011**, *6*, e20063. [CrossRef] [PubMed]

24. Yethon, J.A.; Heinrichs, D.E.; Monteiro, M.A.; Perry, M.B.; Whitfield, C. Involvement of *waaY*, *waaQ*, and *waaP* in the modification of *Escherichia coli* lipopolysaccharide and their role in the formation of a stable outer membrane. *J. Biol. Chem.* **1998**, *273*, 26310–26316. [CrossRef]

25. Reynolds, C.M.; Raetz, C.R.H. Replacement of lipopolysaccharide with free lipid A molecules in *Escherichia coli* mutants lacking all core sugars. *Biochemistry* **2009**, *48*, 9627–9640. [CrossRef]

26. Urban, J.H.; Vogel, J. Two seemingly homologous noncoding RNAs act hierarchically to activate *glmS* mRNA translation. *PLoS Biol.* **2008**, *6*, e64. [CrossRef] [PubMed]

27. Göpel, Y.; Papenfort, K.; Reichenbach, B.; Vogel, J.; Görke, B. Targeted decay of a regulatory small RNA by an adaptor protein for RNase E and counteraction by an anti-adaptor RNA. *Genes Dev.* **2013**, *27*, 552–564. [CrossRef]

28. Reichenbach, B.; Göpel, Y.; Görke, B. Dual control by perfectly overlapping σ^{54}- and σ^{70}-promoters adjusts small RNA GlmY expression to different environmental signals. *Mol. Microbiol.* **2009**, *74*, 1054–1070. [CrossRef] [PubMed]

29. Ogura, T.; Inoue, K.; Tatsuta, T.; Suzaki, T.; Karata, K.; Young, K.; Su, L.H.; Fierke, C.A.; Jackman, J.E.; Raetz, C.R.H.; et al. Balanced biosynthesis of major membrane components through regulated degradation of the committed enzyme of lipid A biosynthesis by the protease FtsH (HflB) in *Escherichia coli*. *Mol. Microbiol.* **1999**, *31*, 833–844. [CrossRef]

30. Führer, F.; Langklotz, S.; Narberhaus, F. The C-terminal end of LpxC is required for degradation by the FtsH protease. *Mol. Microbiol.* **2006**, *59*, 1025–1036. [CrossRef] [PubMed]

31. Mahalakshmi, S.; Sunayana, M.R.; Saisree, L.; Reddy, M. *yciM* is an essential gene required for regulation of lipopolysaccharide synthesis in *Escherichia coli*. *Mol. Microbiol.* **2014**, *91*, 145–157. [CrossRef] [PubMed]

32. Zeng, D.; Zhao, J.; Chung, H.S.; Guan, Z.; Raetz, C.R.H.; Zhou, P. Mutants resistant to LpxC inhibitors by rebalancing cellular homeostasis. *J. Biol. Chem.* **2013**, *288*, 5475–5486. [CrossRef] [PubMed]

33. Guo, M.S.; Updegrove, T.B.; Gogol, E.B.; Shabalina, S.A.; Gross, C.A.; Storz, G. MicL, a new σ^{E}-dependent sRNA, combats envelope stress by repressing synthesis of Lpp, the major outer membrane lipoprotein. *Genes Dev.* **2014**, *28*, 1620–1634. [CrossRef] [PubMed]

34. Metzger IV, L.E.; Raetz, C.R.H. Purification and characterization of the lipid A disaccharide synthase (LpxB) from *Escherichia coli*, a peripheral membrane protein. *Biochemistry* **2009**, *48*, 11559–11571. [CrossRef] [PubMed]

35. Emiola, A.; Andrews, S.S.; Heller, C.; George, J. Crosstalk between the lipopolysaccharide and phospholipid pathways during outer membrane biogenesis in *Escherichia coli*. *Proc. Natl. Acad. Sci. USA* **2016**, *113*, 3108–3113. [CrossRef]

36. Katz, C.; Ron, E.Z. Dual role of FtsH in regulating lipopolysaccharide biosynthesis in *Escherichia coli*. *J. Bacteriol.* **2008**, *190*, 7117–7122. [CrossRef] [PubMed]

37. Prince, C.; Jia, Z. An unexpected duo: Rubredoxin binds nine TPR motifs to form LapB, an essential regulator of lipopolysaccharide synthesis. *Structure* **2015**, *23*, 1500–1506. [CrossRef]

38. Putker, F.; Grutsch, A.; Tommassen, J.; Bos, M.P. Ght protein of *Neisseria meningitidis* is involved in the regulation of lipopolysaccharide biosynthesis. *J. Bacteriol.* **2014**, *196*, 780–789. [CrossRef] [PubMed]

39. Moreau, P.L.; Loiseau, L. Characterization of acetic acid-detoxifying *Escherichia coli* evolved under phosphate starvation conditions. *Microb. Cell Fact.* **2016**, *15*, 42. [CrossRef]

40. Sutterlin, H.A.; Shi, H.; May, K.L.; Miguel, A.; Khare, S.; Huang, K.C.; Silhavy, T.J. Disruption of lipid homeostasis in the Gram-negative cell envelope activates a novel cell death pathway. *Proc. Natl. Acad. Sci. USA* **2016**, *113*, E1565–E1574. [CrossRef]

41. Ward, A.; Reyes, C.L.; Yu, J.; Roth, C.B.; Chang, G. Flexibility in the ABC transporter MsbA: Alternating access with a twist. *Proc. Natl. Acad. Sci. USA* **2007**, *104*, 19005–19010. [CrossRef]

42. Mi, W.; Li, Y.; Yoon, S.H.; Ernst, R.K.; Walz, T.; Liao, M. Structural basis of MsbA-mediated lipopolysaccharide transport. *Nature* **2017**, *549*, 233–237. [CrossRef] [PubMed]

43. Ho, H.; Miu, A.; Alexander, M.K.; Garcia, N.K.; Oh, A.; Zilberleyb, I.; Reichelt, M.; Austin, C.D.; Tam, C.; Shriver, S.; et al. Structural basis for dual-mode inhibition of the ABC transporter MsbA. *Nature* **2018**, *557*, 196–201. [CrossRef]

44. Hicks, G.; Jia, Z. Structural basis for the lipopolysaccharide export activity of the bacterial lipopolysaccharide transport system. *Int. J. Mol. Sci.* **2018**, *19*, 2680. [CrossRef] [PubMed]

45. Sperandeo, P.; Martorana, A.M.; Polissi, A. The lipopolysaccharide transport (Lpt) machinery: A nonconventional transporter for lipopolysaccharide assembly at the outer membrane of Gram-negative bacteria. *J. Biol. Chem.* **2017**, *292*, 17981–17990. [CrossRef] [PubMed]

46. Bohl, T.E.; Aihara, H. Current progress in the structural and biochemical characterization of proteins involved in the assembly of lipopolysaccharide. *Int. J. Microbiol.* **2018**, *2018*, 5319146. [CrossRef] [PubMed]

47. Okuda, S.; Sherman, D.J.; Silhavy, T.J.; Ruiz, N.; Kahne, D. Lipopolysaccharide transport and assembly at the outer membrane: The PEZ model. *Nat. Rev. Microbiol.* **2016**, *14*, 337–345. [CrossRef] [PubMed]

48. Luo, Q.; Yang, X.; Yu, S.; Shi, H.; Wang, K.; Xiao, L.; Zhu, G.; Sun, C.; Li, T.; Li, D.; et al. Structural basis for lipopolysaccharide extraction by ABC transporter LptB$_2$FG. *Nat. Struct. Mol. Biol.* **2017**, *24*, 469–474. [CrossRef]

49. Sherman, D.J.; Xie, R.; Taylor, R.J.; George, A.H.; Okuda, S.; Foster, P.J.; Needleman, D.J.; Kahne, D. Lipopolysaccharide is transported to the cell surface by a membrane-to-membrane protein bridge. *Science* **2018**, *359*, 798–801. [CrossRef]

50. Qiao, S.; Luo, Q.; Zhao, Y.; Zhang, X.C.; Huang, Y. Structural basis for lipopolysaccharide insertion in the bacterial outer membrane. *Nature* **2014**, *511*, 108–111. [CrossRef]

51. Dong, H.; Xiang, Q.; Gu, Y.; Wang, Z.; Paterson, N.G.; Stansfeld, P.J.; He, C.; Zhang, Y.; Wang, W.; Dong, C. Structural basis for outer membrane lipopolysaccharide insertion. *Nature* **2014**, *511*, 52–56. [CrossRef]

52. Missiakas, D.; Raina, S. Protein folding in the bacterial periplasm. *J. Bacteriol.* **1997**, *179*, 2465–2471. [CrossRef] [PubMed]

53. Chng, S.S.; Xue, M.; Garner, R.A.; Kadokura, H.; Boyd, D.; Beckwith, J.; Kahne, D. Disulfide rearrangement triggered by translocon assembly controls lipopolysaccharide export. *Science* **2012**, *337*, 1665–1668. [CrossRef] [PubMed]

54. Narita, S.; Masui, C.; Suzuki, T.; Dohmae, N.; Akiyama, Y. Protease homolog BepA (YfgC) promotes assembly and degradation of β-barrel membrane proteins in *Escherichia coli*. *Proc. Natl. Acad. Sci. USA* **2013**, *110*, E3612–E3621. [CrossRef] [PubMed]

55. Klein, G.; Müller-Loennies, S.; Lindner, B.; Kobylak, N.; Brade, H.; Raina, S. Molecular and structural basis of inner core lipopolysaccharide alterations in *Escherichia coli*: Incorporation of glucuronic acid and phosphoethanolamine in the heptose region. *J. Biol. Chem.* **2013**, *288*, 8111–8127. [CrossRef] [PubMed]

56. Szczesny, M.; Beloin, C.; Ghigo, J.M. Increased osmolarity in biofilm triggers RcsB-dependent lipid A palmitoylation in *Escherichia coli*. *mBio* **2018**, *9*, e01415–e01418. [CrossRef] [PubMed]

57. Raetz, C.R.; Reynolds, C.M.; Trent, M.S.; Bishop, R.E. Lipid A modification systems in Gram-negative bacteria. *Annu. Rev. Biochem.* **2007**, *76*, 295–329. [CrossRef]

58. Rubin, E.J.; Herrera, C.M.; Crofts, A.A.; Trent, M.S. PmrD is required for modifications to *Escherichia coli* endotoxin that promote antimicrobial resistance. *Antimicrob. Agents Chemother.* **2015**, *59*, 2051–2061. [CrossRef]

59. Herrera, C.M.; Hankins, J.V.; Trent, M.S. Activation of PmrA inhibits LpxT-dependent phosphorylation of lipid A promoting resistance to antimicrobial peptides. *Mol. Microbiol.* **2010**, *76*, 1444–1460. [CrossRef]

60. Kato, A.; Chen, H.D.; Latifi, T.; Groisman, E.A. Reciprocal control between a bacterium's regulatory system and the modification status of its lipopolysaccharide. *Mol. Cell* **2012**, *47*, 897–908. [CrossRef]

61. Hong, X.; Chen, H.D.; Groisman, E.A. Gene expression kinetics governs stimulus-specific decoration of the *Salmonella* outer membrane. *Sci. Signal.* **2018**, *11*, eaar7921. [CrossRef]

62. Klein, G.; Lindner, B.; Brade, H.; Raina, S. Molecular basis of lipopolysaccharide heterogeneity in *Escherichia coli*: Envelope stress-responsive regulators control the incorporation of glycoforms with a third 3-deoxy-α-D-*manno*-oct-2-ulosonic acid and rhamnose. *J. Biol. Chem.* **2011**, *286*, 42787–42807. [CrossRef] [PubMed]

63. Coornaert, A.; Lu, A.; Mandin, P.; Springer, M.; Gottesman, S.; Guillier, M. MicA sRNA links the PhoP regulon to cell envelope stress. *Mol. Microbiol.* **2010**, *76*, 467–479. [CrossRef] [PubMed]

64. Gogol, E.B.; Rhodius, V.A.; Papenfort, K.; Vogel, J.; Gross, C.A. Small RNAs endow a transcriptional activator with essential repressor functions for single-tier control of a global stress regulon. *Proc. Natl. Acad. Sci. USA* **2012**, *108*, 12875–12880. [CrossRef] [PubMed]

65. Coornaert, A.; Chiaruttini, C.; Springer, M.; Guillier, M. Post-transcriptional control of the *Escherichia coli* PhoQ-PhoP two-component system by multiple sRNAs involves a novel pairing region of GcvB. *PLoS Genet.* **2013**, *9*, e1003156. [CrossRef]

66. Bishop, R.E.; Gibbons, H.S.; Guina, T.; Trent, M.S.; Miller, S.I.; Raetz, C.R.H. Transfer of palmitate from phospholipids to lipid A in outer membranes of Gram-negative bacteria. *EMBO J.* **2000**, *19*, 5071–5080. [CrossRef]

67. Trent, M.S.; Pabich, W.; Raetz, C.R.H.; Miller, S.I. A PhoP/PhoQ-induced lipase (PagL) that catalyzes 3-*O*-deacylation of lipid A precursors in membranes of *Salmonella typhimurium*. *J. Biol. Chem.* **2001**, *276*, 9083–9092. [CrossRef]

68. Han, K.; Tjaden, B.; Lory, S. GRIL-seq provides a method for identifying direct targets of bacterial small regulatory RNA by in vivo proximity ligation. *Nat. Microbiol.* **2016**, *2*, 16239. [CrossRef] [PubMed]

69. Kawano, M.; Manabe, T.; Kawasaki, K. *Salmonella enterica* serovar Typhimurium lipopolysaccharide deacylation enhances its intracellular growth within macrophages. *FEBS Lett.* **2010**, *584*, 207–212. [CrossRef] [PubMed]

70. Reynolds, C.M.; Ribeiro, A.A.; McGrath, S.C.; Cotter, R.J.; Raetz, C.R.H.; Trent, M.S. An outer membrane enzyme encoded by *Salmonella typhimurium lpxR* that removes the 3′-acyloxyacyl moiety of lipid A. *J. Biol. Chem.* **2006**, *281*, 21974–21987. [CrossRef]

71. Corcoran, C.P.; Podkaminski, D.; Papenfort, K.; Urban, J.H.; Hinton, J.C.D.; Vogel, J. Superfolder GFP reporters validate diverse new mRNA targets of the classic porin regulator, MicF RNA. *Mol. Microbiol.* **2012**, *84*, 428–445. [CrossRef]

72. Moon, K.; Six, D.A.; Lee, H.J.; Lee, C.R.; Raetz, C.R.H.; Gottesman, S. Complex transcriptional and post-transcriptional regulation of an enzyme for lipopolysaccharide modification. *Mol. Microbiol.* **2013**, *89*, 52–64. [CrossRef]

73. Moon, K.; Gottesman, S. A PhoQ/P-regulated small RNA regulates sensitivity of *Escherichia coli* to antimicrobial peptides. *Mol. Microbiol.* **2009**, *74*, 1314–1330. [CrossRef]

74. Müller-Loennies, S.; Holst, O.; Lindner, B.; Brade, H. Isolation and structural analysis of phosphorylated oligosaccharides obtained from *Escherichia coli* J-5 lipopolysaccharides. *Eur. J. Biochem.* **1999**, *260*, 235–249. [CrossRef] [PubMed]

75. Fresno, S.; Jiménez, N.; Canals, R.; Merino, S.; Corsaro, M.M.; Lanzetta, R.; Parrilli, M.; Pieretti, G.; Regué, M.; Tomás, J.M. A second galacturonic acid transferase is required for core lipopolysaccharide biosynthesis and complete capsule association with the cell surface in *Klebsiella pneumoniae*. *J. Bacteriol.* **2007**, *189*, 1128–1137. [CrossRef]

76. Müller-Loennies, S.; Lindner, B.; Brade, H. Structural analysis of oligosaccharides from lipopolysaccharide (LPS) of *Escherichia coli* K12 strain W3100 reveals a link between inner and outer core LPS biosynthesis. *J. Biol. Chem.* **2003**, *278*, 34090–34101. [CrossRef] [PubMed]

77. Reynolds, C.M.; Kalb, S.R.; Cotter, R.J.; Raetz, C.R.H. A phosphoethanolamine transferase specific for the outer 3-deoxy-D-*manno*-octulosonic acid residue of *Escherichia coli* lipopolysaccharide. Identification of the *eptB* gene and Ca^{2+} hypersensitivity of an *eptB* deletion mutant. *J. Biol. Chem.* **2005**, *280*, 21202–21211. [CrossRef]

78. Amar, A.; Pezzoni, M.; Pizarro, R.A.; Costa, C.S. New envelope stress factors involved in σ^E activation and conditional lethality of *rpoE* mutations in *Salmonella enterica*. *Microbiology* **2018**, *164*, 1293–1307. [CrossRef]

79. Burmann, B.M.; Knauer, S.H.; Sevostyanova, A.; Schweimer, K.; Mooney, R.A.; Landick, R.; Artsimovitch, I.; Rösch, P. An α helix to β barrel domain switch transforms the transcription factor RfaH into a translation factor. *Cell* **2012**, *150*, 291–303. [CrossRef] [PubMed]

80. Majdalani, N.; Heck, M.; Stout, V.; Gottesman, S. Role of RcsF in signaling to the Rcs phosphorelay pathway in *Escherichia coli*. *J. Bacteriol.* **2005**, *187*, 6770–6778. [CrossRef] [PubMed]

81. Raina, S.; Missiakas, D.; Georgopoulos, C. The *rpoE* gene encoding the σ^E (σ^{24}) heat shock sigma factor of *Escherichia coli*. *EMBO J.* **1995**, *14*, 1043–1055. [CrossRef] [PubMed]

82. Rouvière, P.E.; De Las Peñas, A.; Mecsas, J.; Lu, C.Z.; Rudd, K.E.; Gross, C.A. *rpoE*, the gene encoding the second heat-shock sigma factor, σ^E, in *Escherichia coli*. *EMBO J.* **1995**, *14*, 1032–1042. [CrossRef] [PubMed]

83. Missiakas, D.; Mayer, M.P.; Lemaire, M.; Georgopoulos, C.; Raina, S. Modulation of the *Escherichia coli* σ^E (RpoE) heat-shock transcription-factor activity by the RseA, RseB and RseC proteins. *Mol. Microbiol.* **1997**, *24*, 355–371. [CrossRef] [PubMed]

84. Lima, S.; Guo, M.S.; Chaba, R.; Gross, C.A.; Sauer, R.T. Dual molecular signals mediate the bacterial response to outer-membrane stress. *Science* **2013**, *340*, 837–841. [CrossRef] [PubMed]

85. Sperandeo, P.; Lau, F.K.; Carpentieri, A.; De Castro, C.; Molinaro, A.; Dehò, G.; Silhavy, T.J.; Polissi, A. Functional analysis of the protein machinery required for transport of lipopolysaccharide to the outer membrane of *Escherichia coli*. *J. Bacteriol.* **2008**, *190*, 4460–4469. [CrossRef]

86. Konovalova, A.; Mitchell, A.M.; Silhavy, T.J. A lipoprotein/β-barrel complex monitors lipopolysaccharide integrity transducing information across the outer membrane. *eLife* **2016**, *5*, e15276. [CrossRef] [PubMed]
87. Noor, R.; Murata, M.; Nagamitsu, H.; Klein, G.; Raina, S.; Yamada, M. Dissection of σ^E-dependent cell lysis in *Escherichia coli*: Roles of RpoE regulators RseA, RseB and periplasmic folding catalysts PpiD. *Genes Cells* **2009**, *14*, 885–899. [CrossRef]

International Journal of
Molecular Sciences

Review

Structural Basis for the Lipopolysaccharide Export Activity of the Bacterial Lipopolysaccharide Transport System

Greg Hicks and Zongchao Jia *

Department of Biomedical and Molecular Sciences, Queen's University, Kingston, ON K7L 3N6, Canada; 8gah@queensu.ca
* Correspondence: jia@queensu.ca; Tel.: +1-613-533-6277

Received: 20 August 2018; Accepted: 5 September 2018; Published: 10 September 2018

Abstract: Gram-negative bacteria have a dense outer membrane (OM) coating of lipopolysaccharides, which is essential to their survival. This coating is assembled by the LPS (lipopolysaccharide) transport (Lpt) system, a coordinated seven-subunit protein complex that spans the cellular envelope. LPS transport is driven by an ATPase-dependent mechanism dubbed the "PEZ" model, whereby a continuous stream of LPS molecules is pushed from subunit to subunit. This review explores recent structural and functional findings that have elucidated the subunit-scale mechanisms of LPS transport, including the novel ABC-like mechanism of the LptB$_2$FG subcomplex and the lateral insertion of LPS into the OM by LptD/E. New questions are also raised about the functional significance of LptA oligomerization and LptC. The tightly regulated interactions between these connected subcomplexes suggest a pathway that can react dynamically to membrane stress and may prove to be a valuable target for new antibiotic therapies for Gram-negative pathogens.

Keywords: lipopolysaccharides; membrane transport; Gram-negative bacteria

---

1. Introduction

Gram-negative bacteria protect themselves from chemical stressors by incorporating hydrophobic lipopolysaccharide (LPS) into their outer membranes. The LPS coating acts as a physical barrier to antibiotics and has antigenic properties that make it critical to understanding and countering pathogenicity. Individual LPS are amphipathic, consisting of a hydrophobic lipid moiety (Lipid A) and a hydrophilic oligosaccharide core (OS). Some bacteria add an O-antigen (Oag), a repeating saccharide moiety that extends out from the OS core. The LPS coating is critical to the integrity of the cellular envelope, and Gram-negative bacteria need to replenish it at an astonishing rate to survive.

LPS is synthesized at the cytoplasmic side of the inner membrane (IM) before it is transported to the outer membrane (OM). Lipid A is synthesized in the cytoplasm via the Raetz pathway and then ligated to the OS core [1]. It is then flipped to the outer leaflet by the ATP-binding cassette (ABC)-type transporter MsbA [2]. The O-antigen is synthesized and transported to the periplasm by a parallel pathway before it is ligated to the OS [3,4]. Producing LPS in the cytoplasm presents the physiologically daunting task of getting the amphipathic LPS across two membranes, the periplasm and a substantial concentration gradient. Gram-negative bacteria accomplish this through the specialized transportation machinery of the LPS transport (Lpt) system, a linear protein bridge that spans the cellular envelope.

The Lpt system is an oligomeric complex consisting of Lpt proteins A through G. The membrane-bound LptB, F, G and C subunits are connected to the LptD/E heterodimer in the OM by periplasmic LptA (Figure 1). The LptB$_2$FG tetramer extracts LPS from the outer leaflet of the IM and provides the energy to drive LPS transport through an ATPase-dependent mechanism. LptC and LptA provide a continuous LPS binding surface that conveys it to the OM. There, LptD/E inserts

LPS laterally into the outer leaflet of the OM where it is dispersed across the extracellular surface. With some species-dependent exceptions, all seven of the Lpt proteins are essential to LPS transport and Gram-negative bacteria survival.

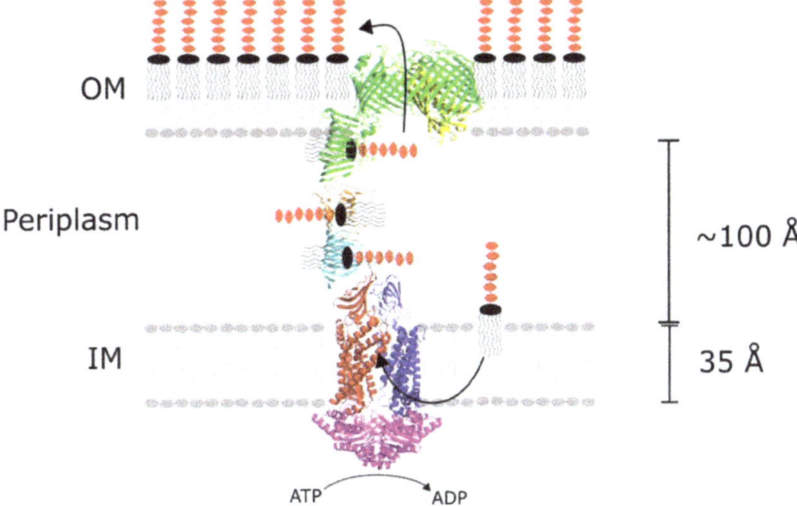

Figure 1. The Lpt (LPS transport) system forms a continuous protein bridge across the inner membrane, periplasm and outer membrane. LptB (purple), LptG (red) and LptF (blue) extract LPS (lipopolysaccharide) from the inner leaflet of the IM (inner membrane) through an ATPase-dependent mechanism. LPS cross the periplasm along a continuous hydrophobic groove formed by LptC (cyan), LptA (orange) and LptD (green). LptD and LptE (yellow) sort LPS to the outer leaflet of the outer membrane (OM).

The LPS coating of the Gram-negative outer membrane is a literal, physical barrier to the development of new antibiotics. Needing to cross the lipid barrier limits the physiochemical properties of antibiotic compounds to narrow ranges of size and lipophilicity [5]. Disrupting the Lpt system can increase cell permeability, and its OM-bound components make it more immediately accessible. The Lpt system is therefore an excellent target for new antibiotics to directly target cell survival or complement other compounds. Further characterization of the Lpt system's subunit-to-subunit interactions and LPS bindings could contribute to the rational design of new antibiotics for the treatment of Gram-negative pathogens.

Until recently, there were major gaps in the experimentally-derived structures of the Lpt subunits. The periplasmic LPS-binding domain of LptD was absent from the available structure, making it difficult to visualize how the Lipid A and OS moieties crossed the membrane together. On the IM side of the complex, the lack of LptF and LptG structures limited the structural characterization of LptB's extraction mechanism to predictions based on other ABC transporter systems. With the publication of the complete LptD/E and LptB$_2$FG structures solved by Botos et al. and Luo et al., respectively, a complete picture of the Lpt system's components is available [6,7]. These structures have answered longstanding questions about the mechanisms of LPS transport and prompted intriguing new ones about its assembly and regulation.

The recent breakthroughs in Lpt structure were accompanied by validation of the protein-bridge "PEZ" model of LPS transport. The model wherein the Lpt subunits form a continuous complex from the IM to the OM and LPS is propelled along it continuously by the ATPase activity of LptB has long been favoured over the periplasmic dissolution or membrane-junction models. It is consistent with the individual interactions between the Lpt subunits and the structural features of the Gram-negative cellular envelope, but direct evidence of the complete bridge and its activity was not observed until a recent study where Sherman et al. demonstrated the Lpt-dependent transmission of LPS between proteoliposomes [8]. Between their findings and the newly-solved Lpt structures, a more complete structural-functional model of LPS transport is now possible.

This review was prompted by the crossing of these two major milestones in Lpt research. The essentiality of the Lpt system and its specific role in preserving the impermeability of the outer membrane make it a promising target for antibiotics, and these new findings could lead to the development of new rationally-designed inhibitors. Questions remain as to how the specific interactions within the Lpt complex facilitate LPS binding and transport. This review will summarize the current research on the structural interactions within the Lpt system and discuss their potential implications regarding function and regulation of OM biogenesis overall.

2. The LptB$_2$FG Complex Drives LPS Extraction from the IM to the Periplasm

The overall mechanism of the Lpt system is often likened to a "PEZ" candy dispenser (Figure 2), wherein a spring at the bottom of a candy-filled chamber slides the contents up as candies are removed from the top one at a time. In vivo, this means that every LPS-binding site in the assembly is continuously passing the molecule from the previous subunit to the next until it reaches the OM. Obviously, this description does not account for how the individual Lpt subunits bind and release LPS, but it is useful for emphasizing the core role of the LptB$_2$FG complex as the "spring" that provides the motivating force for the LPS "candy".

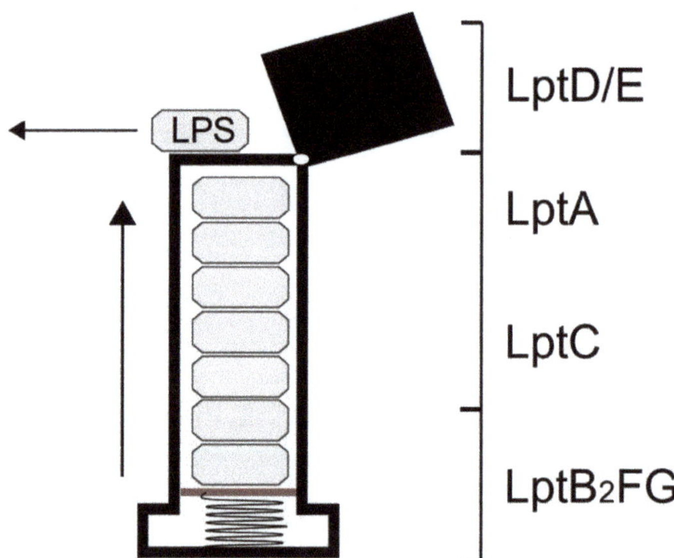

Figure 2. The mechanism of LPS transport to the outer membrane is often referred to as the "PEZ" model in reference to the spring-loaded candy dispenser. In this analogy, the LptB$_2$FG complex is the "spring" that pushes the LPS "candy" through the chamber (LptA and LptC) to the cap (LptD/E), which is lifted to release them one at a time.

The subcomplex of LptF, LptG and dimeric LptB spans the IM and is responsible for extracting LPS from the membrane and propelling it along the periplasmic bridge. The LptB dimer is located in the cytoplasm, where it binds and hydrolyzes ATP. LptB forms a complex with the transmembrane LptF/G dimer. This configuration resembles other ABC transporters: a cytoplasmic nucleotide-binding domain (NBD) hydrolyzes ATP to switch the substrate cavity of the transmembrane domains (TMD) between inward- and outward-facing conformations [9]. In this case, the LptB dimer hydrolyzes ATP to alter the conformation of the transmembrane LptF/G dimer. However, the mechanism of the LptB$_2$FG complex necessarily differs from other ABC transporters because it extracts its substrate directly from the same leaflet of the membrane, rather than switching it from one side of the membrane to another. LptB also differs from other ABC NBDs because it retains some ATPase activity even when purified in vitro, whereas other ABCs are inactive without the full complex [10]. The complete LptB$_2$FG complex displays substantially higher ATPase activity, but curiously, activity decreases by 40% with the inclusion of LptC [7]. Until recently, these differences were difficult to explore as much of LptB$_2$FG's structural mechanism had to be inferred from other ABC complexes.

LptB's catalytic activity couples to the LptF/G heterodimer's extraction of LPS like other ABC transporters, wherein the coupling helices of the TMD interact with the variable Q-loop of the NBD. Structural comparison of ATP-and ADP-bound LptB shows that ATP binding, hydrolysis and release induce conformational changes in the Q-loop region, mediated predominantly by two conserved residues (F90 and R91) [11]. The interaction between LptB and LptF/G evidently has some activating effect on the catalytic domain, as ATPase activity is substantially higher with the full complex. LptB$_2$FG activity and subsequent LPS transport are further enhanced in the presence of the antibiotic novobiocin, a hydrophobic DNA gyrase inhibitor [12]. Novobiocin binds to the Q-loop and directly interacts with F90 and R91, strongly indicating that LptF/G's effect on LptB ATPase activity is mediated directly through their coupling interaction. The actual mechanism of this effect remains unknown, but novobiocin is a potentially promising lead compound for the rational design of antibiotics that target LPS export from the IM.

The crystal structures of the LptB$_2$FG tetramer solved by Luo et al. indicate a mechanism whereby the complex cycles through three conformational states to extract LPS from the IM to the periplasm [7] (Figure 3):

1. Resting: The LptB nucleotide-binding sites are unoccupied, and the LptF/G cavity is oriented inwards.
2. Open: ATP binds LptB, inducing the LptF/G cavity to open away from the IM, and receives the Lipid A moiety of LPS, which is still embedded in the IM.
3. Close: LptB hydrolyzes ATP, inducing the LptF/G cavity to close again. LPS is forced out of the IM into the periplasm.

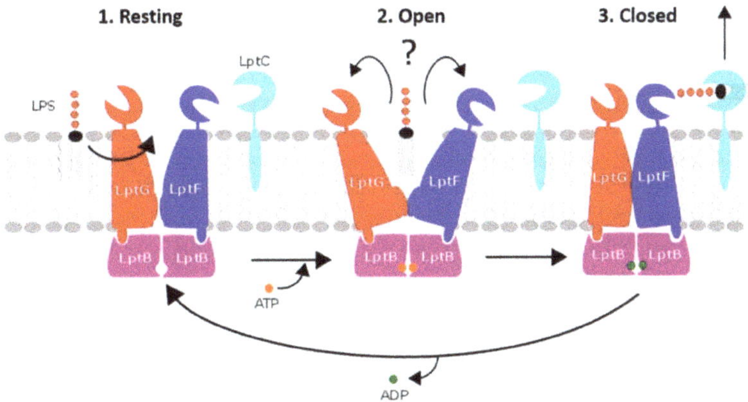

Figure 3. The mechanism of LPS transport to the outer membrane is often referred to as the "PEZ" model in reference to the spring-loaded candy dispenser. In this analogy, the LptB$_2$FG complex is the "spring" that pushes the LPS "candy" through the chamber (LptA and LptC) to the cap (LptD/E), which is lifted to release them one at a time.

Based on this mechanism, Luo et al. have suggested that LptB$_2$FG represents a third distinct type of ABC transporter, deemed type-III.

The overall mechanism of LptB$_2$FG is consistent with structural observations, but LPS's immediate destination following extraction from the IM remains ambiguous. The hydrophobic grooves of the LPS-binding domains of LptF and LptG are oriented away from and towards the IM, respectively, suggesting they may have differing functions in the transmission of LPS. Luo et al. theorized that the two may alternate as the binding site for LPS, although LPS binding to either component has yet to be reported [7]. Induced crosslinking experiments have successfully characterized specific LPS interactions with LptA and LptC, so a similar approach or other in vitro experiments could further elucidate the proposed mechanism. Simpson et al. hypothesized that LptF/G either extracts LPS directly through their LPS-binding domains or triggers the handoff indirectly by stimulating a conformational change in LptC [13]. Now that a general overview of the conformational changes that LptB$_2$FG undergoes is available, further investigation of how LPS interacts with its various domains is needed to understand how it facilitates transport to the other Lpt subunits.

How the LptB$_2$FG complex interacts with the other IM-bound subunit, LptC, and how that influences LPS extraction are still open questions. LptC reduces the ATPase of the activity of the complex in vitro, even though its localization to the periplasmic side of the IM would imply no direct interaction with LptB [7]. Although functional studies have confirmed that LptC is the intermediate between the IM complex and LptA, how or why it influences LptB's activity remains unclear.

3. LptC's Role in IM-Periplasm Transport Is Ambiguous

LptC is essential to LPS transport and cell survival, but the function of several of its structural elements is unclear. LptC consists of a C-terminal periplasmic domain that closely resembles the other LPS-binding domains and a single N-terminal transmembrane helix. Its periplasmic domain complexes with LptB$_2$FG and LptA and has been shown to bind LPS in vivo. Although LptC is bound to the IM, the N-terminal helix is not essential to Lpt assembly, as completely truncating it does not affect cell survival or the formation of the LptB$_2$FG complex [14]. LptC may localize to the IM to improve the efficiency of Lpt assembly, but its specific interactions with the other subunits are sufficient for LPS transport.

LptC's essentiality is dependent on specific interactions with LptF. Benedet et al. found that cells lacking LptC could be made viable by mutating a single LptF residue (R212) [15]. Membrane permeability increases as a result, but some LPS is still able to reach the surface. The likeliest explanation is that the IM complex is able to recruit an additional LptA to receive LPS in the absence of LptC. This conditional dependence also suggests that wildtype LptF forms a specific interaction with LptC that normally excludes LptA. Benedet et al. proposed that LptC could be a late evolutionary addition to what used to be a six-subunit complex [15]. Martorana et al. also showed that overexpressing LptB compensates for truncation of LptC's C-terminus (residues 139–191), possibly by shifting the binding equilibrium in favor of the full complex [16]. If this is the case, LptC may be important to the efficient and stable assembly of the LptB$_2$FG complex, in addition to directly transporting LPS.

LptC's periplasmic domain might serve a role in stabilizing the LptB$_2$FG complex, boosting the efficiency of the Lpt system. Although it has been theorized that the truncated LptC may be able to form an alternative edgewise interaction with LptA, this seems less likely than the alternative suggested by Benedet et al. wherein LptA is able to receive LPS from the IM independent of LptC [15]. The C-term LptC mutation could reduce the stability of the overall LptB$_2$FGC complex, so increased LptB expression could compensate by shifting the binding equilibrium in favor of the LptB2FG complex.

LptC has also been shown to form a homodimer, but no functional significance has been identified. The N-terminal edges of LptC's periplasmic domains form a head-to-head dimer in vitro [17]. Mutations disrupting this interaction have little effect on LptC's ability to bind LPS or LptA [18].

LptC's ability to dimerize may just be a side-effect of the β-jellyroll fold's modularity, but it may be worth accounting for in future investigations of its in vivo function.

4. LPS-Binding Domains Span the Periplasm

LptA and the periplasmic domains of LptC, LptD, LptF and LptG have divergent sequences, but share a common OstA-like twisted β-jellyroll fold that binds the Lipid A moiety of LPS. The 16 antiparallel β-strands form a pair of β-sheets in a "V" shape. The interior of the "V" forms a hydrophobic groove along the interior that widens slightly near the N and C-termini [19]. This groove likely shields Lipid A from the periplasmic environment, while the hydrophilic OS core and O-antigen remain exposed. The β-jellyroll domains oligomerize from head-to-tail with a 90° twist per subunit, forming a continuous multiprotein β-sheet and a spiraling hydrophobic groove. This rod-like oligomer provides the path for LPS from the IM to OM. The relative arrangement of the β-jellyroll domains is well-established, and residues essential to oligomerization have been identified; however, how these interactions contribute to LPS transport beyond providing the shielding Lipid A in the periplasm is still under investigation.

The Lpt β-jellyroll folds are very similar to one another, but the positioning of their N-terminal helices varies. LptA, LptD and LptG have short N-terminal helices that loop around to the C-terminal end of the hydrophobic groove of the jellyroll [6,7,19]. LptC and LptF lack similar helices: the N-terminus of LptC is disordered, and the N-terminal end of the LptF periplasmic domain links directly to helix 5 of its TM domain [7,20]. In LptD, the N-terminal helix forms one of two disulfide bonds that are essential to functional assembly of the LptD/E complex.

The apo structures of LptA, C and D indicate that the β-jellyroll must undergo extensive conformational changes to receive and subsequently hand off LPS. The hydrophobic cavity observed in the crystal structures of the empty LPS-binding domains are too small to accommodate the fatty acid chains of Lipid A [6,7,19]. A recent series of electron paramagnetic resonance (EPR) experiments by Schultz et al. demonstrated that LPS binding induces a substantial rearrangement in the inward-facing side of the hydrophobic groove of LptA and LptC [17,21]. The largest change is a partial unfolding of the N-terminal edge of the pocket, while the C-terminal edge remains relatively immobile.

It was previously thought that the transfer of LPS from the periplasmic domain of LptC to LptA is driven in part by the latter's higher affinity for LPS. In the absence of ATP, LptA has been shown to displace LPS from LptC in solution [20]. However, more recent assessments of the LPS affinity of LptC and LptA suggest the difference may be small or even inverted, with K_d values ranging from 11–28 and 7–35 uM, respectively [17,21]. The transmission of LPS between Lpt β-jellyroll subunits may therefore be a more dynamic process than the "PEZ" model implies alone, whereby the periplasmic bridge undergoes "ripples" of ordered-disorder transitions that shuffle LPS along its length.

5. LptA Oligomerization and Membrane Stress

LptA's function and relative placement in the Lpt system is understood, but its stoichiometry in vivo remains uncertain. LptA is the only component of the Lpt system that lacks a transmembrane component, as it spans the periplasmic space between the IM and OM complexes. LptA co-purifies almost exclusively with membrane fractions when expressed at native levels, indicating that most LptA is bound to other Lpt components rather than floating freely in the periplasm [22]. The first crystal structure of LptA solved by Suits et al. showed it arranged in an end-to-end fibrous tetramer, which forms a continuous hydrophobic groove between the LptA monomers (Figure 4). Mass spectral analysis later confirmed that LptA forms 2–5-member oligomers in a concentration-dependent manner when purified in vitro and that the resultant complexes are stabilized by LPS [23]. UV-dependent dimers of modified LptA have been purified, although these could be experimental artifacts resulting from the presence of tags or altered expression [24].

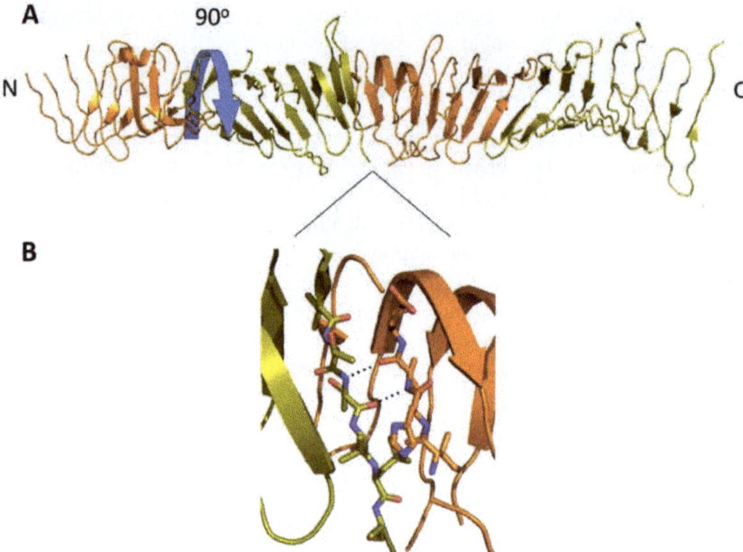

Figure 4. (**A**) The β-jellyroll domains of LptA form a rod-like head-to-tail multimer (alternating orange and gold) in solution with an ~90° rotation per subunit (blue arrow). (**B**) The N- and C-terminal β-strands form a continuous β-sheet along the length of the multimer. This is likely analogous to LptA's interactions with the β-jellyrolls of LptC and LptD.

The LptA-LptA binding interface probably uses the same contacts as LptA-LptC and LptA-LptD, so it is difficult to investigate the in vivo relevance of LptA oligomerization via mutagenesis without also disrupting LPS transport. The recent development of high-resolution in situ imaging techniques like electron cytomography may circumvent this issue by making it possible to directly measure the transenvelope length of the Lpt complex, although it may be necessary to first identify conditions that favour LptA oligomerization [25].

It is tempting to speculate that LptA oligomerization allows the Lpt system to tolerate changes in the intermembrane space resulting from osmotic stress. A single-LptA complex can bridge a typical *Escherichia coli* intermembrane distance of 100 Å, but the width of the periplasm could vary between species and in response to osmotic stress. In *E. coli*, the LptA appears to express at the same level as the other monomeric components of the Lpt system under normal conditions. However, functional analysis of the promoters associated with LptA shows additional responsiveness to changes in LPS biogenesis [26]. It is possible that the Lpt complex varies between single and multiple-LptA states in response to increased LptA availability, influenced either by the disruption of a single-subunit complex or independent increases in expression.

LPS's last stop in the periplasm is the β-jellyroll domain of LptD. The interaction between LptA and LptD appears to be a key regulatory checkpoint in the assembly of the functional Lpt complex. LptA binds LptD/E preferentially to LptC, so during Lpt assembly, it is likely recruited to the OM before connecting to the IM subcomplex [22]. LptA binds the β-jellyroll of LptD in a head-to-tail interaction like the other periplasmic domains and forms no direct interactions with LptE [27]. However, LptA does not bind LptD if its specific disulfide bonds are not formed correctly [24]. It is intriguing that such a small difference would disrupt the otherwise robust β-jellyroll oligomerization process, especially given that the LptD subunit shows a great deal of rotational flexibility relative to the OM [6]. This requirement may be a specific adaptation that prevents malformed LptD from connecting with the rest of the Lpt system.

Recent structural advancements in our understanding of the Lpt complex suggests an ability to dynamically respond to changes to cellular envelopes' dimensions. The Rcs stress response system in enterobacteria was recently shown to respond to changes in periplasmic width, suggesting that essential transenvelope complexes are adapted to transenvelope stress [28]. Given how crucial the Lpt system is to cell survival, it stands to reason that it would be adapted to compensate for expansions and contractions of the periplasmic space resulting from osmotic stress. LptA could allow the system to tolerate changes in intermembrane distance by recruiting newly-detached membrane translocons.

6. LptD/E Inserts LPS into the Outer Membrane Laterally

Until recently, the available structures of the LptD/E dimer were insufficient to fully confirm how the OM complex conveys the Lipid A and OS moieties to the outer leaflet. The C-terminal TM portion of LptD is a massive 26-strand β-barrel, and the LptE lipoprotein sits inside its lumen. The full-length structure solved by Botos et al. shows that the N-terminal domain of LptD is a β-jellyroll that extends from the periplasm and into the hydrophilic core of the OM [6]. This suggests a two-portal mechanism whereby Lipid A is inserted directly into the membrane by the β-jellyroll, while the OS component proceeds through the hydrophilic lumen of the β-barrel (Figure 5).

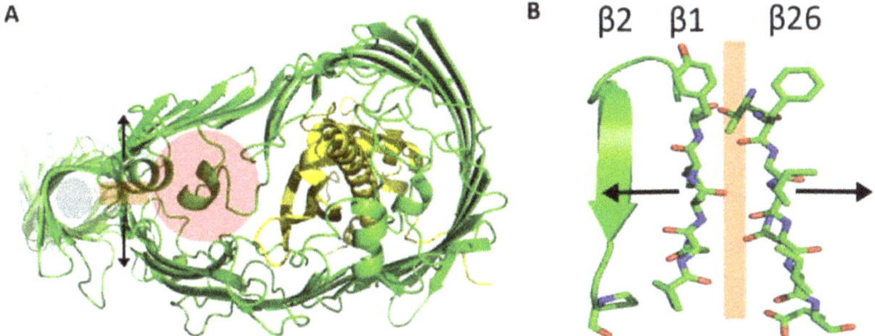

Figure 5. (**A**) The top-down view of the crystal structure of LptD (green) and LptE (gold) shows how the Lipid A and OS moieties of LPS exit the IM through openings in the N-terminal β-jellyroll (grey) and β-barrel domain (pink) respectively. This is made possible by the lateral gate that forms between the two domains. (**B**) Strands β1 and β26 of the transmembrane β-barrel separate perpendicular to the path of LPS. Proline residues on β1 and β2 (P231 and P246) limit the strands' ability to form a sheet, which allows β1 and β2 to separate.

A lateral gate in the β-barrel of LptD allows the Lipid A and OS moieties to proceed to the outer leaflet of the OM. Conserved phenylalanine residues opposite one another on the β1 and β2 strands of the barrel prevent β1 from forming a complete hydrogen bonding network with β26 [6]. This allows β1 and β26 to slide apart, so the OS and O-antigen moieties can move out of the barrel to the cell surface. Functional LptD/E's very specific disulfide bond configuration restricts the size of the lateral gate on the periplasmic side, possibly to prevent incorrect insertion of LPS into the inner leaflet of the OM. The extracellular side of the barrel lumen is rich in negatively-charged residues, which likely help to repel the similarly charged saccharide moieties out of the complex. LptD's two-portal structure gives LPS a clear path from the periplasm to its final destination on the extracellular surface.

7. LptE Is a Multifunction Facilitator of LPS Export

Unlike the other Lpt proteins, LptE's essential functions are not obvious from the overview of the "PEZ" model. LptE is embedded into the LptD lumen, but it is inserted into the OM independently by the Lol pathway [29]. LptE is a lipoprotein with little structural similarity to the other Lpt components, consisting of two α-helices closely associated with a four-strand β-sheet [30]. Rather than act as a direct

LPS transporter, LptE appears to have three have overlapping functions in LPS transport: functional LptD assembly, preserving membrane impermeability and distributing LPS to the cell surface.

Functional assembly of LptD requires LptE as a template. LptD assembly requires the assistance of the BAM complex and extensive reshuffling of disulfide bonds [31,32]. Without the correct disulfide bond configuration, LptD cannot bind LptA or export LPS. LptD is still inserted into the OM when LptE is depleted, but the population of functionally-oxidized protein is depleted, as well. A three-residue LptE mutant (PIS117–119R) induces a similar effect [33]. The affected region of LptE may act as a template for LptD that ensures the cysteine residues are correctly positioned for disulfide bond rearrangement.

LptE also preserves membrane impermeability by plugging LptD's large cavity. LptD's transmembrane β-barrel is exceptionally large for a bacterial protein, consisting of 26 individual β-strands and spanning 50 Å. This large cavity structure is needed to export the diverse, flexible OS moieties of LPS, but also produces a large gap in the membrane. The crystal structure of LptD/E shows that LptE is positioned between the extracellular loops of the LptD β-barrel, blocking part of the extracellular opening [30]. Grabowicz et al. identified an LptE mutant that caused an increase in membrane permeability without affecting LptD assembly or LPS export [34]. No individual change produced the same high-permeability phenotype, but the four most N-terminal mutations all affect residues near the extracellular end of LptE.

LptE has been shown to bind and break up surface-bound aggregates of LPS in vitro [30,35]. It is difficult to disentangle LptE's in vivo interactions with LPS from its other established roles in LptD assembly and plugging, as disrupting either process also increases membrane permeability and hinders LPS transport. However, Malojcic et al. found that mutating the positively-charged residues (R91D, K136D) in a loop homologous to the LPS binding sites in other LPS-binding proteins disrupted binding in vitro and increased in vivo membrane permeability without affecting LptD assembly [30]. Based on this and the observation that LptE will break up surface-bound LPS aggregates, they proposed that LptE facilitates the final transfer of LPS to the outer leaflet by preventing their aggregation at the inner leaflet. Interestingly, one of Malojcic's rationally-designed mutations, K136D, is on the same locus as the K136W mutation in Grabowicz's randomly-generated plugging-deficient LptE14 mutant. This implies a degree of functional overlap between the structural components of LptE.

The LptD/E complex's location at the extracellular surface makes it a more promising antibiotic target than the other Lpt subunits. Antibiotic compounds do not need to permeate the LPS barrier to reach it, and disrupting plugging of the LptD β-barrel could complement other antibiotics by increasing the permeability of the OM. The compound murepavadin and peptide mimetics have been shown to prevent LPS barrier biogenesis in *Pseudomonas aeruginosa* through direct interactions with LptD [36,37]. Crosslinking experiments have demonstrated that the peptide mimetic interacts with an extended region of the β-jellyroll domain found in the pseudomonads [38]. Andolina et al. suggested that this blocks the Lipid A channel, but other mechanisms influencing LptD assembly or LptA binding should not be discounted. Further structural characterization of the *P. aeruginosa* LptD periplasmic domain, which is currently unsolved, could elucidate this mechanism and provide clues for the rational development for more broad-spectrum antibiotics. LptE, meanwhile, may be less promising as a target as at least one species of bacteria, *Neisseria meningitidis*, can survive without LptE [39].

8. Summary

New structures and functional analyses of the Lpt system have made it possible to begin identifying the specific subunit-to-subunit interactions that make the continuous transport of LPS from the cytoplasm to the exterior of the outer membrane possible. The potentially novel type-III ABC mechanism employed by the LptB$_2$FG subcomplex raises new questions about the mechanisms of LPS extraction and the ambiguous role of LptC. Determining how the Lpt system responds to shifts in the periplasmic space may contribute to a broader understanding of how Gram-negative bacteria adapt to environmental stresses affecting the cellular envelope. The LptD/E OM complex has proven a promising target for new antibiotics, which could be further refined through rational design informed by the recent structural characterization of its export mechanism.

Author Contributions: Conceptualization, G.H. and Z.J. Writing, original draft preparation, G.H. Writing, review and editing, G.H. and Z.J. All authors read and approved the final manuscript.

Funding: This research was funded by the Natural Sciences and Engineering Research Council of Canada (RGPIN-2018-04427).

Acknowledgments: We thank all members of the Jia lab who have provided support and guidance.

Conflicts of Interest: The authors declare no conflict of interest.

Abbreviations

ABC	ATP-binding cassette
EPR	electron paramagnetic resonance
IM	inner membrane
LPS	lipopolysaccharide
Lpt	lipopolysaccharide transport
NBD	nucleotide binding domain
OM	outer membrane
OS	oligosaccharide
TMD	transmembrane domain

References

1. Raetz, C.R.H.; Guan, Z.; Ingram, B.O.; Six, D.A.; Song, F.; Wang, X.; Zhao, J. Discovery of new biosynthetic pathways: The lipid A story. *J. Lipid Res.* **2009**, *50*, S103–S108. [CrossRef] [PubMed]

2. Zhou, Z.; White, K.A.; Polissi, A.; Georgopoulos, C.; Raetz, C.R. Function of *Escherichia coli* MsbA, an essential ABC family transporter, in lipid A and phospholipid biosynthesis. *J. Biol. Chem.* **1998**, *273*, 12466–12475. [CrossRef] [PubMed]

3. Rocchetta, H.L.; Burrows, L.L.; Lam, J.S. Genetics of O-antigen biosynthesis in *Pseudomonas aeruginosa*. *Microbiol. Mol. Biol. Rev.* **1999**, *63*, 523–553. [PubMed]

4. Abeyrathne, P.D.; Daniels, C.; Poon, K.K.H.; Matewish, M.J.; Lam, J.S. Functional characterization of WaaL, a ligase associated with linking O-antigen polysaccharide to the core of *Pseudomonas aeruginosa* lipopolysaccharide. *J. Bacteriol.* **2005**, *187*, 3002–3012. [CrossRef] [PubMed]

5. Brown, D.G.; May-Dracka, T.L.; Gagnon, M.M.; Tommasi, R. Trends and exceptions of physical properties on antibacterial activity for gram-positive and gram-negative pathogens. *J. Med. Chem.* **2014**, *57*, 10144–10161. [CrossRef] [PubMed]

6. Botos, I.; Majdalani, N.; Mayclin, S.J.; McCarthy, J.G.; Lundquist, K.; Wojtowicz, D.; Barnard, T.J.; Gumbart, J.C.; Buchanan, S.K. Structural and functional characterization of the LPS transporter LptDE from gram-negative pathogens. *Structure* **2016**, *24*, 965–976. [CrossRef] [PubMed]

7. Luo, Q.; Yang, X.; Yu, S.; Shi, H.; Wang, K.; Xiao, L.; Zhu, G.; Sun, C.; Li, T.; Li, D.; et al. Structural basis for lipopolysaccharide extraction by ABC transporter LptB$_2$FG. *Nat. Struct. Mol. Biol.* **2017**, *24*, 469–474. [CrossRef] [PubMed]

8. Sherman, D.J.; Xie, R.; Taylor, R.J.; George, A.H.; Okuda, S.; Foster, P.J.; Needleman, D.J.; Kahne, D. Lipopolysaccharide is transported to the cell surface by a membrane-to-membrane protein bridge. *Science* **2018**, *359*, 798–801. [CrossRef] [PubMed]

9. Higgins, C.F.; Linton, K.J. The ATP switch model for ABC transporters. *Nat. Struct. Mol. Biol.* **2004**, *11*, 918–926. [CrossRef] [PubMed]

10. Gronenberg, L.S.; Kahne, D. Development of an activity assay for discovery of inhibitors of lipopolysaccharide transport. *J. Am. Chem. Soc.* **2010**, *132*, 2518–2519. [CrossRef] [PubMed]

11. Sherman, D.J.; Lazarus, M.B.; Murphy, L.; Liu, C.; Walker, S.; Ruiz, N.; Kahne, D. Decoupling catalytic activity from biological function of the ATPase that powers lipopolysaccharide transport. *Proc. Natl. Acad. Sci. USA* **2014**, *111*, 4982–4987. [CrossRef] [PubMed]

12. May, J.M.; Owens, T.W.; Mandler, M.D.; Simpson, B.W.; Lazarus, M.B.; Sherman, D.J.; Davis, R.M.; Okuda, S.; Massefski, W.; Ruiz, N.; et al. The antibiotic novobiocin binds and activates the ATPase that powers lipopolysaccharide transport. *J. Am. Chem. Soc.* **2017**, *139*, 17221–17224. [CrossRef] [PubMed]

13. Simpson, B.W.; Owens, T.W.; Orabella, M.J.; Davis, R.M.; May, J.M.; Trauger, S.A.; Kahne, D.; Ruiz, N. Identification of residues in the lipopolysaccharide ABC transporter that coordinate ATPase activity with extractor function. *MBio* **2016**, *7*, e01729-16. [CrossRef] [PubMed]

14. Villa, R.; Martorana, A.M.; Okuda, S.; Gourlay, L.J.; Nardini, M.; Sperandeo, P.; Dehò, G.; Bolognesi, M.; Kahne, D.; Polissi, A. The *E. coli* Lpt transenvelope protein complex for lipopolysaccharide export is assembled via conserved structurally homologous domains. *J. Bacteriol.* **2013**, *195*, 1100–1108. [CrossRef] [PubMed]

15. Benedet, M.; Falchi, F.A.; Puccio, S.; di Benedetto, C.; Peano, C.; Polissi, A.; Dehò, G. The lack of the essential LptC protein in the *trans*-envelope lipopolysaccharide transport machine is circumvented by suppressor mutations in LptF, an inner membrane component of the *E. coli* transporter. *PLoS ONE* **2016**, *11*, e0161354. [CrossRef] [PubMed]

16. Martorana, A.M.; Benedet, M.; Maccagni, E.A.; Sperandeo, P.; Villa, R.; Dehò, G.; Polissi, A. Functional interaction between the cytoplasmic ABC protein LptB and the inner membrane LptC protein, components of the lipopolysaccharide transport machinery in *Escherichia coli*. *J. Bacteriol.* **2016**, *198*, 2192–2203. [CrossRef] [PubMed]

17. Schultz, K.M.; Klug, C.S. Characterization of and lipopolysaccharide binding to the *E. coli* LptC protein dimer. *Protein Sci.* **2018**, *27*, 381–389. [CrossRef] [PubMed]

18. Schultz, K.M.; Fischer, M.A.; Noey, E.L.; Klug, C.S. Disruption of the *E. coli* LptC dimerization interface and characterization of lipopolysaccharide and LptA binding to monomeric LptC. *Protein Sci.* **2018**. [CrossRef] [PubMed]

19. Suits, M.D.L.; Sperandeo, P.; Dehò, G.; Polissi, A.; Jia, Z. novel structure of the conserved gram-negative lipopolysaccharide transport protein A and mutagenesis analysis. *J. Mol. Biol.* **2008**, *380*, 476–488. [CrossRef] [PubMed]

20. Tran, A.X.; Dong, C.; Whitfield, C. Structure and functional analysis of LptC, a conserved membrane protein involved in the lipopolysaccharide export pathway in *Escherichia coli*. *J. Biol. Chem.* **2010**, *285*, 33529–335239. [CrossRef] [PubMed]

21. Schultz, K.M.; Lundquist, T.J.; Klug, C.S. Lipopolysaccharide binding to the periplasmic protein LptA. *Protein Sci.* **2017**, *26*, 1517–1523. [CrossRef] [PubMed]

22. Chng, S.S.; Gronenberg, L.S.; Kahne, D. Proteins required for lipopolysaccharide assembly in *Escherichia coli* form a transenvelope complex. *Biochemistry* **2010**, *49*, 4565–4567. [CrossRef] [PubMed]

23. Santambrogio, C.; Sperandeo, P.; Villa, R.; Sobott, F.; Polissi, A.; Grandori, R. LptA assembles into rod-like oligomers involving disorder-to-order transitions. *J. Am. Soc. Mass Spectrom.* **2013**, *24*, 1593–1602. [CrossRef] [PubMed]

24. Freinkman, E.; Okuda, S.; Ruiz, N.; Kahne, D. Regulated assembly of the transenvelope protein complex required for lipopolysaccharide export. *Biochemistry* **2012**, *51*, 4800–4806. [CrossRef] [PubMed]

25. Oikonomou, C.M.; Jensen, G.J. Cellular electron cryotomography: Toward structural biology in situ. *Annu. Rev. Biochem.* **2017**, *86*, 873–896. [CrossRef] [PubMed]

26. Martorana, A.M.; Sperandeo, P.; Polissi, A.; Dehò, G. Complex transcriptional organization regulates an *Escherichia coli* locus implicated in lipopolysaccharide biogenesis. *Res. Microbiol.* **2011**, *162*, 470–482. [CrossRef] [PubMed]

27. Bowyer, A.; Baardsnes, J.; Ajamian, E.; Zhang, L.; Cygler, M. Characterization of interactions between LPS transport proteins of the Lpt system. *Biochem. Biophys. Res. Commun.* **2011**, *404*, 1093–1098. [CrossRef] [PubMed]

28. Asmar, A.T.; Ferreira, J.L.; Cohen, E.J.; Cho, S.H.; Beeby, M.; Hughes, K.T.; Collet, J.F. Communication across the bacterial cell envelope depends on the size of the periplasm. *PLoS Biol.* **2017**, *15*, e2004303. [CrossRef] [PubMed]

29. Yokota, N.; Kuroda, T.; Matsuyama, S.; Tokuda, H. Characterization of the LolA-LolB system as the general lipoprotein localization mechanism of *Escherichia coli*. *J. Biol. Chem.* **1999**, *274*, 30995–30999. [CrossRef] [PubMed]

30. Malojčić, G.; Andres, D.; Grabowicz, M.; George, A.H.; Ruiz, N.; Silhavy, T.J.; Kahne, D. LptE binds to and alters the physical state of LPS to catalyze its assembly at the cell surface. *Proc. Natl. Acad. Sci. USA* **2014**, *111*, 9467–9472. [CrossRef] [PubMed]

31. Denoncin, K.; Vertommen, D.; Paek, E.; Collet, J.F. The protein-disulfide isomerase DsbC cooperates with SurA and DsbA in the assembly of the essential β-barrel protein LptD. *J. Biol. Chem.* **2010**, *285*, 29425–29433. [CrossRef] [PubMed]

32. Wu, T.; Malinverni, J.; Ruiz, N.; Kim, S.; Silhavy, T.J.; Kahne, D. Identification of a multicomponent complex required for outer membrane biogenesis in *Escherichia coli*. *Cell* **2005**, *121*, 235–245. [CrossRef] [PubMed]

33. Chimalakonda, G.; Ruiz, N.; Chng, S.S.; Garner, R.A.; Kahne, D.; Silhavy, T.J. Lipoprotein LptE is required for the assembly of LptD by the β-barrel assembly machine in the outer membrane of *Escherichia coli*. *Proc. Natl. Acad. Sci. USA* **2011**, *108*, 2492–2497. [CrossRef] [PubMed]

34. Grabowicz, M.; Yeh, J.; Silhavy, T.J. Dominant negative lptE mutation that supports a role for LptE as a plug in the LptD barrel. *J. Bacteriol.* **2013**, *195*, 1327–1334. [CrossRef] [PubMed]

35. Chng, S.S.; Ruiz, N.; Chimalakonda, G.; Silhavy, T.J.; Kahne, D. Characterization of the two-protein complex in *Escherichia coli* responsible for lipopolysaccharide assembly at the outer membrane. *Proc. Natl. Acad. Sci. USA* **2010**, *107*, 5363–5368. [CrossRef] [PubMed]

36. Srinivas, N.; Jetter, P.; Ueberbacher, B.J.; Werneburg, M.; Zerbe, K.; Steinmann, J.; Van der Meijden, B.; Bernardini, F.; Lederer, A.; Dias, R.L.A.; et al. Peptidomimetic antibiotics target outer-membrane biogenesis in Pseudomonas aeruginosa. *Science* **2010**, *327*, 1010–1013. [CrossRef] [PubMed]

37. Werneburg, M.; Zerbe, K.; Juhas, M.; Bigler, L.; Stalder, U.; Kaech, A.; Ziegler, U.; Obrecht, D.; Eberl, L.; Robinson, J.A. Inhibition of lipopolysaccharide transport to the outer membrane in pseudomonas aeruginosa by peptidomimetic antibiotics. *ChemBioChem* **2012**, *13*, 1767–1775. [CrossRef] [PubMed]

38. Andolina, G.; Bencze, L.C.; Zerbe, K.; Müller, M.; Steinmann, J.; Kocherla, H.; Mondal, M.; Sobek, J.; Moehle, K.; Malojčić, G.; et al. A Peptidomimetic antibiotic interacts with the periplasmic domain of LptD from Pseudomonas aeruginosa. *ACS Chem. Biol.* **2018**, *13*, 666–675. [CrossRef] [PubMed]

39. Bos, M.P.; Tommassen, J. The LptD chaperone LptE is not directly involved in lipopolysaccharide transport in Neisseria meningitidis. *J. Biol. Chem.* **2011**, *286*, 28688–28696. [CrossRef] [PubMed]

 International Journal of
Molecular Sciences

Review

Immune Mechanisms Underlying Susceptibility to Endotoxin Shock in Aged Hosts: Implication in Age-Augmented Generalized Shwartzman Reaction

Manabu Kinoshita *, Masahiro Nakashima, Hiroyuki Nakashima and Shuhji Seki *

Department of Immunology and Microbiology, National Defense Medical College, 3-2 Namiki, Tokorozawa, Saitama 359-8513, Japan

* Correspondence: manabu@ndmc.ac.jp (M.K.); btraums@ndmc.ac.jp (S.S.); Tel.: +81-4-2995-1541 (M.K.); Fax: +81-4-2996-5194 (M.K.)

Received: 11 June 2019; Accepted: 1 July 2019; Published: 2 July 2019

Abstract: In recent decades, the elderly population has been rapidly increasing in many countries. Such patients are susceptible to Gram-negative septic shock, namely endotoxin shock. Mortality due to endotoxin shock remains high despite recent advances in medical care. The generalized Shwartzman reaction is well recognized as an experimental endotoxin shock. Aged mice are similarly susceptible to the generalized Shwartzman reaction and show an increased mortality accompanied by the enhanced production of tumor necrosis factor (TNF). Consistent with the findings in the murine model, the in vitro Shwartzman reaction-like response is also age-dependently augmented in human peripheral blood mononuclear cells, as assessed by enhanced TNF production. Interestingly, age-dependently increased innate lymphocytes with T cell receptor-that intermediate expression, such as that of $CD8^+CD122^+$T cells in mice and $CD57^+$T cells in humans, may collaborate with macrophages and induce the exacerbation of the Shwartzman reaction in elderly individuals. However, endotoxin tolerance in mice, which resembles a mirror phenomenon of the generalized Shwartzman reaction, drastically reduces the TNF production of macrophages while strongly activating their bactericidal activity in infection. Importantly, this effect can be induced in aged mice. The safe induction of endotoxin tolerance may be a potential therapeutic strategy for refractory septic shock in elderly patients.

Keywords: LPS; septic shock; elderly; innate lymphocytes; endotoxin tolerance

---

1. Introduction

Severe sepsis, particularly Gram-negative sepsis, remains a grave concern for humans, being recognized as a major leading cause of death in the United States as well as other countries [1]. Gram-negative septic shock, namely endotoxin shock, shows a particularly high mortality rate despite recent advances in intensive care. For several decades, the elderly populations has been rapidly increasing in many countries, and such patients are susceptible to endotoxin shock [2]. Therefore, medical countermeasures for this population are an urgent issue to be addressed. The generalized Shwartzman reaction is well recognized as an experimental endotoxin shock. We investigated the effect of aging on host defenses against endotoxin shock using a mouse model and human peripheral blood mononuclear cells (PBMCs) as an in vitro model of a Shwartzman reaction-like response, focusing on age-related immunological changes. We found that a certain T lymphocyte population that is age-dependently increased in mice as well as in humans collaborates with macrophages and causes the exacerbation of endotoxin shock in elderly subjects. In this review, we demonstrate why elderly hosts are susceptible to endotoxin shock, based on our immunological studies in murine and human PBMCs, and we also address potential therapeutic approaches against endotoxin shock in the elderly.

2. Definition of Septic Shock

Septic shock is the most severe form of sepsis and remains a grave concern because of its high mortality rate [3,4]. In 1991, the consensus conference of American College of Chest Physicians/Society of Critical Care Medicine defined sepsis as a systemic inflammatory response syndrome (SIRS) to infection [5,6]. Severe sepsis was also defined in that consensus conference as instances in which sepsis was complicated by acute organ dysfunctions, and septic shock was defined as sepsis complicated by either hypotension that was refractory to fluid resuscitation or by hyperlactatemia indicating tissue hypoxia [5,6]. At the time, the criteria for SIRS were thought to be clinically useful as a screening tool for the diagnosis of septic patients, as it can easily and simply define the severe infection and then effective therapy can be promptly started for SIRS conditions. In addition, these definitions were thought to be helpful for establishing patient entry criteria in clinical trials on sepsis.

One decade later, a second consensus conference was held in 2001 that endorsed most of these concepts, but some caveats were raised: Signs of SIRS (tachycardia, increased white blood cell count, fever elevation, tachypnea, etc.) may occur in not only infectious but also noninfectious conditions, so these criteria of SIRS may not always be useful for distinguishing sepsis from other critical conditions [7].

In 2016, a third consensus conference was held that defined 'Sepsis-3' [8]. In that conference, 'sepsis' was defined as life-threatening organ dysfunctions caused by a dysregulated host response to infection, similar to the definition of 'severe sepsis' proposed at the first conference in 1991. Septic shock was also defined as a subset of sepsis in which particularly profound circulatory, cellular, and metabolic abnormalities were associated with a greater risk of multi-organ dysfunctions and mortality than with sepsis alone [8].

3. High Mortality and Morbidity of Septic Shock in the Elderly

Aging is one of the most representative risk factors of septic shock [9]. Elderly people tend to have various kinds of chronic comorbid illness, such as ischemic heart disease, cerebrovascular disease, diabetes, and renal or pulmonary disease. These comorbid illnesses are presumably susceptible to sepsis/severe sepsis, leading to shock and multi-organ dysfunctions [2]. Mortality due to septic shock has decreased in the past decade thanks to the prompt initiation of appropriate antimicrobial therapy [10] and sophisticated organ support therapies such as extracorporeal membrane oxygenation (ECMO), continuous hemodiafiltration (CHDF), and direct hemoperfusion with immobilized polymyxin B fiber (PMX-DHP) [11,12]. Nevertheless, mortality due to septic shock remains high at approximately 20% [13–15]. Gram-negative bacteremia induce a more severe inflammatory response in septic patients than Gram-positive bacteremia [16]. Therefore, Gram-negative sepsis is prone to progress to septic shock, and the mortality due to Gram-negative septic shock is still higher than that associated with Gram-positive sepsis at approximately 30–40% [16–18]. Aging is also a risk factor for colonization by Gram-negative bacteria, such as the *Klebsiella*, *Escherichia coli*, and *Enterobacter* species, predisposing elderly patients to Gram-negative bacterial sepsis and septic shock [19]. Taken together, these findings indicate that elderly patients are more susceptible to sepsis than younger ones, in particular Gram-negative sepsis, and may suffer sepsis-induced multi-organ dysfunctions, thereby leading to a higher mortality [2,19–21].

4. Pivotal Role of Tumor Necrosis Factor-α (TNF) in Human Endotoxin Shock

Gram-negative bacteria contain endotoxin, a main component of their bacterial envelope, that is also known as lipopolysaccharide (LPS). Endotoxin/LPS is responsible for the pathogenesis of septic shock induced by Gram-negative bacterial infections [22,23]. Plasma endotoxin/LPS levels are associated with increased severity of illness, as measured by the indicator of total organ dysfunctions [24]. Endotoxin/LPS binds to toll-like receptor 4 (TLR4) expressed on the cellular membrane of macrophages and stimulates the NF-κB-dependent inflammatory cascade to produce tumor necrosis factor (TNF), a strongly representative proinflammatory cytokine [23]. Though TNF is important for host antibacterial resistance [25,26], the excessive production of TNF by the stimulation of bacterium-derived LPS may typically cause endothelial damage, multi-organ dysfunctions, and septic shock [27]. TNF transmits its signals into the cells through TNF-receptor (TNFR) 1 and 2, which are expressed on the surface of various mammalian cells [28]. TNFR1 contains a cytoplasmic 'death domain' that induces cell death of TNFR1-expressing cells [28]. Thus, LPS-induced increased TNF production is likely a major contributor to endotoxin/LPS shock. Indeed, patients with septic shock show high TNF levels in the plasma, and their TNF levels are related to the patients' outcomes [21,29,30]. In addition, the administration of endotoxin/LPS to healthy volunteers increased the plasma TNF levels in the initial phase, suggesting that TNF causes the subsequent inflammatory cytokine response [31–33]. Consistently, the administration of TNF to healthy volunteers resulted in a similar inflammatory reaction in humans to that of endotoxin/LPS administration [31,34]. TNF thus plays pivotal roles in the inflammatory cascade in humans with endotoxemia/Gram-negative sepsis.

5. Generalized Shwartzman Reaction as an Experimental Endotoxin Shock Model

Humans are much more sensitive to LPS than other mammals, such as rodents. The administration of 2–4 ng/kg LPS to healthy volunteers has been shown to evoke various inflammatory responses, including the elevation of plasma TNF [31–33]. In contrast, even when mice were administered 2–4 µg/kg LPS, which was almost 1000 times the concentration administered to humans, they only showed a slight elevation of plasma TNF without any organ damage [35]. We then considered that induction of a hypersensitive response to LPS was important for simulating/mimicking human endotoxin shock in mice. The generalized Shwartzman reaction is well recognized as a hypersensitive innate immune response in experimental animals [36].

Roughly 90 years ago, Shwartzman first described the intradermal injection of the sterile culture filtrate of the Gram-negative bacterium *Bacillus Salmonella typhosus* as a preparatory injection and noted that the subsequent intravenous injection with a provocative dose of the same culture filtrate 24 h later induced severe hemorrhagic necrosis at the first injection site in rabbits [37]. If the provocative challenge was too short (less than 2 h) or too long (beyond 48 h), dermal necrosis did not occur. Shwartzman also performed the experiment using a similar preparation with sterile culture filtrate of Gram-positive streptococcal species but failed to duplicate dermal reactions, suggesting that the heat-stable component of Gram-negative bacteria (later identified as LPS) is important for inducing this reaction [36]. However, Shwartzman described only the localized dermal response and not the systemic response in this phenomenon. According to the review of Chahin et al., the generalized Shwartzman reaction was actually discovered by Sanarelli four years earlier than the report by Shwartzman [36]. The generalized Shwartzman reaction is now considered a potentially lethal endotoxin shock reaction that can be induced by the administration of a sublethal dose of LPS into LPS-primed animals at an interval of 24 h (Figure 1A) [38].

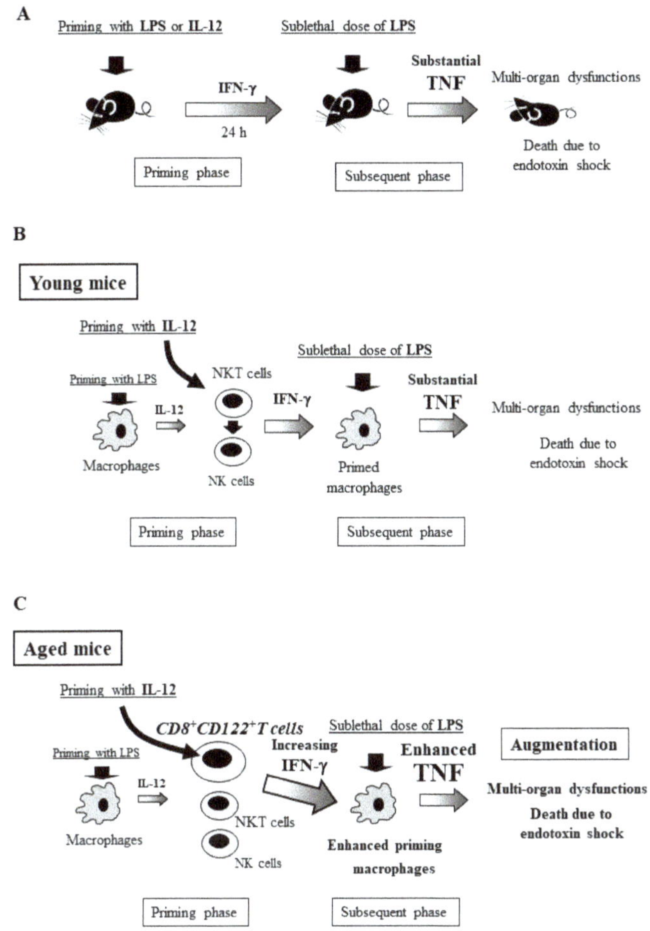

Figure 1. The generalized Shwartzman reaction induced by interleukin (IL)-12 priming and subsequent lipopolysaccharide (LPS) challenge. (**A**) Priming with LPS or IL-12 induces IFN-γ production in mice, and 24 h later, subsequent LPS challenge induces substantial tumor necrosis factor (TNF) production, resulting in the generalized Shwartzman reaction. (**B**) In young mice, IL-12 priming activates natural killer (NK) and natural killer T (NKT) cells to produce IFN-γ, which primes macrophages. Subsequent LPS challenge strongly activates the primed macrophages to produce large amounts of TNF, resulting in the generalized Shwartzman reaction. (**C**) In aged mice, age-dependently increased IL-12-activating CD8⁺CD122⁺T cells augment IFN-γ production, resulting in the further enhancement of TNF production by macrophages and leading to increased lethality of the generalized Shwartzman reaction. Figures are revised from [39].

6. Generalized Shwartzman Reaction Induced by Interleukin (IL)-12 Priming and Sublethal LPS Challenge

Though the generalized Shwartzman reaction is well recognized as an endotoxin shock model, it cannot be induced every time in animals, as this phenomenon is likely an example of a related phenomenon known as endotoxin tolerance [35,36,40]. Through the generalized Shwartzman reaction, injection of interleukin (IL)-12 or IFN-γ, instead of LPS, can induce a lethal reaction in mice if they are further challenged with a sublethal dose of LPS [41]. IL-12, which is mainly produced by macrophages, is a Th 1 cytokine that stimulates natural killer (NK) cells, natural killer T (NKT) cells, and T cells to produce IFN-γ [42,43]. Neutralizing antibody against IL-12, when administered together with the priming agent, prevents the lethal reaction in mice primed with either LPS or IL-12 but not with IFN-γ [41]. Thus, IL-12 and its induced IFN-γ in the priming phase are crucial for the occurrence of the generalized Shwartzman reaction (Figure 1B) [41,44]. Upon subsequent LPS challenge, the lethal Shwartzman reaction is induced by the massive production of inflammatory cytokines, particularly TNF, that act on the target organs already sensitized by IFN-γ and resultant multi-organ dysfunctions in mice [41]. TNF is a key cytokine in the generalized Shwartzman reaction as well as endotoxin shock and results in high mortality [39,41,44]. Notably, IL-12 priming rather than LPS priming can constantly induces the lethal generalized Shwartzman reaction in mice [39,44]. We speculate that although the doses of LPS are different from LPS tolerance, two times LPS injections (LPS priming and subsequent challenge with LPS) may induce a LPS tolerant condition (as described later), leading to failure to effectively induce a generalized Shwartzman reaction [35,40].

7. Crucial Role of NKT Cells and Their Produced IFN-γ for the Murine Generalized Shwartzman Reaction

In the generalized Shwartzman reaction, which is induced by IL-12 priming and subsequent LPS challenge in mice, depletion of both NK cells and NKT cells (by anti-NK1.1 antibody) greatly reduced the elevation of serum IFN-γ after IL-12 priming and reduced the mortality after LPS challenge, whereas depletion of NK cells alone (by anti-asialo GM1 antibody) only partially decreased the serum IFN-γ and did not affect the mortality [44]. NKT cells and their produced IFN-γ may be crucial in the priming phase for the induction of lethality in the generalized Shwartzman reaction. However, NK cells play important roles in protecting their host against sepsis/bacterial infections [45], as they have potent IFN-γ-producing capability in response to bacterial infection and can effectively induce bacterial killing by macrophages, which is crucial for bacterial elimination [45]. We should not underestimate NK cell-producing IFN-γ even in the generalized Shwartzman reaction as well as sepsis/infections.

8. Involvement of Innate Lymphocytes in the Generalized Shwartzman Reaction

The innate immune system serves as the first line of the host defense against invading bacteria. NK cells, NKT cells, and other innate lymphocytes are crucially involved in this front-line defense [46,47]. Severe surgical stress, such as burn injury, drastically reduces the IFN-γ-producing capability of these cells, thereby increasing susceptibility to sepsis/infection [45,48]. In turn, excessive activation of these innate lymphocytes can induce an exaggerated inflammatory response through the substantial production of IFN-γ [45,48]. Thus, dysregulation (including both down- and up-regulation) of these innate lymphocyte functions can be harmful for the host defense [47,48]. In the generalized Shwartzman reaction, these innate lymphocytes also greatly contribute to the lethality of this reaction. The combination of both IFN-γ-producing lymphocytes (in the priming phase) and TNF-producing macrophages (in the subsequent phase) plays a key role in the generalized Shwartzman reaction (Figure 1B).

9. Age-Dependent Increase in Mortality Due to the Generalized Shwartzman Reaction in Mice

Interestingly, the generalized Shwartzman reaction induced by IL-12 priming and subsequent low-dose LPS challenge drastically exacerbates the lethality with increasing age in mice (Figure 2A) [39]. We primed young C57BL/6 mice (4 weeks old) with 0.5 µg/body of IL-12, and, 24 h later, we challenged them with 50 µg/body of LPS [39]. At such doses of IL-12 and LPS, all of the young mice survived the generalized Shwartzman reaction (Figure 2A). However, when older mice (beyond 20 weeks old) were primed/challenged with the same doses of IL-12 and LPS, none survived (Figure 2A), despite showing a mean body weight that was markedly heavier than that of young mice [39]. The serum TNF levels at 1 h after LPS challenge (following IL-12 priming) were also age-dependently increased in the generalized Shwartzman reaction (Figure 2B), indicating TNF-induced mortality, although no age-dependent increase in TNF was observed in the mice challenged with LPS alone.

In contrast, the serum IFN-γ levels were not age-dependently increased in mice. However, interestingly mononuclear cells in the liver (but not spleen) showed significantly increased IFN-γ production by IL-12 priming in middle-aged mice (24 weeks old). NK cells and NKT cells are abundant in the liver and have a potent IFN-γ-producing capability in response to IL-12 priming [46]; we therefore consider that the liver plays crucial roles in the generalized Shwartzman reaction [39,44,46]. However, NK and NKT cells do not increase age-dependently [39]. Regarding age-dependent increases in IFN-γ-producing innate lymphocytes, CD8$^+$CD122$^+$T cells are age-dependently increased and have potent IFN-γ-producing capability, as described later.

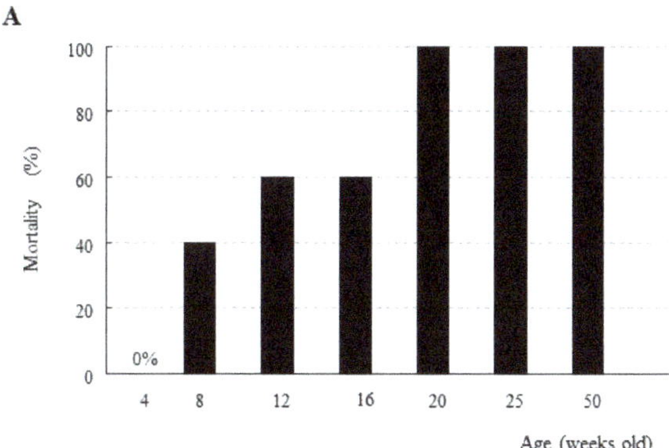

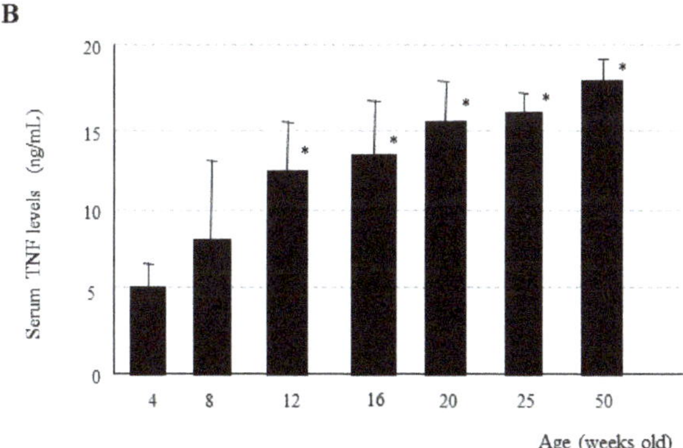

Figure 2. Age-dependent increase in mortality due to the generalized Shwartzman reaction accompanied by elevation of serum TNF. (**A**) Mortality due to the generalized Shwartzman reaction induced by IL-12 priming, and, subsequent LPS challenge was age-dependently increased in mice. (**B**) Serum TNF levels were also age-dependently increased at 1 h after LPS challenge following IL-12 priming. Data are mean (**A**) and means ± SE (**B**) from five mice in each age group. * $p < 0.01$ vs. four weeks old, using an analysis of variance with Scheffe's F test. Figures are revised from [39].

10. Immunosenescence and Thymus-Independent T Cells in Mice

Age-related remodeling of the immune system is termed 'immunosenescence' and profoundly affects changes in the host defense [49,50]. In particular, T cell-mediated immunosenescence may be greatly affected by thymus involution, which drastically curtails the production of thymus-dependent T cells. However, this loss of thymic cell output with aging results in no significant changes to the systemic T cell counts [51]. Reductions in the peripheral T cell numbers in elderly individuals may be compensated by the thymus-independent expansion of mature T cells [46,52]. These thymus-independent T cells have unique characteristics of more primitive lymphocytes, such as intermediate levels of T cell receptor (TCR) expression [46,52,53]. Their TCR intensity is lower than that of regular αβTCR⁺ T cells but higher than that of immature CD4⁺CD8⁺ double-positive thymocytes (TCR dull expression) before selection (Figure 3A) [54] (as described in detail in next paragraph). Though aging also reportedly attenuates the macrophage function [55,56], aged mice showed no marked reduction in cytokine productions [39,57]. It may be possible that age-dependently increased thymus-independent T cells compensates for the reduced macrophage function in the innate immune system in elderly hosts.

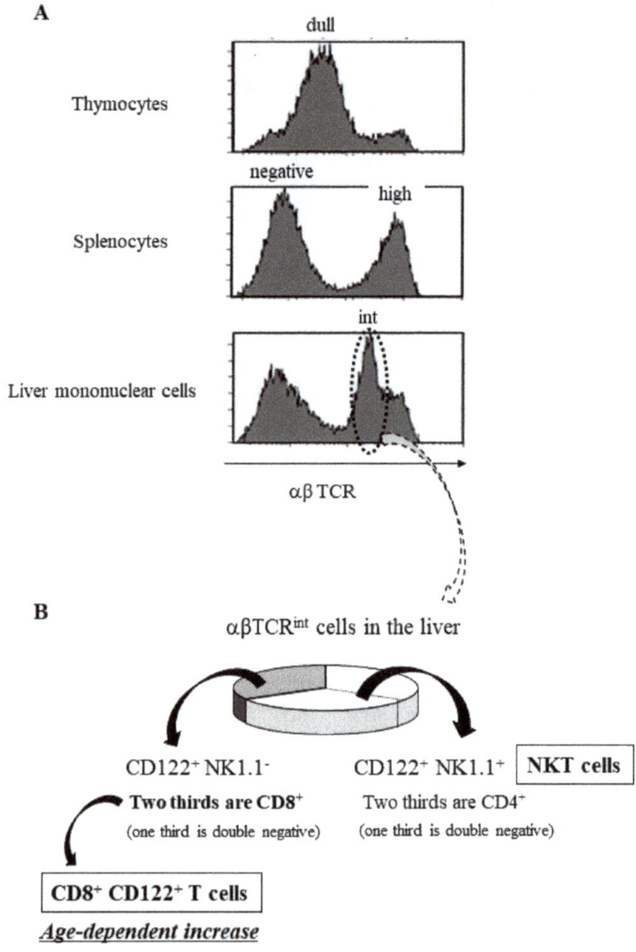

Figure 3. T cell receptor (TCR) intensity in lymphocytes of the thymus, spleen, and liver along with subsets of liver TCR-intermediate cells. (**A**) Immature CD4⁺CD8⁺ double-positive cells show dull TCR intensity in thymocytes, mature T cells show high TCR intensity in the splenocytes, and NKT and CD8⁺CD122⁺T cells show intermediate TCR intensity in the liver mononuclear cells (indicated by dotted circle). (**B**) TCR-intermediate cells in the liver include CD122⁺ NK1.1⁺ cells (NKT cells; two thirds are CD4⁺), and CD122⁺ NK1.1⁻ cells (two thirds are CD8⁺). The cells of this CD8⁺ subset are called CD8⁺CD122⁺T cells and demonstrate an age-dependent increase Figures are revised from [46].

11. Age-Dependent Increases in Murine CD8$^+$CD122$^+$ T Cells

Interestingly, these intermediate TCR-expressing cells are positive for CD122 [46], which is an IL-2 receptor β-chain that contributes to the formation of IL-2 and IL-15 signaling complexes, as both cytokines share this receptor subunit (CD122) [58,59]. CD122 is also highly expressed on NK cells [60,61]. Among CD122$^+$ cells with intermediate TCR (TCR^{int}) in the murine liver, two thirds are NK1.1 marker-positive (NKT cells), and one third are NK1.1-negative (Figure 3B) [62,63]. In these murine NKT cells, two thirds are CD4-positive, but one third are double-negative (CD4$^-$ CD8$^-$) [46,52]. In contrast, in the CD122$^+$NK1.1$^-$ cells, two thirds are CD8-positive, but one third are double-negative (Figure 3B) [46,52]. Interestingly, these CD122$^+$ NK1.1$^-$ CD8$^+$ cells, also known as CD8$^+$CD122$^+$T cells, progressively increase with age and have a potent IFN-γ-producing capability [39,64]. However, there are some caveats, as the antigen-specific memory CD8$^+$T cells with a memory phenotype (CD44^{int}) are also CD122-positive [65], although their CD122 expression is intermediate and distinguished from CD122^{high} NK cells [59]. Antigen-specific memory CD8$^+$T cells (CD44^{int}, CD122^{int}) are generated/primed by various environmental antigens and long-lived [65], while the current CD8$^+$CD122$^+$T cells (CD44^{high}, CD122^{high}) may be generated independently of environmental antigens and are short-lived; these cells are therefore also termed 'memory phenotype' CD8$^+$T cells [66–68]. They are not engaged in chronic responses to environmental antigens but are subject to nonantigen-specific stimulation through contact with cytokines released in response to various stressors, such as bacterial infections and/or LPS [66–68].

12. Crucial Role of CD8$^+$CD122$^+$T Cells in Mortality Due to the Generalized Shwartzman Reaction in Aged Mice

By the generalized Shwartzman reaction in middle-aged mice (24 weeks old), depletion of NK and NKT cells did not reduce the elevation of serum IFN-γ after IL-12 priming, whereas additional depletion of CD8$^+$CD122$^+$T cells to NK/NKT cells (by anti-TMβ1 antibody) markedly reduced the elevation of IFN-γ after IL-12 priming [39]. Consistently, depletion of CD8$^+$CD122$^+$T cells in addition to NK/NKT cells drastically decreased the mortality due to the generalized Shwartzman reaction in middle-aged mice, accompanied by the marked reduction in serum TNF levels after LPS challenge [39]. Adoptive transfer of CD8$^+$CD122$^+$T cells from aged mice markedly increased the mortality due to the generalized Shwartzman reaction in young mice, accompanied by increases in IFN-γ after IL-12 priming and in TNF after subsequent LPS challenge [39]. Thus, age-dependent increases in CD8$^+$CD122$^+$T cells greatly contribute to the high mortality due to the generalized Shwartzman reaction in aged mice (Figure 1C).

13. Beneficial Roles of CD8$^+$CD122$^+$T Cells in Host Defense

To clarify why the numbers of CD8$^+$CD122$^+$T cells in mice increase with age, we pretreated young mice with IL-15, which induces CD8$^+$CD122$^+$T cells [57]. As expected, IL-15-induced CD8$^+$CD122$^+$T cells augmented the generalized Shwartzman reaction, even in young mice, while the depletion of these CD8$^+$CD122$^+$T cells eliminated this harmful effect of IL-15 [57]. Interestingly, IL-15-induced CD8$^+$CD122$^+$T cells also increased the survival after lethal bacterial infection (*Escherichia coli*) in young mice with enhanced IFN-γ production. In addition, IL-15-indcued CD8$^+$CD122$^+$T cells also increased the anti-tumor activity against EL4 cells (murine lymphoma cells) [57]. Consistently, α-galactoceramide, which is a synthetic ligand for NKT cells, induced the proliferation of CD8$^+$CD122$^+$T cells with an anti-tumor function in mouse liver [69]. Thus, age-dependently increased CD8$^+$CD122$^+$T cells may augment antibacterial and anti-tumor immune responses [57]. These beneficial effects of CD8$^+$CD122$^+$T cells on the host defense appear to be rational in aged hosts, as the elderly are susceptible to bacterial infections and tumor progression/invasion. However, there are some questions raised. Gain of CD8$^+$CD122$^+$T cells may be just affected by cytokines such as IL-15, but not aging itself. While IL-15 levels are also increased in the liver of aged mice in comparison to the young mice [57]. Therefore, aging may possibly affect the gain of CD8$^+$CD122$^+$T cells via an increase in IL-15 levels. However, phrase of 'age-dependent immune-alteration' that we used in this review may be termed as 'correlation of immunological status in the hosts at a certain age', because strictly speaking, we just observed the age-correlated alteration of immune system.

14. In Vitro Shwartzman Reaction-Like Response in Human Peripheral Blood Mononuclear Cells (PBMCs)

Is such an age-dependent augmentation of the generalized Shwartzman reaction observed in humans? Five decades ago, Starzl et al. reported three patient cases the suggesting Shwartzman reaction after renal transplantation [70]. However, evaluating the Shwartzman reaction in healthy volunteers is ethically unacceptable. Therefore, little information or evidence of the human Shwartzman reaction has been gathered thus far, although humans are much more sensitive to LPS than animals.

We therefore examined the in vitro Shwartzman reaction-like response using human PBMCs including lymphocytes and macrophages [21]. PBMCs obtained from healthy adult volunteers were cultured with IL-12 (20 µg/mL) for 24 h, followed by culture with LPS (10 ng/mL) for 24 h to examine their TNF production (Figure 4A). CD56$^+$NK cells, CD56$^+$T cells, and CD57$^+$T cells produced IFN-γ in response to IL-12 stimulation, and this IFN-γ primed macrophages to produce large amounts of TNF on subsequent LPS stimulation, suggesting that an in vitro Shwartzman reaction-like response was induced in human PBMCs (Figure 4B) [21]. Nevertheless, we should be careful of the differences in the immune systems between the human and murine, because these immune systems are similar in some parts but not exactly same.

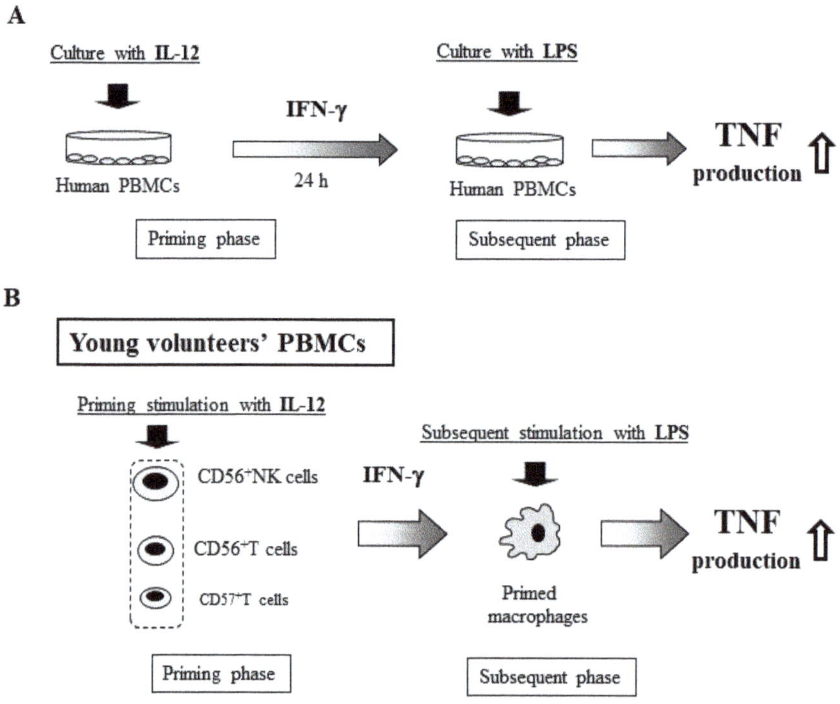

Figure 4. *Cont.*

Figure 4. The in vitro Shwartzman reaction-like response in human peripheral blood mononuclear cells (PBMCs). (**A**) IL-12 stimulation and subsequent LPS stimulation 24 h later induced enhanced TNF production in cultured human PBMCs, suggesting an in vitro Shwartzman reaction-like response. (**B**) In PBMCs from young volunteers, IL-12 priming activates CD56$^+$NK, CD56$^+$T, and CD57$^+$T cells to produce IFN-γ, which primes macrophages. Subsequent LPS stimulation strongly activates the primed macrophages to produce large amounts of TNF, suggesting an in vitro Shwartzman reaction-like response. (**C**) In PBMCs from elderly volunteers, IL-12-priming potently activates CD57$^+$T cells, which show an age-dependent increase, and augments IFN-γ production, which enhances macrophage priming. The subsequent LPS stimulation further augments the TNF production from potently primed macrophages, suggesting an enhancement of the in vitro Shwartzman reaction-like response. Figures are revised from [21].

15. Age-Dependent Augmentation of the In Vitro Shwartzman Reaction-Like Response in Human PBMCs

Interestingly, PBMCs from healthy elderly volunteers (≥70 years old) markedly produced larger amounts of TNF after LPS stimulation following IL-12 stimulation than did PBMCs from younger adult volunteers (20–40 years old) (Figure 5A), suggesting that the in vitro Shwartzman-like reaction was augmented in elderly PMBCs. In contrast, PBMCs from child volunteers showed no enhanced TNF production after IL-12/LPS stimulation (Figure 5A), suggesting that the Shwartzman-like reaction did not occur in child PBMCs [21]. IL-12-induced IFN-γ production from PBMCs was also age-dependently increased. However, there were no marked differences in the TNF production after stimulation with LPS alone (without IL-12 stimulation) among PBMCs from children, adults, and elderly volunteers (Figure 5A). Thus, the age-dependent enhancement of the Shwartzman-like reaction may occur in human PBMCs.

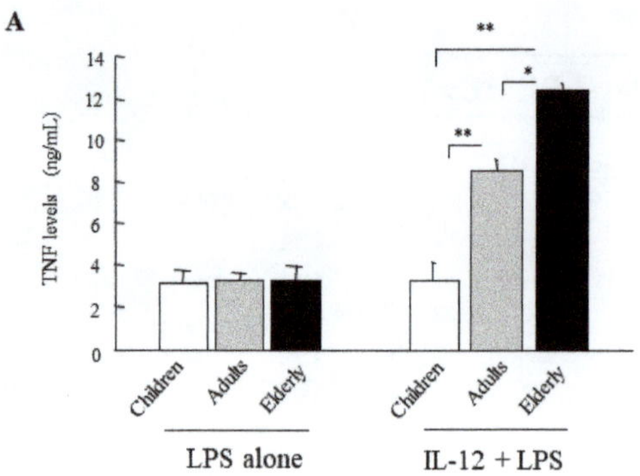

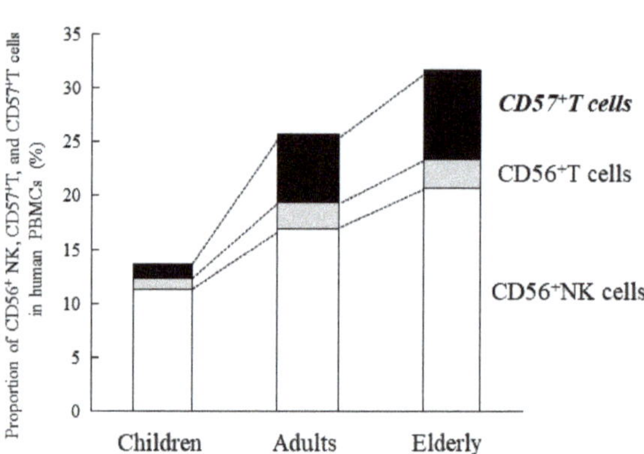

Figure 5. Age-dependent increases in TNF production from human PBMCs by stimulation with IL-12 and LPS and age-dependent increases in CD56⁺NK, CD56⁺T and CD57⁺T cells in human PBMCs. (**A**) Human PBMCs age-dependently increased TNF production by stimulation with LPS following IL-12 stimulation but not by LPS stimulation alone. (**B**) The proportions of CD56⁺NK, CD56⁺T, and CD57⁺T cells were age-dependently increased in human PBMCs. In particular, CD57⁺T cells were markedly increased age-dependently. Data are mean ± SE (**A**) and mean (**B**) from seven, ten, and six healthy volunteers in the child, adult, and elderly groups, respectively. ** $p < 0.01$, * $p < 0.05$, using an analysis of variance with Scheffe's F test. Figures are revised from [21].

The proportion of CD56⁺NK, CD56⁺T, and CD57⁺T cells with potent IFN-γ-producing capabilities also increased with age in human PBMCs (Figure 5B) [21]. In particular, CD57⁺T cells were markedly increased (more than five-fold) in elderly PBMCs compared with child PBMCs (Figure 5B). These age-dependent increases in innate lymphocytes, which have potent IFN-γ-producing capability, may play a key role in the Shwartzman reaction in humans (Figure 4C). Interestingly, both CD56⁺T and CD57⁺T cells show intermediate TCR intensity similar to murine NKT cells and CD8⁺CD122⁺T cells, which are involved in thymus-independent T cells [71].

16. High Mortality and Morbidity of Septic Shock in Elderly Patients Showing Similar Plasma Endotoxin Levels as Adults (Elderly vs. Adults)

We examined the rate of septic shock and its mortality in elderly (≥70 years old) and adult (<70 years old) patients with sepsis/severe sepsis, according to the recent definition [8]. Though both the elderly and adult septic patients showed similar plasma endotoxin levels (7.6 vs. 7.4 pg/mL in average), the elderly patients showed higher serum TNF levels and were more susceptible to septic shock than the younger adult patients, leading to an extremely high mortality due to septic shock compared with younger patients (91% vs. 25%) (Figure 6) [21]. These findings may confirm the characteristic clinical picture in elderly patients with sepsis/septic shock, which have been described before, and are consistent with the experimental results of the age-dependent augmentation of the murine generalized Shwartzman reaction [39] and the in vitro Shwartzman-like reaction using human PBMCs [21].

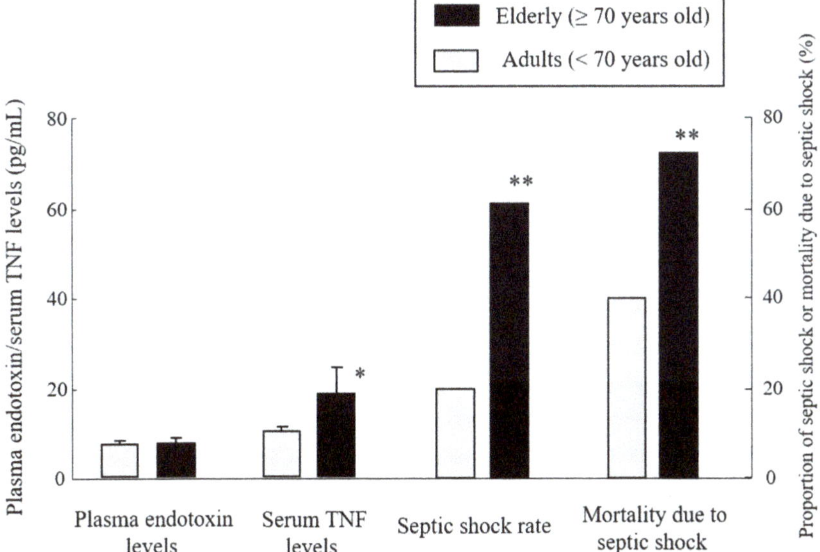

Figure 6. Characteristics of septic shock in elderly patients. Plasma endotoxin levels, serum TNF levels, proportion of septic shock rates, and mortality due to septic shock were compared between elderly and young adult septic patients. Though both patient groups showed similar plasma endotoxin levels, the elderly patients showed markedly higher serum TNF levels and septic shock rates, leading to a high mortality due to septic shock. Data are means ± SE or mean from 18 elderly patients and 20 young adult patients. ** $p < 0.01$, * $p < 0.05$, using the Mann–Whitney U-test and chi-squared test. Figure data are revised from [21].

17. Management of Severe Sepsis/Septic Shock in Elderly Patients: Endotoxin Tolerance as a Potential Therapeutic Strategy in the Future

Elderly patients are susceptible to Gram-negative septic shock with enhanced production of TNF, resulting in a high mortality. When considering effective therapies for elderly patients in such a critical condition, it may be important to reduce the inflammatory cytokine responses, particularly the TNF production by macrophages, and augment the bactericidal activity of these macrophages. However, this seems difficult, as both inflammatory cytokine production and bacterial killing are closely associated with macrophage activation and appear to be synchronized phenomena at a glance. For instance, the induction of CD8$^+$CD122$^+$T cells by IL-15 pretreatment rendered mice susceptible to the generalized Shwartzman reaction (indicating an exaggerated inflammatory response) but resistant to bacterial infection due to deriving an appropriate inflammatory response [57]. Regarding the generalized Shwartzman reaction, aging mainly affects innate lymphocytes and does not directly affect macrophages [21,39].

Interestingly, endotoxin tolerance mainly affects macrophages, drastically reducing the inflammatory cytokine production—particularly TNF production—while strongly augmenting the bactericidal activity [35,40]. These effects can be induced in aged hosts [72,73] and do not directly affect the innate lymphocytes because endotoxin tolerance can be induced in mice even after the elimination of these lymphocytes (our unpublished data) (Figure 7). Though how to clinically induce endotoxin tolerance remains unclear, the mechanisms underlying endotoxin tolerance are quite attractive for the treatment of sepsis in elderly patients, and this therapeutic strategy may become an effective tool for delivering critical care for refractory septic shock in elderly patients.

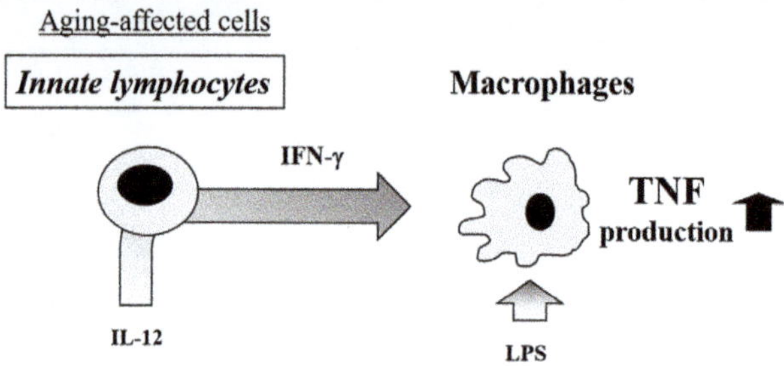

Generalized Shwartzman reaction

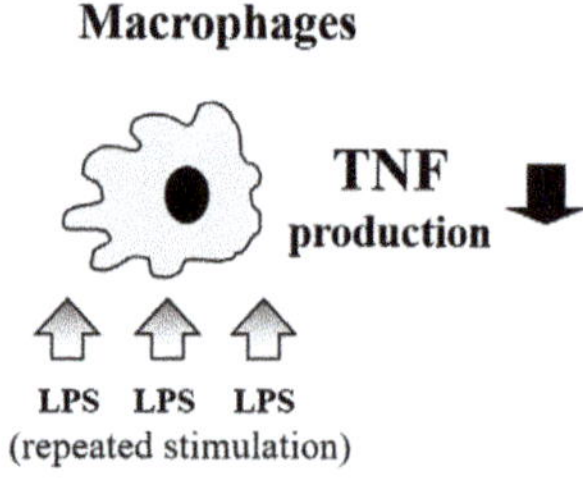

Endotoxin tolerance

Figure 7. Immune mechanisms in the generalized Shwartzman reaction and endotoxin tolerance. In the generalized Shwartzman reaction, innate lymphocytes, which are age-affected cells, are closely involved in the priming phase to stimulate macrophages, leading to an enhanced TNF production. In contrast, endotoxin tolerance directly affects macrophages to induce a reduction in TNF production.

18. Concluding Remarks

Elderly patients are susceptible to septic shock, particularly Gram-negative septic shock (also known as endotoxin shock), and they have an increased associated mortality compared with younger patients. The generalized Shwartzman reaction is well recognized as an experimental endotoxin shock. Aged mice have similarly been shown to be susceptible to the generalized Shwartzman reaction with an increased mortality accompanied by enhanced TNF production. Consistent with this murine model, the in vitro Shwartzman reaction-like response was shown to be age-dependently augmented in human PBMCs, as shown by enhanced TNF production. Interestingly, the age-dependent increase in TCR-intermediate innate lymphocytes, such as CD8$^+$CD122$^+$T cells for mice and CD57$^+$T cells for humans, was found to be closely involved in this age-dependent augmentation of the Shwartzman reaction.

Endotoxin tolerance mainly affects macrophages and drastically reduces their TNF production but strongly activates their bactericidal activity in response to infection. Importantly, endotoxin tolerance can be age-independently induced; therefore, if this tolerant condition can be safely induced in elderly patients, it may be a useful therapeutic strategy for managing refractory septic shock in this population.

Author Contributions: Conceptualization, M.K. and S.S.; investigation, M.K., M.N. and H.N.; data curation, M.K.; writing—original draft preparation, M.K.; writing—review and editing, M.K. and S.S.; visualization, M.K.; supervision, S.S.; funding acquisition, M.K.

Funding: This research was supported in part by the Special Research Grant of National Defense Medical College (M.K.) and JSPS KAKENHI grant No. 25293369 (M.K.).

Conflicts of Interest: The authors declare no conflict of interest.

References

1. Liu, V.; Escobar, G.J.; Greene, J.D.; Soule, J.; Whippy, A.; Angus, D.C.; Iwashyna, T.J. Hospital deaths in patients with sepsis from 2 independent cohorts. *JAMA* **2014**, *312*, 90–92. [CrossRef] [PubMed]
2. Nasa, P.; Juneja, D.; Singh, O. Severe sepsis and septic shock in the elderly: An overview. *World J. Crit. Care Med.* **2012**, *1*, 23–30. [CrossRef] [PubMed]
3. Dellinger, R.P.; Levy, M.M.; Rhodes, A.; Annane, D.; Gerlach, H.; Opal, S.M.; Sevransky, J.E.; Sprung, C.L.; Douglas, I.S.; Jaeschke, R.; et al. Surviving Sepsis Campaign: International guidelines for management of severe sepsis and septic shock, 2012. *Intensive Care Med.* **2013**, *39*, 165–228. [CrossRef] [PubMed]
4. Huang, C.T.; Tsai, Y.J.; Tsai, P.R.; Yu, C.J.; Ko, W.J. Severe Sepsis and Septic Shock: Timing of Septic Shock Onset Matters. *Shock* **2016**, *45*, 518–524. [CrossRef] [PubMed]
5. American College of Chest Physicians; Society of Critical Care Medicine Consensus Conference Committee. American College of Chest Physicians/Society of Critical Care Medicine Consensus Conference: Definitions for sepsis and organ failure and guidelines for the use of innovative therapies in sepsis. *Crit. Care Med.* **1992**, *20*, 864–874.
6. Bone, R.C.; Sibbald, W.J.; Sprung, C.L. The ACCP-SCCM consensus conference on sepsis and organ failure. *Chest* **1992**, *101*, 1481–1483. [CrossRef]
7. Levy, M.M.; Fink, M.P.; Marshall, J.C.; Abraham, E.; Angus, D.; Cook, D.; Cohen, J.; Opal, S.M.; Vincent, J.L.; Ramsay, G. 2001 SCCM/ESICM/ACCP/ATS/SIS International Sepsis Definitions Conference. *Intensive Care Med.* **2003**, *29*, 530–538. [CrossRef] [PubMed]
8. Singer, M.; Deutschman, C.S.; Seymour, C.W.; Shankar-Hari, M.; Annane, D.; Bauer, M.; Bellomo, R.; Bernard, G.R.; Chiche, J.D.; Coopersmith, C.M.; et al. The Third International Consensus Definitions for Sepsis and Septic Shock (Sepsis-3). *JAMA* **2016**, *315*, 801–810. [CrossRef]
9. Angus, D.C.; van der Poll, T. Severe sepsis and septic shock. *N. Engl. J. Med.* **2013**, *369*, 840–851. [CrossRef]
10. Sherwin, R.; Winters, M.E.; Vilke, G.M.; Wardi, G. Does Early and Appropriate Antibiotic Administration Improve Mortality in Emergency Department Patients with Severe Sepsis or Septic Shock? *J. Emerg. Med.* **2017**, *53*, 588–595. [CrossRef]
11. Banjas, N.; Hopf, H.B.; Hanisch, E.; Friedrichson, B.; Fichte, J.; Buia, A. ECMO-treatment in patients with acute lung failure, cardiogenic, and septic shock: Mortality and ECMO-learning curve over a 6-year period. *J. Intensive Care* **2018**, *6*, 84. [CrossRef] [PubMed]
12. Ueno, T.; Ikeda, T.; Yokoyama, T.; Kihara, Y.; Konno, O.; Nakamura, Y.; Iwamoto, H.; Shimizu, T.; McGrath, M.M.; Chandraker, A. Reduction in circulating level of HMGB-1 following continuous renal replacement therapy in sepsis. *Cytokine* **2016**, *83*, 206–209. [CrossRef] [PubMed]
13. Yealy, D.M.; Kellum, J.A.; Huang, D.T.; Barnato, A.E.; Weissfeld, L.A.; Pike, F.; Terndrup, T.; Wang, H.E.; Hou, P.C.; LoVecchio, F.; et al. A randomized trial of protocol-based care for early septic shock. *N. Engl. J. Med.* **2014**, *370*, 1683–1693. [CrossRef] [PubMed]
14. Peake, S.L.; Delaney, A.; Bailey, M.; Bellomo, R.; Cameron, P.A.; Cooper, D.J.; Higgins, A.M.; Holdgate, A.; Howe, B.D.; Webb, S.A.; et al. Goal-directed resuscitation for patients with early septic shock. *N. Engl. J. Med.* **2014**, *371*, 1496–1506. [CrossRef] [PubMed]
15. Mouncey, P.R.; Osborn, T.M.; Power, G.S.; Harrison, D.A.; Sadique, M.Z.; Grieve, R.D.; Jahan, R.; Harvey, S.E.; Bell, D.; Bion, J.F.; et al. Trial of early, goal-directed resuscitation for septic shock. *N. Engl. J. Med.* **2015**, *372*, 1301–1311. [CrossRef]
16. Abe, R.; Oda, S.; Sadahiro, T.; Nakamura, M.; Hirayama, Y.; Tateishi, Y.; Shinozaki, K.; Hirasawa, H. Gram-negative bacteremia induces greater magnitude of inflammatory response than Gram-positive bacteremia. *Crit. Care* **2010**, *14*, R27. [CrossRef]

17. Sligl, W.I.; Dragan, T.; Smith, S.W. Nosocomial Gram-negative bacteremia in intensive care: Epidemiology, antimicrobial susceptibilities, and outcomes. *Int. J. Infect. Dis.* **2015**, *37*, 129–134. [CrossRef]

18. Saito, N.; Sugiyama, K.; Ohnuma, T.; Kanemura, T.; Nasu, M.; Yoshidomi, Y.; Tsujimoto, Y.; Adachi, H.; Koami, H.; Tochiki, A.; et al. Efficacy of polymyxin B-immobilized fiber hemoperfusion for patients with septic shock caused by Gram-negative bacillus infection. *PLoS ONE* **2017**, *12*, e0173633. [CrossRef]

19. Valenti, W.M.; Trudell, R.G.; Bentley, D.W. Factors predisposing to oropharyngeal colonization with gram-negative bacilli in the aged. *N. Engl. J. Med.* **1978**, *298*, 1108–1111. [CrossRef]

20. Martin, G.S.; Mannino, D.M.; Moss, M. The effect of age on the development and outcome of adult sepsis. *Crit. Care Med.* **2006**, *34*, 15–21. [CrossRef]

21. Motegi, A.; Kinoshita, M.; Sato, K.; Shinomiya, N.; Ono, S.; Nonoyama, S.; Hiraide, H.; Seki, S. An In Vitro Shwartzman reaction-like response is augmented age-dependently in human peripheral blood mononuclear cells. *J. Leukoc. Biol.* **2006**, *79*, 463–472. [CrossRef] [PubMed]

22. Opal, S.M.; Scannon, P.J.; Vincent, J.L.; White, M.; Carroll, S.F.; Palardy, J.E.; Parejo, N.A.; Pribble, J.P.; Lemke, J.H. Relationship between plasma levels of lipopolysaccharide (LPS) and LPS-binding protein in patients with severe sepsis and septic shock. *J. Infect. Dis.* **1999**, *180*, 1584–1589. [CrossRef] [PubMed]

23. Munford, R.S. Endotoxemia-menace, marker, or mistake? *J. Leukoc. Biol.* **2016**, *100*, 687–698. [CrossRef] [PubMed]

24. Klein, D.J.; Derzko, A.; Foster, D.; Seely, A.J.; Brunet, F.; Romaschin, A.D.; Marshall, J.C. Daily variation in endotoxin levels is associated with increased organ failure in critically ill patients. *Shock* **2007**, *28*, 524–529. [CrossRef]

25. Laichalk, L.L.; Bucknell, K.A.; Huffnagle, G.B.; Wilkowski, J.M.; Moore, T.A.; Romanelli, R.J.; Standiford, T.J. Intrapulmonary delivery of tumor necrosis factor agonist peptide augments host defense in murine gram-negative bacterial pneumonia. *Infect. Immun.* **1998**, *66*, 2822–2826. [PubMed]

26. Havell, E.A. Evidence that tumor necrosis factor has an important role in antibacterial resistance. *J. Immunol.* **1989**, *143*, 2894–2899.

27. Cauwels, A.; Brouckaert, P. Survival of TNF toxicity: Dependence on caspases and NO. *Arch. Biochem. Biophys.* **2007**, *462*, 132–139. [CrossRef]

28. Brenner, D.; Blaser, H.; Mak, T.W. Regulation of tumour necrosis factor signalling: Live or let die. *Nat. Rev. Immunol.* **2015**, *15*, 362–374. [CrossRef]

29. Oberholzer, A.; Souza, S.M.; Tschoeke, S.K.; Oberholzer, C.; Abouhamze, A.; Pribble, J.P.; Moldawer, L.L. Plasma cytokine measurements augment prognostic scores as indicators of outcome in patients with severe sepsis. *Shock* **2005**, *23*, 488–493.

30. Rice, T.W.; Wheeler, A.P.; Morris, P.E.; Paz, H.L.; Russell, J.A.; Edens, T.R.; Bernard, G.R. Safety and efficacy of affinity-purified, anti-tumor necrosis factor-alpha, ovine fab for injection (CytoFab) in severe sepsis. *Crit. Care Med.* **2006**, *34*, 2271–2281. [CrossRef]

31. Michie, H.R.; Manogue, K.R.; Spriggs, D.R.; Revhaug, A.; O'Dwyer, S.; Dinarello, C.A.; Cerami, A.; Wolff, S.M.; Wilmore, D.W. Detection of circulating tumor necrosis factor after endotoxin administration. *N. Engl. J. Med.* **1988**, *318*, 1481–1486. [CrossRef] [PubMed]

32. Hesse, D.G.; Tracey, K.J.; Fong, Y.; Manogue, K.R.; Palladino, M.A., Jr.; Cerami, A.; Shires, G.T.; Lowry, S.F. Cytokine appearance in human endotoxemia and primate bacteremia. *Surg. Gynecol. Obstet.* **1988**, *166*, 147–153. [PubMed]

33. Van Deventer, S.J.; Buller, H.R.; ten Cate, J.W.; Aarden, L.A.; Hack, C.E.; Sturk, A. Experimental endotoxemia in humans: Analysis of cytokine release and coagulation, fibrinolytic, and complement pathways. *Blood* **1990**, *76*, 2520–2526. [PubMed]

34. Van der Poll, T.; Buller, H.R.; ten Cate, H.; Wortel, C.H.; Bauer, K.A.; van Deventer, S.J.; Hack, C.E.; Sauerwein, H.P.; Rosenberg, R.D.; ten Cate, J.W. Activation of coagulation after administration of tumor necrosis factor to normal subjects. *N. Engl. J. Med.* **1990**, *322*, 1622–1627. [CrossRef] [PubMed]

35. Kinoshita, M.; Miyazaki, H.; Nakashima, H.; Nakashima, M.; Nishikawa, M.; Ishikiriyama, T.; Kato, S.; Iwaya, K.; Hiroi, S.; Shinomiya, N.; et al. In Vivo Lipopolysaccharide Tolerance Recruits CD11b+ Macrophages to the Liver with Enhanced Bactericidal Activity and Low Tumor Necrosis Factor-Releasing Capability, Resulting in Drastic Resistance to Lethal Septicemia. *J. Innate Immun.* **2017**, *9*, 493–510. [CrossRef] [PubMed]

36. Chahin, A.B.; Opal, J.M.; Opal, S.M. Whatever happened to the Shwartzman phenomenon? *Innate Immun.* **2018**, *24*, 466–479. [CrossRef] [PubMed]

37. Shwartzman, G. Studies on bacillus typhosus toxic substances: I. Phenomenon of local skin reactivity to *B. typhosus* culture filtrate. *J. Exp. Med.* **1928**, *48*, 247–268. [CrossRef] [PubMed]

38. Thomas, L.; Good, R.A. Studies on the generalized Shwartzman reaction: I. General observations concerning the phenomenon. *J. Exp. Med.* **1952**, *96*, 605–624. [CrossRef]

39. Sato, K.; Kinoshita, M.; Motegi, A.; Habu, Y.; Takayama, E.; Nonoyama, S.; Hiraide, H.; Seki, S. Critical role of the liver CD8$^+$ CD122$^+$ T cells in the generalized Shwartzman reaction of mice. *Eur. J. Immunol.* **2005**, *35*, 593–602. [CrossRef]

40. Biswas, S.K.; Lopez-Collazo, E. Endotoxin tolerance: New mechanisms, molecules and clinical significance. *Trends Immunol.* **2009**, *30*, 475–487. [CrossRef]

41. Ozmen, L.; Pericin, M.; Hakimi, J.; Chizzonite, R.A.; Wysocka, M.; Trinchieri, G.; Gately, M.; Garotta, G. Interleukin 12, interferon gamma, and tumor necrosis factor alpha are the key cytokines of the generalized Shwartzman reaction. *J. Exp. Med.* **1994**, *180*, 907–915. [CrossRef] [PubMed]

42. Trinchieri, G. Interleukin-12: A proinflammatory cytokine with immunoregulatory functions that bridge innate resistance and antigen-specific adaptive immunity. *Annu. Rev. Immunol.* **1995**, *13*, 251–276. [CrossRef] [PubMed]

43. Tait Wojno, E.D.; Hunter, C.A.; Stumhofer, J.S. The Immunobiology of the Interleukin-12 Family: Room for Discovery. *Immunity* **2019**, *50*, 851–870. [CrossRef] [PubMed]

44. Ogasawara, K.; Takeda, K.; Hashimoto, W.; Satoh, M.; Okuyama, R.; Yanai, N.; Obinata, M.; Kumagai, K.; Takada, H.; Hiraide, H.; et al. Involvement of NK1+ T cells and their IFN-gamma production in the generalized Shwartzman reaction. *J. Immunol.* **1998**, *160*, 3522–3527. [PubMed]

45. Kinoshita, M.; Seki, S.; Ono, S.; Shinomiya, N.; Hiraide, H. Paradoxical effect of IL-18 therapy on the severe and mild Escherichia coli infections in burn-injured mice. *Ann. Surg.* **2004**, *240*, 313–320. [CrossRef] [PubMed]

46. Seki, S.; Habu, Y.; Kawamura, T.; Takeda, K.; Dobashi, H.; Ohkawa, T.; Hiraide, H. The liver as a crucial organ in the first line of host defense: The roles of Kupffer cells, natural killer (NK) cells and NK1.1 Ag+ T cells in T helper 1 immune responses. *Immunol. Rev.* **2000**, *174*, 35–46. [CrossRef] [PubMed]

47. Eberl, G.; Colonna, M.; Di Santo, J.P.; McKenzie, A.N. Innate lymphoid cells. Innate lymphoid cells: A new paradigm in immunology. *Science* **2015**, *348*, aaa6566. [CrossRef] [PubMed]

48. Kinoshita, M.; Miyazaki, H.; Ono, S.; Seki, S. Immunoenhancing therapy with interleukin-18 against bacterial infection in immunocompromised hosts after severe surgical stress. *J. Leukoc. Biol.* **2013**, *93*, 689–698. [CrossRef] [PubMed]

49. Cambier, J. Immunosenescence: A problem of lymphopoiesis, homeostasis, microenvironment, and signaling. *Immunol. Rev.* **2005**, *205*, 5–6. [CrossRef]

50. Miller, R.A.; Berger, S.B.; Burke, D.T.; Galecki, A.; Garcia, G.G.; Harper, J.M.; Sadighi Akha, A.A. T cells in aging mice: Genetic, developmental, and biochemical analyses. *Immunol. Rev.* **2005**, *205*, 94–103. [CrossRef]

51. Berzins, S.P.; Boyd, R.L.; Miller, J.F. The role of the thymus and recent thymic migrants in the maintenance of the adult peripheral lymphocyte pool. *J. Exp. Med.* **1998**, *187*, 1839–1848. [CrossRef] [PubMed]

52. Abo, T. Extrathymic pathways of T cell differentiation. *Arch. Immunol. Ther. Exp.* **2001**, *49*, 81–90.

53. Sato, K.; Ohtsuka, K.; Hasegawa, K.; Yamagiwa, S.; Watanabe, H.; Asakura, H.; Abo, T. Evidence for extrathymic generation of intermediate T cell receptor cells in the liver revealed in thymectomized, irradiated mice subjected to bone marrow transplantation. *J. Exp. Med.* **1995**, *182*, 759–767. [CrossRef] [PubMed]

54. Seki, S.; Abo, T.; Ohteki, T.; Sugiura, K.; Kumagai, K. Unusual alpha beta-T cells expanded in autoimmune lpr mice are probably a counterpart of normal T cells in the liver. *J. Immunol.* **1991**, *147*, 1214–1221. [PubMed]

55. Fei, F.; Lee, K.M.; McCarry, B.E.; Bowdish, D.M. Age-associated metabolic dysregulation in bone marrow-derived macrophages stimulated with lipopolysaccharide. *Sci. Rep.* **2016**, *6*, 22637. [CrossRef]

56. Franceschi, C.; Bonafe, M.; Valensin, S.; Olivieri, F.; De Luca, M.; Ottaviani, E.; De Benedictis, G. Inflamm-aging. An evolutionary perspective on immunosenescence. *Ann. N. Y. Acad. Sci.* **2000**, *908*, 244–254. [CrossRef] [PubMed]

57. Motegi, A.; Kinoshita, M.; Inatsu, A.; Habu, Y.; Saitoh, D.; Seki, S. IL-15-induced CD8$^+$ CD122$^+$ T cells increase antibacterial and anti-tumor immune responses: Implications for immune function in aged mice. *J. Leukoc. Biol.* **2008**, *84*, 1047–1056. [CrossRef]

58. Ikemizu, S.; Chirifu, M.; Davis, S.J. IL-2 and IL-15 signaling complexes: Different but the same. *Nat. Immunol.* **2012**, *13*, 1141–1142. [CrossRef]

59. Waldmann, T.A. The biology of interleukin-2 and interleukin-15: Implications for cancer therapy and vaccine design. *Nat. Rev. Immunol.* **2006**, *6*, 595–601. [CrossRef]

60. Rosmaraki, E.E.; Douagi, I.; Roth, C.; Colucci, F.; Cumano, A.; Di Santo, J.P. Identification of committed NK cell progenitors in adult murine bone marrow. *Eur. J. Immunol.* **2001**, *31*, 1900–1909. [CrossRef]

61. Yoshizawa, K.; Nakajima, S.; Notake, T.; Miyagawa, S.; Hida, S.; Taki, S. IL-15-high-responder developing NK cells bearing Ly49 receptors in IL-15-/-mice. *J. Immunol.* **2011**, *187*, 5162–5169. [CrossRef] [PubMed]

62. Watanabe, H.; Miyaji, C.; Kawachi, Y.; Iiai, T.; Ohtsuka, K.; Iwanage, T.; Takahashi-Iwanaga, H.; Abo, T. Relationships between intermediate TCR cells and NK1.1+ T cells in various immune organs. NK1.1+ T cells are present within a population of intermediate TCR cells. *J. Immunol.* **1995**, *155*, 2972–2983. [PubMed]

63. Tsukahara, A.; Seki, S.; Iiai, T.; Moroda, T.; Watanabe, H.; Suzuki, S.; Tada, T.; Hiraide, H.; Hatakeyama, K.; Abo, T. Mouse liver T cells: Their change with aging and in comparison with peripheral T cells. *Hepatology* **1997**, *26*, 301–309. [CrossRef] [PubMed]

64. Takayama, E.; Seki, S.; Ohkawa, T.; Ami, K.; Habu, Y.; Yamaguchi, T.; Tadakuma, T.; Hiraide, H. Mouse CD8$^+$ CD122$^+$ T cells with intermediate TCR increasing with age provide a source of early IFN-gamma production. *J. Immunol.* **2000**, *164*, 5652–5658. [CrossRef] [PubMed]

65. Mathews, D.V.; Dong, Y.; Higginbotham, L.B.; Kim, S.C.; Breeden, C.P.; Stobert, E.A.; Jenkins, J.; Tso, J.Y.; Larsen, C.P.; Adams, A.B. CD122 signaling in CD8$^+$ memory T cells drives costimulation-independent rejection. *J. Clin. Investig.* **2018**, *128*, 4557–4572. [CrossRef]

66. Judge, A.D.; Zhang, X.; Fujii, H.; Surh, C.D.; Sprent, J. Interleukin 15 controls both proliferation and survival of a subset of memory-phenotype CD8(+) T cells. *J. Exp. Med.* **2002**, *196*, 935–946. [CrossRef]

67. Yamada, H.; Matsuzaki, G.; Chen, Q.; Iwamoto, Y.; Nomoto, K. Reevaluation of the origin of CD44(high) "memory phenotype" CD8 T cells: Comparison between memory CD8 T cells and thymus-independent CD8 T cells. *Eur. J. Immunol.* **2001**, *31*, 1917–1926. [CrossRef]

68. Zhang, X.; Sun, S.; Hwang, I.; Tough, D.F.; Sprent, J. Potent and selective stimulation of memory-phenotype CD8+ T cells in vivo by IL-15. *Immunity* **1998**, *8*, 591–599. [CrossRef]

69. Nakagawa, R.; Inui, T.; Nagafune, I.; Tazunoki, Y.; Motoki, K.; Yamauchi, A.; Hirashima, M.; Habu, Y.; Nakashima, H.; Seki, S. Essential role of bystander cytotoxic CD122$^+$CD8$^+$ T cells for the antitumor immunity induced in the liver of mice by alpha-galactosylceramide. *J. Immunol.* **2004**, *172*, 6550–6557. [CrossRef]

70. Starzl, T.E.; Lerner, R.A.; Dixon, F.J.; Groth, C.G.; Brettschneider, L.; Terasaki, P.I. Shwartzman reaction after human renal homotransplantation. *N. Engl. J. Med.* **1968**, *278*, 642–648. [CrossRef]

71. Takayama, E.; Koike, Y.; Ohkawa, T.; Majima, T.; Fukasawa, M.; Shinomiya, N.; Yamaguchi, T.; Konishi, M.; Hiraide, H.; Tadakuma, T.; et al. Functional and Vbeta repertoire characterization of human CD8$^+$ T-cell subsets with natural killer cell markers, CD56$^+$ CD57$^-$ T cells, CD56$^+$ CD57$^+$ T cells and CD56$^-$ CD57$^+$ T cells. *Immunology* **2003**, *108*, 211–219. [CrossRef] [PubMed]

72. Sun, Y.; Li, H.; Yang, M.F.; Shu, W.; Sun, M.J.; Xu, Y. Effects of aging on endotoxin tolerance induced by lipopolysaccharides derived from *Porphyromonas gingivalis* and *Escherichia coli*. *PLoS ONE* **2012**, *7*, e39224. [CrossRef] [PubMed]

73. Zhang, Z.; Ji, M.; Liao, Y.; Yang, J.; Gao, J. Endotoxin tolerance induced by lipopolysaccharide preconditioning protects against surgeryinduced cognitive impairment in aging mice. *Mol. Med. Rep.* **2018**, *17*, 3845–3852. [CrossRef] [PubMed]

International Journal of
Molecular Sciences

Article

Biophysical Analysis of Lipopolysaccharide Formulations for an Understanding of the Low Endotoxin Recovery (LER) Phenomenon

Wilmar Correa [1], Klaus Brandenburg [1,2,*], Ulrich Zähringer [1], Kishore Ravuri [3], Tarik Khan [3] and Friedrich von Wintzingerode [4]

[1] Forschungszentrum Borstel, Leibniz-Zentrum für Medizin und Biowissenschaften, Parkallee 1-40, D-23845 Borstel, Germany; wcorrea@fz-borstel.de (W.C.); uzaehr@web.de (U.Z.)
[2] Brandenburg Antiinfektiva GmbH, Parkallee 10b, D-23845 Borstel, Germany
[3] Pharmaceutical Development & Supplies, F. Hoffmann-La Roche Ltd., 4070 Basel, Switzerland; satya_krishna_kishore.ravuri@roche.com (K.R.); tarik.khan@roche.com (T.K.)
[4] Roche Diagnostics GmbH, Nonnenwald 2, Roche Diagnostics GmbH, 82377 Penzberg, Germany; friedrich.von_wintzingerode@roche.com
* Correspondence: kbrandenburg@fz-borstel.de; Tel.: +49-4537-1882350

Received: 29 November 2017; Accepted: 13 December 2017; Published: 16 December 2017

Abstract: Lipopolysaccharides (LPS, endotoxin) are complex and indispensable components of the outer membrane of most Gram-negative bacteria. They represent stimuli for many biological effects with pathophysiological character. Recombinant therapeutic proteins that are manufactured using biotechnological processes are prone to LPS contaminations due to their ubiquitous occurrence. The maximum endotoxin load of recombinant therapeutic proteins must be below the pyrogenic threshold. Certain matrices that are commonly used for recombinant therapeutic proteins show a phenomenon called "Low Endotoxin Recovery (LER)". LER is defined as the loss of detectable endotoxin activity over time using compendial *Limulus* amebocyte lysate (LAL) assays when undiluted products are spiked with known amount of endotoxin standards. Because LER poses potential risks that endotoxin contaminations in products may be underestimated or undetected by the LAL assay, the United States (U.S.) Food and Drug Administration's (FDA's) Center for Drug Evaluation and Research (CDER) has recently started requesting that companies conduct endotoxin spike/hold recovery studies to determine whether a given biological product causes LER. Here, we have performed an analysis of different LPS preparations with relevant detergents studying their acyl chain phase transition, their aggregate structures, their size distributions, and binding affinity with a particular anti-endotoxin peptide, and correlating it with the respective data in the macrophage activation test. In this way, we have worked out biophysical parameters that are important for an understanding of LER.

Keywords: endotoxin; lipopolysaccharide; Low Endotoxin Recovery; phase transitions; polysorbate; LPS aggregates; Small Angle X-ray Scattering; MAT; LAL and LER

---

1. Introduction

Lipopolysaccharides (LPS), the endotoxins of most Gram-negative bacteria, belong to the strongest immune-stimulating compounds known in nature. This property may be beneficial at low concentrations, but pathophysiological at high concentrations, leading to severe sepsis and septic shock with high lethality [1]. Since LPS is a constituent of nearly all Gram-negative bacteria, it is a ubiquitous contaminant. It is well-known that the lipid A part of LPS is its "endotoxic principle", which for most relevant bacterial species, such as *Escherichia coli* and *Salmonella* spp. consists of a bisphosphorylated diglucosamine backbone to which six acyl chain residues are linked [1]. There are also lipid A with underacylated lipid A parts (tetra or pentaacyl) with low biological activity [1], which, however, are not relevant in the context of this investigation.

LPS is able to elicit severe safety risks even at very low concentrations. Thus, in clinical studies of Opal and co-workers, it was found that sepsis patients belonging to the survivors had a medium LPS serum concentration of 0.3 ng/mL and the non-survivors 0.7 ng/mL [2]. The reason for this is the fact that LPS induces a "cytokine storm" (interleukins, tumor-necrosis-factor-α (TNF-α) and many others), leading to a septic shock. Therefore, it is of uttermost importance to control LPS-load, in particular, in parenteral pharmaceutical formulations, such as recombinant therapeutic proteins, which are anticipated to be injected.

Low Endotoxin Recovery (LER) poses potential risks that endotoxin contaminations in products may be underestimated or undetected by the *Limulus* amebocyte lysate (LAL) assay.

There are various publications in recent years dealing with this subject, in most cases giving no coherent explanation for the occurrence of the effect [3,4]. As one possible explanation, it was proposed that the presence of certain buffers and detergents, in most cases citrate and polysorbate 20 and 80, leads to a drastic disaggregation of LPS down to a monomeric form, which was found to represent an inactive form in the LAL assay, as well as the macrophage activation test (MAT) [5]. Since there are also papers that are indicative of an active monomeric form of LPS [6], and since the assumption of a monomeric form induced by the detergents could not be verified directly to date, we had a closer biophysical look on the phenomenon. Regarding the biological techniques to prove the presence of LPS, usually the *Limulus* amebocyte lysate (LAL) assays in different modifications (gel clot, chromogenic, turbidimetric) and the recombinant Factor C assay are used, which are all based on the interaction of factor C of the Limulus cascade with LPS [7]. An alternative method is the determination of the LPS-induced stimulation of human cells, such as mononuclear cells (MNCs, monocytes, or macrophages (generally called MAT test)) [8]. It must be noted that both techniques suffer from disadvantages: For activating the LAL test, the structural presence of only a LPS part structure is necessary, i.e., the 4′-phosphate diglucosamine backbone of lipid A [8] in acylated form. Therefore, a LAL signal is already seen with underacylated lipid A structures (tri-, tetra-, and pentaacyl groups), which, in the human immune system, do not or only to a small degree elicit an inflammation reaction [9]. Furthermore, the LAL assay can also be activated by β-D-glucans. The disadvantage of the MAT lies in the fact that it reacts to all of the bacterial immune-stimulating toxins, i.e., also to those from Gram-positive origin, which has been shown to result mainly from lipoproteins and/or their shortened lipopeptide variants [10].

In the present work, we have applied a variety of biophysical techniques to gain more insight into the LER phenomenon and to characterize a possible influence of different detergent formulations on (i) the gel to liquid crystalline phase transition behavior of different LPS, (ii) their aggregate structure, and (iii) aggregate sizes. Furthermore, the action of a recently well-described antimicrobial peptide Aspidasept® [11,12] on the binding to LPS in different formulations was studied.

These data were directly correlated to their activity in the MAT. In this way, we should be able to better understand the influences of the different parameters on LER.

2. Results and Interpretations

2.1. Gel to Liquid Crystalline Phase Transition of the Acyl Chains

All amphiphilic compounds, such as LPS, can adopt two states of order of the acyl chains, one highly ordered (gel) with relatively rigid chains at lower temperatures, and one unordered (liquid-like) with highly fluid chains at higher temperatures, Figure 1A. Typically, the phase transition temperature T_c of enterobacterial LPS is around 30 to 37 °C, i.e., close to the physiological temperature [13]. For the different smooth form LPS, however, due to the heterogeneous LPS mixtures, differing in the degree of acylation and length of the saccharide chains, frequently lower

T_c values may be observed. This parameter might be of importance for the ability of the compounds to interact with target structures. Fourier-infrared spectroscopy (FTIR) is the method of choice, by monitoring the symmetric stretching vibrational band at 2850–2853 cm^{-1} of the methylene groups, with the former value characteristic for the gel and the latter for the liquid crystalline phase. In the following, selected commonly used LPS from wild-type strains from *E. coli* O55:B5 and *E. coli* O111:B4 (Sigma, Deisenhofen, Germany), LPS Rb and Rd mutants from *Salmonella minnesota* R345 and *Salmonella minnesota* R7 (own purified samples), respectively, in different buffers and detergents were analysed. The data are shown in Figure 1B–F. As can be seen, the values of T_c are sensitively dependent on the different formulations, with values of approximately 17 °C for LPS O111:B4 in buffer, and increasing values up to 27 °C for the polysorbate 80 (at concentrations below the critical micellar concentration (CMC)) preparation. Interestingly, when the polysorbate concentrations are increased to values higher than the CMC, then the transition values considerably decrease (Figure 1B).

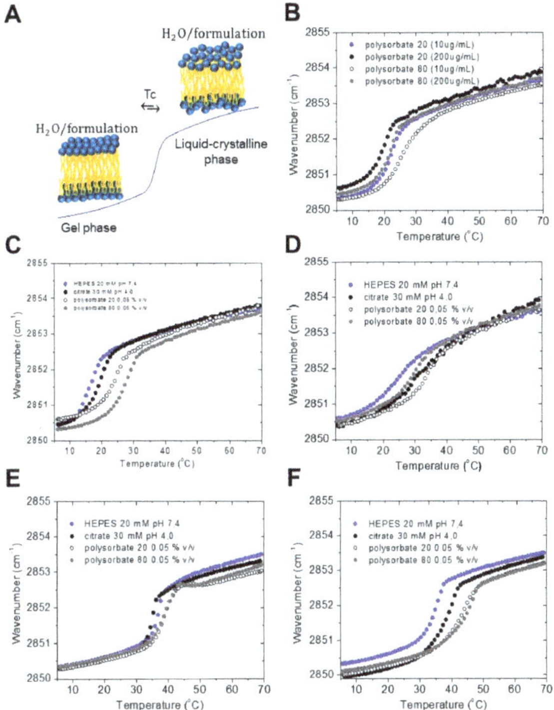

Figure 1. Effect of the formulation on the gel to liquid crystalline phase transition of lipopolysaccharides (LPS). (**A**) Schematic representation of the gel to liquid-crystalline phase transition of a lipid bilayer. Phase transitions of (**B**) LPS from *E. coli* O111:B4 in two different concentration of polysorbate 20 and 80; (**C**) LPS from *E. coli* O111:B4 in different formulations; (**D**) LPS from *E. coli* O55:B5 in different formulations; (**E**) LPS from *S. minnesota* Rb mutant strain R345 in different formulations; and, (**F**) LPS from *S. minnesota* Rd-mutant strain R7 in different formulations.

The measurements for LPS O55:B5 show a very broad phase transition range, which is indicative of a very heterogeneous mixture of this smooth form LPS (Figure 1C). Again, the transition value is lowest (appr. 25 °C) for the buffer system, but in this case, highest for the polysorbate 20 sample. In Figure 1B, the phase transition behavior is measured at two polysorbate concentration (10 and 200 μg/mL), representing values below and above their respective critical micellar concentration (CMC). It can be seen that there is a decrease of the T_c value at the higher polysorbate concentrations. Furthermore, there is an increase of the T_c value in the sequence citrate, polysorbate 20, and polysorbate 80.

Since it has been found that the bioactive form within the heterogeneous wild-type strains corresponds to R-mutants (Ra- or Rb-mutants as found in [14]), two of them were also investigated. The data for the Rb-mutant LPS from *S. minnesota* (Figure 1E) clearly show a much smaller and sharp transition range, according to the fact that this LPS is homogenous and pure. The T_c-values are lowest for the citrate and are highest for the polysorbate formulations, and corresponds to previous findings with values around 35 to 37 °C [15]. The results for a LPS Rd from *S. minnesota* strain R7, which has the lowest T_c (ca. 34 °C) in buffer due to the short oligosaccharide chain, are presented in Figure 1F. Surprisingly, there is a stronger increase in the phase transition temperature, as seen for the other LPS samples.

Summarized, two tendencies are observed: for the two polysorbates, the concentration above their CMCs (200 µg/mL) lead to a lower phase transition temperature of LPS. For the citrate and polysorbate formulation below CMC (10 µg/mL) there is an increase in transition temperature as compared to the HEPES control. The latter effect corresponds to an increase of the rigidity of the hydrocarbon chains, and with that, of the whole LPS assembly.

2.2. Aggregate Structures of LPS Preparations Used in This Study

The aggregate structure of LPS was described as important parameter, which determines the property of these amphiphilic compounds to exhibit biological activity [16,17]. For this, small-angle X-ray scattering (SAXS) via synchrotron radiation was applied, since under near physiological conditions (high water content) laboratory X-ray sources are not sufficient due to lack in brilliance. In the Figure 2, the LPS samples described above were measured in the temperature range 20 (blue line) to 80 °C (red line). Presented are the logarithm of the scattering intensity log I versus the scattering vector s ($s = \frac{1}{d}$, d = spacings of the reflections). The data for LPS from *E. coli* O111:B4 (Figure 2A) exhibit a complex scattering pattern, which is characteristic for a bilayered structure in the s-value range 0.1 to 0.35/nm (reflection centered at 6.0 to 6.3 nm), and a cubic periodicity (reflections at 40 to 56 nm). Interestingly, the polysorbate formulations show changes in the aggregate structure, which is obviously correlated with the phase transitions temperature. This can be deduced from the jump of the reflections at 33.8 and 35.8 nm to values above 50 nm. For LPS *E. coli* O55:B5 (Figure 2B), two main reflections around 5 to 6 nm and 20 to 30 nm are seen. In the case of the polysorbate 20 formulation, there is a complex reflection pattern between 20 and 50 nm. Also, in the case of LPS Rb from *S. minnesota* R345, the situation is similarly complex (Figure 2C). The LPS in HEPES indicates unresolved spectra, the LPS in citrate, and polysorbate indicate a higher degree of order, by showing multilamellar-like reflections at 8.20 and 4.32 nm (citrate), 8.94 and 4.46 nm (polysorbate 20), and 8.92 and 4.48 nm (polysorbate 80). Moreover, the polysorbate formulations exhibit reflections at around 20 to more than 50 nm. The observation is different for LPS Rd mutant (Figure 2D). The patterns for the sample in HEPES already exhibit some weak scattering maxima at 7.66 and 3.97 nm, which can be assigned to a multilamellar arrangement. This is strongly expressed for the LPS in citrate, in which at the lowest temperature peaks are clearly seen at 7.52 and 3.80 nm, 1st and 2nd order of a multilamellar aggregate, which shift to higher values at the higher temperature due to acyl chain melting (Figure 1E). Interestingly, no sharp reflections are seen for this LPS in the polysorbate formulations, but scattering intensity is seen in the s-value range 0.13 to 0.35/nm.

Summarized, for the wild type LPS from *E. coli* O111:B4 as well as Rb-mutant from *Salmonella minnesota* R345, the scattering patterns clearly indicate a complex change of the aggregate structures in the polysorbate containing chelating buffers as compared to HEPES and citrate formulations alone, lacking the detergents. For a forward assessment, it should be noted that the multilamellar structures that are seen here for LPS Rb and LPS Rd correspond to the bio-inactive structures of LPS [15,17,18].

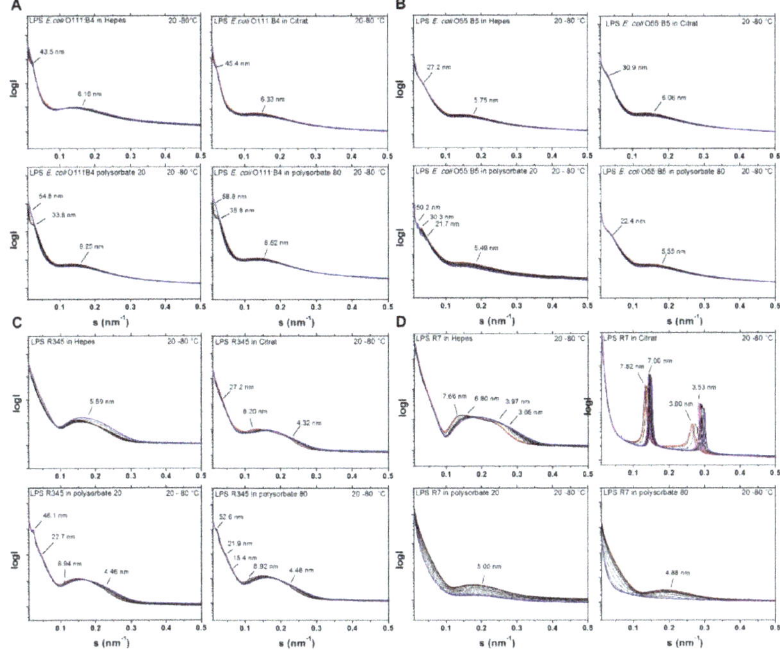

Figure 2. Effect of the formulation on the supramolecular LPS aggregate structure. Synchrotron radiation small-angle X-ray scattering patterns of LPSs from (**A**) *E. coli* O111:B4; (**B**) *E. coli* O55:B5; (**C**) *S. minnesota* Rb strain R345; and, (**D**) *S. minnesota* Rd strain R7. The logarithm of the scattering intensity is plotted versus the scattering vector s (s = $\frac{1}{d}$, with d being the spacings of the reflections).

2.3. Thermodynamics of Binding of the Synthetic Anti-LPS Peptide (SALP) Pep19-2.5 with the Different LPS Preparations

It has been reported that particular antimicrobial peptides (AMP) from the SALP (synthetic anti-LPS peptides) series, compound Pep19-2.5, binds and neutralizes LPS very efficiently [11,19]. This peptide is a 20'mer and consists of a N-terminal region with charged and polar amino acids and a C-terminal region with essentially hydrophobic amino acids. It is scheduled to fight against severe infections, such as sepsis [11]. The binding of the peptide to LPS preparations was tested here because it is known that the lipid A backbones, in particular the lipid A phosphates, are targets for the peptides, which is important with respect to the biological assay: Cellular activation in the MAT runs via the binding of the bisphosphorylated lipid A backbone to the TLR4 receptor.

To test the neutralizing activity of Pep19-2.5 with the different LPS formulations, isothermal titration calorimetry (ITC) was applied. The enthalpy change of this interaction can give information about the kind of binding process, which may be of exothermic or endothermic nature, or a mixture of both, with which the driving force of the interactions can be determined. In the experiments, in a first step all of the compounds were dissolved in the respective formulations, and in a second step the peptide was dissolved in water and was then added to the different formulations.

The data (Figure S1, Table 1) show similar binding characteristics, only for the polysorbate 20 formulation at 200 µg/mL there is an increase of the saturation curve to higher Pep19-2.5:LPS molar ratio values. Furthermore, the data for the peptide dissolved in water and then dispersed into the LPS formulations indicates a lower binding enthalpy of 40–50 kJ/mole at the beginning of the titration.

From the Figure S1A–C, the thermodynamic parameters can be calculated as presented in Table 1. The ITC results show that the basic neutralization mechanisms of LPS by Pep19-2.5 remain similar for all of the different formulations. There are some variations of the initial enthalpy change ΔH and the saturation values as indicated, with the values of LPS in HEPES buffer at −67 kJ/mole and saturation value $n = 0.245$ exhibiting highest affinity and LPS in polysorbate 80 (10 μg/mL) at −43 kJ/mole and $n = 0.4$ exhibiting lowest affinity.

Table 1. Thermodynamic parameters of the interaction of LPS from *E. coli* 055:B5 with the synthetic anti-endotoxin peptide (Pep19-2.5) formulations. PS: polysorbate.

Thermodyamic Parameters	LPS 055:B5 and Pep19-2.5 Dissolved in the Same Medium					LPS 055:B5 Dissolved in Polysorbates Pep19-2.5 Dissolved in Water			
	LPS in HEPES	PS20 10 μg/mL	PS20 200 μg/mL	PS80 10 μg/mL	PS80 200 μg/mL	PS20 10 μg/mL	PS20 200 μg/mL	PS80 10 μg/mL	PS80 200 μg/mL
Mass ratio (Peptide/LPS)	0.25	0.29	0.38	0.39	0.34	0.33	0.32	0.40	0.37
Kd (nM)	862	529	225	104	78	200	218	46	261
ΔH (kJ/mol)	−67.31	−59.82	−45.64	−58.35	−56.58	−48.49	−49.14	−43.00	−48.60
ΔS (kJ/mol·K)	−0.10	−0.07	−0.02	−0.05	−0.05	−0.03	−0.03	0.01	−0.03

To summarize, the findings for the LPS representing different *O*-serotypes and formulations are indicating similar neutralizations mechanisms, which is a matter of fact that the peptide essentially binds to the lipid A part of LPS, its "endotoxic principle". For all bioactive LPS, the lipid A part consists of a hexaacylated diglucosamine moiety phosphorylated in positions 1 and 4′. Surprisingly, the neutralization (saturation) of LPS takes place at higher peptide concentrations for the polysorbate formulations.

2.4. Stimulation of Immune Cells by the LPS in the Different Formulations (MAT)

The immune-stimulating activity of human mononuclear cells by compounds can be tested in an ELISA (MAT), for which TNF-α as sensitive cytokine is selected, which is secreted by the cells already after some hours. In a first step, the LPS from *E. coli* O55:B5 was tested in different formulations, NaCl 0.9%, polysorbate 20 and polysorbate 80, each at two concentrations (10 and 200 μg/mL), see Figure 3. As can be seen, the TNF-α secretion is highest for the sample in NaCl, whereas the highest concentrations of the two polysorbate samples lead to a strong reduction of the activity, with polysorbate 20 having the strongest influence on the reduction.

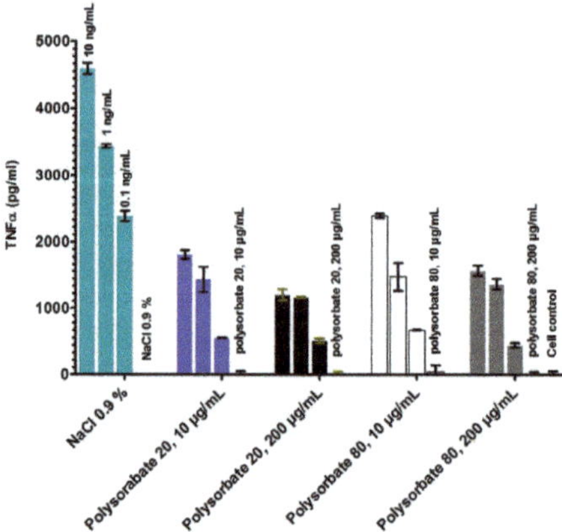

Figure 3. Secretion of tumor-necrosis-factorα (TNF-α) by human mononuclear cells induced by LPS from *E. coli* O55:B5. LPS aggregates were prepared in NaCl 0.9% and polysorbate 20 and 80. Three LPS concentrations 10, 1 and 0.1 ng/mL were tested. The error bar comes from two-fold determination of TNF-α concentration in the ELISA.

To examine also the dependence on the LPS chemotypes, further stimulation data were obtained by investigating LPS O55:B5 and LPS Rb mutant R345 in different formulations (Figure 4A–C). The data indicate differences in particular for the polysorbate formulations.

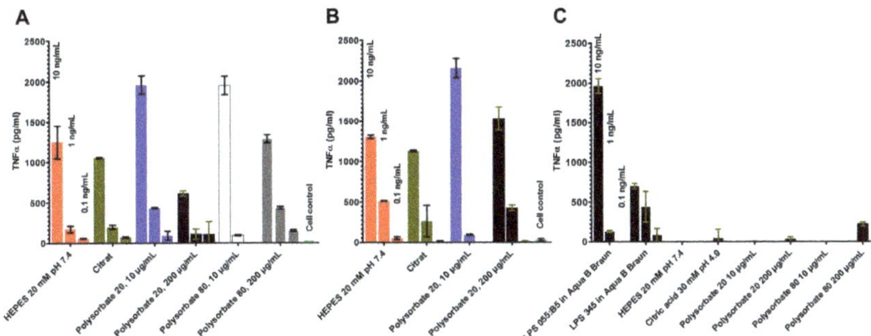

Figure 4. LPS-induced secretion of TNF-α by human mononuclear cells with different LPS formulations. LPS from *E. coli* O55:B5 (**A**), LPS R345 (**B**), and buffers and LPS control dissolved in water (**C**). Stimulation of human mononuclear cells was made at the three concentration: 1.0, 0.1 and 0.01 ng/mL, and the activity is recalculated. The error bar results from twofold measurement of TNF-α in the ELISA.

2.5. Measurements of the Size Distribution of LPS Aggregates by Zeta Sizer

Aggregate sizes and their distributions has been discussed as a parameter, which influences LER [3]. We therefore determined the LPS aggregate sizes and their distributions in a Zeta sizer, by analyzing the diffusion of the aggregates via measurement of the backscatter signals. Again, rough mutant LPS (LPS from *S. minnesota* Re (R595) and Ra (R60), as well as smooth form (O55:B5)) were analysed.

2.5.1. Results for Deep Rough Mutant LPS R595

In the following Figure 5 the results are presented for deep rough mutant LPS from *S. minnesota* R595. On the left-hand side, the size distribution is shown, on the right-hand the side polydispersity, i.e., the respective size distributions (see Table 2).

Table 2. Polydispersity index (PDI) for lipopolysaccharides aggregates in different formulations.

Formulation	Polydispersity Index (PDI)		
	LPS R595	**LPS Ra**	**LPS O55:B5**
HEPES 20 mM pH 7.4	0.449 ± 0.012	0.983 ± 0.029	0.436 ± 0.004
NaCl 0.9%	0.935 ± 0.112	1.000 ± 0.000	0.524 ± 0.030
Citrate 30 mM pH 4.0	1.000 ± 0.000	1.000 ± 0.000	0.536 ± 0.018
Polysorbate 20, 10 µg/mL	1.000 ± 0.000	0.934 ± 0.073	0.514 ± 0.028
Polysorbate 20, 200 µg/mL	0.966 ± 0.058	0.966 ± 0.058	0.506 ± 0.059
Polysorbate 80, 10 µg/mL	0.911 ± 0.083	1.000 ± 0.000	0.463 ± 0.002
Polysorbate 80, 200 µg/mL	0.814 ± 0.050	1.000 ± 0.000	0.469 ± 0.056

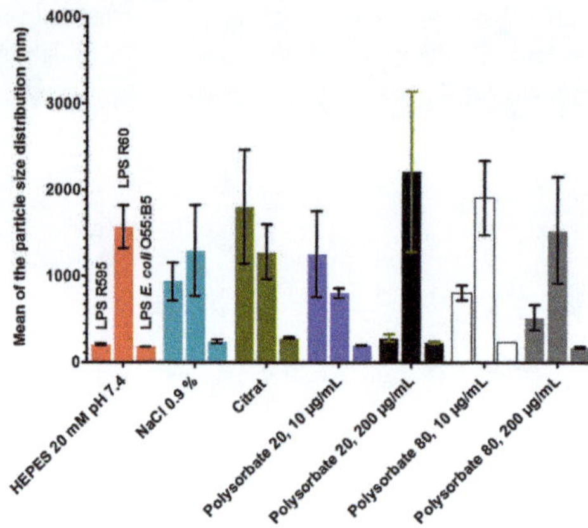

Figure 5. Effect of the formulation on the aggregate size of LPS. The figure shows the particle size for seven different preparations. For each preparation, LPS aggregates are shown as LPS R595 (first bar), LPS R60 (second bar), and LPS *E. coli* O55:B5 (third bar). The error bar results from twofold measurement of TNF-α in the ELISA.

It becomes clear that the LPS sample in HEPES buffer (top left) has lowest values of the peak around 205 nm and a distribution factor of 0.45. Interestingly, the values for the aggregates in NaCl are much higher, and are highest in the citrate formulation. In the two latter samples, also the distributions are broadest. Both polysorbate preparations at 200 µg/mL have rather low sizes, whereas their peak sizes at the smaller polysorbate concentrations are significantly higher.

2.5.2. Results for Rough Mutant LPS Ra with Complete Core Oligosaccharide

In the following, the results are presented for rough mutant LPS from *S. minnesota* R60 (Figure 5 and Table 2).

The results for the rough mutant LPS R60 with complete core oligosaccharide differs considerably from the results for the deep rough mutant LPS. The sizes and their distributions are much more homogenous. In general, the sizes are significantly larger than those from LPS R595. Interestingly, the results for the preparation with polysorbate 20, 200 µg/mL, exhibits the largest sizes, whereas the values for LPS R595 are indicative of very small sizes.

The corresponding data for the wild-type LPS O55:B5 are shown in Figure 5 and Table 2. It is striking that the sizes of the LPS aggregates in the different formulations are considerably lower than for the two rough mutant LPS. Furthermore, similar to LPS R595, the citrate formulation has a highest size (285 nm), which, however, is very low as compared to the former LPS (1811 nm).

Summarized, the data give evidence for a strong dependence on the size distributions from the LPS representing different serotypes. It should be mentioned here, that these results are of course influenced strongly by the facts that the chemical structures of rough mutant LPS are relatively homogenous, whereas wild type forms usually consists of a heterogenous assembly of various part structures, containing an Ra- or Rb-type LPS as bioactive moiety [14].

2.6. Size Distribution in Relation to Cytokine Induction in Human Mononuclear Cells

The same samples, which were analysed in light scattering experiments, were added to human mononuclear cells that were obtained after blood separation form healthy donors, and their ability to induce tumor-necrosis-factorα (TNF-α) was measured in an ELISA (MAT). In Figure S2, the results are shown for deep rough mutant LPS from R595 for two concentrations 10 and 1 ng/mL. At the higher concentration, the stimulation values are rather homogeneous, except for the citrate value, whereas at the lower concentration, only the value for polysorbate 80 (10 µg/mL) deviates to lower values.

The results for the rough mutant LPS R60 are given in Figure S3. It can be seen that the absolute TNF-values tend to be lower than for LPS R595. The observation of lower values for citrate at the higher LPS concentration is also observed here, whereas the pattern of the TNF values is more homogeneous, but significantly lower than for LPS R595.

In a similar way, data are presented for LPS S-form O55:B5 (Figure S4). Also, here, the citrate formulation at 10 ng/mL has the lowest activity. Surprisingly, there is a great difference to the cytokine values at 1 ng/mL. The comparison of the three LPS shows that with an increasing length of the sugar chain, which is shortest for LPS Re, longer for LPS Ra, and longest for LPS S-form, the results become more variable.

In another approach, the MAT was performed with two LPS (LPS R60 and O55:B5) and with two pretreatments. Sonicated LPS should produce small, vortexed LPS large aggregate structures. This was performed according to the findings of Komoro et al. [20], who found better reactivity in the pyrogen test and LAL with sonicated LPS preparations. As can be deduced from Figure S5, there is no significant difference in the response of the MAT at both sonicated and vortexed samples.

Summarized, the data presented here do not indicate a general dependence of the biological activities in the MAT assay of different LPS preparations on the respective sizes and size distributions. It should be noted here that the term aggregate size in a sense of a well-defined spherical form for LPS is not well-defined, in particular, for LPS with long saccharide chains, such as S-form (wild-type) LPS.

3. Discussion

In a comprehensive analysis, we have performed biophysical analyses of different LPS formulations (detergent, chelating buffer) being assumed to represent the major factors that are mediating the LER-effect. In addition, we also investigated different LPS varying in size and structure, i.e., from wild-type (S-form) over various rough-mutants differing in the size of the LPS core-oligosaccharide (Re-, Rd-, and Ra- mutant LPS). In a first step, we have analyzed the single constituents of the complex compositions of the pharmaceutical, formulations, i.e., citrate, polysorbate 20 and polysorbate 80. The data presented here can serve as the basis for further investigations, in which the complete formulation leading to LER will be tested.

We have found in various test systems, that there are clear changes of different parameters, with variations of the formulation. These data give hints with respect to the occurrence of the LER, in which the LPS backbone structure shows reduced LAL reactivity.

We have analyzed the following physical-chemical parameters, which might be responsible for the LER in LPS formulations:

- Fluidity of their hydrocarbon chains;
- Aggregate size and structure;
- Head group conformation and orientation.

Following this line, we have investigated the:

(i)   (i) gel to liquid crystalline phase transition of the hydrocarbon chains of LPS, and with that, the fluidity of the acyl chains, with Fourier-transform infrared spectroscopy (FTIR);

(ii)   three-dimensional aggregate structure of LPS by using synchrotron radiation small-angle X-ray scattering (SAXS);

(iii)   LPS aggregate sizes by dynamic light scattering and have related these data to the biological activities in the MAT;

(iv)   Furthermore, the interaction of LPS with a synthetic anti-LPS peptide Pep19-2.5 was monitored to find out whether differences in head group binding are observed.

It has been shown that the order of the acyl chains (highly ordered = gel phase, less ordered = liquid crystalline phase) influences the bioactivity of LPS and lipid A preparations [21]. Thus, with increasing order (lower fluidity) interaction with target structures such as the factor C in the *Limulus* assay or cell surface receptors, such as CD14 or the TLR4/MD2 complex, are impeded. Therefore, the data for the samples with T_c increases (for example, see Figure 1E) should have lowered biological activity in the MAT, because the acyl chains are more rigid. This relates in first line to the polysorbate formulations, which may influence the biological responses. The SAXS data show only small, but significant, changes of the observed aggregate structures on the different formulations. In particular, the existence of highly ordered phases observed for the smooth, as well as rough, mutant LPS R7 and R345 for the polysorbate formulations may give a hint for a masking process, which will be tested in further experiments with the complete formulation system. It has been shown in previous papers [15,17] for lipid A and rough mutant LPS as well as in a recent paper on wild-type LPS [18] that the aggregate structure of LPS is a determinant for its biological activity in the MAT. Thus, non-lamellar, in most cases cubic structures are the bioactive units of LPS. The observation of a shift of the broad scattering range from 0.1 to 0.25 /nm (Figure 2A–C) to 0.13 to 0.35 (Figure 2D) indicates a new, probably highly ordered, phase, for LPS Rd in the polysorbate formulations (interpretation from unpublished results).

Regarding the data from the size distributions presented here, the results indicate for the different LPS mutants/smooth forms quite diverging results. The data are indicative of medium sizes for LPS Re, high sizes for LPS Ra, and low sizes for LPS S-form. For an understanding, the results from studies of LPS morphologies may be useful. It was found that for most rough mutant LPS spherical-like morphologies were reported, by using cryo- and freeze-fracture electron microscopy [22]. In contrast, in LPS with longer sugar chain, in particular S-LPS, membrane vesicles, bilayer disks, and ribbon-like aggregates are found. These data are in accordance with the size distribution that is obtained via ultracentrufigation, in which R-LPS showed size distribution between 100 to 600 nm, whereas for S-LPS, the values were around 50 to 200 nm [22].

It should be noted that in the evaluation of the Zeta sizer measurements, a simple assumption of spherical-like structures would give directly comparable results for the medium sizes. Therefore, the size values for compounds with long saccharide chains, such as S-form LPS, are not the radius of a sphere, but give only a medium value for its non-spherical morphology. Finally, it should be noted that polysorbates—which are added to drug products to inhibit protein aggregation—do not lead to LPS disaggregation at least when administered solely (see Figure S5).

Regarding the comparison of the results from the biological assay at the selected concentrations with those of the three LPS with differences in the saccharide chain lengths in different buffers do not show any systematic dependence of the MAT response with the aggregate sizes and their distributions. Finally, the ITC data of LPS binding to Pep19-2.5 indicate a significantly higher peptide to LPS ratio for binding saturation for the polysorbate formulations, which is indicative of a change in the LPS head group conformation.

Literature data on the one hand explain the LER by increases in aggregate size and stability [23], and, on the other hand, by a decrease down to monomers.

Reich et al. have proposed 'the supramolecular structure of endotoxin is altered and exhibits only a limited susceptibility in binding of the factor C of Limulus-based detection systems. Although, in our analysis under conditions with reduced complexity (only pure citrate or polysorbates were used, but not a combination therefrom), we observed some changes in the supramolecular assembly and the phase transitions of the acyl chains, in particular when polysorbate is present in the LPS preparations.

Masked endotoxin may adopt a supramolecular conformation not detectable by the LAL test. Schwarz et al. [24] have found for masked endotoxin—as evidenced by the chromogenic endpoint LAL—the expression of pro-inflammatory cytokines and surface activation markers. This is an observation, which we will address in future experiments, in particular by investigating the complete system relevant in LER.

The presented data will form the basis for detailed investigations into the dependence of biophysical parameters of the complete detergent system, in particular on the influence of the 4′-phosphate diglucosamine backbone of the lipid A part of LPS, the recognition structure of LPS by the *Limulus* assay. It is envisaged to continue the investigations by using also the factor C of the *Limulus* assay in recombinant form, and possibly LPS-binding sequences of this, and comparing it with the well-known behavior of anti-endotoxin peptides, such as Pep19-2.5.

From these observations, the following questions seem to be important with respect to the occurrence of the LER: is the lack of endotoxin detection by LAL a problem of LPS in an undetectable, inactive conformation or a failure of the measuring system LAL?

The headgroup conformation, in particular of the 4′-phosphate group in the lipid A part is of central importance. We will perform in a next step an analysis via FTIR by studying the interaction of LPS with rFC and part structures.

Could the change of the LPS conformation into in monomeric form be responsible for the LER? Müller et al. [5] found in the MAT as well as the LAL no biological activity of LPS monomers. There are other publications, however, which come to a completely different conclusion [25]. Therefore, this hypothesis will be in the focus of further studies.

4. Materials and Methods

4.1. Peptides, Reagents and LPS Formulations

Lipopolysaccharides O55:B5 and O111:B4 from *Escherichia coli* wild-type strains (S-form LPS) were purchased from Sigma (Deisenhofen, Germany), rough mutant LPS Ra strain R60, Rb strain R345, Rd strain R7, and Re strain R595 from *Salmonella minnesota* were extracted from bacteria by phenol/chloroform/petrol ether, according to the protocol of Galanos et al. [26]. For wild-type strains, the chemical structure of LPS consists of the lipid A part, which represents the outer leaflet of the bacterial outer membrane, the oligosaccharide core, and the O-antigen, a polysaccharide moiety directing outwards. The chemical structures of the single segments of the LPS molecule from the commonly used wild type strains are—except for the relatively homogenous lipid A moiety ("conservative" motif [1])—not well described, and varies from strain to strain. Usually, the core oligosaccharide, which is bound to the lipid A part, consists of 10 to 12 monosaccharide units, and the subsequent O-antigen has a largely varying polysaccharide chain. Moreover, S-form LPS consists of different fractions, which may have also underacylated lipid A parts [1,14]. The details of these inhomogenities are in most cases unknown except for single analyses as for example performed by Jiao and Galanos [14] for wild-type LPS from *Salmonella abortus equi*.

Rough mutant LPS lack the O-antigen, and have a varying length of the oligosaccharide, Ra with a complete one, and the other mutants having a shorter oligosaccharide in the sequence Rb > Rc > Rd > Re.

The antimicrobial peptide Pep19-2.5 (Aspidasept®) with a sequence of GCKKYRRFRWKF KGKFWFWG was synthesized by BACHEM (Bubendorf, Switzerland) with a purity of >95%. All of the other chemicals were from Merck (Mannheim, Germany). Sodium citrate and polysorbate 20 and 80 was purchased from Merck (Mannheim, Germany) and Sigma (Deisenhofen, Germany).

For all of the applied techniques listed below, the LPS samples were prepared as aqueous dispersions in 20 mM HEPES pH 7.4, 30 mM sodium citrate pH 4.0, polysorbate 20 and 80, the latter each at 10 and 200 µg/mL. The latter concentration corresponds to values below and above the critical micellar concentration, respectively. LPSs were suspended directly in buffer by extensively vortexing, sonicated in a water bath at 60 °C for 30 min, cooled down to 5 °C, and subjected to three cycles of heating and cooling from 60 to 5 °C. After that, the lipid samples were stored for at least 24 h at 4 °C before performing the measurements.

4.2. Acyl Chain Melting Behavior by Fourier-Transform Infrared Spectroscopy

The infrared spectroscopic measurements were performed on a FTIR spectrometer IFS-55, from Bruker (Karlsruhe, Germany). The lipid samples were placed in a CaF_2 cuvette separated by a 12.5 mm thick teflon spacer. Temperature-scans were performed automatically in the range from 10 to 65–80 °C with a heating rate of 0.6 °C min^{-1}. Every 3 °C, 200 interferograms were accumulated, apodized, Fourier transformed, and converted to absorbance spectra. The phase behaviour was monitored by using the peak position of the symmetric stretching vibration ν_s (CH_2) in the wavbenumber range 2850 to 2853 cm^{-1}. The phase transition temperature T_c can be determined by taking the midpoint of the intersection of the tangents of the curve in the gel phase with that of the inflection point of the transition range, and the intersection of the latter with the tangent of the curve in the liquid crystalline phase.

4.3. Aggregate Structure Determined by Small-Angle X-ray Scattering (SAXS)

The X-ray scattering measurements were performed on the X33 beamline of the European Molecular Biology Laboratory (EMBL) outstation at HASYLAB on the storage ring PETRA of the Deutsches Elektronen Synchrotron (DESY) at Hamburg [27].

Briefly, scattering patterns in the range of scattering vector $0.05 < s < 1$ nm^{-1} ($s = 2 \sin \theta / \lambda$, 2θ is the scattering angle and λ the wavelength = 0.15 nm) were recorded, with exposure times of 1min using an image plate detector with online readout (Mar345; Marresearch, Norderstedt, Germany). Further details concerning the data acquisition and evaluation have been described previously [12]. In the diffraction patterns that are presented below, the logarithm of the diffracted intensities I(s) is plotted versus s. The X-ray spectra were evaluated using standard procedures [11], which allow for assigning the spacing ratios of the diffraction maxi-ma to defined three-dimensional structures of the endotoxin: detergent samples.

Structures occuring for endotoxins comprise lamellar (L) phases with spacing ratios lying at equidistant positions and nonlamellar phases like cubic (Q) and inverted hexagonal (HII) that are characterized by square root spacing ratios [28].

4.4. Binding Affinity of LPS to Pep19-2.5 via Isothermal Titration Calorimetry (ITC)

The interaction of the peptide Pep19-2.5 with LPS in various formulations was analyzed by microcalorimetric measurements in the ITC200 (GE Healthcare, Munich, Germany), as recently described [19]. For this, 1 mM (2.71 mg/mL) Pep19-2.5 in different formulations was titrated into 430 μg LPS from *E. coli* O55:B5 and the measured enthalpy changes (ΔH) were recorded versus time and the peptide: LPS concentration ratio.

4.5. Particle Size Measurements by Dynamic Light Scattering on a Zeta Sizer

Dynamic light scattering of the particle sizes of LPS aggregates was performed in different formulations, by measuring the diffusion velocity in a Malvern Zeta sizer Nano (Malvern, Herrenberg, Germany). The method is based on the measurement of the diffusion of small particles according to the Stokes-Einstein equation $D = \mu \times k_B \times T$ (μ = mobility of the particles, k_B = Boltzmann constant), measuring the back-scattering, and calculating the autocorrelation function. Each particle scatters the light to the detector, and the fluctuations of the scattering intensity is smaller for large than for small particles.

In detail, the LPS samples were measured for 3 min in a fixed laser position of 173° (backscattering), relative to the incident laser beam. The measured intensities were correlated over time and analysed by a multiple exponential, non-negative least square fit to obtain relative intensities for the different particle sizes. The LPS samples at concentrations of 10 μM were dispersed in following formulations: 20 mM Hepes buffer at pH 7, 0.9% NaCl, 30 mM citrate, 10 and 200 μg/mL polysorbate 20, respectively, and 10 and 200 μg/mL polysorbate 80. The samples were prepared by sonication and temperature-cycled between 20–60 °C, and were stored at room temperature.

4.6. Stimulation of Human Mononuclear Cells

Mononuclear cells (MNC) were isolated from heparinized blood samples that were obtained from healthy donors, as described previously [15]. The cells were resuspended in medium (RPMI 1640), and their number was equilibrated at 5×10^6 cells/mL. For stimulation, 200 µL MNC (1×10^6 cells) was transferred into each well of a 96-well culture plate. The LPS formulations were preincubated for 30 min at 37 °C and were added to the cultures at 20 µL per well. The cultures were incubated for 4 h at 37 °C with 5% CO_2. Supernatants were collected after centrifugation of the culture plates for 10 min at $400 \times g$ and stored at 20 °C until immunological determination of tumor necrosis factor alpha (TNF-α), carried out with a sandwich enzyme-linked immunosorbent assay (ELISA) using a monoclonal antibody against TNF (clone 6b; Intex AG, Basel, Switzerland), and described previously in detail [19].

5. Conclusions

In conclusion, the current work has laid out a number of analytical approaches to study the LPS system and provide insight into the structural changes that the LPS might be going through, subsequently leading to the LER effect. The next steps to be investigated will be the combination of chelating buffers and polysorbate and study their individual impact on the LER effect, which is most relevant for pharmaceutical preparations. Also, the study in various (chelating) buffers (e.g., histidine, citrate, acetate, succinate, phosphate, etc.) on structural details, in the presence of polysorbate 20 and 80, respectively. In addition, the presence or absence of divalent cations, such as Mg^{2+} and Ca^{2+}, which are necessary for the formation of defined and complex negatively charged LPS aggregates, as well as the pH-value seems to be of outmost importance for the understanding of the LER-effect on a molecular level. Finally, the same holds true for surfactant concentrations from 0.1 to 2 mg/mL, i.e., in a pharmaceutical relevant range.

Supplementary Materials: The following are available online at www.mdpi.com/1422-0067/18/12/2737/s1.

Acknowledgments: The authors are indebted to Sabrina Groth for performing the MAT. The authors would like to thank Nicolas Gisch for helpful comments on the manuscript.

Author Contributions: Wilmar Correa was responsible for the FTIR, SAXS, ITC measurements and writing, Klaus Brandenburg for the cytokine assays and writing. Ulrich Zähringer, Tarik Khan and Kishore Ravuri for details of study concepts and reading, and Friedrich von Wintzingerode for the general study concept, reading, and writing.

Conflicts of Interest: The authors declare no conflict of interest.

References

1. Rietschel, E.T.; Kirikae, T.; Schade, F.U.; Mamat, U.; Schmidt, G.; Loppnow, H.; Ulmer, A.J.; Zähringer, U.; Seydel, U.; di Padova, F. Bacterial endotoxin: Molecular relationships of structure to activity and function. *FASEB J.* **1994**, *8*, 217–225. [PubMed]
2. Opal, S.M.; Yu, R.L. Antiendotoxin Strategies for the Prevention and Treatment of Septic Shock. *Drugs* **1998**, *55*, 497–508. [CrossRef] [PubMed]
3. Reich, J.; Lang, P.; Grallert, H.; Motschmann, H. Masking of endotoxin in surfactant samples: Effects on Limulus-based detection systems. *Biologicals* **2016**, *44*, 417–422. [CrossRef] [PubMed]
4. Bolden, J.S.; Warburton, R.E.; Phelan, R.; Murphy, M.; Smith, K.R.; de Felippis, M.R.; Chen, D. Endotoxin recovery using limulus amebocyte lysate (LAL) assay. *Biologicals* **2016**, *44*, 434–440. [CrossRef] [PubMed]
5. Mueller, M.; Lindner, B.; Kusumoto, S.; Fukase, K.; Schromm, A.B.; Seydel, U. Aggregates are the biologically active units of endotoxin. *J. Biol. Chem.* **2004**, *279*, 26307–26313. [CrossRef] [PubMed]
6. Gioannini, T.L.; Teghanemt, A.; Zhang, D.; Levis, E.N.; Weiss, J.P. Monomeric endotoxin:protein complexes are essential for TLR4-dependent cell activation. *J. Endotoxin Res.* **2005**, *11*, 117–123. [CrossRef] [PubMed]
7. Tan, N.S.; Ho, B.; Ding, J.L. High-affinity LPS binding domain(s) in recombinant factor C of a horseshoe crab neutralizes LPS-induced lethality. *FASEB J.* **2000**, *14*, 859–870. [PubMed]

8. Brandenburg, K.; Howe, J.; Gutsmann, T.; Garidel, P. The Expression of Endotoxic Activity in the Limulus Test as Compared to Cytokine Production in Immune Cells. *Curr. Med. Chem.* **2009**, *16*, 2653–2660. [CrossRef] [PubMed]

9. Howe, J.; Garidel, P.; Roessle, M.; Richter, W.; Alexander, C.; Fournier, K.; Mach, J.P.; Waelli, T.; Gorczynski, R.M.; Ulmer, A.J.; et al. Structural investigations into the interaction of hemoglobin and part structures with bacterial endotoxins. *Innate Immun.* **2008**, *14*, 39–49. [CrossRef] [PubMed]

10. Martinez de Tejada, G.; Heinbockel, L.; Ferrer-Espada, R.; Heine, H.; Alexander, C.; Barcena-Varela, S.; Goldmann, T.; Correa, W.; Wiesmuller, K.H.; Gisch, N.; et al. Lipoproteins/peptides are sepsis-inducing toxins from bacteria that can be neutralized by synthetic anti-endotoxin peptides. *Sci. Rep.* **2015**, *5*, 14292. [CrossRef] [PubMed]

11. Gutsmann, T.; Razquin-Olazaran, I.; Kowalski, I.; Kaconis, Y.; Howe, J.; Bartels, R.; Hornef, M.; Schurholz, T.; Rossle, M.; Sanchez-Gomez, S.; et al. New antiseptic peptides to protect against endotoxin-mediated shock. *Antimicrob. Agents Chemother.* **2010**, *54*, 3817–3824. [CrossRef] [PubMed]

12. Kaconis, Y.; Kowalski, I.; Howe, J.; Brauser, A.; Richter, W.; Razquin-Olazaran, I.; Inigo-Pestana, M.; Garidel, P.; Rossle, M.; Martinez de Tejada, G.; et al. Biophysical mechanisms of endotoxin neutralization by cationic amphiphilic peptides. *Biophys. J.* **2011**, *100*, 2652–2661. [CrossRef] [PubMed]

13. Brandenburg, K.; Funari, S.S.; Koch, M.H.J.; Seydel, U. Investigation into the Acyl Chain Packing of Endotoxins and Phospholipids under Near Physiological Conditions by WAXS and FTIR Spectroscopy. *J. Struct. Biol.* **1999**, *128*, 175–186. [CrossRef] [PubMed]

14. Jiao, B.; Freudenberg, M.; Galanos, C. Characterization of the lipid A component of genuine smooth-form lipopolysaccharide. *Eur. J. Biochem.* **1989**, *180*, 515–518. [CrossRef] [PubMed]

15. Brandenburg, K.; Andrä, J.; Müller, M.; Koch, M.H.J.; Garidel, P. Physicochemical properties of bacterial glycopolymers in relation to bioactivity. *Carbohydr. Res.* **2003**, *338*, 2477–2489. [CrossRef] [PubMed]

16. Brandenburg, K.; Wiese, A. Endotoxins: Relationships between Structure, Function, and Activity. *Curr. Top. Med. Chem.* **2004**, *4*, 1127–1146. [CrossRef] [PubMed]

17. Brandenburg, K.; Mayer, H.; Koch, M.H.J.; Weckesser, J.; Rietschel, E.T.; Seydel, U. Influence of the supramolecular structure of free lipid A on its biological activity. *Eur. J. Biochem.* **1993**, *218*, 555–563. [CrossRef] [PubMed]

18. Brandenburg, K.; Heinbockel, L.; Correa, W.; Fukuoka, S.; Gutsmann, T.; Zähringer, U.; Koch, M.H.J. Supramolecular structure of enterobacterial wild-type lipopolysaccharides (LPS), fractions thereof, and their neutralization by Pep19-2.5. *J. Struct. Biol.* **2016**, *194*, 68–77. [CrossRef] [PubMed]

19. Heinbockel, L.; Sánchez-Gómez, S.; Martinez de Tejada, G.; Dömming, S.; Brandenburg, J.; Kaconis, Y.; Hornef, M.; Dupont, A.; Marwitz, S.; Goldmann, T.; et al. Preclinical Investigations Reveal the Broad-Spectrum Neutralizing Activity of Peptide Pep19-2.5 on Bacterial Pathogenicity Factors. *Antimicrop. Agents Chemother.* **2013**, *57*, 1480–1487. [CrossRef] [PubMed]

20. Komuro, T.; Murai, T.; Kawasaki, H. Effect of sonication on the dispersion state of lipopolysaccharide and its pyrogenicity in rabbits. *Chem. Pharm. Bull.* **1987**, *35*, 4946–4952. [CrossRef] [PubMed]

21. Brandenburg, K.; Schromm, A.B.; Koch, M.H.; Seydel, U. Conformation and fluidity of endotoxins as determinants of biological activity. *Prog. Clin. Biol. Res.* **1995**, *392*, 1671–1682.

22. Richter, W.; Vogel, V.; Howe, J.; Steiniger, F.; Brauser, A.; Koch, M.H.; Roessle, M.; Gutsmann, T.; Garidel, P.; Mäntele, W.; et al. Morphology, size distribution, and aggregate structure of lipopolysaccharide and lipid A dispersions from enterobacterial origin. *Innate Immun.* **2011**, *17*, 427–438. [CrossRef] [PubMed]

23. Tsuchiva, M. *Possible Mechanism of Low Endotoxin Recovery*; American Pharmaceutical Review: Fishers, IN, USA, 2014; pp. 1–5.

24. Schwarz, H.; Gornicec, J.; Neuper, T.; Parigiani, M.A.; Wallner, M.; Duschl, A.; Horejs-Hoeck, J. Biological Activity of Masked Endotoxin. *Sci. Rep.* **2017**, *7*, 44750. [CrossRef] [PubMed]

25. Gioannini, T.L.; Weiss, J.P. Regulation of interactions of Gram-negative bacterial endotoxins with mammalian cells. *Immunol. Res.* **2007**, *39*, 249–260. [CrossRef] [PubMed]

26. Galanos, C.; Lüderitz, O.; Westphal, O. A New Method for the Extraction of R Lipopolysaccharides. *Eur. J. Biochem.* **1969**, *9*, 245–249. [CrossRef] [PubMed]

27. Roessle, M.W.; Klaering, R.; Ristau, U.; Robrahn, B.; Jahn, D.; Gehrmann, T.; Konarev, P.; Round, A.; Fiedler, S.; Hermes, C.; et al. Upgrade of the small-angle X-ray scattering beamline X33 at the European Molecular Biology Laboratory, Hamburg. *J. Appl. Cryst.* **2007**, *40*, s190–s194. [CrossRef]
28. Luzzati, V.; Vargas, R.; Mariani, P.; Gulik, A.; Delacroix, H. Cubic Phases of Lipid-containing Systems: Elements of a Theory and Biological Connotations. *J. Mol. Biol.* **1993**, *229*, 540–551. [CrossRef] [PubMed]

International Journal of
Molecular Sciences

Article

Time Response of Oxidative/Nitrosative Stress and Inflammation in LPS-Induced Endotoxaemia—A Comparative Study of Mice and Rats

Sebastian Steven [1,2], Mobin Dib [1], Siyer Roohani [1], Fatemeh Kashani [1], Thomas Münzel [1] and Andreas Daiber [1,2,*]

[1] Center for Cardiology, Cardiology I, University Medical Center of the Johannes Gutenberg-University, D-55131 Mainz, Germany; sebastiansteven@gmx.de (S.S.); mobindib@yahoo.com (M.D.); SiyerRoohani@gmx.de (S.R.); fatemehkashani_91@yahoo.com (F.K.); tmuenzel@uni-mainz.de (T.M.)

[2] Center for Thrombosis and Hemostasis, University Medical Center of the Johannes Gutenberg-University, D-55131 Mainz, Germany

* Correspondence: daiber@uni-mainz.de; Tel.: +49-(0)6131-17-6280

Received: 6 September 2017; Accepted: 9 October 2017; Published: 18 October 2017

Abstract: Sepsis is a severe and multifactorial disease with a high mortality rate. It represents a strong inflammatory response to an infection and is associated with vascular inflammation and oxidative/nitrosative stress. Here, we studied the underlying time responses in the widely used lipopolysaccharide (LPS)-induced endotoxaemia model in mice and rats. LPS (10 mg/kg; from Salmonella Typhosa) was intraperitoneally injected into mice and rats. Animals of every species were divided into five groups and sacrificed at specific points in time (0, 3, 6, 9, 12 h). White blood cells (WBC) decreased significantly in both species after 3 h and partially recovered with time, whereas platelet decrease did not recover. Oxidative burst and iNOS-derived nitrosyl-iron hemoglobin (HbNO) increased with time (maxima at 9 or 12 h). Immune cell infiltration (CD68 and F4/80 content) showed an increase with time, which was supported by increased vascular mRNA expression of *VCAM-1, P-selectin, IL-6* and *TNF-α*. We characterized the time responses of vascular inflammation and oxidative/nitrosative stress in LPS-induced endotoxaemic mice and rats. The results of this study will help to interpret and compare data from different animal species in LPS-induced endotoxaemia models for the identification of new drug targets.

Keywords: sepsis; time response; inflammation; oxidative stress; endotoxaemia; mouse; rat

---

1. Introduction

Sepsis is a clinical syndrome that is caused by a dysregulated and overshooting response of the inflammatory system to an infection. Sepsis is most frequently caused by bacteria, to a lower extend by fungi, and can rapidly become life threatening. It arises from infections of the skin, lung, abdomen, and urinary tract. Despite aggressive treatment on intensive care unit and ambitious effort in research, sepsis remains a leading cause of death, even in Western countries [1]. In 2008, costs spent on hospitalizations for sepsis in the USA were estimated at \$14.6 billion [2]. The early identification of sepsis is necessary for the sufficient treatment of septic patients and scores like the modified Sepsis-related Organ Failure Assessment score (quickSOFA) are helpful tools for initiation of optimal therapy [3]. Nevertheless, the improvement of survival and reduction of costs can be best achieved by better understanding of the causes and pathophysiology of sepsis.

The basic treatment regimen for sepsis did not change for decades. Early fluid supplementation and antibiotics are known to significantly improve survival in septic shock, but clinical trials targeting the reduction of the inflammatory response failed [4]. Although there are major limitations for the translation of findings of animal studies to the human setting, it cannot be denied that animal models have significantly improved the knowledge of sepsis. In humans, sepsis is characterized by an initial pro-inflammatory phase, which is followed by an anti-inflammatory or immunosuppressive phase [5]. Although several animal models of sepsis were reported to mimic the inflammatory and genomic responses of humans to septic stimuli, this is still under debate for murine sepsis models [6,7]. An important requirement for the translation of data from animals to humans is the detailed pathophysiologic characterization of sepsis in the animal models. However, not only different protocols for the induction of sepsis, such as bacterial infusion model, cecal ligation and puncture (CLP), colon ascendens stent peritonitis (CASP), and lipopolysaccharide (LPS)-induced endotoxaemia, are used to investigate sepsis in animals. Even more problematic is the use of different species and strains in these sepsis models. In the last years, our group used LPS-injection models in mice and rats for three studies and we noticed some differences in inflammatory response and mortality between the species [8–11]. Since it is already challenging to translate findings from animal models to human sepsis, a clear characterization of the inflammatory response in each animal model of different species is highly recommended.

With the present study, we aimed to compare the time response pattern (0, 3, 6, 9, and 12 h) of white blood cell (WBC) derived oxidative and nitrosative stress, as well as vascular inflammation parameters in LPS-induced endotoxaemia in mice and rats.

2. Results

2.1. Time Response of Thrombocyte Count and White Blood Cell Derived Oxidative Burst in Mice and Rats

Thrombocytes in whole blood were significantly higher at the beginning of the experiments. After LPS injection the thrombocyte counts in mice and rats dropped similarly (Figure 1A). Already before LPS injection, rats and mice had significantly different numbers of WBC in whole blood. In both species, the WBC count dropped significantly at 3 h after LPS injection and increased with time showing a partial normalization at 12 h. Of note, the WBC count showed better recovery in rats as compared to mice (Figure 1B). Zymosan A-induced oxidative burst in whole blood was significantly higher in rats at 0, 3, 9, and 12 h after LPS injection as compared to mice. A first significant increase of oxidative burst was detectable at 6 h after LPS injection in mice and 9 h in rats. In both species the peak level of oxidative burst was observed after 9 h (Figure 1C). A quite similar observation was made for the kinetics of PDBu-induced oxidative burst. The first significant increase with time was again detected at 6 vs. 9 h in mice as compared to rats. The maximal oxidative burst in whole blood from mice was at 9 h for zymosan A and PDBu stimulation. In contrast to the maximal zymosan A-induced oxidative burst in rats at 9 h, the peak level of oxidative burst was found at 12 h after LPS injection (Figure 1D). The absolute level of oxidative burst was higher with zymosan A than with PDBu stimulation.

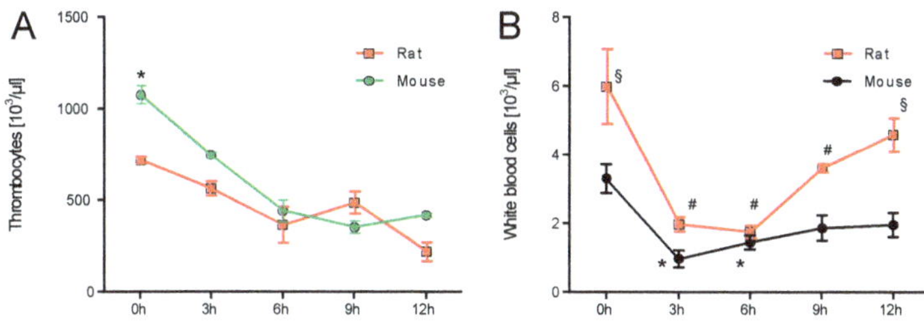

Figure 1. *Cont.*

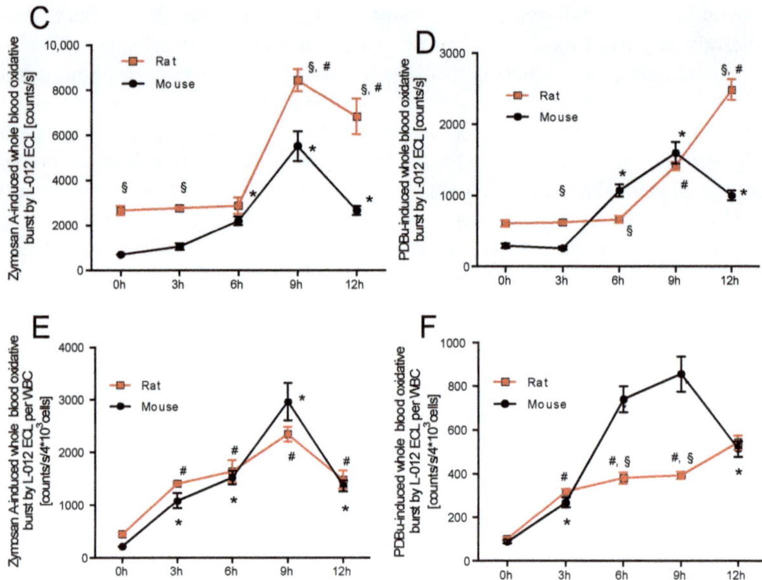

Figure 1. Time response of platelets and white blood cell derived oxidative burst in mice and rats. Thrombocyte count (green line already published in Steven et. al. BJP 2017 [10]) (**A**), white blood cell (WBC) count (**B**) and oxidative burst (nicotinamide adenine dinucleotide phosphate-oxidase (NADPH) oxidase and myeloperoxidase activity) in whole blood after zymosan A (**C**) or PDBu (**D**) stimulation in mice and rats was determined by chemiluminescence (L-012) over a 12 h time response. (**E,F**) Oxidative burst was normalized to the WBC count. The data are mean ± SEM from 6 different animals per group. * $p < 0.05$ vs. 0 h mouse, # $p < 0.05$ vs. 0 h rat and § $p < 0.05$ vs. mouse (same point in time).

Normalization of oxidative burst to WBC count in whole blood changed not only the kinetic pattern but also the relative level in the species significantly for both assays. Whereas, mice and rats showed a similar level of zymosan A-induced oxidative burst at each point in time (Figure 1E), mice showed a significantly higher PDBu-induced oxidative burst level at 6 and 9 h after LPS injection when compared to rats (Figure 1F). Finally, after 12 h burst levels were equal in both species. Of note, although the WBC count is higher in human subjects than in rodents, these numbers are comparable in rats and mice [12,13].

2.2. Time Response of Nitrosyl-Iron Hemoglobin and Inos Expression of Isolated WBC in Mice and Rats

Nitrosyl-iron hemoglobin (HbNO) significantly increased in mice after 9 h and in rats after 6 h of LPS treatment. The peak levels of HbNO were detected at 12 h after LPS injection in both species. HbNO formation was significantly (2–3-fold) higher in rats after 6, 9, and 12 h of LPS treatment as compared to mice (Figure 2A). If corrected to WBC count, the time response of HbNO levels changed. Rats demonstrated a strong peak after 6 h, whereas the time response of HbNO levels in mice showed a similar pattern as compared to the data without normalization to WBC count (Figure 2B).

Vascular mRNA of iNOS increased after 3 h of LPS treatment in both species. A substantial peak level was detected after 6 h in rats, whereas this maximum was much less pronounced in mice. mRNA expression of iNOS decreased with time and was normalized in mice, but still significantly elevated in rats at 12 h after LPS injection (Figure 2C). Immunohistochemical staining of aorta revealed an increase in 3-nitrotyrosine positive proteins throughout the entire vascular wall in both species after 12 h, although the staining was more pronounced in rats as compared to mice (Figure 2D). iNOS expression in isolated WBC of mice and rats was elevated at 12 h after induction of endotoxaemia (Figure 2E). However, rats showed a significantly higher iNOS protein expression as compared to mice.

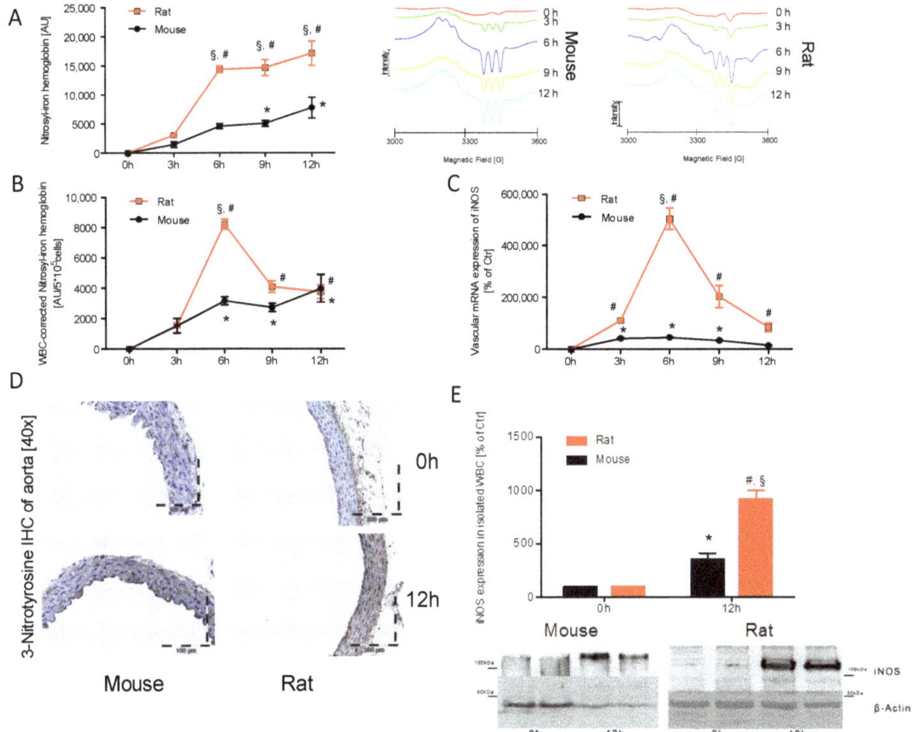

Figure 2. Time response of nitrosyl-iron hemoglobin and iNOS expression in isolated WBC of mice and rats. Whole blood Hb-NO levels were determined by Electron Paramagnetic Resonance (EPR) spectroscopy as a read-out of iNOS activity (**A**) and were normalized to the WBC count (**B**). qRT-PCR was used to determine mRNA expression levels of *iNOS* in aortic tissue (**C**). iNOS protein expression was further visualized by immunohistochemistry of paraffin embedded aortic sections after 12 h (**D**). iNOS protein expression was further investigated in isolated WBC of both species after 12 h using Western-blot technique (**E**). Each lane in the original blot represents a protein sample from 1–2 animals. The data are mean ± SEM from 6 different animals per group. * $p < 0.05$ vs. 0 h mouse, # $p < 0.05$ vs. 0 h rat and § $p < 0.05$ vs. mouse (same point in time).

2.3. Time Response of Vascular Inflammation in Mice and Rats

VCAM-1 mRNA expression increased significantly with time in endotoxaemic mice and rats. *VCAM-1* expression in mice showed a maximum at 3 h after LPS injection (3-fold higher compared to rats) and *VCAM-1* expression declined after the 3 h maximum but was still significantly higher at all points in time compared to 0 h. The maximum *VCAM-1* expression in rats was found after 12 h of LPS treatment with a similar amount as compared to mice and expression levels continuously increased with time in rats (Figure 3A). Vascular mRNA expression of *P-selectin* showed a peak after 3 h of LPS treatment in both species and time-dependently returned to normal levels afterwards. The absolute expression levels and time responses were almost identical for both species (Figure 3B). *IL-6* mRNA expression showed a maximum at 3 h after LPS injection in mice and at 6 h in rats. Afterwards, the *IL-6* levels were normalized in both species with an overall almost identical time response (Figure 3C). *TNF-α* mRNA expression showed a peak at 3 h after LPS injection in both species, but this maximum was less pronounced in rats as compared to mice. Expression levels were higher at all time response compared to 0 h values in mice and rats (Figure 3D).

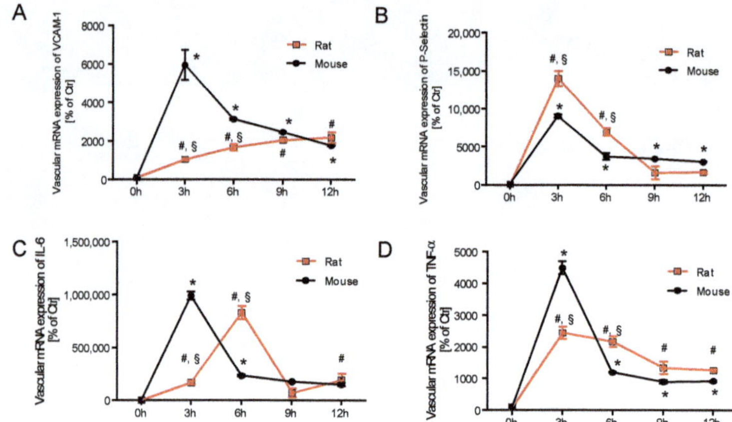

Figure 3. Time response of vascular inflammation in mice and rats. qRT-PCR was used to determine mRNA expression levels of *VCAM-1* (**A**), *P-selectin* (**B**), *IL-6* (**C**) and *TNF-α* (**D**) in aortic tissue over a 12 h time response. The data are mean ± SEM from 6 different animals per group. * $p < 0.05$ vs. 0 h mouse, [#] $p < 0.05$ vs. 0 h rat and [§] $p < 0.05$ vs. mouse (same point in time).

mRNA expression of *CD68*, as a marker for leukocyte infiltration, was increased at the 9 and 12 h point in time (Figure 4A). Increased CD68 positive protein was also detected by immunohistochemical staining of rat aorta at 12 h after LPS injection (Figure 4C). *CD11b* mRNA levels in vascular tissue showed a time-dependent significant increase in mice at the 6, 9, and 12 h points in time (Figure 4B). F4/80 immunohistochemical staining of mouse aorta revealed more F4/80 positive protein throughout the entire vascular wall after 12 h of LPS treatment (Figure 4D). Whereas, CD68 and F4/80 are good markers for infiltrated macrophages, CD11b is a rather unspecific marker that is expressed on all myelomonocytic cells.

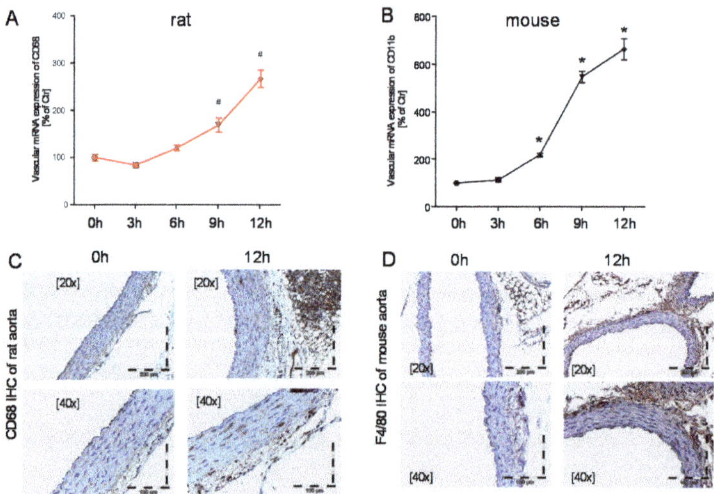

Figure 4. Time response of vascular inflammation in mice and rats. qRT-PCR was used to determine mRNA expression levels of *CD68* (**A**) in rats and *CD11b* in mice (**B**). CD68 protein levels in rats (**C**) and F4/80 protein levels in mice (**D**) were further visualized by immunohistochemistry of paraffin embedded aortic sections after 12 h. The data are mean ± SEM from 6 different animals per group. * $p < 0.05$ vs. 0 h mouse, [#] $p < 0.05$ vs. 0 h rat (same point in time).

3. Discussion

In this study, we provide very detailed time responses (0, 3, 6, 9, and 12 h) of oxidative/nitrosative stress and vascular inflammation in LPS-induced endotoxaemia in mice and rats. We also provide a detailed comparison of these time responses between mice and rats to identify species-independent pathways that might be of relevance for human sepsis as well. LPS-induced endotoxaemia is a frequently used model for preclinical research on sepsis. However, so far, all published studies used different points in time and multiple species, which makes it hard to compare the disease-relevant pathomechanisms and translate the findings to the human setting. Besides species-dependent differences in the inflammatory response, there are even notable differences between strains of the same species. It is known, that C57BL/6j (B6) tend to have a TH1-predominat response to pathogens, whereas A/J, BALB/C, and DBA/2 mice have a TH2-type preference [14]. Furthermore, B6 mice when compared to A/J mice have a significant higher mortality, which was accompanied by higher IL-10 plasma levels and myeloperoxidase activity [15]. The present study underlines the importance of using similar time points, strains and species, and highlights some alarming differences in the time responses and absolute changes across two animal species warranting cautious interpretation of data from experimental studies when translating them to human sepsis.

The endotoxin LPS from gram-negative bacteria triggers inflammation by binding directly via CD14 on monocytes and via toll-like receptors (TLR) on endothelial cells after forming a complex with LPS binding protein (LBP) [16]. The LPS model nicely mirrors the severe inflammatory response of the organism but does not challenge the host with living bacteria. Anti-inflammatory treatment of sepsis should counteract the excessive inflammatory cascade and contribute to improved survival. Statins have such anti-inflammatory properties and a recent trial tested the use of rosuvastatin in patients suffering from acute respiratory distress syndrome (ARDS) but the results were disappointing and statin therapy failed to improve clinical outcome [17]. Several other compounds were tested in clinical and pre-clinical trials to find evidence for improved survival by the reduction of the inflammatory response in sepsis, but only cortisone therapy can be found in recent clinical guidelines [4].

The importance of time responses for our understanding of the septic pathophysiology is underlined by literature reports on different time windows or phases for human sepsis but also LPS-induced endotoxaemia [8,18]: the first phase is between 0 and 2 h after bacterial invasion/LPS exposure, and the second phase between 2 and 12 h. Within these two phases, pathophysiological processes (cell activation, inflammation, and hypotension but also local endothelin-1 formation [8,18,19]) are reversible and can be pharmacologically modified [18]. The upregulation of cell activation markers and inflammation were also observed in the present study (e.g., CD68, F4/80, VCAM-1). After these two phases, a point of no return was postulated, and all later events contribute directly to the high mortality (e.g., massive oxidative damage, cell death, disseminated intravasal coagulation (DIC), and end organ failure), and cannot be easily modified by pharmacotherapy [8]. According to a clinical study of Vargas et al. the decision on survival or death in septic shock develops between 40 and 60 h post infection (probably the decision is already taken somewhat earlier, between 12 and 24 h) and can be predicted from the serum pattern of the inflammation markers IL-6 and IL-8 as well as the adhesion molecules sELAM-1 and sICAM-1 [20]. After 40 h, the level of these inflammation markers returned to normal in the survivor group but escalated in the non-survivor group. Another clinical study showed that the serum levels of the regulator of endothelial cell migration and endothelial permeability VEGF-A as well as its receptor sFlt-1 were significantly upregulated (48 h after the onset of fever) in patients with septic shock but were at normal level in patients who had developed sepsis without shock [21]. Therefore, according to this study, the decision on development of shock or "mild" sepsis was already taken at 48 h post infection. In an animal experimental study on septic sheep, Lange et al. presented data on the time response of different clinical and biochemical markers and revealed a dramatic change in the expression of NOS isoforms between 8 and 12 h post infection, which was followed by severe worsening of the clinical parameters [22].

In addition, a large number of studies investigated the time response of one marker in plasma or serum such as angiopoietin-2, gelsolin, neopterin, C-reactive protein (CRP), and selenium. In summary, despite many efforts it is not clear how the septic situation in a given animal develops to the critical point at which the vascular biochemistry and physiology either return to normal or develop into the lethal situation of septic shock. It is also unclear which animal model should be applied to mimic human sepsis but some data from patients are helpful in pointing out the crucial steps involved. In the first place this applies to the high output of prostacyclin (PGI_2, measured as 6-keto-$PGF_{1\alpha}$) before shock develops [23–26]. Of clinical importance is the fact that high levels of 6-keto-$PGF_{1\alpha}$ are associated with a bad prognosis of septic patients [23]. One may speculate that similar to the inhibition of PGI_2 synthesis in the endothelium after 1 h of LPS exposure the beneficial effects of smooth muscle derived PGI_2 after 4–12 h could be eliminated by the inhibition of its synthesis, which could either be brought about by inhibition of inducible cyclooxygenase (COX-2) (e.g., downregulation or nitration of COX-2 [27]) or even by nitration of PGI_2 synthase after breakdown of the smooth muscle antioxidant potential, followed by a rise of the peroxynitrite levels in this previously resistant cell layer. Of note, despite the presence of severe hypotension, a considerable degree of vascular dysfunction may exist, a paradox that is best explained as follows: the substantial stimulation of the sGC/cGMP signaling cascade with high formation rates of $^\bullet$NO from iNOS (as shown here by EPR-based measurement of nitrosyl-iron-Hb) may result in a desensitization of this signaling pathway, endothelial dysfunction, and vasoconstriction [28,29]. This would provide a reasonable explanation for the impaired vascular function despite systemic hypotension and overproduction of vasodilators. Therefore, the time window between 12 and 24 h may have special significance for the development of lethality.

In the third phase, NOS-2 and COX-2 expression are decreased, PGI_2 synthase is nitrated/inactivated due to huge amounts of ROS and peroxynitrite from infiltrated leukocytes and endothelial dysfunction, at least in the larger vessels, is observed. At this time point (upon loss of vasodilators), enhanced thrombosis is encountered [30], as also observed in our LPS treated mice [10]. The relevance of microthrombi formation (disseminated intravascular coagulation) for the prognosis of septic patients was also demonstrated by an increased mortality of septic patients with reduced numbers of circulating platelets (thrombocytopenia) [31], as also observed in our LPS treated rats and mice [10,11]. This represents another hint for the loss of the potent antiaggregatory compounds PGI_2 and $^\bullet$NO. Therefore, the impaired vascular function observed in our previous studies, in vessels from LPS-treated rats and mice, is in good accordance with the clinical and experimental time response of sepsis [8,10,11]. For LPS-treated rats previous studies have shown substantial hypotension at 6 h after LPS treatment and increase in mean arterial pressure afterwards [19,32]. This would be in good accordance with the here postulated third phase of sepsis, characterized by impaired vascular relaxation and even a mild hypertension (at least restricted to certain vessel areas or organs) as well as the previously reported increased endothelin-1 levels can be observed in the late phase of sepsis.

Previous studies from our group and others show that median survival of C57/Bl6j mice upon i.p. injection of 10 mg/kg LPS is approximately 20 h after injection, whereas the median survival rate of Wistar rats challenged with the same amount of LPS intra peritoneal was shorter (6–10 h) [8,11,33]. Interestingly, mRNA levels of inflammatory makers like *VCAM-1*, *IL-6*, and *TNF-α* were significantly higher and had an earlier peak (*IL-6* and *VCAM-1*) in mice as compared to rats, which reflects a stronger and earlier inflammatory response in mice compared to rats. However, the time response and absolute level of nitrosative/oxidative stress rather showed an opposite pattern. In both species WBC count in whole blood decreased significantly after 3 h of LPS treatment, but only in rats the number of WBC recovered significantly. Interestingly, the absolute WBC-derived oxidative burst was significantly higher in rats when compared to mice at 0, 3, 9, and 12 h points in time.

NO mediates several beneficial effects like vasodilation, inhibition of platelet aggregation, anti-inflammatory and anti-apoptotic effects via activating soluble guanylate cyclase (sGC) [34]. In the setting of septic shock excessive generation of $^\bullet$NO by the iNOS in combination with excessive production of superoxide ($O_2^{\bullet-}$) by NADPH oxidase of WBC leads to generation of peroxynitrite ($ONOO^-$). The latter is capable to affect enzyme function by oxidation or nitration, induces cell damage by lipid peroxidation, and attacks mitochondrial enzymes leading to mitochondrial dysfunction [35]. Our data demonstrate in detail, that $^{\bullet}$NO generation in total is significantly higher in rats 6 h after induction of sepsis, which was due to increased expression of iNOS in WBC and vascular tissue. Together with the high burden of superoxide generation a strong peroxynitrite footprint was detectable by immunohistochemical staining of vascular tissue. Taken together, the burden of peroxynitrite formation in rats was higher than in mice.

4. Materials and Methods

4.1. Materials

The Bradford reagent was obtained from BioRad (Munich, Germany). The QuantiTect probe RT-PCR Kit was purchased from Qiagen (Hilden, Germany). All TaqMan probes were purchased from Applied Biosystems (Darmstadt, Germany). L-012 (8-amino-5-chloro-7-phenylpyrido[3,4-d] pyridazine-1,4-(2*H*,3*H*)dione sodium salt) was purchased from Wako Pure Chemical Industries (Osaka, Japan). For the induction of septic shock, we used LPS from *Salmonella Thyphosa* from Sigma-Aldrich (St. Louis, MO, USA). All of the other chemicals were of analytical grade and were obtained from Sigma-Aldrich, Fluka (St. Louis, MO, USA) or Merck (Darmstadt, Germany).

4.2. Animals and In Vivo Treatment

All of the animal care and experimental procedures were in accordance with the Guide for the Care and Use of Laboratory Animals as adopted by the US National Institutes of Health, and approved by the Ethics Committee of the University Medical Center Mainz and the Landesuntersuchungsamt Koblenz (#23 177-07/G 14-1-039, 12 June 2014). In this study, a total number of 30 C57BL/6j mice and 30 Wistar rats received single injection of LPS intraperitoneally at a dosage of 10 mg/kg for both species. Animals of both species were divided into five groups, six animals each, and sacrificed by exsanguination under isoflurane anesthesia (5% inhalant in room air) at 0, 3, 6, 9, and 12 h after injection of LPS. All C57BL/6j mice were male, and weighed 25–30 g at the time of the experiment. All of the Wistar rats were male, and weighed 270–330 g at the time of the experiment.

4.3. Chemiluminescence-Based Detection of Oxidative Stress in Whole Blood

Whole blood oxidative burst mainly reflects leukocyte NADPH oxidase and myeloperoxidase activity, and was used as a global readout of the burden of oxidant formation as well as of the activation state of inflammatory pathways. To measure whole blood leukocyte-dependent ROS formation, fresh blood (in citrate tubes) was stimulated with phorbol ester PDBu (10 µM) or the fungal endotoxin zymosan A (50 µg/mL) and assessed in PBS containing Ca^{2+}/Mg^{2+} (1 mM) by L-012 (100 µM) enhanced chemiluminescence (ECL) using a Centro chemiluminescence plate reader from Berthold Technologies (Bad Wildbad, Germany) [8].

4.4. Quantification of Nitrosyl-Iron Hemoglobin in Whole Blood by Electron Paramagnetic Resonance (EPR) Spectroscopy

NO synthesis and the burden of nitrosative stress were also assessed by the EPR-based detection of Hb-NO. Samples of venous blood for Hb-NO/EPR studies were obtained by the puncture of the right heart of anesthetized mice and rats; blood samples were immediately frozen and stored in liquid nitrogen. The EPR measurements were carried out at 77 K using an X-band table-top spectrometer MS400 from Magnettech (Berlin, Germany). The instrument settings were as follows: 10 mW microwave power, 7000 mG amplitude modulation, 100 kHz modulation frequency, 3300 G center field, 300 G sweep width, 60 s sweep time and three scans, as described before [11].

4.5. Isolation of White Blood Cells

The procedure was described before [36]. Briefly, erythrocytes in mice and rat blood were separated by sedimentation in 15 mL heparin-supplemented tubes after addition of an equal volume of dextran solution (MW 485,000, T500 9219.1 from Roth). 1.9 g dextran were dissolved in 50 mL 0.9% NaCl solution. The leukocyte-containing supernatant was centrifuged on Histopaque-1083 from Sigma Aldrich for 30 min at $500 \times g$ at 20 °C, resulting in a neutrophil (PMN)-containing pellet and the monocyte/lymphocyte-enriched (WBC) "buffy coat" between the aqueous and Ficoll phases. The WBC fraction was collected and purified by further centrifugation for 10 min at $500 \times g$, followed by resuspension in PBS. The PMN pellet was freed from residual erythrocytes by hypotonic lysis in distilled water and centrifugation at $500 \times g$ (two to three times). The total blood cell count and the purity of the fractions were evaluated using an automated approach using a hematology analyzer KX-21N from Sysmex Europe GmbH (Norderstedt, Germany). Typical content of WBC in each fraction was previously published [37].

4.6. Western Blot Analysis

Isolated white blood cell proteins from mice and rats were separated by 7.5% SDS-PAGE under reducing conditions. After blotting on a nitrocellulose membrane, immunoblotting was accomplished using antibodies against β-actin (rabbit, monoclonal, 1:2500, Sigma-Aldrich, Seelze, Germany), and iNOS (purified anti iNOS, mouse, monoclonal, 1:2500, BD Biosciences, Franklin Lakes, NJ). Detection was accomplished with either Super Signal Substrate (Pierce, Rockford, IL, USA) or ECL Reagent (Amersham, Piscataway, NJ, USA). Bands were evaluated by densitometry.

4.7. Reverse Transcription Real-Time PCR (qRT-PCR)

mRNA expression was analyzed with quantitative real-time RT-PCR, as previously described [38]. RNA from aorta and white blood cells was isolated in both species for the experiments. Briefly, total RNA from mouse and rat aorta, and from isolated whole blood cells of mice and rats was isolated (RNeasy Fibrous Tissue Mini Kit; Qiagen, Hilden, Germany), and 50 ng RNA was used for real-time RT-PCR analysis with the QuantiTect Probe RT-PCR kit (Qiagen). TaqMan® gene expression assays for *vascular cell adhesion molecule 1* (*VCAM-1*), the cytokine *interleukin 6* (*IL-6*), the adhesion molecule *P-selectin*, the immune-signaling protein *tumor necrosis factor-α* (*TNF-α*), as well as the house keeping gene *TATA-box binding protein* (*TBP*) were purchased as probe-and-primer sets (Applied Biosystems, Foster City, CA) for mice and rats. TaqMan® gene expression assays for leukocyte marker *CD68* in mice and *macrophage-1 antigen* also known as *CD11b* in rats from the same provider were also tested. The comparative Delta Ct method was used for relative mRNA quantification. Gene expression was normalized to the endogenous control, *TBP* mRNA, and the amount of target gene mRNA expression in each sample was expressed relative to that of control for every species.

4.8. Immunohistochemistry and Fluorescence Microscopy

For immunohistochemistry aorta segments from rats and mice were fixed in paraformaldehyde (4%) and embedded in paraffin. Aortic segments from both species were stained with primary antibodies against 3-nitrotyrosine (Millipore, Burlington MA, USA). Mouse aortic segments were also stained with primary antibodies against F4/80 (eBioscience) and rat aortic segments were also stained for CD68 (LS Bio, Seattle, WA, USA). Anti-rat, anti-rabbit, and LSAB (Vector; Sigma; DAKO, Glostrup, Denmark, respectively) were used as secondary antibodies. For immunohistochemical detection ABC reagent (Vector, Burlingame, CA, USA) and then DAB reagent (peroxidase substrate Kit, Vector, Burlingame, CA, USA) were used.

4.9. Statistical Analysis

Results are expressed as mean $\pm$ SEM. Two-way ANOVA (with Holm-Sidak's correction for comparison of multiple means) was used for comparisons of time responses of WBC, mRNA expression (*VCAM-1*, *P-selectin*, *IL-6*, *TNF-α*), HbNO-levels and oxidative burst. One-way ANOVA (with Bonferroni's correction for comparison of multiple means) was used for comparisons of mRNA expression (*CD11b*, *CD68*) and iNOS protein levels. p values < 0.05 were considered as statistically significant. All of the statistical analyses were performed using GraphPad Prism 6.0d.

5. Conclusions

Human sepsis is poorly understood and more research is needed to understand the pathomechanisms. Sepsis studies are often performed in different species (mice and rats) and even different strains (C57Bl6/j, BALB/c, A/J) challenged with LPS or CLP-procedure and the results are compared to each other. Nevertheless, the models are poorly characterized, especially with respect to time responses for the entire set of parameters. With the present study, we present species specific, time-dependent changes of regulators of cell activation, inflammation, and oxidative/nitrosative stress in mice and rats after LPS-challenge. Based on our present findings, the survival of endotoxaemic animals might rather be related to suppression of oxidative/nitrosative stress than to complete suppression of inflammatory responses. These data can help to interpret sepsis studies performed in models of LPS-induced endotoxaemia in mice and rats and translate these data to the human setting. Based on this study, further time- and species-dependent characterization of sepsis relevant parameters like hemostasis, vascular function, or immune cell differentiation and recruitment should be conducted.

Acknowledgments: We are indebted to Jessica Rudolph, Bettina Mros, Jörg Schreiner, Angelica Karpi and Nicole Glas for expert technical assistance. This paper contains results that are part of the doctoral thesis of Mobin Dib. Sebastian Steven holds a Virchow-Fellowship from the Center of Thrombosis and Hemostasis (Mainz, Germany) funded by the German Federal Ministry of Education and Research (BMBF 01EO1003).

Author Contributions: Sebastian Steven and Andreas Daiber conceived and designed the experiments; Sebastian Steven, Mobin Dib, Siyer Roohani, Fatemeh Kashani performed the experiments; Sebastian Steven analyzed the data; Sebastian Steven and Andreas Daiber wrote the paper; Thomas Münzel critically revised the paper.

Conflicts of Interest: The authors declare no conflict of interest.

Abbreviations

LPS	Lipopolysaccharide
WBC	White blood cell
HbNO	nitrosyl-haemoglobin
iNOS	inducible NOS (type 2)
IL-6	Interleukin-6
TNF-α	tumor necrosis factor-α
VCAM-1	vascular adhesion molecule-1
PDBu	Phorbol 12,13-dibutyrate

References

1. Rhodes, A.; Phillips, G.; Beale, R.; Cecconi, M.; Chiche, J.D.; De Backer, D.; Divatia, J.; Du, B.; Evans, L.; Ferrer, R.; et al. The surviving sepsis campaign bundles and outcome: Results from the international multicentre prevalence study on sepsis (the impress study). *Intensive Care Med.* **2015**, *41*, 1620–1628. [CrossRef] [PubMed]
2. Hall, M.J.; Williams, S.N.; DeFrances, C.J.; Golosinskiy, A. Inpatient care for septicemia or sepsis: A challenge for patients and hospitals. *NCHS Data Brief* **2011**, *62*, 1–8.
3. Singer, M.; Deutschman, C.S.; Seymour, C.W.; Shankar-Hari, M.; Annane, D.; Bauer, M.; Bellomo, R.; Bernard, G.R.; Chiche, J.D.; Coopersmith, C.M.; et al. The third international consensus definitions for sepsis and septic shock (sepsis-3). *JAMA* **2016**, *315*, 801–810. [CrossRef] [PubMed]

4. Gotts, J.E.; Matthay, M.A. Sepsis: Pathophysiology and clinical management. *BMJ* **2016**, *353*, i1585. [CrossRef] [PubMed]

5. Bone, R.C. Sir isaac newton, sepsis, sirs, and cars. *Crit. Care Med.* **1996**, *24*, 1125–1128. [CrossRef] [PubMed]

6. Takao, K.; Miyakawa, T. Genomic responses in mouse models greatly mimic human inflammatory diseases. *Proc. Natl. Acad. Sci. USA* **2015**, *112*, 1167–1172. [CrossRef] [PubMed]

7. Seok, J.; Warren, H.S.; Cuenca, A.G.; Mindrinos, M.N.; Baker, H.V.; Xu, W.; Richards, D.R.; McDonald-Smith, G.P.; Gao, H.; Hennessy, L.; et al. Genomic responses in mouse models poorly mimic human inflammatory diseases. *Proc. Natl. Acad. Sci. USA* **2013**, *110*, 3507–3512. [CrossRef] [PubMed]

8. Kroller-Schon, S.; Knorr, M.; Hausding, M.; Oelze, M.; Schuff, A.; Schell, R.; Sudowe, S.; Scholz, A.; Daub, S.; Karbach, S.; et al. Glucose-independent improvement of vascular dysfunction in experimental sepsis by dipeptidyl-peptidase 4 inhibition. *Cardiovasc. Res.* **2012**, *96*, 140–149. [CrossRef] [PubMed]

9. Steven, S.; Munzel, T.; Daiber, A. Exploiting the pleiotropic antioxidant effects of established drugs in cardiovascular disease. *Int. J. Mol. Sci.* **2015**, *16*, 18185–18223. [CrossRef] [PubMed]

10. Steven, S.; Jurk, K.; Kopp, M.; Kroller-Schon, S.; Mikhed, Y.; Schwierczek, K.; Roohani, S.; Kashani, F.; Oelze, M.; Klein, T.; et al. Glucagon-like peptide-1 receptor signalling reduces microvascular thrombosis, nitro-oxidative stress and platelet activation in endotoxaemic mice. *Br. J. Pharmacol.* **2017**, *174*, 1620–1632. [CrossRef] [PubMed]

11. Steven, S.; Hausding, M.; Kroller-Schon, S.; Mader, M.; Mikhed, Y.; Stamm, P.; Zinssius, E.; Pfeffer, A.; Welschof, P.; Agdauletova, S.; et al. Gliptin and glp-1 analog treatment improves survival and vascular inflammation/dysfunction in animals with lipopolysaccharide-induced endotoxemia. *Basic Res. Cardiol.* **2015**, *110*, 6. [CrossRef] [PubMed]

12. Mestas, J.; Hughes, C.C. Of mice and not men: Differences between mouse and human immunology. *J. Immunol.* **2004**, *172*, 2731–2738. [CrossRef] [PubMed]

13. Petterino, C.; Argentino-Storino, A. Clinical chemistry and haematology historical data in control sprague-dawley rats from pre-clinical toxicity studies. *Exp. Toxicol. Pathol.* **2006**, *57*, 213–219. [CrossRef] [PubMed]

14. Sellers, R.S.; Clifford, C.B.; Treuting, P.M.; Brayton, C. Immunological variation between inbred laboratory mouse strains: Points to consider in phenotyping genetically immunomodified mice. *Vet. Pathol.* **2012**, *49*, 32–43. [CrossRef] [PubMed]

15. Stewart, D.; Fulton, W.B.; Wilson, C.; Monitto, C.L.; Paidas, C.N.; Reeves, R.H.; De Maio, A. Genetic contribution to the septic response in a mouse model. *Shock* **2002**, *18*, 342–347. [CrossRef] [PubMed]

16. Anas, A.A.; Wiersinga, W.J.; de Vos, A.F.; van der Poll, T. Recent insights into the pathogenesis of bacterial sepsis. *Neth. J. Med.* **2010**, *68*, 147–152. [PubMed]

17. National Heart, L.; Blood Institute, A.C.T.N.; Truwit, J.D.; Bernard, G.R.; Steingrub, J.; Matthay, M.A.; Liu, K.D.; Albertson, T.E.; Brower, R.G.; Shanholtz, C.; et al. Rosuvastatin for sepsis-associated acute respiratory distress syndrome. *N. Engl. J. Med.* **2014**, *370*, 2191–2200.

18. Bachschmid, M.; Ullrich, V. Redox signalling in endothelial cells—Novel mechanisms explaining endothelium function and dysfunction in health, disease and aging. *BIF futura* **2003**, *18*, 223–230.

19. Yamaguchi, N.; Jesmin, S.; Zaedi, S.; Shimojo, N.; Maeda, S.; Gando, S.; Koyama, A.; Miyauchi, T. Time-dependent expression of renal vaso-regulatory molecules in lps-induced endotoxemia in rat. *Peptides* **2006**, *27*, 2258–2270. [CrossRef] [PubMed]

20. Hein, O.V.; Misterek, K.; Tessmann, J.P.; van Dossow, V.; Krimphove, M.; Spies, C. Time course of endothelial damage in septic shock: Prediction of outcome. *Crit. Care* **2005**, *9*, R323–R330. [CrossRef] [PubMed]

21. Alves, B.E.; Montalvao, S.A.; Aranha, F.J.; Lorand-Metze, I.; De Souza, C.A.; Annichino-Bizzacchi, J.M.; De Paula, E.V. Time-course of sFlt-1 and VEGF-A release in neutropenic patients with sepsis and septic shock: A prospective study. *J. Transl. Med.* **2011**, *9*, 23. [CrossRef] [PubMed]

22. Lange, M.; Connelly, R.; Traber, D.L.; Hamahata, A.; Nakano, Y.; Esechie, A.; Jonkam, C.; von Borzyskowski, S.; Traber, L.D.; Schmalstieg, F.C.; et al. Time course of nitric oxide synthases, nitrosative stress, and poly(adp ribosylation) in an ovine sepsis model. *Crit. Care* **2010**, *14*, R129. [CrossRef] [PubMed]

23. Nakae, H.; Endo, S.; Inada, K.; Watanabe, M.; Baba, N.; Yoshida, M. Relationship between leukotriene b4 and prostaglandin i2 in patients with sepsis. *Res. Commun. Mol. Pathol. Pharmacol.* **1994**, *86*, 37–42. [PubMed]

24. Yellin, S.A.; Nguyen, D.; Quinn, J.V.; Burchard, K.W.; Crowley, J.P.; Slotman, G.J. Prostacyclin and thromboxane a2 in septic shock: Species differences. *Circ. Shock* **1986**, *20*, 291–297. [PubMed]

25. Butler, R.R., Jr.; Wise, W.C.; Halushka, P.V.; Cook, J.A. Thromboxane and prostacyclin production during septic shock. *Adv. Shock Res.* **1982**, *7*, 133–145. [PubMed]
26. Cook, J.A.; Wise, W.C.; Butler, R.R.; Reines, H.D.; Rambo, W.; Halushka, P.V. The potential role of thromboxane and prostacyclin in endotoxic and septic shock. *Am. J. Emerg. Med.* **1984**, *2*, 28–37. [CrossRef]
27. Schildknecht, S.; Heinz, K.; Daiber, A.; Hamacher, J.; Kavakli, C.; Ullrich, V.; Bachschmid, M. Autocatalytic tyrosine nitration of prostaglandin endoperoxide synthase-2 in lps-stimulated raw 264.7 macrophages. *Biochem. Biophys. Res. Commun.* **2006**, *340*, 318–325. [CrossRef] [PubMed]
28. August, M.; Wingerter, O.; Oelze, M.; Wenzel, P.; Kleschyov, A.L.; Daiber, A.; Mulsch, A.; Munzel, T.; Tsilimingas, N. Mechanisms underlying dysfunction of carotid arteries in genetically hyperlipidemic rabbits. *Nitric Oxide* **2006**, *15*, 241–251. [CrossRef] [PubMed]
29. Kessler, P.; Bauersachs, J.; Busse, R.; Schini-Kerth, V.B. Inhibition of inducible nitric oxide synthase restores endothelium-dependent relaxations in proinflammatory mediator-induced blood vessels. *Arterioscler. Thromb. Vasc. Biol.* **1997**, *17*, 1746–1755. [CrossRef] [PubMed]
30. Holub, M. Thromboembolic complications of sepsis: What is the incidence and pathophysiological mechanisms involved? *Thromb. Haemost.* **2008**, *99*, 801–802. [CrossRef] [PubMed]
31. Tridente, A.; Clarke, G.M.; Walden, A.; Gordon, A.C.; Hutton, P.; Chiche, J.D.; Holloway, P.A.; Mills, G.H.; Bion, J.; Stuber, F.; et al. Association between trends in clinical variables and outcome in intensive care patients with faecal peritonitis: Analysis of the genosept cohort. *Crit. Care* **2015**, *19*, 210. [CrossRef] [PubMed]
32. Hocherl, K.; Schmidt, C.; Kurt, B.; Bucher, M. Activation of the pgi(2)/ip system contributes to the development of circulatory failure in a rat model of endotoxic shock. *Hypertension* **2008**, *52*, 330–335. [CrossRef] [PubMed]
33. Ku, H.C.; Chen, W.P.; Su, M.J. Glp-1 signaling preserves cardiac function in endotoxemic fischer 344 and dpp4-deficient rats. *Naunyn Schmiedebergs Arch. Pharmacol.* **2010**, *382*, 463–474. [CrossRef] [PubMed]
34. Pacher, P.; Beckman, J.S.; Liaudet, L. Nitric oxide and peroxynitrite in health and disease. *Physiol. Rev.* **2007**, *87*, 315–424. [CrossRef] [PubMed]
35. Szabo, C.; Modis, K. Pathophysiological roles of peroxynitrite in circulatory shock. *Shock* **2010**, *34* (Suppl. S1), 4–14. [CrossRef] [PubMed]
36. Daiber, A.; August, M.; Baldus, S.; Wendt, M.; Oelze, M.; Sydow, K.; Kleschyov, A.L.; Munzel, T. Measurement of nad(p)h oxidase-derived superoxide with the luminol analogue l-012. *Free Radic. Biol. Med.* **2004**, *36*, 101–111. [CrossRef] [PubMed]
37. Wenzel, P.; Schulz, E.; Gori, T.; Ostad, M.A.; Mathner, F.; Schildknecht, S.; Gobel, S.; Oelze, M.; Stalleicken, D.; Warnholtz, A.; et al. Monitoring white blood cell mitochondrial aldehyde dehydrogenase activity: Implications for nitrate therapy in humans. *J. Pharmacol. Exp. Ther.* **2009**, *330*, 63–71. [CrossRef] [PubMed]
38. Hausding, M.; Jurk, K.; Daub, S.; Kroller-Schon, S.; Stein, J.; Schwenk, M.; Oelze, M.; Mikhed, Y.; Kerahrodi, J.G.; Kossmann, S.; et al. Cd40l contributes to angiotensin ii-induced pro-thrombotic state, vascular inflammation, oxidative stress and endothelial dysfunction. *Basic Res. Cardiol.* **2013**, *108*, 386. [CrossRef] [PubMed]

International Journal of
Molecular Sciences

Article

LPS-Induced Low-Grade Inflammation Increases Hypothalamic JNK Expression and Causes Central Insulin Resistance Irrespective of Body Weight Changes

Rodrigo Rorato [1,*,†], Beatriz de Carvalho Borges [1,3,†], Ernane Torres Uchoa [1,2],
José Antunes-Rodrigues [1], Carol Fuzeti Elias [3] and Lucila Leico Kagohara Elias [1,*]

[1] Department of Physiology, Ribeirao Preto Medical School, University of Sao Paulo,
    Sao Paulo 14049-900, Brazil; borgesbc@gmail.com (B.d.C.B.); ernane_uchoa@yahoo.com.br (E.T.U.);
    jantunesr@gmail.com (J.A.-R.)
[2] Department of Physiological Sciences, State University of Londrina, Londrina 86057-970, Brazil
[3] Department of Molecular and Integrative Physiology, University of Michigan,
    Ann Arbor, MI 48109-5622, USA; cfelias@med.umich.edu
*   Correspondence: rcrorato@yahoo.com.br (R.R.); llelias@fmrp.usp.br (L.L.K.E.);
    Tel.: +55-1633-1530-57 (R.R. & L.L.K.E.); Fax: +55-1633-1500-17 (R.R. & L.L.K.E.)
†   These authors contribute equally to this study.

Received: 26 April 2017; Accepted: 27 June 2017; Published: 4 July 2017

Abstract: Metabolic endotoxemia contributes to low-grade inflammation in obesity, which causes insulin resistance due to the activation of intracellular proinflammatory pathways, such as the c-Jun N-terminal Kinase (JNK) cascade in the hypothalamus and other tissues. However, it remains unclear whether the proinflammatory process precedes insulin resistance or it appears because of the development of obesity. Hypothalamic low-grade inflammation was induced by prolonged lipopolysaccharide (LPS) exposure to investigate if central insulin resistance is induced by an inflammatory stimulus regardless of obesity. Male Wistar rats were treated with single (1 LPS) or repeated injections (6 LPS) of LPS (100 µg/kg, IP) to evaluate the phosphorylation of the insulin receptor substrate-1 (IRS1), Protein kinase B (AKT), and JNK in the hypothalamus. Single LPS increased the expression of pIRS1, pAKT, and pJNK, whereas the repeated LPS treatment failed to recruit pIRS1 and pAKT. The 6 LPS treated rats showed increased total JNK and pJNK. The 6 LPS rats became unresponsive to the hypophagic effect induced by central insulin administration (12 µM/5 µL, ICV). Prolonged exposure to LPS (24 h) impaired the insulin-induced AKT phosphorylation and the translocation of the transcription factor forkhead box protein O1 (FoxO1) from the nucleus to the cytoplasm of the cultured hypothalamic GT1-7 cells. Central administration of the JNK inhibitor (20 µM/5 µL, ICV) restored the ability of insulin to phosphorylate IRS1 and AKT in 6 LPS rats. The present data suggest that an increased JNK activity in the hypothalamus underlies the development of insulin resistance during prolonged exposure to endotoxins. Our study reveals that weight gain is not mandatory for the development of hypothalamic insulin resistance and the blockade of proinflammatory pathways could be useful for restoring the insulin signaling during prolonged low-grade inflammation as seen in obesity.

Keywords: LPS tolerance; hypothalamic inflammation; insulin resistance; pJNK

---

1. Introduction

Chronic low-grade inflammation is associated with leptin and insulin resistance, which contributes to the establishment of obesity and its comorbidities, such as type 2 diabetes, cancer, and cardiovascular diseases [1]. We have previously demonstrated that low-grade inflammation induced by repeated injections of the Gram-negative bacterial lipopolysaccharide (6 daily doses of LPS) induces tolerance to the hypophagic effect of the endotoxin [2,3]. The endotoxin failure to reduce food consumption and body weight in LPS-tolerant rats is associated with unresponsiveness to leptin to phosphorylate the signal transduction and activator of transcription 3 (pSTAT3) protein in the hypothalamic nuclei crucial for the control of the energy homeostasis, such as the paraventricular (PVN) and arcuate (ARC) nuclei [2]. Despite the unresponsiveness to leptin in rats treated with six daily doses of LPS, these animals do not show body weight change as opposed to an increase of body weight and fat mass seen in rats after seven days of a high fat diet regimen [4]. Hence, repeated exposure to LPS might be a useful approach to dissociate the impact of the low-grade inflammation that precedes the increased adiposity signals on the development of peripheral and hypothalamic resistance seen in obesity.

The central infusion of insulin induces a negative energy balance comparable to the central administration of leptin [5]. The appetite-suppressive and weight-reducing effects of insulin have been shown in both rodents and primates [6,7]. These effects are supported by the presence of the insulin receptor (IR) and the expression of intracellular components of the insulin signaling, the insulin receptor substrate 1 (IRS-1), and the phosphoinositide 3-kinase (PI3K) pathway-induced proteins in the hypothalamus [8–11]. The PI3K pathway is recruited by both insulin and leptin in the control of energy homeostasis. Pharmacological blockade of PI3K prevents the suppression of food consumption induced by intracerebroventricular (ICV) administration of both leptin and insulin [10,12]. The PI3K pathway also plays a role in the hypophagia acutely induced by LPS [13]. However, the impact of low-grade inflammation induced by prolonged LPS exposure on central insulin signaling has not been addressed.

Hypothalamic insulin resistance may be evoked by inflammatory stimuli [14,15] via the activation of intracellular proinflammatory pathways that increase the expression of intermediate proteins like Jun NH_2-terminal kinase (JNK), protein tyrosine phosphatase 1B (PTP1B), inhibitor of nuclear factor kappa kinase (IKK), and endoplasmic reticulum stress [16–19]. Increased expression of these intermediates of cytokine signaling induces insulin resistance via the alteration of IR activation and expression, as well as by increasing the phosphorylation of the IRS-1 in serine residues, which impairs insulin signal transduction [14,15]. The JNK is a serine kinase activated by cytokines and free fatty acids [15,19,20] and increased JNK activity in the hypothalamus was observed in High fat diet (HFD) fed animals [16,21]. Reinforcing the role of JNK in the development of hypothalamic insulin resistance in HFD fed rodents, mice with JNK1 deficiency in the brain exhibit improved insulin sensitivity in both central and peripheral tissues, preventing adipose tissue dysfunction and hepatic steatosis under HFD feeding [21].

Despite the evident association between obesity and insulin resistance and the activation of the JNK-mediated inflammatory pathway, it remains unclear whether the activation of this pathway precedes the development of obesity in HFD fed animals. To circumvent this challenge, we used prolonged LPS treatment to induce low-grade inflammation and to investigate whether the repeated exposure to endotoxins might increase JNK phosphorylation and cause hypothalamic insulin resistance, independently on the development of obesity. To further address the effect of endotoxins on insulin signaling in neurons, we used cultured GT1-7 cells treated with LPS.

2. Results and Discussion

2.1. Acute, but Not Prolonged Endotoxemia, Promotes Insulin Secretion and Activates Insulin Signaling Cascades. Short and Long-Term LPS Exposure Increases N-Terminal Kinase (JNK) Phosphorylation in the Hypothalamus

To evaluate whether long-term endotoxemia stimulates insulin secretion and insulin signaling in the hypothalamus, rats were treated with repeated LPS injections in comparison with single LPS treatment. Plasma insulin levels were increased 4 h after single, but not repeated LPS treatment

(Figure 1A). Coupled to this response, glucose levels were reduced in single LPS treated rats (Figure 1B) at that time. Insulin secretion following acute LPS administration has been previously described [22]. The release of insulin during the infection process is an adaptive response to the increased production of cytokines [22–24]. Because repeated exposure to endotoxins leads to a desensitization of the neuroendocrine and immunological systems [25], the unaltered insulin levels in the 6 LPS group, compared with the controls, indicates a tolerance to the LPS effects.

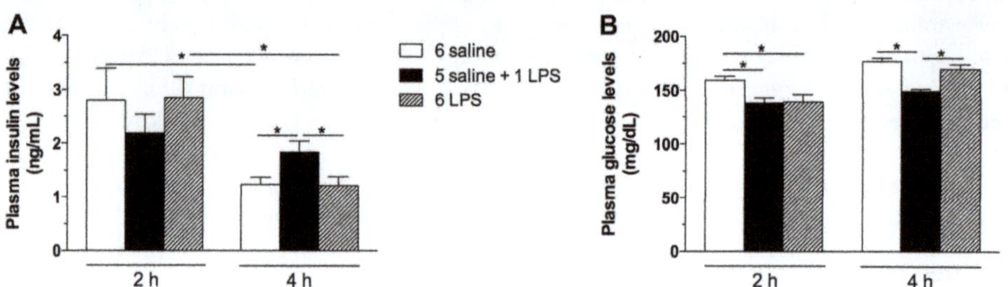

Figure 1. Acute, but not prolonged endotoxemia, increases plasma insulin and reduces glucose levels. Plasma insulin (**A**) and glucose (**B**) levels in saline (6 saline), single (5 saline + 1 LPS, 100 μg/kg ip), or repeated lipopolysaccharide (6 LPS) treated-animals. The representative results of two independent experiments (n = 6–8/group) are shown with the measurements performed with samples from the same animal. One-way ANOVA, followed by the Fisher post hoc test were performed. Data are expressed as means ± SE. Differences were accepted as significant at * $p < 0.05$.

Evaluating the activation of insulin signaling, increased phosphorylation of the IRS-1 at the tyrosine 1222 residue and Protein kinase B (AKT) at the serine 473 and T308 residues in the hypothalamus of the single LPS group were observed, but not in repeated LPS treated rats (Figure 2A,B,E), confirming that acute LPS activates the insulin signaling pathways in the hypothalamus. A previous study from our group [13] reported the phosphorylation of AKT in leptin receptor-expressing neurons in the ARC of mice after acute LPS treatment. As leptin and insulin receptors partly colocalize in the ARC neurons [26], it is reasonable to postulate that an inflammatory stimulus is likely to increase phosphorylation of IRS-1/AKT proteins both in response to leptin and insulin. At present, it was not possible to determine if the single LPS treatment increases insulin signaling per se or if this is an indirect effect mediated by the LPS-induced cytokines, as the expression of Toll-like receptor 4 in neurons is not established [27]. The direct role of endotoxins in the activation of insulin signaling cascades deserves further investigation. Interestingly, as opposed to the 1 LPS group, 6 LPS treated rats did not show increased pIRS-1 and pAKT, supporting our hypothesis that prolonged endotoxemia induces hypothalamic failure to activate insulin signaling proteins.

Insulin resistance can be induced by the increased expression of several kinases that inhibit the eIRS-PI3K-AKT pathways, including JNK. JNK proteins are known to be recruited by LPS as well as during HFD feeding [16,28]. JNK activation increases spontaneous action potential firing of hypothalamic Agouti related peptide (AgRP) expressing neurons and induces both central and peripheral leptin resistance [29]. The phosphorylation of JNK impairs the phosphorylation of tyrosine residues of the IRS-1 and subsequently inhibits AKT activation by insulin in HFD fed rodents [16]. In the present study, prolonged LPS exposure increased both total JNK expression and JNK phosphorylation in the hypothalamus, whereas acute LPS treatment increased only Phospho-c-Jun N-terminal Kinase (pJNK) expression (Figure 2C,D). It is important to highlight that the increased pJNK expression induced by 6 LPS injection was significantly higher than that induced by the single LPS treatment (Figure 2D). Taken together, these findings suggest that hypothalamic unresponsiveness to insulin during a low-grade inflammation challenge could be due to a higher JNK activation.

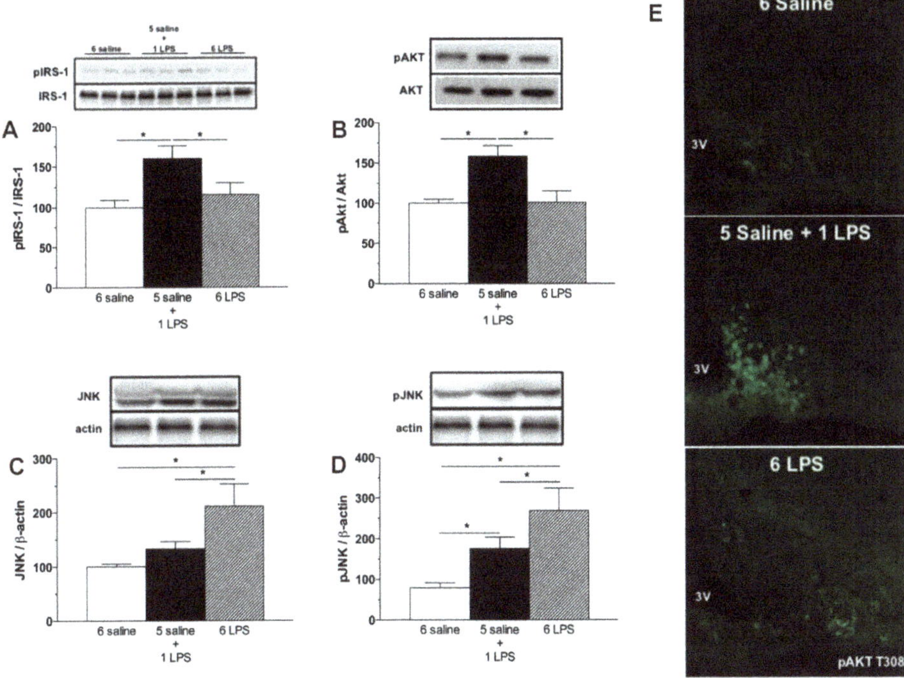

Figure 2. Acute endotoxemia increases phosphorylation of insulin signaling proteins whereas prolonged endotoxemia induces higher expression of the inhibitor of insulin cascade signaling, c-Jun N-terminal Kinase (JNK). Percentage of Phospho-Insulin receptor substrate-1 (pIRS1) (**A**), Phospho-Protein kinase B (pAKT) (**B**), JNK (**C**), and Phospho-c-Jun N-terminal Kinase (pJNK) (**D**) expression in the mediobasal hypothalamus of saline (6 saline), single (5 saline + 1 LPS, 100 ug/kg ip), or repeated LPS (6 LPS) treated-animals. The representative results of two independent experiments (n = 6–8/group) are shown with the measurements performed with samples from the same animal. One-way ANOVA, followed by Fisher post hoc test was performed. Data are expressed as means ± SE. Differences were accepted as significant at * $p < 0.05$; (**E**) representative photomicrographs showing the pAKT T308 expression (green) in neurons from the arcuate (ARC) nucleus of the hypothalamus of animals treated with 6 saline, 5 Saline + 1 LPS, and 6 LPS injections. 3V, third ventricle. Objective: 40×.

2.2. Hypophagic Effect of Central Insulin Injection Is Blunted in Acute and Prolonged Lipopolysaccharide—Treated Rats

Hypothalamic leptin and insulin resistance are important in the pathophysiology of obesity after high fat feeding and they have been associated with hypothalamic inflammation. In HFD fed animals, the hypothalamic inflammation characterized by the production of cytokines and activation of the microglia precedes the weight gain, the peripheral fat accumulation, and the development of insulin resistance [4]. Insulin receptors are expressed in hypothalamic neurons of the melanocortin system [30] and insulin treatment increases proopiomelanocortin (POMC) and decreases Agouti-related peptide (AgRP)/neuropeptide Y (NPY) expression in the hypothalamus through the activation of the IRS-PI3K-AKT pathway [30–32], inhibiting feeding responses. Neuronal IR knockout mice are obese and sensitive to HFD [33], reinforcing the important role of central insulin action in the regulation of energy homeostasis.

We then tested if prolonged endotoxemia could cause central insulin resistance independent of changes in the body weight and in the peripheral sensitivity to insulin. According to previous reports,

ICV insulin administration causes hypophagia [6,7]. As expected, we reproduced these findings given that central insulin administration inhibited food intake in saline treated rats (Figure 3A). The single LPS group displayed a reduced food intake and weight gain (Figure 3B), which were not further reduced by central insulin treatment. This result indicates that acute LPS treatment recruits the signaling cascade common to insulin signaling, as seen by the increased IRS-1 and AKT phosphorylation after the single endotoxin injection. On the other hand, 6 LPS rats, which present no altered food intake and weight gain, were not sensitive to the hypophagic effect of central insulin, suggesting that LPS-tolerant rats are centrally resistant to insulin (Figure 3).

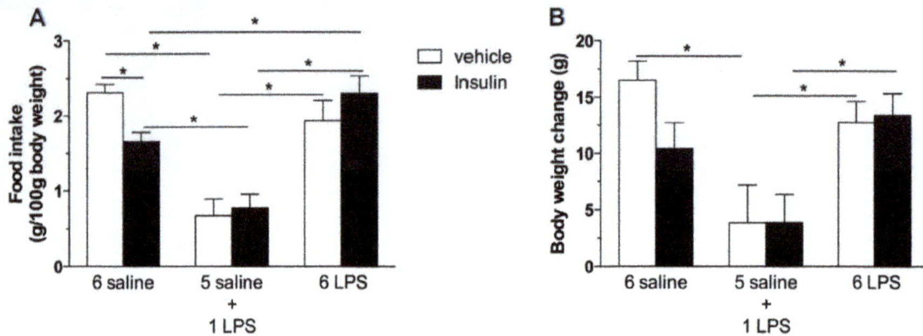

Figure 3. Prolonged endotoxemia induces resistance to the hypophagic effect of icv insulin treatment. Effect of icv injection of vehicle (saline, 5 µL) or insulin (12 µM in 5 µL) on food intake (**A**) and body weight gain (**B**) in saline (6 saline), single (5 saline + 1 LPS, 100 ug/kg ip), or repeated LPS (6 LPS) treated-animals (6–8 animals per group). The representative results of two independent experiments (*n* = 6–8/group) are shown. Two-way ANOVA, followed by the Fisher post hoc test were performed. Data are expressed as means ± SE. Differences were accepted as significant at * *p* < 0.05.

2.3. Prolonged LPS Treatment Blocks Insulin-Induced Protein Kinase B (AKT) Phosphorylation in the Culture of Neuronal Cells

As previously mentioned, the expression of toll-like receptors in neurons is undetermined [27]. However, it was reported that the GT1-7 cell line, derived originally from immortalized hypothalamic neurons, expresses toll like receptors and presents increased interleukins and tumor necrosis factor (TNF)-α mRNA expression after inflammatory stimulus with LPS [34]. To confirm whether long-term exposure to LPS impairs insulin signaling in neurons, we performed in vitro experiments in which we treated GT1-7 cells with LPS for short (30′) or long (24 h) periods, followed by treatment with insulin. LPS at both 30′ and 24 h did not induce pAKT expression in GT1-7 cells. As expected, cultured cells exhibited an increased AKT phosphorylation at both the S473 and T308 residues after insulin treatment. Short-term exposure to LPS (30 min) did not affect insulin-induced pAKT in these cells. Remarkably, prolonged LPS exposure (24 h) impaired the phosphorylation of AKT in both residues after insulin treatment (Figure 4). Activation of the IRS-PI3K-AKT pathway by insulin culminates with the sequestration of the Forkhead box protein O1 (FoxO1) into the cytoplasm, due to FoxO1 phosphorylation. FoxO1 mediates the anorectic effects of leptin and insulin in the ARC by regulating the transcription of POMC and AgRP [35]. We performed an in vitro qualitative study to evaluate the FoxO1 subcellular localization in GT1-7 cells under short- or long-term LPS exposure. In control cells, FoxO1 (green) is widely expressed in the whole cell. After insulin treatment, both control and LPS treated cells (30′) present a qualitatively higher FoxO1 expression predominantly located in the cytoplasm, as evidenced by a clear red DAPI nuclear staining. Short-term LPS treatment did not change FoxO1 expression/distribution in the GT1-7 cells, nor insulin-induced FoxO1 translocation from the nucleus to the cytoplasm. Interestingly, prolonged LPS treatment impaired the ability of insulin to induce the translocation of FoxO1 from the nucleus to the cytoplasm (Figure 5), evidencing the absence of insulin-induced phosphorylation and translocation of this protein to the cytoplasm. These data strongly suggest that low-grade inflammation induced by prolonged exposure to endotoxins impairs insulin signaling and leads to insulin resistant-like phenomena in the neurons.

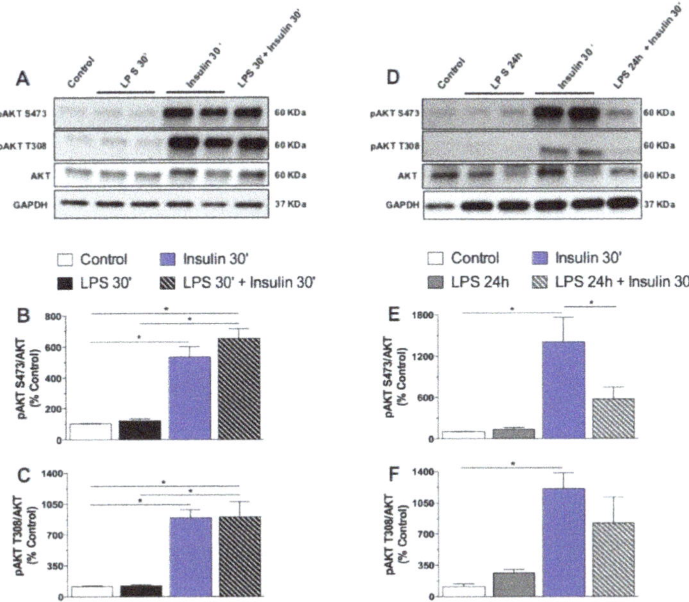

Figure 4. Prolonged exposure to LPS impairs insulin signaling in hypothalamic neurons in vitro. Percentage of pAKT S473 (**A**) and pAKT T308 (**B**) in the mouse GT1–7 cell line after saline (control), short (30′) LPS (1 µg/mL), insulin (100 mM/mL), or LPS (30′) + insulin treatment. Percentage of pAKT S473 (**E**) and pAKT T308 (**F**) in the mouse GT1–7 cell line after saline (control), prolonged (24 h) LPS (1 µg/mL), insulin (100 mM/mL), or LPS (24 h) + insulin treatment. In Vitro experiments were performed in 3–4 different assays. One-way ANOVA, followed by the Fisher post hoc test were performed. Data are expressed as means $\pm$ SE. Differences were accepted as significant at * $p < 0.05$.

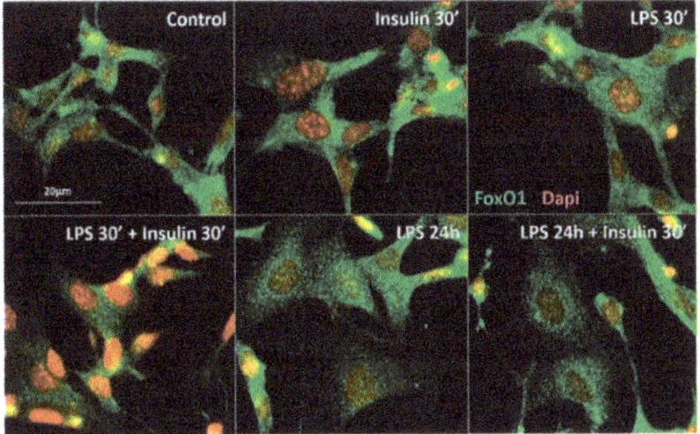

Figure 5. Prolonged LPS treatment impairs the ability of insulin to translocate Forkhead box protein O1 (FoxO1) from the nucleus to the cytoplasm. Representative photomicrograph of the qualitative FoxO1 immunostaning in the mouse GT1-7 cell line after short (30′) LPS (1 µg/mL), insulin (100 mM/mL), LPS (30′) + insulin, prolonged (24 h) LPS, or LPS (24 h) + insulin treatment. FoxO1 expression is in green and DAPI nuclear expression is in red. Scale bar: 20 µm.

2.4. Central JNK Inhibition Restores the Hypothalamic Insulin Responsiveness in Rats Exposed to Repeated LPS Injections

Since JNK causes insulin resistance and we have observed a higher expression of this kinase in the hypothalamus of 6 LPS treated rats, we investigated if we could alleviate the hypothalamic insulin resistance inhibiting the JNK activity. No effect of the selective JNK inhibitor SP600125 was observed in saline or single LPS treated animals (Figure 6A). Remarkably, JNK inhibition restored the hypophagic effect of LPS in 6 LPS tolerant rats. Additionally, there was no further hypophagia in 6 LPS animals treated with SP600125 after stimulation with central insulin (Figure 6A). Body weight was not affected by JNK inhibition (Figure 6B). The data support the hypothesis that JNK plays a role in the tolerance to the hypophagic effect of LPS during prolonged endotoxemia.

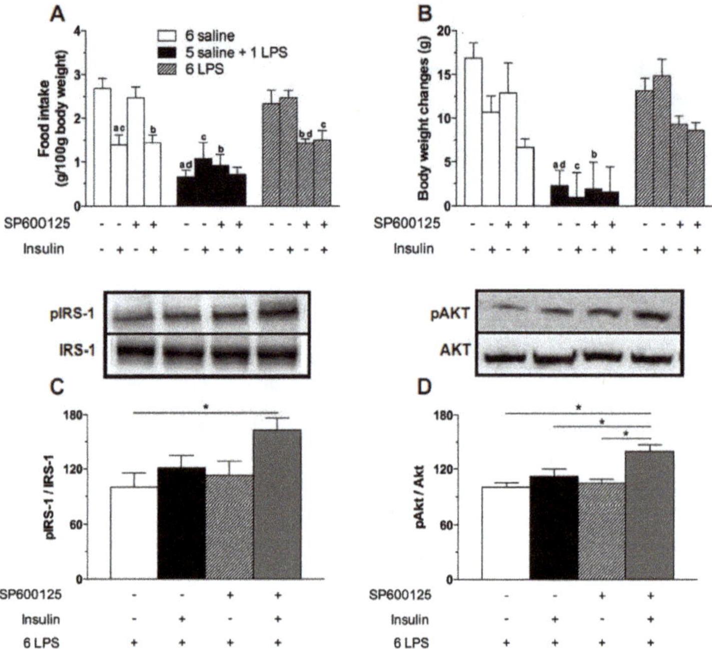

Figure 6. JNK inhibitor treatment restores the hypophagic effect of icv insulin treatment in 6 LPS-treated animals as well as restores the ability of icv insulin to phosphorylate IRS1 and AKT in 6-LPS treated rats. Effect of icv injection of vehicle (saline, 5 μL) or insulin (12 μM in 5 μL) on food intake (**A**) and body weight gain (**B**) in saline (6 saline), single (5 saline + 1 LPS, 100 ug/kg ip), or repeated LPS (6 LPS) treated-animals (6–8 animals per group) previously treated with icv injection of vehicle (saline, 5 μL) or SP600125 (20 μM in 5 μL). The representative results of three independent experiments (n = 6–8/group) are shown. Three-way ANOVA, followed by the Fisher post hoc test were performed. Data are expressed as means ± SE. Differences were accepted as significant at * $p < 0.05$. a vs. 6 saline + vehicle + saline, b vs. 6 saline + SP600125 + saline, c vs. 6 LPS + vehicle + Insulin, d vs. 6 LPS + vehicle + saline and e vs. 1 LPS + vehicle + Insulin. Graphs (**C,D**) show the percentage of pIRS-1 and pAKT expression, respectively, in the mediobasal hypothalamus of repeated LPS (6 LPS) treated-animals that received an icv injection of vehicle (saline, 5 μL) or insulin (12 μM in 5 μL) and were previously treated with an icv injection of vehicle (saline, 5 μL) or SP600125 (20 μM in 5 μL). The representative results of two independent experiments (n = 6–8/group) are shown with the measurements performed with samples from the same animal. One-way ANOVA, followed by the Fisher post hoc test were performed. Data are expressed as means ± SE. Differences were accepted as significant at * $p < 0.05$.

To verify if the reversion of hypophagia in LPS tolerant rats after JNK inhibition was associated with a recovery of the hypothalamic ability to promote insulin signaling, we assessed the expression of pIRS1 and pAKT in the 6 LPS group. Insulin-induced phosphorylation of both IRS1 and AKT was higher in the 6 LPS rats treated with SP600125, compared with the other groups. Interestingly, in the 6 LPS group the JNK inhibitor restored the hypophagic effect of the endotoxin, despite no changes in the activation of insulin signaling proteins. Since it was recently demonstrated that JNK inhibition was able to sensitize leptin's anorectic effect by increasing leptin-induced STAT3 activation and SOCS3 downregulation in the hypothalamus of DIO animals [36], we propose that the activation of alternative pathways not investigated in our study might account for the restoration of the hypophagia. Interestingly, the JNK inhibitor restored the insulin signaling in 6 LPS tolerant rats (Figure 6C,D), reinforcing the role of JNK in the development of insulin resistance during low-grade inflammation. Supporting our results, JNK1 ablation in the CNS improves hypothalamic and systemic insulin sensitivity in mice fed with HFD [21,37]. Administration of the JNK selective inhibitor SP600125 restored the insulin signaling in HFD obese rats [16]. Therefore, our study demonstrates the involvement of the intracellular pro-inflammatory JNK pathway in the development of hypothalamic insulin resistance during prolonged endotoxemia. Our data also suggest that insulin sensitivity can be restored through treatment with drugs that selectively inhibit this intracellular pro-inflammatory pathway.

3. Material and Methods

3.1. Animals

Adult male Wistar rats weighing 220–250 g (Central Animal Facility of the University of Sao Paulo-Campus Ribeirao Preto) were individually housed under controlled light (12:12 h light-dark cycle; lights off at 06:00 p.m.) and temperature conditions ($23 \pm 1\ °C$), with free access to water and food, unless otherwise stated. Food consumption and body weight were recorded daily between 03:30–04:00 p.m. Rats were acclimated to the procedures of drug administration during the experiments by daily handling.

All procedures for the care and use of animals were approved by the Ethical Committee for Animal Use of the Ribeirao Preto Medical School (027/2011, 28 March 2011).

3.2. Experimental Procedures

3.2.1. Effect of Single or Repeated LPS Injections on Plasma Insulin and Glucose Levels, as Well as in the Hypothalamic Content of Insulin Signaling Proteins

Rats were assigned into three groups (n = 6–8/group): (1) saline once daily for 6 days (6 Saline), (2) saline once daily for 5 days and an injection of LPS on the 6th day (5 Saline + 1 LPS), and (3) LPS once daily for 6 days (6 LPS). Rats received saline (0.15 M NaCl, 1 mL/kg) or LPS (100 µg/kg, 1 mL/kg; Serotype 026:B6; Sigma, St. Louis, MO, USA) by an intraperitoneal (IP) injection at 04:00 p.m. On the 6th day of injection, food was withdrawn at 04:00 p.m. One set of animals was decapitated 2 h and another one was decapitated 4 h after treatments, for trunk blood collection for plasma insulin and glucose determination. We also collected the brains 2 h after treatments, for hypothalamic determination of the pIRS-1, pAKT, and pJNK by Western blotting. Another set of rats were submitted to the treatment described above and were perfused 2 h after injections for a qualitative analysis of the pAKT expression in the ARC by immunofluorescence.

3.2.2. Effect of Central Insulin Administration (ICV) Insulin Administration in Rats Treated with Single or Repeated LPS Injections on Food Intake and Body Weight

One week before the experiment, a cannula was placed in the lateral ventricle of the rats that were then treated as previously described (6 Saline, 5 Saline + 1 LPS, and 6 LPS). On the day of the experiment, at 04:00 p.m., food was withdrawn and the animals were weighed and received the last injection (saline or LPS). Forty-five min after the saline or LPS injection, half of each group received an ICV injection of vehicle (saline, 5 μL) or insulin (12 μM in 5 μL; Sigma, St. Louis, MO, USA) and after 30 min food was reoffered ($n = 6$–8/group). Food consumption and body weight changes were assessed 2 and 14 h after food replacement, respectively.

3.2.3. Effect of Short and Prolonged LPS Treatment on Insulin Signaling in GT1-7 Cells

To address whether neuronal insulin signaling is affected by short (30′) or prolonged (24 h) LPS treatment, we performed in vitro experiments using the mouse cell line GT1–7 (kindly provided by Pamela Mellon, University of California, San Diego, CA, USA), derived originally from immortalized GnRH hypothalamic neurons. These cells were shown to be responsive to both insulin and LPS treatment [34]. The cells were cultured in Dulbecco's modified Eagle medium (DMEM; Gibco-Invitrogen, Carlsbad, CA, USA), containing 10% fetal bovine serum and penicillin-streptomycin and maintained at 37 °C in 5% CO_2. In Vitro experiments were performed in 3–4 different assays. GT1-7 cells were subjected to experimental conditions after 3 days in culture, when the cells were 90% confluent. 4 h before treatment, the cells were kept in serum free medium and then treated with vehicle (sterile PBS), LPS (1 μg/mL) for 30 min, insulin (100 nM/mL) and LPS (1 μg/mL) for 30 min, followed by insulin (100 nM/mL) for 30 min. Another set of GT1-7 cells culture was treated with vehicle or LPS (1 μg/mL) for 24 h, followed by treatment with vehicle or insulin (100 nM/mL) for 30 min. At the end of the incubations, cells were harvested and Western blottings for AKT, pAKT, and GAPDH proteins were performed. To qualitatively evaluate the pattern of expression/distribution of the insulin-induced expression of the transcription factor forkhead box protein O1 (FoxO1) in LPS treated cells, GT1-7 cells were cultured overnight on glass coverslips coated with poly-L-Lysine for immunocytochemistry.

3.2.4. Effect of JNK Inhibition on Food Intake and Body Weight Measurements after ICV Insulin Administration in Rats Treated with Single or Repeated LPS Injections.

Rats implanted with a cannula in the lateral ventricle ($n = 6$–8/group) were assigned into the experimental groups (6 Saline, 1 LPS and 6 LPS) as described above. On the day of the experiment, food was removed at 04:00 p.m. and the rats from each group were ICV pre-treated with vehicle (saline, 5 μL) or JNK inhibitor (20 μM in 5 μL; Tocris, Ellisville, MO, USA), followed 105 min after by vehicle (saline, 5 μL) or insulin (12 μM in 5 μL) ICV injection. Food was reoffered 30 min after vehicle or insulin injection. Food consumption and body weight gain were assessed 2 and 14 h after food replacement, respectively.

3.2.5. Effects of JNK Inhibition on Hypothalamic Insulin Signaling in Rats Treated with Repeated LPS Injection

To evaluate whether JNK inhibition might restore the insulin signaling in the hypothalamus of 6 LPS treated rats, animals ($n = 6$–8/group) were implanted with a cannula in the lateral ventricle. On the 6th day, at 04:00 p.m. the rats received the last IP injection of LPS and 30 min after they received an ICV injection of vehicle (0.2% Dimethyl sulfoxide (DMSO) in 5 μL saline) or JNK inhibitor (20 μM in 5 μL; Tocris Co.: Ellisville, MO, USA). Thereafter, the rats received an ICV injection of vehicle or insulin 105 min after saline or LPS injection. Fifteen min after vehicle or insulin injection, the animals were decapitated and the brains were harvested for hypothalamic pIRS-1 and pAkt measurements by Western blotting.

3.3. Cannula Implantation in the Lateral Ventricle

Rats were anesthetized with a mixture of ketamine (60 mg/kg) and xylazine (7.5 mg/kg) at a volume of 0.1 mL/100 g and placed in a stereotaxic instrument (model 900, David Kopf Instruments: Tujunga, CA, USA). A stainless-steel guide cannula (10 mm) was implanted into the right lateral ventricle of the brain. We used the following stereotaxic coordinates: AP = −0.6 mm, LL = −1.5 mm, and depth = −3.6 mm from bregma. The cannula was held in place with two stainless-steel screws and dental acrylic resin on the skull. To prevent occlusion of the guide cannula, a 30-gauge metal wire filled the cannula. After surgery, the rats received a prophylactic injection of penicillin (50,000 U, intramuscular). The placement of the cannula was confirmed at the end of the experiment by histological analysis.

3.4. Western Blotting Analysis of Mediobasal Hypothalamus

The mediobasal hypothalamic protein was extracted using Triton-X 100 (1%), Tris-HCl pH 7.4 (100 mM), sodium pirofosfate (100 mM), sodium fluoride (100 mM), EDTA (10 mM), sodium ortovanadate (10 mM), PMSF (2 mM), aprotinin (0.2 mg/mL), and leupeptin (0.2 mg/mL), at 4 °C, 15,000× g for 40 min. Aliquots of the lysates containing 50 µg of protein were denatured in Laemmli sample buffer (6% SDS, 30% glycerol, 0.02% bromophenol blue, 200 mm Tris-HCl (pH 6.8), and 250 mm mercaptoethanol), at 95 °C for 5 min. Samples were blotted onto a nitrocellulose membrane. Nonspecific binding was prevented by immersing the membranes in blocking buffer (10% BSA in Tris-buffered saline-Tween 20, TBS-T) for 90 min at room temperature. The membranes were then exposed overnight to the primary antibodies: rabbit anti-IRS-1 (1:4000, Cell Signaling # 2390); rabbit anti-phospho IRS-1 tyr1222 (1:4000, Cell Signaling # 3066), rabbit anti β-actin (1:1000, Cell Signaling # 8457), rabbit anti-AKT (1:4000, Cell Signaling # 4691), rabbit anti-phospho AKT S473 (1:4000, Cell signaling # 9271), rabbit anti-JNK (1:1000, Cell Signaling # 9252), and rabbit anti-phospho JNK (1:10,000, Cell Signaling # 9251). The blots were rinsed in TBS-T and then incubated with horseradish peroxidase (HRP) conjugated anti-rabbit antibody (1:5000, Cell Signaling # 7074) for one hour at room temperature. Antibody-antigen complexes were visualized by detecting enhanced chemiluminescence using an ECL detection system (Amersham Biosciences) on the digital images using the Quantity One 4.5.0 software (Bio-Rad: Hercules, CA, USA). To normalize pIRS and pAKT expression, total receptor substrate-1 (IRS1) and total AKT were used, respectively. B-actin was used to normalize JNK and pJNK expression. Because we have used antibodies from a rabbit host, total and phosphorylated protein expression were determined in different blots in the same running.

3.5. Western Blotting Analysis in Cell Culture

Cells were lysed in ice-cold lysis buffer (25 mM Tris pH 8.0; 1.5 mM EGTA, 0.5 mM EDTA, protease inhibitor cocktail, and Triton X-100 1%). The total protein concentration was determined with the Bradford reagent (Bio-Rad: Hercules, CA, USA). Equivalent amounts of proteins (15 µg/well) were then separated by 10% SDS-PAGE and transferred to nitrocellulose membranes (Bio-Rad). The membranes were subsequently immunoblotted with the appropriate primary antibody at 4 °C overnight (rabbit anti-AKT (1:4000, Cell Signaling # 4691); rabbit anti-phospho AKT S473 (1:4000, Cell signaling # 9271); rabbit anti-phospho AKT T308 (1:4000, Cell signaling # 9275)) and then incubated with HRP conjugated secondary anti-mouse or anti-rabbit antibody (1:4000, Jackson Immuno Research: West Grove, PA, USA). Antibody-antigen complexes were visualized by detecting enhanced chemiluminescence using the ECL detection system (Thermo Scientific: Waltham, MA, USA) on digital images (Bio-Rad). Equal protein loading was assessed by the expression of GAPDH (rabbit anti-GAPDH (1:4000, Cell signaling # 5174)). Total and phosphorylated protein expression were determined in different blots in the same running.

3.6. Immunocytochemistry for FoxO1-Dapi

Glass coverslips containing the GT1-7 cells from each experimental group were rinsed with ice-cold PBS and fixed using 4% paraformaldehyde in PBS pH 7.4, at room temperature. After fixation, the cells were rinsed with PBS containing 0.1% Triton-X for permeabilization. After rinsing, the cells were incubated for 30 min in a blocking solution of 1% BSA in PBS containing 0.1% Triton-X. Cells were incubated overnight at 4 °C with the primary antibody (1:5000; rabbit anti-FoxO1, Cell signaling # 2880). After rinsing, the cells were incubated with donkey anti-rabbit IgG conjugated with Alexa 488 secondary antibody (1:500; Jackson Immuno Research: West Grove, PA, USA).

Slides were then mounted using a medium containing DAPI (DNA staining, red). Photomicrographs were acquired using a Leica confocal laser scanning microscope. The immunoreactive structures were excited using argon or helium-neon green lasers with the excitation and barrier filters set for the fluorochrome used (green or red). Images showing the fluorescence were obtained from sequentially acquired images of slices excited by the laser.

3.7. Plasma Insulin and Glucose Determination

Plasma insulin and glucose concentrations were measured by radioimmunoassay (Millipore: Billerica, MA, USA) and the glucose oxidase method (Doles Regents: Goiania, GO, Brazil), respectively, using commercial kits according to the manufacturer's protocol.

3.8. Immunofluorescence for pAKT

Sections were rinsed with 0.1 M PBS and incubated for 48 h at room temperature with rabbit anti-phospho AKT (1:4000, Cell signaling # 9275). After rinsing, sections were incubated with donkey anti-rabbit IgG conjugated with Alexa 488 secondary antibody (1:400; Jackson Immuno Research: West Grove, PA, USA). Finally, the sections were coverslipped with Fluoromont-G mounting medium (Southern Biotechnology Associates: Birmingham, AL, USA). Photomicrographs were acquired using a Leica confocal laser scanning microscope. The immunoreactive structures were excited using argon or helium-neon green lasers with the excitation and barrier filters set for the fluorochrome used (green). Images showing the fluorescence were obtained from sequentially acquired images of slices excited by the laser.

3.9. Statistical Analysis

Results were expressed as means ± SEM and were analyzed using the Software Statistica® (Software Statistica® 10, Tulsa, OK, USA). One-way analysis of variance (ANOVA) followed by the Fisher post hoc test were used to analyze the experiments with single or repeated LPS treatments. Two-way ANOVA, followed by the Fisher post hoc test, were used to analyze the experiment of food intake and body weight gain with single or repeated LPS treatment followed by ICV insulin administration. We used three-way ANOVA, followed by the Fisher post hoc test, to analyze the experiments with single or repeated LPS treatments, JNK inhibitor, and insulin injection. Differences were accepted as significant at $p < 0.05$.

4. Conclusions

The main finding of our study is that hypothalamic insulin resistance, induced by an inflammatory challenge after repeated exposure to endotoxins, can be observed in the absence of increased body weight gain. We used a model of 6 days of prolonged low-grade inflammation to cause hypothalamic insulin resistance without body weight changes. The maintenance of this scenario could advance to obesity and diabetes, as shown by Cani and coworkers, which used 4 weeks of endotoxemia to induce glucose homeostasis disturbance, including peripheral insulin resistance, associated with increased body weight [38]. The additive effects of hypothalamic and peripheral insulin resistance during nutrient overload are the key hallmarks of obesity, which could be caused in part by metabolic endotoxemia. The caveat of the present study is that the peripheral effects of insulin were not investigated to assess if central resistance is coupled to systemic resistance to this hormone. Because both peripheral and central insulin resistance are features of obesity, future studies driven to investigate the time course of peripheral insulin effects on metabolism during prolonged endotoxemia must be performed. Our data indicate that alleviation of low-grade inflammation, as a therapeutic target, may help to restore insulin actions in the hypothalamus and the effects on feeding behavior.

It is also important to point out that the reduced metabolism that leads to the increased body weight during the low-grade inflammation is not observed in the more intense inflammatory process such as that seen in individuals or experimental animal models suffering from different cancers. In fact, it is known that severe inflammation induced by cancers induces an opposite metabolic profile known as cachexia [39,40].

Other central inflammatory processes, such as Alzheimer's disease, also leads to the development of insensitivity to insulin actions and defects in learning and memory [41]. It is crucial to understand the mechanisms underlying impaired insulin signaling during inflammatory processes, to advance in the treatment of obesity and to help alleviate people suffering from degenerative disorders associated with inflammatory markers in the brain. In this context, our study reveals that the inhibition of brain JNK could be used as an intervention approach against obesity and its comorbidities.

Acknowledgments: We would like to thank Maria Valci A. S. Silva and Milene Mantovani for their technical support. This work was supported by the CNPq (Brazilian National Council for Scientific and Technological Development, fellowship to Rodrigo Rorato, 158106/2010-5), FAPESP (Sao Paulo Research Foundation, 2013/09799-1–fellowship to Beatriz Carvalho Borges, 2013/09799-1 to Jose Antunes-Rodrigues and Lucila Leico Kagohara Elias and 2012/18179-4 to Rodrigo Rorato), and the National Institutes of Health (R01-HD-069702 to Carol Fuzeti Elias).

Author Contributions: Rodrigo Rorato and Lucila Leico Kagohara Elias conceived and designed the study; Rodrigo Rorato, Beatriz de Carvalho Borges, and Ernane Torres Uchoa performed the experiments; Rodrigo Rorato, Beatriz de Carvalho Borges and Ernane Torres Uchoa analyzed the data; Jose Antunes-Rodrigues, Carol Fuzeti Elias, and Lucila Leico Kagohara Elias contributed reagents/materials/analysis tools; Rodrigo Rorato, Beatriz de Carvalho Borges, and Lucila Leico Kagohara Elias wrote the paper. All authors read and approved the final manuscript.

Conflicts of Interest: The authors declare no conflict of interest.

Abbreviations

LPS	lipopolysaccharide
IRS-1	insulin receptor substrate 1
PI3K	phosphoinositide 3- kinase
JNK	Jun NH$_2$-terminal kinase
AKT	Protein kinase B
GAPDH	Glyceraldehyde 3-phosphate dehydrogenase
β-actin	Beta-actin
FoxO1	Forkhead box protein O1
SP600125	JNK inhibitor

1. Guh, D.P.; Zhang, W.; Bansback, N.; Amarsi, Z.; Birmingham, C.L.; Anis, A.H. The incidence of co-morbidities related to obesity and overweight: A systematic review and meta-analysis. *BMC Public Health* **2009**, *9*, 88. [CrossRef] [PubMed]

2. Borges, B.C.; Rorato, R.; Avraham, Y.; da Silva, L.E.; Castro, M.; Vorobiav, L.; Berry, E.; Antunes-Rodrigues, J.; Elias, L.L. Leptin resistance and desensitization of hypophagia during prolonged inflammatory challenge. *Am. J. Physiol. Endocrinol. Metab.* **2011**, *300*, E858–E869. [CrossRef] [PubMed]

3. Borges, B.C.; Rorato, R.C.; Uchoa, E.T.; Marangon, P.B.; Elias, C.F.; Antunes-Rodrigues, J.; Elias, L.L. Protein tyrosine phosphatase-1B contributes to LPS-induced leptina resistance in male rats. *Am. J. Physiol. Endocrinol. Metab.* **2015**, *308*, E40–E50. [CrossRef] [PubMed]

4. Thaler, J.P.; Yi, C.X.; Schur, E.A.; Guyenet, S.J.; Hwang, B.H.; Dietrich, M.O.; Zhao, X.; Sarruf, D.A.; Izgur, V.; Maravilla, K.R.; et al. Obesity is associated with hypothalamic injury in rodents and humans. *J. Clin. Investig.* **2012**, *122*, 153–162. [CrossRef] [PubMed]

5. Niswender, K.D.; Schwartz, M.W. Insulin and leptin revisited: Adiposity signals with overlapping physiological and intracellular signaling capabilities. *Front. Neuroendocrinol.* **2003**, *24*, 1–10. [CrossRef]

6. Woods, S.C.; Lotter, E.C.; McKay, L.D.; Porte, J.R.D. Chronic intracerebroventricular infusion of insulin reduces food intake and body weight of baboons. *Nature* **1979**, *282*, 503–505. [CrossRef] [PubMed]

7. Brief, D.J.; Davis, J.D. Reduction of food intake and body weight by chronic intraventricular insulin infusion. *Brain Res. Bull.* **1984**, *12*, 571–575. [CrossRef]

8. Havrankova, J.; Roth, J.; Brownstein, M. Insulin receptors are widely distributed in the central nervous system of the rat. *Nature* **1978**, *272*, 827–829. [CrossRef] [PubMed]

9. Van Houten, M.; Posner, B.I.; Kopriwa, B.M.; Brawer, J.R. Insulin binding sites localized to nerve terminals in rat median eminence and arcuate nucleus. *Science* **1980**, *207*, 1081–1083. [CrossRef] [PubMed]

10. Niswender, K.D.; Morrison, C.D.; Clegg, D.J.; Olson, R.; Baskin, D.G.; Myers, M.G.; Seeley, R.J.; Schwartz, M.W. Insulin activation of phosphatidylinositol 3-kinase in the hypothalamic arcuate nucleus: A key mediator of insulin-induced anorexia. *Diabetes* **2003**, *52*, 227–231. [CrossRef] [PubMed]

11. Plum, L.; Belgardt, B.F.; Brüning, J.C. Central insulin action in energy and glucose homeostasis. *J. Clin. Investig.* **2006**, *116*, 1761–1766. [CrossRef] [PubMed]

12. Zhao, A.Z.; Huan, J.N.; Gupta, S.; Pal, R.; Sahu, A. A phosphatidylinositol 3-kinase phosphodiesterase 3B-cyclic AMP pathway in hypothalamic action of leptin on feeding. *Nat. Neurosci.* **2002**, *5*, 727–728. [CrossRef] [PubMed]

13. Borges, B.C.; Garcia-Galiano, D.; Rorato, R.; Elias, L.L.; Elias, C.F. PI3K p110 β subunit in leptin receptor expressing cells is required for the acute hypophagia induced by endotoxemia. *Mol. Metab.* **2016**, *5*, 379–391. [CrossRef] [PubMed]

14. Hirosumi, J.; Tuncman, G.; Chang, L.; Görgün, C.Z.; Uysal, K.T.; Maeda, K.; Karin, M.; Hotamisligil, G.S. A central role for JNK in obesity and insulin resistance. *Nature* **2002**, *420*, 333–336. [CrossRef] [PubMed]

15. Samuel, V.T.; Shulman, G.I. Mechanisms for insulin resistance: Common threads and missing links. *Cell* **2012**, *148*, 852–871. [CrossRef] [PubMed]

16. De Souza, C.T.; Araujo, E.P.; Bordin, S.; Ashimine, R.; Zollner, R.L.; Boschero, A.C.; Saad, M.J.; Velloso, L.A. Consumption of a fat-rich diet activates a proinflammatory response and induces insulin resistance in the hypothalamus. *Endocrinology* **2005**, *146*, 4192–4199. [CrossRef] [PubMed]

17. Zhang, X.; Zhang, G.; Zhang, H.; Karin, M.; Bai, H.; Cai, D. Hypothalamic IKKβ/NF-κB and ER stress link overnutrition to energy imbalance and obesity. *Cell* **2008**, *135*, 61–73. [CrossRef] [PubMed]

18. Zabolotny, J.M.; Kim, Y.B.; Welsh, L.A.; Kershaw, E.E.; Neel, B.G.; Kahn, B.B. Protein-tyrosine phosphatase 1B expression is induced by inflammation in vivo. *J. Biol. Chem.* **2008**, *283*, 14230–14241. [CrossRef] [PubMed]

19. Milanski, M.; Degasperi, G.; Coope, A.; Morari, J.; Denis, R.; Cintra, D.E.; Tsukumo, D.M.; Anhe, G.; Amaral, M.E.; Takahashi, H.K.; et al. Saturated fatty acids produce an inflammatory response predominantly through the activation of TLR4 signaling in hypothalamus: Implications for the pathogenesis of obesity. *J. Neurosci.* **2009**, *29*, 359–370. [CrossRef] [PubMed]

20. Benoit, S.C.; Air, E.L.; Coolen, L.M.; Strauss, R.; Jackman, A.; Clegg, D.J.; Seeley, R.J.; Woods, S.C. The catabolic action of insulin in the brain is mediated by melanocortins. *J. Neurosci.* **2002**, *22*, 9048–9052. [PubMed]

21. Belgardt, B.F.; Mauer, J.; Wunderlich, F.T.; Ernst, M.B.; Pal, M.; Spohn, G.; Brönneke, H.S.; Brodesser, S.; Hampel, B.; Schauss, A.C.; et al. Hypothalamic and pituitary c-Jun N-terminal kinase 1 signaling coordinately regulates glucose metabolism. *Proc. Natl. Acad. Sci. USA* **2010**, *107*, 6028–6033. [CrossRef] [PubMed]

22. Kim, Y.W.; Kim, K.H.; Ahn, D.K.; Kim, H.S.; Kim, J.Y.; Lee, D.C.; Park, S.Y. Time-course changes of hormones and cytokines by lipopolysaccharide and its relation with anorexia. *J. Physiol. Sci.* **2007**, *57*, 159–165. [CrossRef] [PubMed]

23. Sato, T.; Laviano, A.; Meguid, M.M.; Chen, C.; Rossi-Fanelli, F.; Hatakeyama, K. Involvement of plasma leptin, insulin and free tryptophan in cytokine-induced anorexia. *Clin. Nutr.* **2003**, *22*, 139–146. [CrossRef] [PubMed]

24. Zawalich, W.S.; Zawalich, K.C. Interleukin 1 is a potent stimulator of islet insulin secretion and phosphoinositide hydrolysis. *Am. J. Physiol.* **1989**, *256*, E19–E24. [PubMed]

25. Cavaillon, J.M.; Adrie, C.; Fitting, C.; Adib-Conquy, M. Endotoxin tolerance: Is there a clinical relevance? *J. Endotoxin Res.* **2003**, *9*, 101–107. [CrossRef] [PubMed]

26. Hill, J.W.; Elias, C.F.; Fukuda, M.; Williams, K.W.; Berglund, E.D.; Holland, W.L.; Cho, Y.R.; Chuang, J.C.; Xu, Y.; Choi, M.; et al. Direct insulin and leptin action on pro-opiomelanocortin neurons is required for normal glucose homeostasis and fertility. *Cell Metab.* **2010**, *11*, 286–297. [CrossRef] [PubMed]

27. Lehnardt, S.; Massillon, L.; Follett, P.; Jensen, F.E.; Ratan, R.; Rosenberg, P.A.; Volpe, J.J.; Vartanian, T. Activation of innate immunity in the CNS triggers neurodegeneration through a Toll-like receptor 4-dependent pathway. *Proc. Natl. Acad. Sci. USA* **2003**, *100*, 8514–8519. [CrossRef] [PubMed]

28. Shibata, H.; Katsuki, H.; Okawara, M.; Kume, T.; Akaike, A. c-Jun N-terminal kinase inhibition and alpha-tocopherol protect midbrain dopaminergic neurons from interferon-gamma/lipopolysaccharide-induced injury without affecting nitric oxide production. *J. Neurosci. Res.* **2006**, *83*, 102–109. [CrossRef] [PubMed]

29. Tsaousidou, E.; Paeger, L.; Belgardt, B.F.; Pal, M.; Wunderlich, C.M.; Brönneke, H.; Collienne, U.; Hampel, B.; Wunderlich, F.T.; et al. Distinct Roles for JNK and IKK Activation in Agouti-Related Peptide Neurons in the Development of Obesity and Insulin Resistance. *Cell Rep.* **2014**, *9*, 1495–1506. [CrossRef] [PubMed]

30. Benoit, S.C.; Kemp, C.J.; Elias, C.F.; Abplanalp, W.; Herman, J.P.; Migrenne, S.; Lefevre, A.L.; Cruciani-Guglielmacci, C.; Magnan, C.; Yu, F.; et al. Palmitic acid mediates hypothalamic insulin resistance by altering PKC-theta subcellular localization in rodents. *J. Clin. Investig.* **2009**, *119*, 2577–2589. [CrossRef] [PubMed]

31. Schwartz, M.W.; Marks, J.L.; Sipols, A.J.; Baskin, D.G.; Woods, S.C.; Kahn, S.E.; Porte, D. Central insulin administration reduces neuropeptide Y mRNA expression in the arcuate nucleus of food-deprived lean (Fa/Fa) but not obese (fa/fa) Zucker rats. *Endocrinology* **1991**, *128*, 2645–2647. [CrossRef] [PubMed]

32. Obici, S.; Feng, Z.; Karkanias, G.; Baskin, D.G.; Rossetti, L. Decreasing hypothalamic insulin receptors causes hyperphagia and insulin resistance in rats. *Nat. Neurosci.* **2002**, *5*, 566–572. [CrossRef] [PubMed]

33. Brüning, J.C.; Gautam, D.; Burks, D.J.; Gillette, J.; Schubert, M.; Orban, P.C.; Klein, R.; Krone, W.; Müller-Wieland, D.; Kahn, C.R. Role of brain insulin receptor in control of body weight and reproduction. *Science* **2000**, *289*, 2122–2125. [CrossRef] [PubMed]

34. Choi, S.J.; Kim, F.; Schwartz, M.W.; Wisse, B.E. Cultured hypothalamic neurons are resistant to inflammation and insulin resistance induced by saturated fatty acids. *Am. J. Physiol. Endocrinol. Metab.* **2010**, *298*, E1122–E1130. [CrossRef] [PubMed]

35. Kwon, O.; Kim, K.W.; Kim, M.S. Leptin signalling pathways in hypothalamic neurons. *Cell. Mol. Life Sci.* **2016**, *73*, 1457–1477. [CrossRef] [PubMed]

36. Gao, S.; Howard, S.; LoGrass, P.V. Pharmacological Inhibition of c-Jun N-terminal Kinase Reduces Food Intake and Sensitizes Leptin's Anorectic Signaling Actions. *Sci. Rep.* **2017**, *7*, 41795. [CrossRef] [PubMed]

37. Unger, E.K.; Piper, M.L.; Olofsson, L.E.; Xu, A.W. Functional role of c-Jun-N-terminal kinase in feeding regulation. *Endocrinology* **2010**, *151*, 671–682. [CrossRef] [PubMed]

38. Cani, P.D.; Amar, J.; Iglesias, M.A.; Poggi, M.; Knauf, C.; Bastelica, D.; Neyrinck, A.M.; Fava, F.; Tuohy, K.M.; Chabo, C.; et al. Metabolic endotoxemia initiates obesity and insulin resistance. *Diabetes* **2007**, *56*, 1761–1772. [CrossRef] [PubMed]

39. Arruda, A.P.; Milanski, M.; Coope, A.; Torsoni, A.S.; Ropelle, E.; Carvalho, D.P.; Carvalheira, J.B.; Velloso, L.A. Low-grade hypothalamic inflammation leads to defective thermogenesis, insulin resistance, and impaired insulin secretion. *Endocrinology* **2011**, *152*, 1314–1326. [CrossRef] [PubMed]

40. Morley, J.E.; Thomas, D.R.; Wilson, M.M. Cachexia: Pathophysiology and clinical relevance. *Am. J. Clin. Nutr.* **2006**, *83*, 735–743. [PubMed]

41. De Felice, F.G.; Lourenco, M.V.; Ferreira, S.T. How does brain insulin resistance develop in Alzheimer's disease? *Alzheimers Dement.* **2014**, *10*, S26–S32. [CrossRef] [PubMed]

 International Journal of
Molecular Sciences

Article

Identification and Characterization of Lipopolysaccharide Induced TNFα Factor from Blunt Snout Bream, *Megalobrama amblycephala*

Yina Lv [1,†], Xinying Xiang [2,†], Yuhong Jiang [1,†], Leilei Tang [1], Yi Zhou [3], Huan Zhong [3], Jun Xiao [3] and Jinpeng Yan [1,*]

[1] Department of Cell Biology, School of Life Sciences, Central South University, Changsha 410017, China; lvyina@csu.edu.cn (Y.L.); jiangyuhong@csu.edu.cn (Y.J.); tangleilei142511021@163.com (L.T.)

[2] Center of Biological Experiments, School of Life Sciences, Central South University, Changsha 410017, China; bestxxy@126.com

[3] Guangxi Key Laboratory of Aquatic Genetic Breeding and Healthy Aquaculture, Guangxi Academy of Fishery Sciences, Nanning 530021, China; zhouyi1982cn@126.com (Y.Z.); zhonghuanzh@126.com (H.Z.); dreamshaw@hotmail.com (J.X.)

* Correspondence: andy_yan@csu.edu.cn; Tel./Fax: +86-731-8265-0230

† These authors contributed equally to this work.

Academic Editor: Juan M. Tomás
Received: 19 October 2016; Accepted: 18 January 2017; Published: 15 February 2017

Abstract: Lipopolysaccharide induced TNFα factor (LITAF) is an important transcription factor responsible for regulation of tumor necrosis factor α. In this study, a novel *litaf* gene (designated as *Malitaf*) was identified and characterized from blunt snout bream, *Megalobrama amblycephala*. The full-length cDNA of *Malitaf* was of 956 bp, encoding a polypeptide of 161 amino acids with high similarity to other known LITAFs. A phylogenetic tree also showed that *Malitaf* significantly clustered with those of other teleost, indicating that *Malitaf* was a new member of fish LITAF family. The putative maLITAF protein possessed a highly conserved LITAF domain with two CXXC motifs. The mRNA transcripts of *Malitaf* were detected in all examined tissues of healthy *M. amblycephala*, including kidney, head kidney, muscle, liver, spleen, gill, and heart, and with the highest expression in immune organs: spleen and head kidney. The expression level of *Malitaf* in spleen was rapidly up-regulated and peaked (1.29-fold, $p < 0.05$) at 2 h after lipopolysaccharide (LPS) stimulation. Followed the stimulation of *Malitaf*, *Matnfα* transcriptional level was also transiently induced to a high level (51.74-fold, $p < 0.001$) at 4 h after LPS stimulation. Taken together, we have identified a putative fish LITAF ortholog, which was a constitutive and inducible immune response gene involved in *M. amblycephala* innate immunity during the course of a pathogenic infection.

Keywords: *Megalobrama amblycephala*; lipopolysaccharide induced TNFα factor; lipopolysaccharide stimulation; innate immune

---

1. Introduction

Blunt snout bream (*Megalobrama amblycephala*) is one of the major economically important species in freshwater polyculture fish aquaculture in China [1]. It has been widely cultured because of its herbivorous habit, faster growth rate, and delicate flesh quality, as well as increasing demand in China during the last few decades [2,3]. In 2013, its production has reached 0.73 million tons, ranking seventh in Chinese freshwater fish production [4]. Associated with intensive farming, however, diseases caused by infectious bacteria, mainly *Aeromonas hydrophila*, frequently occur [5]. The infectious disease outbreak with quick spreading has led to serious economic losses in *M. amblycephala* culture industry. Innate immunity plays crucial roles in defense against bacterial infections in fish [6]. The innate immune response to a bacterial pathogen is characterized by the immediate release of pro-inflammatory cytokines, which act as key mediators of the immune system to eliminate the pathogen [7,8]. Finding more molecular components involved in *M. amblycephala*'s innate immunity, therefore, will facilitate our understandings in the largely unveiled complex immunity in fish.

Among the pro-inflammatory cytokines, tumor necrosis factor α (TNFα) has been confirmed to significantly trigger host immunity, increase phagocytic activity, and provoke the induction of inflammatory cytokines [9,10]. TNFα is one of the most well-known pleiotropic cytokines and is secreted by various cell types and can be regulated by different transcription factors, such as nuclear factor κB (NF-κB) [11], nuclear factor of activated T-cells (NF-AT) [12], activator protein 1 (AP-1) [13], and lipopolysaccharide induced TNFα factor (LITAF) [14]. Lipopolysaccharide induced TNFα factor (LITAF) is an important transcription factor mediating transcription of various inflammatory cytokines, especially TNFα [15]. It has been demonstrated that LITAF can directly interact with the signal transducer and activator of transcription (STAT) 6B and translocates into the nucleus where it binds to the promoter regions of TNFα and other cytokines to modulate their transcription [16]. LITAF was initially identified and characterized in the human macrophage cell line, THP-1 [17]. Since then, a large amount of LITAF homologues have been obtained in several aquatic animals including mollusk [18–23], arthropod [24,25], sea cucumber (*Apostichopus japonicus*) [26], and amphioxus (*Branchiostoma belcheri*) [27], suggesting a conserved function in innate immunity. Although the *litaf* gene has been characterized in several fish species, the knowledge of the LITAF orthologs in most teleosts is still limited [28–31].

Therefore, in the present study, we identified a novel *litaf* homolog cDNA (designated as *Malitaf*) in *M. amblycephala*, analyzed its phylogenetic relationship, and characterized its expression pattern in response to LPS stimulation. Considering that *litaf* is a vital regulator for *tnfα* expression, we therefore subsequently investigated the expression profile of *tnfα* (*Matnfα*) in *M. amblycephala*. To our knowledge, this is the first study in the LPS-induced response of *Malitaf*. The achieved results will provide a better understanding of the immune defense mechanisms and further improve the healthy management efficiency in this species.

2. Results

2.1. Isolation and Characterization of Malitaf

The complete *Malitaf* cDNA sequence was 956 bp, which was composed of an 88-bp length 5′-untranslated region (5′-UTR), a 486-bp open reading frame encoding a protein comprising 161 amino acids, and a 358-bp 3′-UTR followed by a poly (A) tail (Figure 1). One putative polyadenylation signal (AATAAA) was recognized at the nucleotide position 906, which was 21 nucleotides upstream of the poly (A) tail. Furthermore, there were two cytokine RNA instability motifs (ATTTA) at the 3′-UTR of *Malitaf*, which were also presented in the *litaf* gene of *Paralichthys olivaceus* [28]. This cDNA sequence has been deposited in the GenBank database under accession number KX421367.

The deduced protein from *Malitaf* gene (MaLITAF) possessed with an estimated molecular mass of 17.2 kDa and an isoelectric point of 6.00. The protein analysis by Basic Local Alignment Search Tool (BLAST) showed that MaLITAF protein shared the highest identity (91.3%) with that of grass carp (*Ctenopharyngodon idellus*). Multiple alignment of amino acid sequences of LITAFs from different species revealed that the C-terminal region of the MaLITAF showed much higher homology than that of other regions (Figure 2). Furthermore, similar to other LITAF homologues, the maLITAF protein possessed the typical LITAF domain (91–160 aa) that contained an N-terminal CXXC "knuckle" followed by a long hydrophobic region, and a C-terminal (H)XCXXC knuckle (Figure 2). The CXXC and HXCXXC motifs are highly conserved among invertebrates, mammals, and fish.

A phylogenetic tree was constructed by the neighbor-joining method based on entire amino acid sequences of LITAFs from *M. amblycephala* and other species. As shown in Figure 3, all LITAFs were split into five categories, including mammalian, avian, amphibian, teleost, and invertebrate LITAFs.

Obviously, *Malitaf* was located into the fish LITAF group which was distinct from the mammalian cluster. As expected, *Malitaf* showed the closest relationship to *C. idellus*. The results indicated that *Malitaf* represents a new member of fish LITAF family.

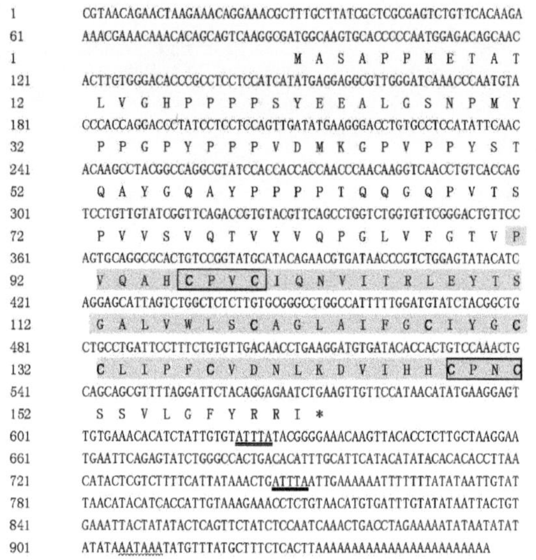

```
  1   CGTAACAGAACTAAGAAACAGGAAACGCTTTGCTTATCGCTCGCGAGTCTGTTCACAAGA
 61   AAACGAAACAAACACAGCAGTCAAGGCGATGGCAAGTGCACCCCCAATGGAGACAGCAAC
  1                             M  A  S  A  P  P  M  E  T  A  T
121   ACTTGTGGGACACCCGCCTCCTCCATCATATGAGGAGGCGTTGGGATCAAACCCAATGTA
 12    L  V  G  H  P  P  P  P  S  Y  E  E  A  L  G  S  N  P  M  Y
181   CCCACCAGGACCCTATCCTCCTCCAGTTGATATGAAGGGACCTGTGCCTCCATATTCAAC
 32    P  P  G  P  Y  P  P  P  V  D  M  K  G  P  V  P  P  Y  S  T
241   ACAAGCCTACGGCCAGGCGTATCCACCACCACCAACCCAACAAGGTCAACCTGTCACCAG
 52    Q  A  Y  G  Q  A  Y  P  P  P  P  T  Q  Q  G  Q  P  V  T  S
301   TCCTGTTGTATCGGTTCAGACCGTGTACGTTCAGCCTGGTCTGGTGTTCGGGACTGTTCC
 72    P  V  V  S  V  Q  T  V  Y  V  Q  P  G  L  V  F  G  T  V  P
361   AGTGCAGGCGCACTGTCCGGTATGCATACAGAACGTGATAACCCGTCTGGAGTATACATC
 92    V  Q  A  H  C  P  V  C  I  Q  N  V  I  T  R  L  E  Y  T  S
421   AGGAGCATTAGTCTGGCTCTCTTGTGCGGGCCTGGCCATTTTTGGATGTATCTACGGCTG
112    G  A  L  V  W  L  S  C  A  G  L  A  I  F  G  C  I  Y  G  C
481   CTGCCTGATTCCTTTCTGTGTTGACAACCTGAAGGATGTGATACACCACTGTCCAAACTG
132    C  L  I  P  F  C  V  D  N  L  K  D  V  I  H  H  C  P  N  C
541   CAGCAGCGTTTTAGGATTCTACAGGAGAATCTGAAGTTGTTCCATAACATATGAAGGAGT
152    S  S  V  L  G  F  Y  R  R  I  *
601   TGTGAAACACATCTATTGTGTATTTATACGGGGAAACAAGTTACACCTCTTGCTAAGGAA
661   TGAATTCAGAGTATCTGGGCCACTGACACATTTGCATTCATACATATACACACACCTTAA
721   CATACTCGTCTTTTCATTATAAACTGATTTAATTGAAAAAATTTTTTTATATAATTGTAT
781   TAACATACATCACCATTGTAAAGAAAACCTCTGTAACATGTGATTTGTATATAATTACTGT
841   GAAATTACTATATACTCAGTTCTATCTCCAATCAAACTGACCTAGAAAAATATAATATAT
901   ATATAAATAAATATGTTTATGCTTTCTCACTTAAAAAAAAAAAAAAAAAAAAAAAAAAAA
```

Figure 1. The nucleotide and deduced amino acid sequences of *Malitaf*. The predicted lipopolysaccharide induced TNFα factor (LITAF) domain is shaded. The two CXXC motifs are in shaded boxes, and the cysteine residues are indicated in shaded bold face. The ATTTA and AATAAA are double and wavy underlined, respectively. "*" shows stop codon.

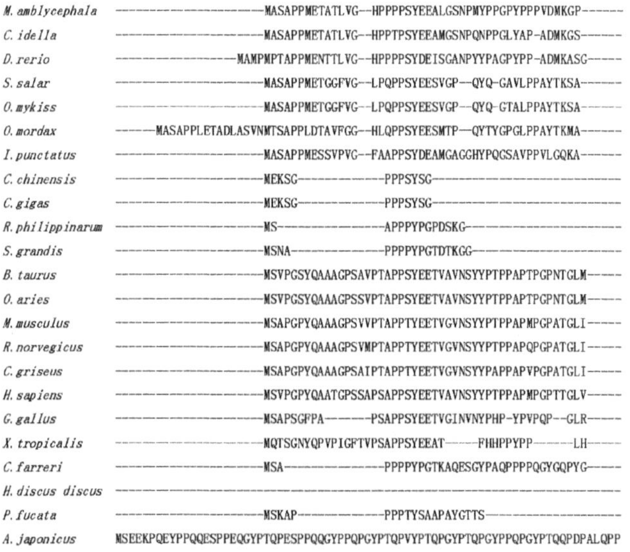

```
M. amblycephala   ------------------------MASAPPMETATLVG--HPPPPSYEEALGSNPMYPPGPYPPPVDMKGP------
C. idella         ------------------------MASAPPMETATLVG--HPPTPSYEEAMGSNPQNPPGLYAP-ADMKGS------
D. rerio          ------------------------MAMPMPTAPPMENTTLVG--HPPPPSYDEISGANPYYPAGPYPP-ADMKASG-----
S. salar          ------------------------MASAPPMETGGFVG--LPQPPSYEESVGP--QYQ-GAVLPPAYTKSA-------
O. mykiss         ------------------------MASAPPMETGGFVG--LPQPPSYEESVGP--QYQ-GTALPPAYTKSA-------
O. mordax         -------MASAPPLETADLASVNMTSAPPLDTAVFGG--HLQPPSYEESMTP--QYTYGPGLPPAYTKMA-------
I. punctatus      ------------------------MASAPPMESSVPVG--FAAPPSYDEAMGAGGHYPQGSAVPPVLGQKA-------
C. chinensis      ----------------MEKSG-----------PPPSYSG-----------------------
C. gigas          ----------------MEKSG-----------PPPSYSG-----------------------
R. philippinarum  ----------------MS-------------APPPYPGPDSKG------------------
S. grandis        ----------------MSNA-----------PPPPYPGTDTKGG------------------
B. taurus         ---------------MSVPGSYQAAAGPSAVPTAPPSYEETVAVNSYYPTPPAPTPGPNTGLM-----
O. aries          ---------------MSVPGSYQAAAGPSSVPTAPPSYEETVAVNSYYPTPPAPTPGPNTGLM-----
M. musculus       ---------------MSAPGPYQAAAGPSVVPTAPPTYEETVGVNSYYPTPPAPMPGPATGLI-----
R. norvegicus     ---------------MSAPGPYQAAAGPSVMPTAPPTYEETVGVNSYYPTPPAPQPGPATGLI-----
C. griseus        ---------------MSAPGPYQAAAGPSAIPTAPPTYEETVGVNSYYPAPPAPVPGPATGLI-----
H. sapiens        ---------------MSVPGPYQAATGPSSAPSAPPSYEETVAVNSYYPTPPAPMPGPTTGLV-----
G. gallus         ---------------MSAPSGFPA------PSAPPSYEETVGINVNYPHP-YPVQP--GLR-----
X. tropicalis     ---------------MQTSGNYQPVPIGFTVPSAPPSYEEAT------FHHPPYPP-------LH-----
C. farreri        ---------------MSA------------PPPPYPGTKAQESGYPAQPPPPQGYGQPYG-----
H. discus discus  -------------------------------------------------------------
P. fucata         ---------------MSKAP----------PPPTYSAAPAYGTTS-----------------
A. japonicus      MSEEKPQEYPPQQESPPEQGYPTQPESPPQQGYPPQPGYPTQPVYPTQPGYPTQPGYPPQPGYPTQQPDPALQPP
```

Figure 2. *Cont.*

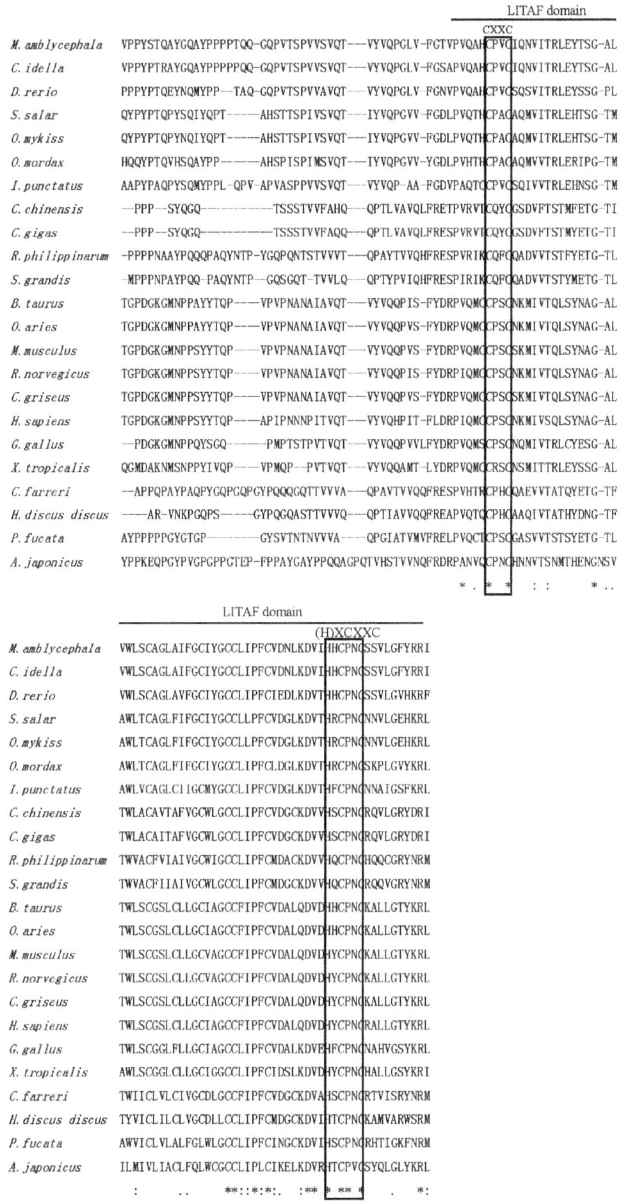

Figure 2. Multiple sequence alignment of LITAFs. The identical, highly conserved, and less conserved amino acid residues are indicated by "*", ":", and ".", respectively. The gaps in the alignment are indicated by "-". The LITAF domain was labeled above the sequences, and two motifs were indicated with rectangles.

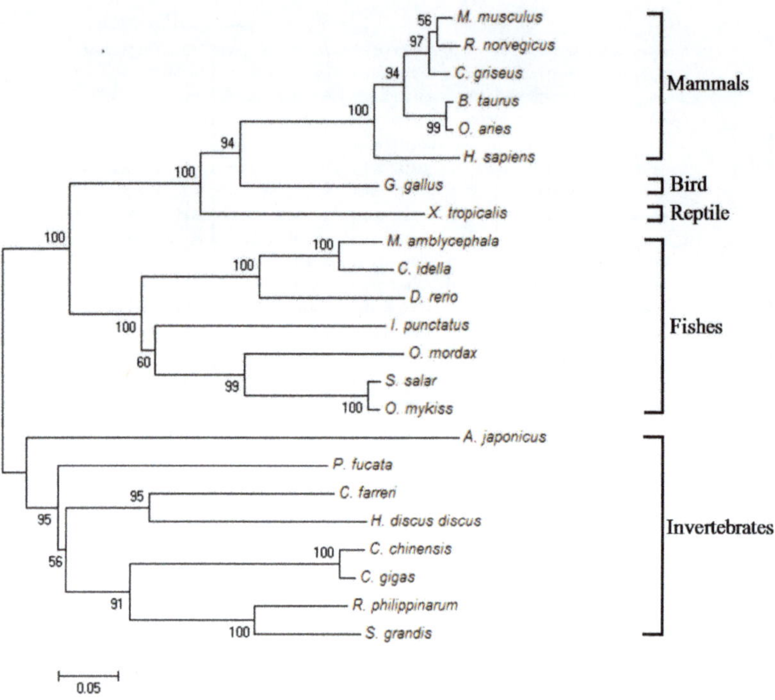

Figure 3. The neighbor-joining phylogenetic tree of MaLITAF protein and other known homologues from GenBank registered. The numbers at the nodes represent bootstrap values for 1000 replications and the bar (0.1) indicates genetic distance. The sequences of LITAFs from other species were downloaded from GenBank: *C. idella* (ACT68335), *Danio rerio* (NP_001002184), *Ictalurus punctatus* (NP_001187935), *Oncorhynchus mykiss* (NP_001158593), *Osmerus mordax* (ACO09164), *Salmo salar* (ACI67257), *Gallus gallus* (NP_989598), *Xenopus tropicalis* (NP_988970), *Mus musculus* (NP_064364), *Homo sapiens* (NP_004853), *Bos taurus* (NP_001039717), *Cricetulus griseus* (XP_003496739), *Ovis aries* (XP_004020821), *Rattus norvegicus* (NP_001099205), *Solen grandis* (AEW43450), *Ruditapes philippinarum* (ADX31291), *Crassostrea gigas* (ABO70331), *Cipangopaludina chinensis* (AEX08893), *Haliotis discus discus* (ADI72430), *Chlamys farreri* (ABI79459), *Pinctada fucata* (ACN70008), and *A. japonicus* (AIB51692).

2.2. Tissues Distribution of Malitaf

To further understand the potential function of this new gene, the presence of *Malitaf* mRNA transcript in different tissues from healthy *M. amblycephala* was examined by quantitative real-time PCR (qRT-PCR) analysis, which can accurately quantify transcripts at a low copy number [32]. Expression was normalized to the tissue with the lowest observed mRNA level-kidney (set as 1). As shown in Figure 4, the mRNA transcript of *Malitaf* was ubiquitously detected in a wide range of tissues examined from healthy fish. However, the relative gene expression of *Malitaf* to kidney was the highest in spleen (15.21-fold, $p < 0.001$), which was also observed in spleen of rock bream (*Oplegnathus fasciatus*) [30], followed by head kidney (12.45-fold). The transcriptional level of *Malitaf* decreased gradually in liver (11.12-fold), heart (4.9-fold), gill (2.29-fold), and muscle (1.75-fold).

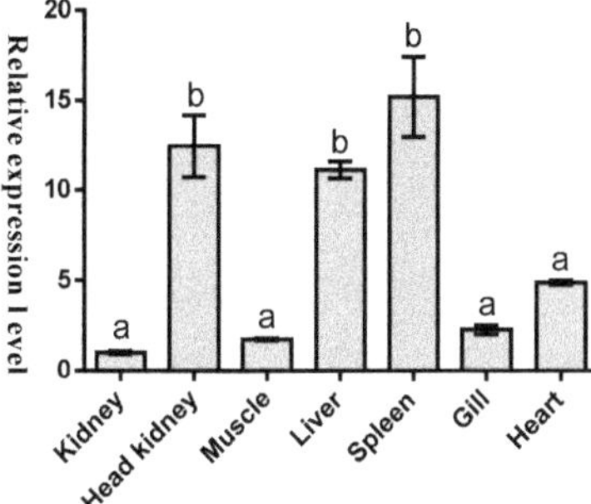

Figure 4. The expression of *Malitaf* mRNA in different tissues from healthy *M. amblycephala*. Significant pairwise expression-level differences between tissues are indicated by different letters above the bars.

2.3. Temporal Expression of Malitaf and Matnfα after LPS Stimulation

To preliminarily unravel the potential role of *Malitaf* in innate immunity, we characterized the temporal expression pattern of *Malitaf* in spleen in response to stimulation with LPS. As shown in Figure 5A, the mRNA expression of *Malitaf* was rapidly up-regulated after LPS stimulation. The expression reached a peak level at 2 h post-stimulation, which was about 1.29-fold higher than control group ($p < 0.05$). Subsequently, it was found that its expression decreased gradually to the control level as time elapsed. However, a significant decrease of *Malitaf* expression was observed at 12 h after LPS stimulation. Considering that *litaf* is an important regulator for *tnfα* expression, we therefore tested the expression profile of *Matnfα* following the elevated expression of *Malitaf* under LPS stimulations at different time points. As shown in Figure 5B, the transcription of *Matnf* maintained the control level at 2 h post LPS stimulation. However, a sharp increase of *Matnfα* mRNA to peak level was detected at 4 h post LPS stimulation (51.74-fold, $p < 0.001$), and then dropped to the original level at 8 h post LPS stimulation. Interestingly, the transcription of *Matnfα* surged again to a higher level at 12 and 24 h post LPS stimulation (3.66- and 1.71-fold, respectively) compared with that of the control group.

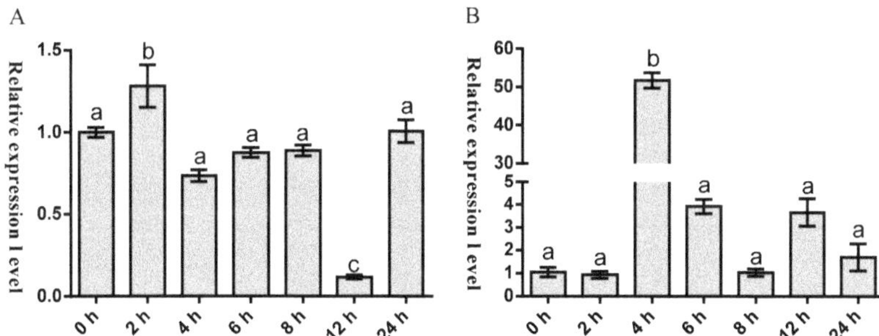

Figure 5. Temporal expression analysis of *Malitaf* (**A**); and *Matnfα* (**B**) mRNA in spleen from *M. amblycephala* after LPS stimulation. Significant pairwise expression-level differences among different time-points are indicated by different letters above the bars.

3. Discussion

Cytokines are the key regulators of innate immunity against pathogens [33]. Until now, *litaf* gene had been cloned and characterized from many organisms including both vertebrate and invertebrate animals, which suggested a conserved function in innate immunity. In mammals, the *litaf* has been reported as an important transcript factor for the regulation of TNFα and transcription of various inflammatory cytokines in mammals [34]. However, the information on the systemic reaction of fish *litaf*s during bacterial infection was rather rare. Therefore, we cloned the novel *litaf* gene from blunt snout bream and examined their expression profiles to understand its potential role in innate immunity.

The LITAF domain, which is a key feature for the LITAF family, exists in viruses, fungi, plants, and Metazoa, [35]. In the present study, the amino acid alignment indicated a highly conserved LITAF domain with N-terminal CXXC and C-terminal HXCXXC knuckles that formed a compact Zn^{2+}-binding structure. These characteristics are a key feature of intracellular Zn^{2+}-binding domains that the N-terminal region binds to the intracellular molecule and the hydrophobic region does not span the membrane [36]. These observations indicated that the LITAF domain of *Malitaf* appears to be capable of responding similarly to the mammalian LITAF. In a phylogenetic tree of selected vertebrate and invertebrate LITAF amino acid sequences, the MaLITAF was obviously separated from the mammalian cluster and formed one distinct cluster with other teleost LITAFs, suggesting that it is a fish-specific LITAF.

We performed qRT-PCR to monitor the transcriptional level of *Malitaf* in various tissues from the healthy fish. The results showed the constitutive distribution of *Malitaf*, indicating that its important role in immune defense against invaders and its immune responses could be occurring in the whole fish body. In teleost fish, spleen and head kidney could be considered as the important central immune organs synthesizing proteins involved in fish immune defense [37]. Therefore, the higher expression level of *Malitaf* in spleen and head kidney also particularly suggested its crucial role in immune defense against potential pathogens. In other teleost fish, the constitutive expression of *litaf* mRNA has been reported in a wide range of tissues, but with varying expression levels. For example, the relative gene expression of *litaf1* of rock bream was the highest in spleen. In Japanese flounder, the *litaf* mRNA was detected in all examined tissues with the greatest amount in gill, followed by blood, skin, gonad, hepatopancreas, head kidney, heart, brain, trunk kidney, and spleen in a descending order [28]. The grouper *litaf* gene was also widely expressed in different tissues analyzed, including liver, spleen, kidney, head kidney, intestine, skin, gill, brain, muscle, heart, and stomach [29]. However, the relatively low expression levels were detected in muscle and liver. In grass carp, the *litaf* gene was found to express in various tissues but with a high expression level in gill [31]. The inconsistent constitutive expression of *litaf* gene among different fish species may be a reflection of physiological differences among these species, or even environmental influences. Furthermore, the differences in *litaf* gene expression are thought to be possibly due to different expressing cells and functions of *litaf* gene. Certainly, further studies on functional differentiation of *litaf* gene between fish and other species may yield more novel information on the immune regulatory response of fish.

The *Malitaf* expression was significantly induced in spleen by LPS, a compound that mimics a Gram-negative bacteria infection, which was similar to the previous reports on human THP-1 cells [17], mouse [38], chicken macrophages [39], Pacific oyster [22], and scallop hemocytes [23]. The results strongly suggested its significantly responsive to LPS and involvement in innate immune response against Gram-negative bacteria. After 12 h of LPS stimulation, the *Malitaf* mRNA has significantly decreased, which might possibly prevent the excess production of cytokines, and then the innate immune response to pathogens could be efficiently controlled [33]. The expression pattern of *Malitaf* appeared to be similar to Japanese flounder *litaf* with respect to the expression profile in response to LPS stimulation [28]. TNFα effects can be both beneficial and detrimental to the host [40]. TNFα must be tightly regulated because over-production can be lethal to the host as in septic shock syndrome. The mouse *litaf* gene induces activation of *tnfα* gene expression by itself [38]. Inhibiting human *litaf* expression in a human monocytic cell line leads to a reduction in the *tnfα* transcript [15]. Therefore, in the present study, we tested the expression profile of *Matnfα* following the elevated expression of *Malitaf* under LPS stimulation. The *Matnfα* was also induced and rapidly peaked at 4 h, not at 2 h, post LPS administration indicating that *Malitaf* may significantly contribute to the up-regulated *Matnfα* expression in the spleen. In addition, the associated up-regulation of *Malitaf* and

Matnfα in spleen during the early phase of LPS stimulation process indicated that *Malitaf* may be an essential regulator involved in *M. amblycephala* innate immune response, probably through regulation of *Matnfα* expression.

4. Materials and Methods

4.1. Fish, LPS Stimulation, and RNA Isolation

Adult *M. amblycephala* (weight: 405 ± 12.4 g) were obtained from a fish farm (Changsha, China). Fish were maintained with a flow-through water supply at room temperature. After acclimating for one week, the normal fish were used for the stimulation experiments. LPS isolated from *Escherichia coli* (L2880, Sigma, St. Louis, MO, USA) was suspended into sterilized phosphate buffered saline (PBS), and then was intraperitoneally injected into fish at a dose of 0.1 mg/100 g fish. Afterwards, fish were anesthetized with MS-222 (3-aminobenzoic acid ethyl ester; Sigma). Various tissues from three healthy individuals, including kidney, head kidney, muscle, liver, spleen, gill, and heart, were collected. Similarly, the spleen from three individuals was collected at different time points (0, 2, 4, 6, 8, 12, and 24 h) after stimulation. Total RNA from above tissues was extracted using TRIzol reagent (Invitrogen, Carlsbad, CA, USA), and quantified based on the absorbance at 260 nm. The integrity of RNA was checked by agarose gel electrophoresis. The animal experiments were approved by the Ethics Committee of School of Life Sciences of Central South University with the following reference number (SLSEC0028) in 10 March 2015. The tissues collected at 0 h were from fish injected with the same volume of PBS.

4.2. Cloning and Characterization of Malitaf

To obtain the partial cDNA of *Malitaf*, the degenerate primers were designed based on an alignment of its counterparts from other teleost fish (Table 1). Gene-specific primers were used to amplify each end according to the manufacturer's instructions. All resulting PCR products were purified and then cloned into the pMD18-T vector (TaKaRa, Dalian, China) and sequenced bi-directionally at Sangon (Shanghai, China). The open reading frame (ORF) of *Malitaf* cDNA was detected using the ORF finder (available on: http://www.ncbi.nlm.nih.gov/project/gorf). Protein domains were predicted by the Simple Modular Architecture Research Tool (SMART) [41]. Multiple sequence alignments were created using the Clustal W program [42]. Phylogenetic and molecular evolutionary analyses were constructed by the Neighbor-Joining method in Molecular Evolutionary Genetics Analysis (MEGA) software (version 7.01, Tokyo Metropolitan University, Tokyo, Japan), and support for each node was bootstrapped with 1000 replicates [43].

4.3. Spatial and Temporal Expression Analysis of Malitaf and Matnfα

The quantitative real-time PCR (qRT-PCR) was performed to investigate *Malitaf* mRNA expression levels in different tissues of healthy *M. amblycephala*. In addition, the mRNA expression pattern of *Malitaf* and *Matnfα* was determined in spleen after LPS stimulation by qRT-PCR. The β-actin and 18S rRNA was selected as an internal control to verify the successful reverse transcription and to calibrate the cDNA template in spatial and temporal expression analysis, respectively. The qRT-PCR was implemented using an ABI 7500 Real-time PCR system (Applied Biosystems, Foster, CA, USA) in a total volume of 20 μL, including 10 μL SYBR® Premix Ex Taq™ II (2×) (TaKaRa, Dalian, China), 0.4 μL ROX Reference Dye II (50×), 0.4 μL of each primer (10 μmol·L^{-1}), 2 μL 1:5 diluted cDNA, and 6.8 μL of PCR-grade water. The thermal profile was 95 °C for 30 s followed by 40 cycles of 95 °C for 5 s, 60 °C for 34 s, and 72 °C for 30 s. Melting curve analysis of the amplified products was performed at the end of each PCR to confirm that a single PCR product was generated. The $2^{-\Delta\Delta Ct}$ method was used to analyze the expression levels of *Malitaf* and *Matnfα* genes [32]. The data obtained from three independent biological replicates were subjected to statistical analysis and the values represented the *n*-fold difference relative to the references (kidney and 0 h).

Table 1. Sequences of primers used in this study.

Primer	Sequence(5′–3′)	Comment
LITAF-F	AARCGCTTTGYTTRTCGCTC	Gene cloning
LITAF-R	AGGTGYAAMTTGTTTCCCCG	
3′-Adaptor primer	GCTGTCAACGATACGCTACGTAACGGCATGACAGTG(T)$_{18}$	3′RACE
3′-Primer	GCTGTCAACGATACGCTACGTAACG	
3′-Nested primer	CGCTACGTAACGGCATGACAGTG	
LITAF-3′-GSP	ACAGCAACACTTGTGGGACACCCGCCT	
LITAF-3′-NGSP	AGCCTGGTCTGGTGTTCGGGACTGTT	
AAP	GGCCACGCGTCGACTAGTACGGGIIGGGIIGGGIIG	5′RACE
AUAP	GGCCACGCGTCGACTAGTAC	
LITAF-5′-GSP	CTGTGCCTCCATATTCAACACAAG	
LITAF-5′-NGSP	GTTGGGATCAAACCCAATGTACC	
LITAF-qF	CACCAGTCCTGTTGTATCGG	Real-time PCR
LITAF-qR	CGCACAAGAGAGCCAGACTA	
TNFα-qF	CTGCTGTCTGCTTCACGCTC	
TNFα-qR	TAAATGGATGGCTGCCTTGG	
β-actin-qF	CGGACAGGTCATCACCATTG	
β-actin-qR	CGCAAGACTCCATACCCAAGA	
18S rRNA-qF	CAAGACGGACGAGAGCGAAA	
18S rRNA-qR	GCGGGTTGGCATAGTTTACG	

4.4. Statistical Analysis

All data of qRT-PCR were presented as means $\pm$ SD and checked for homogeneity of variances and normality. Statistical analysis was performed using GraphPad Prism 5.0 (GraphPad Software Inc., San Diego, CA, USA). Significant differences among samples were determined by one-way analysis of variance (one-way ANOVA) followed by a Tukey's multiple comparison test. Differences were considered significant at $p < 0.05$ and extremely significant at $p < 0.01$.

5. Conclusions

We identified and characterized a new member of LITAF family, *Malitaf*, in *M. amblycephala*. We confirmed that *Malitaf* mRNA constitutively expresses in all examined tissues and displays a higher expression level in spleen and head kidney than in other tissues. We also demonstrated that *Malitaf* expression could be induced by bacterial endotoxin LPS stimulation. Furthermore, we showed that the expression of *Matnfα*, a pleiotropic cytokine regulated by *Malitaf*, was up-regulated associated with the enhanced expression of *Malitaf* in vivo. Our results significantly suggested that *Malitaf* may play a key role in blunt snout bream innate immunity. Further studies on the functions of *Malitaf* will contribute to a better understanding of the fish immune system and may help elucidate fish immunoregulatory pathways.

Acknowledgments: This work was supported by the National Natural Science Foundation of China (31201984 and 81500231), the Natural Science Foundation of Hunan Province, China (14JJ3047), the Fundamental Research Funds for the Central Universities (2011QNZT129), the Student Innovation Program for Central South University (Grant no. CX2015508), and the Student Free Exploration Program for Central South University (Grant no. ZY2015785).

Author Contributions: Yina Lv and Yuhong Jiang conducted the gene clone and wrote the manuscript; Leilei Tang conducted the expression analysis of *Malitaf* and *Matnfα*; Jun Xiao, Yi Zhou, and Huan Zhong analyzed the data; Jinpeng Yan and Xinying Xiang designed the experiments and modified the manuscript; All authors reviewed and approved the final manuscript.

Conflicts of Interest: The authors declare no conflict of interest.

1. Li, S.F.; Cai, W.Q.; Zhou, B.Y. Variation in morphology and biochemical genetic markers among populations of blunt snout bream (*Megalobrama amblycephala*). *Aquaculture* **1993**, *111*, 117–127. [CrossRef]

2. Tang, L.L.; Liang, Y.H.; Jiang, Y.H.; Liu, S.J.; Zhang, F.Y.; He, X.; Wang, T.Y.; Zhou, Y.; Zhong, H.; Yan, J.P. Identification and expression analysis on bactericidal permeability-increasing protein/lipopolysaccharide-binding protein of blunt snout bream, *Megalobrama amblycephala*. *Fish Shellfish Immunol.* **2015**, *45*, 630–640. [CrossRef] [PubMed]

3. Jiang, Y.H.; Tang, L.L.; Zhang, F.Y.; Jiang, H.Y.; Li, X.W.; Yang, L.Y.; Zhang, L.; Mao, J.R.; Yang, J.P. Identification and characterization of immune-related microRNAs in blunt snout bream, *Megalobrama amblycephala*. *Fish Shellfish Immunol.* **2016**, *49*, 470–492.

4. Bureau, C.F. *China Fisheries Yearbook*; Chinese Agriculture Press: Beijing, China, 2013.

5. Nielsen, M.E.; Hoi, L.; Schmidt, A.S.; Qian, D.; Shimada, T.; Shen, J.Y.; Larsen, J.L. Is *Aeromonas hydrophila* the dominant motile *Aeromonas* species that causes disease outbreaks in aquaculture production in the Zhejiang Province of China? *Dis. Aquat. Organ.* **2001**, *46*, 23–29. [CrossRef] [PubMed]

6. Ellis, A.E. Innate host defense mechanisms of fish against viruses and bacteria. *Dev. Comp. Immunol.* **2001**, *25*, 827–839. [CrossRef]

7. Zhu, L.Y.; Nie, L.; Zhu, G.; Xiang, L.X.; Shao, J.Z. Advances in research of fish immune-relevant genes: A comparative overview of innate and adaptive immunity in teleosts. *Dev. Comp. Immunol.* **2013**, *39*, 39–62. [CrossRef] [PubMed]

8. Tang, L.L.; Xiang, X.Y.; Jiang, Y.H.; Lv, Y.N.; Zhou, Y.; Zhong, H.; Xiao, J.; Zhang, F.Y.; Jiang, H.Y.; Yan, J.P. Identification and characterization of a novel Toll-like receptor 4 homologue in blunt snout bream, *Megalobrama amblycephala*. *Fish Shellfish Immunol.* **2016**, *57*, 25–34. [CrossRef] [PubMed]

9. Grayfer, L.; Walsh, J.G.; Belosevic, M. Characterization and functional analysis of goldfish (*Carassius auratus* L.) tumor necrosis factor-α. *Dev. Comp. Immunol.* **2008**, *32*, 532–543. [CrossRef] [PubMed]

10. Li, M.F.; Zhang, J. CsTNF1, a teleost tumor necrosis factor that promotes antibacterial and antiviral immune defense in a manner that depends on the conserved receptor binding site. *Dev. Comp. Immunol.* **2016**, *55*, 65–75. [CrossRef] [PubMed]

11. Collart, M.A.; Baeuerle, P.; Vassalli, P. Regulation of tumor necrosis factor α transcription in macrophages: Involvement of four κB-like motifs and of constitutive and inducible forms of NF-κB. *Mol. Cell Biol.* **1990**, *10*, 1498–1506. [CrossRef] [PubMed]

12. Lee, M.H.; Park, J.; Chung, S.W.; Kang, B.Y.; Kim, S.H.; Kim, T.S. Enhancement of interleukin-4 production in activated CD4⁺ T cells by diphthalate plasticizers via increased NF-AT binding activity. *Int. Arch. Allergy Immunol.* **2004**, *134*, 213–222. [CrossRef] [PubMed]

13. Vitiello, M.; D'Isanto, M.; Galdiero, M.; Raieta, K.; Tortora, A.; Rotondo, P.; Peluso, L.; Galdiero, M. Interleukin-8 production by THP-1 cells stimulated by Salmonella enterica serovar Typhimurium porins is mediated by AP-1, NF-κB and MAPK pathways. *Cytokine* **2004**, *27*, 15–24. [CrossRef]

14. Stucchi, A.; Reed, K.; O'Brien, M.; Cerda, S.; Andrews, C.; Gower, A.; Bushell, K.; Amar, S.; Leeman, S.; Becker, J. A new transcription factor that regulates *TNF-α* gene expression, LITAF, is increased in intestinal tissues from patients with CD and UC. *Inflamm. Bowel Dis.* **2006**, *12*, 581–587. [CrossRef] [PubMed]

15. Tang, X.; Metzger, D.; Leeman, S.; Amar, S. LPS-induced TNF-α factor (LITAF)-deficient mice express reduced LPS-induced cytokine: Evidence for LITAF-dependent LPS signaling pathways. *Proc. Natl. Acad. Sci. USA* **2006**, *103*, 13777–13782. [CrossRef] [PubMed]

16. Tang, X.; Marciano, D.L.; Leeman, S.E.; Amar, S. LPS induces the interaction of a transcription factor, LPS-induced TNF-α factor, and STAT6(B) with effects on multiple cytokines. *Proc. Natl. Acad. Sci. USA* **2005**, *102*, 5132–5137. [CrossRef] [PubMed]

17. Myokai, F.; Takashiba, S.; Lebo, R.; Amar, S. A novel lipopolysaccharide-induced transcription factor regulating tumor necrosis factor α gene expression: Molecular cloning, sequencing, characterization, and chromosomal assignment. *Proc. Natl. Acad. Sci. USA* **1999**, *96*, 4518–4523. [CrossRef]

18. De Zoysa, M.; Nikapitiya, C.; Oh, C.; Whang, I.; Lee, J.S.; Jung, S.J.; Choi, C.Y.; Lee, J. Molecular evidence for the existence of lipopolysaccharide-induced TNF-α factor (LITAF) and Rel/NF-κB pathways in disk abalone (*Haliotis discus discus*). *Fish Shellfish Immunol.* **2010**, *28*, 754–763. [CrossRef]

19. Li, H.J.; Yang, Q.; Gao, X.G.; Su, H.; Wang, J.; He, C.B. Identification and expression of a putative LPS-induced TNF-α factor from Asiatic hard clam *Meretrix meretrix*. *Mol. Biol. Rep.* **2012**, *39*, 865–871. [CrossRef] [PubMed]

20. Zhang, D.; Jiang, J.; Jiang, S.; Ma, J.; Su, T.; Qiu, L.; Zhu, C.; Xu, X. Molecular characterization and expression analysis of a putative LPS-induced TNF-α factor (LITAF) from pearl oyster *Pinctada fucata*. *Fish Shellfish Immunol.* **2009**, *27*, 391–396. [CrossRef] [PubMed]

21. Yang, D.; Wei, X.; Yang, J.; Yang, J.; Xu, J.; Fang, J.; Wang, S.; Liu, X. Identification of a LPS-induced TNF-α factor (LITAF) from mollusk Solen grandis and its expression pattern towards PAMPs stimulation. *Fish Shellfish Immunol.* **2013**, *35*, 1325–1328. [CrossRef] [PubMed]

22. Yu, F.; Zhang, Y.; Yu, Z. Characteristics and expression patterns of the lipopolysaccharide-induced TNF-α factor (LITAF) gene family in the Pacific oyster, *Crassostrea gigas*. *Fish Shellfish Immunol.* **2012**, *33*, 899–908. [CrossRef] [PubMed]

23. Yu, Y.; Qiu, L.; Song, L.; Zhao, J.; Ni, D.; Zhang, Y.; Xu, W. Molecular cloning and characterization of a putative lipopolysaccharide-induced TNF-α factor (LITAF) gene homologue from Zhikong scallop *Chlamys farreri*. *Fish Shellfish Immunol.* **2007**, *23*, 419–429. [CrossRef] [PubMed]

24. Li, S.; Jia, Z.; Li, X.; Geng, X.; Sun, J. Identification and expression analysis of lipopolysaccharide-induced TNF-α factor gene in Chinese mitten crab *Eriocheir sinensis*. *Fish Shellfish Immunol.* **2014**, *38*, 190–195. [CrossRef] [PubMed]

25. Wang, P.H.; Wan, D.H.; Pang, L.R.; Gu, Z.H.; Qiu, W.; Weng, S.P.; Yu, X.Q.; He, J.G. Molecular cloning, characterization and expression analysis of the tumor necrosis factor (TNF) superfamily gene, TNF receptor superfamily gene and lipopolysaccharide-induced TNF-α factor (LITAF) gene from *Litopenaeus vannamei*. *Dev. Comp. Immunol.* **2012**, *36*, 39–50. [CrossRef] [PubMed]

26. Zhang, X.; Zhang, P.; Li, C.; Li, Y.; Jin, C.; Zhang, W. Characterization of two regulators of the TNF-α signaling pathway in *Apostichopus japonicus*: LPS-induced TNF-α factor and baculoviral inhibitor of apoptosis repeat-containing 2. *Dev. Comp. Immunol.* **2015**, *48*, 138–142. [CrossRef] [PubMed]

27. Jin, P.; Hu, J.; Qian, J.; Chen, L.; Xu, X.; Ma, F. Identification and characterization of a putative lipopolysaccharide-induced TNF-α factor (LITAF) gene from Amphioxus (*Branchiostoma belcheri*): An insight into the innate immunity of Amphioxus and the evolution of LITAF. *Fish Shellfish Immunol.* **2012**, *32*, 1223–1228. [CrossRef] [PubMed]

28. Li, S.; Li, X.; Gen, X.; Chen, Y.; Wei, J.; Sun, J. Identification and characterization of lipopolysaccharide-induced TNF-α factor gene from Japanese flounder *Paralichthys olivaceus*. *Vet. Immunol. Immunopathol.* **2014**, *157*, 182–189. [CrossRef] [PubMed]

29. Cai, J.; Huang, Y.; Wei, S.; Ouyang, Z.; Huang, X.; Qin, Q. Characterization of LPS-induced TNF-α factor (LITAF) from orange-spotted grouper, *Epinephelus coioides*. *Fish Shellfish Immunol.* **2013**, *35*, 1858–1866. [CrossRef] [PubMed]

30. Hwang, S.D.; Sang, H.S.; Kwon, M.G.; Chae, Y.S.; Shim, W.J.; Jung, J.H.; Kim, J.W.; Park, C.I. Molecular cloning and expression analysis of two lipopolysaccharide-induced TNF-α factors (LITAFs) from rock bream, *Oplegnathus fasciatus*. *Fish Shellfish Immunol.* **2014**, *36*, 467–474. [CrossRef] [PubMed]

31. Wang, H.; Shen, X.; Xu, D.; Lu, L. Lipopolysaccharide-induced TNF-α factor in grass carp (*Ctenopharyngodon idella*): Evidence for its involvement in antiviral innate immunity. *Fish Shellfish Immunol.* **2013**, *34*, 538–545. [CrossRef] [PubMed]

32. Livak, K.J.; Schmittgen, T.D. Analysis of relative gene expression data using real-time quantitative PCR and the $2^{-\Delta\Delta Ct}$ method. *Methods* **2001**, *25*, 402–408. [CrossRef] [PubMed]

33. Bonecchi, R.; Garlanda, C.; Mantovani, A.; Riva, F. Cytokine decoy and scavenger receptors as key regulators of immunity and inflammation. *Cytokine* **2016**, *87*, 37–45. [CrossRef] [PubMed]

34. Merrill, J.C.; You, J.; Constable, C.; Leeman, S.E.; Amar, S. Whole-body deletion of LPS-induced TNF-α factor (LITAF) markedly improves experimental endotoxic shock and inflammatory arthritis. *Proc. Natl. Acad. Sci. USA* **2011**, *108*, 21247–21252. [CrossRef] [PubMed]

35. Moriwaki, Y.; Begum, N.A.; Kobayashi, M.; Matsumoto, M.; Toyoshima, K.; Seya, T. Mycobacterium bovis Bacillus Calmette-Guerin and its cell wall complex induce a novel lysosomal membrane protein, SIMPLE, that bridges the missing link between lipopolysaccharide and p53-inducible gene, LITAF(PIG7), and estrogen-inducible gene, EET-1. *J. Biol. Chem.* **2001**, *276*, 23065–23076. [CrossRef] [PubMed]

36. Ponting, C.P.; Mott, R.; Bork, P.; Copley, R.R. Novel protein domains and repeats in Drosophila melanogaster: Insights into structure, function, and evolution. *Genome Res.* **2001**, *11*, 1996–2008. [CrossRef] [PubMed]

37. Bromage, E.S.; Kaattari, I.M.; Zwollo, P.; Kaattari, S.L. Plasmablast and plasma cell production and distribution in trout immune tissues. *J. Immunol.* **2004**, *173*, 7317–7323. [CrossRef] [PubMed]

38. Bolcatobellemin, A.L.; Mattei, M.G.; Fenton, M.; Amar, S. Molecular cloning and characterization of mouse LITAF cDNA: Role in the regulation of tumor necrosis factor-α (TNF-α) gene expression. *J. Endotoxin Res.* **2004**, *10*, 15–23.

39. Hong, Y.H.; Lillehoj, H.S.; Lee, S.H.; Park, D.W.; Lillehoj, E.P. Molecular cloning and characterization of chicken lipopolysaccharide-induced TNF-α factor (LITAF). *Dev. Comp. Immunol.* **2006**, *30*, 919–929. [CrossRef] [PubMed]

40. Beutler, B.; Cerami, A. Cachectin and tumour necrosis factor as two sides of the same biological coin. *Nature* **1986**, *320*, 584–588. [CrossRef] [PubMed]

41. Letunic, I.; Copley, R.R.; Pils, B.; Pinkert, S.; Schultz, J.; Bork, P. SMART 5: Domains in the context of genomes and networks. *Nucleic Acids Res.* **2006**, *34*, D257–D260. [CrossRef] [PubMed]

42. Thompson, J.D.; Higgins, D.G.; Gibson, T.J. CLUSTAL W: Improving the sensitivity of progressive multiple sequence alignment through sequence weighting, position-specific gap penalties and weight matrix choice. *Nucleic Acids Res.* **1994**, *22*, 4673–4680. [CrossRef] [PubMed]

43. Kumar, S.; Stecher, G.; Tamura, K. MEGA7: Molecular evolutionary genetics analysis version 7.0 for bigger datasets. *Mol. Biol. Evol.* **2016**, *33*, 1870–1874. [CrossRef] [PubMed]

 International Journal of
Molecular Sciences

Review

The Interactive Roles of Lipopolysaccharides and dsRNA/Viruses on Respiratory Epithelial Cells and Dendritic Cells in Allergic Respiratory Disorders: The Hygiene Hypothesis

Tsang-Hsiung Lin [1], Hsing-Hao Su [2], Hong-Yo Kang [1,3,* and Tsung-Hsien Chang [4,5,*

[1]   Graduate Institute of Clinical Medical Sciences, College of Medicine, Chang Gung University, Kaohsiung 81362, Taiwan; joanne.chiou@msa.hinet.net

[2]   Department of Otorhinolaryngology—Head & Neck Surgery, Kaohsiung Veterans General Hospital, Kaohsiung 81362, Taiwan; shsu@vghks.gov.tw

[3]   Hormone Research Center and Department of Obstetrics and Gynecology, Kaohsiung Chang Gung Memorial Hospital, Kaohsiung 83301, Taiwan

[4]   Department of Medical Education and Research, Kaohsiung Veterans General Hospital, Kaohsiung 81362, Taiwan

[5]   Department of Medical Laboratory Science and Biotechnology, Chung Hwa University of Medical Technology, Tainan 71703, Taiwan

*   Correspondence: hkang3@mail.cgu.edu.tw (H.-Y.K.); changth@vghks.gov.tw (T.-H.C.)

Received: 26 September 2017; Accepted: 19 October 2017; Published: 23 October 2017

Abstract: The original hygiene hypothesis declares "more infections in early childhood protect against later atopy". According to the hygiene hypothesis, the increased incidence of allergic disorders in developed countries is explained by the decrease of infections. Epithelial cells and dendritic cells play key roles in bridging the innate and adaptive immune systems. Among the various pattern-recognition receptor systems of epithelial cells and dendritic cells, including toll-like receptors (TLRs), nucleotide-binding oligomerization domain (NOD)-like receptors (NLRs) and others, TLRs are the key systems of immune response regulation. In humans, TLRs consist of TLR1 to TLR10. They regulate cellular responses through engagement with TLR ligands, e.g., lipopolysaccharides (LPS) acts through TLR4 and dsRNA acts through TLR3, but there are certain common components between these two TLR pathways. dsRNA activates epithelial cells and dendritic cells in different directions, resulting in allergy-related Th2-skewing tendency in epithelial cells, and Th1-skewing tendency in dendritic cells. The Th2-skewing effect by stimulation of dsRNA on epithelial cells could be suppressed by the presence of LPS above some threshold. When LPS level decreases, the Th2-skewing effect increases. It may be via these interrelated networks and related factors that LPS modifies the allergic responses and provides a plausible mechanism of the hygiene hypothesis. Several hygiene hypothesis-related phenomena, seemingly conflicting, are also discussed in this review, along with their proposed mechanisms.

Keywords: lipopolysaccharide; double-stranded RNA; epithelial cell; dendritic cell; allergic respiratory disorder; hygiene hypothesis; rhinovirus; respiratory syncytial virus; toll-like receptor

---

1. Introduction

According to the hygiene hypothesis, the increased incidence of allergic disorders in developed countries is explained by the decrease in infections [1,2]. Several studies have shown that exposure to more LPS (lipopolysaccharide and endotoxin), a major component of the outer membrane of gram-negative bacteria, in early childhood protects against the later development of allergic disorders [3–5]. However, these studies were done in rural areas in Europe. An urban study examined a birth cohort in the inner-city environment, and declared that exposure to specific Firmicutes and Bacteriodetes in house dust during children's first year of life was associated with decreased atopy and atopic wheeze. Exposure to high levels of both allergens and this subset of bacteria in infancy was also inversely related to the incidence of atopy or wheeze [6]. Another study focusing on the common cold also concluded that more runny nose episodes in infancy protected against later atopy [7]. Some questions, however, remained to be answered, especially the mechanisms of protection, the nature of protective infections, and why, after sensitization, LPS seems to exacerbate the conditions, rather than protect the organisms [8]. The mechanism by which LPS might downregulate allergic airway inflammation and subsequently suppress airway hyperreactivity was not clear, but two recently published studies proposed a plausible model that clearly elucidates the role of LPS in downregulating allergic inflammation [9,10]. In this review, we propose a model that includes four major players: LPS, dsRNA or viruses, epithelial cells and dendritic cells. This model is used to explore a possible protective mechanism based on more frequent occurrences of the common cold and more LPS exposure in early childhood (i.e., before sensitization) leading to less development of allergy.

2. Four Major Players in the Proposed Simplified Model of Hygiene Hypothesis: Epithelial Cells (ECs), Dendritic Cells (DCs), dsRNA and LPS

2.1. Epithelial Cells Play Key Roles in Bridging the Innate and Adaptive Immune System

The airway epithelium contributes significantly to the barrier function of airway tract, which has three active components: the mucociliary escalator, the intercellular apical junctional complexes and the secreted antimicrobial peptides. An impaired barrier function would increase susceptibility to infection and sensitization as well as chronic inflammation [11]. However, beyond the barrier function, which was once thought to be the only function of epithelial cells, in recent years epithelial cells were found to play key roles in bridging the innate and adaptive immune system [12–16]. Epithelial cells can activate DCs, B cells, and T cells, thereby stimulating their differentiation, or modifying the above effects [12]. Epithelial-derived cytokines can also activate basophils, eosinophils, mast cells and nuocytes [17].

2.2. DCs Interact Closely with ECs to Orchestrate the Immune Responses

DCs are responsible for initiating all antigen-specific immune responses. They capture and process antigens, express lymphocyte co-stimulatory molecules, regulate the functions of B and T lymphocytes and secrete cytokines to initiate immune responses. In addition, they are also responsible for inducing tolerance of T cells to innate antigens. Before recognition of the pivotal role of ECs in immune response, most researches emphasized the DCs [18–20]. Allergens, microbial compounds and various environmental and genetic risk factors for allergic disorders, however, often interfere with the immune functions of airway ECs and DCs [17,21,22]. At the 2014 International DC Symposium, at least 28 different DC subsets were described using various surface markers and nomenclature systems in distinct species [23]. However, these numerous species are generally classified into type 1 conventional DC (cCD1), type2 conventional DC (cCD2) and plasmacytoid DC (pDC) [16]. Conventional DCs, also referred to as myeloid DCs (mDCs) [24], play important roles in the pathogenesis of allergic airway inflammation. By contrast, pDCs are related to immune tolerance, and host defense against viral infections at the mucosal site, thereby modulating the extent of inflammation and tissue damage [24]. Although DCs cultured in vitro from monocytes, called moDCs, do not show the same behavior or capability as their ex vivo isolated counterpart, they are often used for research for their easier availability [23,25].

2.3. Most Bridging Effects Start from Activation via TLRs and Other Receptors of Epithelial Cells and DCs

Among the various pattern-recognition receptor systems of ECs and DCs, including toll-like receptors (TLRs) and NOD-like receptors (NLRs) and others, which could respond to the pathogen-associated molecular pattern (PAMP) and the danger-associated molecular pattern (DAMP),

TLRs are the key systems. In humans, TLRs consist of TLR1 to TLR10 and could respond to various microbial products. However, this review discusses only the most relevant TLR3 and TLR4. Lipopolysaccharide (LPS) acts via TLR4 and dsRNA acts via TLR3. However, there are many common components in these two signaling pathways [26,27], which provide the mechanisms for interactive regulation. Furthermore, TLRs may cooperate with NLRs and PARs (protease-activated receptors) in mediating the immune response [28,29], which can be upregulated or down regulated by cytokines and chemokines in the context [30–32].

Below, we will discuss these two TLRs of ECs and DCs in the order of TLR3/ECs, TLR4/ECs, TLR3/DCs and TLR4/DCs.

2.4. dsRNA or Viruses Can Activate TLR3 of Respiratory Epithelial Cells and Stimulate the Production of Various Proallergic Cytokines

dsRNA or viruses can activate the TLR3 of epithelial cells (ECs) and stimulate the production of various proallergic cytokines [9], including thymic stromal lymphopoietin (TSLP), interleukin 33 (IL33) and IL25 [33]. Of all the TLR ligands tested, only polyI:C, representing dsRNA, the TLR3 ligand stimulates a high level of TSLP expression in keratinocytes and human bronchial ECs. Ligands of other TLRs induce only minimal expression of TSLP [31,32]. In the human epithelium IL33 is induced mainly by polyI:C and flagellin, the ligands to TLR3 and TLR5, respectively [34], following parasitic or viral infections, but also by other non-TLR factors, such as exposure to allergens. IL25 is induced by microflora, allergens, helminth or particle (TiO2) [35]. PolyI:C induces only modest expression of IL25 in respiratory ECs [9]. Downstream Th2 cytokines, such as IL4 and IL13, of the allergic triad consisting of TSLP, IL33 and IL25, would augment the stimulatory effect of polyI:C on ECs synergistically to produce more Th2 cytokines, thus promoting a positive feedback cycle [31].

Although some respiratory viruses such as influenza and respiratory syncytial virus (RSV) destroy the airway epithelial barrier, rhinovirus (RV) by itself does not cause cytopathology. The RV infection only disrupts the epithelial barrier function, specifically disrupting tight junctions, as well as increasing vascular leakage and mucus secretion [36]. In cultured human nasal ECs, RV infections showed decreased zona occluden-1, claudin-1, and E-cadherin levels, consistent with the effect of polyI:C and could expose basolateral epithelial receptors, where TLRs and other pattern recognition receptors (PRRs) are prominently located [37]. polyI:C activates TLR3 and induces apoptosis in cells [38]. Fortunately, LPS could suppress the pathogenic effect of polyI:C on ECs, and may protect the epithelial integrity [9].

2.5. LPS Activate TLR4 of Respiratory Epithelial Cell, Using Bidirectional Capacity to Modulate Allergic Disorders through Multiple Pathways

LPS is composed of lipid A (a hydrophobic domain), a core oligosaccharide, and a distal polysaccharide (or O-antigen) [39]. Core and O-antigen sugars protect bacteria from antibiotics, the complement system, and other environmental stresses. Lipid A is the moiety that activates TLR4.

Epidemiological studies have repeatedly confirmed that exposure to environmental LPS in early childhood would reduce the incidence of allergic disorders in later life [3,5,40]. Animal studies confirmed LPS exposure before or shortly after sensitization protects against the development of allergy [41], but after that critical time point LPS worsens the inflammation by attracting neutrophils and eosinophils, possibly via dendritic cells and B cells [42] (Table 1). Neutrophilia are discussed further in Section 2.5.3.

Table 1. Studies demonstrating the biphasic capacity of lipopolysaccharides (LPS).

Authors	Model	Origin of LPS	LPS Dose and Pathway Used	Allergen or Antigen	Allergen Dose and Pathway Used for Sensitization	Allergen Dose and Pathway Used for Challenging	Protocol	Result	Note
Tulic et al., 2000 [41]	male PVG rat	*Salmonella typhimurium*	50 µg/mL inhaled	OVA	100 µg/mL i.p.	Aerosolized 1% OVA at Day 11 after sensitization	1. Sensitized rats exposed 1 d before or 1, 2, 4, 6, 8, 10, or 12 d after sensitization 2. A second group of sensitized rats were exposed to LPS 18 h after OVA challenge.	1. Single aerosol exposure to LPS—1 d, and up to 4 d after i.p. OVA protected against allergy. LPS exposure ≥6, 8, or 10 d after sensitization exacerbated the allergy with cellular influx. 2. Exposure of sensitized rats to LPS on Day 12, 18 h after allergen challenge further potentiated the allergen induced inflammatory cell influx, predominantly due to a 20-fold increase in neutrophil influx, making up ≥80% of the cellular content.	Timing of LPS exposure determines protection or exacerbation of allergy.
Eisenbarth et al., 2002 [43]	BALB/cJ mice and BALB/cAnNCr	*Escherichia coli*	Concomitant use of 100 µg (high dose) or 0.1 µg LPS with OVA in sensitizing period	OVA	100 µg OVA in 50 µL PBS intranasally, or 100 µg OVA in 2 mg Al(OH)$_3$ intraperitoneally, with LPS depletion	25 µg OVA intranasally	Sensitized on Days 0, 1 and 2, challenged on Days 14,15,18 and 19, killed on Day 21	Mice exposed to LPS-depleted OVA showed no airway inflammatory responses after challenge; those sensitized with OVA containing low dose LPS demonstrated significant Th2 lung infiltrates; those exposed to PBS or low dose LPS alone did not generate pulmonary inflammation after challenge; those sensitized with OVA containing high dose LPS resulted in a Th1 associated response.	TLR4 signaling Is required for Th2 priming to inhaled antigens, and the dose of LPS during sensitizing period regulates the predominance of Th1 or Th2 response.
Lowe et al., 2015 [44]	Male Dunkin-Hartley guinea pigs (GPs)	LPS source not mentioned	Inhaled 30 µg/mL	OVA	Bil. i.p. injection of OVA,150 µg/mL and Al(OH)$_3$ 100 mg/mL normal saline	Sensitized GPs were exposed to inhaled OVA (300 µg/mL) on Day 21.	LPS (30 µg/mL) exposure was by two protocols: 72 and 24 h pre-OVA exposure, 48 h pre-OVA and co-administered with OVA by nebulizer, at rate of 0.3 mL/min for 1 h	LPS exposure 24 h before allergen challenge attenuates the early asthmatic response (EAR), whereas co-administered LPS does not influence the EAR. The addition of a second LPS exposure co-administered with OVA prolonged the EAR. Similarly, LPS exposure 24 h before allergen challenge diminished airway hyperreactivity (AHR) to histamine, whereas co-administered LPS prolonged the AHR	Emphasizing to the timing of LPS application
Langenkamp et al., 2000 [45]	Dendritic cells	1. *Salmonella abortus equi.* 2. Toxic shock syndrome toxin-1 (TSST-1)	1. 20–100 ng/mL LPS 2. 0.1 or 10 ng/mL TSST-1	TSST-1 as antigen			Pretreatment of dendritic cells with LPS, then after 8 or 48 h, 0.1 or 10 ng/mL TSST-1 was added to culture medium	1. Soon after LPS stimulation DCs primed strong Th1 responses, but later favored Th2 responses. 2. High dose antigen favored Th1 response, yet low dose antigen favored Th2 response. 3. Exogenous IL12 favored Th1 tendency, yet IL4 favored Th2.	This study explored DC's priming tendency after LPS pre-stimulation, in response to superantigen TSST-1, which was an inflammatory response, but not allergic response.

d: days.

2.5.1. Protective Role of LPS against Allergic Disorders

Focusing on the Role of LPS

The protective mechanism of LPS was not previously clear because LPS has little or no direct effect of stimulating ECs to produce cytokines [31,46] (Table 2). Although LPS cause a 10-fold increase in cytokine expression when serum is present [47], under general condition, when without other coexisting factors, ECs are relatively resistant to LPS in producing cytokines [48]. Thus, previously via what pathway LPS exerts its protective function was a confusing issue. However, recently, pretreatment of LPS was found to induce the A20 protein, which is a ubiquitin-modifying enzyme, and in this way reduce the inflammation caused by house dust mite in a mouse model [10]. Pretreated LPS also attenuates the induction of proallergic cytokines, including TSLP and IL33 in respiratory ECs stimulated with polyI:C and human parechovirus by attenuating TANK-binding kinase 1, IRF3, and NF-κB activation [9]. Thus, LPS does not directly antagonize the inflammation; instead, when pretreated before sensitization, it inhibits the effect of proallergic ligands of TLR3 and simultaneously induces A20, which is a potent NF-κB inhibitor [49]. Pretreatment of LPS was also shown to be protective against inflammation in other kinds of cells, including macrophages, cardiac myocytes [50], and neurons [51], under certain conditions (Table 2).

Table 2. Studies demonstrating the protective capacity of LPS.

Authors	Model	Origin of LPS	LPS Dose and Pathway Used	Allergen, Antigen or Stimulus	Allergen Dose and Pathway Used for Sensitization	Allergen Dose and Pathway Used for Challenging	Protocol	Result	Note
Carlsten et al., 2011 [52]	Human of age 7	Home dust	Inhaled from environment	Dog allergen	Inhaled from environment	Inhaled from environment	Correlation study	Endotoxin was associated with decreased risk of sensitization to dog allergen.	HDM was also associated with decreased risk of sensitization to dog allergen, which needs further confirmatory studies. However, in Schuijs' study below, HDM was noted to induce A20 also, though less pronounced than LPS.
Braun-Fahrlander et al., 2002 [3]	Human of age 6-13	Home dust	Inhaled from environment	unspecified	Inhaled from environment	Inhaled from environment	Correlation study	Endotoxin levels in dust were inversely related to the incidence of hay fever, atopic asthma, and atopic sensitization.	
Schuijs et al., 2015 [10]	1. Female C57Bl/6 wild-type 2. Human normal bronchial epithelial cells 3. Human epithelial cells from asthma patients	Ultrapure LPS purchased from Invivogen. Strain and species not specified.	1. A single intranasal injection of 1 µg LPS on Day 14, or i.n 100 ng every other day (starting on Day 14). 2. 100 ng overnight 3. 100 ng every other day for 1 week	HDM	1. 1 µg HDM extracts 2. No sensitization 3. Accurate time point of sensitization could not be traced.	1. 10 µg HDM extracts 2. HDM extracts (dose not mentioned) 3. HDM extracts (dose not mentioned)	1. Mice sensitized on Day 0 with 1 µg HDM, and challenged on Days 7–11 with 10 µg HDM extracts 2. Cells exposed overnight to 100 ng LPS. After 2 weeks, cells stimulated with HDM extract. 3. cell cultures exposed to 100 ng of LPS for 1 week before stimulation with HDMs	1. Protective LPS led to decreasing of IL5 and IL13 in mediastinal lymph node cells and downregulation of HDM-induced recruitment of cDCs, with moDCs unaffected. 2. LPS pretreatment reduced allergic cytokines production.	TLR4 signaling in ECs induces attenuators of signaling such as A20.
Ganesh et al., 2014 [53]	BALB/c and DO11.10 mice	*Salmonella enterica serovar typhimurium aroA strain SL 7207*	Intragastric inoculation with 0.5~1 × 10⁹ CFU of whole *S. typhimurium* (SL 7207)	OVA	10 µg OVA + adjuvant, i.p.	intranasal use of 30 µg of OVA	Mice were sensitized with OVA i.p. on Days 7, 8, 9, and 20, infected intragastrically with *S. typhimurium* on Days 0, 7, 20, and 27, challenged on Days 20, 24, 27, 30, and 34 by intranasal administration of of OVA	*S. typhimurium* infection in mice results in attenuation of cellular airway inflammation, reduced pathology and mucus production in the lungs, expansion of CD11b⁺Gr1⁺ myeloid cells, with no apparent diversion toward Th1.	This study used whole bacteria for experiment, instead of LPS only.
Rodriguez et al., 2003 [54]	C57BL/6J, BALB/c and C3H/HeJ mice	*Salmonella abortus equi*	LPS at a dose of 20 µg/animal was delivered intravenously concomitantly with a second OVA challenge	OVA	4 µg OVA/1.6 mg aluminum hydroxide	10 µg OVA/50 µL saline intranasally	Mice were immunized on Days 0 and 7, and challenged on Days 14 and 21 intranasally	1. LPS administration suppresses allergic airway inflammation and cytokine production through a mechanism independent of IL12 or IFNγ. 2. Systemic LPS inhibited airway inflammation. 3. Systemic LPS reduced airway hyperreactivity (AHR).	Thus, systemic LPS displayed protective effect, while local LPS displayed pro-inflammatory effect with neutrophilia reaction.
Lin et al., 2016 [9]	H292 cell line	*Escherichia coli*	0.3 to 30 µg/mL co-culture	polyI:C, HPeV1			LPS pretreatment 2 h before polyI:C or HPeV1 co-culture with H292 cells.	The downstream production of TSLP and IL33 by stimulating H292 cells with polyI:C or HPeV1 was reduced with 30 µg/mL LPS pretreatment, but not 0.3 µg/mL LPS	

Focusing on the Role of LPS-Related Actions in Farming Households

The LPS levels of exposure for children from farming households are significantly higher than those from non-farming households, and an inverse association was also noted between the LPS exposure level and the prevalence of hay fever and atopic sensitization [3]. More than 20 studies focusing on the issue of protection conferred by farm environment exposure against atopy development revealed that the protective farm effect was related to microbial exposure [55]. An inverse relationship between exposure to LPS in the mattress dust of children and the occurrence of atopic diseases was also shown in rural environments [4]. The same pattern was noted with the diversity of microbial exposure inversely related to the risk of asthma [56]. The unstimulated peripheral blood mononuclear cells of farm children produced more IL10, IL12 and IFNγ than those of non-farm children, indicating increased spontaneous production of Th1 and regulatory cytokines. Decreased TNF responses to short-term LPS stimulation in farm-exposed children may imply tolerance [57]. The mRNA expression of Th1/Th2/Th17-associated cell markers decreased between ages of 4.5 and 6 years, and, at the age of six, regulatory T cells (Treg) were decreased with farm exposure and increased within asthmatics, compared with levels at age of 4.5, implicating a critical "time window" for Treg-mediated asthma protection via environmental exposure before age of six years [58,59]. Finally, it must be emphasized that not all farming environments protect against the development of asthma and wheeze in children, e.g., the keeping of hares and rabbits, using pressed hay and the presence of sheep were positively associated [59,60].

2.5.2. The Proallergic Role of LPS

In asymptomatic one-month-old neonates colonization of the airways with one or more of the pathogens *S. pneumoniae*, *H. influenzae*, *M. catarrhalis*, was associated with increased risk of a first wheezy episode, persistent wheeze, acute exacerbation of wheeze, increased blood eosinophil counts and total IgE and increased risk of developing asthma by the age of five years. However, the study did not specify the mechanism of increased asthma risk, whether via ECs, DCs or their interaction [61].

Using irradiated chimeric mice, Hammad et al. demonstrated that TLR4 expression on lung ECs is required and sufficient for house dust mite (HDM) to activate DCs and prime for Th2 responses. Moreover, LPS binding to TLR4 of ECs in the presence of HDM led to the production of proallergic cytokines, including TSLP, GM-CSF, IL25 and IL33. Knockout of TLR4 on ECs, but not on hematopoietic cells, abolished HDM driven allergic airway inflammation. A TLR4 antagonist targeting exposed ECs suppressed the airway hyperreactive features of asthma [62]. Thus, the role of LPS-TLR4 binding in HDM-induced allergic disorders was confirmed.

2.5.3. The Pro-Inflammatory Non-Allergic Role of LPS

When the allergic state was already established, further LPS application in the airway might induce a state of worsened inflammation with neutrophilia predominant with eosinophilia persistence or abolished [41,54]. The airway neutrophilia could be caused by two pathways: first, direct activation of the airway epithelial cells by LPS with production of IL8 [47], which is a strong neutrophil attractant chemokines [63] and second, indirect pathway with Th17 cells activated first, and then the released IL17 activates the airway epithelial cells to produce IL8 and other related cytokines [64]. Here we have to emphasize that normal airway epithelial cells are relatively resistant to common type LPS stimulation [47], yet they will become inflamed in the presence of blood or serum, but by contrast, not plasma [65,66], indicating some factors in blood or serum critically influence their responses. Besides, unusually high dose or even to the toxic level or sublethal dose as in Table 3 would activate another non-allergic inflammatory pathway with neutrophils predominant [67–69]. LPS of strong pathogens, such as *Pseudomonas aeruginosa* would cause robust airway inflammation even at a much less dose [70]. The role of LPS in pro-inflammatory or non-allergic effect is summarized in Table 3. In addition, the dual roles of LPS is discussed in more detail in Sections 2.7, 3 and 4.

Table 3. Studies demonstrating the pro-inflammatory capacity of LPS.

Authors	Model	Origin of LPS	LPS Dose and Pathway Used	Allergen or Antigen	Allergen Dose and Pathway Used for Sensitization	Allergen Dose and Pathway Used for Challenging	Protocol	Result	Note
Rittirsch et al., 2008 [67]	1. C57BL/6 mice 2. C57BL/6 mice with neutrophil depleted by antibody	*Escherichia coli* (*serotype O111:B4*)	50 µg LPS in 40 µL PBS intratracheally, total 2,550,100 µg	nil	nil	nil	Permeability index checked from bronchoalveolar lavage at 0, 2, 4, 6, 8 h.	1. Maximal permeability at 50 µg, no different from 100 µg; at 6 h, no different from 8 h. 2. When neutrophils depleted, no permeability change. 3. Pathology includes: interstitial and intraalveolar deposits of neutrophils and fibrin, prominence of alveolar macrophages, and intraalveolar hemorrhage.	The LPS concentration used is 1250 µg/mL, as compared to the 0.3 and 30 µg/mL in cell line model [9], and the total LPS used is 50 µg, as compared with total 100 ng to 1 µg LPS in Schuijs' study [10].
Eutamene et al., 2005 [70]	1. Male Wistar rats 2. NCI-H292 human airway epithelial cells	1. *Pseudomonas aeruginosa* 2. *Escherichia coli* (*SO55:B5*)	1. 1 µg LPS per rat via intra-tracheal instillate 2. 2 µg/mL LPS for co-culture	nil	nil	nil	1. LPS from *P. aeruginosa* instilled in the trachea at a constant rate of 10 µL/min for 15 min. 2. LPS from *E. coli* for 15 and 30 min and 1, 2, 3 and 6 h.	1. Airway epithelial paracellular permeability was increased , Leukocytes number in BAL fluid was sixfold higher, with macrophage, neutrophil and lymphocyte numbers significantly increased. 2. Myosin light chain (MLC) phosphorylation occurs after *E. coli* co-culture, and tight junction permeability increased.	*P. aeruginosa* is a strong pathogen for airway [71], so total amount of LPS used is less, as compared with studies above.
Rojas et al., 2005 [68]	C57BL/6 male mice	*Escherichia coli* O111:B6	Intraperitoneally with 1 mg/kg LPS	nil	nil	nil	Mice were inoculated intraperitoneally with 1 mg/kg of LPS from *E. coli* O111:B6.	Sublethal dose of i.p. LPS to mice caused rapid onset of interstitial pulmonary edema, inflammatory cell accumulation, and deposition of fibronectin and collagen in the lungs.	The scale of mg/kg is sublethal, compared to the protective dose scale of ng/mL to µg/mL.
Yao et al., 2017 [69]	1. Male C57BL/6J mice 2. male Wistar rats	LPS, source not specified	1. i.p. LPS at the doses of 8 mg/kg 2. i.p. LPS at the doses of 5 mg/kg	nil	nil	nil	Lung injury in mice and rats were induced by i.p. LPS.	Lung tissues revealed interstitial edema and hemorrhage, alveolar wall thickening, increased infiltration of neutrophils and macrophages in the lung parenchyma and alveolar spaces.	Again, the dose of causing acute lung injury is on the scale of mg/kg.

214

Table 3. *Cont.*

Authors	Model	Origin of LPS	LPS Dose and Pathway Used	Allergen or Antigen	Allergen Dose and Pathway Used for Sensitization	Allergen Dose and Pathway Used for Challenging	Protocol	Result	Note
Taveira da Silva et al., 1993 [72]	Human	*Salmonella minnesota*	i.v. LPS	nil	nil	nil	The patient administered i.v. 1 mg of *S. minnesota* LPS, in sterile water, in an attempt to treat a tumor.	Septic shock syndrome induced, including a high-cardiac-output hypotension, disseminated intravascular coagulation, abnormalities of hepatic and renal function, and non-cardiogenic pulmonary edema.	1 mg of purified LPS is equivalent to 15,000 ng/kg, thousands times higher than the usual dose of 4 ng/kg given to normal volunteers in experimental studies. Endothelial cells are much more sensitive to LPS than epithelial cells, with pg/mL level LPS activating endothelial cells in the presence of blood, compared to the relative resistance of respiratory epithelial cells to µg/mL level LPS [66].
Pugin et al., 1993 [66]	Human umbilical vein endothelial cells (HUVEC)	1. *Escherichia coli* 0111:B4 2. *Salmonella minnesota*	Incubated with different dilutions of *E. coli 0111:B4* or *S. minnesota* wild-type LPS, from 10^{-1} to 10^4 pg/mL	nil	nil	nil	HUVECs incubated with different dilutions of LPS for 6 h	In the presence of whole blood, 1000-fold less LPS was required to achieve the level of HUVEC activation (assessed by VCAM-1 upregulation) observed with plasma alone.	Endothelial cells are sensitive to ng/mL LPS in the absence of blood, but much more sensitive even to pg/mL LPS in the presence of blood.
Rodriguez et al., 2003 [54]	C57BL/6J, BALB/c and C3H/HeJ mice	*Salmonella abortus equi*	LPS at a dose of 20 µg/animal was delivered intranasally concomitantly with a second OVA challenge	OVA	4 µg OVA/1.6 mg aluminum hydroxide	10 µg OVA/50 µL saline intranasally	Mice were immunized on Days 0 and 7, and challenged on Days14 and 21 intranasally	1. LPS administration suppresses allergic airway inflammation and cytokine production through a mechanism independent of IL12 or IFNγ 2. Local LPS switched the airway inflammation from eosinophilia to neutrophilia. 3. Local LPS increased AHR by neutrophilic inflammation.	Systemic LPS displayed protective effect, while local LPS displayed pro-inflammatory effect with neutrophilia reaction.
Hammad et al., 2009 [62]	Radiation-induced chimeric *Tlr4*-deficient mice with DCs deficient or ECs-like cells	*Rhodobacter sphaeroides*	10 µg or 100 ng per mouse, in 80 µL PBS, intratracheal	HDM	nil	Intratracheal 100 µg HDM	80 µL PBS intratracheal with HDM and LPS	TLR4 expression on lung structural cells, but not on DCs, is necessary and sufficient for lung DC activation and for priming of effector T helper responses to HDM.	TLR4 triggering on structural cells in the presence of HDM caused production of TSLP, GM-CSF, IL25 and IL33. The absence of TLR4 on structural cells, but not on hematopoietic cells, abolished HDM-driven allergic airway inflammation.

nil: not in list.

2.6. dsRNA or Many Viruses Activate TLR3 of Dendritic Cells, Thus, Induce DCs with Th1-Promoting Capacity with Some Exceptions

2.6.1. dsRNA Activates the TLR3 of DCs, and Cause Them to Become DCs with Th1-Promoting Capacity

dsRNA can facilitate the development of Th1 cells even after the active period [73]. The Th1-promoting effect is mediated by IL12 and some unknown factor because anti-IL12 Abs could only partially block it [73], and IL12p40/p70-deficient mice could still mount a strong Th1 response [74]. DCs pretreated with polyI:C expand T cells with high Th1 polarity (70–90%), and this pattern is also confirmed in influenza [75].

2.6.2. Respiratory Syncytial Virus (RSV) Is Probably an Exception, Which Likely Skew DC towards DC with Th2-Promoting Capacity

RSV infection is known to be associated with higher risk of later allergic disorders [76,77]. Clinical studies in RSV bronchiolitis revealed low Th1 responses with reduced IFNγ production and robust Th2 cytokines production [78]. In predisposed individuals, RSV infection cause the release of tissue alarmins and promote a cytokine microenvironment that is Th2 prone [79]. DCs, mainly myeloid DCs (mDCs), are likely to play pro-inflammatory roles in RSV infection and induce a Th2 response and may increase the later risk of developing allergic asthma [80]. The Th2-biased response is quite different from Th1 response to other respiratory viruses, such as influenza virus and adenovirus with strong IFNγ production. Upon RSV infection, mDCs mature, strongly activate naïve T cells and promote Th2 responses. In contrast, in healthy lungs, most of these cells are immature and unable to activate naïve T cells [80]. However, there remain some controversies. In Stein's series, RSV lower respiratory infections (LRTIs) were associated with an increased infrequent and frequent wheeze by age six. Risk decreased markedly with age and was not significant by age 13. No association between RSV LRTIs and subsequent atopic status was noted. In addition, it was concluded that RSV LRTIs in early childhood is an independent risk factor for the later occurrence of wheezing up to age 11 years but not at age 13. This association is not caused by an increased risk of allergic sensitization [81]. In Kotaniemi-Syrjänen's cohort series, RSV infection was found to be associated with a relatively low risk of later childhood asthma among young children hospitalized for wheezing. However, when compared with non-selected school-aged children, the risk of asthma for RSV-positive children is increased, though only less than four folds [82], not as high as reported by Sigurs et al. (12 folds) [76,77].

2.6.3. Is Rhinovirus (RV) Another Exception?

Severe rhinovirus infection in infancy is closely related to later asthma development [83]. The risk of asthma at age six years in children who wheezed during the first three years of life with RV infection is 9.8-fold greater in comparison with those who wheezed without RV or RSV infection [84], with RSV-infected children displaying a 2.6-fold increase. However, in contrast to infancy, RV infection in non-asthmatic adult generally only causes runny nose, stuffy nose and sore throat, limiting symptoms to the upper respiratory tract only [85,86]. Whether RV infection predisposes the children to allergy or just reveals the children who already have an allergic predisposition is an issue to be resolved [87].

Related Mechanisms of Rhinovirus Infection

dsRNA and many virus infections generally would facilitate the development of Th1 cells via skewing DCs [73], but some strains of human rhinoviruses, e.g., HRV14, which belongs to the major group human RV (HRV), can efficiently inhibit the T cell stimulatory capacity of DCs through binding to its cellular receptor human intercellular adhesion molecule-1 (ICAM-1), inducing inhibitory cell surface receptors, and induce a promiscuous and deep anergic state in co-cultured T cells, despite high levels of MHC molecules as well as co-stimulatory molecules in in vitro study. This effect is independent of inhibitory soluble factors such as IL10 [88]. In this way, RV induces a hypoproliferative state in co-cultured T cells if they can come to contact DCs, but the Th1 predominant pattern is still preserved, only with decreased intensity. However, this study was performed by co-culturing HRV14 and DCs for more than 24 h in the laboratory. It is not certain if the circumstances could be reproduced in infected non-allergic individuals. Furthermore, even though the adaptive immune response was inhibited, the net response still was Th1, with the IFNγ level much higher than the IL4 level. In addition, another strain HRV2, belonging to a minor group, could bind to low-density lipoprotein receptor (LDLR) and stimulate much more production of IFNγ than HRV14 stimulation without causing anergy at all. Thusm both groups support the Th1 response, but maybe with different strengths [88].

Question: Under Normal Circumstances, Will Rhinoviruses Easily Approach DCs, as Shown in the In Vitro Study Above?

In Arruda's series done on nasal and nasopharyngeal biopsy tissue infected by HRV, only low numbers of ciliated cells were infected by HRV. Infected non-ciliated epithelial cells were also detected in the nasopharynx, indicating that only a very small proportion of cells (usually ≤10%) in the nasal epithelium and in both ciliated and non-ciliated cells in the nasopharynx were infected [89,90]. In the lower airway epithelium, the frequency of rhinovirus-infectable cells was noted to be similar to that in the upper airway [90]. The infected epithelium revealed no cytopathic change, and in studies of both natural and experimentally-induce colds, no viral RNA can be detected in the subepithelial layer. Thus, the viral infection is confined to the epithelial layers, with the degradation product of RV possibly via lysosome or proteasome pathways activating the epithelial cells to release various cytokines, chemokines etc. and attracting inflammatory cells [91]. Another study concluded that RV viremia was rare in children with respiratory infections, with 25.0% (7/28) children viremic with asthma exacerbations, 7.7% (2/26) viremic with common cold, 4.0% (1/25) viremic with bronchiolitis, and 0% (0/9) viremic with pneumonia [92]. Further experiments should be done to elucidate the possibilities of direct stimulation of RV on DCs. At present, there is limited data because most HRV strains (major groups) do not bind to murine receptor ICAM-1, so no proper murine model of HRV can be used.

In Transgenic Mouse Model

In the ICAM-1 transgenic mouse model, the minor group RV-1B caused neutrophilia and lymphocytosis without detectable thymic stromal lymphopoietin (TSLP), IL4, IL13 and IL17. IFNγ production was also increased in murine lung leukocytes, thus favoring Th1 response. This study proposed that allergic airway inflammation could be exacerbated by RV-induced Th1 response [93]. However, there are approximately 150 strains of rhinovirus, and different RV strains may have different stimulating capacities in the induction of allergic disease.

Conclusion on the Protective or Pro-Inflammatory Role of RV

For non-sensitized individuals, whether infants or adults, RV infection causes a minimal epithelial reaction, with a minimal Th2 response, which could possibly be easily suppressed by LPS when LPS of non-pathogenic bacteria are present. RV infections then stimulate DCs to produce a minor Th1 response (Figure 1a), because most of the infections are localized to the epithelial layers, so the viral load on DCs is minor, and would not cause the severe neutrophilia and lymphocytosis as seen in mouse model due to RV-1B [93]. However, in sensitized individuals, RV infections would stimulate the ECs with junctions of epithelial layers disrupted, proallergic cytokines such as TSLP and IL33 would be released and stimulate DCs towards Th2-promoting development. In addition, more viruses may penetrate the epithelial layers and stimulate DCs towards Th1-promoting development. The net result is that both Th1- and Th2-promoting activities of DCs would be increased, as noted in clinical cases of the exacerbation of the asthmatic patients after RV infection [94]. More studies are necessary to prove this deduction, although it is compatible with the already published data.

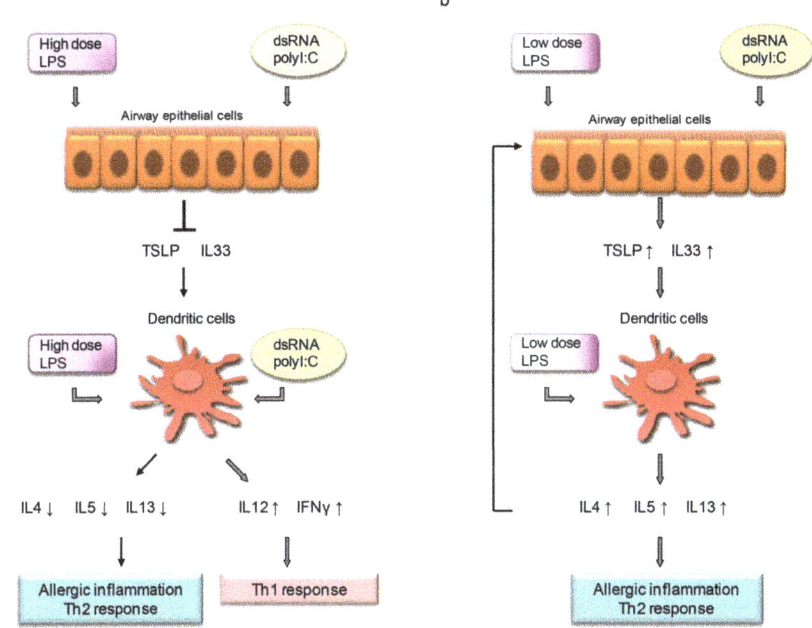

Figure 1. (**a**) Before sensitization, when the LPS level in the environment is high, each episode of common cold infection (dsRNA) cause no effect on epithelial cells (ECs), but stimulate dendritic cells (DCs) to skew towards Th1 prone except RSV infection, which is Th2 prone. (**b**) When the LPS level in the environment is low, each episode of common cold infection (dsRNA) then stimulate the epithelial cells to produce proallergic cytokines, such as thymic stromal lymphopoietin (TSLP) and IL33, which would facilitate T cells to produce more IL4, IL5 and IL13, and these Th2 cytokines would work synergistically with dsRNA to stimulate ECs to produce more IL33 and TSLP, forming a positive feedback cycle, and thus shaping the allergic predisposition seen clinically.

2.7. LPS Has the Potential to Activate Immature Dendritic Cells (DC) into Mature DCs with Th1- or Th2-Promoting Capacity

LPS stimulates DCs to produce IL12 transiently and thus induces strong Th1 polarization (the so-called "active period"), but shifts to induce Th2 polarization afterwards in the "exhausted" period [45]. During the immune response, there is a dynamic regulation of the Th1/Th2 cells balance. The polarity can also be influenced by the dose of superantigen toxic shock syndrome toxin-2 with a high dosage (10 ng/mL) favoring Th1 skewing, and a low dosage (0.1 ng/mL) favoring Th2 skewing. Exogenous IL12 or IL4 is the third factor that would move the total profile towards Th1 or Th2 polarity respectively [45]. In this model, the predominance of IL12 or IL4 determines the central axis of Th1/Th2 polarity, with minor modifications by the high/low dose of antigens and active/exhausted factors. Therefore, active DCs with high dose antigen in the presence of IL12 would yield the maximal Th1 response, and exhausted DCs with low dose antigen in the presence of IL4 would yield the maximal Th2 response. In another mouse model in which ovalbumin (OVA) was used as allergen at fixed dose, the concomitant use of 100 µg (high dose) or 0.1 µg LPS in sensitizing period led to Th1 and Th2 responses later respectively. Again the dose of LPS determines the final Th1/Th2 skewing pattern [43]. When the allergen was changed to HDM, in a mouse model, increasing doses of LPS (0.001–10 µg) dose-dependently inhibited HDM-induced eosinophil recruitment and the production of Th2 cytokines in the lungs, revealing a shift toward Th1 inflammation with predominant neutrophilia [95]. However, the local switching of the airway inflammation from eosinophilia to neutrophilia did not quench the airway inflammation; instead, airway hyperreactivity was increased [54] in accordance with the clinical observation [96] (Table 1).

3. The Protective Role and Mechanism of LPS: Pre-Exposure to LPS Protects the Respiratory Epithelial Cells and Downregulates the Effect of dsRNA or Allergen in Producing Proallergic Cytokines, Indicating a Delicate Cross-Regulation Mechanism Exists between dsRNA (TLR3 Pathway) or Allergen and LPS (TLR4 Pathway), at Least at Epithelial Level

3.1. Pretreatment with Lps Attenuates Induction of Proallergic Cytokines, TSLP and IL33 in Respiratory Epithelial Cells Stimulated with polyI:C and Human Parechovirus.

In H292 cells, polyI:C activated the TLR3 signaling pathway with activation of the transcription factors IRF3 and NF-κB p65/50 by I kappa B kinase (IKK) and IKK-related kinases, such as TBK-1, IKKε, IKKα, and IKKβ. IRF3 phosphorylation was detected at 3 h after stimulation [26]. Phospho-TBK1, -IRF3, and -NF-κB p65 (Ser456 and Ser536) was markedly increased at 12 h. The protein level of NF-κB p65 and IκBα degradation was increased with downstream production of allergic cytokines, such as TSLP and IL33 increased as well [9]. The polyI:C-induced IRF3 phosphorylation was inhibited by pretreatment of high dose (30 μg/mL) LPS, but by contrast, lower dose of LPS (0.3 μg/mL) enhanced polyI:C-mediated IRF3 phosphorylation as compared with polyI:C stimulation alone. High dose LPS pretreatment significantly decreased IRF3 phosphorylation induced by polyI:C stimulation. The total NF-κB p65 level was not significantly changed by LPS pretreatment, but 3 and 30 μg/mL LPS treatment inhibited the polyI:C-induced IκBα degradation, which suggests that the polyI:C-mediated activation of the NF-κB pathway was downregulated with high dose LPS [9].

3.2. Pretreatment with LPS Protects against Allergy through A20 Induction in Lung Epithelial Cells

After pretreatment with LPS in the mouse model, the lung levels of granulocyte macrophage colony-stimulating factor (GM-CSF, the maturation factor of recruited lung DCs) and CCL20 protein (the chemokine of attracting DCs [62]) induced by house dust mites were reduced through A20 induction. A20 is a negative regulator induced by allergen stimulation via NF-κB activation to avoid deleterious effect due to overstimulation and maintains homeostasis [97]. IL33 mRNA was noted to be decreased with Th2 cytokine levels also downregulated, including IL5 and IL13 [10].

3.3. Pretreatment with E. coli in Mice Models Protects against Allergy via Two Pathways

Pretreatment with *E. coli* in mice models via intranasal inoculation leads to suppression of allergic airway inflammation by recruited γδ T cells (a subset of T cells with potent cytotoxicity and interferon-γ production [98]), and dampening of DC function in the lung, thus decreasing the effectiveness of presenting antigen to effector T cells [99]. These findings may be the associated events of Sections 3.1 and 3.2. Though Th1 and Treg responses do not play a role in the context, γδ T cells are noted for their potent cytotoxicity and interferon-γ production.

3.4. Pretreatment with Salmonella enterica Serovar Typhimurium Protects against Allergic Airway Inflammation in Mice

Intragastric pretreatment with *Salmonella enterica serovar typhimurium* in murine models leads to protection from induced allergic airway inflammation via expansion of a $CD11b^+$ $Gr1^+$ myeloid cell populations, which reduce airway inflammation by influencing Th2 cells. These groups of myeloid cells consist of macrophages, immature granulocytes, early myeloid progenitors, and DCs, and exhibit their inhibitory effect by altering GATA-3 expression and IL4 production by Th2 cells [53].

3.5. LPS Suppresses Asthma-Like Responses via Nitric Oxide Synthase (NOS2) Activity

The regulatory role of NOS2 in airway allergy was revealed. Mice were immunized with OVA on Days 0 and 7, then on Day 14, challenged intranasally with OVA/saline to induce an allergic airway response, and on Day 21, rechallenged with intranasal OVA concomitantly with either intranasal or intravenous delivery of LPS. LPS via both routes completely suppressed airway eosinophilia, but in the intranasal group, the total cell number in the bronchoalveolar fluid was not reduced. The suppression effect was lost in NOS2$^{-/-}$ mice, indicating the LPS suppressed the allergic inflammation via NOS2 activity. However, in this experiment, LPS was given after the establishment of allergic state [54]. Thus, whether pretreatment with LPS would also work via this way must be confirmed by further study.

4. The Pro-Inflammatory Role of LPS: Why Does LPS Induce Inflammation, Instead of Protecting against Inflammation on Many Occasions?

Although LPS acts like a protector of respiratory epithelial cells against allergic inflammation, on many occasions, it creates difficulties by inducing a large amount of inflammatory responses, even to the extent of shock or death of the host. There are probably several reasons for this. Table 3 summarized some concepts shaped by already published studies.

4.1. First, the Timing of Delivering LPS

In rat models, LPS protects the animal only when it is delivered before or shortly after allergen sensitization. After that, LPS exacerbates the inflammatory responses, often with neutrophils predominant [41]. In asthmatic children, a significant correlation was found between levels of LPS and airway neutrophils in bronchoalveolar lavage [100].

4.2. Second, the Dose of LPS Delivered

In a mouse model, 100 ng LPS every other day for two weeks before house dust mite (HDM) sensitization or 1 µg two weeks before HDM sensitization suppress all of the key asthma feature [10]. In contrast, when 8 mL of 100 µg/mL LPS suspension was nebulized and delivered in the mouse model, inhalation of LPS results in acute neutrophilic inflammation of the distal air spaces of the lungs [48]. Another mice experiment indicated that allergen sensitization with low dose LPS (0.1 µg) and OVA induced type 2 responses with airway hyperresponsiveness, eosinophilic inflammation, and allergen-specific IgE up-regulation, but, again, sensitization with high dose LPS (10 µg) and OVA induced asthma phenotypes with noneosinophilic airway inflammation [43]. In an epithelial cell line model, 30 µg/mL LPS in culture medium attenuates induction of proallergic cytokines, TSLP and IL33 in response to polyI:C or human parechovirus, but 0.3 µg/mL LPS enhanced the induction effect with increased phosphorylation of IRF3 and decreased inhibitors of NF-κB, the IκBα on the contrary [9].

4.3. Third, the Monocytes/Macrophage or Dendritic Cells Which Are Also Activated by LPS

As little as 10 ng/mL LPS would induce the release of inflammatory and chemotactic chemokines from these cells [101], in contrast to the 10,000 ng/mL LPS required for activation of respiratory epithelial cells [47]. If the monocytes/ macrophage or dendritic cells were activated, they would release several cytokines, such as tumor necrosis factor-α and IL1, and enhance epithelial barrier dysfunction [70].

4.4. Fourth, the Synergistic Effect between LPS and Environmental Cofactors

The synergistic effect between LPS and environmental cofactors, such as concomitant ozone exposure, causes worsened inflammation. In rat model, 100 µg LPS intranasally instilled cause slight neutrophilia in nasal epithelia in comparison to saline (relative value 19/14), but pretreatment with ozone before instillation of same amount of LPS cause significant neutrophilia (relative value 33/1) [102].

4.5. Fifth, the Resource of LPS Delivered

In the Copenhagen Birth Cohort Study, the presence of potentially pathogenic species, including *M. catarrhalis*, *H. influenzae*, or *S. pneumoniae*, in the oropharynx of one-month-old infants significantly correlates with increased risk of developing asthma in later childhood, but *S. aureus* does not [61]. Thus, it is reasonably deduced that different LPS may have a different effect, either protective or pro-inflammatory.

4.6. Sixth, the Presence of Serum or Whole Blood on Lung Alveolar Cells and Bronchial Epithelial Cells

The presence of serum or whole blood makes the human umbilical vein endothelial cells become very sensitive to even 0.1 ng/mL LPS [65]. When endothelial cells were incubated with whole blood or serum, they were fully activated with picomolar doses of LPS. In contrast, one thousand fold of LPS were required for the same level of activation when plasma only was used. When anti-CD14 mAbs were used, the endothelial responses to LPS in the presence of blood could be almost completely blocked. However, this enhancement effect by blood was not observed in lung epithelial cells A549 at this minimal dose [65] as reported by Pugin. However, when the LPS level was increased to 10 ng/mL, the serum would cause a much greater increase in cytokine expression in both alveolar epithelial cells (A549) and bronchial epithelial cells (BEAS-2B), perhaps through different mechanisms [47], with CD14-dependence in A549 and CD14-independence in BEAS-2B. In contrast, in the absence of serum, LPS has to reach 10,000 ng/mL (here we use ng instead of μg to emphasize the contrast) to activate respiratory epithelial cells. Currently, the significance of the presence of serum or blood in respiratory epithelial cells has not yet been fully elucidated, but the contents in inflammatory exudate may play roles in enhancing the activation of respiratory epithelial cells.

4.7. Seventh, the Type of Epithelial Cells Tested

As pointed out in Section 4.6, alveolar epithelial cells and bronchial epithelial cells react to LPS via different mechanisms. H292 cells, which are bronchial epithelial cells, grow well even in 30,000 ng/mL LPS up to eight days, demonstrating the relative resistance of bronchial epithelial cells [9]. Here, we preserve the nanogram (ng) scale to emphasize the difference of LPS dosage used. By contrast, if endothelial cells are tested, they are thousands or even millions times sensitive to LPS activation, with pg/mL level of LPS capable of activating endothelial cells in the presence of blood, whose concentration is only 1/1000 of the ng/mL level [66].

5. Proposed Mechanism Supporting Hygiene Hypothesis

5.1. First, Why Early Exposure to Environmental LPS, Such as Farm Dust, Would Protect against the Development of Allergic Disorder in Later Life? Two Mechanisms May Possibly Explain the Observed Phenomenon

5.1.1. Pre-Exposure to LPS Attenuates the Signaling Pathway Necessary for Allergic Cytokines Production, but Spares the dsRNA/DCs Route

After a baby is exposed to more environmental LPS, which is relatively nonpathogenic, then when it later encounters a common viral infection, the dsRNA cannot activate the epithelial cells to produce allergic cytokines, such as TSLP, IL33 and IL25 [9]. Instead, it activates DCs to produce more Th1-skewing cytokines [73], and thus more Th1 cells, and more IL12. In the context of more IL12, more LPS would stimulate DCs to produce more Th1 cytokines in the presence of antigens [43,45]. Thus, each time a baby gets a common cold, it becomes more prone to Th1. Even when an allergen is present, a high dosage of LPS still favors the Th1 response to a later challenge [43] (Table 4) (Figure 1a).

5.1.2. Pre-Exposure to LPS Suppresses Responsiveness of Airway Epithelial Cells via Increased Synthesis of A20

Pre-exposure of airway epithelial cells to LPS suppresses their allergic responsiveness to house dust mite (HDM) via increasing synthesis of A20, encoded by the Tnfaip3 gene. Ex vivo cultures of human bronchial epithelial cells revealed a similar result [10]. Other mechanisms may also be involved, as discussed in Sections 3.3–3.5.

Table 4. Profile of four major players of the hygiene hypothesis model in the presence of high dose LPS before sensitization.

Cell Type	High Dose LPS	polyI:C or Virus
Epithelial cells	Minimal or no effect [31,46].	Neutral because allergic inflammation due to TLRs pathway activation was blocked by pre-exposure to high dose LPS [9].
Dendritic cells	Basically slight Th1 skewing due to high dose LPS with no IL12/IL4 skewing in the context [43,45] and less Th2-promoting mDC2s [57]. Unstimulated peripheral blood mononuclear cells produced more IL10, IL12 and IFNγ,indicating increased spontaneous production of Th1 and regulatory cytokines [103].	Th1 predominant except RSV infection, which displays Th2 pattern [73,75].
Net result: Th1 predominant [#]		

[#] Here what we call high dose is relative to low dose, but not to the extent of toxic dose as in Table 3. High dose LPS has minimal or no effect on epithelial cells, but will skew the DCs slightly toward Th1 response when no IL12/IL4 skewing in the context, and produce less Th2-promoting mDC2s.On the other hand, the Th2-prone allergic inflammation of polyI:C or virus was blocked by pre-exposure to high dose LPS, and polyI:C or virus will stimulate DCs to produce Th1 predominant cytokines with the exception of RSV infection, which displays Th2 pattern. The net result of polyI:C or virus stimulation in the presence of high dose LPS on DCs and epithelial cells would thus be Th1 predominant.

5.2. Second, Why Is More Common Cold in Early Life Associated with Less Allergy in Later Life?

A German birth cohort multicenter allergy study (MAS) group confirmed that children with ≤1 episode of runny nose before the age of one year were more likely to be diagnosed as asthmatic at seven years old or to have wheeze at seven years old, compared with those with ≥2 episodes, and were more likely to be atopic before the age of five years. Similarly, having ≥1 viral infection of the herpes type in the first three years of life was inversely associated with asthma at age seven [7]. The conclusion excludes the lower respiratory infection, which instead showed a positive association with later wheeze in that study. No significant associations were found between bacterial, fungal, or gastrointestinal infections and later asthma [7]. More instances of the common cold are closely related to poorer hygiene due to more crowded living or poorer economic conditions, so on average there would be more environmental LPS. This was described in studies by Giovannangelo et al., who found that having more than four persons living in the home were consistently associated with up to 1.7-fold higher endotoxin concentrations in the mattress and floor dust [104]. Similarly, in Brazil's study, LPS levels in day care centers and preschools were three times higher than in elementary schools [105]. Again for each episode of URI, more LPS would block the effect of dsRNA to induce allergic cytokines at epithelial level [9], and the common cold virus would stimulate the DCs towards Th1 skewing, because dsRNA stimulates both the maturation and resistance of DCs, and makes them capable of trigger naïve T cells and drives polarized Th1 responses [75] (Table 4, Figure 1a). The Tucson Children's Respiratory Study proposed a conclusion that indirectly supports the protective effect of 'poor hygiene'. It states that more exposure of young children to older children at home or longer stays at day care center confers more protection against the development of asthma and frequent wheezing in late childhood [106]. Maier etc. confirmed that there is a significant difference of the bacterial community structure in house dust between families with children attending day care and those without [107].

A complex question then arises: is rhinovirus beneficial or detrimental in allergic response? According to the German MAS study above, the infection of common cold virus of course is protective [7], and since around 50 to 83% of episodes of the common cold are caused by rhinoviruses [85,108], so rhinovirus should be protective. However, there is a pitfall in the study:

because in that study more lower respiratory tract infections (LRTIs) before age three is positively associated with more wheeze at age seven [7]. Unfortunately except for the six months group, in which RSV accounted for most of the LRTIs, rhinovirus accounts for the largest portion of pathogens causing LRTIs alone or in combination with other viruses [82]. The other viral pathogens include RSV, parainfluenza viruses, adenoviruses and enteroviruses etc. Thus, in those children with no wheezing predisposition, according to Illis's study, rhinovirus should be protective; but for those with wheezing tendency, rhinovirus seems to play a proallergic role. Thus, again, as with the dual role of LPS [41], rhinovirus seems to have a bidirectional capacity of enhancing or reducing the allergic tendency of children depending on their already preexisting allergic predisposition. Another indication of the effect of predisposition is that in children less than four years old, rates of asymptomatic infection range from 12 to 32% [94,109–111]. Readers are encouraged to refer back to Sections 2.4 and 2.6.3 for the protective mechanism of rhinovirus.

5.3. Summary of the Four Major Players in Hygiene Hypothesis

Usually LPS does not directly act on epithelial cells, instead, it protects the epithelial cells against the action of dsRNA. LPS stimulates DCs to become more Th1-prone with a large dose, and more Th2-prone with a small dose, more Th1-prone in an active period, and more Th2-prone in an exhausted period, and the presence of IL12 or IL4 is the third factor, which would facilitate the Th1/Th2 skewing in the presence of antigen [45]. dsRNA differentially acts on epithelial cells and DCs, with Th2-prone on epithelial cells, and Th1-prone on DCs. Thus, before sensitization, when the LPS level in the environment is high, each episode of common cold infection (dsRNA) has no effect on ECs, and stimulate DCs to be more Th1-prone, except RSV infection, which is Th2-prone (Table 4 and Figure 1a). However, when the LPS level in the environment is low, each episode of common cold infection (dsRNA) then stimulates the epithelial cells to produce proallergic cytokines, like TSLP and IL33, which would facilitate T cells to produce more IL4, IL5 and IL13, and these Th2 cytokines would work synergistically with dsRNA to stimulate ECs to produce more IL33 and TSLP, forming a positive feedback cycle, and thus shaping the allergic predisposition seen clinically (Table 5 and Figure 1b). These four players basically regulate the key networks of immune responses related to hygiene hypothesis.

However, after sensitization has occurred, the protective effects of LPS and dsRNA may be lost, as discussed below.

Table 5. Profile of four major players of the hygiene hypothesis model in the presence of low dose LPS before sensitization.

Cell Type	Low Dose LPS	polyI:C or Virus
Epithelial cells	Minimal or no effect.	Th2 predominant due to TLR3 pathway activation with production of TSLP, IL33, IL25 etc. [31–35], even slightly enhanced by pre-exposure to low dose LPS [9].
Dendritic cells	Basically slight Th2 skewing due to low dose LPS when no IL12/IL4 skewing in the context [43,45,57].	Th1 or Th2 skewing, depending on the relative stimulatory force between Th2-prone allergic cytokines, such as TSLP and IL33 [112,113], and Th1-prone polyI:C or virus [73,75], except RSV, which displays Th2 pattern.

Net result:Th2 predominant [#]

[#] Here what we call high dose is relative to low dose, but not to the extent of toxic dose as in Table 3. Low dose LPS has minimal or no effect on epithelial cells, but will skew the DCs slightly toward Th2 response when no IL12/IL4 skewing in the context. On the other hand, the Th2-prone allergic inflammation of polyI:C or virus was not blocked, or even slightly enhanced by pre-exposure to low dose LPS. polyI:C or virus will stimulate DCs to produce Th1 predominant cytokines with the exception of RSV infection, which displays Th2 pattern. However, Th2-prone signals coming from epithelial cells, such as TSLP and IL33 will skew the DCs toward Th2. The net result of polyI:C or virus stimulation in the presence of low dose LPS on DCs and epithelial cells would thus be Th2 predominant.

6. Why Does the Hygiene Hypothesis Work Only before One Critical Time Point in Early life? After Allergy Is Established, Why Do the dsRNA and LPS No More Play Protective Roles?

6.1. Why, in Established Atopic Patients, Does Exposure to More Environmental LPS No Longer Protect Them against Allergy?

Tulic et al. proposed a mechanism, which stated that before or shortly after sensitization, LPS exposure causes the production of IFNγ and/or IL12 and drives the B cells toward IgG antibody production but inhibits class switching to IgE. Subsequently, plasma cells committed to producing IgE are already present, and these committed IgE-producing cells are resistant to regulation by IL12 and/or IFNγ. Therefore, LPS must be present before the isotype switch to IgE has occurred in order to exert its inhibitory effect [41]. In his study, LPS exposure on Days 6, 8 and 10 with challenging OVA on Day 11 resulted in more cellular influx, with increases in both neutrophils and eosinophils. Lowe performed his study on guinea pigs and concluded LPS exposure 24 h before allergen challenge diminished airway hyperreactivity (AHR) to histamine, whereas co-administered LPS prolonged the AHR [44]. The above rule seems to be applicable to DCs and in whole rat or mouse models, where LPS enhances both Th1/Th2 responses after sensitization, with Th2 response seemingly resistant to LPS treatment.

However, in Rodriguez's study, LPS via either intranasal or intravenous route could still completely suppress airway eosinophilia [54]. The disparity may be due to the difference of experimental animals (rat vs. mice), the dose of LPS (50 µg/mL vs. 20 µg in one animal) and the pathway (inhaled by aerosol in a chamber vs. intranasally or intravenously). Similarly, in Schuijs' study, when bronchial epithelial cells were removed from asthmatic patients via endobronchial biopsy and grown to confluence, and then cultured in air-liquid interface model, 100 ng LPS pretreatment every other day for one week still was able to reduce the production of IL1α and GM-CSF [10]. Thus, LPS seems protective still when considering respiratory epithelial cells only.

To reconcile these seemingly conflicting data, more studies are needed to elucidate the delicate and complex underlying mechanisms.

6.2. Why in Established Atopic Patients, Does Exposure to Viral Infections No Longer Protect Them against Allergies, but, Instead, Worsen the Allergic Disorders

TSLP production is increased in allergic individuals, and further synergistically enhanced by a combination of IL4 and dsRNA [31]. This implies that in asthmatic airway, respiratory viral infection and the recruitment of Th2 cytokine producing cells may increase the production of TSLP and amplify Th2 inflammation. When mice challenged with the already sensitized allergen, rhinovirus infection exacerbated neutrophilic, eosinophilic and lymphocytic airway inflammation, airway hyperresponsiveness, mucus secretion and production of both Th1 and Th2 cytokines [93]. Thus, the mechanisms of the synergistic interaction between virus infection and allergen exposure to increase the risk of asthma exacerbations are elucidated. Allergen exposure causes increased allergic cytokines, which would be further increased by the concomitant activation of TLR3 of epithelial cells by viral dsRNA. However, since most viral infections stimulate DCs to produce Th1 cytokines, so both Th1 and Th2 cytokines would both be increased.

In addition, in healthy humans, plasmacytoid dendritic cells (pDCs) and the IFN$\alpha\beta$ they secrete selectively negatively regulate Th2 cytokine synthesis following RV exposure in vitro. In asthmatic patients, this important regulatory mechanism may be lost and contribute to asthma exacerbations during RV infections [114,115]. The pDCs may play key roles in healthy humans to negate the Th2-prone effect of rhinovirus. For allergic patients, the environmental LPS concentration is on average less than their normal counterpart [116], and thus confers less protection at epithelial level. Thus, when these patients contract the common cold, dsRNA would activate the epithelial cells and induce the production of allergic cytokines. In addition, increased Th2 cytokines in this context would augment the dsRNA effect in stimulating epithelial cells to produce more proallergic cytokines [31].

7. What Factors Initiate the Disruption of Immune Balance towards the Allergic Predisposition?

As mentioned above in Section 2.6.3, for some individuals allergic predisposition exists before rhinovirus infection. In a similar manner, some risk factors are already present before the allergy-inciting event, which make the individuals more susceptible to the development of allergic disorders. The proposed mechanisms leading to allergic predisposition include mostly genetic deficiency [10], airway and gut microbial dysbiosis (i.e., deviations from healthy microbial compositions) [79], and possibly environmental hazard factors.

7.1. Genetic Deficiency

Defects in TLR7/interferon regulatory factor 7 (IRF7) signaling predisposes to severe viral bronchiolitis and subsequent asthma [117,118]. Levels of plasmacytoid DCs (pDCs) during infancy were inversely correlated with childhood respiratory tract infections and wheezing up to age five years [24]. Reduced circulating pDCs during early life predisposes young children to more frequent and more severe respiratory tract viral infections and more wheezing. Recently several SNPs in the TNFAIP3 interacting protein (TNIP-1) were identified as associated with asthma, further supporting the importance of the A20 protective pathway [10]. Furthermore, common mutations in TLR4 are associated with differences in LPS responsiveness in humans, and these people are hyporesponsive to inhaled LPS [119]. Children sensitized to any allergen early in life and sensitized to inhalant allergens by the age of seven years were found to be at a significantly increased risk of being asthmatic at this age if a positive parental history of asthma or atopy was present, with the effect being strongest for maternal asthma, indicating the existence of an underlying determining genetic factor [120].

7.2. Microbial Dysbiosis

7.2.1. Airway Microbial Dysbiosis

The microbes inhabiting the lower airways show substantial differences between asthmatics and healthy subjects. Though there are some variations, generally speaking, in asthmatics, more *Hemophilus*, *Moraxella*, *Neisseria* and *Streptococcus* spp. were found, but less *Bacteroides* [79,121]. However, once again, whether the difference of microbiome precedes and leads to the development of allergic disorders or just develops after sensitization begins needs more study to reach a firm conclusion. However, just as we have discussed above, the amount of protective LPS in the environment (which represent the amount of LPS inhaled) or the type of LPS (pathogenic or non-pathogenic) in the airways are closely related with later atopy development [61,122]. Here, we have to point out the importance of organ specificity of LPS in determining its role of protection or pro-inflammation. For example, *S. typhimurium* is an intestinal pathogen of mice, but it could protect against airway inflammation when applied before or shortly after sensitization [41].

7.2.2. Gut Microbial Dysbiosis

Human guts are colonized with many trillions of bacteria, viruses and eukaryotes. Their colonization in early life plays an important role in the development of our immune system [123]. During a critical period in early life, disruption of the optimal host-commensal interactions, i.e., dysbiosis of these microbes might cause allergic disorders [124]. For example, babies born by cesarean section harbor more *Staphylococcus*, *Corynebacterium* and *Propionibacterium* than those born vaginally, but less *Lactobacillus*, *Prevotella* and *Sneathia* species [125]. The resulting change of microbiota in the intestine may lead to aberrant long-term colonization and subsequent altering of immune development [126]. Cesarean delivery is associated with a significantly increased rate of allergic disorders [127], supporting this conclusion.

A modified hygiene hypothesis suggests that an altered normal intestinal colonization pattern in infancy, which fails to induce immunological tolerance, could be responsible for the increase in allergies [128]. Two other studies concluded that bacterial diversity in the intestinal flora of infancy was inversely associated with the risk of later allergic sensitization [129,130]. Therapeutic trials with various strains of oral probiotics achieved success to varying degrees by restoring intestinal homeostasis and preventing or alleviating allergy, at least in part by interacting with the intestinal immune cells [131–133], although this conclusion is not agreed with by all [134,135].

Microbial products and metabolites can also affect allergic inflammation. Dietary fermentable fiber content changes the gut and lung microbiota, and the metabolized fibers consequently increase the concentration of circulating short-chain fatty acids (SCFAs), which protect against allergic lung inflammation [136], possibly via promoting peripheral regulatory T-cell generation [137].

7.3. Environmental Hazard Factors

Exposure to air pollutants, including higher levels of O_3 and others, can cause acute exacerbations in those who already have allergic respiratory disorders, but its role in the initiation of new cases of asthma is not yet confirmed [138]. A cohort study of 3863 children confirmed the association between traffic-related air pollution and the development of asthma and allergies during the first eight years of life. According to the study, PM2.5 levels and NO_2 were associated with a significant increase in incidence and prevalence of asthma [139]. In another study, infants at high familial risk for asthma were recruited and the birth year home exposures to NO, NO_2 and PM2.5 were found to be associated with a markedly increased risk of asthma at age of seven with an odds ratio around 3 [140]. Indoor dampness and mold are also determinants for developing asthma. Chemical emissions from damp structures and surface materials may be other causal agents related to the development of asthma [141,142].

8. Conclusions

8.1. LPS Has a High Potential for Prevention Modality; However, Application of LPS as Treatment Modality Should Be Considered Cautiously

LPS, or maybe farm dust, attenuates the induction of proallergic cytokines, including TSLP, IL33, and others in respiratory epithelial cells in response to viral infection, and it does not disturb the Th1-prone effect of viruses on DCs. Thus, it should be a high potential candidate for applications of prevention modality against allergic development. However, although LPS is promising as a prevention modality against allergic disorders, the application of LPS is questionable after the establishment of allergic disorders, and may exacerbate preexisting disorders. The original Th1/Th2 balance model is not applicable since many studies have confirmed the presence of both increased Th1 and Th2 cells in asthmatic airways [143], and specifically, both Th1 and Th17 cells are crucial for the development of neutrophilic inflammation in the airways [144–146].

8.2. Limitations of the Proposed Model

In our simplified model, the role of regulatory T cells (Treg), Th17, innate lymphoid cells group 2 (ILC2) and signaling via other pathways, such as NOD-like receptors were not presented due to insufficient studies to propose an incorporating mechanism. Allergens also were not mentioned, both because allergens act via different pathways and because their relationship with the hygiene hypothesis is not yet clear.

8.3. The Content of "Hygiene Hypothesis" Could Be Modified

The original statement pointing out "more infections in early childhood protect against later atopy" could be modified into "more non-epithelium-damaging viral infections in the presence of organ-specific non-pathogenic bacteria (or certain bacterial products) in early childhood protect against later atopy" to accommodate most of the exceptions.

Acknowledgments: This work was supported in part by grants from Ministry of Science and Technology (MOST 105-2321-B-075B-001), Kaohsiung Chang Gung Hospital (CMRPD8G0081-3 and CMRPD8F0171-3) and Kaohsiung

Veterans General Hospital (VGHKS106-150). The funders had no role in preparation of the manuscript or decision to publish.

Author Contributions: Tsang-Hsiung Lin and Tsung-Hsien Chang prepared the manuscript. Hsing-Hao Su and Hong-Yo Kang provided critical feedback and helped shape the manuscript. All authors reviewed the manuscript.

Conflicts of Interest: The authors declare no competing financial interests.

References

1. Strachan, D.P. Hay fever, hygiene, and household size. *BMJ* **1989**, *299*, 1259–1260. [CrossRef] [PubMed]
2. Strachan, D.P. Family size, infection and atopy: The first decade of the "hygiene hypothesis". *Thorax* **2000**, *55* (Suppl. 1), S2–S10. [CrossRef] [PubMed]
3. Braun-Fahrlander, C.; Riedler, J.; Herz, U.; Eder, W.; Waser, M.; Grize, L.; Maisch, S.; Carr, D.; Gerlach, F.; Bufe, A.; et al. Environmental exposure to endotoxin and its relation to asthma in school-age children. *N. Engl. J. Med.* **2002**, *347*, 869–877. [CrossRef] [PubMed]
4. Braun-Fahrlander, C. Environmental exposure to endotoxin and other microbial products and the decreased risk of childhood atopy: Evaluating developments since April 2002. *Curr. Opin. Allergy Clin. Immunol.* **2003**, *3*, 325–329. [CrossRef] [PubMed]
5. Riedler, J.; Braun-Fahrlander, C.; Eder, W.; Schreuer, M.; Waser, M.; Maisch, S.; Carr, D.; Schierl, R.; Nowak, D.; von Mutius, E. Exposure to farming in early life and development of asthma and allergy: A cross-sectional survey. *Lancet* **2001**, *358*, 1129–1133. [CrossRef]
6. Lynch, S.V.; Wood, R.A.; Boushey, H.; Bacharier, L.B.; Bloomberg, G.R.; Kattan, M.; O'Connor, G.T.; Sandel, M.T.; Calatroni, A.; Matsui, E.; et al. Effects of early-life exposure to allergens and bacteria on recurrent wheeze and atopy in urban children. *J. Allergy Clin. Immunol.* **2014**, *134*, 593–601. [CrossRef] [PubMed]
7. Illi, S.; von Mutius, E.; Lau, S.; Bergmann, R.; Niggemann, B.; Sommerfeld, C.; Wahn, U. Early childhood infectious diseases and the development of asthma up to school age: A birth cohort study. *BMJ* **2001**, *322*, 390–395. [CrossRef] [PubMed]
8. Bach, J.F. Six questions about the hygiene hypothesis. *Cell. Immunol.* **2005**, *233*, 158–161. [CrossRef] [PubMed]
9. Lin, T.-H.; Cheng, C.-C.; Su, H.-H.; Huang, N.-C.; Chen, J.-J.; Kang, H.-Y.; Chang, T.-H. Lipopolysaccharide Attenuates Induction of Proallergic Cytokines, Thymic Stromal Lymphopoietin, and Interleukin 33 in Respiratory Epithelial Cells Stimulated with PolyI:C and Human Parechovirus. *Front. Immunol.* **2016**, *7*, 440. [CrossRef] [PubMed]
10. Schuijs, M.J.; Willart, M.A.; Vergote, K.; Gras, D.; Deswarte, K.; Ege, M.J.; Madeira, F.B.; Beyaert, R.; van Loo, G.; Bracher, F.; et al. Farm dust and endotoxin protect against allergy through A20 induction in lung epithelial cells. *Science* **2015**, *349*, 1106–1110. [CrossRef] [PubMed]
11. Ganesan, S.; Comstock, A.T.; Sajjan, U.S. Barrier function of airway tract epithelium. *Tissue Barriers* **2013**, *1*, e24997. [CrossRef] [PubMed]
12. Schleimer, R.P.; Kato, A.; Kern, R.; Kuperman, D.; Avila, P.C. Epithelium: At the interface of innate and adaptive immune responses. *J. Allergy Clin. Immunol.* **2007**, *120*, 1279–1284. [CrossRef] [PubMed]
13. Whitsett, J.A.; Alenghat, T. Respiratory epithelial cells orchestrate pulmonary innate immunity. *Nat. Immunol.* **2015**, *16*, 27–35. [CrossRef] [PubMed]
14. Lambrecht, B.N.; Hammad, H. The airway epithelium in asthma. *Nat. Med.* **2012**, *18*, 684–692. [CrossRef] [PubMed]
15. Vroling, A.B.; Fokkens, W.J.; van Drunen, C.M. How epithelial cells detect danger: Aiding the immune response. *Allergy* **2008**, *63*, 1110–1123. [CrossRef] [PubMed]
16. Deckers, J.; De Bosscher, K.; Lambrecht, B.N.; Hammad, H. Interplay between barrier epithelial cells and dendritic cells in allergic sensitization through the lung and the skin. *Immunol. Rev.* **2017**, *278*, 131–144. [CrossRef] [PubMed]
17. Lambrecht, B.N.; Hammad, H. The role of dendritic and epithelial cells as master regulators of allergic airway inflammation. *Lancet* **2010**, *376*, 835–843. [CrossRef]
18. Mellman, I. Dendritic cells: Master regulators of the immune response. *Cancer Immunol. Res.* **2013**, *1*, 145–149. [CrossRef] [PubMed]

19. Mildner, A.; Jung, S. Development and Function of Dendritic Cell Subsets. *Immunity* **2014**, *40*, 642–656. [CrossRef] [PubMed]

20. Swiecki, M.; Colonna, M. The multifaceted biology of plasmacytoid dendritic cells. *Nat. Rev. Immunol.* **2015**, *15*, 471–485. [CrossRef] [PubMed]

21. Hammad, H.; Lambrecht, B.N. Dendritic cells and epithelial cells: Linking innate and adaptive immunity in asthma. *Nat. Rev. Immunol.* **2008**, *8*, 193–204. [CrossRef] [PubMed]

22. Guilliams, M.; Ginhoux, F.; Jakubzick, C.; Naik, S.H.; Onai, N.; Schraml, B.U.; Segura, E.; Tussiwand, R.; Yona, S. Dendritic cells, monocytes and macrophages: A unified nomenclature based on ontogeny. *Nat. Rev. Immunol.* **2014**, *14*, 571–578. [CrossRef] [PubMed]

23. Guilliams, M.; van de Laar, L. A Hitchhiker's Guide to Myeloid Cell Subsets: Practical Implementation of a Novel Mononuclear Phagocyte Classification System. *Front. Immunol.* **2015**, *6*, 406. [CrossRef] [PubMed]

24. Upham, J.W.; Zhang, G.; Rate, A.; Yerkovich, S.T.; Kusel, M.; Sly, P.D.; Holt, P.G. Plasmacytoid dendritic cells during infancy are inversely associated with childhood respiratory tract infections and wheezing. *J. Allergy Clin. Immunol.* **2009**, *124*, 707–713. [CrossRef] [PubMed]

25. Worbs, T.; Hammerschmidt, S.I.; Forster, R. Dendritic cell migration in health and disease. *Nat. Rev. Immunol.* **2017**, *17*, 30–48. [CrossRef] [PubMed]

26. Kawai, T.; Akira, S. Toll-like receptors and their crosstalk with other innate receptors in infection and immunity. *Immunity* **2011**, *34*, 637–650. [CrossRef] [PubMed]

27. Kawai, T.; Akira, S. The role of pattern-recognition receptors in innate immunity: Update on Toll-like receptors. *Nat. Immunol.* **2010**, *11*, 373–384. [CrossRef] [PubMed]

28. Nhu, Q.M.; Shirey, K.; Teijaro, J.R.; Farber, D.L.; Netzel-Arnett, S.; Antalis, T.M.; Fasano, A.; Vogel, S.N. Novel signaling interactions between proteinase-activated receptor 2 and Toll-like receptors in vitro and in vivo. *Mucosal Immunol.* **2010**, *3*, 29–39. [CrossRef] [PubMed]

29. Gieseler, F.; Ungefroren, H.; Settmacher, U.; Hollenberg, M.D.; Kaufmann, R. Proteinase-activated receptors (PARs)—Focus on receptor-receptor-interactions and their physiological and pathophysiological impact. *Cell Commun. Signal. CCS* **2013**, *11*, 86. [CrossRef] [PubMed]

30. Bogiatzi, S.I.; Fernandez, I.; Bichet, J.C.; Marloie-Provost, M.A.; Volpe, E.; Sastre, X.; Soumelis, V. Cutting Edge: Proinflammatory and Th2 cytokines synergize to induce thymic stromal lymphopoietin production by human skin keratinocytes. *J. Immunol.* **2007**, *178*, 3373–3377. [CrossRef] [PubMed]

31. Kato, A.; Favoreto, S., Jr.; Avila, P.C.; Schleimer, R.P. TLR3- and Th2 cytokine-dependent production of thymic stromal lymphopoietin in human airway epithelial cells. *J. Immunol.* **2007**, *179*, 1080–1087. [CrossRef] [PubMed]

32. Kinoshita, H.; Takai, T.; Le, T.A.; Kamijo, S.; Wang, X.L.; Ushio, H.; Hara, M.; Kawasaki, J.; Vu, A.T.; Ogawa, T.; et al. Cytokine milieu modulates release of thymic stromal lymphopoietin from human keratinocytes stimulated with double-stranded RNA. *J. Allergy Clin. Immunol.* **2009**, *123*, 179–186. [CrossRef] [PubMed]

33. Saenz, S.A.; Taylor, B.C.; Artis, D. Welcome to the neighborhood: Epithelial cell-derived cytokines license innate and adaptive immune responses at mucosal sites. *Immunol. Rev.* **2008**, *226*, 172–190. [CrossRef] [PubMed]

34. Zhang, L.; Lu, R.; Zhao, G.; Pflugfelder, S.C.; Li, D.Q. TLR-mediated induction of pro-allergic cytokine IL-33 in ocular mucosal epithelium. *Int. J. Biochem. Cell Biol.* **2011**, *43*, 1383–1391. [CrossRef] [PubMed]

35. Bulek, K.; Swaidani, S.; Aronica, M.; Li, X. Epithelium: The interplay between innate and Th2 immunity. *Immunol. Cell Biol.* **2010**, *88*, 257–268. [CrossRef] [PubMed]

36. Sajjan, U.; Wang, Q.; Zhao, Y.; Gruenert, D.C.; Hershenson, M.B. Rhinovirus disrupts the barrier function of polarized airway epithelial cells. *Am. J. Respir. Crit. Care Med.* **2008**, *178*, 1271–1281. [CrossRef] [PubMed]

37. Rezaee, F.; Meednu, N.; Emo, J.A.; Saatian, B.; Chapman, T.J.; Naydenov, N.G.; De Benedetto, A.; Beck, L.A.; Ivanov, A.I.; Georas, S.N. Polyinosinic:polycytidylic acid induces protein kinase D-dependent disassembly of apical junctions and barrier dysfunction in airway epithelial cells. *J. Allergy Clin. Immunol.* **2011**, *128*, 1216–1224. [CrossRef] [PubMed]

38. Sun, R.; Zhang, Y.; Lv, Q.; Liu, B.; Jin, M.; Zhang, W.; He, Q.; Deng, M.; Liu, X.; Li, G.; et al. Toll-like receptor 3 (TLR3) induces apoptosis via death receptors and mitochondria by up-regulating the transactivating p63 isoform alpha (TAP63alpha). *J. Biol. Chem.* **2011**, *286*, 15918–15928. [CrossRef] [PubMed]

39. Raetz, C.R.; Whitfield, C. Lipopolysaccharide endotoxins. *Annu. Rev. Biochem.* **2002**, *71*, 635–700. [CrossRef] [PubMed]

40. Gereda, J.E.; Leung, D.Y.; Liu, A.H. Levels of environmental endotoxin and prevalence of atopic disease. *JAMA* **2000**, *284*, 1652–1653. [CrossRef] [PubMed]

41. Tulic, M.K.; Wale, J.L.; Holt, P.G.; Sly, P.D. Modification of the inflammatory response to allergen challenge after exposure to bacterial lipopolysaccharide. *Am. J. Respir. Cell Mol. Biol.* **2000**, *22*, 604–612. [PubMed]

42. Verhasselt, V.; Buelens, C.; Willems, F.; De Groote, D.; Haeffner-Cavaillon, N.; Goldman, M. Bacterial lipopolysaccharide stimulates the production of cytokines and the expression of costimulatory molecules by human peripheral blood dendritic cells: Evidence for a soluble CD14-dependent pathway. *J. Immunol.* **1997**, *158*, 2919–2925. [PubMed]

43. Eisenbarth, S.C.; Piggott, D.A.; Huleatt, J.W.; Visintin, I.; Herrick, C.A.; Bottomly, K. Lipopolysaccharide-enhanced, toll-like receptor 4-dependent T helper cell type 2 responses to inhaled antigen. *J. Exp. Med.* **2002**, *196*, 1645–1651. [CrossRef] [PubMed]

44. Lowe, A.P.; Thomas, R.S.; Nials, A.T.; Kidd, E.J.; Broadley, K.J.; Ford, W.R. LPS exacerbates functional and inflammatory responses to ovalbumin and decreases sensitivity to inhaled fluticasone propionate in a guinea pig model of asthma. *Br. J. Pharmacol.* **2015**, *172*, 2588–2603. [CrossRef] [PubMed]

45. Langenkamp, A.; Messi, M.; Lanzavecchia, A.; Sallusto, F. Kinetics of dendritic cell activation: Impact on priming of TH1, TH2 and nonpolarized T cells. *Nat. Immunol.* **2000**, *1*, 311–316. [CrossRef] [PubMed]

46. Nakamura, H.; Yoshimura, K.; Jaffe, H.A.; Crystal, R.G. Interleukin-8 gene expression in human bronchial epithelial cells. *J. Biol. Chem.* **1991**, *266*, 19611–19617. [PubMed]

47. Schulz, C.; Farkas, L.; Wolf, K.; Kratzel, K.; Eissner, G.; Pfeifer, M. Differences in LPS-induced activation of bronchial epithelial cells (BEAS-2B) and type II-like pneumocytes (A-549). *Scand. J. Immunol.* **2002**, *56*, 294–302. [CrossRef] [PubMed]

48. Skerrett, S.J.; Liggitt, H.D.; Hajjar, A.M.; Ernst, R.K.; Miller, S.I.; Wilson, C.B. Respiratory epithelial cells regulate lung inflammation in response to inhaled endotoxin. *Am. J. Physiol. Lung Cell. Mol. Physiol.* **2004**, *287*, L143–L152. [CrossRef] [PubMed]

49. Coornaert, B.; Carpentier, I.; Beyaert, R. A20: Central gatekeeper in inflammation and immunity. *J. Biol. Chem.* **2009**, *284*, 8217–8221. [CrossRef] [PubMed]

50. Ha, T.; Hua, F.; Liu, X.; Ma, J.; McMullen, J.R.; Shioi, T.; Izumo, S.; Kelley, J.; Gao, X.; Browder, W.; et al. Lipopolysaccharide-induced myocardial protection against ischaemia/reperfusion injury is mediated through a PI3K/Akt-dependent mechanism. *Cardiovasc. Res.* **2008**, *78*, 546–553. [CrossRef] [PubMed]

51. Ding, Y.; Li, L. Lipopolysaccharide preconditioning induces protection against lipopolysaccharide-induced neurotoxicity in organotypic midbrain slice culture. *Neurosci. Bull.* **2008**, *24*, 209–218. [CrossRef] [PubMed]

52. Carlsten, C.; Ferguson, A.; Dimich-Ward, H.; Chan, H.; DyBuncio, A.; Rousseau, R.; Becker, A.; Chan-Yeung, M. Association between endotoxin and mite allergen exposure with asthma and specific sensitization at age 7 in high-risk children. *Pediatr. Allergy Immunol.* **2011**, *22*, 320–326. [CrossRef] [PubMed]

53. Ganesh, V.; Baru, A.M.; Hesse, C.; Friedrich, C.; Glage, S.; Gohmert, M.; Janke, C.; Sparwasser, T. Salmonella enterica serovar Typhimurium infection-induced CD11b+ Gr1+ cells ameliorate allergic airway inflammation. *Infect. Immun.* **2014**, *82*, 1052–1063. [CrossRef] [PubMed]

54. Rodriguez, D.; Keller, A.C.; Faquim-Mauro, E.L.; de Macedo, M.S.; Cunha, F.Q.; Lefort, J.; Vargaftig, B.B.; Russo, M. Bacterial lipopolysaccharide signaling through Toll-like receptor 4 suppresses asthma-like responses via nitric oxide synthase 2 activity. *J. Immunol.* **2003**, *171*, 1001–1008. [CrossRef] [PubMed]

55. Von Mutius, E.; Vercelli, D. Farm living: Effects on childhood asthma and allergy. *Nat. Rev. Immunol.* **2010**, *10*, 861–868. [CrossRef] [PubMed]

56. Ege, M.J.; Mayer, M.; Normand, A.C.; Genuneit, J.; Cookson, W.O.; Braun-Fahrlander, C.; Heederik, D.; Piarroux, R.; von Mutius, E. Exposure to environmental microorganisms and childhood asthma. *N. Engl. J. Med.* **2011**, *364*, 701–709. [CrossRef] [PubMed]

57. Martikainen, M.V.; Kaario, H.; Karvonen, A.; Schroder, P.C.; Renz, H.; Kaulek, V.; Dalphin, J.C.; von Mutius, E.; Schaub, B.; Pekkanen, J.; et al. Farm exposures are associated with lower percentage of circulating myeloid dendritic cell subtype 2 at age 6. *Allergy* **2015**, *70*, 1278–1287. [CrossRef] [PubMed]

58. Schroder, P.C.; Illi, S.; Casaca, V.I.; Lluis, A.; Bock, A.; Roduit, C.; Depner, M.; Frei, R.; Genuneit, J.; Pfefferle, P.I.; et al. A switch in regulatory T cells through farm exposure during immune maturation in childhood. *Allergy* **2017**, *72*, 604–615. [CrossRef] [PubMed]

59. Lluis, A.; Depner, M.; Gaugler, B.; Saas, P.; Casaca, V.I.; Raedler, D.; Michel, S.; Tost, J.; Liu, J.; Genuneit, J.; et al. Increased regulatory T-cell numbers are associated with farm milk exposure and lower atopic sensitization and asthma in childhood. *J. Allergy Clin. Immunol.* **2014**, *133*, 551–559. [CrossRef] [PubMed]

60. Wickens, K.; Lane, J.M.; Fitzharris, P.; Siebers, R.; Riley, G.; Douwes, J.; Smith, T.; Crane, J. Farm residence and exposures and the risk of allergic diseases in New Zealand children. *Allergy* **2002**, *57*, 1171–1179. [CrossRef] [PubMed]

61. Bisgaard, H.; Hermansen, M.N.; Buchvald, F.; Loland, L.; Halkjaer, L.B.; Bonnelykke, K.; Brasholt, M.; Heltberg, A.; Vissing, N.H.; Thorsen, S.V.; et al. Childhood asthma after bacterial colonization of the airway in neonates. *N. Engl. J. Med.* **2007**, *357*, 1487–1495. [CrossRef] [PubMed]

62. Hammad, H.; Chieppa, M.; Perros, F.; Willart, M.A.; Germain, R.N.; Lambrecht, B.N. House dust mite allergen induces asthma via Toll-like receptor 4 triggering of airway structural cells. *Nat. Med.* **2009**, *15*, 410–416. [CrossRef] [PubMed]

63. Harada, A.; Sekido, N.; Akahoshi, T.; Wada, T.; Mukaida, N.; Matsushima, K. Essential involvement of interleukin-8 (IL-8) in acute inflammation. *J. Leukoc. Biol.* **1994**, *56*, 559–564. [PubMed]

64. Fogli, L.K.; Sundrud, M.S.; Goel, S.; Bajwa, S.; Jensen, K.; Derudder, E.; Sun, A.; Coffre, M.; Uyttenhove, C.; Van Snick, J.; et al. T cell-derived IL-17 mediates epithelial changes in the airway and drives pulmonary neutrophilia. *J. Immunol.* **2013**, *191*, 3100–3111. [CrossRef] [PubMed]

65. Pugin, J.; Schurer-Maly, C.C.; Leturcq, D.; Moriarty, A.; Ulevitch, R.J.; Tobias, P.S. Lipopolysaccharide activation of human endothelial and epithelial cells is mediated by lipopolysaccharide-binding protein and soluble CD14. *Proc. Natl. Acad. Sci. USA* **1993**, *90*, 2744–2748. [CrossRef] [PubMed]

66. Pugin, J.; Ulevitch, R.J.; Tobias, P.S. A critical role for monocytes and CD14 in endotoxin-induced endothelial cell activation. *J. Exp. Med.* **1993**, *178*, 2193–2200. [CrossRef] [PubMed]

67. Rittirsch, D.; Flierl, M.A.; Day, D.E.; Nadeau, B.A.; McGuire, S.R.; Hoesel, L.M.; Ipaktchi, K.; Zetoune, F.S.; Sarma, J.V.; Leng, L.; et al. Acute lung injury induced by lipopolysaccharide is independent of complement activation. *J. Immunol.* **2008**, *180*, 7664–7672. [CrossRef] [PubMed]

68. Rojas, M.; Woods, C.R.; Mora, A.L.; Xu, J.; Brigham, K.L. Endotoxin-induced lung injury in mice: Structural, functional, and biochemical responses. *Am. J. Physiol. Lung Cell. Mol. Physiol.* **2005**, *288*, L333–L341. [CrossRef] [PubMed]

69. Yao, H.; Sun, Y.; Song, S.; Qi, Y.; Tao, X.; Xu, L.; Yin, L.; Han, X.; Xu, Y.; Li, H.; et al. Protective Effects of Dioscin against Lipopolysaccharide-Induced Acute Lung Injury through Inhibition of Oxidative Stress and Inflammation. *Front. Pharmacol.* **2017**, *8*, 120. [CrossRef] [PubMed]

70. Eutamene, H.; Theodorou, V.; Schmidlin, F.; Tondereau, V.; Garcia-Villar, R.; Salvador-Cartier, C.; Chovet, M.; Bertrand, C.; Bueno, L. LPS-induced lung inflammation is linked to increased epithelial permeability: Role of MLCK. *Eur. Respir. J.* **2005**, *25*, 789–796. [CrossRef] [PubMed]

71. Bodey, G.P.; Bolivar, R.; Fainstein, V.; Jadeja, L. Infections Caused by Pseudomonas aeruginosa. *Rev. Infect. Dis.* **1983**, *5*, 279–313. [CrossRef] [PubMed]

72. Da Silva, A.M.T.; Kaulbach, H.C.; Chuidian, F.S.; Lambert, D.R.; Suffredini, A.F.; Danner, R.L. Shock and Multiple-Organ Dysfunction after Self-Administration of Salmonella Endotoxin. *N. Engl. J. Med.* **1993**, *328*, 1457–1460. [CrossRef] [PubMed]

73. De Jong, E.C.; Vieira, P.L.; Kalinski, P.; Schuitemaker, J.H.; Tanaka, Y.; Wierenga, E.A.; Yazdanbakhsh, M.; Kapsenberg, M.L. Microbial compounds selectively induce Th1 cell-promoting or Th2 cell-promoting dendritic cells in vitro with diverse th cell-polarizing signals. *J. Immunol.* **2002**, *168*, 1704–1709. [CrossRef] [PubMed]

74. Schijns, V.E.; Haagmans, B.L.; Wierda, C.M.; Kruithof, B.; Heijnen, I.A.; Alber, G.; Horzinek, M.C. Mice lacking IL-12 develop polarized Th1 cells during viral infection. *J. Immunol.* **1998**, *160*, 3958–3964. [PubMed]

75. Cella, M.; Salio, M.; Sakakibara, Y.; Langen, H.; Julkunen, I.; Lanzavecchia, A. Maturation, activation, and protection of dendritic cells induced by double-stranded RNA. *J. Exp. Med.* **1999**, *189*, 821–829. [CrossRef] [PubMed]

76. Sigurs, N.; Gustafsson, P.M.; Bjarnason, R.; Lundberg, F.; Schmidt, S.; Sigurbergsson, F.; Kjellman, B. Severe respiratory syncytial virus bronchiolitis in infancy and asthma and allergy at age 13. *Am. J. Respir. Crit. Care Med.* **2005**, *171*, 137–141. [CrossRef] [PubMed]

77. Sigurs, N.; Aljassim, F.; Kjellman, B.; Robinson, P.D.; Sigurbergsson, F.; Bjarnason, R.; Gustafsson, P.M. Asthma and allergy patterns over 18 years after severe RSV bronchiolitis in the first year of life. *Thorax* **2010**, *65*, 1045–1052. [CrossRef] [PubMed]

78. Legg, J.P.; Hussain, I.R.; Warner, J.A.; Johnston, S.L.; Warner, J.O. Type 1 and type 2 cytokine imbalance in acute respiratory syncytial virus bronchiolitis. *Am. J. Respir. Crit. Care Med.* **2003**, *168*, 633–639. [CrossRef] [PubMed]

79. Lynch, J.P.; Sikder, M.A.A.; Curren, B.F.; Werder, R.B.; Simpson, J.; Cuív, P.Ó.; Dennis, P.G.; Everard, M.L.; Phipps, S. The Influence of the Microbiome on Early-Life Severe Viral Lower Respiratory Infections and Asthma—Food for Thought? *Front. Immunol.* **2017**, *8*, 156. [CrossRef] [PubMed]

80. Schwarze, J. Lung dendritic cells in respiratory syncytial virus bronchiolitis. *Pediatr. Infect. Dis. J.* **2008**, *27*, S89–S91. [CrossRef] [PubMed]

81. Stein, R.T.; Sherrill, D.; Morgan, W.J.; Holberg, C.J.; Halonen, M.; Taussig, L.M.; Wright, A.L.; Martinez, F.D. Respiratory syncytial virus in early life and risk of wheeze and allergy by age 13 years. *Lancet* **1999**, *354*, 541–545. [CrossRef]

82. Kotaniemi-Syrjänen, A.; Vainionpää, R.; Reijonen, T.M.; Waris, M.; Korhonen, K.; Korppi, M. Rhinovirus-induced wheezing in infancy-the first sign of childhood asthma? *J. Allergy Clin. Immunol.* **2003**, *111*, 66–71. [CrossRef] [PubMed]

83. Kusel, M.M.; de Klerk, N.H.; Kebadze, T.; Vohma, V.; Holt, P.G.; Johnston, S.L.; Sly, P.D. Early-life respiratory viral infections, atopic sensitization, and risk of subsequent development of persistent asthma. *J. Allergy Clin. Immunol.* **2007**, *119*, 1105–1110. [CrossRef] [PubMed]

84. Jackson, D.J.; Gangnon, R.E.; Evans, M.D.; Roberg, K.A.; Anderson, E.L.; Pappas, T.E.; Printz, M.C.; Lee, W.M.; Shult, P.A.; Reisdorf, E.; et al. Wheezing rhinovirus illnesses in early life predict asthma development in high-risk children. *Am. J. Respir. Crit. Care Med.* **2008**, *178*, 667–672. [CrossRef] [PubMed]

85. Arruda, E.; Pitkäranta, A.; Witek, T.J.; Doyle, C.A.; Hayden, F.G. Frequency and natural history of rhinovirus infections in adults during autumn. *J. Clin. Microbiol.* **1997**, *35*, 2864–2868. [PubMed]

86. Kennedy, J.L.; Turner, R.B.; Braciale, T.; Heymann, P.W.; Borish, L. Pathogenesis of Rhinovirus Infection. *Curr. Opin. Virol.* **2012**, *2*, 287–293. [CrossRef] [PubMed]

87. Yoo, J.K.; Kim, T.S.; Hufford, M.M.; Braciale, T.J. Viral infection of the lung: Host response and sequelae. *J. Allergy Clin. Immunol.* **2013**, *132*, 1263–1276. [CrossRef] [PubMed]

88. Kirchberger, S.; Majdic, O.; Steinberger, P.; Bluml, S.; Pfistershammer, K.; Zlabinger, G.; Deszcz, L.; Kuechler, E.; Knapp, W.; Stockl, J. Human rhinoviruses inhibit the accessory function of dendritic cells by inducing sialoadhesin and B7-H1 expression. *J. Immunol.* **2005**, *175*, 1145–1152. [CrossRef] [PubMed]

89. Arruda, E.; Boyle, T.R.; Winther, B.; Pevear, D.C.; Gwaltney, J.M., Jr.; Hayden, F.G. Localization of human rhinovirus replication in the upper respiratory tract by in situ hybridization. *J. Infect. Dis.* **1995**, *171*, 1329–1333. [CrossRef] [PubMed]

90. Mosser, A.G.; Brockman-Schneider, R.; Amineva, S.; Burchell, L.; Sedgwick, J.B.; Busse, W.W.; Gern, J.E. Similar frequency of rhinovirus-infectible cells in upper and lower airway epithelium. *J. Infect. Dis.* **2002**, *185*, 734–743. [CrossRef] [PubMed]

91. Blaas, D.; Fuchs, R. Mechanism of human rhinovirus infections. *Mol. Cell. Pediatr.* **2016**, *3*, 21. [CrossRef] [PubMed]

92. Xatzipsalti, M.; Kyrana, S.; Tsolia, M.; Psarras, S.; Bossios, A.; Laza-Stanca, V.; Johnston, S.L.; Papadopoulos, N.G. Rhinovirus viremia in children with respiratory infections. *Am. J. Respir. Crit. Care Med.* **2005**, *172*, 1037–1040. [CrossRef] [PubMed]

93. Bartlett, N.W.; Walton, R.P.; Edwards, M.R.; Aniscenko, J.; Caramori, G.; Zhu, J.; Glanville, N.; Choy, K.J.; Jourdan, P.; Burnet, J.; et al. Mouse models of rhinovirus-induced disease and exacerbation of allergic airway inflammation. *Nat. Med.* **2008**, *14*, 199–204. [CrossRef] [PubMed]

94. Jacobs, S.E.; Soave, R.; Shore, T.B.; Satlin, M.J.; Schuetz, A.N.; Magro, C.; Jenkins, S.G.; Walsh, T.J. Human rhinovirus infections of the lower respiratory tract in hematopoietic stem cell transplant recipients. *Transpl. Infect. Dis.* **2013**, *15*, 474–486. [CrossRef] [PubMed]

95. De Boer, J.D.; Yang, J.; van den Boogaard, F.E.; Hoogendijk, A.J.; de Beer, R.; van der Zee, J.S.; Roelofs, J.J.; van't Veer, C.; de Vos, A.F.; van der Poll, T. Mast cell-deficient kit mice develop house dust mite-induced lung inflammation despite impaired eosinophil recruitment. *J. Innate Immun.* **2014**, *6*, 219–226. [CrossRef] [PubMed]

96. Boehlecke, B.; Hazucha, M.; Alexis, N.E.; Jacobs, R.; Reist, P.; Bromberg, P.A.; Peden, D.B. Low-dose airborne endotoxin exposure enhances bronchial responsiveness to inhaled allergen in atopic asthmatics. *J. Allergy Clin. Immunol.* **2003**, *112*, 1241–1243. [CrossRef] [PubMed]

97. Catrysse, L.; Vereecke, L.; Beyaert, R.; van Loo, G. A20 in inflammation and autoimmunity. *Trends Immunol.* **2014**, *35*, 22–31. [CrossRef] [PubMed]

98. Silva-Santos, B.; Serre, K.; Norell, H. γδ T cells in cancer. *Nat. Rev. Immunol.* **2015**, *15*, 683–691. [CrossRef] [PubMed]

99. Nembrini, C.; Sichelstiel, A.; Kisielow, J.; Kurrer, M.; Kopf, M.; Marsland, B.J. Bacterial-induced protection against allergic inflammation through a multicomponent immunoregulatory mechanism. *Thorax* **2011**, *66*, 755–763. [CrossRef] [PubMed]

100. Hauk, P.J.; Krawiec, M.; Murphy, J.; Boguniewicz, J.; Schiltz, A.; Goleva, E.; Liu, A.H.; Leung, D.Y. Neutrophilic airway inflammation and association with bacterial lipopolysaccharide in children with asthma and wheezing. *Pediatr. Pulmonol.* **2008**, *43*, 916–923. [CrossRef] [PubMed]

101. Kaufmann, A.; Gemsa, D.; Sprenger, H. Differential desensitization of lipopolysaccharide-inducible chemokine gene expression in human monocytes and macrophages. *Eur. J. Immunol.* **2000**, *30*, 1562–1567. [CrossRef]

102. Wagner, J.G.; Van Dyken, S.J.; Hotchkiss, J.A.; Harkema, J.R. Endotoxin enhancement of ozone-induced mucous cell metaplasia is neutrophil-dependent in rat nasal epithelium. *Toxicol. Sci.* **2001**, *60*, 338–347. [CrossRef] [PubMed]

103. Kääriö, H.; Huttunen, K.; Karvonen, A.M.; Schaub, B.; von Mutius, E.; Pekkanen, J.; Hirvonen, M.R.; Roponen, M. Exposure to a farm environment is associated with T helper 1 and regulatory cytokines at age 4.5 years. *Clin. Exp. Allergy* **2016**, *46*, 71–77. [CrossRef] [PubMed]

104. Giovannangelo, M.; Gehring, U.; Nordling, E.; Oldenwening, M.; Terpstra, G.; Bellander, T.; Hoek, G.; Heinrich, J.; Brunekreef, B. Determinants of house dust endotoxin in three European countries—The AIRALLERG study. *Indoor Air* **2007**, *17*, 70–79. [CrossRef] [PubMed]

105. Rullo, V.E.; Rizzo, M.C.; Arruda, L.K.; Sole, D.; Naspitz, C.K. Daycare centers and schools as sources of exposure to mites, cockroach, and endotoxin in the city of Sao Paulo, Brazil. *J. Allergy Clin. Immunol.* **2002**, *110*, 582–588. [CrossRef] [PubMed]

106. Ball, T.M.; Castro-Rodriguez, J.A.; Griffith, K.A.; Holberg, C.J.; Martinez, F.D.; Wright, A.L. Siblings, day-care attendance, and the risk of asthma and wheezing during childhood. *N. Engl. J. Med.* **2000**, *343*, 538–543. [CrossRef] [PubMed]

107. Maier, R.M.; Palmer, M.W.; Andersen, G.L.; Halonen, M.J.; Josephson, K.C.; Maier, R.S.; Martinez, F.D.; Neilson, J.W.; Stern, D.A.; Vercelli, D.; et al. Environmental determinants of and impact on childhood asthma by the bacterial community in household dust. *Appl. Environ. Microbiol.* **2010**, *76*, 2663–2667. [CrossRef] [PubMed]

108. Makela, M.J.; Puhakka, T.; Ruuskanen, O.; Leinonen, M.; Saikku, P.; Kimpimaki, M.; Blomqvist, S.; Hyypia, T.; Arstila, P. Viruses and bacteria in the etiology of the common cold. *J. Clin. Microbiol.* **1998**, *36*, 539–542. [PubMed]

109. Fry, A.M.; Lu, X.; Olsen, S.J.; Chittaganpitch, M.; Sawatwong, P.; Chantra, S.; Baggett, H.C.; Erdman, D. Human rhinovirus infections in rural Thailand: Epidemiological evidence for rhinovirus as both pathogen and bystander. *PLoS ONE* **2011**, *6*, e17780. [CrossRef] [PubMed]

110. Iwane, M.K.; Prill, M.M.; Lu, X.; Miller, E.K.; Edwards, K.M.; Hall, C.B.; Griffin, M.R.; Staat, M.A.; Anderson, L.J.; Williams, J.V.; et al. Human rhinovirus species associated with hospitalizations for acute respiratory illness in young US children. *J. Infect. Dis.* **2011**, *204*, 1702–1710. [CrossRef] [PubMed]

111. Van Benten, I.; Koopman, L.; Niesters, B.; Hop, W.; van Middelkoop, B.; de Waal, L.; van Drunen, K.; Osterhaus, A.; Neijens, H.; Fokkens, W. Predominance of rhinovirus in the nose of symptomatic and asymptomatic infants. *Pediatr. Allergy Immunol.* **2003**, *14*, 363–370. [CrossRef] [PubMed]

112. Liu, Y.J. Thymic stromal lymphopoietin: Master switch for allergic inflammation. *J. Exp. Med.* **2006**, *203*, 269–273. [CrossRef] [PubMed]

113. Liu, Y.J.; Soumelis, V.; Watanabe, N.; Ito, T.; Wang, Y.H.; Malefyt Rde, W.; Omori, M.; Zhou, B.; Ziegler, S.F. TSLP: An epithelial cell cytokine that regulates T cell differentiation by conditioning dendritic cell maturation. *Annu. Rev. Immunol.* **2007**, *25*, 193–219. [CrossRef] [PubMed]

114. Pritchard, A.L.; Carroll, M.L.; Burel, J.G.; White, O.J.; Phipps, S.; Upham, J.W. Innate IFNs and plasmacytoid dendritic cells constrain Th2 cytokine responses to rhinovirus: A regulatory mechanism with relevance to asthma. *J. Immunol.* **2012**, *188*, 5898–5905. [CrossRef] [PubMed]

115. Wark, P.A.B.; Johnston, S.L.; Bucchieri, F.; Powell, R.; Puddicombe, S.; Laza-Stanca, V.; Holgate, S.T.; Davies, D.E. Asthmatic bronchial epithelial cells have a deficient innate immune response to infection with rhinovirus. *J. Exp. Med.* **2005**, *201*, 937–947. [CrossRef] [PubMed]

116. Braun-Fahrlander, C. Does the 'hygiene hypothesis' provide an explanation for the relatively low prevalence of asthma in Bangladesh? *Int. J. Epidemiol.* **2002**, *31*, 488–489. [CrossRef] [PubMed]

117. Kaiko, G.E.; Loh, Z.; Spann, K.; Lynch, J.P.; Lalwani, A.; Zheng, Z.; Davidson, S.; Uematsu, S.; Akira, S.; Hayball, J.; et al. Toll-like receptor 7 gene deficiency and early-life *Pneumovirus* infection interact to predispose toward the development of asthma-like pathology in mice. *J. Allergy Clin. Immunol.* **2013**, *131*, 1331–1339. [CrossRef] [PubMed]

118. Davidson, S.; Kaiko, G.; Loh, Z.; Lalwani, A.; Zhang, V.; Spann, K.; Foo, S.Y.; Hansbro, N.; Uematsu, S.; Akira, S.; et al. Plasmacytoid Dendritic Cells Promote Host Defense against Acute Pneumovirus Infection via the TLR7–MyD88-Dependent Signaling Pathway. *J. Immunol.* **2011**, *186*, 5938–5948. [CrossRef] [PubMed]

119. Arbour, N.C.; Lorenz, E.; Schutte, B.C.; Zabner, J.; Kline, J.N.; Jones, M.; Frees, K.; Watt, J.L.; Schwartz, D.A. TLR4 mutations are associated with endotoxin hyporesponsiveness in humans. *Nat. Genet.* **2000**, *25*, 187–191. [PubMed]

120. Illi, S.; von Mutius, E.; Lau, S.; Nickel, R.; Niggemann, B.; Sommerfeld, C.; Wahn, U. The pattern of atopic sensitization is associated with the development of asthma in childhood. *J. Allergy Clin. Immunol.* **2001**, *108*, 709–714. [CrossRef] [PubMed]

121. Kloepfer, K.M.; Lee, W.M.; Pappas, T.E.; Kang, T.J.; Vrtis, R.F.; Evans, M.D.; Gangnon, R.E.; Bochkov, Y.A.; Jackson, D.J.; Lemanske, R.F., Jr.; et al. Detection of pathogenic bacteria during rhinovirus infection is associated with increased respiratory symptoms and asthma exacerbations. *J. Allergy Clin. Immunol.* **2014**, *133*, 1301–1307. [CrossRef] [PubMed]

122. Gereda, J.E.; Leung, D.Y.; Thatayatikom, A.; Streib, J.E.; Price, M.R.; Klinnert, M.D.; Liu, A.H. Relation between house-dust endotoxin exposure, type 1 T-cell development, and allergen sensitisation in infants at high risk of asthma. *Lancet* **2000**, *355*, 1680–1683. [CrossRef]

123. Gensollen, T.; Iyer, S.S.; Kasper, D.L.; Blumberg, R.S. How colonization by microbiota in early life shapes the immune system. *Science* **2016**, *352*, 539–544. [CrossRef] [PubMed]

124. Clemente, J.C.; Ursell, L.K.; Parfrey, L.W.; Knight, R. The impact of the gut microbiota on human health: An integrative view. *Cell* **2012**, *148*, 1258–1270. [CrossRef] [PubMed]

125. Dominguez-Bello, M.G.; Costello, E.K.; Contreras, M.; Magris, M.; Hidalgo, G.; Fierer, N.; Knight, R. Delivery mode shapes the acquisition and structure of the initial microbiota across multiple body habitats in newborns. *Proc. Natl. Acad. Sci. USA* **2010**, *107*, 11971–11975. [CrossRef] [PubMed]

126. Neu, J.; Rushing, J. Cesarean versus vaginal delivery: Long-term infant outcomes and the hygiene hypothesis. *Clin. Perinatol.* **2011**, *38*, 321–331. [CrossRef] [PubMed]

127. Penders, J.; Gerhold, K.; Thijs, C.; Zimmermann, K.; Wahn, U.; Lau, S.; Hamelmann, E. New insights into the hygiene hypothesis in allergic diseases: Mediation of sibling and birth mode effects by the gut microbiota. *Gut Microbes* **2014**, *5*, 239–244. [CrossRef] [PubMed]

128. Penders, J.; Gerhold, K.; Stobberingh, E.E.; Thijs, C.; Zimmermann, K.; Lau, S.; Hamelmann, E. Establishment of the intestinal microbiota and its role for atopic dermatitis in early childhood. *J. Allergy Clin. Immunol.* **2013**, *132*, 601–607. [CrossRef] [PubMed]

129. Bisgaard, H.; Li, N.; Bonnelykke, K.; Chawes, B.L.; Skov, T.; Paludan-Muller, G.; Stokholm, J.; Smith, B.; Krogfelt, K.A. Reduced diversity of the intestinal microbiota during infancy is associated with increased risk of allergic disease at school age. *J. Allergy Clin. Immunol.* **2011**, *128*, 646–652. [CrossRef] [PubMed]

130. Abrahamsson, T.R.; Jakobsson, H.E.; Andersson, A.F.; Bjorksten, B.; Engstrand, L.; Jenmalm, M.C. Low gut microbiota diversity in early infancy precedes asthma at school age. *Clin. Exp. Allergy* **2014**, *44*, 842–850. [CrossRef] [PubMed]

131. Kukkonen, K.; Savilahti, E.; Haahtela, T.; Juntunen-Backman, K.; Korpela, R.; Poussa, T.; Tuure, T.; Kuitunen, M. Probiotics and prebiotic galacto-oligosaccharides in the prevention of allergic diseases: A randomized, double-blind, placebo-controlled trial. *J. Allergy Clin. Immunol.* **2007**, *119*, 192–198. [CrossRef] [PubMed]

132. Kalliomäki, M.; Salminen, S.; Arvilommi, H.; Kero, P.; Koskinen, P.; Isolauri, E. Probiotics in primary prevention of atopic disease: A randomised placebo-controlled trial. *Lancet* **2001**, *357*, 1076–1079. [CrossRef]

133. Elazab, N.; Mendy, A.; Gasana, J.; Vieira, E.R.; Quizon, A.; Forno, E. Probiotic Administration in Early Life, Atopy, and Asthma: A Meta-analysis of Clinical Trials. *Pediatrics* **2013**, *132*, e666–e676. [CrossRef] [PubMed]

134. Kopp, M.V.; Hennemuth, I.; Heinzmann, A.; Urbanek, R. Randomized, Double-Blind, Placebo-Controlled Trial of Probiotics for Primary Prevention: No Clinical Effects of *Lactobacillus* GG Supplementation. *Pediatrics* **2008**, *121*, e850–e856. [CrossRef] [PubMed]

135. Brouwer, M.L.; Wolt-Plompen, S.A.; Dubois, A.E.; van der Heide, S.; Jansen, D.F.; Hoijer, M.A.; Kauffman, H.F.; Duiverman, E.J. No effects of probiotics on atopic dermatitis in infancy: A randomized placebo-controlled trial. *Clin. Exp. Allergy* **2006**, *36*, 899–906. [CrossRef] [PubMed]

136. Trompette, A.; Gollwitzer, E.S.; Yadava, K.; Sichelstiel, A.K.; Sprenger, N.; Ngom-Bru, C.; Blanchard, C.; Junt, T.; Nicod, L.P.; Harris, N.L.; et al. Gut microbiota metabolism of dietary fiber influences allergic airway disease and hematopoiesis. *Nat. Med.* **2014**, *20*, 159–166. [CrossRef] [PubMed]

137. Arpaia, N.; Campbell, C.; Fan, X.; Dikiy, S.; van der Veeken, J.; deRoos, P.; Liu, H.; Cross, J.R.; Pfeffer, K.; Coffer, P.J.; et al. Metabolites produced by commensal bacteria promote peripheral regulatory T-cell generation. *Nature* **2013**, *504*, 451–455. [CrossRef] [PubMed]

138. Marino, E.; Caruso, M.; Campagna, D.; Polosa, R. Impact of air quality on lung health: Myth or reality? *Ther. Adv. Chronic Dis.* **2015**, *6*, 286–298. [CrossRef] [PubMed]

139. Gehring, U.; Wijga, A.H.; Brauer, M.; Fischer, P.; de Jongste, J.C.; Kerkhof, M.; Oldenwening, M.; Smit, H.A.; Brunekreef, B. Traffic-related air pollution and the development of asthma and allergies during the first 8 years of life. *Am. J. Respir. Crit. Care Med.* **2010**, *181*, 596–603. [CrossRef] [PubMed]

140. Carlsten, C.; Dybuncio, A.; Becker, A.; Chan-Yeung, M.; Brauer, M. Traffic-related air pollution and incident asthma in a high-risk birth cohort. *Occup. Environ. Med.* **2011**, *68*, 291–295. [CrossRef] [PubMed]

141. Quansah, R.; Jaakkola, M.S.; Hugg, T.T.; Heikkinen, S.A.; Jaakkola, J.J. Residential dampness and molds and the risk of developing asthma: A systematic review and meta-analysis. *PLoS ONE* **2012**, *7*, e47526. [CrossRef] [PubMed]

142. Mendell, M.J.; Mirer, A.G.; Cheung, K.; Tong, M.; Douwes, J. Respiratory and allergic health effects of dampness, mold, and dampness-related agents: A review of the epidemiologic evidence. *Environ. Health Perspect.* **2011**, *119*, 748–756. [CrossRef] [PubMed]

143. Liu, A.H.; Murphy, J.R. Hygiene hypothesis: Fact or fiction? *J. Allergy Clin. Immunol.* **2003**, *111*, 471–478. [CrossRef] [PubMed]

144. Salvi, S.S.; Babu, K.S.; Holgate, S.T. Is asthma really due to a polarized T cell response toward a helper T cell type 2 phenotype? *Am. J. Respir. Crit. Care Med.* **2001**, *164*, 1343–1346. [CrossRef] [PubMed]

145. Kim, Y.-M.; Kim, Y.-S.; Jeon, S.G.; Kim, Y.-K. Immunopathogenesis of Allergic Asthma: More than the Th2 Hypothesis. *Allergy Asthma Immunol. Res.* **2013**, *5*, 189–196. [CrossRef] [PubMed]

146. Hellings, P.W.; Kasran, A.; Liu, Z.; Vandekerckhove, P.; Wuyts, A.; Overbergh, L.; Mathieu, C.; Ceuppens, J.L. Interleukin-17 orchestrates the granulocyte influx into airways after allergen inhalation in a mouse model of allergic asthma. *Am. J. Respir. Cell Mol. Biol.* **2003**, *28*, 42–50. [CrossRef] [PubMed]

International Journal of
Molecular Sciences

Article

Neuroimmunological Implications of Subclinical Lipopolysaccharide from *Salmonella* Enteritidis

Anita Mikołajczyk [1],* and Dagmara Złotkowska [2]

1   Department of Public Health, Faculty of Health Sciences, Collegium Medicum,
    University of Warmia and Mazury in Olsztyn, 10-082 Olsztyn, Poland
2   Department of Food Immunology and Microbiology, Institute of Animal Reproduction and Food Research,
    Polish Academy of Sciences in Olsztyn, 10-748 Olsztyn, Poland; d.zlotkowska@pan.olsztyn.pl
*   Correspondence: anita.mikolajczyk@uwm.edu.pl; Tel.: +48-89-524-53-28

Received: 2 September 2018; Accepted: 18 October 2018; Published: 22 October 2018

Abstract: Mounting evidence has indicated that lipopolysaccharide (LPS) is implicated in neuroimmunological responses, but the body's response to subclinical doses of bacterial endotoxin remains poorly understood. The influence of a low single dose of LPS from *Salmonella* Enteritidis, which does not result in any clinical symptoms of intoxication (subclinical lipopolysaccharide), on selected cells and signal molecules of the neuroimmune system was tested. Five juvenile crossbred female pigs were intravenously injected with LPS from *S.* Enteritidis (5 µg/kg body weight (b.w.)), while five pigs from the control group received sodium chloride in the same way. Our data demonstrated that subclinical LPS from *S.* Enteritidis increased levels of dopamine in the brain and neuropeptides such as substance P (SP), galanin (GAL), neuropeptide Y (NPY), and active intestinal peptide (VIP) in the cervical lymph nodes with serum hyperhaptoglobinaemia and reduction of plasma CD4 and CD8 T-lymphocytes seven days after lipopolysaccharide administration. CD4 and CD8 T-lymphocytes from the cervical lymph node and serum interleukin-6 and tumour necrosis factor α showed no significant differences between the control and lipopolysaccharide groups. Subclinical lipopolysaccharide from *S.* Enteritidis can affect cells and signal molecules of the neuroimmune system. The presence of subclinical lipopolysaccharide from S. Enteritidis is associated with unknown prolonged consequences and may require eradication and a deeper search into the asymptomatic carrier state of *Salmonella* spp.

Keywords: LPS from *S.* Enteritidis; neuropeptides; dopamine; CD4 T-lymphocytes; CD8 T-lymphocytes; haptoglobin; cytokines; cervical lymph nodes; prefrontal cortex; substantia nigra

1. Introduction

Our growing knowledge of the role of viruses and bacteria as a cause of mental disorders, cancer, and neurodegenerative and metabolic diseases may help in preventing certain human chronic diseases that pose serious public health problems [1–5]. Endotoxin LPS (lipopolysaccharide) is one of the most important bacterial components contributing to many chronic diseases and sepsis [6]. Lipopolysaccharide animal models for induction of Parkinson's disease have been used by many researchers [7–10]. The mechanisms underlying Parkinson's disease (PD) models indicate that LPS induces microglial activation. LPS inflammation by the activation of glial cells and a series of inflammatory mediators, including proinflammatory cytokines, chemokines, reactive oxygen species, and nitric oxide, plays an important role in neurodegenerative diseases, including PD, Alzheimer's disease (AD), amyotrophic lateral sclerosis, psychiatric disorders, cognitive impairment, and depression [11–14]. Recent evidence suggests that LPS may play an important role in some neurodegeneration and metabolic diseases, not only in rodents but also in humans [15,16]. Pretorius et al. [17] point to the potentially important role of even very low LPS concentrations (0.2 ng/L) in healthy individuals in the etiology of PD, and they hypothesized that lipopolysaccharide-binding protein may have a protective role in the context of PD. LPS is probably associated with AD neuropathology in humans because it is able to transit physiological barriers to access the brain [18].

How LPS given peripherally induces its effects on the brain is still not clear. Banks et al. [19] suggested that high doses of LPS, in contrast to low doses, can disrupt the blood brain barrier (BBB). They suggested that at low peripheral doses of LPS, the amount entering the brain is below that needed to directly affect the brain. Only about 0.025% of an intravenously administered dose of LPS from *Salmonella* enterica reaches the brain parenchyma [20]. Despite the low passage of LPS across the BBB, LPS could potentially affect brain function by a release of substances from the periphery that can either cross the BBB; interact with immune cells; or alter BBB permeability and functions of the BBB, for example, through dysfunction of vascular endothelial cells or induction of the synthesis of mediators by the cells of the blood–cerebrospinal fluid barrier (BCSFB) and the BBB. Both the BCSFB and the BBB seem equipped to convey signals to the brain parenchyma in response to a low dose of LPS [21,22]. On the other hand, it is known that LPS or its constituents can persist in various tissues and organs for years [23,24]. A large part of LPS is rapidly cleared from the circulation, but the remaining ~20% of LPS could be bound to immune cells such as monocytes, tissue macrophages, neutrophils, or platelets, and thus could potentially be involved in signalling [25].

Knowledge of asymptomatic *Salmonella* infection and latent carriers is limited, but it is well known that LPS causing symptoms of disease can induce an acute phase reaction in humans and animals and release pro-inflammatory cytokines such as interleukin-6 (IL-6) and tumour necrosis factor α (TNF-α) [26–28]. Additionally, haptoglobin (Hp), a positive acute-phase protein in response to IL-6, can modulate the inflammatory response induced by LPS. Hp interacts with both resting and activated CD4 and CD8 T-lymphocytes and can play a modulating role in type 1 and 2 T-helper cells, balancing immune responses. [29,30].

LPS can also influence neuropeptides, which are neuroimmune modulators in the communication between the nervous system and the immune system. A single low dose of LPS of *S.* Enteritidis can modulate main enteric neuropeptides [31]. Neuropeptides are widely distributed and involved in many physiological and pathological processes. Neuropeptides of neural and non-neural origin are released in the lymphoid organs such as lymph nodes (LN) and contribute to the modulation of the function of many immunological cells, including lymphocytes and monocytes. Galanin (GAL), neuropeptide Y (NPY), substance P (SP), and vasoactive intestinal peptide (VIP) represent the neuropeptides most involved in neuroimmune modulation. [32–34]. Recent studies have illustrated the importance of immune regulation by neuropeptides through direct effects on and CD4 and CD8 lymphocytes [35].

One of the more important catecholamines in health and disease is dopamine (DA). A growing body of research now suggests that not only neuropeptides, but also catecholamines, serve as a link between the nervous and immune systems during physiological and pathological processes [36]. Among its many roles, DA also plays an important role in the immune system. In the immune system, DA acts upon receptors present in immune cells, especially lymphocytes and can be synthesized and released by immune cells themselves [37]. The presence of DA receptors in immune cells suggests that DA plays a physiological role in the regulation of the immune response, and that its deregulation could be involved in many pathological processes such as autoimmune disorders, schizophrenia, and Parkinson's disease [38,39]. In addition, a decrease of DA in substantia nigra (SN) can lead to PD, and dysregulation of DA in the prefrontal cortex (PFC) is associated with schizophrenia and other psychiatric disorders [40,41].

DA interactions between the nervous and immune systems may be very important in connection with the discovery of meningeal lymphatic vessels [42,43]. Louveau et al. [43] demonstrated that meningeal lymphatic vessels were capable of carrying leukocytes to the deep cervical lymph nodes (dcLN) and, at later time points, to the superficial cervical lymph nodes (scLN). Additionally, cerebrospinal fluid (CSF) drained from the subarachnoid space along the olfactory nerves to nasal lymphatic vessels and subsequently migrated to the scLN [44,45]. Recent studies have also indicated the critical role of the cervical lymph nodes (cLN) on influencing neuroimmunological reactions [46].

In summary, mounting evidence has highlighted that LPS is implicated in neuroimmunological response, but the knowledge of the body's responses to subclinical doses of bacterial endotoxin remains poorly understood. Taking into consideration all of the facts mentioned above, we are the first to test the influence of a low single dose of LPS from *S*. Enteritidis, which does not result in any clinical symptoms of intoxication (subclinical LPS), and which can hypothetically take place during, for example, the asymptomatic carrier state of *Salmonella* spp. on selected cells and signal molecules of the neuroimmune system:

(1)   Peripheral blood levels of CD4 T-lymphocytes, CD8 T-lymphocytes, IL-6, TNF-α, and Hp
(2)   cLN levels of CD4 T-lymphocytes, CD8 T-lymphocytes and neuropeptides such as GAL, NPY, SP and VIP
(3)   SN and PFC levels of DA

In our study, animal experiments remain essential, and appropriate animal models are irreplaceable because to expose humans to even such low doses of LPS with unknown prolonged consequences is not ethical in scientific studies. In addition, the study of human active substance is hampered by tissue inaccessibility for biopsy. It should be noted that using the pig as a biomedical model plays a critical role in understanding the physiological and pathophysiological processes in the human body. Experiments that use the pig as a biomedical model have very good repeatability of results and recapitulation of human conditions, because pigs are phylogenetically closer to humans than rodents [47,48].

2. Results

On each day during the experiment, all animals were assessed by a veterinary surgeon as being clinically healthy. All animals in the LPS and control groups were without any symptoms of disease during this investigation. The low dose of LPS *S*. Enteritidis used in this experiment did not evoke any differences in health status, appearance, temperature, or body weight in animals of the LPS group compared with the control group during the entire period of the experiment. Similar body weights with no significant statistically changes were observed between LPS and control group (Figure 1A). Both in the LPS and in the control group, each measurement of the rectal temperature in the morning (Figure 1B) and in the afternoon (Figure 1C) fluctuated in the normal temperature range of pigs. There were no statistically significant differences in the rectal temperature levels between LPS and the control group.

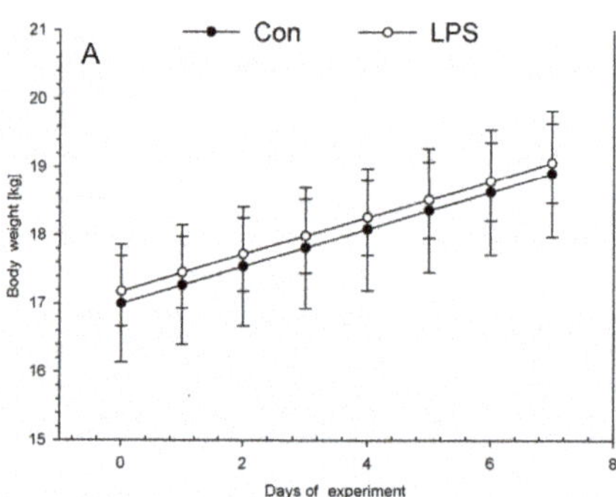

Figure 1. *Cont.*

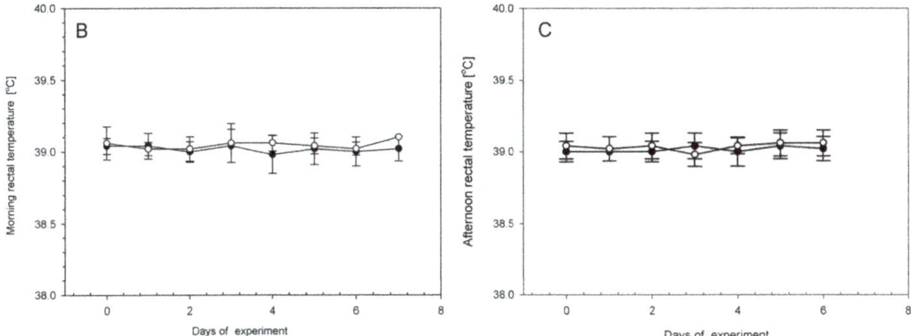

Figure 1. The daily body weights and rectal temperatures in the control group (Con), which received saline, and in the treatment group (LPS), which received lipopolysaccharide (LPS) from *S.* Enteritidis at a dose of 5 µg/kg body weight (b.w.); *n* = 5 pigs/group. (**A**) The body weights were determined once a day in the morning (between 07:00 and 07:30). The first measurement was before premedication and LPS or NaCl administration (day 0), and the final measurement was on the last day of the experiment before antecedent sample collection (day 7). (**B**) The rectal temperatures were measured in the morning (between 07:00 and 07:30). The first measurement was before premedication and LPS or NaCl administration (day 0), and the final measurement was on the last day of the experiment before antecedent sample collection (day 7). (**C**) The rectal temperatures in the afternoon (between 17:00 and 17:30). The first measurement was day 0 (in the day of LPS or NaCl administration), the final measurement was on day 6 of the experiment. The body weight per day and rectal temperatures in the morning and in the afternoon were presented as the mean from group ± SD. The results of rectal temperature and body weight were compared between the control and the LPS group using Student's *t*-test. The result of rectal temperature and body weight was not statistically significant.

The effects of LPS on studied levels of parameters and active substances from the blood, scLN, and brain seven days after administration of LPS *S.* Enteritidis are shown in Figures 2–5. Figure 2 depicts the levels of Hp, IL-6, and TNF-α in the peripheral blood of animals from the control and LPS group. After the administration of LPS *S.* Enteritidis, pigs had significantly enhanced serum levels of Hp. The level of Hp was almost three-fold higher compared with the control group and amounted to 3599.71 ± 135.69 µg/mL (an increase from 1330.97 ± 344.67 to 3599.71 ± 135.69) (Figure 2A). Although LPS increases serum Hp, no statistically significant changes in serum levels of IL-6 and TNF-α were observed (Figure 2B,C).

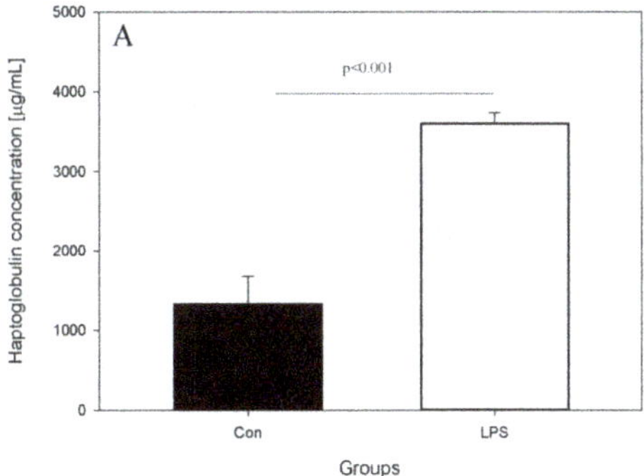

Figure 2. *Cont.*

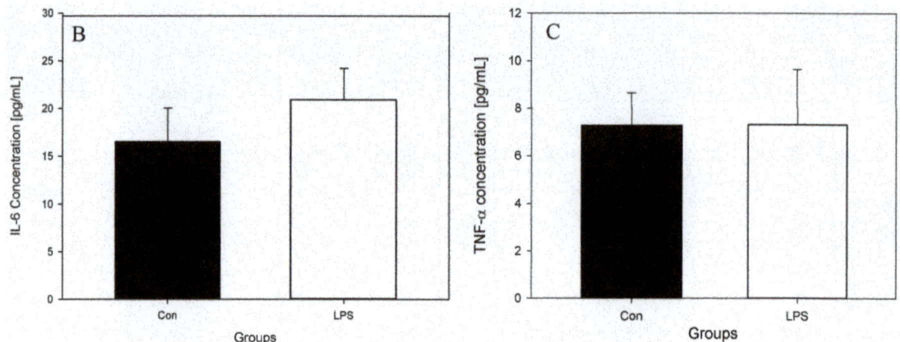

Figure 2. Concentrations of acute phase protein and cytokines in the peripheral blood in the control group (Con), which received saline, and in the treatment group (LPS), which received LPS from *S.* Enteritidis at a dose of 5 μg/kg b.w.; $n = 5$ pigs/group. Bars represent mean $\pm$ SD (standard deviation). The statistical analysis was performed by one-way analysis of variance (ANOVA) and Tukey's tests. (**A**) Serum concentrations of haptoglobin (Hp). The result was statistically different at $p < 0.001$ as compared with the control group (**B**) Serum concentrations of interleukin-6 (IL-6). The result was not statistically significant. (**C**) Serum concentrations of tumour necrosis factor α (TNF-α). The result was not statistically significant.

Figure 3 depicts the percentage of CD4 and CD8 T-lymphocytes in peripheral blood of animals from the control and LPS group. A decrease in the mean percentage of both plasma CD4 T-lymphocytes from 26.13 ± 1.49 (in control group) to 18.54 ± 1.16 (in LPS group) and plasma CD8 T-lymphocytes from 42.01 ± 5.46 (in control group) to 33.63 ± 5.64 (in LPS group) was noted (Figure 3).

DA levels were determined in the brain tissue. Figure 4 depicts the DA levels in SN and PFC in the control group and the LPS group. Compared with the control, the values of DA increased significantly following the injection of LPS from 91.86 ± 3.05 to 138.58 ± 10.07 ng/g in SN and from 15.41 ± 2.46 to 53.59 ± 3.90 ng/g in PFC (Figure 4).

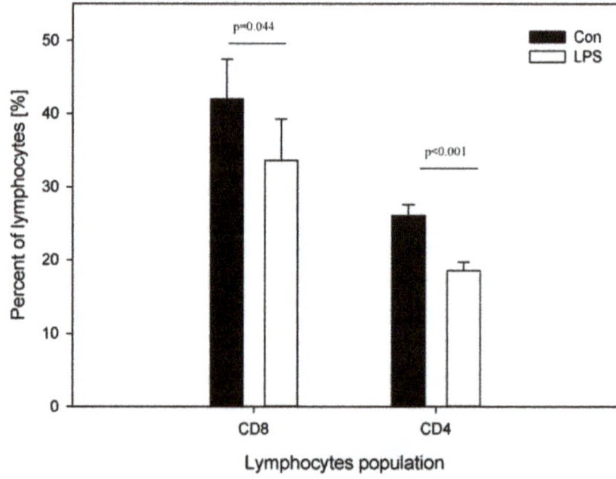

Figure 3. *Cont.*

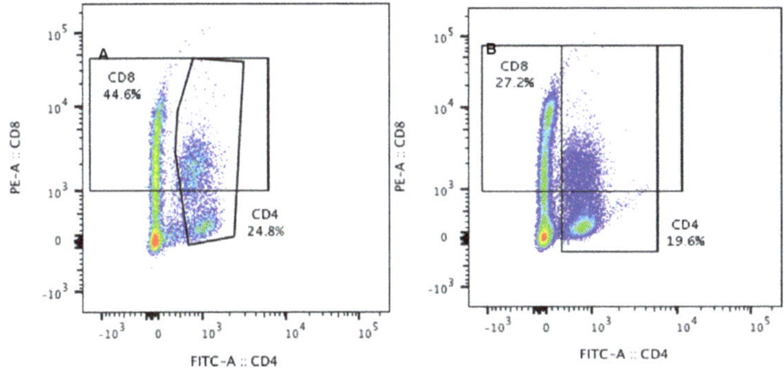

Figure 3. Percentages of CD4 and CD8 T-lymphocytes in the peripheral blood in the control group (Con), which received saline, and in the treatment group (LPS), which received LPS from *S.* Enteritidis at a dose of 5 µg/kg b.w.; *n* = 5 pigs/group. Bars represent mean ± SD (standard deviation). The statistical analysis was performed by one-way ANOVA and Tukey's tests. Statistically different at *p* = 0.044 (for the percentage of plasma CD8 T-lymphocytes) and *p* < 0.001 (for the percentage of plasma CD4 T-lymphocytes) as compared with the control group. Panels A and B present exemplary dot plots with distribution of CD4 and CD8 T-lymphocytes of the control and treatment group. Samples were stained with FITC Anti-Pig CD4a (BD) and PE Anti-Pig CD8a (BD) and analysed on flow cytometer (Beckman Coulter).

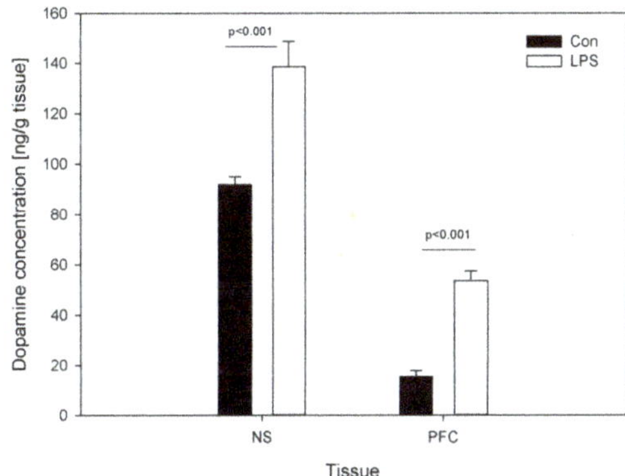

Figure 4. Concentrations of dopamine (DA) in substantia nigra (SN) and in the prefrontal cortex (PFC) in control groups (Con), which received saline, and treatment group (LPS) following the administration of subclinical LPS from *S.* Enteritidis. Bars represent mean ± SD (standard deviation); *n* = 5 pigs/group. The statistical analysis was performed by one-way ANOVA and Tukey's tests. In both SN and PFC, statistically significant differences of DA concentration for *p* < 0.001 as compared with the control group were observed.

The results of the concentration of neuropeptides and the measurements of the percentage of CD4 and CD8 T-lymphocytes in scLN after LPS administration are depicted in Figures 5 and 6. The levels of all studied neuropeptides reached with LPS were significantly greater than in the control group. LPS induced an increased level of GAL (from 1.76 ± 0.21 to 3.17 ± 0.43), NPY (from 2.09 ± 0.14 to 3.04 ± 0.46), SP (from 2.36 ± 0.26 to 2.97 ± 0.33), and VIP (from 0.58 ± 0.08 to 1.03 ± 0.14) (Figure 5). An analysis of the percentages of CD4 T-lymphocytes and CD8 T-lymphocytes in scSN showed no significant differences between the control and the LPS group (Figure 6).

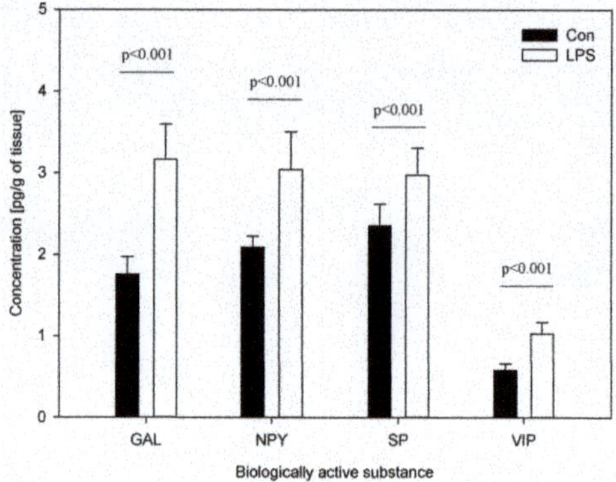

Figure 5. Concentrations of neuropeptides: galanin (GAL), neuropeptide Y (NPY), substance P (SP), and vasoactive intestinal peptide (VIP) in the superficial cervical lymph nodes (scLN). Bars represent mean ± SD (standard deviation); $n = 5$ pigs/group; Con—the control group, which received saline; LPS—the treatment group, which received LPS from *S*. Enteritidis at a dose of 5 µg/kg b.w. (in saline solution). The statistical analysis was performed by one-way ANOVA and Tukey's tests. The results were statistically different at $p < 0.001$ as compared with the control group.

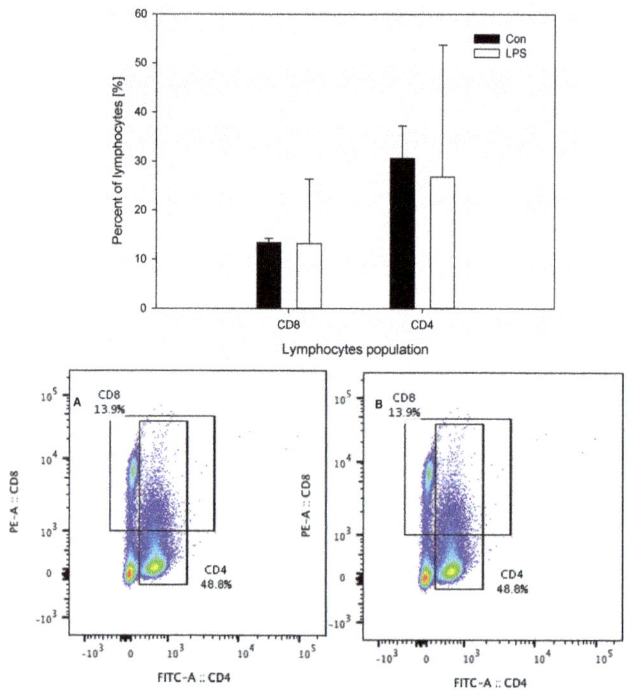

Figure 6. Percentages of CD4 and CD8 T-lymphocytes in the superficial cervical lymph nodes (scLN) in the control group (Con), which received saline, and in the treatment group (LPS), which received LPS from *S*. Enteritidis at a dose of 5 µg/kg b.w.; $n = 5$ pigs/group. Bars represent mean ± SD (standard deviation). The statistical analysis was performed by one-way ANOVA and Tukey's tests. The result was not statistically significant. Panels A and B present exemplary dot plots with distribution of CD4 and CD8 T-lymphocytes of the control and treatment group. Samples were stained with FITC Anti-Pig CD4a (BD) and PE Anti-Pig CD8a (BD) and analysed on flow cytometer (Beckman Coulter).

3. Discussion

Although the role of LPS as a major pro-inflammatory component released by gram-negative bacteria has been widely established, knowledge of asymptomatic LPS infections remains limited. It may be the case that some of the pathogens or their components, which are able to persist in organisms without causing any symptoms of the disease, can interfere with neural or/and immunological cell functions. To understand the mechanisms of the ability of *Salmonella* spp. to survive in various host cells, the activity of their lipopolysaccharides may be very important for both eradication strategies and the prevention of various disease processes, including neurodegenerative diseases. Additionally, the intra-species and inter-species LPS influence on the immune response control of bacterial infections and LPS biological activity varies [49–51]. Even LPS derived from two related species of G-bacteria impacted the regulation of Th-cell responses and T-cell cytokine balances in different ways [52]. Our previous in vitro observations confirmed that LPS shows differences in activity to SP and GAL even within particular serotypes of *Salmonella* spp. [53]. Therefore, considerations of different LPS biological activity depending on various bacterial sources and long-term consequences of LPS different doses must be considered in a research strategy. The ability of LPS to impact cell responses may be attributed to the different bacterial source of LPS, the dosage of LPS, and the time of LPS or the presence of its metabolites in the body. Niehaus [24] reported that 14 years after a laboratory worker developed polyneuropathy, encephalopathy, and parkinsonism after accidental exposure to *Salmonella* Minnesota LPS—the lipopolysaccharides had not been detoxified by the body. Thus, considering the probable neuroimmunological effects of subclinical *S.* Enteritidis LPS with unknown prolonged consequences—it can be seen that asymptomatic *Salmonella* infection and latent carriers are more serious problems than had been assumed [54]. In this study, we investigated the effects of subclinical *S.* Enteritidis LPS in vivo, focusing on the neuroimmune response of Hp, IL-6, TNF-α, and CD4 and CD8 T-lymphocytes in peripheral blood, selected neuropeptides and CD4 and CD8 T-lymphocytes in cLN, and DA in PFC and SN.

Our data demonstrated that seven days after the administration of LPS *S.* Enteritidis, pigs had significantly increased serum levels of Hp, but did not have increased serum levels of TNF-α or IL-6 compared with the control group (Figure 1A–C). Previous studies have used the administration of single or repeated doses of LPS, leading to clinical symptoms of intoxication, which induced short-term increases in serum IL-6 and TNF-α. Qin et al. [55] indicated that a single high dose of systemic LPS in mice led to a decline in the short-term increase of TNF-α to return to the base level by 9 h post treatment, and induced a significant chronic loss of DA neurons beginning at seven months post-treatment. Calvano and Coyle [56] suggested (using the human endotoxin model generated using LPS from *E. coli* O:113 in normal human volunteers) that serum TNF-α and IL-6 peaked at 1.5–2 h after endotoxin administration and again became undetectable after 4 and 6 h, respectively. Similarly, previous studies [57,58] have shown that 12 h after the injection of LPS from *E. coli*, pro-inflammatory cytokines such as IL-6 and TNF-α maintained the same levels as in the control pigs. Despite the lack of differences in the proinflammatory cytokines' levels in our study, the Hp level was significantly increased. It may be the case that serum levels of Hp were altered in response to subclinical LPS challenges because asymptomatic LPS results in an inflammatory process. It seems likely that subclinical inflammatory activity takes place in the case of such a low dose of LPS with an absence of clinical symptoms of a disease. It is worth noting that, because of the relationship of inflammatory processes with psychiatric and neurodegenerative diseases, Hp is the most frequently studied acute phase protein in major depression and has provided the most consistent results. Hyperhaptoglobinaemia in major depression is significantly related to the activation of cell-mediated immunity from activated T-lymphocytes to monocytes [59]. Major depression patients have abnormal levels of IL-6 and TNF-α, and increased Hp [60]. Robertson et al. [61] found alterations in lymphocyte T subsets without differences in the total numbers of T-cells in patients with multiple sclerosis and with major depression. We have observed a decrease in both plasma CD4 and CD8 T-lymphocytes after subclinical *S.* Enteritidis LPS administration (Figure 3). Juffermans et al. [62]

observed a decrease in the number of the fraction of CD4 T-lymphocytes in peripheral blood in healthy volunteers four hours after intravenous administration of LPS from *E. coli*. This is in line with our observations of decreased CD4 T-lymphocytes (Figure 3). However, Juffermans et al. [62] assessed the effects of a single high dose of LPS that caused clinical symptoms of intoxication, rather than the subclinical dose of LPS as was observed in our study. Palmer et al. [63] reported decreased CD4/CD8 T-cell ratios in human subjects without any symptoms of intoxication, but with the presence of subclinical LPS in serum; unfortunately, they did not investigate the source of LPS from the specific bacterial groups that were present in the body.

To our knowledge, there are no prior studies that describe the effect of a single subclinical dose of LPS *S.* Enteritidis on the changes of DA in pig SN and PCF. Our studies have shown that levels of DA in the SN and PFC were significantly greater in the group that received asymptomatic LPS from *S.* Enteritidis than in the control group (Figure 4). Previous studies have described DA levels as a consequence of the inflammation response after *S.* Typhimurium [64] or LPS from *E. coli* [64,65] administration, and that is probably the main reason why they are not in accordance with our results. We used a dose that does not result in any clinical symptoms, as well as LPS from different gram-negative bacteria than that used by the above-mentioned researchers. Guzmán et al. [64] demonstrated that DA levels decreased in hemisphere regions and did not change in cortex regions, the cerebellum, or the medulla oblongata five days after oral administration of *S.* Typhimurium in rats. Another study [65] showed that intrastriatal injection of LPS from *E. coli* in a high dose decreased the content of DA in the rat SN 72 h after an injection. Noworyta-Sokołowska et al. [66] suggested that a high single i.p. LPS from *E. coli* administration does not affect the striatal DA level 180 min after LPS administration, but LPS given repeatedly for five days decreased the DA level. The data indicate that repeated doses and a long period of time are necessary for the progression of inflammation symptoms induced by LPS. Considering the differences between the previous studies and our study, we conclude that subclinical LPS affect DA in a different way than LPS in inducing symptoms of intoxication. It is known that higher levels of DA released in the ventromedial prefrontal cortex play an important role in reward and motivation, and drugs like cocaine, amphetamines, and heroin are highly addictive because they increase the release of DA and then act as DA re-uptake inhibitors. Repeated stressors may also influence the onset of, or relapse to, a number of DA-related disorders [67]. The latest findings [68] show DA dysregulation in different regions of the brain in schizophrenia. Enhancing DA levels in SN during DA treatment for Parkinson's disease can cause psychotic side effects mimicking the symptoms of schizophrenia. On the other hand, reducing dopaminergic transmission in the treatment of schizophrenia and psychosis can cause Parkinson-like symptoms. Furthermore, data from previous studies [40] have provided evidence for brain DA dysregulation in the schizophrenic brain and confirmed a deficit in DA release in PFC in schizophrenia. Moreover, pathophysiology of the DA synthesis and/or release in mania and in bipolar depression is still inconsistent. Both DA agonists and anti-dopaminergics can improve bipolar depressive symptoms [41]. Hence, finding the hypothetical relationships that may increase DA levels in SN and PFC seven days after subclinical LPS administration with psychiatric or/and neurodegenerative disorders is difficult and requires further research.

Neuropeptides (like DA), which can be called neuroimmune transmitters, also play neuroimmunological roles as modulatory molecules between immune and nervous cells [69]. Both neuropeptides and DA can be released by neuronal and immunological cells. Under specific stimuli, these cells may release modulatory molecules into the extracellular compartment, thus enabling communications between other cells. As an example, T lymphocytes express neuropeptide receptors for SP and VIP. SP and VIP are released from the lymphoid organs from neuronal and immune cells. At the same time, neural cells express receptors for cytokines, which are released from the immune system and affect neural differentiation. The nervous and immune systems may produce and respond to mediators of immune–neuronal interaction, such as neurotransmitters and cytokines [69]. As we can see, the nervous system is necessary for full immune function and vice-versa, but the abnormal activity of immune or nervous cells may result in disease. In the present study, subclinical LPS increase all examined neuropeptide concentrations in cLN. SP and VIP are the best-studied immunomodulatory neuropeptides. Specific high-affinity receptors for both have been identified on lymphocytes and monocytes, which suggests that they are involved in the direct interaction between immune cells. SP is an immunostimulatory neuropeptide and modulates a number of important immunological functions, including direct effects on T-cell activation [22]. VIP is a well-known anti-inflammatory mediator and may have a dual role in the neuromodulatory system. VIP is generally a suppressive neuropeptide for T-cell proliferation, but can also enhance certain lymphocyte functions by interacting with different VIP receptors and can inhibit activation-induced apoptosis in T lymphocytes [70]. NPY modulates differentiation of T-helper cells and plays an important homeostatic role in balancing disturbances of neuroimmune systems. In addition, NPY is also able to modulate the immunomodulatory effects of other neuropeptides and acts as a neuroimmune transmitter and co-transmitter in neuroimmune crosstalk [33,71]. NPY infusion improves survival during septic shock induced by LPS in mice [72]. The beneficial effects of NPY result from its vasoconstrictor abilities and potentiation of catecholaminergic vasopressor effects, as well as improved critical immune functions [73]. Although the influence of GAL on lymphocytes has not been thoroughly studied, it was found that this neuropeptide has a potent anti-proliferative influence on (at least) certain lymphocyte subpopulations [74]. Considering the above information and our findings of the high level of GAL, NPY, SP, and VIP in cLN seven days after subclinical LPS administration, we hypothesize that dysregulation of neuropeptides is connected with their influence on immune cells. This hypothesis may particularly be confirmed by the fact that in the present study, despite increasing the selected neuropeptides, LPS did not increase either CD4 or CD8 T-lymphocyte levels in cLN. It should be noted that changes in immune cell functions can also influence the expression of neuropeptides in lymphoid organs [75]. Therefore, it is possible that neuropeptides act as a negative feedback inhibitor to decrease CD4 and CD8 T-lymphocyte levels, which have been upregulated in response to LPS.

The observable occurrence of functional interconnections between the nervous and immune systems is inextricably interlinked by recent studies revealing the meningeal lymphatic vessels. The cLN appear to be critical for this neuroimmunological connection [41–43,76]. Ligation of the collecting vessels draining to the dcLN resulted in the distension of the dural lymphatic vessels and the accumulation of T-lymphocytes, which suggests that the meningeal lymphatic vessels' pathway may play a role in antigen presentation and in the movement of peripheral immune cells out of the brain to cLN [42]. Surprisingly, CD4 T-lymphocytes are required for normal learning and memory in the brain, because removal of the dcLN disrupted T-cells circulation and induced learning and memory impairments in mice [77]. Additional experiments have shown that whereas CD8 T-lymphocytes deficiency is negligible, the participation of CD4 T-lymphocytes is important for promoting neurodegeneration of DAergic neurons in the SN of mice undergoing PD [78,79]. The presence of infiltrating immune cells in the central nervous system parenchyma has been detected in most studied neurodegenerative diseases [38,80,81]. Recent evidence suggests that exposure to a high dose of LPS in vitro leads to suppression of CD8 T-lymphocyte proliferation in the mice spleen but not in cLN [82]. The suppression of T-cells in the spleen, but not in the cLN, makes it possible for the continuation of T-cell activation in cLN. Similarly, in the current study, subclinical LPS did not make any direct (or indirect, probably thanks to neuropeptides) changes in the CD4 and CD8 T-lymphocyte levels in comparison with the control group (Figure 6). The presence of receptors for DA, neuropeptides, cytokines, and LPS on T cells in the LN [83] and the possibility of interaction in cLN between CD4 and CD8 lymphocytes and neuropeptides, together with the discovery of meningeal lymphatic vessels, may indicate the existence of relationships that until now have not been considered. A biological factor such as LPS and a pathological process from Alzheimer's to Parkinson's disease, schizophrenia, and depression may be linked to a dysfunctional neuroimmune system. Moreover, our study is very interesting in reference to a recent study by Kozina et al. [84], which suggested that peripheral immune signalling plays an unexpected and very important role in the neurodegeneration process.

In our in vivo studies, in comparison with in vitro systems, an account was taken of the normal anatomical distribution of CD 4 and CD8 T-lymphocytes and neuropeptides in LN and DA in the brain and of the integration of the effects of those cells and active substances with cytokines and Hp in blood. Numerous studies investigating the neuroimmunological system have focused on in vitro systems in which concentrations of neuroimmune-active substances could be outside the physiological range and in which no account was taken of the normal anatomical distribution of lymphocytes and neural innervation of lymphoid organs and peripheral and central tissues. Our in vivo study integrated the effect of LPS on the levels of neuropeptides, DA, CD4 T-lymphocytes, CD8 T-lymphocytes, acute phase cytokines, and Hp derived from different tissue and cells following asymptomatic LPS stimulation.

4. Materials and Methods

4.1. Animal Housing, Health Status, and LPS Administration

All experimental procedures were approved by the Local Ethical Committee for Animal Experimentation in Olsztyn, Poland (decision no.: 73/2015, from 29 September 2015). Animals were kept and treated in accordance with all institutional and national guidelines applicable within the Republic of Poland, as per the Federal Law of 15 January 2015 on Animal Welfare for Science and Education (Dz.U.2015.0.266).

The study was performed on ten juvenile crossbred female pigs (Pietrain × Duroc) and was conducted when the pigs were between eight and nine weeks of age with body weights of 16–18 kg. The animals were maintained for two weeks prior to the experiment in order to allow adaptation to the new environment. The pigs were kept under standard laboratory conditions and fed commercial feed for pigs of this age group. Following assessment by a veterinary surgeon (DVM, Ph.D.), only clinically healthy animals with negative results of analyses of *Salmonella* spp. in faecal samples qualified for the experiment.

After a two-week adaptive period, pigs included in the experiment were randomly divided into two groups (five pigs each): a control and a treatment group (LPS group). In the morning, animals of both groups were subjected to premedication, according to the method previously described [85], with an intramuscular injection of atropine (Atropinum Sulfuricum Polfa Warszawa S.A., Poland, 0.035 mg/kg b.w.), ketamine (Bioketan, Vetoquinol Biowet Sp. z o.o., Poland and Vetoquinol S.A., France, 7.0 mg/kg b.w.), and medetomidine (Cepetor, CP-Pharma Handelsges mbH, Germany, 0.063 mg/kg b.w.). The premedication of animals allowed accurate and safe (for investigators) injections of LPS. Under premedication, the control animals were injected with 10 mL of 0.9% NaCl (sodium chloride 0.9% WET Baxter, 9 g/1000 mL, Baxter Sp. z o.o. Poland) saline solution into the marginal ear vein, while pigs of the treatment group received lipopolysaccharides from *Salmonella enterica* serotype Enteritidis in the same way (i.e., intravenously) (L7770 Sigma, Aldrich, Germany) at a dose of 5 µg/kg b.w. (in 10 mL saline solution). Such a dose has been previously described as a "low single dose", which does not result in any clinical symptoms of disease [31,56]. All procedures and drugs were managed and administered by a veterinary surgeon (DVM, Ph.D.).

During the clinical assessment of the pigs' health status, conducted by a veterinary surgeon, the observations of the animal care staff were always considered. In addition, during the physical examination, the measurements of temperature and body weight, both in the control and LPS group, were taken into accounts. Pigs were individually weighed once a day in the morning (between 07:00 and 07:30). The body weight data were presented as the mean from group ± SD (standard deviation). Rectal temperature was determined using an animal digital thermometer (model SC 12, SCALA Electronic GmbH, Stahnsdorf, Germany). Temperatures were determined twice daily: in the morning (between 07:00 and 07:30) and in the afternoon (between 17:00 and 17:30) during the seven days after LPS administration. The first measurement was taken before premedication and LPS or NaCl administration, and the final measurement was taken on the last day of the experiment before the antecedent sample collection. The temperature data were presented as the mean from group ± SD.

4.2. Sample Collection

After a seven-day period, which has been described as sufficient for the emergence of changes in the nervous system in previous studies [31,86,87], all animals were pre-medicated (in the above-described manner) and subjected to general anaesthesia using propofol (Scanofol, NORBROOK, Northern Ireland, IRL.PN, 4.5 mg/kg b.w. given intravenously into the marginal ear vein). Blood from the LPS-treated animals and the control group was taken from the vena orbital sinus. Serum was harvested after centrifugation and stored at $-80\,^{\circ}$C for further analyses of IL-6, TNF-α, and Hp. Heparinised blood samples were collected to isolate peripheral blood mononuclear cells (PBMCs). After euthanasia, with pentobarbital (Morbital, Biowet Puławy Sp. z o.o, Poland, 60–70 mg/kg b.w., given intravenously), the scLN were then isolated [88]. Next, the brains were rapidly removed from the skulls. The SN and PFC were isolated as described by Jelsing et al. [89]. The right dorsal scLN, the SN, and the fragment of PFC were frozen immediately after collection in liquid nitrogen and were stored at $-80\,^{\circ}$C until processing for further analysis. The left dorsal scLN was used for lymphocyte isolation.

4.3. Lymphocytes Isolation Phenotyping

4.3.1. Isolation of Peripheral Blood Mononuclear Cells (PBMCs) from Blood

PBMCs were isolated by gradient centrifugation [90]. An equal volume of phosphate buffered saline (PBS) was mixed with blood. The mixture was layered on Histopaque-1077 (10771, Sigma) at a 3:1 ratio (v/v) and centrifuged at 512× g/18 $^{\circ}$C for 35 min (Eppendorf 5804R, Hamburg, Germany). Mononuclear cells were aspired and washed in incomplete medium (RPMI 1640 cat. no. 11875093; Thermo Fisher Scientific supplemented with 10 mM HEPES (4-(2-hydroxyethyl)-1-piperazineethanesulfonic acid) and 10 units/mL penicillin–streptomycin solution, Waltham, MA, USA), centrifuged at 413× g/10 $^{\circ}$C for 10 min, and suspended in 1 mL of incomplete medium. The number of lymphocytes was calculated using a Burker Cell Counter after Trypan Blue staining.

4.3.2. Lymphocytes Isolation from the scLN

The scLN were dounced in incomplete medium and filtered through an 80 µm nylon filter to remove cell debris. The cells were then washed and centrifuged at 413× g (Eppendorf 5418R, Hamburg, Germany) at 10 $^{\circ}$C for 10 min and suspended in 1 mL of incomplete medium [91]. The number of lymphocytes was calculated using a Burker Cell Counter after Trypan Blue staining.

4.3.3. Lymphocyte Phenotyping

Lymphocytes were stained with fluorescein isothiocyanate (FITC) Mouse Anti-Pig CD4a (cat. no. 559585; BD Biosecinesces, Clone 74-12-4) and phycoerythrin (PE)Mouse Anti-Pig CD8a (cat. no. 559584; BD Biosciences, Clone 76-2-11). Cells were incubated at 4 $^{\circ}$C for 15 min, washed with FACS buffer (PBS: 0.1 M phosphate buffered saline, pH = 7.2 supplemented with 5% foetal bovine serum), and centrifuged at 10 $^{\circ}$C for 10 min and 413× g (Eppendorf 5418R). The supernatant was removed and the cells were fixed in 2% paraformaldehyde and analysed using a MoFloTM XDP flow cytometer (Beckman Coulter Inc., Miami, Fullerton, CA, USA) equipped with a 488 nm air-cooled argon laser. Forward scatter, side scatter, and green and red fluorescence channels were used to collect specific multi-parameter data from the cells. A total of 50,000 events were collected from each sample. The data were analysed with the Summit 5.2 (Beckman-Coulter Inc., Miami, FL, USA) software package. The gating tree was set as follows: forward-scattered light (FSC)/side-scattered light (SSC) (represent distribution cells on size and intracellular composition) lymphocytes were gated in the range 100–150 kDa, followed by CD4+ and CD8+ [92].

4.4. Measurement of Serum Concentration of Hp, IL-6, and TNF-α

The serum concentrations of Hp, IL-6, and TNF-α were quantified by commercial the enzyme-linked immunosorbent assay (ELISA) kits:

- Pig Haptoglobin ELISA (HAPT-9, Life Diagnostic Inc., West Chester, PA, USA); sensitivity range from 18.75 to 300 ng/mL
- Quantizing® ELISA Porcine IL-6 (P6000B, R&D Systems, Minneapolis, MN, USA); sensitivity range from 0.68 to 4.30 pg/mL
- Quantizing® ELISA Porcine TNF-α (PTA00, R&D Systems, Minneapolis, MN, USA); sensitivity range from 2.8 to 5.0 pg/mL

For all analyses, serum samples were tested in duplicate, according to the manufacturer's recommendations. These assays employed the quantitative sandwich enzyme immunoassay technique. A monoclonal antibody specific for a particular protein was pre-coated onto a microplate. The standards, control, and samples were pipetted into the wells and, if it was present, it was bound by the immobilized antibody. After washing away any unbound substances, an enzyme-linked monoclonal antibody specific for peptides was added to the wells. Following a wash step, a substrate solution was added to the wells. After stopping the enzyme reaction, the intensity of the colour was measured. The sample values were then read off the standard curve.

4.5. Brain Sample Preparation and Determination of DA Concentrations

Quantitative determination of DA in SN and in PFC was carried out using a commercially available Dopamine ELISA kit (RE59161 IBL International; Hamburg, Germany) according to previously described protocols [93,94]. Briefly, brain samples were weighed and homogenized with an extract solution containing acetonitrile (0.5 mL/100 mg brain tissue), 0.1 M HCl (0.4 mL/100 mg brain tissue), 27 mM ethylenedinitrilotetraacetic acid (EDTA) water solution (0.1 mL/100 mg tissue), and SIGMAFAST Protease Inhibitor Cocktail Tablet EDTA free (cat. no. S8830; Sigma-Aldrich, St. Louis, MO, USA). The sample was then centrifuged at $4500\times$ *g* (Eppendorf 5804) and the supernatant was filtered on syringe filters without pre-filtering (Millex-HV Filter, 0.45 μm, PVDF, Millipore, Burlington, MA, USA). Samples were concentrated on a miVac centrifugal vacuum concentrator, model DNA-23050-800, with SpeedTrap (Genevac Limited, Ipswich, UK) for 2 h, and then lyophilized using an ALPHA 1-4 LSC freeze dryer (MARTIN CHRIST Gefriertrocknungsanlagen GmbH Germany, Osterode am Harz, Germany). Lyophilized samples were stored at -80 °C until analysis. A dopamine ELISA kit was then used according to the manufacturer's recommendations. First, DA extraction from all samples (unknown, standards, and controls) was done on 24 wells plates with extraction, shaking, and releasing steps (all reagents provided in the kit) followed by ELISA (protocol provided in manual). An Infinite 200 spectrophotometer (Tecan Group, Männedorf, Switzerland) with Magellan software was used to read samples absorbance at a wavelength of 450 nm and calculate dopamine concentration in unknown samples. The results were presented as the average per group $\pm$ SD.

4.6. Purification and Determination of Neuropeptide Levels from scLN

The levels of GAL, NPY, SP, and VIP were determined in the scLN in the three-step procedure described below.

4.6.1. Sample Preparation and High-Temperature Extraction

Brain peptide extracts from tissues were prepared according to the Conlon procedure [95]. Briefly, frozen tissue was cut into small pieces and 10 mL of hot 1 M acetic acid was added per gram tissue and boiled for 5 min. The samples were then homogenized using Ultra Turax IKA T-25 (Jankel & Kunkel IKA, Germany) at RT for 5 min and centrifuged at 4 °C for 40 min at $4500\times$ *g* (Eppendorf 5804). The supernatant was subject to solid-phase extraction (SPE) step.

4.6.2. Solid-Phase Extraction (SPE) and Concentration

The supernatants were filtered through syringe filters with a graduated glass fibre pre-filter (Millex-HPF HV Filter, 0.45 μm, PVDF, Millipore). Trifluoroacetic acid (TFA) was added to filtrates to obtain a final concentration of 0.1% (v/v). A Sep-Pak Plus Light Cartridge (130 mg of C18 sorbent per cartridge, Waters, Milford, MA, USA) was used according to the producer's protocol using a Baker Vacuum Manifold SPE-12G unit (J.T.Baker, Germany). Samples were concentrated on a miVac centrifugal vacuum concentrator, model DNA-23050-800, with SpeedTrap (Genevac Limited, UK) for 2h, and then lyophilized using an ALPHA 1-4 LSC freeze dryer (MARTIN CHRIST Gefriertrocknungsanlagen GmbH Germany). Lyophilized samples were stored at $-80\,^{\circ}$C until analysis.

All chemicals used were of commercial origin with high performance liquid chromatography (HPLC) grade purity: glacial acetic acid (cat. no. 951503, J.T. Baker), trifluoroacetic acid–TFA (cat. no. 9470, J.T. Baker) and acetonitrile–LC-MS reagent (cat. no. 9821.1000, J.T. Baker, Germany).

4.6.3. Enzyme-Linked Immunosorbent Assay for Quantitative Determination of GAL, NPY, SP, and VIP in scLN

For quantitative determination of GAL, NPY, SP, and VIP in scLN, the following commercial ELISA tests were used: Galanin (GAL) EIA kit, 0–10 ng/mL (S-1210; Peninsula Laboratories International, Inc., San Carlos, CA, USA); Neuropeptide Y (NPY) EIA kit, 0–100 ng/mL (EK-049-03CE; Phoenix Pharmaceuticals, Inc., Burlingame, CA, USA); Substance P EIA kit, 0–5 ng/mL (S-1180; Peninsula Laboratories International, Inc., San Carlos, CA, USA); and Vasoactive Intestinal Peptide (VIP) EIA kit, 0–25 ng/mL (EK-064-16CE; Phoenix Pharmaceuticals, Inc., Burlingame, CA, USA). The manufacturers provided reagents for each assay and protocols (temperatures of incubation, sample volume, reagent volumes) and we followed them. All tests were based on the standard sandwich ELISA method. Briefly, samples together with primary antibodies were put on plates and incubated at room temperature. The plate was then washed and conjugated antibodies were added, and the plates were incubated. After washing, a one-step substrate reagent was added to each well and the plates were incubated again at room temperature. The reaction was terminated with a stop reagent. The absorbance was read at $\lambda = 450$ nm on a microplate spectrophotometer Infinite 200 (Tecan) for each sample.

A four-parameter ELISA curve was prepared for each determined neuropeptide (an Excel sheet was provided by Peninsula Laboratories service). Each sample was assayed in duplicate and the peptide concentration was read from the curve. Each peptide concentration was presented as the mean from group $\pm$ SD per g of tissue.

4.7. Statistical Analysis

The results of body weight and rectal temperature were compared between the control and LPS group using Student's *t*-test at a significance level of $p < 0.05$. The other results were analysed statistically using a one-way analysis of variance (ANOVA) and the significance of differences between groups was determined using Tukey's test at a significance level of $p < 0.05$. The data were expressed as mean values $\pm$ SD and the calculations were performed with SigmaPlot 12 (Systat Software Inc., Cracow, Poland)

5. Conclusions

In conclusion, the results of this study indicate that subclinical LPS from *S*. Enteritidis can affect cells and signal molecules involved in the neuroimmune interaction. We demonstrated increasing levels of DA in the brain and neuropeptides in cLN with a decrease in plasma CD4 and CD8 T-lymphocyte levels after subclinical *S*. Enteritidis LPS administration. Moreover, our data indicate that LPS increases serum Hp levels seven days after *S*. Enteritidis LPS administration. It is possible that even such low doses of LPS from *S*. Enteritidis that do not result in any clinical symptoms of the disease may require eradication, which may be very important, especially in connection with the asymptomatic carrier state of *Salmonella* spp. Understanding how asymptomatic LPS from *S*. Enteritidis impact neuroimmune transmitters and neuroimmune T-lymphocytes responses may be useful for the prevention of pathological processes. Additionally, the results of our study and our considerations in this paper may be helpful in assessing the long-term consequences of low doses of LPS from *S*. Enteritidis. This is especially important in the era of using LPS in developing drugs ranging from vaccines and cancer therapy to immunostimulants, which are becoming increasingly common in use.

Author Contributions: A.M. conceived and designed the study, analysed the data, and wrote the paper. A.M. and D.Z. performed the experimental procedures.

Funding: This research was supported by statutory grant No. 25.610.001-300 from the Faculty of Medical Sciences, the University of Warmia and Mazury in Olsztyn in Poland.

Acknowledgments: This study was supported by statutory grant No. 25.610.001-300, Faculty of Medical Sciences, the University of Warmia and Mazury in Olsztyn, Poland.

Conflicts of Interest: The authors declare no conflict of interest.

Abbreviations

AD	Alzheimer's disease
BBB	blood brain barrier
BCSFB	blood-cerebrospinal fluid barrier
cLN	cervical lymph nodes
CSF	cerebrospinal fluid
DA	dopamine
dcLN	deep cervical lymph nodes
GAL	galanin
Hp	haptoglobin
IL-6	interleukin-6
LN	lymph nodes
LPS	lipopolysaccharide
NPY	neuropeptide Y
PD	Parkinson's disease
PFC	prefrontal cortex
scLN	superficial cervical lymph nodes
SN	substantia nigra
SP	substance P
TNF-α	tumour necrosis factor α
VIP	vasoactive intestinal peptide

References

1. De Martel, C.; Ferlay, J.; Franceschi, S.; Vignat, J.; Bray, F.; Forman, D.; Plummer, M. Global burden of cancers attributable to infections in 2008: A review and synthetic analysis. *Lancet Oncol.* **2012**, *13*, 607–615. [CrossRef]
2. Mesri, E.A.; Feitelson, M.A.; Munger, K. Human viral oncogenesis: A cancer hallmarks analysis. *Cell Host Microbe* **2014**, *15*, 266–282. [CrossRef] [PubMed]
3. Vedham, V.; Divi, R.L.; Starks, V.L.; Verma, M. Multiple Infections and Cancer: Implications in Epidemiology. *Technol. Cancer Res. Treat.* **2014**, *13*, 177–194. [CrossRef] [PubMed]
4. Mikołajczyk, A. Invited Brief Commentary on the Article "Breast Cancer Association with Cytomegalo Virus—A Tertiary Center Case-Control Study" Is Cytomegalo Virus a Breast Cancer Etiologic Risk Factor? *J. Investig. Surg.* **2017**, *30*, 1–2. [CrossRef] [PubMed]
5. Zloza, A. Viruses, bacteria, and parasites—Oh my! A resurgence of interest in microbial-based therapy for cancer. *J. Immunother. Cancer* **2018**, *6*, 4–6. [CrossRef] [PubMed]
6. Ramachandran, G. Gram-positive and gram-negative bacterial toxins in sepsis. *Virulence* **2014**, *5*, 213–218. [CrossRef] [PubMed]

7.   Liu, M.; Bing, G. Lipopolysaccharide animal models for Parkinson's disease. *Parkinsons. Dis.* **2011**, *2011*, 1–7. [CrossRef] [PubMed]

8.   Hoban, D.B.; Connaughton, E.; Connaughton, C.; Hogan, G.; Thornton, C.; Mulcahy, P.; Moloney, T.C.; Dowd, E. Further characterisation of the LPS model of Parkinson's disease: A comparison of intra-nigral and intra-striatal lipopolysaccharide administration on motor function, microgliosis and nigrostriatal neurodegeneration in the rat. *Brain Behav. Immun.* **2013**, *27*, 91–100. [CrossRef] [PubMed]

9.   Sharma, N.; Nehru, B. Characterization of the lipopolysaccharide induced model of Parkinson's disease: Role of oxidative stress and neuroinflammation. *Neurochem. Int.* **2015**, *87*, 92–105. [CrossRef] [PubMed]

10.   Huang, B.; Liu, J.; Ju, C.; Yang, D.; Chen, G.; Xu, S.; Zeng, Y.; Yan, X.; Wang, W.; Liu, D.; et al. Licochalcone A prevents the loss of dopaminergic neurons by inhibiting microglial activation in lipopolysaccharide (LPS)-induced Parkinson's disease models. *Int. J. Mol. Sci.* **2017**, *18*, 1–20. [CrossRef] [PubMed]

11.   Nguyen, M.D. Exacerbation of Motor Neuron Disease by Chronic Stimulation of Innate Immunity in a Mouse Model of Amyotrophic Lateral Sclerosis. *J. Neurosci.* **2004**, *24*, 1340–1349. [CrossRef] [PubMed]

12.   Choi, D.; Liu, M.; Hunter, R.L.; Cass, W.A.; Pandya, J.D.; Patrick, G.; Shin, E.; Kim, H.; Gash, D.M.; Bing, G. Striatal Neuroinflammation Promotes Parkinsonism in Rats. *PLoS ONE* **2009**, *4*, e5482. [CrossRef] [PubMed]

13.   Ma, M.; Ren, Q.; Yang, C.; Zhang, J.C.; Yao, W.; Dong, C.; Ohgi, Y.; Futamura, T.; Hashimoto, K. Antidepressant effects of combination of brexpiprazole and fluoxetine on depression-like behavior and dendritic changes in mice after inflammation. *Psychopharmacology* **2017**, *234*, 525–533. [CrossRef] [PubMed]

14.   Zhang, X.Y.; Cao, J.B.; Zhang, L.M.; Li, Y.F.; Mi, W.D. Deferoxamine attenuates lipopolysaccharide-induced neuroinflammation and memory impairment in mice. *J. Neuroinflamm.* **2015**, *12*, 1–13. [CrossRef] [PubMed]

15.   Hawkesworth, S.; Moore, S.E.; Fulford, A.J.C.; Barclay, G.R.; Darboe, A.A.; Mark, H.; Nyan, O.A.; Prentice, A.M. Evidence for metabolic endotoxemia in obese and diabetic Gambian women. *Nutr. Diabetes* **2013**, *3*, e83–e86. [CrossRef] [PubMed]

16.   Zhan, X.; Stamova, B.; Jin, L.W.; DeCarli, C.; Phinney, B.; Sharp, F.R. Gram-negative bacterial molecules associate with Alzheimer disease pathology. *Neurology* **2016**, *87*, 2324–2332. [CrossRef] [PubMed]

17.   Pretorius, E.; Page, M.J.; Mbotwe, S.; Kell, D.B. Lipopolysaccharide-binding protein (LBP) can reverse the amyloid state of fibrin seen or induced in Parkinson's disease. *PLoS ONE* **2018**, *13*, 1–16. [CrossRef] [PubMed]

18.   Zhao, Y.; Jaber, V.; Lukiw, W.J. Secretory Products of the Human GI Tract Microbiome and Their Potential Impact on Alzheimer's Disease (AD): Detection of Lipopolysaccharide (LPS) in AD Hippocampus. *Front. Cell. Infect. Microbiol.* **2017**, *7*, 1–9. [CrossRef] [PubMed]

19.   Banks, W.A.; Gray, A.M.; Erickson, M.A.; Salameh, T.S.; Damodarasamy, M.; Sheibani, N.; Meabon, J.S.; Wing, E.E.; Morofuji, Y.; Cook, D.G.; et al. Lipopolysaccharide-induced blood-brain barrier disruption: Roles of cyclooxygenase, oxidative stress, neuroinflammation, and elements of the neurovascular unit. *J. Neuroinflamm.* **2015**, *12*, 1–15. [CrossRef] [PubMed]

20.   Banks, W.A.; Robinson, S.M. Minimal penetration of lipopolysaccharide across the murine blood-brain barrier. *Brain Behav. Immun.* **2010**, *24*, 102–109. [CrossRef] [PubMed]

21.   Marques, F.; Sousa, J.C.; Coppola, G.; Falcao, A.M.; Rodrigues, A.J.; Geschwind, D.H.; Sousa, N.; Correia-Neves, M.; Palha, J.A. Kinetic profile of the transcriptome changes induced in the choroid plexus by peripheral inflammation. *J. Cereb. Blood Flow Metab.* **2009**, *29*, 921–932. [CrossRef] [PubMed]

22.   Marques, F.; Sousa, J.C.; Coppola, G.; Geschwind, D.H.; Sousa, N.; Palha, J.A.; Correia-Neves, M. The choroid plexus response to peripheral inflammatory stimulus. *BMC Neurosci.* **2009**, *10*, 135. [CrossRef] [PubMed]

23.   Niehaus, I.; Lange, J.H. Endotoxin: Is it an environmental factor in the cause of Parkinson's disease? *Occup. Environ. Med.* **2003**, *60*, 378. [CrossRef] [PubMed]

24.   Niehaus, I. In vivo Radiodetoxification of Salmonella minnesota Lipopolysaccharides with radio-labeled Leucine Enkephalin cures sensory polyneuropathy: A Case report. *Niger. Health J.* **2010**, *10*, 26–33.

25.   Yao, Z.; Mates, J.M.; Cheplowitz, A.M.; Hammer, L.P.; Phillips, G.S.; Wewers, M.D.; Rajaram, M.V.S.; John, M.; Anderson, C.L.; Ganesan, L.P.; et al. Blood-Borne Lipopolysaccharide Is Rapidly Eliminated by Liver Sinusoidal Endothelial Cells via High-Density Lipoprotein. *J. Immunol.* **2016**, *197*, 2390–2399. [CrossRef] [PubMed]

26.   Franco, R.F.; de Jonge, E.; Dekkers, P.E.; Timmerman, J.J.; Spek, C.A.; van Deventer, S.J.; van Deursen, P.; van Kerkhoff, L.; van Gemen, B.; ten Cate, H.; et al. The in vivo kinetics of tissue factor messenger RNA expression during human endotoxemia: Relationship with activation of coagulation. *Blood* **2000**, *96*, 554–559. [PubMed]

27. Maxwell, J.R.; Ruby, C.; Kerkvliet, N.I.; Vella, A.T. Contrasting the Roles of Costimulation and the Natural Adjuvant Lipopolysaccharide during the Induction of T Cell Immunity. *J. Immunol.* **2002**, *168*, 4372–4381. [CrossRef] [PubMed]

28. Haudek, S.B.; Natmessnig, B.E.; Fürst, W.; Bahrami, S.; Schlag, G.; Redl, H. Lipopolysaccharide dose response in baboons. *Shock* **2003**, *20*, 431–436. [CrossRef] [PubMed]

29. Arredouani, M.; Matthijs, P.; Van Hoeyveld, E.; Kasran, A.; Baumann, H.; Ceuppens, J.L.; Stevens, E. Haptoglobin directly affects T cells and suppresses T helper cell type 2 cytokine release. *Immunology* **2003**, *108*, 144–151. [CrossRef] [PubMed]

30. Arredouani, M.S.; Kasran, A.; Vanoirbeek, J.A.; Berger, F.G.; Baumann, H.; Ceuppens, J.L. Haptoglobin dampens endotoxin-induced inflammatory effects both in vitro and in vivo. *Immunology* **2005**, *114*, 263–271. [CrossRef] [PubMed]

31. Mikołajczyk, A.; Gonkowski, S.; Złotkowska, D. Modulation of the main porcine enteric neuropeptides by a single low-dose of lipopolysaccharide (LPS) *Salmonella* Enteritidis. *Gut Pathog.* **2017**, *9*, 1–10. [CrossRef] [PubMed]

32. Lambrecht, B.N. Immunologists getting nervous: Neuropeptides, dendritic cells and T cell activation. *Respir. Res.* **2001**, *2*, 133–138. [CrossRef] [PubMed]

33. Farzi, A.; Reichmann, F.; Holzer, P. The homeostatic role of neuropeptide Y in immune function and its impact on mood and behaviour. *Acta Physiol.* **2015**, *213*, 603–627. [CrossRef] [PubMed]

34. Wasowicz, K.; Winnicka, A.; Kaleczyc, J.; Zalecki, M.; Podlasz, P.; Pidsudko, Z. Neuropeptides and lymphocyte populations in the porcine ileum and ileocecal lymph nodes during postnatal life. *PLoS ONE* **2018**, *13*, 1–14. [CrossRef] [PubMed]

35. Ganea, D.; Hooper, K.M.; Kong, W. The neuropeptide VIP: Direct effects on immune cells and involvement in inflammatory and autoimmune diseases. *Acta Physiol.* **2015**, *213*, 442–452. [CrossRef] [PubMed]

36. Madva, E.N.; Granstein, R.D. Nerve-derived Transmitters Including Peptides Influence Cutaneous Immunology. *Brain Behav. Immun.* **2013**, *34*, 1–10. [CrossRef] [PubMed]

37. Huang, Y.; Qiu, A.W.; Peng, Y.P.; Liu, Y.; Huang, H.W.; Qiu, Y.H. Roles of dopamine receptor subtypes in mediating modulation of T lymphocyte function. *Neuroendocrinol. Lett.* **2010**, *31*, 782–791. [PubMed]

38. Sarkar, C.; Basu, B.; Chakroborty, D.; Dasgupta, P.S.; Basu, S. The immunoregulatory role of dopamine: An update and. *Brain Behav. Immun.* **2010**, *24*, 1–8. [CrossRef] [PubMed]

39. Pacheco, R.; Contreras, F.; Zouali, M. The dopaminergic system in autoimmune diseases. *Front. Immunol.* **2014**, *5*, 1–17. [CrossRef] [PubMed]

40. Slifstein, M.; Van De Giessen, E.; Van Snellenberg, J.; Thompson, J.L.; Narendran, R.; Gil, R.; Hackett, E.; Girgis, R.; Ojeil, N.; Moore, H.; et al. Deficits in prefrontal cortical and extrastriatal dopamine release in schizophrenia a positron emission tomographic functional magnetic resonance imaging study. *JAMA Psychiatry* **2015**, *72*, 316–324. [CrossRef] [PubMed]

41. Ashok, A.H.; Marques, T.R.; Jauhar, S.; Nour, M.M.; Goodwin, G.M.; Young, A.H.; Howes, O.D. The dopamine hypothesis of bipolar affective disorder: The state of the art and implications for treatment. *Mol. Psychiatry* **2017**, *22*, 666–679. [CrossRef] [PubMed]

42. Aspelund, A.; Antila, S.; Proulx, S.T.; Karlsen, T.V.; Karaman, S.; Detmar, M.; Wiig, H.; Alitalo, K. A dural lymphatic vascular system that drains brain interstitial fluid and macromolecules. *J. Exp. Med.* **2015**, *212*, 991–999. [CrossRef] [PubMed]

43. Louveau, A.; Harris, T.H.; Kipnis, J. Revisiting the concept of CNS immune privilege Antoine. *Trends Immunol.* **2015**, *36*, 569–577. [CrossRef] [PubMed]

44. Liu, H.; Ni, Z.; Chen, Y.; Wang, D.; Qi, Y.; Zhang, Q.; Wang, S. Olfactory route for cerebrospinal fluid drainage into the cervical lymphatic system in a rabbit experimental model. *Neural Regen. Res.* **2012**, *7*, 766–771. [CrossRef] [PubMed]

45. Chen, L.; Elias, G.; Yostos, M.P.; Stimec, B.; Fasel, J.; Murphy, K. Pathways of cerebrospinal fluid outflow: A deeper understanding of resorption. *Neuroradiology* **2015**, *57*, 139–147. [CrossRef] [PubMed]

46. Laman, J.D.; Weller, R.O. Drainage of cells and soluble antigen from the CNS to regional lymph nodes. *J. Neuroimmune Pharmacol.* **2013**, *8*, 840–856. [CrossRef] [PubMed]

47. Swindle, M.M.; Makin, A.; Herron, A.J.; Clubb, F.J., Jr.; Frazier, K.S. Swine as models in biomedical research and toxicology testing. *Vet. Pathol.* **2012**, *49*, 344–356. [CrossRef] [PubMed]

48. Bassols, A.; Costa, C.; Eckersall, P.D.; Osada, J.; Sabrià, J.; Tibau, J. The pig as an animal model for human pathologies: A proteomics perspective. *Proteom. Clin. Appl.* **2014**, *8*, 715–731. [CrossRef] [PubMed]

49. Pulendran, B.; Kumar, P.; Cutler, C.W.; Mohamadzadeh, M.; Van Dyke, T.; Banchereau, J. Lipopolysaccharides from Distinct Pathogens Induce Different Classes of Immune Responses In Vivo. *J. Immunol.* **2001**, *167*, 5067–5076. [CrossRef] [PubMed]

50. Nedrebø, T.; Reed, R.K. Different serotypes of endotoxin (lipopolysaccharide) cause different increases in albumin extravasation in rats. *Shock* **2002**, *18*, 138–141. [CrossRef] [PubMed]

51. Bryant, C.E.; Spring, D.R.; Gangloff, M.; Gay, N.J. The molecular basis of the host response to lipopolysaccharide. *Nat. Rev. Microbiol.* **2010**, *8*, 8–14. [CrossRef] [PubMed]

52. Fedele, G.; Nasso, M.; Spensieri, F.; Palazzo, R.; Frasca, L.; Watanabe, M.; Ausiello, C.M. Lipopolysaccharides from *Bordetella pertussis* and *Bordetella parapertussis* Differently Modulate Human Dendritic Cell Functions Resulting in Divergent Prevalence of Th17-Polarized Responses. *J. Immunol.* **2008**, *181*, 208–216. [CrossRef] [PubMed]

53. Mikołajczyk, A.; Kozłowska, A.; Gonkowski, S. Distribution and Neurochemistry of the Porcine Ileocaecal Valve Projecting Sensory Neurons in the Dorsal Root Ganglia and the Influence of Lipopolysaccharide from Different Serotypes of *Salmonella* spp. on the Chemical Coding of DRG Neurons in the Cell Cultures. *Int. J. Mol. Sci.* **2018**, *19*, 2551. [CrossRef]

54. Maciel, B.M.; Rezende, R.P.; Sriranganathan, N.K. Salmonella enterica: Latency. In *Current Topics in Salmonella and Salmonellosis*; Mares, M., Ed.; InTechOpen: Rijeka, Croatia, 2017; pp. 41–58.

55. Qin, L.; Wu, X.; Block, M.L.; Liu, Y.; Breese, G.R.; Knapp, D.J.; Crews, F.T.; Hill, C.; Carolina, N.; Park, T. Systemic LPS Causes Chronic Neuroinflammation and Progressive Neurodegeneration. *Glia* **2007**, *55*, 453–462. [CrossRef] [PubMed]

56. Calvano, S.E.; Coyle, S.M. Experimental Human Endotoxemia: A Model of the Systemic Inflammatory Response Syndrome? *Surg. Infect.* **2012**, *13*, 293–299. [CrossRef] [PubMed]

57. Webel, D.M.; Finck, B.N.; Baker, D.H.; Johnson, R.W. Time course of increased plasma cytokines, cortisol, and urea nitrogen in pigs following intraperitoneal injection of lipopolysaccharide. *J. Anim. Sci.* **1997**, *75*, 1514–1520. [CrossRef] [PubMed]

58. Llamas Moya, S.; Boyle, L.; Lynch, P.B.; Arkins, S. Pro-inflammatory cytokine and acute phase protein responses to low-dose lipopolysaccharide (LPS) challenge in pigs. *Anim. Sci.* **2006**, *82*, 527–534. [CrossRef]

59. Maes, M. Depression is an inflammatory disease, but cell-mediated immune activation is the key component of depression. *Prog. Neuro-Psychopharmacol. Biol. Psychiatry* **2011**, *35*, 664–675. [CrossRef] [PubMed]

60. Liu, R.H.; Pan, J.Q.; Tang, X.E.; Li, B.; Liu, S.F.; Ma, W.L. The role of immune abnormality in depression and cardiovascular disease. *J. Geriatr. Cardiol.* **2017**, *14*, 703–710. [PubMed]

61. Robertson, M.J.; Schacterle, R.S.; Mackin, G.A.; Wilson, S.N.; Bloomingdale, K.L.; Ritz, J.; Komaroff, A.L. Lymphocyte subset differences in patients with chronic fatigue syndrome, multiple sclerosis and major depression. *Clin. Exp. Immunol.* **2005**, *141*, 326–332. [CrossRef] [PubMed]

62. Juffermans, N.P.; Paxton, W.A.; Dekkers, P.E.; Verbon, A.; de Jonge, E.; Speelman, P.; van Deventer, S.J.; van der Poll, T. Up-regulation of HIV coreceptors CXCR4 and CCR5 on CD4(+) T cells during human endotoxemia and after stimulation with (myco)bacterial antigens: The role of cytokines. *Blood* **2000**, *96*, 2649–2654. [PubMed]

63. Palmer, C.D.; Tomassilli, J.; Sirignano, M.; Tejeda, M.R.; Arnold, B.; Che, D.; Lauffenburger, D.A.; Jost, S.; Allen, T.; Mayer, K.H.; et al. Enhanced Immune Activation Linked to Endotoxemia in HIV-1 Seronegative Men who have Sex with Men. *AIDS* **2014**, *28*, 2162–2166. [CrossRef] [PubMed]

64. Guzmán, D.C.; Herrera, M.O.; Brizuela, N.O.; Mejía, G.B.; Jiménez, F.T.; García, E.H.; Olguín, H.J. Assessment of the effects of oseltamivir and indomethacin on dopamine, 5-HIAA, and some oxidative stress markers in stomach and brain of Salmonella typhimurium-infected rats. *Neuroendocrinol. Lett. Vol.* **2016**, *37*, 129–136.

65. Gołembiowska, K.; Wardas, J.; Noworyta-Sokołowska, K.; Kamińska, K.; Górska, A. Effects of adenosine receptor antagonists on the in vivo lps-induced inflammation model of parkinson's disease. *Neurotox. Res.* **2013**, *24*, 29–40. [CrossRef] [PubMed]

66. Noworyta-Sokolowska, K.; Gorska, A.; Golembiowska, K. LPS-induced oxidative stress and inflammatory reaction in the rat striatum. *Pharmacol. Rep.* **2013**, *65*, 863–869. [CrossRef]

67. Booij, L.; Welfeld, K.; Leyton, M.; Dagher, A.; Boileau, I.; Sibon, I.; Baker, G.B.; Diksic, M.; Soucy, J.P.; Pruessner, J.C.; et al. Dopamine cross-sensitization between psychostimulant drugs and stress in healthy male volunteers. *Transl. Psychiatry* **2016**, *6*, e740. [CrossRef] [PubMed]

68. Weinstein, J.J.; Weinstein, J.J.; Chohan, M.O.; Slifstein, M.; Kegeles, L.S.; Moore, H.; Abi-dargham, A. Pathway-Specific Dopamine Abnormalities in Schizophrenia Review Pathway-Speci fi c Dopamine Abnormalities in Schizophrenia. *Biol. Psychiatry* **2017**, *81*, 31–42. [CrossRef] [PubMed]

69. Pacheco, R.; Contreras, F.; Prado, C. Cells, molecules and mechanisms involved the neuro-immune interaction. In *Cell Interaction*; Gowder, S., Ed.; InTechOpen: Rijeka, Croatia, 2012; pp. 139–166.

70. Delgado, M.; Ganea, D. VIP and PACAP inhibit activation induced apoptosis in T lymphocytes. *Ann. N. Y. Acad. Sci.* **2000**, *921*, 55–67. [CrossRef] [PubMed]

71. Bedoui, S.; Kawamura, N.; Straub, R.H.; Pabst, R.; Yamamura, T.; Von Hörsten, S. Relevance of neuropeptide Y for the neuroimmune crosstalk. *J. Neuroimmunol.* **2003**, *134*, 1–11. [CrossRef]

72. Hauser, G.J.; Myers, A.K.; Dayao, E.K.; Zukowska-Grojec, Z. Neuropeptide Y infusion improves hemodynamics and survival in rat endotoxic shock. *Am. J. Physiol.* **1993**, *265*, H1416–H1423. [CrossRef] [PubMed]

73. Bedoui, S.; von Hörsten, S.; Gebhardt, T. A role for neuropeptide Y (NPY) in phagocytosis: Implications for innate and adaptive immunity. *Peptides* **2007**, *28*, 373–376. [CrossRef] [PubMed]

74. Trejter, M.; Brelinska, R.; Warchol, J.B.; Butowska, W.; Neri, G.; Rebuffat, P.; Gottardo, L.; Malendowicz, L.K. Effects of galanin on proliferation and apoptosis of immature rat thymocytes. *Int. J. Mol. Med.* **2002**, *10*, 183–186. [CrossRef] [PubMed]

75. Mignini, F.; Streccioni, V.; Amenta, F. Autonomic innervation of immune organs and neuroimmune modulation. *Auton. Autacoid Pharmacol.* **2003**, *23*, 1–25. [CrossRef] [PubMed]

76. Johnston, M.; Zakharov, A.; Papaiconomou, C.; Salmasi, G.; Armstrong, D. Evidence of connections between cerebrospinal fluid and nasal lymphatic vessels in humans, non-human primates and other mammalian species. *Cerebrospinal Fluid Res.* **2004**, *1*, 1–13. [CrossRef] [PubMed]

77. Radjavi, A.; Smirnov, I.; Derecki, N.; Kipnis, J. Dynamics of the Meningeal CD4+ T-cell repertoire are defined by the cervical lymph nodes and facilitate cognitive task performance in mice. *Mol. Psychiatry* **2014**, *19*, 531–533. [CrossRef] [PubMed]

78. Brochard, V.; Combadière, B.; Prigent, A.; Laouar, Y.; Perrin, A.; Beray-Berthat, V.; Bonduelle, O.; Alvarez-Fischer, D.; Callebert, J.; Launay, J.M.; et al. Infiltration of CD4+ lymphocytes into the brain contributes to neurodegeneration in a mouse model of Parkinson disease. *J. Clin. Investig.* **2009**, *119*, 182–192. [CrossRef] [PubMed]

79. Schetters, S.T.T.; Gomez-Nicola, D.; Garcia-Vallejo, J.J.; Van Kooyk, Y. Neuroinflammation: Microglia and T cells get ready to tango. *Front. Immunol.* **2018**, *8*, 1–11. [CrossRef] [PubMed]

80. McKenna, F.; McLaughlin, P.J.; Lewis, B.J.; Sibbring, G.C.; Cummerson, J.A.; Bowen-Jones, D.; Moots, R.J. Dopamine receptor expression on human T- and B-lymphocytes, monocytes, neutrophils, eosinophils and NK cells: A flow cytometric study. *J. Neuroimmunol.* **2002**, *132*, 34–40. [CrossRef]

81. Lucin, K.; Wyss-Coray, T. Immune activation in brain aging and neurodegeneration: Too much or too little? *Neuron* **2009**, *64*, 110–122. [CrossRef] [PubMed]

82. Greifenberg, V.; Ribechini, E.; Rößner, S.; Lutz, M.B. Myeloid-derived suppressor cell activation by combined LPS and IFN-γ treatment impairs DC development. *Eur. J. Immunol.* **2009**, *39*, 2865–2876. [CrossRef] [PubMed]

83. Arreola, R.; Alvarez-Herrera, S.; Pérez-Sánchez, G.; Becerril-Villanueva, E.; Cruz-Fuentes, C.; Flores-Gutierrez, E.O.; Garcés-Alvarez, M.E.; De La Cruz-Aguilera, D.L.; Medina-Rivero, E.; Hurtado-Alvarado, G.; et al. Immunomodulatory Effects Mediated by Dopamine. *J. Immunol. Res.* **2016**, *2016*, 1–31. [CrossRef] [PubMed]

84. Kozina, E.; Sadasivan, S.; Jiao, Y.; Dou, Y.; Ma, Z.; Tan, H.; Kodali, K.; Shaw, T.; Peng, J.; Smeyne, R.J. Mutant LRRK2 mediates peripheral and central immune responses leading to neurodegeneration in vivo. *Brain* **2018**, *141*, 1753–1769. [CrossRef] [PubMed]

85. Mikołajczyk, A. Safe and effective anaesthesiological protocols in domestic pig. *Ann. Warsaw Univ. Life Sci. SGGW Anim. Sci.* **2016**, *55*, 219–227.

86. Fu, H.Q.; Yang, T.; Xiao, W.; Fan, L.; Wu, Y.; Terrando, N.; Wang, T.L. Prolonged neuroinflammation after lipopolysaccharide exposure in aged rats. *PLoS ONE* **2014**, *9*. [CrossRef] [PubMed]

87. Lopes, P.C. LPS and neuroinflammation: A matter of timing. *Inflammopharmacology* **2016**, *24*, 291–293. [CrossRef] [PubMed]

88. Saar, L.I. Lymph Nodes of the Head, Neck and Shoulder Region of Swine. *Iowa State Univ. Vet. Digit. Respir.* **1962**, *25*, 120–134.

89. Jelsing, J.; Hay-Schmidt, A.; Dyrby, T.; Hemmingsen, R.; Uylings, H.B.M.; Pakkenberg, B. The prefrontal cortex in the Göttingen minipig brain defined by neural projection criteria and cytoarchitecture. *Brain Res. Bull.* **2006**, *70*, 322–336. [CrossRef] [PubMed]

90. Waters, W.R.; Sacco, R.E.; Dorn, A.D.; Hontecillas, R.; Zuckermann, F.A.; Wannemuehler, M.J. Systemic and mucosal immune responses of pigs to parenteral immunization with a pepsin-digested Serpulina hyodysenteriae bacterin. *Vet. Immunol. Immunopathol.* **1999**, *69*, 75–87. [CrossRef]

91. Mierzejewska, D.; Mitrowska, P.; Rudnicka, B.; Kubicka, E.; Kostyra, H. Effect of non-enzymatic glycosylation of pea albumins on their immunoreactive properties. *Food Chem.* **2008**, *111*, 127–131. [CrossRef]

92. Jun, S.M.; Ochoa-Repáraz, J.; Zlotkowska, D.; Hoyt, T.; Pascual, D.W. Bystander-mediated stimulation of proteolipid protein-specific regulatory T (Treg) cells confers protection against experimental autoimmune encephalomyelitis (EAE) via TGF-β. *J. Neuroimmunol.* **2012**, *245*, 39–47. [CrossRef] [PubMed]

93. Li, Q.; Wong, J.H.; Lu, G.; Antonio, G.E.; Yeung, D.K.; Ng, T.B.; Forster, L.E.; Yew, D.T. Gene expression of synaptosomal-associated protein 25 (SNAP-25) in the prefrontal cortex of the spontaneously hypertensive rat (SHR). *Biochim. Biophys. Acta* **2009**, *1792*, 766–776. [CrossRef] [PubMed]

94. Najmanová, V.; Rambousek, L.; Syslová, K.; Bubeníková, V.; Šlamberová, R.; Valeš, K.; Kačer, P. LC-ESI-MS-MS method for monitoring dopamine, serotonin and their metabolites in brain tissue. *Chromatographia* **2011**, *73*, 143–149. [CrossRef]

95. Conlon, J.M. Purification of naturally occurring peptides by reversed-phase HPLC. *Nat. Protoc.* **2007**, *2*, 191–197. [CrossRef] [PubMed]

International Journal of
Molecular Sciences

Review

Corneal Fibroblasts as Sentinel Cells and Local Immune Modulators in Infectious Keratitis

Ken Fukuda [1,*], Waka Ishida [1], Atsuki Fukushima [1] and Teruo Nishida [2,3]

1   Department of Ophthalmology, Kochi Medical School, Nankoku City 783-8505, Japan;
    wakai@kochi-u.ac.jp (W.I.); fukusima@kochi-u.ac.jp (A.F.)
2   Department of Ophthalmology, Yamaguchi University Graduate School of Medicine, Ube City,
    Yamaguchi 755-8505, Japan; tnishida@yamaguchi-u.ac.jp
3   Ohshima Eye Hospital, Fukuoka City 812-0036, Japan
*   Correspondence: k.fukuda@kochi-u.ac.jp; Tel.: +81-88-880-2391

Received: 31 July 2017; Accepted: 21 August 2017; Published: 23 August 2017

Abstract: The cornea serves as a barrier to protect the eye against external insults including microbial pathogens and antigens. Bacterial infection of the cornea often results in corneal melting and scarring that can lead to severe visual impairment. Not only live bacteria but also their components such as lipopolysaccharide (LPS) of Gram-negative bacteria contribute to the development of inflammation and subsequent corneal damage in infectious keratitis. We describe the important role played by corneal stromal fibroblasts (activated keratocytes) as sentinel cells, immune modulators, and effector cells in infectious keratitis. Corneal fibroblasts sense bacterial infection through Toll-like receptor (TLR)–mediated detection of a complex of LPS with soluble cluster of differentiation 14 (CD14) and LPS binding protein present in tear fluid. The cells then initiate innate immune responses including the expression of chemokines and adhesion molecules that promote the recruitment of inflammatory cells necessary for elimination of the infecting bacteria. Infiltrated neutrophils are activated by corneal stromal collagen and release mediators that stimulate the production of pro–matrix metalloproteinases by corneal fibroblasts. Elastase produced by *Pseudomonas aeruginosa* (*P. aeruginosa*) activates these released metalloproteinases, resulting in the degradation of stromal collagen. The modulation of corneal fibroblast activation and of the interaction of these cells with inflammatory cells and bacteria is thus important to minimize corneal scarring during treatment of infectious keratitis. Pharmacological agents that are able to restrain such activities of corneal fibroblasts without allowing bacterial growth represent a potential novel treatment option for prevention of excessive scarring and tissue destruction in the cornea.

Keywords: fibroblast; keratocyte; cornea; lipopolysaccharide; bacteria; chemokine; adhesion molecule; collagen; tear fluid

---

1. Introduction

The cornea is located on the external surface at the front of the eyeball and differs from most other tissues in that it is transparent and avascular, properties that allow it to contribute to ocular refraction. The cornea has a relatively simple structure consisting of three layers: the epithelium, stroma, and endothelium with each layer consisting of a different type of structural cell—epithelial cells, keratocytes, and endothelial cells, respectively. Given its location, the cornea is frequently exposed to external insults including microbes, antigens, and inflammatory mediators.

Worldwide, bacterial keratitis is a major cause of visual disturbance and blindness as a result of the corneal melting and scarring that occur if the infection is not treated promptly and appropriately. The Gram-negative bacterium, *Pseudomonas aeruginosa* is one of the most common isolates from individuals with microbial keratitis, especially those who use extended-wear contact lenses [1–3].

Not only live bacteria but also their components or products, including the lipopolysaccharide (LPS, also known as lipoglycan or endotoxin) of Gram-negative bacteria, are able to initiate keratitis and subsequent corneal damage [4]. Given that the cornea is an avascular tissue and contains few immune cells, corneal resident cells function as sentinel cells as well as immune modulators during corneal inflammation [5–7]. Whereas the corneal epithelium serves as an effective barrier to protect the eye from external agents, corneal stromal fibroblasts (activated keratocytes) play a key role in the recruitment of inflammatory cells into the cornea during acquired or innate immune responses. We previously showed that corneal fibroblasts, but not corneal epithelial cells, recognize the T helper 2 (Th2) cytokines interleukin (IL)-4 and IL-13 in tear fluid [8] and express the chemokines eotaxin (CCL11) [9,10] and thymus- and activation-regulated chemokine (CCL17) [11] as well as vascular cell adhesion molecule-1 in response to such recognition [9,12]. Corneal fibroblasts may thus play an important role in the corneal inflammation associated with severe ocular allergic diseases by promoting the recruitment of eosinophils (via CCL11) and Th2 cells (via CCL17) to the cornea [5,6].

Corneal fibroblasts also recognize bacterial components such as LPS through Toll-like receptors (TLRs) expressed on their surface and thereby activate appropriate innate immune responses in bacterial keratitis. In this review, we will address the role of corneal fibroblasts in LPS-related ocular surface inflammation.

2. Innate Immune Responses to LPS in Corneal Cells

LPS is a component of the cell membrane in Gram-negative bacteria and is a potent secretagogue for various cytokines produced by inflammatory cells. Among various TLRs, TLR4 recognizes LPS of Gram-negative bacteria [13]. Although human corneal epithelial cells have been shown to express TLR4, its role in such cells has been unclear. Whereas some studies found this receptor to be functional in corneal epithelial cells [14], others reported that it is expressed intracellularly and therefore does not confer sensitivity to LPS, with the consequent "immunosilent environment" preventing unnecessary responses to commensal bacteria [15]. Although LPS is anchored in the outer membrane of bacteria, it is released spontaneously during bacterial growth. LPS released into tear fluid is not able to stimulate stromal fibroblasts unless the barrier function of the corneal epithelium is compromised. Thus, it is thought to enter the corneal stroma by diffusion only at sites of epithelial abrasion [16].

Injection of LPS into the corneal stroma induced rapid infiltration of inflammatory cells including neutrophils and monocytic cells into the stroma and led to the development of corneal ulceration in rabbits [17,18]. On the basis of these observations, we hypothesized that stromal resident fibroblasts recognize the presence of LPS and trigger inflammatory cell infiltration through expression of chemokines and adhesion molecules. We found that human corneal fibroblasts express TLR4, cluster of differentiation (CD)14, and MD-2, that these proteins form functional LPS receptors [19], and that the cells initiate innate immune responses to LPS through activation of this receptor complex. Corneal fibroblasts produce cytokines and chemokines including IL-6, monocyte chemotactic protein (MCP) 1 (CCL2), and IL-8 (CXCL8), but not IL-1β or tumor necrosis factor-α (TNF-α), in response to LPS (Figure 1). LPS-stimulated corneal fibroblasts also express intercellular adhesion molecule (ICAM) 1 (also known as CD54) [19,20]. Given that the chemotactic activities of MCP-1 and IL-8 as well as adhesion to ICAM-1 expressed on the surface of structural cells mediate the local infiltration and activation of monocytes and neutrophils, the activation of corneal fibroblasts by direct stimulation with LPS may be an important step in the pathogenesis of bacterial infection in the cornea. The cytokine IL-1β plays an essential role in bacterial clearance as well as in the recruitment of neutrophils into the cornea during infectious keratitis [21,22]. However, IL-1β is secreted predominantly by infiltrated inflammatory cells—in particular, by neutrophils themselves—rather than by resident cells. The mechanisms of cleavage and activation of pro–IL-1β are thought to differ between murine models of *P. aeruginosa* keratitis and *Streptococcus pneumoniae* keratitis [21,22]. Infection with *P. aeruginosa* stimulates the production of IL-1β by cultured corneal fibroblasts in a manner that is largely dependent on the sensing of extracellular flagellin of the bacteria by TLR5 [23].

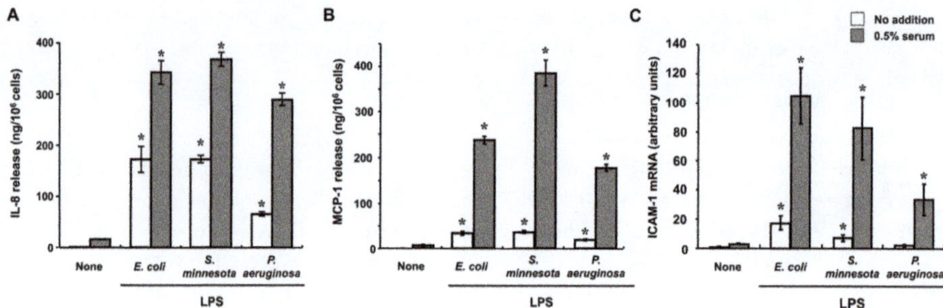

Figure 1. Effects of lipopolysaccharide (LPS) on chemokine release and the abundance of ICAM-1 mRNA in human corneal fibroblasts. Cells were incubated for 24 h (**A**,**B**) or 6 h (**C**) with or without LPS (10 ng/mL) from *Escherichia coli* (*E. coli*), *Salmonella minnesota* (*S. minnesota*), or *Pseudomonas aeruginosa* (*P. aeruginosa*) as well as in the absence (open bars) or presence (closed bars) of 0.5% human serum. The amounts of IL-8 (**A**) and MCP-1 (**B**) released into the culture medium were then determined by enzyme-linked immunosorbent assays, and the amount of ICAM-1 mRNA in the cells (**C**) was determined by reverse transcription and real-time polymerase chain reaction analysis. Data are means ± standard error of the mean (SEM) of quadruplicates from representative experiments. * $p < 0.05$ versus the corresponding value for cells incubated in the absence of LPS. Reprinted with permission from [19].

The various effects of LPS on corneal fibroblasts were found to be potentiated by the presence of a low concentration of human serum (Figure 1) [19], suggesting that factors in serum may contribute to the activation of these cells by LPS. Several serum proteins and lipids bind to LPS, and we found that two serum-derived soluble factors—LPS binding protein (LBP) and soluble CD14 (sCD14)—potentiate LPS-induced innate immune responses in corneal fibroblasts [24]. LPS is an amphipathic molecule with a small hydrophilic domain and a large hydrophobic component [25]. Lipid A, the lipophilic portion of LPS, is necessary for endotoxic activity and is highly conserved structurally [26]. In an aqueous environment such as tear fluid, LPS forms polymeric aggregates with the lipid A region facing inward and the hydrophilic polysaccharide component facing outward [25]. Polymeric forms of LPS bind poorly to cells and fail to provoke responses at low concentrations [27]. LBP is produced mostly by the liver and was initially identified as an acute-phase reactant in serum. LBP binds LPS and renders it monomeric, thereby exposing the active lipid A moiety. LBP thus facilitates detection of LPS by its receptors expressed on corneal fibroblasts [18,28,29]. CD14 exists in two forms: a glycophosphatidylinositol-anchored membrane-bound form (mCD14), and a soluble form. Various cell types including inflammatory cells as well as human corneal fibroblasts constitutively express mCD14 at the cell surface, whereas sCD14 forms a complex with LPS that is thought to bind to mCD14-negative cells and thereby to confer sensitivity to LPS. LPS-induced innate immune responses including the expression of chemokines and adhesion molecules in corneal fibroblasts are enhanced by the addition of either LBP or sCD14 (Figure 2) [24]. Although the cornea lacks blood vessels, soluble serum factors are present in the tear fluid that covers the ocular surface. We and others have shown that the tear fluid of healthy adults contains both sCD14 and LBP (Figure 3) [30,31] and our in vitro experiments suggest that they are present at concentrations sufficient to support maximal effects of LPS on innate immune responses in corneal fibroblasts.

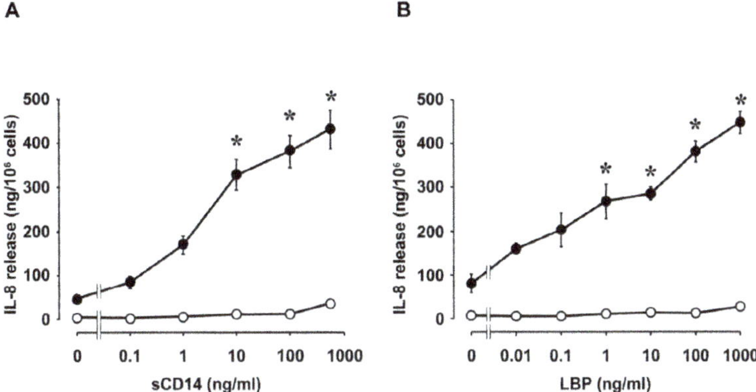

Figure 2. Concentration-dependent effects of soluble cluster of differentiation 14 (sCD14) and LPS binding protein (LBP) on IL-8 release by human corneal fibroblasts. Cells deprived of serum for 24 h were incubated for 24 h in the absence (open circles) or presence (closed circles) of LPS (10 ng/mL) and with the indicated concentrations of sCD14 (**A**) or LBP (**B**), after which the amount of IL-8 released into the culture medium was determined. Data are means ± SEM of triplicates from representative experiments. * $p < 0.01$ versus the corresponding value for cells incubated with LPS in the absence of sCD14 or LBP. Reprinted with permission from [24].

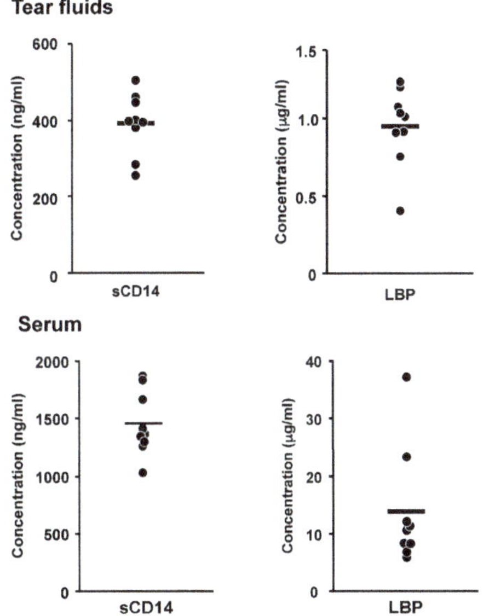

Figure 3. Concentrations of sCD14 and LBP in human tear fluid and serum from the same individuals. Circles, individual values; bars, mean values. Reprinted with permission from [31].

Taken together, these various observations suggest that sCD14 and LBP in tear fluid bind LPS and enhance the perception of LPS by corneal fibroblasts, thereby contributing to the first line of immune defense of the cornea against microorganisms.

The innate immune recognition of intracellular LPS by a TLR4-independent mechanism was recently uncovered [32,33]. The lipid A moiety of LPS, when present in the cytoplasm, was thus found to trigger noncanonical inflammasome activation that results in the activation of caspase-11, pyroptosis, and the proteolytic processing of pro-IL-1β and pro-IL-18. However, the presence of this intracellular LPS-sensing pathway and its potential role in corneal fibroblasts remains to be elucidated.

3. Therapeutic Intervention Targeting the Role of Corneal Fibroblasts in Infectious Keratitis

The treatment for individuals with bacterial keratitis is administration of appropriate antibiotics. However, LPS is rapidly released from bacteria as a consequence of antibiotic therapy [34], and once corneal fibroblasts have been activated by such released LPS, antibiotics are not able to influence the inflammatory responses of these cells. Although such responses by corneal fibroblasts are the first line of immune defense in the avascular cornea and are important for protection of the host from pathogens at the early stage of infection, persistent and excessive inflammatory responses result in the destruction of corneal tissue.

Keratocytes contribute to maintenance of corneal stromal structure by synthesizing and degrading stromal extracellular matrix proteins including collagen under physiological conditions. Such degradation of the stromal matrix is mediated by matrix metalloproteinases (MMPs) derived from the cells [35,36]. Under pathological conditions such as bacterial infection, the interaction of corneal fibroblasts with invading bacteria and infiltrated neutrophils leads to excessive degradation of stromal collagen. The results of in vitro experiments in which corneal fibroblasts are cultured in a three-dimensional collagen gel suggest that factors including IL-1 secreted by collagen-stimulated neutrophils augment collagen degradation by corneal fibroblasts through a stimulatory effect on pro-MMP synthesis [37]. During infection with *P. aeruginosa*, pseudomonal elastase both degrades type I collagen directly and promotes collagen degradation by corneal fibroblasts through the activation of pro-MMPs released from the fibroblasts [38]. The uncontrolled and prolonged activation of corneal fibroblasts by inflammatory cells or infectious pathogens may therefore lead to destruction of the corneal stroma and corneal scarring. Given the important role of the cornea in ocular refraction, such scarring of the cornea directly results in a loss of vision. It is therefore important that, regardless of the causative factors, corneal inflammation be treated in such a manner as to ensure minimal scarring. Given that infiltration of inflammatory cells into the avascular cornea is regulated by chemokines released by corneal fibroblasts and that stromal collagen degradation is mediated by MMPs produced by these cells, the targeting of stromal fibroblast function is a potential approach to the treatment of corneal inflammation [7]. Although corticosteroids are potent immunosuppressants and attenuate both collagen degradation by corneal fibroblasts as well as the infiltration of mononuclear cells into the cornea in a rabbit model of LPS-induced keratitis [17,39], steroids also promote the proliferation of infecting pathogens. There are currently no eyedrops clinically available for the treatment of infectious keratitis that are able to suppress an excessive inflammatory response without adverse effects.

Given that activation of nuclear factor–κB (NF-κB) is a key step in the LPS-induced expression of chemokines and adhesion molecules in corneal fibroblasts [19], drugs that are able to inhibit the NF-κB signaling pathway in these cells might be expected to limit the infiltration of immune cells into the cornea. Triptolide, which is present in extracts of the Chinese herb *Tripterygium wilfordii* hook f, possesses anti-inflammatory activity for various cell types including immune cells and tissue resident cells. Triptolide inhibits the activation of NF-κB and thereby attenuates both LPS-induced chemokine and adhesion molecule expression in as well as collagen degradation by human corneal fibroblasts [20,40].

Rebamipide eyedrops were recently introduced to the Japanese market for the treatment of dry eye on the basis of the mucin secretagogue activity of this drug. Rebamipide also manifests various anti-inflammatory effects on corneal epithelial cells and gastric epithelial cells. We recently showed that rebamipide increases the barrier function of human corneal epithelial cells, attenuates the loss of such barrier function induced by the proinflammatory cytokine TNF-α, and inhibits the TNF-α–induced expression of IL-6 and IL-8 in these cells [41]. Rebamipide also suppresses the LPS-induced synthesis of IL-8 through inhibition of NF-κB signaling in human corneal fibroblasts (Figure 4) [42].

Rebamipide may therefore prove effective for the treatment of not only dry eye–related epitheliopathy, but also corneal stromal inflammation associated with bacterial infection or allergy.

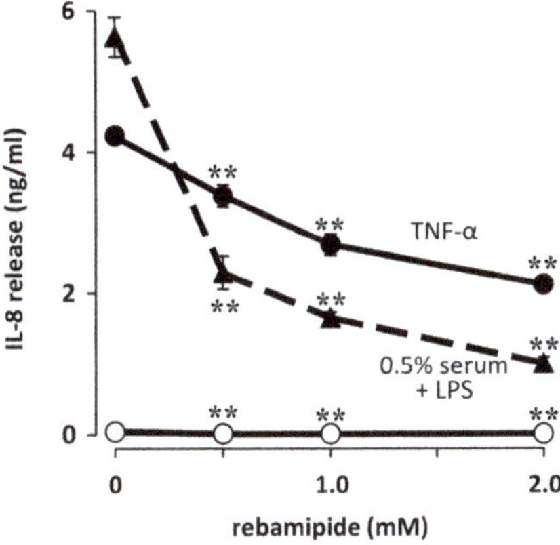

Figure 4. Effect of rebamipide on IL-8 release induced by LPS in human corneal fibroblasts. Cells were incubated first for 1 h with the indicated concentrations of rebamipide and then for 48 h in the additional absence (open circles) or presence either of LPS (100 ng/mL, triangles) plus 0.5% human serum or of tumor necrosis factor-α (TNF-α) (10 ng/mL, closed circles), after which the concentration of IL-8 in culture supernatants was determined. Data are means ± SEM of quadruplicates from a representative experiment. ** $p < 0.01$ versus the corresponding value for cells incubated without rebamipide. Reprinted with permission from [42].

A small peptide derived from human pancreatitis-associated protein [43] and an 11-amino acid peptide (RNPRGEEGGPW) derived from hepatocyte growth factor [44] were both found to inhibit the LPS-induced expression of chemokines such as IL-8 and MCP-1 as well as the adhesion molecule ICAM-1 in corneal fibroblasts by preventing NF-κB activation. These peptides also attenuated the corneal inflammation associated with LPS-induced keratitis in mice. In addition, an inhibitor of hyaluronic acid synthesis, 4-methylumbelliferone, as well as hyaluronic acid of high molecular weight were each shown to attenuate the LPS-induced up-regulation of the expression of inflammatory cytokines including IL-6 and IL-8 in rabbit corneal fibroblasts [45]. Furthermore, leukocyte infiltration into the cornea associated with LPS-induced keratitis was found to be restrained in mice deficient in urokinase-type plasminogen activator (uPA) compared with wild-type mice, and the LPS-induced production of both chemokines and MMP-9 was attenuated in corneal fibroblasts from the u-PA-deficient mice compared with those from wild-type mice. These results suggest that targeting of uPA in corneal fibroblasts may inhibit LPS-induced corneal inflammation through down-regulation of chemokine production [46].

Endogenous antimicrobial peptides play a key role in defense against infection at the ocular surface. These molecules are essentially small cationic peptides with broad-spectrum antimicrobial activity against bacteria, fungi, and viruses. Defensins and cathelicidins are two major categories of mammalian antimicrobial peptides and are present at the ocular surface [47,48]. At the ocular surface, α-defensins including human neutrophil peptides 1 to 3 are derived mostly from infiltrating neutrophils, whereas β-defensins such as human β-defensins 1 to 3 are synthesized and secreted by corneal or conjunctival epithelial cells. LL37 is the only member of the cathelicidin family in humans and is secreted by corneal epithelial cells and fibroblasts [49]. The expression of antimicrobial peptides in epithelial cells at the ocular surface is up-regulated in response to bacterial infection through TLRs. TLR2 activation by peptidoglycan or lipopeptide from *Staphylococcus aureus* enhanced the production of LL37 and human β-defensin-2 in corneal epithelial cells [50,51]. LPS from *P. aeruginosa* was also found to stimulate the expression of human β-defensin–2 in corneal and conjunctival epithelial cells via activation of TLR4 [52]. In addition, *P. aeruginosa* flagellin induced the production of human β-defensin–2 and LL37 by corneal epithelial cells through the activation of TLR5 [53]. Several peptides with antimicrobial properties have been tested in vitro as well as in in vivo models for their potential either alone or in combination with antibiotics to treat infectious keratitis [54].

Drugs that are able to suppress immune responses as well as kill bacteria are also good candidates for the treatment of infectious keratitis. In addition to their direct antimicrobial action, recent studies have revealed that some antimicrobial peptides have pleiotropic effects on various cell types [55]. Among such antimicrobial peptides present at the ocular surface in humans, LL37 and angiogenin have been found to have anti-inflammatory effects on corneal fibroblasts. LL37 reportedly stimulates corneal epithelial wound healing as well as cytokine synthesis by corneal epithelial cells [55]. We have shown that LL37 directly suppresses LPS-induced innate immune responses in corneal fibroblasts (Figure 5), in contrast to its action in corneal epithelial cells. LL37 thus inhibited the expression of IL-6, IL-8, and ICAM-1, as well as the activation of NF-κB, induced by LPS in corneal fibroblasts, whereas it did not attenuate such effects elicited by TNF-α. These inhibitory effects of LL37 on cytokine and adhesion molecule expression thus appeared not to be attributable to nonspecific suppression of NF-κB signaling but rather to be mediated by the binding of LL37 to LPS or to CD14 at the cell surface [56]. We have also shown that exogenous LL37 did not induce corneal inflammation but instead significantly suppressed LPS-induced keratitis in mice (Figure 6). Taken together, these observations suggest that administration of LL37 may be beneficial for the treatment of infectious keratitis on the basis of both its inhibitory effects on corneal fibroblasts and its antimicrobial and pro-wound healing activities.

Angiogenin is a pro-angiogenic molecule but also acts as an antimicrobial peptide in tear fluid [57,58]. In addition, angiogenin was shown to inhibit the LPS-induced production of IL-6, IL-8, MCP-1, MCP-2, and TNF-α by corneal fibroblasts [59]. Angiogenin may therefore also have a beneficial action in infectious keratitis as a result of its antimicrobial activity and its attenuation of innate immune responses of corneal fibroblasts.

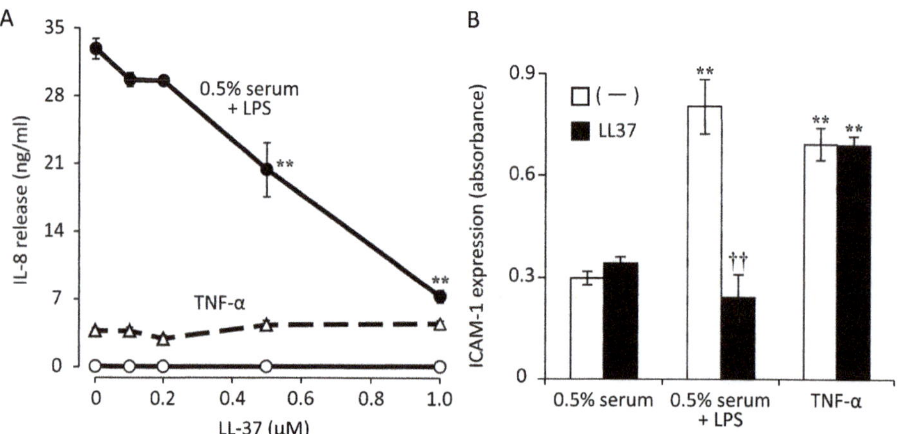

Figure 5. Effects of LL37 on LPS- or TNF-α–induced IL-8 and ICAM-1 expression in human corneal fibroblasts. Cells were incubated first for 2 h with the indicated concentrations of (**A**) or 1 μM (**B**) LL37 and then for 48 h (**A**) or 24 h (**B**) in the additional presence of LPS (100 ng/mL) plus 0.5% human serum (closed circles), of 0.5% human serum alone (open circles), or of TNF-α (10 ng/mL, triangles). The concentration of IL-8 in culture supernatants (**A**) and the cell surface expression of ICAM-1 (**B**) were then determined. ** $p < 0.01$ versus the corresponding value for cells incubated without LL37 (**A**). ** $p < 0.01$ versus the corresponding value for cells incubated with 0.5% serum alone; †† $p < 0.01$ versus the corresponding value for cells incubated without LL37 (**B**). Reprinted with permission from [56].

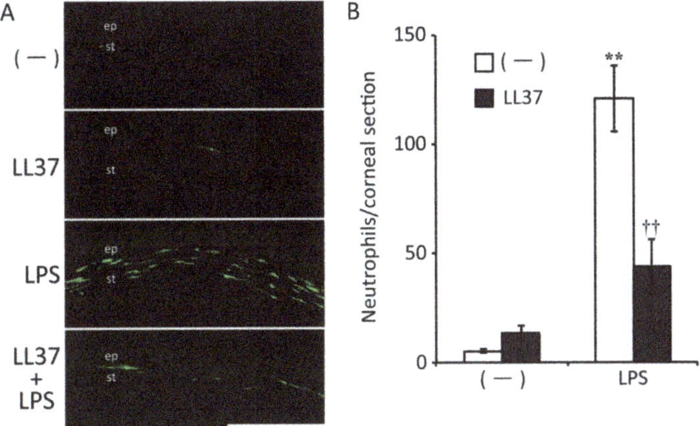

Figure 6. Effects of LL37 in a mouse model of LPS-induced keratitis. (**A**) The cornea was scratched and either LL37, LPS, both LPS and LL37, or phosphate-buffered saline (PBS) vehicle was applied. After 24 h, the eye was enucleated for immunohistofluorescence staining of neutrophils in the cornea. The corneal epithelium (ep) and stroma (st) are indicated. Scale bar, 200 μm; (**B**) the number of infiltrating neutrophils in the corneal stroma in images similar to those in (**A**) was counted. Data are means ± SEM for one section examined for each of four eyes. ** $p < 0.01$ versus the value for PBS alone; †† $p < 0.01$ versus the value for LPS alone. Reprinted with permission from [56].

4. Conclusions

The studies described in this review highlight the central role of corneal fibroblasts in the development of corneal inflammation during infectious keratitis (Figure 7). Corneal fibroblasts sense bacterial infection through the detection of LPS by TLR with the assistance of sCD14 and LBP in tear fluid. Such detection of LPS by corneal fibroblasts triggers innate immune responses including the expression of chemokines and adhesion molecules that promote the recruitment of inflammatory cells, mostly neutrophils, and the consequent elimination of infecting bacteria. Appropriate resolution of infectious inflammation may be prevented, however, by the prolonged overproduction of such inflammatory mediators by corneal fibroblasts, leading to tissue remodeling or destruction and, eventually, to corneal stromal scarring. Infiltrated neutrophils are activated by corneal stromal collagen and release factors including IL-1 that stimulate the production of pro-MMPs by corneal fibroblasts. Certain proteases such as elastase produced by *P. aeruginosa* activate these released pro-MMPs and thereby promote stromal collagen degradation. Corneal fibroblasts thus act as sentinel cells, immune modulators, and effector cells in infectious keratitis. Modulation of corneal fibroblast activation and of the interaction of these cells with inflammatory cells or infecting bacteria is therefore critical to minimize corneal scarring during treatment of infectious keratitis. Drugs that are able to restrain these activities of corneal fibroblasts are needed to expand and improve the treatment options available for infectious keratitis. Pharmacological agents described in this review may provide such novel treatment options to prevent excessive scarring and tissue destruction in the cornea by regulating corneal fibroblast function.

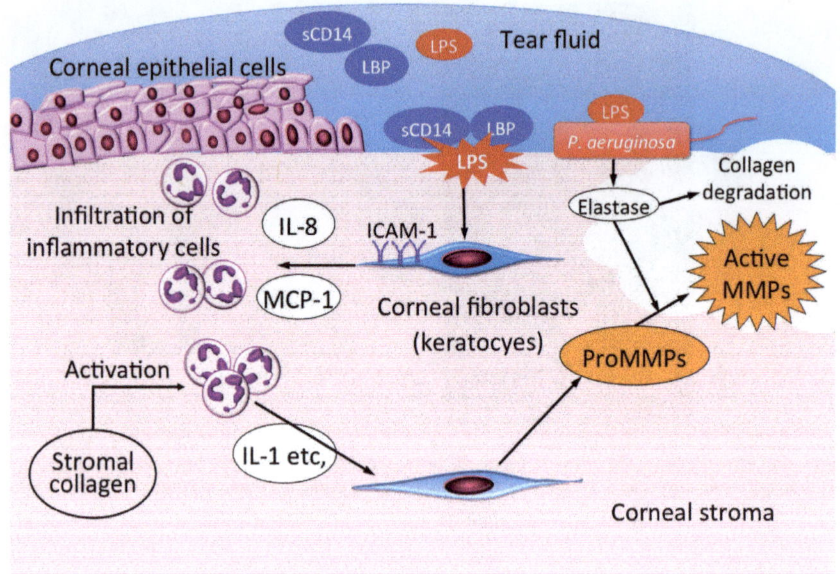

Figure 7. Role of corneal fibroblasts in bacterial keratitis. LPS released from bacteria binds to LBP and sCD14 in tear fluid, and the LPS-LBP-sCD14 complex then activates corneal fibroblasts. The activated fibroblasts promote inflammatory cell recruitment through expression of chemokines and adhesion molecules. Infiltrated neutrophils are activated by corneal stromal collagen and secrete inflammatory mediators including IL-1 that stimulate pro-MMP secretion by fibroblasts. The released pro-MMPs are activated by bacterial proteases such as *Pseudomonas aeruginosa* elastase, and the active MMPs then degrade stromal collagen, leading to corneal melting.

Conflicts of Interest: The authors declare no conflict of interest.

Abbreviations

LPS	Lipopolysaccharide
Th2	T helper 2
IL	Interleukin
TLR	Toll-like receptor
CD	Cluster of differentiation
MCP	Monocyte chemotactic protein
TNF-α	Tumor necrosis factor-α
ICAM-1	Intercellular adhesion molecule-1
LBP	Lipopolysaccharide binding protein
sCD14	Soluble CD14
mCD14	Membrane-bound CD14
MMP	Matrix metalloproteinase
NF-κB	Nuclear factor-κB
uPA	Urokinase-type plasminogen activator

References

1. Green, M.; Apel, A.; Stapleton, F. Risk factors and causative organisms in microbial keratitis. *Cornea* **2008**, *27*, 22–27. [CrossRef] [PubMed]
2. Hazlett, L.D. Bacterial infections of the cornea (*Pseudomonas aeruginosa*). *Chem. Immunol. Allergy* **2007**, *92*, 185–194. [CrossRef] [PubMed]

3. Pachigolla, G.; Blomquist, P.; Cavanagh, H.D. Microbial keratitis pathogens and antibiotic susceptibilities: A 5-year review of cases at an urban county hospital in north Texas. *Eye Contact Lens* **2007**, *33*, 45–49. [CrossRef] [PubMed]

4. Pearlman, E.; Sun, Y.; Roy, S.; Karmakar, M.; Hise, A.G.; Szczotka-Flynn, L.; Ghannoum, M.; Chinnery, H.R.; McMenamin, P.G.; Rietsch, A. Host defense at the ocular surface. *Int. Rev. Immunol.* **2013**, *32*, 4–18. [CrossRef] [PubMed]

5. Fukuda, K.; Kumagai, N.; Fujitsu, Y.; Nishida, T. Fibroblasts as local immune modulators in ocular allergic disease. *Allergol. Int.* **2006**, *55*, 121–129. [CrossRef] [PubMed]

6. Kumagai, N.; Fukuda, K.; Fujitsu, Y.; Yamamoto, K.; Nishida, T. Role of structural cells of the cornea and conjunctiva in the pathogenesis of vernal keratoconjunctivitis. *Prog. Retin. Eye Res.* **2006**, *25*, 165–187. [CrossRef] [PubMed]

7. Nishida, T. Commanding roles of keratocytes in health and disease. *Cornea* **2010**, *29* (Suppl. S1), S3–S6. [CrossRef] [PubMed]

8. Fukuda, K.; Fujitsu, Y.; Kumagai, N.; Nishida, T. Characterization of the interleukin-4 receptor complex in human corneal fibroblasts. *Investig. Ophthalmol. Vis. Sci.* **2002**, *43*, 183–188.

9. Fukuda, K.; Nishida, T.; Fukushima, A. Synergistic induction of eotaxin and VCAM-1 expression in human corneal fibroblasts by staphylococcal peptidoglycan and either IL-4 or IL-13. *Allergol. Int.* **2011**, *60*, 355–363. [CrossRef] [PubMed]

10. Kumagai, N.; Fukuda, K.; Ishimura, Y.; Nishida, T. Synergistic induction of eotaxin expression in human keratocytes by TNF-α and IL-4 or IL-13. *Investig. Ophthalmol. Vis. Sci.* **2000**, *41*, 1448–1453.

11. Fukuda, K.; Fujitsu, Y.; Seki, K.; Kumagai, N.; Nishida, T. Differential expression of thymus- and activation-regulated chemokine (CCL17) and macrophage-derived chemokine (CCL22) by human fibroblasts from cornea, skin, and lung. *J. Allergy Clin. Immunol.* **2003**, *111*, 520–526. [CrossRef] [PubMed]

12. Kumagai, N.; Fukuda, K.; Fujitsu, Y.; Nishida, T. Synergistic effect of TNF-α and either IL-4 or IL-13 on VCAM-1 expression by cultured human corneal fibroblasts. *Cornea* **2003**, *22*, 557–561. [CrossRef] [PubMed]

13. Beutler, B. Tlr4: Central component of the sole mammalian LPS sensor. *Curr. Opin. Immunol.* **2000**, *12*, 20–26. [CrossRef]

14. Song, P.I.; Abraham, T.A.; Park, Y.; Zivony, A.S.; Harten, B.; Edelhauser, H.F.; Ward, S.L.; Armstrong, C.A.; Ansel, J.C. The expression of functional LPS receptor proteins CD14 and toll-like receptor 4 in human corneal cells. *Investig. Ophthalmol. Vis. Sci.* **2001**, *42*, 2867–2877.

15. Ueta, M.; Nochi, T.; Jang, M.H.; Park, E.J.; Igarashi, O.; Hino, A.; Kawasaki, S.; Shikina, T.; Hiroi, T.; Kinoshita, S.; et al. Intracellularly expressed TLR2s and TLR4s contribution to an immunosilent environment at the ocular mucosal epithelium. *J. Immunol.* **2004**, *173*, 3337–3347. [CrossRef] [PubMed]

16. Schultz, C.L.; Buret, A.G.; Olson, M.E.; Ceri, H.; Read, R.R.; Morck, D.W. Lipopolysaccharide entry in the damaged cornea and specific uptake by polymorphonuclear neutrophils. *Infect. Immun.* **2000**, *68*, 1731–1734. [CrossRef] [PubMed]

17. Howes, E.L.; Cruse, V.K.; Kwok, M.T. Mononuclear cells in the corneal response to endotoxin. *Investig. Ophthalmol. Vis. Sci.* **1982**, *22*, 494–501.

18. Schultz, C.L.; Morck, D.W.; McKay, S.G.; Olson, M.E.; Buret, A. Lipopolysaccharide induced acute red eye and corneal ulcers. *Exp. Eye Res.* **1997**, *64*, 3–9. [CrossRef] [PubMed]

19. Kumagai, N.; Fukuda, K.; Fujitsu, Y.; Lu, Y.; Chikamoto, N.; Nishida, T. Lipopolysaccharide-induced expression of intercellular adhesion molecule-1 and chemokines in cultured human corneal fibroblasts. *Investig. Ophthalmol. Vis. Sci.* **2005**, *46*, 114–120. [CrossRef] [PubMed]

20. Lu, Y.; Liu, Y.; Fukuda, K.; Nakamura, Y.; Kumagai, N.; Nishida, T. Inhibition by triptolide of chemokine, proinflammatory cytokine, and adhesion molecule expression induced by lipopolysaccharide in corneal fibroblasts. *Investig. Ophthalmol. Vis. Sci.* **2006**, *47*, 3796–3800. [CrossRef] [PubMed]

21. Karmakar, M.; Katsnelson, M.; Malak, H.A.; Greene, N.G.; Howell, S.J.; Hise, A.G.; Camilli, A.; Kadioglu, A.; Dubyak, G.R.; Pearlman, E. Neutrophil IL-1β processing induced by pneumolysin is mediated by the NLRP3/ASC inflammasome and caspase-1 activation and is dependent on K⁺ efflux. *J. Immunol.* **2015**, *194*, 1763–1775. [CrossRef] [PubMed]

22. Karmakar, M.; Sun, Y.; Hise, A.G.; Rietsch, A.; Pearlman, E. L-1β processing during Pseudomonas aeruginosa infection is mediated by neutrophil serine proteases and is independent of NLRC4 and caspase-1. *J. Immunol.* **2012**, *189*, 4231–4235. [CrossRef] [PubMed]

23. Cendra, M.D.M.; Christodoulides, M.; Hossain, P. Signaling Mediated by Toll-Like Receptor 5 Sensing of Pseudomonas aeruginosa Flagellin Influences IL-1β and IL-18 Production by Primary Fibroblasts Derived from the Human Cornea. *Front. Cell. Infect. Microbiol.* **2017**, *7*, 130. [CrossRef] [PubMed]

24. Fukuda, K.; Kumagai, N.; Yamamoto, K.; Fujitsu, Y.; Chikamoto, N.; Nishida, T. Potentiation of lipopolysaccharide-induced chemokine and adhesion molecule expression in corneal fibroblasts by soluble CD14 or LPS-binding protein. *Investig. Ophthalmol. Vis. Sci.* **2005**, *46*, 3095–3101. [CrossRef] [PubMed]

25. Seydel, U.; Labischinski, H.; Kastowsky, M.; Brandenburg, K. Phase behavior, supramolecular structure, and molecular conformation of lipopolysaccharide. *Immunobiology* **1993**, *187*, 191–211. [CrossRef]

26. Galanos, C.; Luderitz, O.; Rietschel, E.T.; Westphal, O.; Brade, H.; Brade, L.; Freudenberg, M.; Schade, U.; Imoto, M.; Yoshimura, H.; et al. Synthetic and natural *Escherichia coli* free lipid A express identical endotoxic activities. *Eur. J. Biochem.* **1985**, *148*, 1–5. [CrossRef] [PubMed]

27. Hailman, E.; Lichenstein, H.S.; Wurfel, M.M.; Miller, D.S.; Johnson, D.A.; Kelley, M.; Busse, L.A.; Zukowski, M.M.; Wright, S.D. Lipopolysaccharide (LPS)-binding protein accelerates the binding of LPS to CD14. *J. Exp. Med.* **1994**, *179*, 269–277. [CrossRef] [PubMed]

28. Gioannini, T.L.; Teghanemt, A.; Zhang, D.; Levis, E.N.; Weiss, J.P. Monomeric endotoxin: Protein complexes are essential for TLR4-dependent cell activation. *J. Endotoxin Res.* **2005**, *11*, 117–123. [CrossRef] [PubMed]

29. Wright, S.D.; Ramos, R.A.; Tobias, P.S.; Ulevitch, R.J.; Mathison, J.C. CD14, a receptor for complexes of lipopolysaccharide (LPS) and LPS binding protein. *Science* **1990**, *249*, 1431–1433. [CrossRef] [PubMed]

30. Blais, D.R.; Vascotto, S.G.; Griffith, M.; Altosaar, I. LBP and CD14 secreted in tears by the lacrimal glands modulate the LPS response of corneal epithelial cells. *Investig. Ophthalmol. Vis. Sci.* **2005**, *46*, 4235–4244. [CrossRef] [PubMed]

31. Fukuda, K.; Kumagai, N.; Nishida, T. Levels of soluble CD14 and lipopolysaccharide-binding protein in human basal tears. *Jpn. J. Ophthalmol.* **2010**, *54*, 241–242. [CrossRef] [PubMed]

32. Kayagaki, N.; Wong, M.T.; Stowe, I.B.; Ramani, S.R.; Gonzalez, L.C.; Akashi-Takamura, S.; Miyake, K.; Zhang, J.; Lee, W.P.; Muszynski, A.; et al. Noncanonical inflammasome activation by intracellular LPS independent of TLR4. *Science* **2013**, *341*, 1246–1249. [CrossRef] [PubMed]

33. Hagar, J.A.; Powell, D.A.; Aachoui, Y.; Ernst, R.K.; Miao, E.A. Cytoplasmic LPS activates caspase-11: Implications in TLR4-independent endotoxic shock. *Science* **2013**, *341*, 1250–1253. [CrossRef] [PubMed]

34. Hurley, J.C. Antibiotic-induced release of endotoxin: A reappraisal. *Clin. Infect. Dis.* **1992**, *15*, 840–854. [CrossRef] [PubMed]

35. Hao, J.L.; Nagano, T.; Nakamura, M.; Kumagai, N.; Mishima, H.; Nishida, T. Galardin inhibits collagen degradation by rabbit keratocytes by inhibiting the activation of pro-matrix metalloproteinases. *Exp. Eye Res.* **1999**, *68*, 565–572. [CrossRef] [PubMed]

36. Yamamoto, K.; Kumagai, N.; Fukuda, K.; Fujitsu, Y.; Nishida, T. Activation of corneal fibroblast-derived matrix metalloproteinase-2 by tryptase. *Curr. Eye Res.* **2006**, *31*, 313–317. [CrossRef] [PubMed]

37. Li, Q.; Fukuda, K.; Lu, Y.; Nakamura, Y.; Chikama, T.; Kumagai, N.; Nishida, T. Enhancement by neutrophils of collagen degradation by corneal fibroblasts. *J. Leukoc. Biol.* **2003**, *74*, 412–419. [CrossRef] [PubMed]

38. Nagano, T.; Hao, J.L.; Nakamura, M.; Kumagai, N.; Abe, M.; Nakazawa, T.; Nishida, T. Stimulatory effect of pseudomonal elastase on collagen degradation by cultured keratocytes. *Investig. Ophthalmol. Vis. Sci.* **2001**, *42*, 1247–1253.

39. Lu, Y.; Fukuda, K.; Liu, Y.; Kumagai, N.; Nishida, T. Dexamethasone inhibition of IL-1-induced collagen degradation by corneal fibroblasts in three-dimensional culture. *Investig. Ophthalmol. Vis. Sci.* **2004**, *45*, 2998–3004. [CrossRef] [PubMed]

40. Lu, Y.; Fukuda, K.; Seki, K.; Nakamura, Y.; Kumagai, N.; Nishida, T. Inhibition by triptolide of IL-1-induced collagen degradation by corneal fibroblasts. *Investig. Ophthalmol. Vis. Sci.* **2003**, *44*, 5082–5088. [CrossRef]

41. Tanaka, H.; Fukuda, K.; Ishida, W.; Harada, Y.; Sumi, T.; Fukushima, A. Rebamipide increases barrier function and attenuates TNFα-induced barrier disruption and cytokine expression in human corneal epithelial cells. *Br. J. Ophthalmol.* **2013**, *97*, 912–916. [CrossRef] [PubMed]

42. Fukuda, K.; Ishida, W.; Tanaka, H.; Harada, Y.; Fukushima, A. Inhibition by rebamipide of cytokine-induced or lipopolysaccharide-induced chemokine synthesis in human corneal fibroblasts. *Br. J. Ophthalmol.* **2014**, *98*, 1751–1755. [CrossRef] [PubMed]

43. Zhu, S.; Xu, X.; Liu, K.; Gu, Q.; Yang, X. PAPep, a small peptide derived from human pancreatitis-associated protein, attenuates corneal inflammation in vivo and in vitro through the IKKα/β/IκBα/NF-κB signaling pathway. *Pharmacol. Res.* **2015**, *102*, 113–122. [CrossRef] [PubMed]

44. Zhu, S.; Xu, X.; Wang, L.; Su, L.; Gu, Q.; Wei, F.; Liu, K. Inhibitory effect of a novel peptide, H-RN, on keratitis induced by LPS or poly(I:C) in vitro and in vivo via suppressing NF-κB and MAPK activation. *J. Transl. Med.* **2017**, *15*, 20. [CrossRef] [PubMed]

45. Li, F.; Hao, P.; Liu, G.; Wang, W.; Han, R.; Jiang, Z.; Li, X. Effects of 4-methylumbelliferone and high molecular weight hyaluronic acid on the inflammation of corneal stromal cells induced by LPS. *Graefes Arch. Clin. Exp. Ophthalmol.* **2017**, *255*, 559–566. [CrossRef] [PubMed]

46. Sugioka, K.; Kodama, A.; Yoshida, K.; Okada, K.; Mishima, H.; Aomatsu, K.; Matsuo, O.; Shimomura, Y. The roles of urokinase-type plasminogen activator in leukocyte infiltration and inflammatory responses in mice corneas treated with lipopolysaccharide. *Investig. Ophthalmol. Vis. Sci.* **2014**, *55*, 5338–5350. [CrossRef] [PubMed]

47. McDermott, A.M. Defensins and other antimicrobial peptides at the ocular surface. *Ocul. Surf.* **2004**, *2*, 229–247. [CrossRef]

48. Mohammed, I.; Said, D.G.; Dua, H.S. Human antimicrobial peptides in ocular surface defense. *Prog. Retin. Eye Res.* **2017**. [CrossRef] [PubMed]

49. Castaneda-Sanchez, J.I.; Garcia-Perez, B.E.; Munoz-Duarte, A.R.; Baltierra-Uribe, S.L.; Mejia-Lopez, H.; Lopez-Lopez, C.; Bautista-De Lucio, V.M.; Robles-Contreras, A.; Luna-Herrera, J. Defensin production by human limbo-corneal fibroblasts infected with mycobacteria. *Pathogens* **2013**, *2*, 13–32. [CrossRef] [PubMed]

50. Kumar, A.; Zhang, J.; Yu, F.S. Toll-like receptor 2-mediated expression of β-defensin-2 in human corneal epithelial cells. *Microbes Infect.* **2006**, *8*, 380–389. [CrossRef] [PubMed]

51. Li, Q.; Kumar, A.; Gui, J.F.; Yu, F.S. Staphylococcus aureus lipoproteins trigger human corneal epithelial innate response through toll-like receptor-2. *Microb. Pathog.* **2008**, *44*, 426–434. [CrossRef] [PubMed]

52. Mc, N.N.; Van, R.; Tuchin, O.S.; Fleiszig, S.M. Ocular surface epithelia express mRNA for human β defensin-2. *Exp. Eye Res.* **1999**, *69*, 483–490. [CrossRef]

53. Zhang, J.; Xu, K.; Ambati, B.; Yu, F.S. Toll-like receptor 5-mediated corneal epithelial inflammatory responses to Pseudomonas aeruginosa flagellin. *Investig. Ophthalmol. Vis. Sci.* **2003**, *44*, 4247–4254. [CrossRef]

54. Brandt, C.R. Peptide therapeutics for treating ocular surface infections. *J. Ocul. Pharmacol. Ther.* **2014**, *30*, 691–699. [CrossRef] [PubMed]

55. Huang, L.C.; Petkova, T.D.; Reins, R.Y.; Proske, R.J.; McDermott, A.M. Multifunctional roles of human cathelicidin (LL-37) at the ocular surface. *Investig. Ophthalmol. Vis. Sci.* **2006**, *47*, 2369–2380. [CrossRef] [PubMed]

56. Ishida, W.; Harada, Y.; Fukuda, K.; Fukushima, A. Inhibition by the Antimicrobial Peptide LL37 of Lipopolysaccharide-Induced Innate Immune Responses in Human Corneal Fibroblasts. *Investig. Ophthalmol. Vis. Sci.* **2016**, *57*, 30–39. [CrossRef]

57. Sack, R.A.; Conradi, L.; Krumholz, D.; Beaton, A.; Sathe, S.; Morris, C. Membrane array characterization of 80 chemokines, cytokines, and growth factors in open- and closed-eye tears: Angiogenin and other defense system constituents. *Investig. Ophthalmol. Vis. Sci.* **2005**, *46*, 1228–1238. [CrossRef] [PubMed]

58. Hooper, L.V.; Stappenbeck, T.S.; Hong, C.V.; Gordon, J.I. Angiogenins: A new class of microbicidal proteins involved in innate immunity. *Nat. Immunol.* **2003**, *4*, 269–273. [CrossRef] [PubMed]

59. Lee, S.H.; Kim, K.W.; Joo, K.; Kim, J.C. Angiogenin ameliorates corneal opacity and neovascularization via regulating immune response in corneal fibroblasts. *BMC Ophthalmol.* **2016**, *16*, 57. [CrossRef] [PubMed]

International Journal of
Molecular Sciences

Article

Distribution and Neurochemistry of the Porcine Ileocaecal Valve Projecting Sensory Neurons in the Dorsal Root Ganglia and the Influence of Lipopolysaccharide from Different Serotypes of *Salmonella* spp. on the Chemical Coding of DRG Neurons in the Cell Cultures

Anita Mikołajczyk [1,*]**, Anna Kozłowska** [2] **and Sławomir Gonkowski** [3]

[1]   Department of Public Health, Faculty of Health Sciences, Collegium Medicum, University of Warmia and Mazury in Olsztyn, Warszawska 30 Str., 10-082 Olsztyn, Poland

[2]   Department of Human Physiology, School of Medicine, Collegium Medicum, University of Warmia and Mazury in Olsztyn, Warszawska 30 Str., 10-082 Olsztyn, Poland; kozlowska.anna@uwm.edu.pl

[3]   Department of Clinical Physiology, Faculty of Veterinary Medicine, University of Warmia and Mazury in Olsztyn, Oczapowskiego 13 Str., 10-718 Olsztyn, Poland; slawekg@uwm.edu.pl

*   Correspondence: anita.mikolajczyk@uwm.edu.pl

Received: 3 August 2018; Accepted: 25 August 2018; Published: 28 August 2018

Abstract: The ileocecal valve (ICV)—a sphincter muscle between small and large intestine—plays important roles in the physiology of the gastrointestinal (GI) tract, but many aspects connected with the innervation of the ICV remain unknown. Thus, the aim of this study was to investigate the localization and neurochemical characterization of neurons located in the dorsal root ganglia and supplying the ICV of the domestic pig. The results have shown that such neurons mainly located in the dorsal root ganglia (DRG) of thoracic and lumbar neuromers show the presence of substance P (SP), calcitonin gene-related peptide (CGRP), and galanin (GAL). The second part of the experiment consisted of a study on the influence of a low dose of lipopolysaccharide (LPS) from *Salmonella* serotypes Enteritidis Minnesota and Typhimurium on DRG neurons. It has been shown that the LPS of these serotypes in studied doses does not change the number of DRG neurons in the cell cultures, but influences the immunoreactivity to SP and GAL. The observed changes in neurochemical characterization depend on the bacterial serotype. The results show that DRG neurons take part in the innervation of the ICV and may change their neurochemical characterization under the impact of LPS, which is probably connected with direct actions of this substance on the nervous tissue and/or its pro-inflammatory activity.

Keywords: ileocecal valve (ICV); LPS from *S.* Enteritidis; LPS from *S.* Minnesota; LPS from *S.* Typhimurium; neuropeptides of DRG

---

1. Introduction

The innervation of the gastrointestinal (GI) tract consists of two parts: the intestinal nervous system (ENS) and extrinsic innervation [1–4]. The ENS, located in the wall of the oesophagus, stomach, and intestine, is built of millions of neuronal cells divided into ganglionated plexuses, whose quantity depends on animal species [5]. Enteric neurons are very diverse in their morphology, physiology, and neurochemical characteristics and regulate the majority of the functions of the GI tract [1,5]. Due to the significant degree of the independence from the brain, the ENS is called the "second" or "intestinal" brain [6]. However, despite a high degree of autonomy, the ENS is, to some extent, controlled by the central nervous system by the extrinsic innervation of the GI tract. This innervation consists of three fundamental parts, including parasympathetic efferent innervation, sympathetic efferent innervation and afferent innervation [7,8]. The precise localization of neurons participating in the extrinsic innervation of the stomach and intestine clearly depends on innervated segment of the GI tract. In the case of the parasympathetic nervous system, the major part of the GI tract (from oesophagus to transverse colon) is innervated by fibres which are branches of the vagal nerve [7]. Only caudal fragments of the GI tract (descending colon, rectum and anus) are supplied by nerves originating from the parasympathetic nuclei within the intermediolateral column of the sacral spinal cord [9]. In turn, sympathetic neurons innervating the GI tract may be located in the sympathetic chain ganglia and prevertebral ganglia, including the celiac, superior and inferior mesenteric, or pelvic ganglia [2,3,7]. Apart from sympathetic and parasympathetic extrinsic innervation, the GI tract is also supplied by afferent nerves which conduct sensory and pain stimuli from the stomach and intestine to the central nervous system. These nerves are the processes of neuronal cells situated in the nodose ganglia of the vagal nerve or dorsal root ganglia [8]. Previous studies have shown that sensory neurons supplying the gut may contain a wide range of neuronal active substances [8,10]. Among them, the most important factors in sensory and pain stimuli conduction seem to be substance P (SP) and calcitonin gene-related peptide (CGRP). Moreover, it is known that sensory neuronal cells innervating the GI tract may undergo neurochemical changes during pathological processes taking part in the stomach and intestine [8–10], but knowledge concerning these aspects is extremely limited.

The ileocecal valve (ICV), a sphincter muscle between small and large intestine, prevents the reflux of colonic content to the ileum and serves as a barrier to the entering of colonic microbial flora to the small intestine [11–13]. It should be underlined that previous studies on the innervation of the ICV are few [12] and the sensory neurons innervating this part of the GI tract have not been studied at all. On the other hand, it is known that the organization of the ICV innervation is specific [12,14] and disturbances within it may play important roles in disorders of this part of the GI tract, including constipation, feeling bloated, and/or diarrhoea [12,15]. The knowledge of why the ICV can become impaired, which can lead to many pathological processes such as small intestine bacterial overgrowth (SIBO), is still unclear. This assumes that it is rather the composition of the bacterial species that is more crucial than the number of bacteria. Maybe interaction of some bacteria or some bacterial toxins with the nervous system can also play a role in ICV dysfunction. Due to the above, the study of the interaction between sensory neurons supplying ICV and lipopolysaccharide (LPS) from *Salmonella* spp. seems very interesting.

It should be pointed out that the ileum and cecum were the main sites of *Salmonella* growth in a latent carrier mouse. Thus, parts of the intestine can be strategic places for *Salmonella* proliferation in animals without showing any clinical symptoms of disease during latent infection [16]. Additionally, LPS can be derived from the dead *Salmonella* existing in this part of the intestine. It is known that LPS is present within the cellular membrane of all Gram-negative bacteria and demonstrates negative activity on living organisms [17]. Namely, this substance damages various internal organs, which is connected with the release of free radicals. Moreover, LPS acts on the immunological system, causing fever and septic shock [18]. Previous studies have also shown that LPS may affect the nervous system and is involved in, or connected with, neurodegenerative diseases. However, apart from the fact that LPS may change the expression of the neuronal factor in the internal organs [19–21], knowledge concerning changes in neurochemical characterization of neuronal cells under the influence of this substance is lacking. Moreover, it is known that LPS is not a homogeneous substance [22–26]. Additionally, LPS from different species are characterized by various activities [23,27,28] but, until now, the studies on differences in LPS activity derived from various serotypes connected with the influence of this substance on neuronal neurochemical characterization have not yet been studied.

It should be also underlined that the selection of the domestic pig as an experimental animal during this investigation was not accidental. It is relatively well known that the domestic pig is very good animal model of processes occurring in the human organism, due to well confirmed biochemical, neurochemical, and physiological similarities between these species [29]. Thus, the results obtained during this study may reflect the mechanisms connected with the influence of LPS on the human nervous system. The aim of these studies was to investigate the distribution and neurochemistry of the DRG sensory neurons supplying the ICV under physiological conditions Moreover, the influence of a low dose of LPS from various serotypes of *Salmonella* on the neurochemical characterization of neurons located in DRG during the cell cultures has been also studied.

2. Results

During the present study, neuronal cells supplying the ICV were observed in bilateral DRG from neuromers Th7 to L4, and the differences between right and left DRG were not very clear (Figure 1).

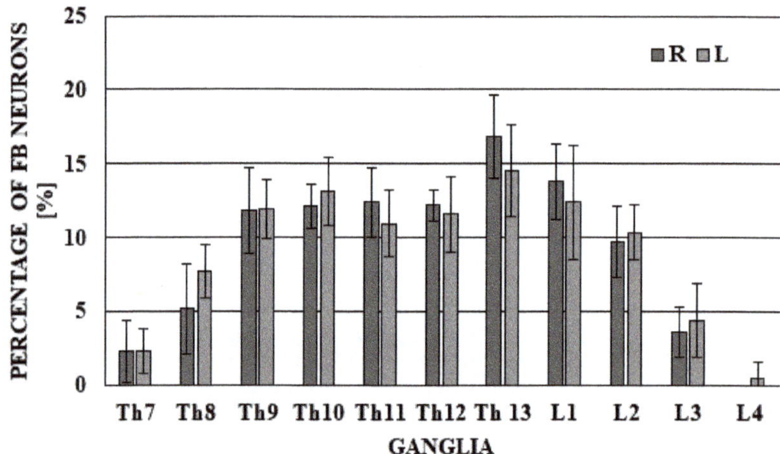

Figure 1. The distribution of fast blue (FB)-labelled neurons supplying the porcine ileocecal valve located in dorsal root ganglia (DRG). The total number of FB-positive neurons located in left and right DRG was considered as 100% (separately for left and right DRG).

The largest number of FB-positive cells was located in neuromers Th13 (16.88 ± 2.81% of all FB+ neurons) in the right DRG and 14.54 ± 3.07% in the left DRG). A slightly lower number of neuronal cells supplying the ICV was noted in neuromers Th9–Th12 (from 11.82 ± 2.90% to 12.41 ± 2.36% in right DRG and from 10.95 ± 2.26% to 13.16 ± 2.32% in left DRG) and L1–L2, respectively, 13.81 ± 2.58% and 9.70 ± 2.41% in right DRG, and 12.42 ± 3.85% and 10.37 ± 1.87% in left DRG). Significantly fewer neurons were present in neuromers Th7–Th8 (2.29 ± 2.10% and 5.18 ± 3.10% in right DRG and 2.34 ± 1.51% and 7.76 ± 1.81% in left DRG) and L3 (3.62 ± 1.71% in right DRG and 4.45 ± 2.51% in left DRG), and within the neuromer L4, FB+ neuronal cells were observed only sporadically (FB+ neurons were observed only in left DRG in quantities of 0.51 ± 1.15%). The total average number of FB+ neurons investigated in one animal amounted to 219.00 ± 67.30 (108.40 ± 35.98 in right DRG and 110.60 ± 32.24 in left DRG)

The highest number of sensory neurons supplying the ICV showed immunoreactivity to CGRP. The percentage of CGRP-positive cells amounted to 57.33 ± 6.89% of all FB+ cells (Figure 2A). The presence of SP was observed in a slightly lower percentage of the ICV projecting neurons (Figure 2B). This value reached 50.10 ± 8.71%. In turn, neurons immunoreactive to GAL were the least numerous. GAL was noted in 39.52 ± 5.33% of all FB-positive cells (Figure 3C).

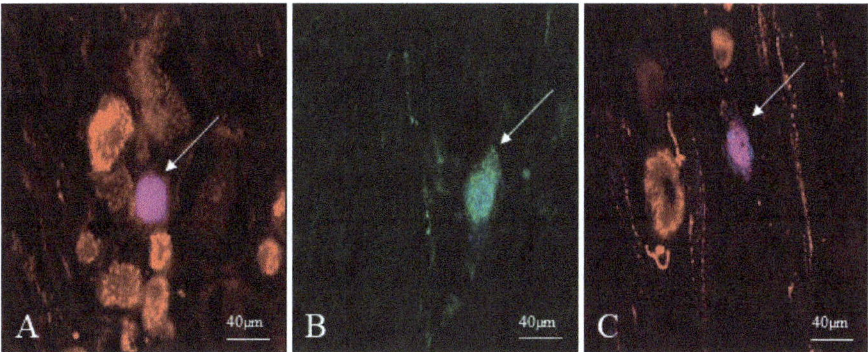

Figure 2. Fast blue-positive neurons in the dorsal root ganglia (DRG) supplying the ileocecal valve (ICV) (indicated with arrows) immunoreactive to calcitonin gene related peptide (CGRP) (**A**), substance P (SP) (**B**) and galanin (GAL) (**C**).

Regarding in vitro culture, it should be underlined that we tried to count both all FB-labelled neurons projecting to the ICV and unlabelled neurons projecting to other parts of the intestine and other organs. Unfortunately, the number of FB-labelled neurons was too low to get reliable statistically results and, therefore, the results obtained from all the population of labelled and un-labelled neurons are presented in the present study.

Studies conducted on cell cultures have shown that LPS derived from all bacterial serotypes studied did not change the average number of neurons on a glass coverslip. The number of neurons in the control group amounted to 48.07 ± 7.62, whereas under the influence of LPS, they achieved 48.69 ± 7.85, 48.66 ± 8.92, and 48.10 ± 8.73 under the impact of LPS *S.* Enteritidis, *LPS S.* Minnesota and *LPS S.* Typhimurium, respectively.

Moreover, during the present study, the influence of LPS on neurons immunoreactive to SP and/or GAL was observed and changes clearly depended on bacterial serotype. In control animals, SP-positive cells amounted to 60.90 ± 3.34% of all neurons. LPS *S.* Enteritidis caused the increase of the number of SP + neurons to 82.09 ± 4.43%, whereas LPS *S.* Minnesota and LPS *S.* Typhimurium resulted in a decrease in the percentage of such cells (to 41.48 ± 3.67% and 30.83 ± 3.71%, respectively) (Figure 3).

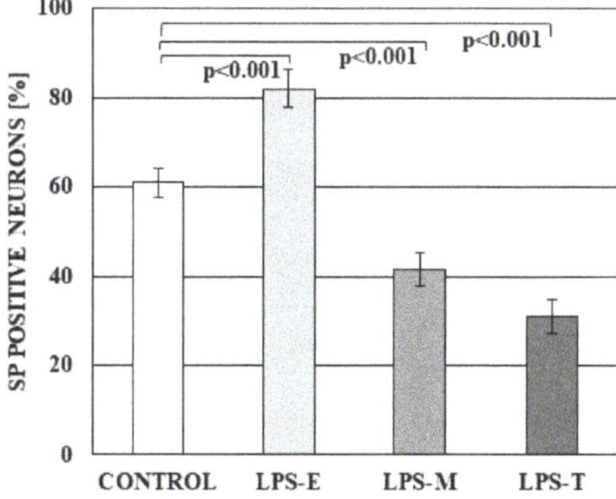

Figure 3. The percentage of neurons in dorsal root ganglia (DRG) immunoreactive to substance P (SP) in the control group and under the impact of LPS from *S.* Enteritidis (LPS-E), *S.* Minnesota (LPS-M), and *S.* Typhimurium (LPS-T). Statistically different at $p < 0.001$ as compared to the control group.

A different situation was observed in the case of GAL-positive neurons. In control animals, the percentage of such cells amounted to 55.44 ± 4.16%. LPS *S.* Enteritidis and LPS *S.* Minnesota caused the decrease in the number of GAL + neurons (to 27.51 ± 1.40% and 26.82 ± 7.08%, respectively), whereas under the impact of LPS *S.* Typhimurium, this value achieved 52.81 ± 6.80% and was not significantly statistically different from the percentage observed in the control group (Figure 4).

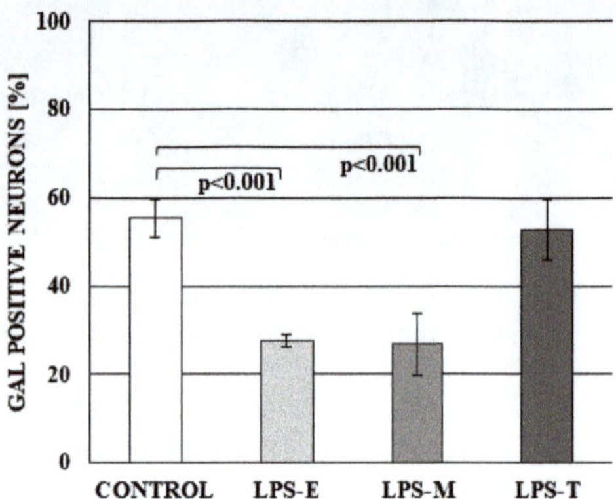

Figure 4. The percentage of neurons in dorsal root ganglia (DRG) immunoreactive to galanin (GAL) in control group and under the impact of LPS from S. Enteritidis (LPS-E), S. Minnesota (LPS-M), and *S.* Typhimurium (LPS-T). Statistically different at *p* < 0.001 as compared to the control group.

During the present investigation, the influence of LPS derived from all bacterial serotypes studied on the percentage of CGRP-positive neurons was not observed. The percentage of such neurons amounted to 67.19 ± 3.49% in control group, 67.52 ± 3.96% in LPS *S.* Enteritidis group, 69.08 ± 4.51% in LPS *S.* Minnesota group and 67.53 ± 5.35% in the LPS *S.* Typhimurium group. Differences between the mentioned above values were not statistically significant (Figure 5).

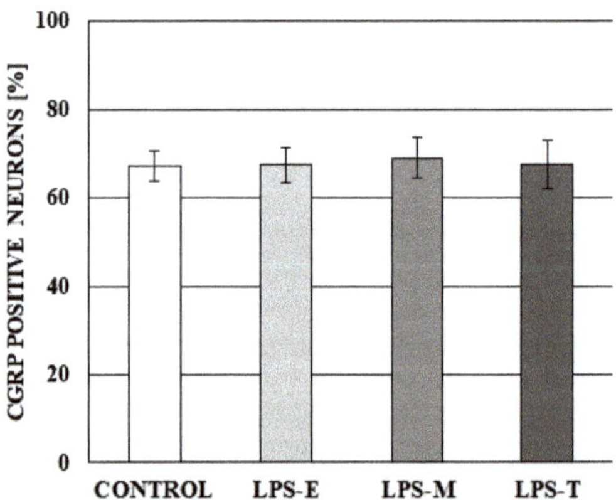

Figure 5. The percentage of neurons in dorsal root ganglia immunoreactive to calcitonin gene-related peptide (CGRP) in control group and under the impact of LPS from S. Enteritidis (LPS-E), S. Minnesota (LPS-M), and S. Typhimurium (LPS-T). Statistically significant differences for *p* < 0.001 as compared to the control group.

The analysis of the cell viability (MTT assay) demonstrated that a 24 h exposure to the low dose of LPS from studied serotypes of *Salmonella* spp. did not exert any toxic or proliferative effect on neuronal and non-neuronal DRG cells viability of DRG (Figure 6).

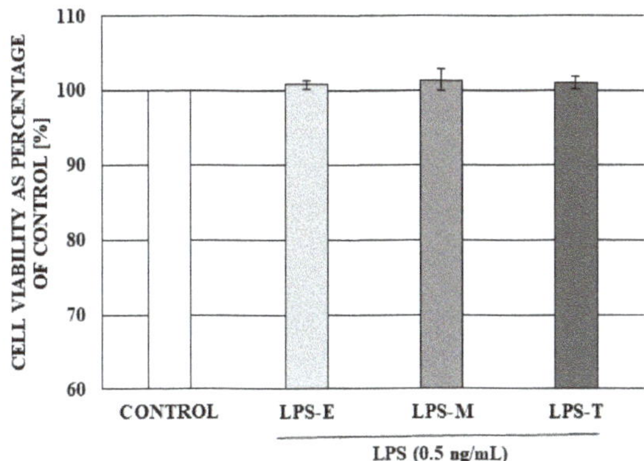

Figure 6. The effect of a low dose (0.5 ng/mL) of LPS from different serotypes of *Salmonella* spp: LPS from *Salmonella* Enteritidis (LPS-E), LPS from *Salmonella* Minnesota (LPS-M), LPS from *Salmonella* Typhimurium (LPS-T) on DRG cell viability. Cell viability was assessed using the MTT method. Data are presented as means of the percentage of the untreated control cells ± SE (*n* = 4).

Representative examples of dorsal root ganglia (DRG) neurons supplying the ileocecal valve observed during the present study in the cell culture are visualised in Figures 7 and 8.

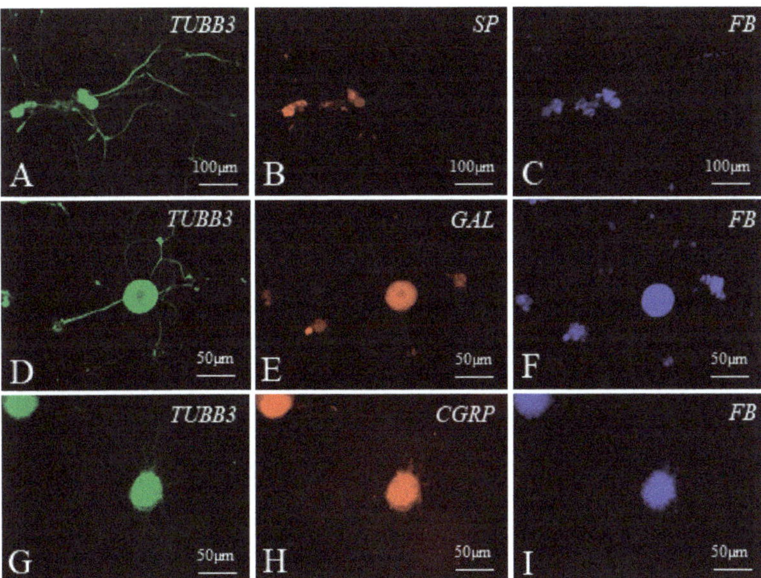

Figure 7. Examples of dorsal root ganglia (DRG) neurons supplying the ileocecal valve (FB-positive, blue) (**C,F,I**), immunoreactive to tubulin (TUBB3 used here as a neuronal marker, green) (**A,D,G**) and substance P (SP) (**B**), galanin (GAL) (**E**) or calcitonin gene-related peptide (CGRP) (**H**) in cell cultures of the control group.

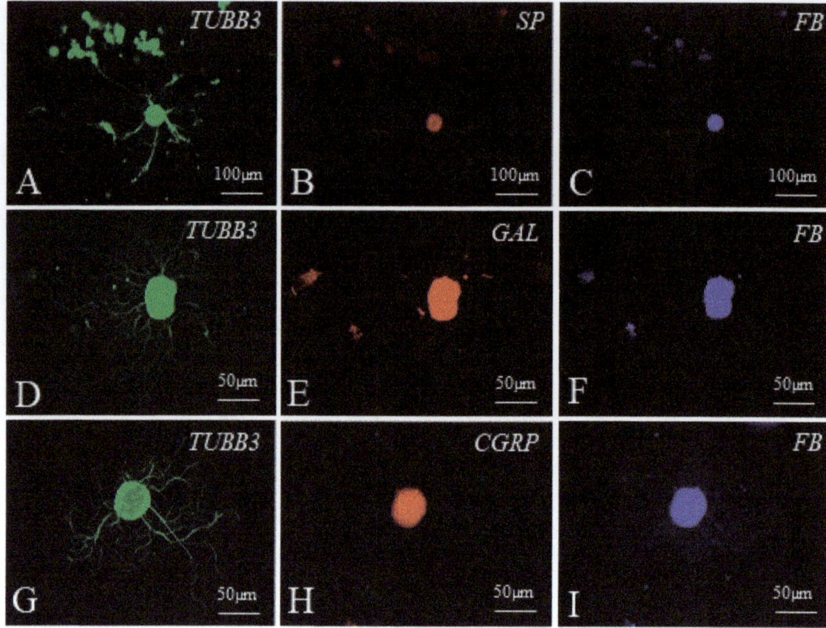

Figure 8. Examples of dorsal root ganglia (DRG) neurons supplying the ileocecal valve (FB-positive blue) (**C,F,I**) immunoreactive to tubulin (TUBB3 used here as a neuronal marker, green) (**A,D,G**) and substance P (SP) (**B**), galanin (GAL) (**E**), and calcitonin gene-related peptide (CGRP) (**H**) (all these substances in red) in cell cultures treated with lipopolysaccharides from *Salmonella* Enteritidis (**A–C**), *Salmonella* Typhimurium (**D–F**) and *Salmonella* Minnesota (**G–I**).

3. Discussion

The results obtained in this study indicate that the ICV, similar to other parts of the gastrointestinal tract, is supplied by neurons located in DRG [30]. Similar to previous studies, where DRG neuronal cells supplying the stomach, duodenum, ileum, and colon were described in various mammal species [30–33], observations made during the present investigations show that DRG plays a role in ICV sensory innervation in the domestic pig. On the other hand, the relatively low number of FB+ neurons observed in this study suggest that sensory components of the vagal nerve may play more important functions in the innervation of the ICV. This is according to the previous studies, where it was found that the network of processes derived from DRG neurons in the colon is more extensive than in the stomach and jejunum [30,32]. This is connected with the fact that branches of the vagal nerve supply the GI tract from the oesophagus to the proximal colon [34]. The majority of FB+ neuronal cells innervating the ICV investigated in this study were located in the thoracic and lumbar DRG, which is generally consistent with the distribution of previously described DRG neuronal cells supplying the small intestine.

Previous studies have shown that neuronal cells located in the DRG may contain (apart from typical substances involved in the sensory and pain stimuli conductions, such as SP and CGRP) a wide range of active neuronal factors, including GAL, CART peptide, vasoactive intestinal polypeptide, nitric oxide, somatostatin, and many others [35,36]. During the present study, the majority of DRG neurons supplying the ICV also contain SP and/or CGRP. SP is a member of the tachykinin family. It is a substance participating in sensory and pain stimuli conduction and has been observed in sensory neuronal cells and fibres in the central and peripheral nervous system of numerous species, including human [37–41]. It should be underlined that SP, besides sensory functions, may also participate in other various regulatory processes within the GI tract. In particular, it is known that SP regulates the intestinal motility and secretory activity [37], blood flow within the intestine and mesentery [42], as well immunological processes [43].

The second important substance participating in sensory and pain conduction, which has been observed in sensory neurons supplying the ICV, is CGRP. This peptide, similarly to SP, has been described in various central and peripheral sensory nervous structures of a wide range of mammals [44,45]. Within the digestive tract of some species, CGRP is considered to be a marker of intrinsic primary afferent neurons, which belong to the enteric nervous system and are a component of short intramural reflexes taking place without the central nervous system [46]. CGRP within the GI tract is also involved in regulatory processes connected with blood flow, secretory activity, and the absorption of nutrients [47,48]. Moreover, this peptide takes part in the protection of the intestinal mucosal layer against injuries and regulates intestinal motility [49].

The third substance studied in this investigation—GAL—is a not classical factor taking part in sensory or pain stimuli conduction. Nevertheless, the presence of GAL has been described in sensory neuronal structures supplying various internal organs [50,51]. It is known that GAL takes part in the regulation of other neurotransmitters by acting on the ion channels within the membrane of neuronal cells [52,53]. Besides sensory conduction, GAL in the GI system regulates the intestinal motility and secretory activity and the character of processes regulated by GAL clearly depends on the animal species and the fragment of the intestine [54,55].

In the light of the previous investigations, it is known that all neuronal factors studied during this experiment participate in regulatory processes not only under physiological conditions, but are also involved in pathological mechanisms of various diseases and intoxications with a wide range substances. Evidence of this includes changes in the expression of SP, CGRP, and GAL both in the enteric nervous system and extrinsic neuronal cells supplying the intestine under the influence of inflammatory processes, neuronal damage, intoxication with mycotoxins, and other toxicological substances and many other pathological stimuli [38,56]. For SP, these changes may be connected with neuroprotective properties and the participation of this peptide in inflammatory processes [38]. Namely, it is known that SP acting on lymphocytes and macrophages stimulates the secretion of pro-inflammatory cytokines, including IL-1 and TNF-α—the most important inflammatory mediator [37,43]. In turn, GAL is known from neuroprotective functions in the central and peripheral nervous system, which have primarily been described during brain injuries and neurodegenerative diseases [57]. Moreover, GAL (contrary to SP) plays an anti-inflammatory role by enhancing synthesis and secretion of IFN-γ and IL-12/23 while simultaneously decreasing the levels of TNF-α and IL-1β [58]. However, CGRP, whose role in intestinal diseases has not been fully elucidated, may inhibit the expression of TNF-α and IL-1β and take part in mechanisms connected with the development of diarrhoea [59]. On the other hand, the participation of CGRP in neuroprotective processes within the innervation of the intestine, despite changes in CGRP expression under various pathological stimuli [60], has not yet been confirmed.

Changes in the immunoreactivity of DRG neurons observed during the present study probably result from the above-mentioned functions of neuronal factors under the impact of LPS. These changes may be connected with neuroprotective and/or adaptive processes used to maintain homeostasis within the nervous tissues These may result from the pro-inflammatory activity of LPS, described in the previous investigations. In particular, it is known that one of the components of LPS —lipid A—strongly induces synthesis and the secretion of pro-inflammatory substances [61]. However, a more likely cause of the observed changes is the direct impact of LPS on neuronal cells. It should be underlined that this impact in the light of the previous studies is not clear and depends on LPS dose. On the one hand, the neurotoxic activity of LPS is relatively well-known. Previous investigations have described the involvement of this substance in neurodegenerative processes in various parts of the nervous system [62–64]. Moreover, LPS is used as a factor inducing experimental Parkinson's disease in rodents [65–68]. On the other hand, it has been shown that low doses of LPS are essential to promote the survival of enteric and hippocampal neurons. Anitaha et al. [69] found that for enteric neuronal survival, microbial-neuronal interaction is essential and a low dose of LPS is essential to maintain neuronal survival, although at higher doses LPS results in neuronal toxicity. Low dose LPS treatment (10 ng/mL) promoted enteric neuronal survival through the activation of NF-κB and TLR4. None of our studied serotypes of *Salmonella* spp. influenced the neuron counts and furthermore, we saw that neurons treated with 0.5 µg/mL LPS have longer and denser neurites compared to the control group. This is only a subjective observation which has not been supported by any analysis of neuron morphology. Perhaps, hypothetically, it was in connection with the role of glial cells, but we did not study it because it was not the intended goal of our study. Additionally, LPS in a low dose has been shown to promote the survival of hippocampal neurons through increased expression of a granulocyte colony-stimulating factor [70]. But, in the central nervous system (CNS), hippocampal neurogenesis and neurological functions were attenuated by lipopolysaccharide-induced TLR4 activation [71]. Moreover, some studies have shown the protective effect of LPS pre-treatment [72] and endotoxin tolerance provides (enhances) the ischemic resistance of neuronal cells [73]. In contrast, Chen et al. [74] reported that pre-conditioning with a super-low or low dose of LPS exacerbates sepsis mortality. Further studies may be required to find consistent issues to combine the different results of various studies.

4. Materials and Methods

The present study consisted of two experiments conducted on nine immature sows of the Pietrain × Duroc breed. All animals were kept in standard laboratory conditions and fed with complete feeding stuff appropriate to the age and species of animals for two weeks prior to the experiment in order to allow adaptation to the new environment. The experiments took place when the pigs were 8–9 weeks of age with body weights of 16–18 kg. Both administration of the drugs and the performance of surgical procedures were performed by a veterinary doctor (DVM, Ph.D.). All procedures in the experiment were approved by the Local Ethical Commission of Experiments on Animals in Olsztyn (decision number 73/2015 from 29 September 2015).

4.1. Experiment No. 1: Localization and Neurochemical Characterization of Neurons Located in the Dorsal Root Ganglia and Supplying the ICV

The first experiment was conducted to study the sensory neurons located in dorsal root ganglia and supplying the ileocecal valve (ICV). For this part of the experiment, five pigs were used.

4.1.1. Surgical Procedures Surgery

Before surgical procedures, five animals were pre-medicated using the method previously described by Mikolajczyk [75] with intramuscular injection of atropine (Atropinum Sulfuricum Polfa Warszawa S.A., Warszawa, Poland, 0.035 mg/kg b.w.), ketamine (Bioketan, Vetoquinol Biowet Sp. z o.o., Poland and Vetoquinol S.A., Lure, France, 7.0 mg/kg b.w.), and medetomidine (Cepetor, CP-Pharma Handelsges mbH, Burgdorf, Germany, 0.063 mg/kg b.w.). After 15 min. the animals were subjected to the general anaesthesia with propofol (Scanofol, NORBROOK, Newry, Northern Ireland, IRL.PN, 4.5 mg/kg b.w. given intravenously) and median laparotomy. During the transaction surgery procedure, a conventional midline incision of the abdominal wall was made. The cecum and ileum were identified and the ICV was isolated from the abdominal cavity. The ICV was injected with 50 µL of a 5% aqueous solution of the fluorescence retrograde neuronal tracer fast blue (FB; EMSChemie GmbH, Groß-Umstadt, Germany, ten injections, 5 µL each) using a Hamilton syringe equipped with a 26-gauge needle. Close attention was paid to avoiding any contamination of the surrounding tissues with FB due to the hydrostatic leakage from the injection canal. To avoid leakage, the needle was left in each site of FB injection for up to a minute. The peritoneum with the transverse abdominal muscles, the internal and external abdominal oblique muscles, and the cutaneous muscle with subcutaneous fascia were closed in a simple continuous pattern. The skin was closed in a subcuticular pattern.

4.1.2. Sample Collection and Processing

After three weeks, the animals were again pre-medicated (as described above) and euthanized with pentobarbital (Morbital, Biowet Puławy Sp. z o.o, Poland, 60–70 mg/kg b.w., given intravenously). After death, the thoracic, lumbar, and sacral dorsal root ganglia were collected. Ganglia were fixed in 4% buffered paraformaldehyde (pH 7.4) for 30 min. and rinsed in phosphate buffer for three days (at 40 °C). They then added a 18% sucrose solution and stored it for least three weeks at 40 °C. After this period, the ganglia were frozen at −20 °C and cut into 10 μm-thick sections using a cryostat (HM 525, Microm International, Germany).

4.1.3. Immunofluorescence Procedures with Counting Neurons

The sections were subjected to examination for the presence of neurons containing FB using a fluorescence Olympus BX51 microscope equipped with an appropriate filter set. Sections with FB-positive cells were subjected to typical single immunofluorescence technique by the method described previously by Gonkowski et al. [76]. Basically, during this method, fragments of DRG were subjected to (1) drying at room temperature (rt) for 1 h; (2) "blocking" in the solution containing 10% normal goat serum, 0.1% bovine serum albumin, 0.01% NaN3, Triton X-100 and thimerozal in PBS for 1 h (rt); (3) incubation with antibodies directed towards SP, CGRP or galanin (GAL) (overnight; rt, in a humid chamber); (4) incubation with secondary antibodies conjugated with appropriate fluorochromes (alexa fluor 594 or 488) to visualise the complexes "antigen—primary antibody". The specifications and working dilution of primary and secondary antisera are presented in Table 1.

Table 1. List of antisera and reagents used in immunohistochemical investigations.

Primary antibodies				
Antigen	Code	Species	Working dilution	Supplier
CGRP	AB5920	Rabbit	1:1600	Chemicon Int Temecula, OH, USA
GAL	T-5036	Guinea Pig	1:2000	Peninsula San Carlos, CA, USA
SP	8450-0505	Rat	1:1000	Bio-Rad (AbD Serotec), Kidlington, UK

Secondary antibodies		
Reagents	Working dilution	Supplier
Alexa fluor 546 donkey anti-rabbit IgG	1:1000	Invitrogen Carlsbad, CA, USA
Alexa fluor 546 donkey anti-guinea pig IgG	1:1000	Invitrogen
Alexa fluor 488 goat anti-rat IgG	1:1000	Invitrogen

To confirm the specificity of the method routine standard controls, such as pre-absorption of the neuropeptide antisera with appropriate antigen, omission and replacement of primary antisera by non-immune sera were performed. To determine the percentage of neurons supplying the ICV immunoreactive to SP, CGRP, or GAL, at least 50 FB+ neuronal cells from each animal were evaluated for the presence of the particular neuronal factors. This relatively low number of neurons included in the experiment was caused by the relatively small number of all neuronal cells supplying the ICV. The obtained data was pooled, expressed as means ± SD.

4.2. Experiment No. 2: Culturing of Primary Sensory Neurons and Various Serotypes of Salmonella spp.

The second experiment conducted during this investigation consisted of a study of the influence of a low dose (0.5 ng/mL [77,78] of LPS from various types of *Salmonella* spp. serotypes: LPS from *Salmonella enterica* subsp. *enterica* serotype Enteritidis (L7770 Sigma), LPS from *Salmonella enterica* subsp. *enterica* serotype Minnesota (L4641 Sigma), LPS from *Salmonella enterica* subsp. *enterica* serotype Typhimurium (L6143 Sigma) on the neurochemical characterization of neurons located in DRG. Thus, during the in vitro experiment, four groups of cultures were used: control, Enteritidis (LPS-E), Minnesota (LPS-M), and Typhimurium (LPS-T). For this part of the experiment, four pigs were used.

4.2.1. Sample Collection and Processing

After three weeks, all animals were again pre-medicated and subjected to general anaesthesia conducted as described above. During anaesthesia, the left and right thoracic (Th7–Th13) ganglia and lumbar (L1–L4) ganglia were exposed. For DRG exposure, an incision was made in the skin of the dorsal midline. The superficial muscular fascia was incised and the paraspinal muscles separated by a combination of sharp and blunt dissection, exposing the lumbar and thoracic vertebrae. A rongeur was used to remove bone fragments of spinal nerves as well as the intervertebral foramina from which they emerge, which were exposed along with particular ganglia. The animals were then euthanized with pentobarbital (in the above-described manner).

DRG were removed and transferred into cold RPMI 1640 W/HEPES W/GLUTAMAX-I medium (cat. no. 72400021, Life Technologies Polska Sp. z.o.o.; Warszawa, Poland) and antibiotic and antimycotic (cat. no. 15240062, Life Technologies Polska Sp. z.o.o.; Poland).

4.2.2. Cell Culture and Treatments

Cell cultures were prepared as described previously [79,80]. Briefly, following removal of connective tissue, DRG were incubated in collagenase (cat. no. 17100017, Life Technologies Polska Sp. z.o.o.; Poland) for 60 min followed by 0.5% trypsin/EDTA (15400054, Life Technologies Polska Sp. z.o.o.; Poland). Neurons were dissociated by passages through a fire-polished Pasteur pipette and centrifuged at a low speed (10 min, 700 rpm). After final centrifugation, the pellet was suspended with TNB 100TM medium (cat. no. F8023; Biochrom AG, Berlin, Germany) containing: antibiotic and antimycotic and protein-lipid complex TM (cat. no. F8820; Biochrom AG, Germany). Equal volumes (50 µL) containing DRG neurons were seeded on glass coverslip (12 mm in diameter) coated with poly-D-lysine (0.01% solution; cat. no. P7280; Aldrich)/laminin (1 mg/mL L-2020, Sigma) placed in six-well multi-dishes (cat. no. 353046; BD Biosciences; Franklin Lakes, NJ, USA) at a density of 50–70 per glass. Cultures were cultivated in TNB medium 1 mL per well with 5 ng/mL nerve growth factor beta (NGF; N1408; Sigma Aldrich; Berlin, Germany) to maintain the survival of neurons [81]. After 36 h incubation at 37 °C in 5% CO_2, the medium was replaced with fresh medium in the control and with fresh medium with the addition of 0.5 ng/mL LPS from *S.* Enteritidis, LPS from *S.* Minnesota (L4641 Sigma), LPS from *S.* Typhimurium (L6143 Sigma) in the LPS-treated group. We decided to use a 36 h cell primary neuron culture because extensive neurite outgrowth was observed after 24 h of culture [82].

4.2.3. Immunocytochemical Labelling

After 60 h (including 24 h of culture with the addition of LPS to the media) in culture, neurons were fixed with 4% paraformaldehyde for 20 min, permeabilised with 0.01% Triton X-100 (X100-100ML; Sigma Aldrich; Germany) in PBS (P5493; Sigma Aldrich; Germany) for 5 min and blocked with blocking buffer (10% goat serum in PBS) for 30 min. DRG neurons were incubated with primary antibodies against substance P (8450-0004; AbD Serotec; Regensburg, Germany; 1:3000), calcitonin gene-related peptide (AB43873; Abcam; Cambridge, UK; 1:18,000), galanin (AB5909; Merck Millipore; Burlington, MA, USA; 1:16,400) and neuron-specific β-III tubulin (MAB1195; R&D Systems; Minneapolis, MN, USA; 1:1000) diluted in blocking buffer for 1 h at room temperature and then incubated with secondary Alexa-555- and Alexa-488-conjugated antibody (A31572; A-11001, Invitrogen, Carlsbad, CA, USA; 1:1000) for 60 min at RT. We stained all DRG neurons (FB-labelled and FB-unlabelled) using neuron-specific β-III tubulin [83] in co-localization with CGRP, SP, and GAL [84].

4.2.4. Counting Cultured Neurons

Both FB-labelled neurons projecting to the ICV and unlabelled neurons projecting to other organs (also large extent, the colon) [85] were counted.

The experiment in vitro was performed in duplicate with two replicate wells (four coverslips in one well, i.e., eight coverslips per one animal) for each group of cultures (control, Enteritidis, Minnesota and Typhimurium). The number of neurons in the control and after LPS treatment group were counted and then expressed in percentage. FB-traced neurons were identified in the cell cultures by their blue fluorescence under the UV illumination using fluorescence Olympus BX51 microscope (V1 module, excitation range 330–385 nm and barrier filter at 420 nm). The number of β-III tubulin neurons in all studied groups (control and with the addition of LPS to the medium) was considered as the number of total neurons (100%). The number of neurons containing studied substances (β-III tubulin, CGRP, GAL, and SP) in intact and LPS-treated group was quantified using a fluorescent Olympus BX51 microscope (Shinjuku, Tokyo, Japan) equipped with an appropriate filter sets for Alexa 488 (B1 module, excitation filter 450–480 nm) and Alexa 555 (G1, excitation filter 510–550 nm). Microphotographs were acquired using 20× objectives and a PC equipped with a CCD camera operated by Cell Sens Dimension image analysis software (Olympus, Warsaw, Poland).

4.2.5. MTT Assay

To determine the cell viability by MTT assay, DRG cells were prepared as described before using six well multi-dishes with glass coverslips and seeded in a 96-well plate (CytoOne cat no: CC7682-7596). After 36 h incubation at 37 °C in 5% CO_2 the cells were treated with 0.5 ng/mL LPS from *S.* Enteritidis, LPS from *S.* Minnesota, and LPS from *S.* Typhimurium or tested without LPS (control group) and incubated at 37 °C in 5% CO_2 for 24 h. After incubation, 50 µL MTT (Thiazolyl Blue Tetrazolium Bromide 5 mg/mL, M5655 Sigma) was added and the plates were incubated at 37 °C in 5% CO_2 for 4 h. At the end of the incubation period, the supernatants were removed, and 100 µL of dimethyl sulfoxide (DMSO 34943, Sigma) was added to each well to enable the release of the blue reaction product-formazan. The absorbance at 570 nm was read on a microplate reader Infinite 200 (Tecan) and the results were expressed as a percentage of the absorbance measured in control cells and in the LPS-treated culture.

4.3. Statistical Analysis

All data were expressed as the mean ± standard deviation (SD) from independent experiments performed using five animals in the experiment 1 ($n = 5$) and four animals in experiment 2 ($n = 4$). Statistical analysis was determined using one-way analysis of variance (ANOVA) followed by Tukey's test for multiple comparisons (Statistica software /version 13.1 /StatSoft, Cracow, Poland). *P* values less than 0.05 were considered significant.

5. Conclusions

To sum up, the results obtained during the present study have shown that sensory processes supplying the ICV may derive from neurons located in DRG from neuromers Th7 to L4. The ICV sensory neurons contained SP, CGRP, and GAL. Moreover, it has been shown that the dose of LPS used in the in vitro study did not change the number of DRG neurons from neuromers Th7 to L4 supplying the ICV and other organs and tissues, but influenced their neurochemical characterization. The observed changes clearly depended on the bacterial serotype of LPS. This is proof that structural differences of LPS—not only between bacterial species, but also within particular serotypes of one species—result in varied impact on the nervous system. Moreover, the obtained results have shown that SP and GAL, contrary to CGRP, are involved in the processes connected with the impact of LPS. Changes in the expression of these substances may result from the direct influence of LPS on the nervous structures and/or pro-inflammatory activity of LPS. However, the explanation of the exact mechanisms connected with LPS-induced fluctuations in neurochemical characterization of DRG neurons requires further study.

Author Contributions: Conceptualization, A.M. (A.M. conceived and designed the study); Methodology and investigation in the experiment no. 1, A.M. and S.G.; Methodology and investigation in the experiment no. 2, A.M., S.G. and A.K.; Data Analysis, A.M.; Interpretation of the results, A.M.; Resources, A.M.; Writing the manuscript, A.M.; Visualization and analysis of microscopic images A.M. and S.G.

Funding: This research was supported by the statutory grant No. 25.610.001-300 from the Faculty of Medical Sciences, the University of Warmia and Mazury in Olsztyn in Poland.

Acknowledgments: This study was supported by the statutory grant No. 25.610.001-300, Faculty of Medical Sciences, the University of Warmia and Mazury in Olsztyn, Poland.

Conflicts of Interest: The authors declare no conflict of interest.

References

1. Furness, J.B. Types of neurons in the enteric nervous system. *J. Auton. Nerv. Syst.* **2000**, *81*, 87–96. [CrossRef]
2. Majewski, M.; Bossowska, A.; Gonkowski, S.; Wojtkiewicz, J.; Brouns, I.; Scheuermann, D.W.; Adriaensen, A.; Timmermans, J.P. Neither axotomy nor target-tissue inflammation changes the NOS- or VIP-synthesis rate in distal bowel-projecting neurons of the porcine inferior mesenteric ganglion (IMG). *Folia Histochemica et Cytobiologica* **2002**, *40*, 151–152. [PubMed]
3. Skobowiat, C.; Gonkowski, S.; Calka, J. Phenotyping of sympathetic chain ganglia (SChG) neurons in porcine colitis. *J. Vet. Med. Sci.* **2010**, *72*, 1269–1274. [CrossRef] [PubMed]
4. Makowska, K.; Gonkowski, S. The Influence of Inflammation and Nerve Damage on the Neurochemical Characterization of Calcitonin Gene-Related Peptide-Like Immunoreactive (CGRP-LI) Neurons in the Enteric Nervous System of the Porcine Descending Colon. *Int. J. Mol. Sci.* **2018**, *19*, 548. [CrossRef] [PubMed]
5. Furness, J.B.; Callaghan, B.P.; Rivera, L.R.; Cho, H.J. The enteric nervous system and gastrointestinal innervation: Integrated local and central control. *Adv. Exp. Med. Biol.* **2014**, *817*, 39–71. [PubMed]
6. Avetisyan, M.; Schill, E.M.; Heuckeroth, R.O. Building a second brain in the bowel. *J. Clin. Investig.* **2015**, *125*, 899–907. [CrossRef] [PubMed]
7. Browning, K.N.; Travagli, R.A. Central nervous system control of gastrointestinal motility and secretion and modulation of gastrointestinal functions. *Compr. Physiol.* **2014**, *4*, 1339–1368. [PubMed]
8. Rytel, L.; Calka, J. Acetylsalicylic acid-induced changes in the chemical coding of extrinsic sensory neurons supplying the prepyloric area of the porcine stomach. *Neurosci. Lett.* **2016**, *617*, 218–224. [CrossRef] [PubMed]
9. Dorofeeva, A.A.; Panteleev, S.S.; Makarov, F.N. Involvement of the sacral parasympathetic nucleus in the innervation of the descending colon and rectum in cats. *Neurosci. Behav. Physiol.* **2009**, *39*, 207–210. [CrossRef] [PubMed]
10. Rytel, L.; Całka, J. Neuropeptide profile changes in sensory neurones after partial prepyloric resection in pigs. *Ann. Anat.* **2016**, *206*, 48–56. [CrossRef] [PubMed]
11. Phillips, S.F.; Camilleri, M. The ileocecal area and the irritable bowel syndrome. *Gastroenterol. Clin. N. Am.* **1991**, *20*, 297–311.
12. Cserni, T.; Paran, S.; Kanyari, Z.; O'Donnell, A.M.; Kutasy, B.; Nemeth, N.; Puri, P. New insights into the neuromuscular anatomy of the ileocecal valve. *Anat. Rec.* **2009**, *292*, 254–261. [CrossRef] [PubMed]
13. Palmisano, S.; Silvestri, M.; Troian, M.; Germani, P.; Giudici, F.; de Manzini, N. Ileocaecal valve syndrome after surgery in adult patients: Myth or reality? *Color. Dis.* **2017**, *19*, e288–e295. [CrossRef] [PubMed]
14. Kaur, U.; Goyal, N.; Gupta, M. Disposition of muscularis propria and nerve elements in the lip of the ileocaecal valve. *Nepal Med. Coll. J.* **2005**, *7*, 125–128. [PubMed]
15. Cserni, T.; Paran, S.; Puri, P. New hypothesis on the pathogenesis of ileocecal intussusception. *J. Pediatr. Surg.* **2007**, *42*, 1515–1519. [CrossRef] [PubMed]
16. Maciel, B.M.; Passos, R.; Sriranganathan, N. Salmonella enterica: Latency. In *Current Topics in Salmonella and Salmonellosis*; Mares, M., Ed.; IntechOpen: Rijeka, Croatia, 2017; pp. 43–58, ISBN 978-953-51-3066-6.
17. Guo, S.; Al-Sadi, R.; Said, H.M.; Ma, T.Y. Lipopolysaccharide causes an increase in intestinal tight junction permeability in vitro and in vivo by inducing enterocyte membrane expression and localization of TLR-4 and CD14. *Am. J. Pathol.* **2013**, *182*, 375–387. [CrossRef] [PubMed]

18. Ramachandran, G. Gram-positive and gram-negative bacterial toxins in sepsis: A brief review. *Virulence* **2014**, *5*, 213–218. [CrossRef] [PubMed]

19. Mikołajczyk, A.; Makowska, K. Cocaine- and amphetamine-regulated transcript peptide (CART) in the nerve fibers of the porcine gallbladder wall under physiological conditions and after Salmonella Enteritidis lipopolysaccharides administration. *Folia Morphol.* **2017**, *76*, 596–602. [CrossRef] [PubMed]

20. Mikołajczyk, A.; Gonkowski, S.; Złotkowska, D. Modulation of the Main Porcine Enteric Neuropeptides by a Single Low-Dose of Lipopolysaccharide (LPS) Salmonella Enteritidis. *Gut Pathog.* **2017**, *9*, 73. [CrossRef] [PubMed]

21. Makowska, K.; Mikolajczyk, A.; Calka, J.; Gonkowski, S. Neurochemical characterization of nerve fibers in the porcine gallbladder wall under physiological conditions and after the administration of Salmonella Enteritidis lipopolysaccharides (LPS). *Toxicol. Res.* **2018**, *7*, 73–83. [CrossRef] [PubMed]

22. Reyes, R.E.; González, C.; Jiménez, R.C.; Ortiz, M.C.; Andrade, A.A. Mechanisms of O-Antigen Structural Variation of Bacterial Lipopolysaccharide (LPS). In *The Complex World of Polysaccharides*; Karunaratne, D.N., Ed.; INTECH Open Access: Kelaniya, Sri Lanka, 2012; pp. 71–98, ISBN 978-953-51-0819-1. [CrossRef]

23. Pieterse, E.; Rother, N.; Yanginlar, C.; Hilbrands, L.B.; van der Vlag, J. Neutrophils Discriminate between Lipopolysaccharides of Different Bacterial Sources and Selectively Release Neutrophil Extracellular Traps. *Front. Immunol.* **2016**, *7*, 484. [CrossRef] [PubMed]

24. Steimle, A.; Autenrieth, I.B.; Frick, J.S. Structure and function: Lipid A modifications in commensals and pathogens. *Int. J. Med. Microbiol.* **2016**, *306*, 290–301. [CrossRef] [PubMed]

25. Vatanen, T.; Kostic, A.D.; D'Hennezel, E.; Siljander, H.; Franzosa, E.A.; Yassour, M.; Kolde, R.; Vlamakis, H.; Arthur, T.D.; Hämäläinen, A.M.; et al. Variation in Microbiome LPS Immunogenicity Contributes to Autoimmunity in Humans. *Cell* **2016**, *165*, 842–853. [CrossRef] [PubMed]

26. Khan, M.M.; Ernst, O.; Sun, J.; Fraser, I.D.C.; Ernst, R.K.; Goodlett, D.R.; Nita-Lazar, A. Mass Spectrometry-based Structural Analysis and Systems Immunoproteomics Strategies for Deciphering the Host Response to Endotoxin. *J. Mol. Biol.* **2018**, *430*, 2641–2660. [CrossRef] [PubMed]

27. Pulendran, B.; Kumar, P.; Cutler, C.W.; Mohamadzadeh, M.; Van Dyke, T.; Banchereau, J. Lipopolysaccharides from distinct pathogens induce different classes of immune responses in vivo. *J. Immunol.* **2001**, *167*, 5067–5076. [CrossRef] [PubMed]

28. Nedrebø, T.; Reed, R.K. Different serotypes of endotoxin (lipopolysaccharide) cause different increases in albumin extravasation in rats. *Shock* **2002**, *18*, 138–141. [CrossRef] [PubMed]

29. Verma, N.; Rettenmeier, A.W.; Schmitz-Spanke, S. Recent advances in the use of *Sus scrofa* (pig) as a model system for proteomic studies. *Proteomics* **2011**, *11*, 776–793. [CrossRef] [PubMed]

30. Chen, B.N.; Olsson, C.; Sharrad, D.F.; Brookes, S.J. Sensory innervation of the guinea pig colon and rectum compared using retrograde tracing and immunohistochemistry. *Neurogastroenterol. Motil.* **2016**, *28*, 1306–1316. [CrossRef] [PubMed]

31. Ohmori, Y.; Atoji, Y.; Saito, S.; Ueno, H.; Inoshima, Y.; Ishiguro, N. Differences in extrinsic innervation patterns of the small intestine in the cattle and sheep. *Auton. Neurosci.* **2012**, *167*, 39–44. [CrossRef] [PubMed]

32. Zalecki, M. Extrinsic primary afferent neurons projecting to the pylorus in the domestic pig—Localization and neurochemical characteristics. *J. Mol. Neurosci.* **2014**, *52*, 82–89. [CrossRef] [PubMed]

33. Spencer, N.J.; Zagorodnyuk, V.; Brookes, S.J.; Hibberd, T. Spinal afferent nerve endings in visceral organs: Recent advances. *Am. J. Physiol. Gastrointest. Liver Physiol.* **2016**, *311*, 1056–1063. [CrossRef] [PubMed]

34. Matteoli, G.; Boeckxstaens, G.E. The vagal innervation of the gut and immune homeostasis. *Gut* **2013**, *62*, 1214–1222. [CrossRef] [PubMed]

35. Bossowska, A.; Crayton, R.; Radziszewski, P.; Kmiec, Z.; Majewski, M. Distribution and neurochemical characterization of sensory dorsal root ganglia neurons supplying porcine urinary bladder. *J. Physiol. Pharmacol.* **2009**, *4*, 77–81.

36. Guić, M.M.; Kosta, V.; Aljinović, J.; Sapunar, D.; Grković, I. Characterization of spinal afferent neurons projecting to different chambers of the rat heart. *Neurosci. Lett.* **2010**, *469*, 314–318. [CrossRef] [PubMed]

37. Shimizu, Y.; Matsuyama, H.; Shiina, T.; Takewaki, T.; Furness, J.B. Tachykinins and their functions in the gastrointestinal tract. *Cell. Mol. Life Sci.* **2008**, *65*, 295–311. [CrossRef] [PubMed]

38. Gonkowski, S. Substance P as a neuronal factor in the enteric nervous system of the porcine descending colon in physiological conditions and during selected pathogenic processes. *Biofactors* **2013**, *39*, 542–551. [CrossRef] [PubMed]

39. Mistrova, E.; Kruzliak, P.; Chottova Dvorakova, M. Role of substance P in the cardiovascular system. *Neuropeptides* **2016**, *58*, 41–51. [CrossRef] [PubMed]

40. Chi, G.; Huang, Z.; Li, X.; Zhang, K.; Li, G. Substance P Regulation in Epilepsy. *Curr. Neuropharmacol.* **2018**, *16*, 43–50. [CrossRef] [PubMed]

41. Lénárd, L.; László, K.; Kertes, E.; Ollmann, T.; Péczely, L.; Kovács, A.; Kállai, V.; Zagorácz, O.; Gálosi, R.; Karádi, Z. Substance P and neurotensin in the limbic system: Their roles in reinforcement and memory consolidation. *Neurosci. Biobehav. Rev.* **2018**, *85*, 1–20. [CrossRef] [PubMed]

42. De Fontgalland, D.; Wattchow, D.A.; Costa, M.; Brookes, S.J. Immunohistochemical characterization of the innervation of human colonic mesenteric and submucosal blood vessels. *Neurogastroenterol. Motil.* **2008**, *20*, 1212–1226. [CrossRef] [PubMed]

43. Zhao, D.; Kuhnt-Moore, S.; Zeng, H.; Pan, A.; Wu, J.S.; Simeonidis, S.; Moyer, M.P.; Pothoulakis, C. Substance P-stimulated interleukin-8 expression in human colonic epithelial cells involves Rho family small GTPases. *Biochem. J.* **2002**, *368*, 665–672. [CrossRef] [PubMed]

44. Eftekhari, S.; Salvatore, C.A.; Johansson, S.; Chen, T.B.; Zeng, Z.; Edvinsson, L. Localization of CGRP, CGRP receptor, PACAP and glutamate in trigeminal ganglion. Relation to the blood-brain barrier. *Brain Res.* **2015**, *1600*, 93–109. [CrossRef] [PubMed]

45. Kestell, G.R.; Anderson, R.L.; Clarke, J.N.; Haberberger, R.V.; Gibbins, I.L. Primary afferent neurons containing calcitonin gene-related peptide but not substance P in forepaw skin, dorsal root ganglia, and spinal cord of mice. *J. Comp. Neurol.* **2015**, *523*, 2555–2569. [CrossRef] [PubMed]

46. Brehmer, A.; Croner, R.; Dimmler, A.; Papadopoulos, T.; Schrödl, F.; Neuhuber, W. Immunohistochemical characterization of putative primary afferent (sensory) myenteric neurons in human small intestine. *Auton. Neurosci.* **2004**, *112*, 49–59. [CrossRef] [PubMed]

47. Nuki, C.; Kawasaki, H.; Kitamura, K.; Takenaga, M.; Kangawa, K.; Eto, T.; Wada, A. Vasodilator effect of an adrenomeullin and calcitonin gene-related peptide receptors in rat mesenteric vascular beds. *Biochem. Biophys. Res. Commun.* **1993**, *196*, 245–251. [CrossRef] [PubMed]

48. Barada, K.A.; Saade, N.E.; Atweh, S.F.; Khoury, C.I.; Nassar, C.F. Calcitonin gene-related peptide regulates amino acid absorption across rat jejunum. *Regul. Pept.* **2000**, *90*, 39–45. [CrossRef]

49. Lambrecht, N.; Burchert, M.; Respondek, M.; Muller, K.M.; Peskar, B.M. Role of calcitonin gene-related peptide and nitric oxide in thegastroprotective effect of capsaicin in the rat. *Gastroenterology* **1993**, *104*, 1371–1380. [CrossRef]

50. Wynick, D.; Thompson, S.W.; McMahon, S.B. The role of galanin as a multi-functional neuropeptide in the nervous system. *Curr. Opin. Pharmacol.* **2001**, *1*, 73–77. [CrossRef]

51. Pidsudko, Z. Immunohistochemical characteristics and distribution of sensory dorsal root Ganglia neurons supplying the urinary bladder in the male pig. *J. Mol. Neurosci.* **2014**, *52*, 71–81. [CrossRef] [PubMed]

52. Sarnelli, G.; Vanden Berghe, P.; Raeymaekers, P.; Janssens, J.; Tack, J. Inhibitory effects of galanin on evoked [Ca^{2+}]I responses in cultured myenteric neurons. *Am. J. Physiol. Gastrointest. Liver Physiol.* **2004**, *286*, 1009–1014. [CrossRef] [PubMed]

53. Piqueras, L.; Tache, Y.; Martinez, V. Galanin inhibits gastric acid secretion through a somatostatin-independent mechanism in mice. *Peptides* **2005**, *25*, 1287–1295. [CrossRef] [PubMed]

54. Fox-Threlkeld, J.E.T.; McDonald, T.J.; Cipris, S.; Woskowska, Z.; Daniel, E.E. Galanin inhibition of vasoactive intestinal polypeptide release and circular muscle motility in the isolated perfused canine ileum. *Gastroenterology* **1991**, *101*, 1471–1476. [CrossRef]

55. Botella, A.; Delvaux, M.; Frexinos, J.; Bueno, L. Comparative effects of galanin on isolated smooth muscle cells from ileum in five mammalian species. *Life Sci.* **1992**, *50*, 1253–1261. [CrossRef]

56. Vasina, V.; Barbara, G.; Talamonti, L.; Stanghellini, V.; Corinaldesi, R.; Tonini, M.; De Ponti, F.; De Georgio, R. Enteric neuroplasticy evoked by inflammation. *Auton. Neurosci.* **2006**, *126*, 264–272. [CrossRef] [PubMed]

57. Kawa, L.; Barde, S.; Arborelius, U.P.; Theodorsson, E.; Agoston, D.; Risling, M.; Hökfelt, T. Expression of galanin and its receptors are perturbed in a rodent model of mild, blast-induced traumatic brain injury. *Exp. Neurol.* **2016**, *279*, 159–167. [CrossRef] [PubMed]

58. Lang, R.; Kofler, B. The galanin peptide family in inflammation. *Neuropeptides* **2011**, *45*, 1–8. [CrossRef] [PubMed]

59. Holzmann, B. Modulation of immune responses by the neuropeptide CGRP. *Amino Acids* **2013**, *45*, 1–7. [CrossRef] [PubMed]

60. Makowska, K.; Obremski, K.; Gonkowski, S. The Impact of T-2 Toxin on Vasoactive Intestinal Polypeptide-Like Immunoreactive (VIP-LI) Nerve Structures in the Wall of the Porcine Stomach and Duodenum. *Toxins* **2018**, *10*, 138. [CrossRef] [PubMed]

61. Korneev, K.V.; Arbatsky, N.P.; Molinaro, A.; Palmigiano, A.; Shaikhutdinova, R.Z.; Shneider, M.M.; Pier, G.B.; Kondakova, A.N.; Sviriaeva, E.N.; Sturiale, L.; et al. Structural Relationship of the Lipid A Acyl Groups to Activation of Murine Toll-Like Receptor 4 by Lipopolysaccharides from Pathogenic Strains of Burkholderia mallei, Acinetobacter baumannii, and Pseudomonas aeruginosa. *Front. Immunol.* **2015**, *6*, 595. [CrossRef] [PubMed]

62. Nguyen, M.D.; D'Aigle, T.; Gowing, G.; Julien, J.P.; Rivest, S. Exacerbation of motor neuron disease by chronic stimulation of innate immunity in a mouse model of amyotrophic lateral sclerosis. *J. Neurosci.* **2004**, *24*, 1340–1349. [CrossRef] [PubMed]

63. Choi, D.Y.; Liu, M.; Hunter, R.; Cass, W.A.; Pandya, J.D.; Sullivan, P.G.; Shin, E.J.; Kim, H.C.; Gash, D.M.; Bing, G. Striatal neuroinflammation promotes Parkinsonism in rats. *PLoS ONE* **2009**, *4*, e5482. [CrossRef] [PubMed]

64. Zhan, X.; Stamova, B.; Jin, L.W.; DeCarli, C.; Phinney, B.; Sharp, F.R. Gram-negative bacterial molecules associate with Alzheimer disease pathology. *Neurology* **2016**, *87*, 2324–2332. [CrossRef] [PubMed]

65. Liu, M.; Bing, G. Lipopolysaccharide Animal Models for Parkinson's Disease. *Parkinson's Dis.* **2011**, *2001*, 327089. [CrossRef] [PubMed]

66. Hoban, D.B.; Connaughton, E.; Connaughton, C.; Hogan, G.; Thornton, C.; Mulcahy, P.; Moloney, T.C.; Dowd, E. Further characterisation of the LPS model of Parkinson's disease: A comparison of intra-nigral and intra-striatal lipopolysaccharide administration on motor function, microgliosis and nigrostriatal neurodegeneration in the rat. *Brain. Behav. Immun.* **2013**, *27*, 91–100. [CrossRef] [PubMed]

67. Sharma, N.; Nehru, B. Characterization of the lipopolysaccharide induced model of Parkinson's disease: Role of oxidative stress and neuroinflammation. *Neurochem. Int.* **2015**, *87*, 92–105. [CrossRef] [PubMed]

68. Huang, B.; Liu, J.; Ju, C.; Yang, D.; Chen, G.; Xu, S.; Zeng, Y.; Yan, X.; Wang, W.; Liu, D.; et al. Licochalcone A Prevents the Loss of Dopaminergic Neurons by Inhibiting Microglial Activation in Lipopolysaccharide (LPS)-Induced Parkinson's Disease Models. *Int. J. Mol. Sci.* **2017**, *18*, 2043. [CrossRef] [PubMed]

69. Anitha, M.; Vijay-Kumar, M.; Sitaraman, S.V.; Gewirtz, A.T.; Srinivasan, S. Gut Microbial Products Regulate Murine Gastrointestinal Motility via Toll-like Receptor 4 Signaling. *Gastroenterology* **2012**, *143*, 1006–1016. [CrossRef] [PubMed]

70. Li, Z.X.; Li, Q.Y.; Qiao, J.; Lu, C.Z.; Xiao, B.G. Granulocyte-colony stimulating factor is involved in low-dose LPS-induced neuroprotection. *Neurosci. Lett.* **2009**, *465*, 128–132. [CrossRef] [PubMed]

71. Ye, Y.; Yang, Y.; Chen, C.; Li, Z.; Jia, Y.; Su, X.; Wang, C.; He, X. Electroacupuncture Improved Hippocampal Neurogenesis Following Traumatic Brain Injury in Mice through Inhibition of TLR4 Signaling Pathway. *Stem Cells Int.* **2017**, *2017*, 5841814. [CrossRef] [PubMed]

72. Nakasone, M.; Nakaso, K.; Horikoshi, Y.; Hanaki, T.; Kitagawa, Y.; Takahashi, T.; Inagaki, Y.; Matsura, T. Preconditioning by low dose LPS prevents subsequent LPS-induced severe liver injury via Nrf2 activation in mice. *Yonago Acta Medica* **2016**, *59*, 223–231. [PubMed]

73. Orio, M.; Kunz, A.; Kawano, T.; Anrather, J.; Zhou, P.; Iadecola, C. Lipopolysaccharide Induces Early Tolerance to Excitotoxicity via Nitric Oxide and CGMP. *Stroke* **2007**, *38*, 2812–2817. [CrossRef] [PubMed]

74. Chen, K.; Geng, S.; Yuan, R.; Diao, N.; Upchurch, Z.; Li, L. Super-Low Dose Endotoxin Pre-Conditioning Exacerbates Sepsis Mortality. *EBioMedicine* **2015**, *2*, 324–333. [CrossRef] [PubMed]

75. Mikołajczyk, A. Safe and Effective Anaesthesiological Protocols in Domestic Pig. *Ann. Warsaw. Univ. Life Sci SGGW Anim. Sci.* **2016**, *55*, 219–227.

76. Gonkowski, S.; Rychlik, A.; Nowicki, M.; Nieradka, R.; Bulc, M.; Całka, J. A population of nesfatin 1-like immunoreactive (LI) cells in the mucosal layer of the canine digestive tract. *Res. Vet. Sci.* **2012**, *93*, 1119–1121. [CrossRef] [PubMed]

Int. J. Mol. Sci. **2018**, *19*, 2551

77. Gao, H.M.; Jiang, J.; Wilson, B.; Zhang, W.; Hong, J.S.; Liu, B. Microglial Activation-Mediated Delayed and Progressive Degeneration of Rat Nigral Dopaminergic Neurons: Relevance to Parkinson's Disease. *J. Neurochem.* **2002**, *81*, 1285–1297. [CrossRef] [PubMed]

78. Gao, H.M.; Liu, B.; Zhang, W.; Hong, J.S. Synergistic Dopaminergic Neurotoxicity of MPTP and Inflammogen Lipopolysaccharide: Relevance to the Etiology of Parkinson's Disease. *J. Neurosci.* **2003**, *23*, 1228–1236. [CrossRef] [PubMed]

79. Marvaldi, L.; Thongrong, S.; Kozłowska, A.; Irschick, R.; Pritz, C.O.; Bäumer, B.; Ronchi, G.; Geuna, S.; Hausott, B.; Klimaschewski, L. Enhanced axon outgrowth and improved long-distance axon regeneration in sprouty2 deficient mice. *Dev. Neurobiol.* **2015**, *75*, 217–231. [CrossRef] [PubMed]

80. Kozłowska, A.; Mikołajczyk, A.; Klimaschewski, L.; Majewski, M. Fibroblast Growth Factor (FGF) Promotes Axon Elongation in Galanin Expressing Dorsal Root Ganglia (DRG) Neurons Supplying Urinary Bladder of the Pig-an in Vitro Study. *Eur. Urol. Suppl.* **2016**, *15*, e1233. [CrossRef]

81. Loy, B.; Apostolova, G.; Dorn, R.; McGuire, V.A.; Arthur, J.S.; Dechant, G. P38α and P38β Mitogen-Activated Protein Kinases Determine Cholinergic Transdifferentiation of Sympathetic Neurons. *J. Neurosci.* **2011**, *31*, 12059–12067. [CrossRef] [PubMed]

82. Lee, S.; Levine, J. Isolation and Growth of Adult Mouse Dorsal Root Ganglia Neurons. *Bio-Protocol* **2015**, *5*, e1601. [CrossRef]

83. Heinrich, T.; Hübner, C.A.; Kurth, I. Isolation and Primary Cell Culture of Mouse Dorsal Root Ganglion Neurons. *Bio-Protocol* **2016**, *6*, e1785. [CrossRef]

84. Ohtori, S.; Takahashi, K.; Moriya, H. Existence of brain-derived neurotrophic factor and vanilloid receptor subtype 1 immunoreactive sensory DRG neurons innervating L5/6 intervertebral discs in rats. *J. Orthop. Sci.* **2003**, *8*, 84–87. [CrossRef] [PubMed]

85. Christianson, J.A.; Liang, R.; Ustinova, E.E.; Davis, B.M.; Fraser, M.O.; Pezzonea, M.A. Convergence of Bladder and Colon Sensory Innervation Occurs at the Primary Afferent Level. *Pain* **2007**, *128*, 235–243. [CrossRef] [PubMed]

International Journal of
Molecular Sciences

Article

Effects of 1-Methyltryptophan on Immune Responses and the Kynurenine Pathway after Lipopolysaccharide Challenge in Pigs

Elisa Wirthgen [1,2], Winfried Otten [1,*], Margret Tuchscherer [1], Armin Tuchscherer [3], Grazyna Domanska [4], Julia Brenmoehl [5], Juliane Günther [5], Daniela Ohde [5], Werner Weitschies [6], Anne Seidlitz [6], Eberhard Scheuch [7] and Ellen Kanitz [1,*]

[1] Institute of Behavioural Physiology, Leibniz Institute for Farm Animal Biology (FBN),
D-18196 Dummerstorf, Germany; elisa.wirthgen@googlemail.com (E.W.);
mtuchsch@fbn-dummerstorf.de (M.T.)

[2] Department of Paediatrics, Institute University Children's Hospital Rostock, D-18057 Rostock, Germany

[3] Institute of Genetics and Biometry, Leibniz Institute for Farm Animal Biology (FBN),
D-18196 Dummerstorf, Germany; atuchsch@fbn-dummerstorf.de

[4] Institute of Immunology and Transfusion Medicine, University of Greifswald,
D-17475 Greifswald, Germany; grazyna.domanska@uni-greifswald.de

[5] Institute of Genome Biology, Leibniz Institute for Farm Animal Biology (FBN),
D-18196 Dummerstorf, Germany; brenmoehl@fbn-dummerstorf.de (J.B.);
j.guenther@fbn-dummerstorf.de (J.G.); ohde@fbn-dummerstorf.de (D.O.)

[6] Institute of Pharmacy, University of Greifswald, D-17489 Greifswald, Germany;
werner.weitschies@uni-greifswald.de (W.W.); anne.seidlitz@uni-greifswald.de (A.S.)

[7] Institute of Pharmacology, University of Greifswald, D-17489 Greifswald, Germany;
scheuch@uni-greifswald.de

[*] Correspondence: otten@fbn-dummerstorf.de (W.O.); ellen.kanitz@fbn-dummerstorf.de
(E.K.);Tel.: +49-38208-68809(W.O.); +49-38208-68807(E.K.)

Received: 9 August 2018; Accepted: 26 September 2018; Published: 2 October 2018

Abstract: An enhanced indoleamine 2,3-dioxygenase 1 (IDO1) activity is associated with an increased mortality risk in sepsis patients. Thus, the preventive inhibition of IDO1 activity may be a promising strategy to attenuate the severity of septic shock. 1-methyltryptophan (1-MT) is currently in the interest of research due to its potential inhibitory effects on IDO1 and immunomodulatory properties. The present study aims to investigate the protective and immunomodulatory effects of 1-methyltryptophan against endotoxin-induced shock in a porcine in vivo model. Effects of 1-MT were determined on lipopolysaccharide (LPS)-induced tryptophan (TRP) degradation, immune response and sickness behaviour. 1-MT increased TRP and its metabolite kynurenic acid (KYNA) in plasma and tissues, suppressed the LPS-induced maturation of neutrophils and increased inactivity of the animals. 1-MT did not inhibit the LPS-induced degradation of TRP to kynurenine (KYN)—a marker for IDO1 activity—although the increase in KYNA indicates that degradation to one branch of the KYN pathway is facilitated. In conclusion, our findings provide no evidence for IDO1 inhibition but reveal the side effects of 1-MT that may result from the proven interference of KYNA and 1-MT with aryl hydrocarbon receptor signalling. These effects should be considered for therapeutic applications of 1-MT.

Keywords: indoleamine 2,3-dioxygenase; kynurenine pathway; methyltryptophan; LPS; pig

---

1. Introduction

Indoleamine 2,3-dioxygenase 1 (IDO1)-mediated degradation of tryptophan (TRP) along the kynurenine pathway plays a significant role as a counter-regulatory mechanism within the inflammatory immune response [1]. Endotoxin-induced IDO1 activation results in the depletion of TRP and the generation of biologically active TRP metabolites such as kynurenine (KYN), kynurenic acid (KYNA) and quinolinic acid (QUIN), which have important immuno- and neuromodulatory properties [2–5]. In sepsis, it is proposed that an early hyper-inflammatory phase is followed or overlapped by a prolonged state of immunosuppression, which are potentially life-threatening by the development of septic shock or sepsis-induced immunoparalysis, respectively [6]. Clinical studies have shown an association of enhanced IDO1 activity in the early pro-inflammatory stage of sepsis with an increased disease severity [7] and increased hypotension, which contributes to septic shock due to acute circulatory failure [8]. However, it remains unclear whether the increased IDO1 activity reflects a compensatory response or contributes to overshooting inflammation. During ongoing sepsis, there are indices that a prolonged elevated IDO1 activity increases the risk for poor outcome by provoking the maintenance of a protracted immunosuppressive phase, diminishing immune responsiveness and host defence [7,9,10]. In these clinical studies, the enhanced IDO1 activity was characterized by an accelerated TRP degradation, resulting in an increased KYN/TRP ratio or increased levels of KYN or KYNA. The finding that IDO1 knockout mice are protected against lipopolysaccharide (LPS)-induced septic shock [11] also supports the assumption that an enhanced endotoxin-induced activation of IDO1 provokes detrimental effects within conditions of sepsis. Compared with wild-type mice, IDO1 knockout mice had a higher survival rate within 48 h after an endotoxin shock accompanied by lower levels of the pro-inflammatory cytokines TNF-α and IL-12 and simultaneously higher levels of the anti-inflammatory cytokine IL-10, 4 h after LPS. These results indicate that IDO1 may elicit detrimental pro-inflammatory effects in the early stage of sepsis in contrast to the mainly immunosuppressive effects assumed during ongoing sepsis. In view of these findings, the inhibition of IDO1 activity may be a promising strategy for the prevention of endotoxin-induced shock e.g., after surgery. In patients, who expected a surgery, psychological stress enhanced IDO1 activity already in the pre- and also in the postoperative period, which may contribute to postoperative modulation of immune response [3]. Therefore, the inhibition of IDO1 before a scheduled surgery may reduce the risk for poor outcome associated with an enhanced IDO1 activity during sepsis. Further studies, e.g., on cancer in humans, are investigating the inhibition of IDO1 activity using applications of 1-methyltryptophan (1-MT) as a possible therapeutic approach [12]. 1-MT has been shown to inhibit IDO1 activation in mice, concomitant with a restoration of T-cell proliferation in vitro [13] and of antibacterial defence concurrent with a reduction of depression-like behaviour [3]. Furthermore, the application of 1-MT reduced TNF-α and increased IL-10 in response to endotoxin shock, supporting the findings in IDO1 knockout mice [11].

In a previous study of our group, a pig model for systemic IDO1 activation by intraperitoneal (i.p.) LPS stimulation was established. It resulted in the LPS-induced increase of inflammatory cytokines such as TNF-α and IL-10 followed by the depletion of TRP and an increased generation of TRP metabolites such as KYN and KYNA in plasma, and IDO1 protein expression in blood and various organs [14]. In a subsequent study, the pharmacokinetic properties of 1-MT after subcutaneous (s.c.) application were evaluated and a model for systemic accumulation of 1-MT in plasma and tissues was developed [15]. It is the aim of the present study to examine the protective and immunomodulatory effects of 1-MT against LPS-induced shock in a porcine model, because pigs (*Sus scrofa domestica*) provide obvious advantages with respect to their relevance to human pathophysiology compared with rodents. This species closely resembles humans in anatomy, genetics and physiology and is increasingly used as a model for humans in biomedical research [16–18]. We hypothesize that 1-MT may inhibit IDO1 activity, resulting in reduced catalytic degradation of TRP and altered metabolite profile along the KYN pathway after LPS challenge. We suppose that this could result in an attenuation of the LPS-induced innate immune response due to the complex immunomodulatory properties of KYN pathway. Hence, the immune response was characterized by cytokines, haematopoiesis and sickness behaviour. An altered activity of IDO1 by 1-MT and its effects on KYN pathway were validated by the measurement of TRP and the downstream metabolites KYN, KYNA and QUIN.

2. Results

2.1. 1-MT Modifies TRP and Its Metabolites in Plasma

According to the experimental design shown in Figure 6, 1-MT was applied 84, 60, 36 and 12 h before the LPS injection (0 h) and 12 h after LPS. Pairwise comparisons revealed no differences in the basal concentrations of TRP, KYN, KYNA, QUIN and the KYN/TRP ratio immediately before the first 1-MT application at the time point −84 h (Figure 1A–E). The repeated s.c. administrations of 1-MT significantly increased plasma concentrations of TRP and KYNA at 60, 36 and 12 h before LPS, whereas KYN, the KYN/TRP ratio and QUIN were not affected by 1-MT.

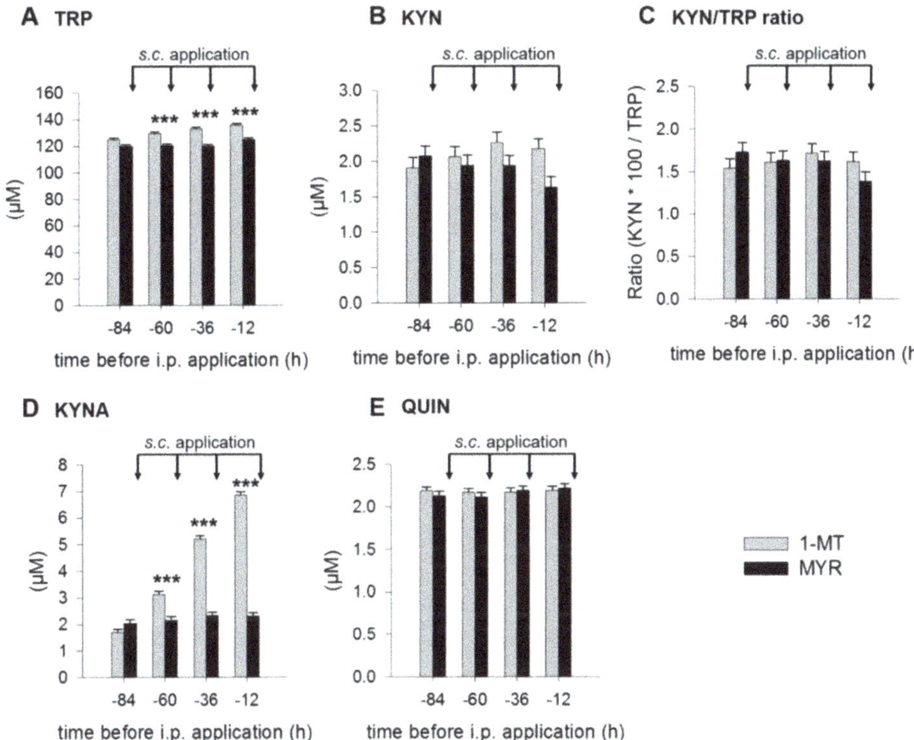

Figure 1. Effects of repeated s.c. 1-MT applications on plasma TRP (**A**), KYN (**B**), KYN/TRP ratio (**C**), KYNA (**D**) and QUIN (**E**) before LPS application. Pre-treatment with MYR was used as a control. TRP and its metabolites were measured using MS/MS and the KYN/TRP ratio was used as a marker for IDO1 activity. The s.c. 1-MT or MYR applications (indicated by black arrows) were injected directly after each blood sampling. The results are presented as LS-means ± SE. Significant differences between the 1-MT- and MYR-pre-treated groups were calculated using the Tukey–Kramer test and are indicated for each time point. 1-MT: $n = 24$; MYR: $n = 23$; *** $p < 0.001$.

Immediately before the LPS application (0 h), KYNA concentrations were significantly increased by 1-MT (Figure 2D, Figure A2D). Likewise, 1-MT generally increased plasma TRP and KYNA after application of LPS or NaCl and reduced plasma concentrations of QUIN (Table 1). Administration of LPS induced a reduction of TRP and an increase of KYN, KYN/TRP ratio, KYNA and QUIN compared to NaCl (Table 1). Significant interactions between 1-MT and LPS effects were found for TRP and KYNA. LPS significantly decreased TRP in Myritol (MYR)-treated controls but not in 1-MT animals and significantly increased KYNA in 1-MT but not in MYR animals (Table 1).

Table 1. Pairwise comparisons of the pre-treatment 1-MT vs. MYR and the treatment LPS vs. NaCl in plasma using a Tukey–Kramer test.

	1-MT		MYR		SE	p Value		
	LPS LSM	NaCl LSM	LPS LSM	NaCl LSM		1-MT vs. MYR	LPS vs. NaCl	Interaction
TRP (μM)	127.38 [A]	129.67 [A]	115.63 [B]	122.29 [C]	1.10	<0.001	<0.001	<0.05
KYN (μM)	1.64 [A]	0.99 [B]	1.54 [A,B]	1.17 [A,B]	0.16	0.78	<0.01	0.40
KYN/TRP ratio	1.22 [A]	0.83 [B]	1.30 [A]	0.98 [A,B]	0.11	0.22	<0.001	0.69
KYNA (μM)	11.9 [A]	11.5 [B]	2.1 [C]	2.1 [C]	0.07	<0.001	<0.001	<0.01
QUIN (μM)	1.98 [A]	1.83 [A]	2.18 [B]	1.98 [A]	0.05	<0.01	<0.001	0.36

Concentrations of TRP, KYN, KYNA, Ratio (KYN $\times$ 100/TRP) are presented as LS-means (LSM) and SE. Within a row, values with different superscript letters differ with $p < 0.05$. 1-MT: $n = 24$, MYR: $n = 23$; LPS: $n = 23$; NaCl: $n = 24$.

Detailed pairwise comparisons between the different sampling times show that LPS induced a significant decrease of plasma TRP in both 1-MT and MYR pre-treated animals between 3 and 12 h after i.p. application (Figure 2A). However, only in MYR treated animals, TRP was also reduced 12 and 24 h after LPS compared to 0 h. LPS caused a significant increase of KYN and the KYN/TRP ratio at 6 h after administration, which was not affected by 1-MT, respectively (Figure 2B,C). KYNA concentrations were significantly elevated in 1-MT animals at all sampling times (Figure 2D). Further, KYNA concentrations increased in 1-MT animals between 0 to 3 h after LPS, whereas no changes were found in the MYR group. Plasma levels of QUIN increased from 1 to 12 h after LPS in the MYR-treated but not in 1-MT-treated animals (Figure 2E). There was a significant decrease of QUIN from 12 to 24 h after LPS, which was not affected by 1-MT.

For a better evaluation of the LPS effect on KYNA in 1-MT animals, differences compared with basal levels were calculated and shown as Figure A1. In 1-MT animals, LPS application induced an additional increase in KYNA compared with NaCl-treated 1-MT animals. With the exception of KYNA, the application of NaCl in control animals induced no time-dependent changes of TRP metabolites (Figure A2).

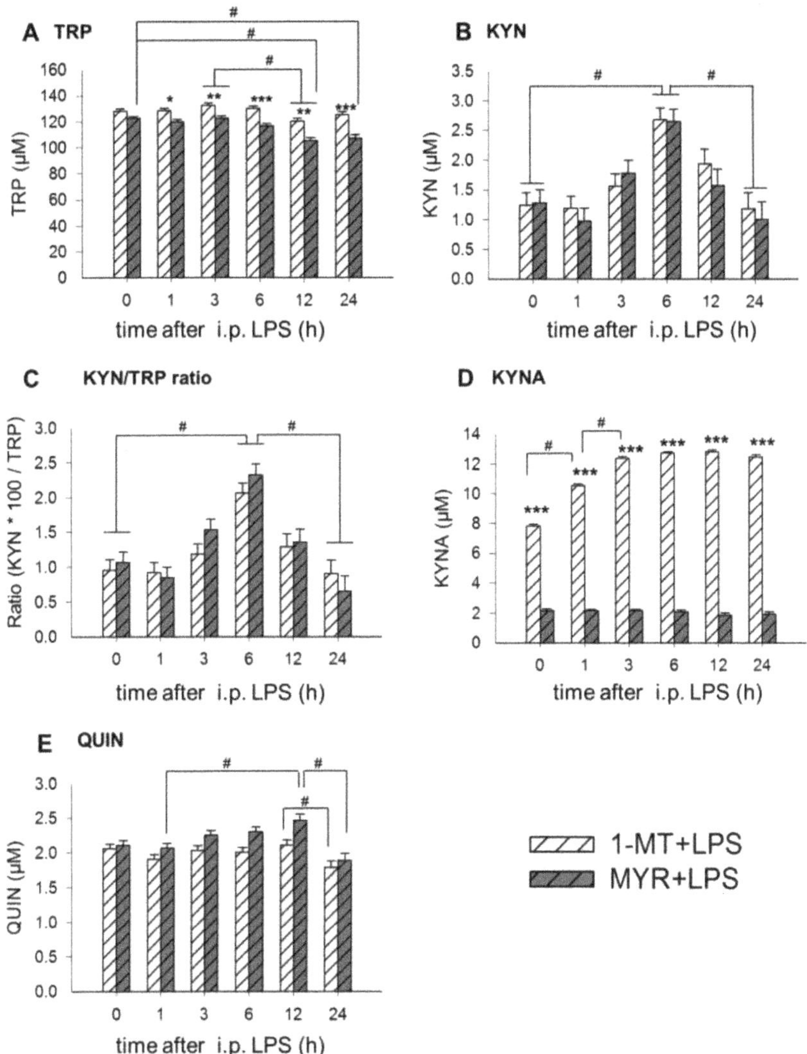

Figure 2. Effects of 1-MT on LPS-induced changes of TRP (**A**), KYN (**B**), KYN/TRP ratio (**C**), KYNA (**D**) and QUIN (**E**) in plasma. As a control for 1-MT-induced effects the excipient MYR was used. TRP and its metabolites were measured using MS/MS and the KYN/TRP ratio was used as a marker for IDO1 activity. LPS was injected i.p. directly after blood sampling at 0 h. The animals received the fifth s.c. 1-MT or MYR injection at 12 h after LPS. The results are presented as LS-means ± SE. Tukey–Kramer procedure was used for all pairwise comparisons. Significant differences are indicated between 1-MT and MYR groups for each time point by asterisks, and between the sampling times by hashes. For the sampling times 0, 1, 3, 6 h: $n = 11$-12/group (total $n = 47$); for time point 24 h: $n = 5$-6/group (total $n = 23$); *** $p < 0.001$, ** $p < 0.01$, * $p < 0.05$, # $p < 0.01$.

2.2. 1-MT Modifies TRP Metabolites in Tissues

TRP, KYN and KYNA were detectable in several tissues and pairwise comparisons of 1-MT vs. MYR and LPS vs. NaCl groups are shown in Table 2. 1-MT significantly increased TRP concentrations in the prefrontal cortex, adrenal gland, muscle, lung, liver, heart, hippocampus, hypothalamus and spleen. KYN was detectable in liver, lung, spleen and hippocampus, but was not affected by 1-MT pre-treatment (Table 2). KYNA was detectable in lung, kidney and liver, and was increased in lung and kidney by 1-MT pre-treatment (Table 2).

Table 2. Pairwise comparisons of 1-MT vs. MYR and of LPS vs. NaCl in peripheral and brain tissues using Tukey–Kramer test.

	1-MT LSM	MYR LSM	SE	*p* Value	LPS LSM	NaCl LSM	SE	*p* Value
TRP (nmol/g protein)								
Amygdala	1094	1067	88	0.83	892	1269	88	<0.01
Prefrontal cortex	563	416	29	<0.001	477	502	29	0.47
Adrenal gland	1062	855	46	<0.001	926	992	46	0.22
Muscle	484	385	19	<0.001	397	472	18	<0.01
Lung	489	393	32	<0.05	396	486	31	<0.05
Liver	715	619	33	<0.05	597	737	32	<0.01
Heart	685	569	41	<0.05	569	685	39	<0.05
Hippocampus	224	178	11	<0.01	189	213	11	0.12
Hypothalamus	79	67	4	<0.05	70	76	4	0.19
Spleen	746	636	31	<0.05	687	696	32	0.84
Kidney	1337	1249	42	0.10	1211	1375	42	<0.01
Thyroid gland	459	495	30	0.31	469	485	29	0.67
KYN (nmol/g protein)								
Liver	604.86	590.86	35.62	0.64	586.74	608.98	34.99	0.48
Lung	123.16	156.84	21.97	0.29	180.15	99.85	22.08	<0.05
Spleen	12.83	11.45	1.79	0.58	18.28	6.00	1.80	<0.001
Hippocampus	19.69	19.74	1.78	0.98	19.50	19.93	1.78	0.87
KYNA (nmol/g protein)								
Lung	3.83	2.4	0.34	<0.01	3.34	2.90	0.005	0.30
Kidney	0.71	0.57	0.05	<0.05	0.65	0.63	0.05	0.80
Liver	5.84	4.75	0.46	0.06	5.64	4.95	0.47	0.23

Results of TRP, KYN and KYNA concentrations are presented as LS-means (LSM) and SE. 1-MT: *n* = 24; MYR: *n* = 23; LPS: *n* = 23; NaCl: *n* = 24.

Application of LPS induced a significant decrease of TRP concentrations in the amygdala, muscle, lung, liver, heart and kidney and an increase of KYN in lung and spleen (Table 2). A significant interaction between 1-MT and LPS effects was only found for TRP in the adrenal gland. LPS significantly decreased TRP in the adrenal gland of MYR but not of 1-MT animals ($p < 0.001$).

2.3. Application of LPS Induces A Reduction of 1-MT in Plasma and Tissue

1-MT was not detectable either in plasma or in tissues of MYR animals. In 1-MT animals, plasma 1-MT concentrations decreased 6, 12 and 24 h after LPS compared to concentrations at 0 h (Figure 3A). In the NaCl group, 1-MT was decreased after 12 h, the time of the next 1-MT application, and increased until 24 h. Comparison of both treatments revealed that after 24 h plasma 1-MT was significantly lower in LPS than in NaCl animals. In tissues, the application of LPS also induced a reduction of 1-MT in liver, lung, kidney, spleen and heart (Figure 3B).

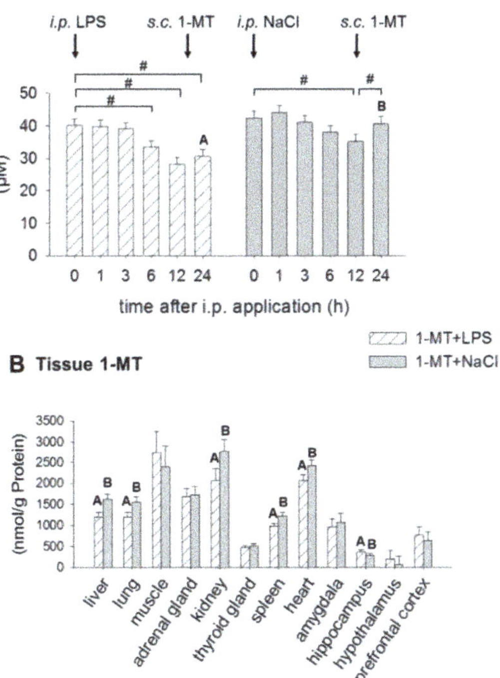

Figure 3. Effects of LPS on 1-MT concentrations in plasma (**A**) and tissues (**B**). Treatment with NaCl was used as a control. MYR-pre-treated animals are not shown due to the absence of 1-MT. 1-MT was quantified using MS/MS. LPS or NaCl (indicated by black arrows) were injected i.p. directly after blood sampling at 0 h. The animals received the fifth s.c. 1-MT injection at 12 h after LPS/NaCl treatment as indicated. The data are presented as LS-means ± SE. In plasma (**A**), significant differences between the time points were calculated using the Tukey–Kramer test and indicated by hashes within each treatment group. Significant differences between LPS and NaCl groups ($p < 0.05$) are indicated using differing letters (A, B). For time points 0, 1, 3, 6 h: $n = 12$/group. For time point 24 h: $n = 6$/group. The data for 1-MT in tissues (**B**) include both sampling times (6 and 24 h, $n = 12$/group) after i.p. application. Significant differences between LPS and NaCl treatment were calculated using the Tukey–Kramer test and are indicated for each tissue, respectively. # $p < 0.01$.

2.4. 1-MT Modulates Inflammatory Response after LPS Application

White blood cells, including lymphocytes and neutrophils were measured before (0 h) and 6 and 24 h after LPS. Summarized over all sampling times, number of lymphocytes was significantly increased in 1-MT animals, irrespective of LPS or NaCl administration (1-MT: 8.3 ± 0.6, MYR: 7.1 ± 0.6 cells × 10^6/mL, $p < 0.05$). In contrast, the number of neutrophils was generally reduced in 1-MT animals (1-MT: 7.7 ± 0.4, MYR 9.1 ± 0.4 cells × 10^6/mL, $p < 0.05$). LPS affected the cell number of lymphocytes, neutrophils, myelocytes and band neutrophils and the N/L ratio 6 h after i.p. application. In contrast, NaCl induced no significant changes 6 and 24 h after application. Pairwise comparisons showed no significant differences between 1-MT and MYR pre-treated groups immediately before LPS at time 0 h. LPS induced a decrease of lymphocytes 6 h after i.p. application in both 1-MT and MYR pre-treated animals followed by an increase at 24 h (Figure 4A). An increase of neutrophils was observed in MYR, but not in 1-MT treated animals 6 h after LPS (Figure 4B). The LPS-induced reduction of lymphocytes and increase of neutrophils resulted in an increased neutrophil to lymphocyte (N/L) ratio at 6 h after LPS (Figure 4C) in both 1-MT and MYR pre-treated groups. However, this increase was significantly reduced in 1-MT animals. This was mainly caused by a reduced increase of myelocytes (Figure 4D) and band neutrophils (Figure 4E) in 1-MT animals 6 h after LPS application. Cell number of segmented neutrophils was not affected by 1-MT or LPS (Figure 4F).

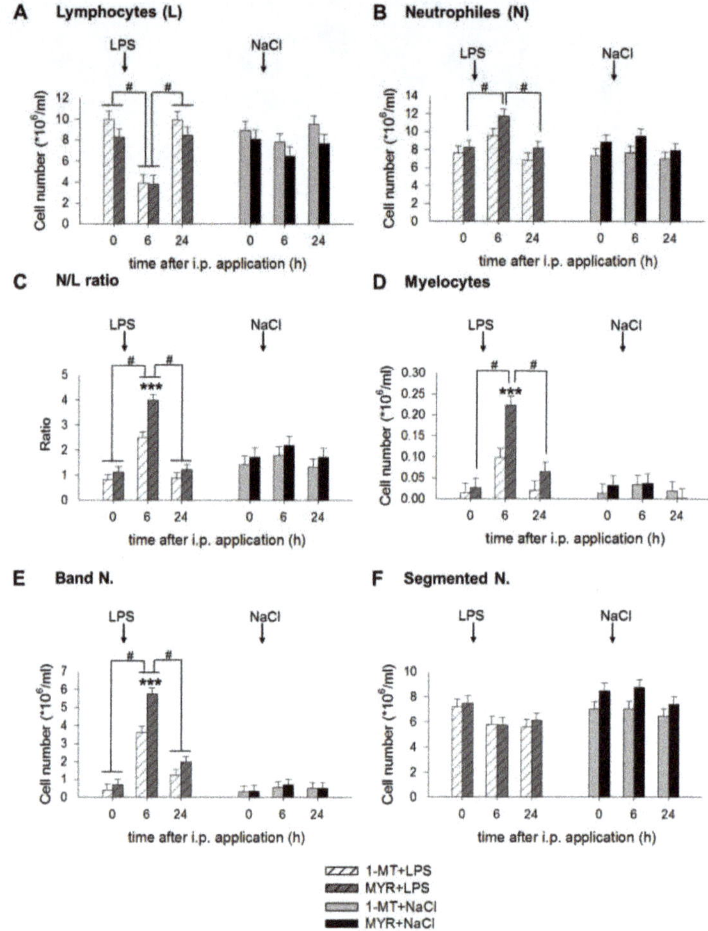

Figure 4. Effects of LPS on the number of lymphocytes (**A**), neutrophils (**B**), N/L ratio (**C**), myelocytes (**D**), band neutrophils (**E**) and segmented neutrophils (**F**) and its interference by 1-MT. Treatment with NaCl and pre-treatment with MYR were used as control groups. LPS or NaCl were injected i.p. directly after blood sampling at 0 h. The animals received the fifth s.c. 1-MT injection at 12 h after LPS/NaCl treatment to prevent a decline in 1-MT levels during the period of the LPS response. The cell number was measured by differential leukocyte counts using a total of 200 leukocytes per animal and time point. All data are presented as LS-means ± SE. Significant differences between 1-MT- and MYR-pre-treated groups were calculated using the Tukey–Kramer test and are indicated for each time point/period by asterisks. Significant differences between the time points are indicated by hashes within each treatment group. For time points 0 and 6 h: $n = 11$-12/group (total $n = 47$); for time point 24 h: $n = 5$-6/group (total $n = 23$); *** $p < 0.001$, # $p < 0.01$.

The concentrations of TNF-α and IL-10, measured immediately before the LPS application (0 h), were not affected by 1-MT. Treatment with LPS induced a significant increase of plasma TNF-α and

IL-10 with the highest concentrations 1 h after LPS and an increase of skin temperature (Figure 5A–C). Pairwise comparisons show that these increases were not affected by 1-MT. Administration of LPS induced sickness behaviour within the 5-h period after administration as shown by inactivity (relative frequency: LPS: 99.97 ± 0.008%, NaCl: 83.36 ± 1.07%, $p < 0.01$) and the occurrence of sickness symptoms (mean number: LPS: 0.69 ± 0.03, NaCl: 0.006 ± 0.01, $p < 0.001$). The severity of sickness after LPS was not affected by 1-MT. However, the pre-treatment with 1-MT increased the inactivity of NaCl-treated control pigs (relative frequency: 1-MT + NaCl: 87.34 ± 1.29%, MYR + NaCl: 78.43 ± 1.67%, $p < 0.01$).

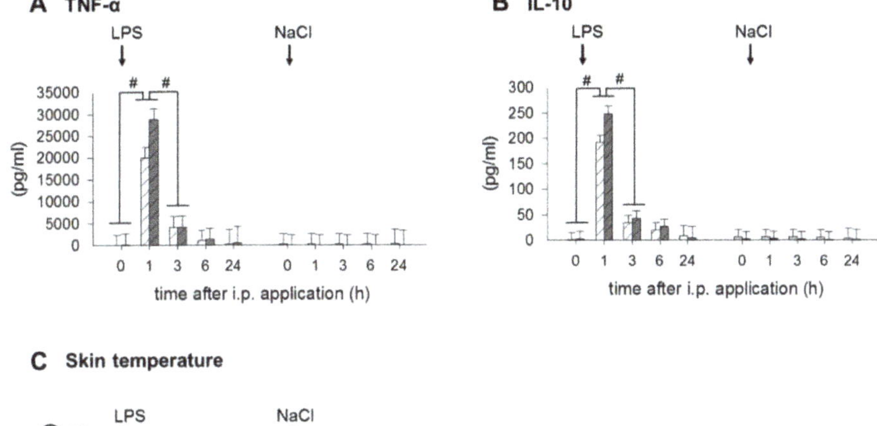

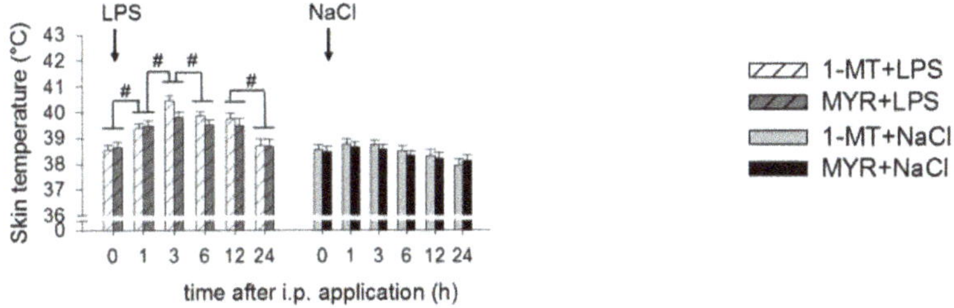

Figure 5. Effects of LPS on plasma TNF-α (**A**), plasma IL-10 (**B**) and skin temperature (**C**) and its interference by 1-MT. Treatment with NaCl and pre-treatment with MYR were used as control groups. LPS or NaCl were injected i.p. directly after blood sampling at 0 h, respectively. The animals received the fifth s.c. 1-MT injection at 12 h after LPS/NaCl treatment to prevent a decline in 1-MT levels during the period of the LPS response. The cytokines TNF-α and IL-10 were quantified by ELISA. Skin temperature was measured using an infrared thermometer directly before blood sampling. All data are presented as LS-means ± SE. For multiple pair-wise comparisons Tukey–Kramer procedure was used. Significant differences between time points are indicated by hashes. For time points 0, 1, 3, 6 h: $n = 11$-12/group (total $n = 47$); for time points 12 and 24 h: $n = 5$-6/group (total $n = 23$); # $p < 0.01$.

2.5. 1-MT Modifies mRNA Expression of Porcine Lung Fibroblasts

A supplementary experiment was conducted to evaluate potential interactions of 1-MT or TRP with aryl hydrocarbon receptor (AhR) or inflammatory signalling pathways in a primary cell culture of porcine lung fibroblasts (See Appendix **??**). The results show that 1-MT significantly increased the mRNA expression of cytochrome P450, family 1, subfamily A, polypeptide 1 (CYP1A1), a marker for AhR activation, compared to the TRP and medium control. The expression of IL-8, granulocyte-macrophage colony-stimulating factor (GM-CSF) and inducible nitric oxide synthase (iNOS) was not affected neither by the 1-MT nor the TRP treatment (Figure A3). TNF-α was expressed on very low levels near or below the detection limit and was not increased, neither by stimulation with 1-MT nor with TRP (data not shown).

3. Discussion

In this study, immunomodulatory effects of 1-MT on LPS-induced IDO1 activation were investigated in a porcine model to reveal new insights into the IDO1-mediated effects during the early stage of inflammatory immune response. Our results show that administrations of 1-MT did not diminish the LPS-induced conversion of TRP to KYN, indicating no inhibition of IDO1 activity in vivo. However, 1-MT increased TRP and enhanced the formation of KYNA, a metabolite in the KYN pathway with various immuno- and neuromodulatory functions. In addition, 1-MT caused a modulation of the LPS-induced increase of N/L ratio.

3.1. 1-MT-Induced Increases in TRP and KYNA

The results of this study show that 1-MT increased the concentrations of TRP in plasma and several tissues, independently of LPS treatment. This may be the result of reduced TRP degradation or increased availability associated with the administration of 1-MT. In vivo, plasma TRP concentrations are affected by both IDO1 and tryptophan 2,3-dioxygenase 2 (TDO2). Homeostasis of TRP is mainly regulated by TDO2 [19], which is not inhibited by 1-MT [20,21], whereas IDO1 expression is predominantly induced by inflammatory stimuli, such as the cytokines TNF-α or IFN-γ [19]. The increased levels of TRP in our study may reflect a reduced degradation as a result of IDO inhibition due to 1-MT. However, the finding that the LPS-induced increase of KYN was not diminished by 1-MT does not support the assumption of IDO inhibition. According to the manufacturer's information, 1-MT has a purity of at least 95%. As confirmed by our own HPLC analyses, the L-1-MT contains approximately 5% TRP (data not shown), which corresponds to an s.c. TRP uptake of 0.05 g TRP per day, in addition to a daily dietary uptake of approximately 1.2 to 1.6 g TRP from the digestive tract. Studies in human glioblastoma cells indicate that TRP contamination of commercially available 1-MT (L-isomer) is converted to KYN in vitro [22]. In addition to TRP contamination of 1-MT, a removal of the methyl group of 1-MT by enzymes or chemical processes would result in an increase in TRP concentrations that could not be distinguished from endogenous TRP. Whether TRP contamination of 1-MT and/or demethylation caused the increase in TRP could not be clarified in this study.

Our results show that the production of KYNA, but not KYN, was increased in plasma and tissues in response to 1-MT, independently of LPS. This indicates an enhanced degradation of TRP to one branch of the KYN pathway, which may be a result of increased levels of available TRP. This finding is supported by studies in rats, in which dietary supplementation of TRP induced an increase in urinary excretion of TRP metabolites including KYNA [23]. Furthermore, studies in mice have shown that a repeated application of 1-MT, which also contains low amounts of TRP, results in increased plasma concentrations of KYNA [3]. In recent years, numerous in vivo and in vitro studies have been directed toward the immunomodulatory functions of KYNA. It has been shown that KYNA mediates immunosuppressive effects such as a reduced cytokine response during inflammation in mice [3,5] or acts as a mediator of early recruitment of human monocytes [2]. Moreover, KYNA has been described as a potent natural AhR ligand [24]. AhR signalling is involved in counter-regulatory effects such as reprogramming of the LPS-induced cytokine response and suppression of adaptive immune response [25,26]. Furthermore, AhR is an important regulator of cell physiology and contributes to the proper function of the hepatic, haematopoietic, cardiovascular and immune system [27].

3.2. 1-MT Does Not Inhibit the Production of KYN But Is Diminished after LPS Treatment

Previous studies in rodents and in cell cultures showed that 1-MT (L-isomer) reduced the production of metabolites in the KYN pathway and prevented the depletion of TRP [3,13,28,29]. In studies using a recombinant IDO1 enzyme in cell-free assay systems, the L-isomer of 1-MT ($K_i = 19 \ \mu M/L$) was found to be a more potent inhibitor of IDO1 than the D-isomer

(K_i > 100 μM/L) [13]. However, the results of this study provide no evidence of an inhibitory effect of the L-isomer of 1-MT on the LPS-induced KYN production in plasma, lung and spleen. Nevertheless, the LPS-induced depletion of TRP was attenuated by the 1-MT-induced general increase in TRP. In vitro studies have shown that the 1-MT-induced increase in TRP may impede the antimicrobial and immunoregulatory functions of LPS-induced TRP depletion [22], facilitating chronic infections due to impaired pathogen growth arrest [12].

One reason for the lack of IDO1 inhibition could be that the concentration of 1-MT was too low to inhibit IDO1 activity in vivo. A phase I trial of tumour patients using 1-MT (D-isomer) as an IDO1 inhibitor has shown that doses higher than 1200 mg 1-MT/patient do not increase peak serum levels [30], indicating a limited accumulation of the applied 1-MT. This finding is in accordance with our previous findings showing that a steady-state 1-MT concentration is already reached after the second 1-MT injection of 1000 mg/animal/day [15], increasing 1-MT to plasma levels similar to those of TRP. Even in lung and spleen, in which 1-MT was approximately 2- to 3-fold higher than the corresponding TRP levels [15], no inhibition of LPS-induced KYN production was detected, assuming that 1-MT did not inhibit IDO1 activity.

It has been shown that 1-MT binds to the ferrous IDO1 enzyme, but the additional methyl group prevents degradation along the KYN pathway due to steric effects [20]. The results of the present study reveal that LPS induces a reduction of 1-MT in plasma and several tissues, which might be a result of an increased urinary clearance and/or of catalytic degradation. There are few indices showing that both IDO1 and TDO2 are able to catalyse 1-MT (L-isomer) with low affinity (K_m = 150 μM) in comparison to TRP (K_m = 7 μM), resulting in the generation of KYN or methyl-KYN [31,32]. However, the results of our study show no evidence of additional production of KYN after LPS treatment. Furthermore, in the subsequent HPLC analysis, using α-methyl-KYN as a reference standard, no methyl-KYN was detectable in samples after LPS treatment (data not shown).

3.3. 1-MT Mediates Immunomodulatory Effects

In the experimental design of this study, IDO1 activation was induced by LPS resulting in an inflammatory response that provoked an increase in pro- and anti-inflammatory cytokines such as TNF-α or IL-10. Contrary to findings in mice [11], no significant effect of 1-MT on the plasma cytokine response or sickness severity was found. However, the results show that 1-MT induced a modulation of LPS-induced immune response. 1-MT reduced the increase in the immature myelocytes and band neutrophils in response to LPS, indicating a suppression of adequate neutrophil maturation. This may also explain the generally reduced number of neutrophils by 1-MT. Furthermore, 1-MT reduced the LPS-induced increase of the N/L ratio, which is used as an early marker of acute inflammation [33,34]. These results support findings in human cancer and dendritic cells, revealing that 1-MT has transcriptional effects that may promote immunosuppressive effects [35,36].

1-MT enhanced the inactivity of the NaCl-treated animals and elevated number of lymphocytes, which might reflect a response to a repeated inflammatory stimulation [37,38]. In vitro, there are indications that 1-MT interferes with TLR signalling in dendritic cells, independently of IDO1 activity [35]. Furthermore, there are indications in human and mouse cells that 1-MT itself acts as an agonist for AhR, enabling the transcriptional expression of AhR-specific target genes [39], which may affect inflammatory pathways. Results of the supplementary experiment show that 1-MT significantly increased the mRNA expression of CYP1A1, which is described as a sensitive marker for AhR activation [40], thus confirming the findings in mouse and human cell lines [39]. In addition, no increased expression of inflammatory mediators such as IL-8, GM-CSF, TNF-α and iNOS was detected on transcription levels after 1-MT incubation. Further, elevated TRP levels, simulating the 5% of TRP contained in 1-MT, had no significant effects on gene expression. Our findings that 1-MT itself acts as an AhR ligand and increases plasma and tissue concentrations of the endogenous AhR ligand KYNA in vivo should be taken into account when using 1-MT for therapeutic applications and require further investigation.

4. Materials and Methods

4.1. Animals

Male German Landrace pigs (n = 96), bred and raised in the experimental pig unit of the Leibniz Institute for Farm Animal Biology, were used in two experiments. All pigs received standard processing (oral iron supplementation and castration) within the first three days of life. At the beginning of the experiments, the pigs were seven weeks old and weighed between 12 and 18 kg. The pigs were fed a commercial diet and had free access to water. All procedures involving animal handling and treatment were in accordance with the German animal protection law and were approved by the responsible authority (Landesamt für Landwirtschaft, Lebensmittelsicherheit und Fischerei, Mecklenburg-Vorpommern; Rostock; Germany; LALLF M-V/TSD/7221.3-1.1-027/10; 02 Juni 2010).

4.2. Experimental Design

The effects of 1-MT application on TRP metabolism and endotoxin-induced immune response were investigated in two experiments (Figure 6). The first experiment focused on 1-MT-induced effects on blood and tissue parameters before and during the 24 h period after a single endotoxin application (n = 48). In the second experiment, the influence of 1-MT on LPS-induced sickness behaviour and haematopoiesis was investigated (n = 48). Both experiments were conducted in three replicates with 16 animals each. One week prior to the commencement of the experiments, animals were housed in single pens. In each experiment, half of the animals received daily administrations of 1 g 1-MT (L-isomer, purity 95%; Sigma-Aldrich, Deisenhofen, Germany) over a period of five days. The dose was selected according to a preceding experiment. At the time of LPS injection (0 h) steady state plasma levels of 1-MT were achieved and 1-MT was accumulated in tissues at a level equal to or higher than TRP (for details and pharmacokinetics see Wirthgen et al., 2016 [15]). Injections were given s.c. in the popliteal fossa of the hind legs and split in two doses of 0.5 g of 1-MT in 4.0 mL of triglyceride solution Myritol®318 (MYR) (Caesar und Loretz GmbH, Hilden, Germany), which was used as an excipient. Control animals received an equivalent volume of the MYR solution. Injections of MYR or 1-MT were given at 08:00 p.m. on five consecutive days (84, 60, 36 and 12 h before, and 12 h after the i.p. LPS/NaCl application). Health status was continuously checked by visual inspection twice daily, and the daily feed uptake was measured. Repetitive s.c. administration of MYR or 1-MT (suspended in MYR) caused local swelling around the puncture sites. However, no fever response or significant changes in feed uptake or body weight were observed.

At 8:00 a.m. on the fifth day, half of the pre-treated and control animals received a single i.p. administration of LPS with a dose of 50 µg/kg live weight (*Escherichia coli* O111:B4; Sigma-Aldrich, Deisenhofen, Germany) dissolved in 3 mL sterile endotoxin-free 0.9% NaCl according to a previous description [41]. The other animals were treated with an equivalent volume of NaCl. Twelve hours before LPS/NaCl application, the feed was removed to avoid an interference of feed uptake with TRP metabolism during the sampling period immediately after endotoxin challenge.

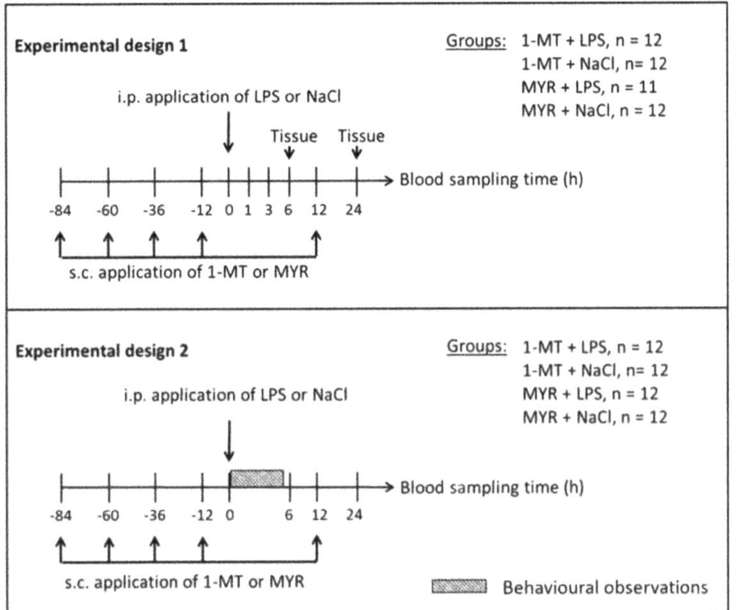

Figure 6. The effects of 1-MT application on TRP metabolism and LPS-induced immune response were investigated in two experiments. In both experiments, 1-MT and its control MYR were repeatedly injected subcutaneously (s.c.) as indicated, aiming at maximum levels of 1-MT at the time of LPS or NaCl injections at 0 h according to the 1-MT kinetics. In experiment 1 (n = 47), blood samples were taken before and at 0, 1, 3, 6, 12 and 24 h after LPS/NaCl application and tissue was sampled after 6 h (n = 24) and after 24 h (n = 23). In experiment 2 (n = 48), blood samples were taken before and at 0, 6, 12 and 24 h after LPS/NaCl application and behaviour was observed in the period from 0 to 5 h. Both experiments were conducted in three replicates with 15 or 16 animals each.

4.2.1. Experiment 1: Effects of 1-MT on Inflammatory Response and KYN Pathway

Because one animal had to be excluded due to health problems, 47 animals were used in total in this experiment, distributed in the following treatment groups as described above: 1-MT + LPS (n = 12), 1-MT + NaCl (n = 12), MYR + LPS (n = 11) and MYR + NaCl (n = 12). Before every s.c. 1-MT/MYR application, blood was collected to analyse plasma concentrations of 1-MT and TRP metabolites. In addition, blood samples were collected immediately before i.p. LPS/NaCl administration (0 h) and at 1, 3, 6, 12 and 24 h after administration for analyses of 1-MT, TRP metabolites and cytokines (TNF-α, IL-10) in plasma. Blood sampling was carried out while pigs were in a supine position by anterior vena cava puncture with the whole procedure lasting less than one minute. Blood samples were collected in ice-cooled tubes containing EDTA or heparin and centrifuged at $2000 \times g$ for 10 min at 4 °C. The blood plasma was then stored at -80 °C until analysis. Before s.c. 1-MT injections and before blood samplings during LPS challenge, skin temperature was measured in the inguinal region using an infrared thermometer (ThermoScan IRT 4020, Braun, Kronberg, Germany).

For collecting tissues, six animals each from every treatment group were euthanized by an injection of T61 (Intervet, Unterschleißheim, Germany) at 6 and 24 h after LPS/NaCl. These times of tissue collection were chosen because they represent the temporal dynamics of IDO1 activity based on alterations of TRP metabolite concentrations as shown previously [14]. After euthanasia, liver, lung, muscle (Musculus deltoideus), adrenal gland, spleen, thyroid gland, heart, kidney and brain tissues were quickly removed. The hypothalamus, hippocampus, amygdala and prefrontal cortex were dissected from the brain, frozen in liquid nitrogen and stored at −80 °C. The stereotaxic atlas of the pig brain served as a reference [42]. Tissues were analysed for 1-MT and TRP metabolite concentrations.

4.2.2. Experiment 2: Effects of 1-MT on Behaviour and Haematopoiesis

In experiment 2, a total of 48 animals were used. Animals were pre-treated with 1-MT or MYR and received either an LPS or NaCl administration in accordance with experiment 1. Blood samples were collected before every s.c. application of 1-MT/MYR, immediately before (0 h), and 6, 12 and 24 h after LPS/NaCl administration. Evaluation of sickness behaviour was conducted from 0 to 5 h after LPS/NaCl treatment. Differential leukocyte counts were evaluated at 0, 6 and 24 h after i.p. LPS/NaCl application.

4.3. Analyses

4.3.1. Quantification of 1-MT and TRP Metabolites

The determination of 1-MT, TRP, KYN, KYNA and QUIN in EDTA plasma and tissues was performed using methods that have been previously described in detail [14,15] using an HPLC-system (Perkin Elmer, series 200, Darmstadt, Germany) and an API2000 tandem mass spectrometer equipped with an electrospray ion source (ABSciex, Darmstadt, Germany). The main quality parameters of between-day and within-day accuracy and precision as well as the analytical ranges and linearity (correlation coefficient, R) are provided in Table 3. The obtained results adhere to international recommendations [43,44]. As an indicator for IDO1 activation, the ratio of KYN and TRP (KYN × 100/TRP) was calculated [45].

Table 3. Analytic quality parameters for 1-MT, TRP and its metabolites.

Analyte	Accuracy (%)	Precision (%)	Range (μM)	Linearity (R)
1-MT	−7.8–4.0	2.3–7.0	2.3–69	0.9966–0.9992
TRP	−5.9–3.3	1.7–9.9	1.2–49	0.9986–0.9995
KYN	−9.5–13.5	3.9–14.5	0.24–24	0.9969–0.9990
KYNA	−9.9–15.7	6.8–14.9	0.01–0.32	0.9931–0.9946
QUIN	−19.8–18.4	7.6–19.8	18–359	0.9590–0.9957

4.3.2. TNF-α and IL-10 Assays

The concentrations of TNF-α and IL-10 were analysed in duplicate in blood plasma using a commercially available pig ELISA kit (Invitrogen, Frederick, MD, USA) according to the manufacturer's instructions. The sensitivities of the TNF-α and IL-10 assays were 3 pg/mL. The intra- and inter-assay coefficients of variation (CV) of TNF-α were 6.2% and 8.2%, respectively. The intra- and inter-assay CVs of IL-10 were 6.3% and 9.4%, respectively [46].

4.3.3. Differential Leukocyte Counts

A blood smear was prepared followed by air drying. The object slide was then incubated for 2 min in May–Grünwald solution and washed with aqua dest, followed by an incubation in Giemsa solution (1:20) for 30 min. After washing with aqua dest, the slide was dried. To calculate the leukocyte distribution, a total of 200 leukocytes was counted using microscopy. The cell types were differentiated as lymphocytes, monocytes, basophiles, eosinophils and neutrophils with subdivision into neutrophilic myelocytes, band neutrophils and segmented neutrophils [47]. As a marker for early acute inflammation and physiological stress, the neutrophil to lymphocyte count ratio (N/L ratio) was calculated [33]. The N/L ratio is a biomarker reacting very early in the course of acute inflammation [34].

4.3.4. Behavioural Observations

The behaviour of each animal was observed using scan sampling [48] every 5 min over a period of 5 h after i.p. LPS/NaCl application, resulting in 60 observations per animal. Thus, at the time of observation, individual behaviour categorized as symptoms of sickness, activity or inactivity was assessed for each animal (Table 4). A value of 1 was assigned to indicate the occurrence of a specific behaviour, whereas a value 0 was assigned to denote its absence. To evaluate the severity of sickness, the values of all sickness symptoms were summed for each observation point for each animal (minimum number = 0, maximum number = 5) and calculated as the mean number. Furthermore, an animal was characterized as active or inactive if the behaviour of the category was recorded at the time of observation. Then, the activity level was calculated as the percentage of the whole observation period of 5 h for each animal.

Table 4. Categories and their related behaviours used for the assessment of sickness severity and activity.

Category	Behaviour
Symptoms of Sickness	Shivering; impeded respiration; vomiting; diarrhoea; circulatory insufficiency
Activity	Walking; drinking; employment with feed, bedding, trough, toy or piglet from neighbouring pen
Inactivity	Lying; sitting; standing without movement

4.4. Statistics

Statistical analyses were performed using SAS software, version 9.4 for Windows (SAS Institute Inc., Cary, NC, USA). Descriptive statistics and tests for normality were calculated with the UNIVARIATE procedure provided with the Base SAS software.

Blood and tissue parameters and skin temperature were evaluated by ANOVA using the MIXED procedure with SAS/STAT software. In experiment 1, plasma TRP metabolites were evaluated before LPS/NaCl application. This model comprised the fixed effects pre-treatment (1-MT, MYR), sampling time (-84, -60, -36, -12 h), replicate (1, 2, 3) and their multiple interactions. The repeated statement in the MIXED procedure and a compound symmetry block diagonal structure of the residual covariance matrix were used with respect to repeated measurements on the same animal. For the plasma parameters and skin temperature from experiment 1 after LPS/NaCl application, ANOVA comprised the fixed effects pre-treatment (1-MT, MYR), treatment (LPS, NaCl), sampling time (0, 1, 3, 6, 24 h) and replicate (1, 2, 3), their multiple interactions, and the random effect sow and repeated statement (as described above). For evaluation of tissue parameters, data of both sampling times (6 and 24 h) were combined and ANOVA consisted of the fixed effects pre-treatment (1-MT, MYR), treatment (LPS, NaCl) and replicate (1, 2, 3), their multiple interactions and the random sow effect. For blood parameters in experiment 2, ANOVA comprised the fixed effects pre-treatment (1-MT, MYR), treatment (LPS, NaCl), sampling time (0, 6, 12, 24 h) and replicate (4, 5, 6), their multiple interactions and the random effect sow. The repeated statement in the MIXED procedure was used as described above.

To evaluate sickness severity during the 5 h observation period after LPS, the GLIMMIX procedure was applied using a Poisson model (model statement: distribution = Poisson, link = log) comprising the fixed effects pre-treatment (1-MT, MYR), treatment (LPS, NaCl) replicate (4, 5, 6) and their interactions. For the evaluation of activity in the 5 h observation period, the GLIMMIX procedure was applied using a logistic model (model statement: distribution = binomial, link = logit) comprising the fixed effects pre-treatment (1-MT, MYR), treatment (LPS, NaCl), replicate (4, 5, 6) and their interactions. Repeated measurements on the same animal were taken into account by the "_RESIDUAL_" keyword in the "random statement" of the GLIMMIX procedure using the autoregressive structure of the first order (type = AR(1)) for the block diagonal residual covariance matrix.

For the presentation of the results, the least square means (LS-means) and their standard errors (SE) were calculated and tested for each fixed effect in the models described earlier using the Tukey–Kramer procedure for all multiple pair-wise comparisons. Effects and differences were considered significant at $p < 0.05$.

5. Conclusions

The results from the present study in pigs indicate that repeated administration of 1-MT does not inhibit the LPS-induced increase of KYN/TRP ratio, which is used as a marker for IDO1 activation in vivo. Therefore, no evidence is provided that the 1-MT-induced effects in this study are related to a reduced IDO activity. Indeed, 1-MT increased the levels of KYNA prior to the LPS challenge, indicating a pre-treatment effect of 1-MT facilitating the degradation to one branch of the KYN pathway. The 1-MT-associated modulation of the LPS-induced haematopoiesis indicates an interference of 1-MT with inflammatory signalling pathways. This might be due to the activation of AhR by 1-MT and KYNA resulting in the transcriptional activation of several inflammation-associated target genes. Furthermore, the general increase in TRP by 1-MT may cause an impaired antimicrobial defence if applied in sepsis patients. These adverse effects of 1-MT should be considered in therapeutic applications of 1-MT, which is used in clinical studies in tumour patients with the goal of preventing IDO1-induced immune escape of cancer cells.

Author Contributions: E.W., W.O., E.K., M.T., A.T., and W.W. designed the study; E.W., E.K., M.T., W.O. and A.S. performed the experiments; E.W., W.O., E.K., M.T. A.T., G.D., E.S., J.B., J.G. and D.O. analysed the data; E.W., W.O., E.K. and M.T. wrote the main manuscript. All authors reviewed the manuscript.

Funding: This work was funded by the Deutsche Forschungsgemeinschaft (www.dfg.de; grant numbers: KA 1266/5-1, SCHU 853/7-1). The publication of this article was funded by the Open Access Fund of the Leibniz Association and the Open Access Fund of the Leibniz Institute for Farm Animal Biology (FBN).

Acknowledgments: The authors are grateful for the excellent technical assistance of Silvia Langhoff, Dagmar Mähling, Petra Müntzel, Martina Pohlmann, Birgit Sobczak and Regina Wal from the Institute of Behavioural Physiology, Leibniz Institute for Farm Animal Biology (FBN), and Thomas Brand and Sabine Ristow from the Institute of Pharmacy, University of Greifswald. The authors are also grateful to Cornelia Müller from the Institute of Clinical Chemistry and Laboratory Medicine, University of Greifswald, for assisting with the quantification of 1-MT and TRP metabolites. In addition, the authors thank Evelin Normann, Heidi Sievert and colleagues from the experimental pig unit for excellent animal care, Eduard Murani for providing porcine primers and Ulrike Gimsa for critically reviewing the manuscript.

Conflicts of Interest: The authors declare no conflict of interest.

Abbreviations

IDO1	Indoleamine 2,3-dioxygenase
1-MT	1-Methyltryptophan
LPS	Lipopolysaccharide
TRP	Tryptophan
KYNA	Kynurenic acid
KYN	Kynurenine
QUIN	Quinolinic acid
MYR	Myritol
N/L ratio	Neutrophil to lymphocyte count ratio
TDO2	Tryptophan 2,3-dioxygenase
AhR	Aryl hydrocarbon receptor
CYP1A1	Cytochrome P450, family 1, subfamily A, polypeptide 1
GM-CSF	Granulocyte-macrophage colony-stimulating factor
iNOS	Inducible nitric oxide synthase

Appendix A

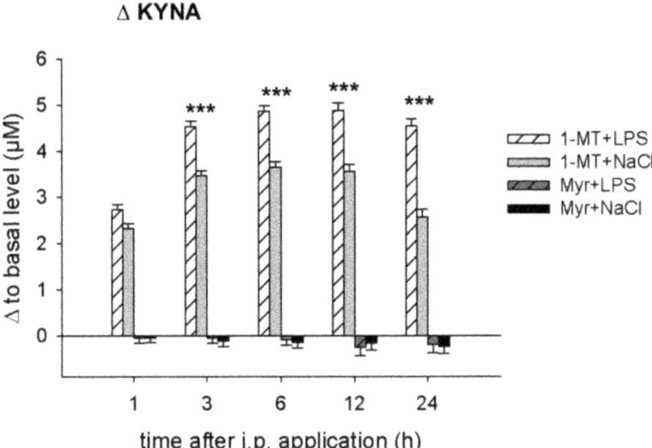

Figure A1. Effect of LPS on the 1-MT-dependent increase of KYNA concentrations in plasma. Treatment with NaCl and pre-treatment with MYR were used as controls. KYNA was measured using MS/MS, and differences in comparison to basal levels (0 h) were calculated to evaluate the LPS effect. The i.p. applications of LPS or NaCl were injected directly after blood sampling at 0 h, respectively. The animals received the fifth s.c. 1-MT or MYR injection at 12 h after LPS/NaCl treatment. The results are presented as LS-means ± SE. Significant differences were calculated using the Tukey–Kramer test and are indicated between LPS and NaCl groups for each time point by asterisks. For time points 0, 1, 3, 6 h: n = 11-12/group (total n = 47); for time point 24 h: n = 5-6/group (total n = 23); *** $p < 0.001$.

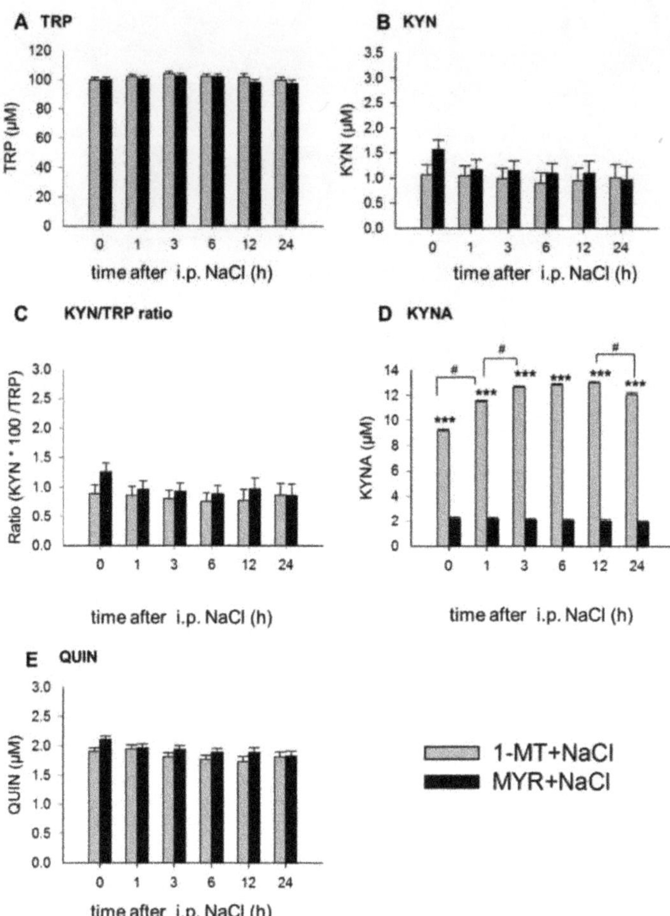

Figure A2. Effect of 1-MT on TRP (**A**), KYN (**B**), KYN/TRP ratio (**C**), KYNA (**D**) and QUIN (**E**) in plasma after i.p. NaCl administration. As a control for 1-MT-induced effects the excipient MYR was used. TRP and its metabolites were measured using MS/MS and the KYN/TRP ratio was used as a marker for IDO1 activity. NaCl was injected i.p. directly after blood sampling at 0 h. The animals received the fifth s.c. 1-MT or MYR injection at 12 h after NaCl. The results are presented as LS-means ± SE. Tukey–Kramer procedure was used for all pairwise comparisons. Significant differences are indicated between 1-MT and MYR groups for each time point by asterisks, and between the sampling times by hashes. For the sampling times 0, 1, 3, 6 h: $n = 11$-12/group (total $n = 47$); for time point 24 h: $n = 5$-6/group (total $n = 23$); *** $p < 0.001$, # $p < 0.01$.

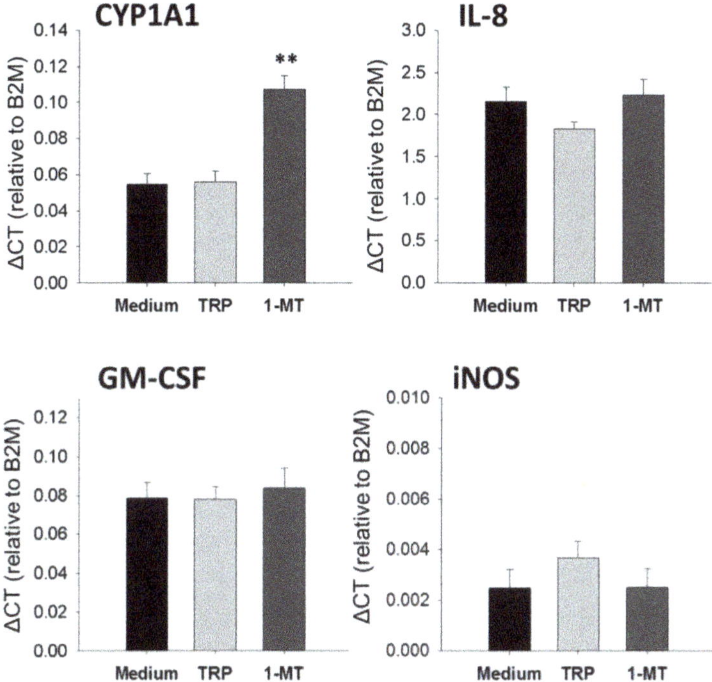

Figure A3. Effects of 1-MT and TRP contained in 1-MT on the mRNA expression of CYP1-A1, IL-8, GM-CSF and iNOS. Porcine lung fibroblasts were incubated with medium (unstimulated) or stimulated with TRP (5 μM) or L-1-MT (100 μM) for 3 h. Data are presented as means ± SE. $n = 6$, ** $p < 0.01$.

Appendix B

Appendix B.1. Supplementary Experiment

The potential interactions of 1-methyltryptophan (1-MT) and the amount of tryptophan (TRP) contained in 1-MT with aryl hydrocarbon receptor (AhR) and inflammatory pathways were evaluated in a primary cell culture of porcine lung fibroblasts. As a sensitive marker for AhR activation, the mRNA expression of cytochrome P450, family 1, subfamily A, polypeptide 1 (CYP1A1) was evaluated [40]. To examine the potential activation of inflammatory pathways by 1-MT, the mRNA expression of TNF-α, IL-8, granulocyte-macrophage colony-stimulating factor (GM-CSF) and inducible nitric oxide synthase (iNOS) was analysed.

Appendix B.2. Methods

Lung fibroblasts were isolated from nine-week-old German Landrace pigs ($n = 6$) directly after slaughter according to the isolation method of human colonic lamina propria fibroblasts as described earlier [49]. Briefly, lung tissue was cut into 1-mm pieces and incubated for 30 min in Hank's Balanced Salt Solution without Ca^{2+} and Mg^{2+} (Lonza, Basel, Switzerland) with 2 mmol/L EDTA (Sigma-Aldrich, Deisenhofen, Germany). The remaining tissue was rinsed and then digested for 1 hour at 37 °C with 1 mg/mL collagenase 1 (Merck, Darmstadt, Germany), 0.3 mg/mL DNase I (Roche, Heidelberg, Germany) and 2 mg/mL hyaluronidase (Sigma-Aldrich, Deisenhofen, Germany) in PBS. The isolated cells were cultured in DMEM (Lonza, Basel, Switzerland) containing penicillin (100 IE/mL), streptomycin (100 μg/mL), and amphotericin B (1 μg/mL) (Lonza, Basel, Switzerland) and 10% FCS (Biochrom, Berlin, Germany). Porcine lung fibroblasts were used between passages 3 and 4. 1-MT (L-isomer, purity 95%; Sigma-Aldrich, Deisenhofen, Germany) was dissolved in 1 N NaOH to a stock concentration of 1M. Since it is known that 1-MT contain 5% L-TRP, a TRP control was prepared by dissolving 50 mM L-TRP (Sigma Aldrich, Deisenhofen, Germany) in 1 N NaOH. For porcine lung fibroblast challenge, 1-MT was diluted in fibroblast cell culture medium to a concentration of 100 μM. For TRP, 50 mM L-TRP stock was diluted to a concentration of 5 μM. A NaOH solvent control was prepared by adding 1 N NaOH to cell culture medium to the same volume as actually used for 1-MT and L-TRP challenge. After a cultivation time of 3 h, fibroblasts were dissolved in 900 μL Trizol (Thermo Fisher Scientific, Waltham, MA, USA) per well and RNA was isolated using Direct-zol™ RNA MiniPrep (Zymo Research,

Irvine, CA, USA) with subsequent quantification on NanoDrop ND-1000 spectrophotometer (NanoDrop, Peqlab, Erlangen, Germany). The amounts of mRNA transcripts in cultured fibroblasts were determined by real-time PCR using LightCycler® 480 instrumentation (Roche Diagnostics, Mannheim, Germany). Reverse transcription was performed with 500 ng of total RNA using SuperScript III First-Strand Synthesis System (Thermo Fisher Scientific) with random hexamers and oligo (dT) primers. For gene-specific polymerase chain reaction (PCR), 12.5 ng cDNA were used in reaction mixtures containing 20 μM each forward and reverse primers (Table A1) and GoTaq® qPCR Master Mix (Promega, Madison, WI, USA), according to the manufacturer's instructions. Templates were amplified in duplicate after 2 min at 95 °C by 45 cycles of the following program: 15 s at 95 °C for denaturation and 1 min at 60 °C for annealing and extension. B2M was not affected by the applied treatments (variance 1.9%) between the analysed groups and therefore used as a housekeeping gene. The crossing points were analysed using LightCycler® 480 software (Roche Diagnostics) and the $2^{-\Delta\Delta Ct}$ method [50] was used to calculate transcript levels of each target gene.

Table A1. Primer pairs for real-time PCR.

Primer	Sequence Forward	Sequence Reverse
CYP1A1	GATCTCTTCAAGGACCTGAATCA	GCTGGATATTGGCATTCTCGTC
TNF-α	CACCACGCTCTTCTGCCTACT	GCTGTCCCTCGGCTTTGACAT
IL-8	GCTGTTGCCTTCTTGGCAGT	CTGCACCCACTTTTCCTTGG
GM-CSF	GCAGACTCGCCTGAACCTGT	CAGCAGTCAAAGGGGATGGT
iNOS	CTGCGTTATGCCACCAACAA	TTTCCAGCCCAGGTCGATAC
B2M	CGTGACTCTCGATAAGCCCAAG	GATTCATCCAACCCAGATGCAG

References

1. Wirthgen, E.; Hoeflich, A. Endotoxin-induced tryptophan degradation along the kynurenine pathway: The role of indolamine 2,3-dioxygenase and aryl hydrocarbon receptor-mediated immunosuppressive effects in endotoxin tolerance and cancer and its implications for immunoparalysis. *J. Amino Acids* **2015**, *2015*, 973548. [CrossRef] [PubMed]

2. Barth, M.C.; Ahluwalia, N.; Anderson, T.J.T.; Hardy, G.J.; Sinha, S.; Alvarez-Cardona, J.A.; Pruitt, I.E.; Rhee, E.P.; Colvin, R.A.; Gerszten, R.E. Kynurenic acid triggers firm arrest of leukocytes to vascular endothelium under flow conditions. *J. Biol. Chem.* **2009**, *284*, 19189–19195. [CrossRef] [PubMed]

3. Kiank, C.; Zeden, J.-P.; Drude, S.; Domanska, G.; Fusch, G.; Otten, W.; Schuett, C. Psychological stress-induced, IDO1-dependent tryptophan catabolism: implications on immunosuppression in mice and humans. *PLoS ONE* **2010**, *5*, e11825. [CrossRef] [PubMed]

4. Terness, P.; Bauer, T.M.; Röse, L.; Dufter, C.; Watzlik, A.; Simon, H.; Opelz, G. Inhibition of allogeneic T cell proliferation by indoleamine 2,3-dioxygenase–expressing dendritic cells mediation of suppression by tryptophan metabolites. *J. Exp. Med.* **2002**, *196*, 447–457. [CrossRef] [PubMed]

5. Wang, J.; Simonavicius, N.; Wu, X.; Swaminath, G.; Reagan, J.; Tian, H.; Ling, L. Kynurenic acid as a ligand for orphan G protein-coupled receptor GPR35. *J. Biol. Chem.* **2006**, *281*, 22021–22028. [CrossRef] [PubMed]

6. Schulte, W.; Bernhagen, J.; Bucala, R. Cytokines in sepsis: potent immunoregulators and potential therapeutic targets—An updated view. *Mediators Inflamm.* **2013**, *2013*, 165974. [CrossRef] [PubMed]

7. Tattevin, P.; Monnier, D.; Tribut, O.; Dulong, J.; Bescher, N.; Mourcin, F.; Uhel, F.; Le Tulzo, Y.; Tarte, K. Enhanced indoleamine 2,3-dioxygenase activity in patients with severe sepsis and septic shock. *J. Infect. Dis.* **2010**, *201*, 956–966. [CrossRef] [PubMed]

8. Changsirivathanathamrong, D.; Wang, Y.; Rajbhandari, D.; Maghzal, G.J.; Mak, W.M.; Woolfe, C.; Duflou, J.; Gebski, V.; dos Remedios, C.G.; Celermajer, D.S. Tryptophan metabolism to kynurenine is a potential novel contributor to hypotension in human sepsis. *Crit. Care Med.* **2011**, *39*, 2678–2683. [CrossRef] [PubMed]

9. Ploder, M.; Spittler, A.; Kurz, K.; Neurauter, G.; Pelinka, L.E.; Roth, E.; Fuchs, D. Accelerated tryptophan degradation predicts poor survival in trauma and sepsis patients. *Int. J. Tryptophan Res.* **2010**, *3*, 61–67. [CrossRef] [PubMed]

10. Zeden, J.; Fusch, G.; Holtfreter, B.; Schefold, J.; Reinke, P.; Domanska, G.; Haas, J.; Gruendling, M.; Westerholt, A.; Schuett, C. Excessive tryptophan catabolism along the kynurenine pathway precedes ongoing sepsis in critically ill patients. *Anaesth. Intensive Care* **2010**, *38*, 307–316. [PubMed]

11. Jung, I.D.; Lee, M.G.; Chang, J.H.; Lee, J.S.; Jeong, Y.I.; Lee, C.M.; Park, W.S.; Han, J.; Seo, S.K.; Lee, S.Y.; et al. Blockade of indoleamine 2,3-dioxygenase protects mice against lipopolysaccharide-induced endotoxin shock. *J. Immunol.* **2009**, *182*, 3146–3154. [CrossRef] [PubMed]

12. Greco, F.A.; Coletti, A.; Camaioni, E.; Carotti, A.; Marinozzi, M.; Gioiello, A.; Macchiarulo, A. The Janus-faced nature of IDO1 in infectious diseases: challenges and therapeutic opportunities. *Future Med. Chem.* **2016**, *8*, 39–54. [CrossRef] [PubMed]

13. Hou, D.Y.; Muller, A.J.; Sharma, M.D.; DuHadaway, J.; Banerjee, T.; Johnson, M.; Mellor, A.L.; Prendergast, G.C.; Munn, D.H. Inhibition of indoleamine 2,3-dioxygenase in dendritic cells by stereoisomers of 1-methyl-tryptophan correlates with antitumor responses. *Cancer Res.* **2007**, *67*, 792–801. [CrossRef] [PubMed]

14. Wirthgen, E.; Tuchscherer, M.; Otten, W.; Domanska, G.; Wollenhaupt, K.; Tuchscherer, A.; Kanitz, E. Activation of indoleamine 2, 3-dioxygenase by LPS in a porcine model. *Innate Immun.* **2014**, *20*, 30–39. [CrossRef] [PubMed]

15. Wirthgen, E.; Kanitz, E.; Tuchscherer, M.; Tuchscherer, A.; Domanska, G.; Weitschies, W.; Seidlitz, A.; Scheuch, E.; Otten, W. Pharmacokinetics of 1-methyl-L-tryptophan after single and repeated subcutaneous application in a porcine model. *Exp. Anim.* **2016**, *65*, 147–155. [CrossRef] [PubMed]

16. Bendixen, E.; Danielsen, M.; Larsen, K.; Bendixen, C. Advances in porcine genomics and proteomics—A toolbox for developing the pig as a model organism for molecular biomedical research. *Brief. Funct. Genom.* **2010**, *9*, 208–219. [CrossRef] [PubMed]

17. Martinez, M.N. Factors influencing the use and interpretation of animal models in the development of parenteral drug delivery systems. *AAPS J.* **2011**, *13*, 632–649. [CrossRef] [PubMed]

18. Roth, W.J.; Kissinger, C.B.; McCain, R.R.; Cooper, B.R.; Marchant-Forde, J.N.; Vreeman, R.C.; Hannou, S.; Knipp, G.T. Assessment of juvenile pigs to serve as human pediatric surrogates for preclinical formulation pharmacokinetic testing. *AAPS J.* **2013**, *15*, 763–774. [CrossRef] [PubMed]

19. Le Floc'h, N.; Otten, W.; Merlot, E. Tryptophan metabolism, from nutrition to potential therapeutic applications. *Amino Acids* **2011**, *41*, 1195–1205. [CrossRef] [PubMed]

20. Cady, S.G.; Sono, M. 1-Methyl-dl-tryptophan, β-(3-benzofuranyl)-dl-alanine (the oxygen analog of tryptophan), and β-[3-benzo (b) thienyl]-dl-alanine (the sulfur analog of tryptophan) are competitive inhibitors for indoleamine 2, 3-dioxygenase. *Arch. Biochem. Biophys.* **1991**, *291*, 326–333. [CrossRef]

21. Forouhar, F.; Anderson, J.R.; Mowat, C.G.; Vorobiev, S.M.; Hussain, A.; Abashidze, M.; Bruckmann, C.; Thackray, S.J.; Seetharaman, J.; Tucker, T. Molecular insights into substrate recognition and catalysis by tryptophan 2, 3-dioxygenase. *Proc. Natl. Acad. Sci. USA* **2007**, *104*, 473–478. [CrossRef] [PubMed]

22. Schmidt, S.K.; Siepmann, S.; Kuhlmann, K.; Meyer, H.E.; Metzger, S.; Pudelko, S.; Leineweber, M.; Daubener, W. Influence of tryptophan contained in 1-Methyl-Tryptophan on antimicrobial and immunoregulatory functions of indoleamine 2,3-dioxygenase. *PLoS ONE* **2012**, *7*, e44797. [CrossRef] [PubMed]

23. Okuno, A.; Fukuwatari, T.; Shibata, K. Urinary excretory ratio of anthranilic acid/kynurenic acid as an index of the tolerable amount of tryptophan. *Biosci. Biotechnol. Biochem.* **2008**, *72*, 1667–1672. [CrossRef] [PubMed]

24. DiNatale, B.C.; Murray, I.A.; Schroeder, J.C.; Flaveny, C.A.; Lahoti, T.S.; Laurenzana, E.M.; Omiecinski, C.J.; Perdew, G.H. Kynurenic acid is a potent endogenous aryl hydrocarbon receptor ligand that synergistically induces interleukin-6 in the presence of inflammatory signaling. *Toxicol. Sci.* **2010**, *115*, 89–97. [CrossRef] [PubMed]

25. De Luca, A.; Montagnoli, C.; Zelante, T.; Bonifazi, P.; Bozza, S.; Moretti, S.; D'Angelo, C.; Vacca, C.; Boon, L.; Bistoni, F. Functional yet balanced reactivity to Candida albicans requires TRIF, MyD88, and IDO-dependent inhibition of Rorc. *J. Immunol.* **2007**, *179*, 5999–6008. [CrossRef] [PubMed]

26. Kerkvliet, N.I. AHR-mediated immunomodulation: the role of altered gene transcription. *Biochem. Pharmacol.* **2009**, *77*, 746–760. [CrossRef] [PubMed]

27. Mulero-Navarro, S.; Fernandez-Salguero, P.M. New trends in aryl hydrocarbon receptor biology. *Front. Cell Dev. Biol.* **2016**, *4*, 45. [CrossRef] [PubMed]

28. Okamoto, T.; Toné, S.; Kanouchi, H.; Miyawaki, C.; Ono, S.; Minatogawa, Y. Transcriptional regulation of indoleamine 2, 3-dioxygenase (IDO) by tryptophan and its analogue. *Cytotechnology* **2007**, *54*, 107–113. [CrossRef] [PubMed]

29. Qian, F.; Villella, J.; Wallace, P.K.; Mhawech-Fauceglia, P.; Tario, J.D.; Andrews, C.; Matsuzaki, J.; Valmori, D.; Ayyoub, M.; Frederick, P.J. Efficacy of levo-1-methyl tryptophan and dextro-1-methyl tryptophan in reversing indoleamine-2, 3-dioxygenase–mediated arrest of T-cell proliferation in human epithelial ovarian cancer. *Cancer Res.* **2009**, *69*, 5498–5504. [CrossRef] [PubMed]

30. Soliman, H.H.; Jackson, E.; Neuger, T.; Dees, E.C.; Harvey, R.D.; Han, H.; Ismail-Khan, R.; Minton, S.; Vahanian, N.N.; Link, C. A first in man phase I trial of the oral immunomodulator, indoximod, combined with docetaxel in patients with metastatic solid tumors. *Oncotarget* **2014**, *5*, 8136. [CrossRef] [PubMed]

31. Basran, J.; Rafice, S.A.; Chauhan, N.; Efimov, I.; Cheesman, M.R.; Ghamsari, L.; Raven, E.L. A kinetic, spectroscopic, and redox study of human tryptophan 2,3-dioxygenase. *Biochemistry* **2008**, *47*, 4752–4760. [CrossRef] [PubMed]

32. Chauhan, N.; Thackray, S.J.; Rafice, S.A.; Eaton, G.; Lee, M.; Efimov, I.; Basran, J.; Jenkins, P.R.; Mowat, C.G.; Chapman, S.K. Reassessment of the reaction mechanism in the heme dioxygenases. *J. Am. Chem. Soc.* **2009**, *131*, 4186–4187. [CrossRef] [PubMed]

33. Bilandžić, N.; Žurić, M.; Lojkić, M.; Šimić, B.; Milić, D.; Barač, I. Cortisol and immune measures in boars exposed to three-day administration of exogenous adrenocorticotropic hormone. *Vet. Res. Commun.* **2006**, *30*, 433–444. [CrossRef] [PubMed]

34. Ljungstrom, L.R.; Jacobsson, G.; Andersson, R. Neutrophil-lymphocyte count ratio as a biomarker of severe sepsis in Escherichia coli infections in adults. *Crit. Care* **2013**, *17*, P25. [CrossRef]

35. Agaugué, S.; Perrin-Cocon, L.; Coutant, F.; André, P.; Lotteau, V. 1-Methyl-tryptophan can interfere with TLR signaling in dendritic cells independently of IDO activity. *J. Immunol.* **2006**, *177*, 2061–2071. [CrossRef] [PubMed]

36. Opitz, C.A.; Litzenburger, U.M.; Opitz, U.; Sahm, F.; Ochs, K.; Lutz, C.; Wick, W.; Platten, M. The indoleamine-2, 3-dioxygenase (IDO) inhibitor 1-methyl-D-tryptophan upregulates IDO1 in human cancer cells. *PLoS ONE* **2011**, *6*, e19823. [CrossRef] [PubMed]

37. Dantzer, R.; O'Connor, J.C.; Freund, G.G.; Johnson, R.W.; Kelley, K.W. From inflammation to sickness and depression: when the immune system subjugates the brain. *Nat. Rev. Neurosci.* **2008**, *9*, 46–56. [CrossRef] [PubMed]

38. Schettler, G.; Andrassy, K. *Innere Medizin: Verstehen-Lernen-Anwenden*, 9th ed.; Thieme: Stuttgart, Germany, 1998; ISBN 3135522091.

39. Moyer, B.J.; Rojas, I.Y.; Murray, I.A.; Lee, S.; Hazlett, H.F.; Perdew, G.H.; Tomlinson, C.R. Indoleamine 2,3-dioxygenase 1 (IDO1) inhibitors activate the aryl hydrocarbon receptor. *Toxicol. Appl. Pharmacol.* **2017**, *323*, 74–80. [CrossRef] [PubMed]

40. Hu, W.; Sorrentino, C.; Denison, M.S.; Kolaja, K.; Fielden, M.R. Induction of cyp1a1 is a nonspecific biomarker of aryl hydrocarbon receptor activation: results of large scale screening of pharmaceuticals and toxicants in vivo and in vitro. *Mol. Pharmacol.* **2007**, *71*, 1475–1486. [CrossRef] [PubMed]

41. Tuchscherer, M.; Kanitz, E.; Puppe, B.; Tuchscherer, A.; Stabenow, B. Effects of postnatal social isolation on hormonal and immune responses of pigs to an acute endotoxin challenge. *Physiol. Behav.* **2004**, *82*, 503–511. [CrossRef] [PubMed]

42. Félix, B.; Léger, M.-E.; Albe-Fessard, D.; Marcilloux, J.-C.; Rampin, O.; Laplace, J.-P.; Duclos, A.; Fort, F.; Gougis, S.; Costa, M. Stereotaxic atlas of the pig brain. *Brain Res. Bull.* **1999**, *49*, 1–137. [CrossRef]

43. EMA, Committee for Medicinal Products for Human Use (CHMP). *Guideline on Bioanalytical Method Validation*; European Medicines Agency: London, UK, 2011.

44. US Food and Drug Administration. *Guidance for Industry: Bioanalytical Method Validation*; U.S. Department of Health and Human Services: Silver Spring, MD, USA, September 2013; pp. 4–10.

45. Suzuki, Y.; Suda, T.; Asada, K.; Miwa, S.; Suzuki, M.; Fujie, M.; Furuhashi, K.; Nakamura, Y.; Inui, N.; Shirai, T.; et al. Serum indoleamine 2,3-dioxygenase activity predicts prognosis of pulmonary tuberculosis. *Clin. Vaccine Immunol.* **2012**, *19*, 436–442. [CrossRef] [PubMed]

46. Tuchscherer, M.; Otten, W.; Kanitz, E.; Gräbner, M.; Tuchscherer, A.; Bellmann, O.; Rehfeldt, C.; Metges, C.C. Effects of inadequate maternal dietary protein: carbohydrate ratios during pregnancy on offspring immunity in pigs. *BMC Vet. Res.* **2012**, *8*, 232. [CrossRef] [PubMed]

47. Niepage, H. *Methoden der praktischen Hämatologie für Tierärzte*, 2nd ed.; Verlag Paul Parey: Berlin, Germany, 1989.

48. Lehner, P.N. Sampling methods in behavior research. *Poult. Sci.* **1992**, *71*, 643–649. [CrossRef] [PubMed]

49. Brenmoehl, J.; Lang, M.; Hausmann, M.; Leeb, S.N.; Falk, W.; Schölmerich, J.; Göke, M.; Rogler, G. Evidence for a differential expression of fibronectin splice forms ED-A and ED-B in Crohn's disease (CD) mucosa. *Int. J. Colorectal Dis.* **2007**, *22*, 611–623. [CrossRef] [PubMed]
50. Livak, K.J.; Schmittgen, T.D. Analysis of relative gene expression data using real-time quantitative PCR and the $2^{-\Delta\Delta Ct}$ method. *Methods* **2001**, *25*, 402–408. [CrossRef] [PubMed]

International Journal of
Molecular Sciences

Article

Aspirin down Regulates Hepcidin by Inhibiting NF-κB and IL6/JAK2/STAT3 Pathways in BV-2 Microglial Cells Treated with Lipopolysaccharide

Wan-Ying Li [1,2,†], Fei-Mi Li [1,†], Yu-Fu Zhou [1], Zhong-Min Wen [2,*], Juan Ma [1,3], Ke Ya [3,*] and Zhong-Ming Qian [1,*]

1   Laboratory of Neuropharmacology, Fudan University School of Pharmacy, Shanghai 201203, China; lj19840208@126.com (W.-Y.L.); lfm19910916@sina.com (F.-M.L.); zyf19850722@126.com (Y.-F.Z.); mj19890412@126.com (J.M.)
2   Department of Neurology, The Second Affiliated Hospital of Soochow University, Suzhou 215004, China
3   School of Biomedical Sciences, Faculty of Medicine, The Chinese University of Hong Kong, Shatin, NT, Hong Kong, China
*   Correspondence: sy19871104@yahoo.com (Z.-M.W.); yake@cuhk.edu.hk (K.Y.); qianzhongming@fudan.edu.cn (Z.-M.Q.); Tel.: +86-21-5198-0178 (Z.-M.W. & Z.-M.Q.); +852-3943-6780 (K.Y.)
†   These authors contributed equally to this work.

Academic Editor: Juan M. Tomás
Received: 30 September 2016; Accepted: 9 November 2016; Published: 16 December 2016

Abstract: Aspirin down regulates transferrin receptor 1 (TfR1) and up regulates ferroportin 1 (Fpn1) and ferritin expression in BV-2 microglial cells treated without lipopolysaccharides (LPS), as well as down regulates hepcidin and interleukin 6 (IL-6) in cells treated with LPS. However, the relevant mechanisms are unknown. Here, we investigate the effects of aspirin on expression of hepcidin and iron regulatory protein 1 (IRP1), phosphorylation of Janus kinase 2 (JAK2), signal transducer and activator of transcription 3 (STAT3) and P65 (nuclear factor-κB), and the production of nitric oxide (NO) in BV-2 microglial cells treated with and without LPS. We demonstrated that aspirin inhibited hepcidin mRNA as well as NO production in cells treated with LPS, but not in cells without LPS, suppresses IL-6, JAK2, STAT3, and P65 (nuclear factor-κB) phosphorylation and has no effect on IRP1 in cells treated with or without LPS. These findings provide evidence that aspirin down regulates hepcidin by inhibiting IL6/JAK2/STAT3 and P65 (nuclear factor-κB) pathways in the cells under inflammatory conditions, and imply that an aspirin-induced reduction in TfR1 and an increase in ferritin are not associated with IRP1 and NO.

Keywords: aspirin;   hepcidin;   P65 (nuclear   factor-κB);   IL-6/JAK2/STAT3   pathway; lipopolysaccharide (LPS); nitric oxide (NO); iron regulatory protein 1 (IRP1)

---

1. Introduction

Aspirin is a non-steroidal anti-inflammatory drug (NSAID) and has been used for many years to treat a wide range of maladies, including pain and inflammation [1]. Preclinical and clinical studies have evidenced that aspirin has beneficial effects on mood disorders and schizophrenia and that high-dose aspirin is associated with a reduced risk of Alzheimer's disease (AD) [2]. This oldest agent in medicine has also been considered to be a potential new therapy for a range of neuropsychiatric disorders [2].

Studies   have   demonstrated   that   aspirin   and   sodium   salicylate   have   a   significant neuro-protective role in 1-methyl-4-phenyl-1,2,3,6-tetrahydropyridine (MPTP) [3,4], rotenone [5], 1-methyl-4-phenylpyridiniumion (MPP$^+$), and 6-hydroxydopamine (6-OHDA) [6] animal models in vivo and in neurons exposed to 6-OHDA and MPP$^+$ in vitro [7]. MPTP, rotenone, 6-OHDA, and MPP$^+$ are all neurotoxins known to trigger oxidative stress [4,7]. The neuro-protective effects of aspirin against oxidative stress induced by these neurotoxins have therefore been considered to be related to its ability to scavenger free radicals [3,4,6].

Abnormally high levels of iron and oxidative stress have been observed in a number of neurodegenerative disorders [8–11]. Iron is a major generator of reactive oxygen species (ROS), and oxidative stress resulting from increased iron in the brain has been widely considered to be an initial causes of neuronal death in some neurodegenerative diseases [12,13]. In addition, it has been demonstrated that aspirin can affect iron metabolism by increasing ferritin synthesis in the cultured bovine pulmonary artery endothelial cells [14] and reducing serum ferritin (SF) in humans [15].

The well-established association of inflammatory and expression of iron regulatory hormone hepcidin [16,17], the anti-inflammatory character of aspirin [2], and the findings as discussed above prompted us to speculate that aspirin might have the ability to affect iron metabolism. Recently, we therefore investigated the effects of aspirin on the expression of three major iron metabolism proteins, transferrin receptor 1 (TfR1), ferroportin 1 (Fpn1), and ferritin, as well as hepcidin and interleukin 6 (IL-6) in BV-2 microglial cells. We found that aspirin significantly down regulates TfR1 and up regulates Fpn1 and ferritin expressions in cells treated without lipopolysaccharides (LPS) in vitro, as well as down regulates hepcidin and IL-6 levels in cells treated with LPS [18]. However, the relevant mechanisms are unknown. In the present study, we investigate the effects of aspirin on expression of hepcidin mRNA and regulating molecules of hepcidin, including IL-6 mRNA, iron regulatory protein 1 (IRP1) protein, phosphorylation of Janus kinase 2 (JAK2), signal transducer and activator of transcription 3 (STAT3), P65 (nuclear factor-κB, NF-κB), and nitric oxide (NO) in BV-2 microglial cells treated with and without LPS.

2. Results

2.1. Aspirin Protects BV-2 Microglial Cells from Lipopolysaccharides (LPS)-Induced Damage

We first investigated the effects of aspirin (ASA) on the cell viability by treating BV-2 microglial cells with a vehicle (0.1% ethanol) for 24 h (The Control), 0.1 mM aspirin for 24 h (ASA), 0.1% ethanol for 18 h + 1 μg/mL of LPS for 6 h (LPS), or aspirin for 18 h + 1 μg/mL of LPS for 6 h (LPS + 0.1 mM ASA). We used 0.1 mM aspirin because this concentration was found to have a significant effect on hepcidin mRNA expression in LPS-treated BV-2 microglial cells in another recent study [18]. The 3-(4,5-dimethylthiazol-2-yl)-2,5-diphenyltetrazolium bromide (MTT) assay showed that there was no significant difference in cell viability between cells treated with the vehicle (control) or with 0.1 mM aspirin (Figure 1). It was also found that the cell viability in cells treated with LPS alone were significantly lower than those in the cells treated with the vehicle or the 0.1 mM aspirin, implying that LPS could induce cell-damage under our in vitro experimental conditions. However, the viability of the cells treated with 0.1 mM aspirin plus LPS was significantly higher than that of the cells treated with LPS alone and almost the same as that of control cells (Figure 1). This result is consistent with our recent findings [18] and re-confirmed that aspirin has a role in protecting BV-2 microglial cells from LPS-induced damage in vitro.

2.2. Aspirin Inhibits Hepcidin mRNA Expression But Has No Effect on IRP1 Protein Expression in BV-2 Microglial Cells Treated with LPS

We then investigated the effects of aspirin on hepcidin mRNA and IRP1 protein expression in BV-2 microglial cells by treating with a vehicle (0.1% ethanol) for 24 h (The Control), 0.1 mM aspirin for 24 h (ASA), 0.1% ethanol for 18 h + 1 μg/mL of LPS for 6 h (LPS), or aspirin for 18 h + 1 μg/mL of LPS for 6 h (LPS + 0.1 mM ASA). Treatment with LPS induced a significant increase in hepcidin mRNA expression, the levels of hepcidin mRNA in the cells treated with LPS being markedly higher than those in the controls (Figure 2A). However, hepcidin mRNA expression in the cells treated with aspirin plus LPS was significantly lower than that in the cells treated with LPS only, but there was no difference between the cells treated with aspirin or with the vehicle. These results demonstrated that aspirin could inhibit hepcidin mRNA expression in BV-2 microglial cells treated with LPS but not in the cells treated without LPS. Western blot analysis showed that there were no differences in IRP1 protein content between cells treated with aspirin or the vehicle, or with LPS or aspirin plus LPS (Figure 2B), evidencing that aspirin has no effect on IRP1 protein expression in BV-2 microglial cells treated with or without LPS.

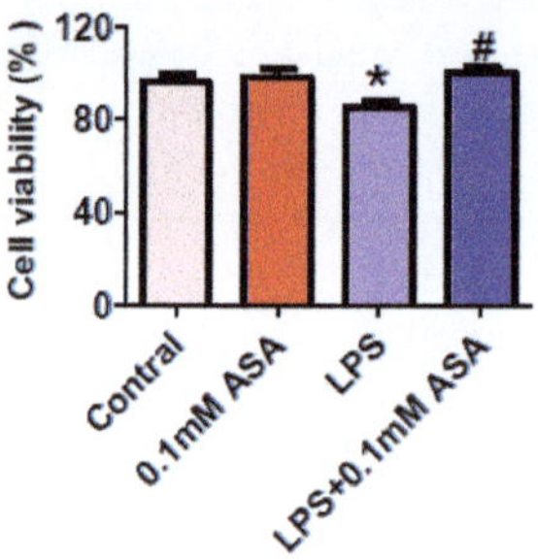

Figure 1. Aspirin protects BV-2 microglial cells from LPS-induced damage. BV-2 microglial cells were treated with 0.1% ethanol (Control) or aspirin (ASA) for 24 h—0.1% ethanol for 18 h and then 1 µg/mL of LPS for another 6 h (LPS) or aspirin for 18 h and then 1 µg/mL of LPS for another 6 h (LPS + 0.1 mM ASA). Cell viability was then conducted as described in Materials and Methods. Data were represented as means $\pm$ SEM ($n = 4$). * $p < 0.05$ vs. the control; # $p < 0.05$ vs. the LPS-treated group.

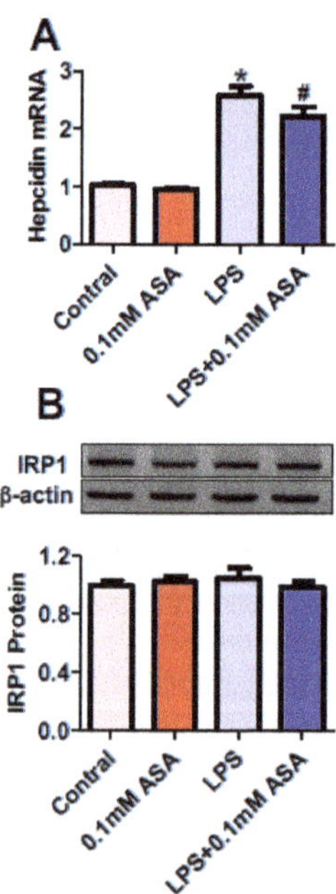

Figure 2. Aspirin inhibits hepcidin mRNA expression but has no effect on IRP1 protein expression in BV-2 microglial cells treated with LPS. BV-2 microglial cells were treated with 0.1% ethanol (Control) or aspirin (ASA) for 24 h—0.1% ethanol for 18 h and then 1 µg/mL of LPS for another 6 h (LPS) or aspirin for 18 h and then 1 µg/mL of LPS for another 6 h (LPS + 0.1 mM ASA). Expression of hepcidin mRNA (**A**) and IRP1 protein (**B**) were measured by RT-PCR and Western blot analysis, respectively. Data were presented as mean $\pm$ SEM ($n = 5$). * $p < 0.05$ vs. the control; # $p < 0.05$ vs. the LPS-treated group.

2.3. Aspirin Inhibits Phosphorylation of JAK2, STAT3, and P65(NF-κB) and Expression of IL-6 mRNA in BV-2 Microglial Cells Treated with or without LPS

To understand why aspirin could inhibit hepcidin expression under inflammatory conditions, we investigated the effects of aspirin on phosphorylation of JAK2, STAT3, and P65 by incubating BV-2 microglial cells with a vehicle (0.1% ethanol) for 24 h (The Control), 0.1 mM aspirin for 24 h (ASA), 0.1% ethanol for 18 h + 1 μg/mL of LPS for 6 h (LPS), or aspirin for 18 h + 1 μg/mL of LPS for 6 h (LPS + 0.1 mM ASA). It was found that the contents of p-JAK2 (Figure 3A), p-STAT3 (Figure 3B), p-P65 (Figure 3C), and IL-6 mRNA (Figure 4A) in the cells treated with LPS were significantly higher than those in the control cells as well as in the cells treated with aspirin plus LPS. This implied that LPS could dramatically increase JAK2, STAT3, and P65(NF-κB) phosphorylation and IL-6 mRNA expression, while aspirin was able to attenuate the LPS-induced increase in phosphorylation and expression. In addition, our findings showed that the levels of IL-6, p-JAK2, p-STAT3, and p-P65(NF-κB) in the cells treated with aspirin only were significantly lower than those in the control cells, suggesting that aspirin was able to inhibit IL-6 mRNA expression, JAK2, STAT3, and P65(NF-κB) phosphorylation under 'normal' conditions in vitro, not only under inflammatory conditions in vitro.

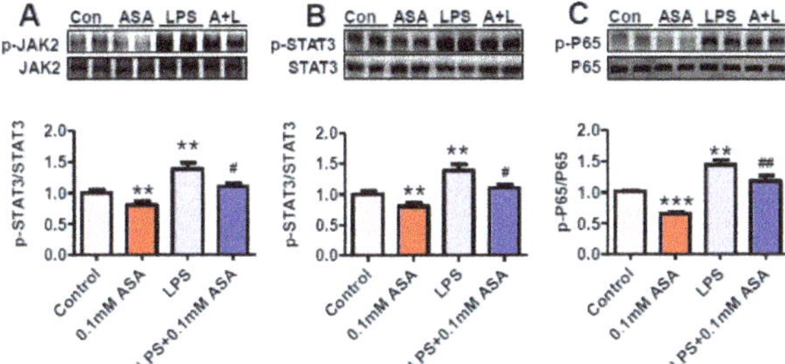

Figure 3. Aspirin inhibits phosphorylation of JAK2, STAT3, and P65(NF-κB) in BV-2 microglial cells treated with or without LPS. BV-2 microglial cells were treated with 0.1% ethanol (Control) or aspirin (ASA) for 24 h—0.1% ethanol for 18 h and then 1 μg/mL of LPS for another 6 h (LPS) or aspirin for 18 h and then 1 μg/mL of LPS for another 6 h (LPS + 0.1 mM ASA). Phosphorylation of JAK2 (**A**); STAT3 (**B**); and P65(NF-κB) (**C**) was detected by Western blot analysis, as described in Materials and Methods. Data were represented as mean ± SEM (n = 5). ** $p < 0.01$; *** $p < 0.001$ vs. the control; # $p < 0.05$; ## $p < 0.01$ vs. the LPS-treated group.

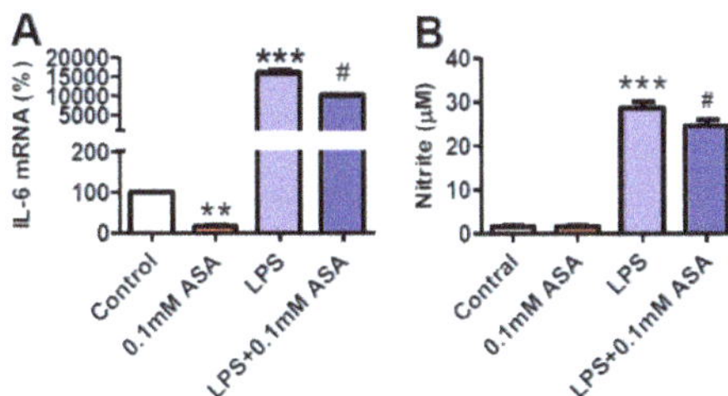

Figure 4. Aspirin inhibits expression of IL-6 mRNA in BV-2 microglial cells treated with or without LPS and NO production in BV-2 microglial cells treated with LPS but not in the cell treated without LPS. BV-2 microglial cells were treated with 0.1% ethanol (Control) or aspirin (ASA) for 24 h—0.1% ethanol for 18 h and then 1 μg/mL of LPS for another 6 h (LPS) or aspirin for 18 h and then 1 μg/mL of LPS for another 6 h (LPS + 0.1 mM ASA). Expression of IL-6 mRNA (**A**) was measured by RT-PCR and production of NO (**B**) was assayed by measuring the levels of nitrite (a metabolite of NO) in culture medium as described in Materials and Methods. Data were represented as mean ± SEM (n = 5). ** $p < 0.01$; *** $p < 0.001$ vs. the control; # $p < 0.05$ vs. the LPS-treated group.

2.4. Aspirin Inhibits NO Production in BV-2 Microglial Cells Treated with LPS But Not in the Cell Treated without LPS

In a previous study, we demonstrated that aspirin down regulates TfR1 and up regulates Fpn1 and ferritin expression in BV-2 microglial cells in vitro; however, the mechanisms are unknown. It has been documented that NO can regulate the expression of TfR1 and ferritin by interacting with IRP1 [19–21]. To find out whether NO is involved in the effects of ASA on TfR1 and ferritin expression, we also investigated the effects of ASA on NO production in the cells treated with or without LPS. Treatment with LPS was found to induce a significant increase in NO levels, while pre-treatment with ASA displayed a marked inhibition on the LPS-induced increase in NO production, the levels of NO being significantly lower in ASA + LPS-treated cells than in LPS-treated cells (Figure 4B). There were no differences in NO content between the ASA-treated and the control cells.

3. Discussion

It has been well-demonstrated that LPS is able to up regulate hepcidin expression [22] via the IL-6/STAT3 signaling pathway and then down regulate expression of TfR1 and Fpn1 in the brain [23]. In our recent study, we [18] demonstrated that aspirin significantly inhibits the LPS-induced increase in IL-6 and hepcidin mRNA expression and revises the LPS-evoked reduction in TfR1 and Fpn1 expression in BV-2 microglial cells. The inhibition of aspirin on IL-6 and hepcidin mRNA expression suggests that aspirin might have play a role in suppressing the activated IL-6/ JAK2/STAT3 signaling pathway in LPS-treated BV-2 microglial cells. To test this hypothesis, we investigated the effects of aspirin on the contents of IL-6 mRNA, p-JAK2, and p-STAT3 in LPS-treated BV-2 microglial cells. We demonstrated that the significant increase in expression of IL-6 mRNA as well as phosphorylation of JAK2 and STAT3 induced by LPS could be largely suppressed by pre-incubation of the cells with aspirin. These findings provide evidence that aspirin down regulates hepcidin at least partly by inhibiting the IL6/JAK2/STAT3 pathway and then alleviates the LPS-induced reduction in TfR1 and Fpn1 expression in LPS-treated BV-2 microglial cells.

The transcription factor NF-κB is critical for the inducible expression of multiple cellular and viral genes involved in inflammation and infection including IL-6 [24,25]. Studies have demonstrated that aspirin and its metabolite sodium salicylate (another anti-inflammatory drug) are both able to inhibit the activation of NF-κB by inhibiting the activity of IκB kinase-beta (IκB-β) to preventing the translocation of NF-κB to the nucleus [24,26,27]. Evidence also shows that the toll-like receptor 4 (TLR4) recognition of LPS (TLR4 ligand), a pathogen-associated molecular pattern, results in the triggering of downstream signaling cascades leading to the activation of NF-κB [25]. These led us to speculate that the activation of NF-κB might play a role in the LPS-induced increase in hepcidin mRNA expression, while the inhibition of aspirin on hepcidin mRNA expression might be partly associated with its role in inhibiting the activation of NF-κB, as has been found in human peripheral blood leukocytes [28]. We therefore investigated the effects of aspirin on P65(NF-κB) phosphorylation in BV-2 microglial cells treated with LPS. P65(NF-κB) was examined here because the RelA(p65)–p50 heterodimer is the most frequently activated form of NF-κB in TLR signaling [29]. Our findings show that LPS can induce a marked increase in phosphorylation of P65(NF-κB), which can be significantly suppressed by the pre-incubation of cells with aspirin. This implies that the down regulation of hepcidin might also be partly associated with the inhibiting role of aspirin on NF-κB phosphorylation in LPS-treated BV-2 microglial cells.

It was noticed that aspirin was able to inhibit levels of phosphorylation of JAK2, STAT3, and P65(NF-κB) not only under in vitro inflammatory conditions but also in vitro 'normal' conditions. The levels of p-JAK2, p-STAT3, and p-P65 were found to be significantly lower in cells treated with ASA than those in the control cells. In theory, this could lead to a reduction in hepcidin mRNA in cells treated with ASA only. However, no difference was found in the contents of hepcidin mRNA between the cells treated with ASA and control cells. Under in vitro 'normal' conditions, the base-line level of JAK2, STAT3, and P65 is relatively lower, as compared with in vitro inflammatory conditions.

Although aspirin can induce a reduction in the content of these mediators, the reduction might not be enough to result in significant changes in hepcidin mRNA expression. This is probably one of causes for the inconformity in the response of mediators and hepcidin to aspirin. Further studies on this possibility and other relevant causes are needed.

In a previous study, we demonstrated that aspirin down regulates TfR1 and up regulates Fpn1 and ferritin expression in BV-2 microglial cells treated without LPS [16]; however, the mechanisms are unknown. The responses of TfR1, Fpn1, and ferritin to aspirin are absolutely unrelated to hepcidin because the peptide has no response to aspirin under in vitro "normal" conditions. Like hepcidin, IRP1 is a key protein involved in the regulation of iron homeostasis. In most types of cells, the coordinated control of TfR1 and ferritin by cellular iron is mediated by IRP1 [30,31]. In addition, it has been documented that NO regulates expression of TfR1 and ferritin by interacting with IRP1 [19–21]. To find out why aspirin is able to down regulate TfR1 and up regulate Fpn1 and ferritin expression in BV-2 microglial cells under in vitro "normal" conditions, we examined the effects of aspirin on IRP1 expression and NO production in BV-2 microglial cells. We found that the levels of IRP1 as well as the NO in the cells treated with aspirin are no significantly different from those in the control cells. This finding implies that an aspirin-induced reduction in TfR1 and an increase in ferritin are not associated with IRP1 and NO and suggest that there might be some unknown mechanism by which aspirin regulates TfR1 and ferritin expression under in vitro "normal" conditions.

Aspirin inhibits inflammation mainly through its ability to suppress cyclooxygenase (COX) activity [32]. There are two isoforms of this enzyme—COX1 and COX2 [33]. COX-1 is constitutively expressed in most tissues and produces prostanoids responsible for normal physiological functions. COX-2 is sparsely present in most healthy tissues [33] and functions as a key enzyme for prostaglandin biosynthesis. In addition, COX-2 has been shown to contribute to the LPS-induced inflammatory process [34,35]. Cellular responses to inflammatory stimuli, including LPS, mainly involve the activation of mitogen-activated protein kinase (MAPK) signaling cascades [36–38]. The p38MAPK and JNK subfamilies play critical roles in regulating expression of pro-inflammatory mediators such as COX-2 and interleukins such as IL-6 [39–42]. These data imply that aspirin should be able to inhibit a LPS-induced increase in COX2 expression in BV-2 microglial cells, probably via inhibiting MAPK/JNK pathway, although the content of COX2 was not measured in this study. In addition, we found that the tendency in the effect of aspirin, LPS, or both on IL-6, JAK2, and P65 is very similar to that on COX2 reported by others [34]. This might suggest that COX2 may play a role in the down regulation of hepcidin induced by aspirin in LPS-treated BV-2 microglial cells. Further studies on this possibility are needed.

4. Materials and Methods

4.1. Chemicals

Unless otherwise stated, all chemicals, including aspirin (ASA), LPS (*Escherichia coli* 055:B5), MTT (3-(4,5-dimethylthiazol-2-yl)-2,5-diphenyltetrazolium bromide), and mouse monoclonal anti-β-actin, were obtained from Sigma Chemical Co., St. Louis, MO, USA. A BCA protein assay kit and a Revert Aid First Strand cDNA Synthesis Kit were purchased from Thermo Scientific, Waltham, MA, USA, and TRIZOL Reagent from Life technologies, Carlsbad, CA, USA. Rabbit monoclonal anti-phospho-JAK2, rabbit monoclonal anti-JAK2, rabbit polyclonal anti-phospho-STAT3, mouse monoclonal anti-STAT3, rabbit monoclonal anti-phospho-P65, and rabbit monoclonal anti-P65 antibodies were supplied by Cell Signaling Technology, Inc., Danvers, MA, USA, and mouse monoclonal anti-IRP1 was from Abcam, San Francisco, CA, USA. Goat anti-rabbit and anti-mouse IRDye 800CW secondary antibodies were bought from LI-COR bio sciences, Lincoln, NE, USA. The Health Department of Hong Kong and Shanghai Government and the Animal Research Ethics Committee of The Chinese University of Hong Kong (the project identification code: GRF14106914, 1 January 2015) and Fudan University (the project identification code: NSFC31271132, 1 January 2013; NSFC31330035, 1 January 2014) approved the experimental procedures of this study.

4.2. BV-2 Microglia Cells

BV-2 microglia cells (a murine microglia cell line) were grown in a 5% CO_2 incubator at 37 °C in Dulbecco's modified Eagle's medium (DMEM) supplemented with 10% FBS (PAN Biotech, Aidenbach Bavaria, Germany) and antibiotics (penicillin 100 U/mL, streptomycin 100 mg/mL), and culture medium was changed every 2 days [24]. Then, BV-2 cells were seeded in 96-well plates (6×10^3 cells/well) for cell viability assay, and 6-well plates (4×10^5 cells/well) for RT-PCR and (8×10^5 cells/well) for Western blotting analysis [18]. The cells were treated with different concentrations of 0.1 mM aspirin (in 0.1% ethanol) and/or LPS (1 mg/mL in PBS), which were dissolved in fresh DMEM without serum.

4.3. Assessment of Cell Viability

The cell viability were measured using an MTT assay as described previously [43]. Briefly, a total of 25 µL of MTT (1 g/L in PBS) was added to each well before the conduction of incubation at 37 °C for 4 h. The assay was stopped by the addition of a 100 µL of lysis buffer (20% SDS in 50% $N'N$-dimethylformamide, pH 4.7). Optical density (OD) was measured at the 570 nm wavelength by the use of an ELX-800 microplate assay reader (Bio-tek, Winooski, VT, USA), and the results were expressed as a percentage of absorbance measured in the control cells.

4.4. NO Production Assay

Production of NO was assayed by measuring the levels of nitrite (a metabolite of NO) in the culture medium using a colorimetric assay with Griess reagent according to Kim et al. [44], After 24 h of treatment with LPS with or without ASA, the culture media were collected and reacted with an equal volume of Griess reagent in 96-well culture plates and were incubated at room temperature for 10 min in the dark. The absorbance was measured at 540 nm using a microplate reader, and nitrite concentrations were calculated by reference to a standard curve generated by known concentrations of sodium nitrite.

4.5. Quantitative Real-Time PCR

Extraction of total RNA and preparation of cDNA were performed using a TRIZOL reagent andreverse transcription kit (Thermo Scientific) in accordance with the instruction of the manufacturers, respectively. The specific primers used for PCR are as follows: hepcidin forward, 5′-GAAGGCAAGATGGCACTAAGCA-3′; hepcidin reverse, 5′-TCTCGTCTGTTGCCGGAGATAG-3′; IL-6 forward, 5′-GAGGATACCACTCCCAACAGACC-3′; IL-6 reverse, 5′-AAGTGCATCATCGT TGTTCATACA-3′; β-actin forward, 5′-AAATCGTGCGTGACATCAAAGA-3′; β-actin reverse, 5′-GCCATCTCCTGCTCGAAGTC-3′ [19,45]. Quantitative real-time PCR was conducted with CFX96 PCR instrument (Bio-Rad, Hercules, CA, USA) using specific primers and SYBR Premix II kit (Takara, Dalian, China). The C_t values of each target gene were normalized to that of the β-actin mRNA. Relative gene expression was calculated by the $2^{-\Delta\Delta Ct}$ method [46].

4.6. Western Blot Analysis

The cells were washed and lysed as described previously [47,48]. After centrifugation at $13,200 \times g$ for 15 min at 4 °C, the supernatant was collected, and protein content was determined using the BCA protein assay kit. Aliquots of the extract containing about 20 µg of protein were loaded and run on a single track of 10% SDS-PAGE under reducing conditions and subsequently transferred to a pure nitrocellulose membrane (Bio-Rad, Hercules, CA, USA). The blots were blocked and then incubated with primary antibodies—rabbit monoclonal anti-phospho-JAK2 (1:1000), rabbit monoclonal anti-JAK2

(1:1000), rabbit polyclonal anti-phospho-STAT3 (1:1000), mouse monoclonal anti-STAT3 (1:1000), rabbit monoclonal anti-phospho-P65 (1:1000), rabbit monoclonal anti-P65 (1:1000), and mouse monoclonal anti-IRP1 (1:1000) antibodies—overnight at 4 °C. After the incubation, the blots were washed three times and then incubated with goat anti-rabbit (1:1000) or anti-mouse IRDye 800 CW secondary antibodies (1:5000) for 2 h at room temperature. The intensity of the specific bands was detected and analyzed by the Odyssey infrared image system (Li-Cor, Lincoln, NE, USA). To ensure even loading of the samples, the same membrane was probed with a mouse monoclonal anti-β-actin antibody at a 1:5000 dilution.

4.7. Statistical Analysis

Statistical analyses were performed using Graphpad Prism. Data were presented as mean ± SEM. The differences between the means were all determined by a two-way analysis of variance (ANOVA). A probability value of $p < 0.05$ was considered statistically significant.

5. Conclusions

We demonstrated that aspirin inhibits hepcidin mRNA as well as NO production in cells treated with LPS, but not in cells without LPS, suppresses IL-6, JAK2, STAT3, and P65(NF-κB) phosphorylation, and has no effect on IRP1 protein in cells treated with or without LPS. The findings provided evidence that aspirin down regulates hepcidin by inhibiting IL6/JAK2/STAT3 as well as P65(NF-κB) pathways in cells under inflammatory conditions and implies that an aspirin-induced reduction in TfR1 and an increase in ferritin were not associated with IRP1 and NO.

Acknowledgments: We thank Christopher Qian for his assistance with preparation and English revision of the manuscript. The studies in our laboratories were supported by the National Natural Science Foundation of China (31330035, 31271132, 31371092; 31571195), the Competitive Earmarked Grants of The Hong Kong Research Grants Council (GRF14106914, GRF14111815), and the Hong Kong Health and Medical Research Fund (01120146).

Author Contributions: Ke Ya and Zhong-Ming Qian conceived, organized, and supervised the study; Wan-Ying Li, Fei-Mi Li, Yu-Fu Zhou, and Juan Ma performed the experiments; Zhong-Min Wen contributed to the analysis of data; Ke Ya and Zhong-Ming Qian prepared, wrote, and revised the manuscript.

Conflicts of Interest: The authors declare no conflict of interest.

Abbreviations

ASA	aspirin
Fpn1	ferroportin 1
IL-6	interleukin 6
IRP1	iron regulatory protein 1
JAK2	Janus kinase 2
LPS	lipopolysaccharides
NO	nitric oxide
NSAID	non-steroidal anti-inflammatory drug
P65(NF-κB)	rela (p65) nuclear factor-κB
SF	serum ferritin
STAT3	signal transducer and activator of transcription 3
TfR1	transferrin receptor 1

1. Mulvihill, M.M.; Nomura, D.K. Therapeutic potential of monoacylglycerol lipase inhibitors. *Life Sci.* **2013**, *92*, 492–497. [CrossRef] [PubMed]
2. Berk, M.; Dean, O.; Drexhage, H.; McNeil, J.J.; Moylan, S.; O'Neil, A.; Sanna, L.; Maes, M. Aspirin: A review of its neurobiological properties and therapeutic potential for mental illness. *BMC Med.* **2013**, *11*, 74. [CrossRef] [PubMed]

3. Aubin, N.; Curet, O.; Deffois, A.; Carter, C. Aspirin and salicylate protect against MPTP-induced dopamine depletion in mice. *J. Neurochem.* **1998**, *71*, 1635–1642. [CrossRef] [PubMed]

4. Mohanakumar, K.P.; Muralikrishnan, D.; Thomas, B. Neuroprotection by sodium salicylate against 1-methyl-4-phenyl-1,2,3, 6-tetrahydropyridine-induced neurotoxicity. *Brain Res.* **2000**, *864*, 281–290. [CrossRef]

5. Madathil, S.K.; Karuppagounder, S.S.; Mohanakumar, K.P. Sodium salicylate protects against rotenone-induced Parkinsonism in rats. *Synapse* **2013**, *67*, 502–514. [CrossRef] [PubMed]

6. Di Matteo, V.; Pierucci, M.; di Giovanni, G.; di Santo, A.; Poggi, A.; Benigno, A.; Esposito, E. Aspirin protects striatal dopaminergic neurons from neurotoxin-induced degeneration: An in vivo microdialysis study. *Brain Res.* **2006**, *1095*, 167–177. [CrossRef] [PubMed]

7. Carrasco, E.; Werner, P. Selective destruction of dopaminergic neurons by low concentrations of 6-OHDA and MPP$^+$: Protection by acetylsalicylic acid aspirin. *Parkinsonism Relat. Disord.* **2002**, *8*, 407–411. [CrossRef]

8. Riederer, P.; Dirr, A.; Goetz, E.; Sofic, E.; Jellinger, K.; Youdim, M.B.H. Distribution of iron in different regions and subcellular compartments in Parkinson's disease. *Ann. Neurol.* **1992**, *32*, s101–s104. [CrossRef] [PubMed]

9. Jenner, P. Oxidative damage in neurodegenerative disease. *Lancet* **1994**, *344*, 796–798. [CrossRef]

10. Gorell, J.M.; Ordidge, R.J.; Brown, G.G.; Deniau, J.C.; Buderer, N.M.; Helpern, J.A. Increased iron-related MRI contrast in the substantianigra in Parkinson's disease. *Neurology* **1995**, *45*, 1138–1143. [CrossRef] [PubMed]

11. Qian, Z.M.; Wang, Q. Expression of iron transport proteins and excessive iron accumulation in the brain in neurodegenerative disorders. *Brain Res. Rev.* **1998**, *27*, 257–267. [CrossRef]

12. Ke, Y.; Qian, Z.M. Iron misregulation in the brain: A primary cause of neurodegenerative disorders. *Lancet Neurol.* **2003**, *2*, 246–253. [CrossRef]

13. Ke, Y.; Qian, Z.M. Brain iron metabolism: Neurobiology and neurochemistry. *Prog. Neurobiol.* **2007**, *83*, 149–173. [CrossRef] [PubMed]

14. Oberle, S.; Polte, T.; Abate, A.; Podhaisky, H.P.; Schröder, H. Aspirin increases ferritin synthesis in endothelial cells: A novel antioxidant pathway. *Circ. Res.* **1998**, *82*, 1016–1020. [CrossRef] [PubMed]

15. Fleming, D.J.; Jacques, P.F.; Massaro, J.M.; D'Agostino, R.B., Sr.; Wilson, P.W.; Wood, R.J. Aspirin intake and the use of serum ferritin as a measure of iron status. *Am. J. Clin. Nutr.* **2001**, *74*, 219–226. [PubMed]

16. Pietrangelo, A.; Dierssen, U.; Valli, L.; Garuti, C.; Rump, A.; Corradini, E.; Ernst, M.; Klein, C.; Trautwein, C. STAT3 Is required for IL-6-gp130-dependent activation of hepcidin in vivo. *Gastroenterology* **2007**, *132*, 294–300. [CrossRef] [PubMed]

17. Wrighting, D.M.; Andrews, N.C. Interleukin-6 induces hepcidin expression through STAT3. *Blood* **2006**, *108*, 3204–3209. [CrossRef] [PubMed]

18. Xu, Y.X.; Du, F.; Jiang, L.R.; Gong, J.; Zhou, Y.F.; Luo, Q.Q.; Qian, Z.M.; Ke, K. Effects of aspirin on expression of iron transport and storage proteins in BV-2 microglial cells. *Neurochem. Int.* **2015**, *91*, 72–77. [CrossRef] [PubMed]

19. Casey, J.L.; Hentze, M.W.; Koeller, D.M.; Caughman, S.W.; Rouault, T.A.; Klausner, R.D.; Harford, J.B. Iron-responsiveelements: Regulatory RNA sequences that control mRNA levelsand translation. *Science* **1988**, *240*, 924–928. [CrossRef] [PubMed]

20. Klausner, R.D.; Rouault, T.; Harford, J.B. Regulating the fate of mRNA: The control of cellular iron metabolism. *Cell* **1993**, *72*, 19–28. [CrossRef]

21. Weiss, G.; Goossen, B.; Doppler, W.; Fuchs, D.; Pantopoulos, K.; Werner-Felmayer, G.; Wachter, H.; Hentze, M.W. Translationalregulation via iron responsive elements by the nitricoxide/NO-synthase pathway. *EMBO J.* **1993**, *12*, 3651–3657. [PubMed]

22. Wang, Q.; Du, F.; Qian, Z.M.; Ge, X.H.; Zhu, L.; Yung, W.H.; Yang, L.; Ke, Y. Lipopolysaccharide induces a significant increase in expression of iron regulatory hormone hepcidin in the cortex and substantianigra in rat brain. *Endocrinology* **2008**, *149*, 3920–3925. [CrossRef] [PubMed]

23. Qian, Z.M.; He, X.; Liang, T.; Wu, K.C.; Yan, Y.C.; Lu, L.N.; Yang, G.; Luo, Q.Q.; Yung, W.H.; Ke, Y. Lipopolysaccharides up regulate hepcidin expression in the neurons via microglia and the IL-6/STAT3 signaling pathway. *Mol. Neurobiol.* **2014**, *50*, 811–820. [CrossRef] [PubMed]

24. Kopp, E.; Ghosh, S. Inhibition of NF-κB by sodium salicylate and aspirin. *Science* **1994**, *265*, 956–969. [CrossRef] [PubMed]

25. Kawai, T.; Akira, S. Signaling to NF-κB by Toll-like receptors. *Trends Mol. Med.* **2007**, *13*, 460–469. [CrossRef] [PubMed]

26. Grilli, M.; Pizzi, M.; Memo, M.; Spano, P. Neuroprotection by aspirin and sodium salicylate through blockade of NF-κB activation. *Science* **1996**, *274*, 1383–1385. [CrossRef] [PubMed]

27. Yin, M.J.; Yamamoto, Y.; Gaynor, R.B. The anti-inflammatory agents aspirin and salicylate inhibit the activity of IκB kinase-β. *Nature* **1998**, *396*, 77–80. [PubMed]

28. Wu, S.; Zhang, K.; Lv, C.; Wang, H.; Cheng, B.; Jin, Y.; Chen, Q.; Lian, Q.; Fang, X. Nuclear Factor-κB mediated lipopolysaccharide-induced mRNA expression of hepcidin in human peripheral blood leukocytes. *Innate Immun.* **2012**, *18*, 318–324. [CrossRef] [PubMed]

29. Hayden, M.S.; West, A.P.; Ghosh, S. NF-κB and the immune response. *Oncogene* **2006**, *25*, 6758–6780. [CrossRef] [PubMed]

30. Kühn, L.C. Iron regulatory proteins and their role in controlling iron metabolism. *Metallomics* **2015**, *7*, 232–243. [CrossRef] [PubMed]

31. Anderson, C.P.; Shen, M.; Eisenstein, R.S.; Leibold, E.A. Mammalian iron metabolism and its control by iron regulatory proteins. *Biochim. Biophys. Acta* **2012**, *1823*, 1468–1483. [CrossRef] [PubMed]

32. Jana, N.R. NSAIDs and apoptosis. *Cell. Mol. Life Sci.* **2008**, *65*, 1295–1301. [CrossRef] [PubMed]

33. Warner, T.D.; Mitchell, J.A. Cyclooxygenases: New forms, new inhibitors, and lessons from the clinic. *FASEB J.* **2004**, *18*, 790–804. [CrossRef] [PubMed]

34. Williams, C.S.; Mann, M.; DuBois, R.N. The role of cyclooxygenases in inflammation, cancer, and development. *Oncogene* **1999**, *18*, 7908–7916. [CrossRef] [PubMed]

35. Ejima, K.; Layne, M.D.; Carvajal, I.M.; Kritek, P.A.; Baron, R.M.; Chen, Y.H.; vom Saal, J.; Levy, B.D.; Yet, S.F.; Perrella, M.A. Cyclooxygenase-2-deficient mice are resistant to endotoxin-induced inflammation and death. *FASEB J.* **2003**, *17*, 1325–1327. [CrossRef] [PubMed]

36. Dong, C.; Davis, R.J.; Flavell, R.A. MAP kinases in the immune response. *Annu. Rev. Immunol.* **2002**, *20*, 55–72. [CrossRef] [PubMed]

37. Takeda, K.; Akira, S. TLR signaling pathways. *Semin. Immunol.* **2004**, *16*, 3–9. [CrossRef] [PubMed]

38. Chuang, Y.F.; Yang, H.Y.; Ko, T.L.; Hsu, Y.F.; Sheu, J.R.; Ou, G.; Hsu, M.J. Valproic acid suppresses lipopolysaccharide-induced cyclooxygenase-2 expression via MKP-1 in murine brain microvascular endothelial cells. *Biochem. Pharmacol.* **2014**, *88*, 372–383. [CrossRef] [PubMed]

39. Jin, J.; Samuvel, D.J.; Zhang, X.; Li, Y.; Lu, Z.; Lopes-Virella, M.F.; Huang, Y. Coactivation of TLR4 and TLR2/6 coordinates an additive augmentation on IL-6 gene transcription via p38MAPK pathway in U937 mononuclear cells. *Mol. Immunol.* **2011**, *49*, 423–432. [CrossRef] [PubMed]

40. Hsu, Y.F.; Sheu, J.R.; Lin, C.H.; Chen, W.C.; Hsiao, G.; Ou, G.; Chiu, P.T.; Hsu, M.J. MAPK phosphatase-1 contributes to trichostatin A inhibition of cyclooxygenase-2 expression in human umbilical vascular endothelial cells exposed to lipopolysaccharide. *Biochim. Biophys. Acta* **2011**, *1810*, 1160–1169. [CrossRef] [PubMed]

41. Hsu, M.J.; Chang, C.K.; Chen, M.C.; Chen, B.C.; Ma, H.P.; Hong, C.Y.; Lin, C.H. Apoptosis signalregulating kinase 1 in peptidoglycan-induced COX-2 expression in macrophages. *J. Leukoc. Biol.* **2010**, *87*, 1069–1082. [CrossRef] [PubMed]

42. Turpeinen, T.; Nieminen, R.; Moilanen, E.; Korhonen, R. Mitogen-activated protein kinase phosphatase-1 negatively regulates the expression of interleukin-6, interleukin-8, and cyclooxygenase-2 in A549 human lung epithelial cells. *J. Pharmacol. Exp. Ther.* **2010**, *333*, 310–318. [CrossRef] [PubMed]

43. He, W.; Qian, Z.M.; Zhu, L.; Christopher, Q.; Du, F.; Yung, W.H.; Ke, Y. Ginkgolides mimic the effects of hypoxic preconditioning to protect C6 cells against ischemic injury by up-regulation of hypoxia-inducible factor-1 α and erythropoietin. *Int. J. Biochem. Cell Biol.* **2008**, *40*, 651–662. [CrossRef] [PubMed]

44. Kim, B.W.; Koppula, S.; Park, S.Y.; Hwang, J.W.; Park, P.J.; Lim, J.H. Attenuation of inflammatory-mediated neurotoxicity by Saururuschinensis extract in LPS-induced BV-2 microgliacells via regulation of NF-κB signaling and anti-oxidant properties. *BMC Complement. Altern. Med.* **2014**, *14*, 502. [CrossRef] [PubMed]

45. Du, F.; Qian, Z.M.; Luo, Q.; Yung, W.H.; Ke, Y. Hepcidin suppresses brain iron accumulation by downregulating iron transport proteins in iron-overloaded rats. *Mol. Neurobiol.* **2015**, *52*, 101–114. [CrossRef] [PubMed]

46. Du, F.; Qian, Z.M.; Gong, Q.; Zhu, Z.J.; Lu, L.; Ke, Y. The iron regulatory hormone hepcidin inhibits expression of iron release as well as iron uptake proteins in J774 cells. *J. Nutr. Biochem.* **2012**, *23*, 1694–1700. [CrossRef] [PubMed]

47. Chang, Y.Z.; Qian, Z.M.; Wang, K.; Zhu, L.; Yang, X.D.; Du, J.R.; Jiang, L.; Ho, K.P.; Wang, Q.; Ke, Y. Effects of development and iron status on ceruloplasmin expression in rat brain. *J. Cell. Physiol.* **2005**, *204*, 623–633. [CrossRef] [PubMed]

48. Du, F.; Qian, C.; Qian, Z.M.; Wu, X.M.; Xie, H.; Yung, W.H.; Ke, Y. Hepcidin directly inhibits transferrin receptor 1 expression in astrocytes via a cyclic AMP-protein kinase A pathway. *Glia* **2011**, *59*, 936–945. [CrossRef] [PubMed]

International Journal of
Molecular Sciences

Article

Glycine Relieves Intestinal Injury by Maintaining mTOR Signaling and Suppressing AMPK, TLR4, and NOD Signaling in Weaned Piglets after Lipopolysaccharide Challenge

Xiao Xu [1], Xiuying Wang [1], Huanting Wu [1], Huiling Zhu [1], Congcong Liu [1], Yongqing Hou [1], Bing Dai [2], Xiuting Liu [2] and Yulan Liu [1,*]

[1]   Hubei Key Laboratory of Animal Nutrition and Feed Science, Hubei Collaborative Innovation Center for Animal Nutrition and Feed Safety, Wuhan Polytechnic University, Wuhan 430023, China; xuxiao200315@163.com (X.X.); xiuyingdk@foxmail.com (X.W.); wht18842@126.com (H.W.); zhuhuiling2004@sina.com (H.Z.); 13545906457@163.com (C.L.); houyq@aliyun.com (Y.H.)

[2]   Zhe Jiang Goshine Test Technologies Co., Ltd., Hangzhou 310030, China; db@gojue.com (B.D.); lxt@gojue.com (X.L.)

*   Correspondence: yulanflower@126.com; Tel./Fax: +86-027-8395-6175

Received: 13 April 2018; Accepted: 3 July 2018; Published: 6 July 2018

Abstract: This study was conducted to envaluate whether glycine could alleviate *Escherichia coli* lipopolysaccharide (LPS)-induced intestinal injury by regulating intestinal epithelial energy status, protein synthesis, and inflammatory response via AMPK, mTOR, TLR4, and NOD signaling pathways. A total of 24 weanling piglets were randomly allotted to 1 of 4 treatments: (1) non-challenged control; (2) LPS-challenged control; (3) LPS + 1% glycine; (4) LPS + 2% glycine. After 28 days feeding, piglets were injected intraperitoneally with saline or LPS. The pigs were slaughtered and intestinal samples were collected at 4 h postinjection. The mRNA expression of key genes in these signaling pathways was measured by real-time PCR. The protein abundance was measured by Western blot analysis. Supplementation with glycine increased jejunal villus height/crypt depth ratio. Glycine also increased the jejunal and ileal protein content, RNA/DNA ratio, and jejunal protein/DNA ratio. The activities of citroyl synthetase in ileum, and α-ketoglutarate dehydrogenase complex in jejunum, were increased in the piglets fed diets supplemented with glycine. In addition, glycine decreased the jejunal and ileal phosphorylation of AMPKα, and increased ileal phosphorylation of mTOR. Furthermore, glycine downregulated the mRNA expression of key genes in inflammatory signaling. Meanwhile, glycine increased the mRNA expression of negative regulators of inflammatory signaling. These results indicate that glycine supplementation could improve energy status and protein synthesis by regulating AMPK and mTOR signaling pathways, and relieve inflammation by inhibiting of TLR4 and NOD signaling pathways to alleviate intestinal injury in LPS-challenged piglets.

Keywords: glycine; inflammatory response; intestine; LPS; weanling piglets

---

1. Introduction

The intestinal epithelium not only plays a key role in digestion and absorption of nutrients, but also has an important function in preventing pathogen invasion and dissemination of commensals [1]. However, the intestinal epithelial health status, especially in young animals, could be easily injured by many factors, such as inflammation and infection [2]. Inflammation often results in intestinal mucosal damage and dysfunction, which negatively affects animal performance and health [3]. In order to alleviate the inflammation and maintain health and function, the intestine needs a high level of energy and amino acids (AAs) [4,5].

Glycine, whose structure is the simplest of all AAs, is regarded as a conditionally essential AA for young mammals, and a nutritionally essential amino acid for fetal and neonatal development of poultry [6,7]. In addition, glycine is the most abundant AA in the body [8], and it is highly required for neonatal growth and development [7,9]. Recently, some reports showed that glycine could alleviate colitis induced by chemicals, small intestine injury induced by endotoxins, and inhibit overproduction of pro-inflammatory cytokines in rats [10–12].

Activation of inflammatory signaling pathways can lead to intestinal damage [13]. Toll-like receptors (TLRs) and nucleotide-binding oligomerization domain proteins (NODs) are important protein families of inflammatory signaling pathways [14,15]. These proteins are expressed in many tissues, including the intestine [16–18], and play key roles in induction of inflammatory responses by recognition of pathogen-associated molecular patterns (PAMPs) [16,19]. Interactions of TLRs or NODs with their specific PAMPs trigger downstream signaling events that lead to activation of nuclear factor-κB (NF-κB), which could further induce the expression of genes related to pro-inflammatory cytokines, such as interleukin-1β (IL-1β), IL-6 and tumor necrosis factor-α (TNF-α) [20]. As a consequence, these pro-inflammatory cytokines adjust the host's defense against invading pathogens. However, the intestine can be easily injured by overproduction of these cytokines, especially TNF-α. Adenosine monophosphate-activated protein kinase (AMPK) is a serine/threonine protein kinase, which widely exists in eukaryotic cells [21]. AMPK can directly mediate metabolic adaptations to a change of energy status [22]. The overproduction of cytokines, such as TNF-α, increases energy consumption which could activate AMPK [23]. The activated AMPK could further inhibit the mammalian target of rapamycin (mTOR) signaling pathway to reduce the synthesis of protein in tissues, including the intestine [24].

The aim of this study was to investigate whether glycine could mitigate lipopolysaccharide (LPS)-induced intestinal injury, and to explore its molecular mechanism(s). We hypothesized that dietary glycine addition could enhance energy status and protein synthesis by suppressing AMPK activation and activating the mTOR signaling pathway, and reduce the production of pro-inflammatory cytokines in the intestine through regulating inflammatory signaling pathways to maintain intestinal integrity. LPS, a component of Gram-negative bacteria, is responsible for neonatal mortality and sepsis, but low concentrations of LPS resulted in tissue protection in some studies [25]. In the present study, *Escherichia coli* (*E. coli*) lipopolysaccharide (LPS; *E. coli* serotype 055:B5; potency $\geq$ 5,000,000 EU/mg) was intraperitoneally injected at 100 µg/kg BW (body weight), aimed to establish the model of endotoxemia [26]. Furthermore, we used the weanling piglet model, which is a suitable animal model for human nutrition research [27,28].

2. Results

2.1. Growth Performance

During the 28 days feeding period (before LPS or saline injection), there was no difference in average daily gain (510 $\pm$ 38 g), average feed intake (785 $\pm$ 76 g), and feed/gain ratio (1.54 $\pm$ 0.08) among the four groups.

2.2. Intestinal Morphology

The intestinal mucosa in control group was in good condition (Figures 1 and 2). However, piglets challenged with LPS exhibited intestinal mucosal damage. Supplementation with Gly alleviated intestinal mucosal injury, to some extent. Compared to the CONTR (control group) piglets, the intestinal morphology in LPS-challenged piglets had no significant difference ($p > 0.05$). Among the LPS-challenged piglets, glycine supplementation decreased jejunal crypt depth (linear, $p < 0.05$; Table 1), and increased jejunal villus height/crypt depth ratio (VCR; linear, $p < 0.05$; quadratic, $p < 0.05$). However, there was no significant effect on intestinal morphology among the LPS-challenged piglets fed diets supplemented with glycine ($p > 0.05$).

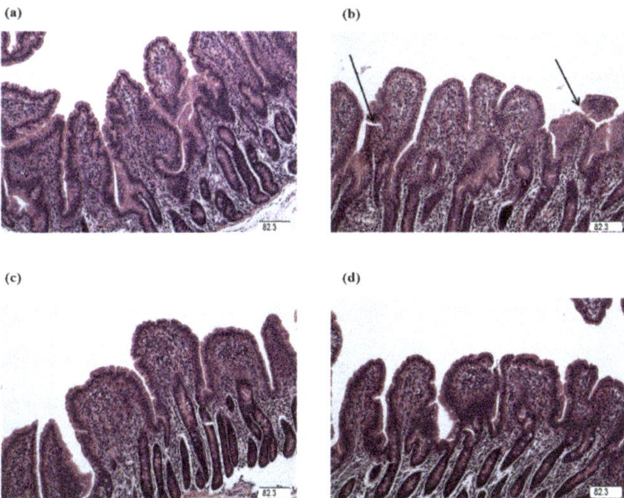

Figure 1. Jejunal mucosal histological appearance (hematoxylin and eosin). (**a**) Pigs fed a control diet and injected with sterile saline. No obvious damage was observed. (**b**) Pigs fed a control diet and injected with LPS. Intestinal mucosa was seriously damaged by LPS. Arrow represents the damaged intestinal mucosa in the piglets. (**c**) Pigs fed a 1.0% Gly diet and injected with LPS. Intestinal damage was alleviated. (**d**) Pigs fed a 2.0% Gly diet and injected with LPS. Intestinal damage was alleviated. Original magnification 100×. Scale bars = 82.3 μm.

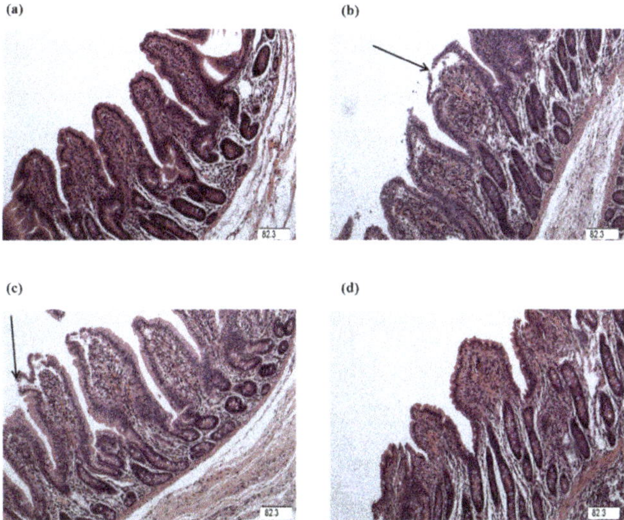

Figure 2. Ileal mucosal histological appearance (hematoxylin and eosin). (**a**) Pigs fed a control diet and injected with sterile saline. No obvious damage was observed. (**b**) Pigs fed a control diet and injected with LPS. Intestinal mucosa was seriously damaged by LPS. Arrow represents the damaged intestinal mucosa in the piglets. (**c**) Pigs fed a 1.0% Gly diet and injected with LPS. Intestinal damage was still existed. (**d**) Pigs fed a 2.0% Gly diet and injected with LPS. Intestinal damage was alleviated. Arrow represents the damaged intestinal mucosa in the piglets. Original magnification 100×. Scale bars = 82.3 μm.

Table 1. Effect of glycine supplementation on intestinal morphology after 4 h lipopolysaccharide (LPS) challenge in piglets.

Item	Treatment [1]				SEM	*p* Value [2]		
	CONTR	LPS	LPS + 1.0% Gly	LPS + 2.0% Gly		CONTR vs. LPS	Linear	Quadratic
Jejunum								
Villus height (μm)	286	286	268	265	9.7	0.973	0.111	0.201
Crypt depth (μm)	106	103	88.6	88.1	5.0	0.688	0.037	0.058
VCR	2.76	2.82	2.91	2.94	0.03	0.356	0.012	0.038
Ileum								
Villus height (μm)	244	241	255	254	10.0	0.902	0.368	0.560
Crypt depth (μm)	83.0	81.0	85.9	86.2	3.5	0.758	0.230	0.412
VCR	2.95	2.93	2.97	2.94	0.03	0.740	0.739	0.467

[1] CONTR (non-challenged control), piglets fed a basal diet as well as injected with 0.9% NaCl solution; LPS (LPS-challenged control), piglets fed the same basal diet as well as injected with *E. coli* LPS; LPS + 1.0% Gly, piglets fed a 1.0% glycine-supplemented diet and injected with LPS; LPS + 2.0% Gly, piglets fed a 2.0% glycine-supplemented diet and injected with LPS. VCR, villus height/crypt depth ratio. SEM, standard error of mean. [2] CONTR vs. LPS was used to obtain the response of LPS challenge. Linear and quadratic polynomial contrasts were used to obtain the response of glycine supplementation in LPS-challenged piglets.

2.3. Mucosal Protein, DNA, and RNA Content

Among the LPS-challenged piglets, supplementation with glycine increased mucosal protein content (linear, $p < 0.05$; Table 2), RNA/DNA ratio (linear, $p < 0.05$), and protein/DNA ratio (linear, $p < 0.05$) in jejunum. Compared with piglets in the CONTR group, the piglets in LPS group had a reduced RNA/DNA ratio in the ileum ($p < 0.05$). Furthermore, the piglets challenged with LPS had increased mucosal protein content (linear, $p < 0.05$; quadratic, $p < 0.05$), and RNA/DNA ratio (linear, $p < 0.05$) in the ileum, when they were fed diets supplemented with glycine.

Table 2. Effect of glycine supplementation on intestinal protein, DNA and RNA contents after 4 h LPS challenge in piglets.

Item	Treatment [1]				SEM	*p* Value [2]		
	CONTR	LPS	LPS + 1.0% Gly	LPS + 2.0% Gly		CONTR vs. LPS	Linear	Quadratic
Jejunum								
Protein (mg/g tissue)	84.4	81.3	89.7	90.8	3.7	0.614	0.049	0.099
RNA/DNA	3.30	3.35	3.77	3.82	0.20	0.888	0.049	0.097
Protein/DNA (mg/μg)	0.20	0.18	0.21	0.21	0.01	0.365	0.021	0.055
Ileum								
Protein (mg/g tissue)	56.3	61.7	68.8	72.3	2.3	0.231	<0.001	0.001
RNA/DNA	11.4	8.29	9.64	9.65	0.59	0.011	0.048	0.125
Protein/DNA (mg/μg)	0.31	0.27	0.32	0.34	0.03	0.395	0.069	0.180

[1] CONTR (non-challenged control), piglets fed a basal diet as well as injected with 0.9% NaCl solution; LPS (LPS-challenged control), piglets fed the same basal diet as well as injected with *E. coli* LPS; LPS + 1.0% Gly, piglets fed a 1.0% glycine-supplemented diet and injected with LPS; LPS + 2.0% Gly, piglets fed a 2.0% glycine-supplemented diet and injected with LPS. SEM, standard error of mean. [2] CONTR vs. LPS was used to obtain the response of LPS challenge. Linear and quadratic polynomial contrasts were used to obtain the response of glycine supplementation in LPS-challenged piglets.

2.4. Intestinal Claudin-1 Protein Expression

As shown in Table 3 and Figure 3, the piglets challenged with LPS had decreased claudin-1 protein abundance in the ileum, compared with the piglets in CONTR group ($p < 0.05$). However, there was no significant difference in claudin-1 protein abundance in the jejunum between CONTR and LPS group ($p > 0.05$). In addition, there was no significant effect on the intestinal claudin-1 protein abundance in the piglets fed diets supplemented with glycine ($p > 0.05$).

Table 3. Effects of glycine supplementation on intestinal protein abundance of claudin-1 after 4 h LPS challenge in piglets.

Item	Treatment [1]				SEM	p Value [2]		
	CONTR	LPS	LPS + 1.0% Gly	LPS + 2.0% Gly		CONTR vs. LPS	Linear	Quadratic
Jejunum claudin-1/β-actin	0.43	0.50	0.52	0.47	0.07	0.282	0.772	0.902
Ileum claudin-1/β-actin	1.03	0.60	0.80	0.64	0.13	0.032	0.814	0.482

[1] CONTR (non-challenged control), piglets fed a basal diet as well as injected with 0.9% NaCl solution; LPS (LPS-challenged control), piglets fed the same basal diet as well as injected with *E. coli* LPS; LPS + 1.0% Gly, piglets fed a 1.0% glycine-supplemented diet and injected with LPS; LPS + 2.0% Gly, piglets fed a 2.0% glycine-supplemented diet and injected with LPS. SEM, standard error of mean. [2] CONTR vs. LPS was used to obtain the response of LPS challenge. Linear and quadratic polynomial contrasts were used to obtain the response of glycine supplementation in LPS-challenged piglets.

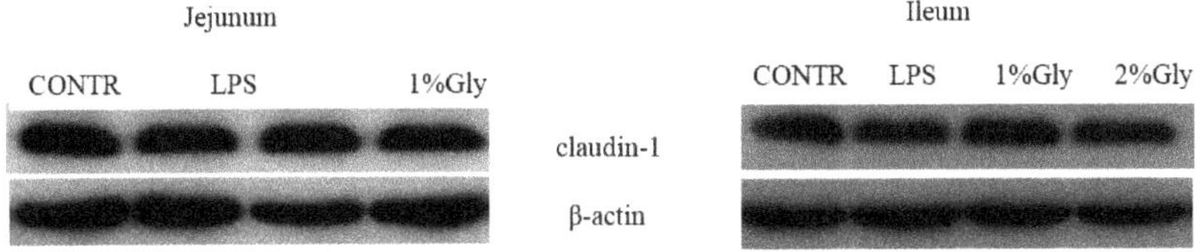

Figure 3. Effects of glycine supplementation on intestinal protein abundance of claudin-1 after 4 h LPS challenge in piglets.

2.5. Intestinal Key Enzyme Activities of the Tricarboxylic Acid (TCA) Cycle

Compared with the piglets in CONTR group, the piglets in LPS group had decreased ileal citrate synthase (CS) and α-ketoglutarate dehydrogenase complex (α-KGDHC) activities ($p < 0.05$; Table 4). Among the piglets challenged with LPS, supplementation with glycine in the diets increased the activities of α-KGDHC (quadratic, $p < 0.05$) in jejunum, and the activity of CS (quadratic, $p < 0.05$) in ileum.

Table 4. Effects of glycine supplementation on intestinal tricarboxylic acid cycle key enzyme activities after 4 h LPS challenge in weaned piglets.

Item	Treatment [1]				SEM	p Value [2]		
	CONTR	LPS	LPS + 1.0% Gly	LPS + 2.0% Gly		CONTR vs. LPS	Linear	Quadratic
Jejunum								
CS (U·g protein^{-1})	1.88	1.66	1.68	1.57	0.10	0.255	0.388	0.560
ICD (mIU·g protein^{-1})	187	168	173	190	7.82	0.088	0.064	0.708
α-KGDHC (µg·g protein^{-1})	3146	2706	2171	2844	169	0.140	0.615	0.015
Ileum								
CS (U·g protein^{-1})	1.37	1.07	0.93	1.21	0.05	0.002	0.192	0.015
ICD (mIU·g protein^{-1})	127	109	110	108	7.62	0.160	0.941	0.979
α-KGDHC (µg·g protein^{-1})	2566	1638	1783	1966	182	0.019	0.196	0.443

CS: citrate synthase, ICD: isocitrate dehydrogenase, α-KGDHC: α-ketoglutarate dehydrogenase complex. SEM, standard error of mean. [1] CONTR (non-challenged control), piglets fed a basal diet as well as injected with 0.9% NaCl solution; LPS (LPS-challenged control), piglets fed the same basal diet as well as injected with *E. coli* LPS; LPS + 1.0% Gly, piglets fed a 1.0% glycine-supplemented diet and injected with LPS; LPS + 2.0% Gly, piglets fed a 2.0% glycine-supplemented diet and injected with LPS. [2] CONTR vs. LPS was used to obtain the response of LPS challenge. Linear and quadratic polynomial contrasts were used to obtain the response of glycine supplementation in LPS-challenged piglets.

2.6. Intestinal Protein Expression of the Key Protein in AMPKα and mTOR Pathways

Among the LPS-challenged piglets, supplementation with glycine decreased the ratio of p-AMPKα/t-AMPKα both in jejunum and ileum (linear, $p < 0.05$; quadratic, $p < 0.05$; Figure 4). Relative to CONTR piglets, LPS challenge decreased the ratio of p-mTOR/t-mTOR in ileum ($p < 0.05$; Figure 5), while there was no difference in the ratio of p-mTOR/t-mTOR in jejunum between CONTR and LPS group ($p > 0.05$). As supplementation with glycine in the diets, the piglets challenged with LPS increased ratio of p-mTOR/t-mTOR in ileum (linear, $p < 0.05$). However, there was no effect of glycine or LPS on the ratio of t-AMPKα, t-mTOR, p-4EBP1, and t-4EBP1 in jejunum and ileum ($p > 0.05$; Figures 4–6).

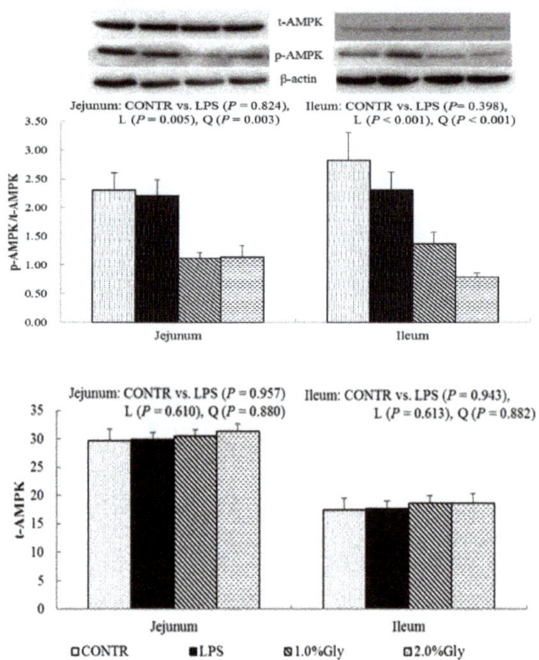

Figure 4. Effects of glycine supplementation on intestinal protein abundance of t-AMPKα and p-AMPKα after 4 h LPS challenge in weanling piglets. Note: L (linear), Q (quadratic).

2.7. Intestinal mRNA Expression of the Key Genes in TLR4 and NOD Pathways

Relative to CONTR piglets, the LPS pigs had higher mRNA expression of TLR4, MyD88, NOD2, RIPK2, and NF-κB ($p < 0.05$; Table 5) in jejunum. Among LPS challenged piglets, supplementation with glycine decreased mRNA expression of TLR4, LBP, MyD88, TRAF6, NOD2, and NF-κB (linear, $p < 0.05$; quadratic, $p < 0.05$) in jejunum. Compared to the piglets in CONTR group, the piglets in LPS group had increased mRNA expression of NOD2, and RIPK2 ($p < 0.05$) in ileum. However, the mRNA expression of IRAK1 in piglets challenged with LPS was reduced compared with the piglets in CONTR group ($p < 0.05$). Glycine supplementation reduced mRNA expression of NOD2 and RIPK2 (quadratic, $p < 0.05$) in ileum among the LPS-challenged pigs.

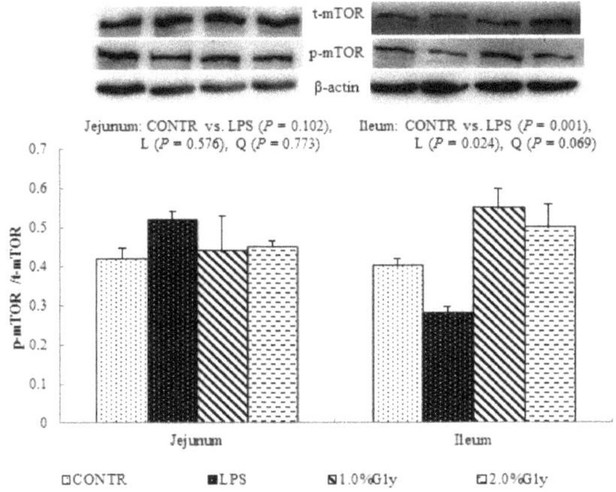

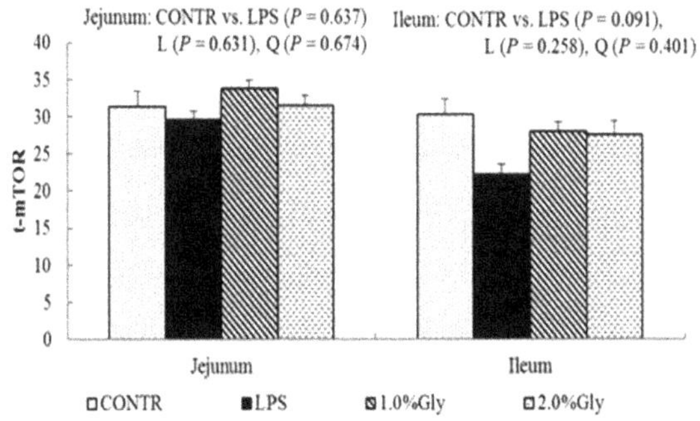

Figure 5. Effects of glycine supplementation on intestinal protein abundance of t-mTOR and p-mTOR after 4 h LPS challenge in weanling piglets. Note: L (linear), Q (quadratic).

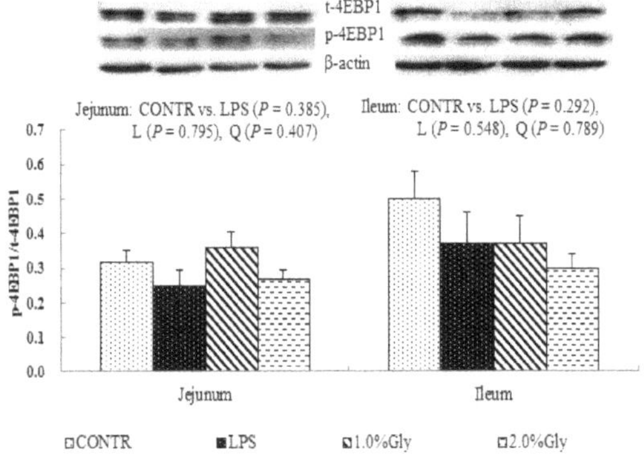

Figure 6. *Cont.*

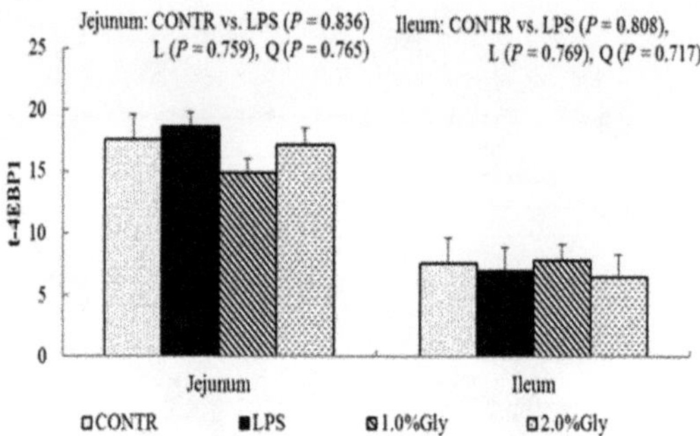

Figure 6. Effects of glycine supplementation on intestinal protein abundance of t-4EBP1 and p-4EBP1 after 4 h LPS challenge in weanling piglets. Note: L (linear), Q (quadratic).

Table 5. Effect of glycine supplementation on intestinal mRNA expression of TLR4 and NOD and their downstream signals after 4 h LPS challenge in piglets.

Item	Treatment [1]				SEM	*p* Value [2]		
	CONTR	LPS	LPS + 1.0% Gly	LPS + 2.0% Gly		CONTR vs. LPS	Linear	Quadratic
Jejunum								
TLR4	1.00	1.67	1.17	1.22	0.09	<0.001	0.005	0.001
LBP	1.00	0.82	0.67	0.44	0.11	0.439	0.005	0.022
MyD88	1.00	1.17	0.92	0.93	0.04	0.017	0.007	0.003
IRAK1	1.00	1.00	0.89	0.95	0.06	0.943	0.579	0.509
TRAF6	1.00	1.22	0.84	0.93	0.07	0.079	0.034	0.007
NOD1	1.00	0.66	0.59	0.53	0.11	0.103	0.249	0.524
NOD2	1.00	2.45	1.42	1.53	0.20	0.006	0.019	0.011
RIPK2	1.00	1.84	1.39	1.68	0.09	<0.001	0.331	0.010
NF-κB	1.00	1.29	1.02	1.10	0.05	0.001	0.039	0.006
Ileum								
TLR4	1.00	1.13	1.22	1.20	0.08	0.215	0.557	0.713
LBP	1.00	0.44	0.27	0.33	0.21	0.198	0.418	0.474
MyD88	1.00	1.04	0.95	0.98	0.06	0.646	0.403	0.431
IRAK1	1.00	0.83	0.69	0.92	0.08	0.021	0.490	0.199
TRAF6	1.00	1.15	1.07	1.16	0.06	0.125	0.972	0.583
NOD1	1.00	0.86	0.81	0.85	0.10	0.415	0.926	0.890
NOD2	1.00	1.97	1.07	1.54	0.20	0.006	0.255	0.039
RIPK2	1.00	1.85	1.37	1.82	0.11	<0.001	0.900	0.019
NF-κB	1.00	1.08	1.02	1.09	0.06	0.381	0.886	0.729

IRAK1: IL-1 receptor-associated kinase, LBP: lipopolysaccharide-binding protein, MyD88: myeloid differentiation factor 88, NF-κB: nuclear factor-κB, NOD1: nucleotide-binding oligomerization domain protein 1, NOD2: nucleotide-binding oligomerization domain protein 2, RIPK2: receptor-interacting protein kinase 2, TLR4: toll-like receptor 4, TRAF6: TNF receptor-associated factor 6, SEM: standard error of mean. [1] CONTR (non-challenged control), piglets fed a basal diet as well as injected with 0.9% NaCl solution; LPS (LPS-challenged control), piglets fed the same basal diet as well as injected with *E. coli* LPS; LPS + 1.0% Gly, piglets fed a 1.0% glycine-supplemented diet and injected with LPS; LPS + 2.0% Gly, piglets fed a 2.0% glycine-supplemented diet and injected with LPS. [2] CONTR vs. LPS was used to obtain the response of LPS challenge. Linear and quadratic polynomial contrasts were used to obtain the response of glycine supplementation in LPS-challenged piglets.

2.8. Intestinal mRNA Expression of Negative Regulators of TLR4 and NOD Pathways

Compared with the pigs in CONTR group, LPS challenge decreased jejunal mRNA expression of Tollip (*p* < 0.05; Table 6), and ileal mRNA expression of Tollip, ERBB2IP, and centaurin β1 (*p* < 0.05) in the piglets. However, LPS challenge increased mRNA expression of SOCS1 in jejunum (*p* < 0.05) and SIGIRR in ileum ($p < 0.05$). Among the LPS challenged piglets, supplementation with glycine increased mRNA expression of Tollip (quadratic, $p < 0.05$) and ERBB2IP (linear, $p < 0.05$) in ileum.

Table 6. Effects of glycine supplementation on intestinal mRNA expression of negative regulators of TLR4 and NOD signal pathway after 4 h LPS challenge in piglets.

Item	Treatment [1]				SEM	p Value [2]		
	CONTR	LPS	LPS + 1.0% Gly	LPS + 2.0% Gly		CONTR vs. LPS	Linear	Quadratic
Jejunum								
RP105	1.00	1.94	0.95	1.59	0.34	0.190	0.558	0.216
SOCS1	1.00	2.74	1.69	1.94	0.30	0.019	0.128	0.103
Tollip	1.00	0.52	0.93	0.59	0.12	0.016	0.702	0.062
SIGIRR	1.00	0.96	1.01	1.00	0.05	0.524	0.525	0.736
ERBB2IP	1.00	0.88	0.76	0.79	0.04	0.074	0.136	0.125
centaurin β1	1.00	0.82	0.56	0.58	0.08	0.202	0.029	0.027
Ileum								
RP105	1.00	0.73	0.79	0.84	0.10	0.116	0.355	0.661
SOCS1	1.00	0.87	1.03	1.10	0.08	0.242	0.063	0.170
Tollip	1.00	0.62	1.16	0.64	0.12	0.007	0.942	0.008
SIGIRR	1.00	1.23	1.13	1.26	0.06	0.001	0.797	0.328
ERBB2IP	1.00	0.77	0.82	0.97	0.06	0.028	0.029	0.082
centaurin β1	1.00	0.74	0.75	0.80	0.07	0.041	0.431	0.696

ERBB2IP: Erbb2 interacting protein, RP105: radioprotective 105, SIGIRR: single immunoglobulin IL-1 related receptor, SOCS1: suppressor of cytokine signaling 1, Tollip: toll-interacting protein, SEM: standard error of mean. [1] CONTR (non-challenged control), piglets fed a basal diet, as well as injected with 0.9% NaCl solution; LPS (LPS-challenged control), piglets fed the same basal diet as well as injected with *E. coli* LPS; LPS + 1.0% Gly, piglets fed a 1.0% glycine-supplemented diet and injected with LPS; LPS + 2.0% Gly, piglets fed a 2.0% glycine-supplemented diet and injected with LPS. [2] CONTR vs. LPS was used to obtain the response of LPS challenge. Linear and quadratic polynomial contrasts were used to obtain the response of glycine supplementation in LPS-challenged piglets.

3. Discussion

In the present study, the effect of glycine on intestinal integrity after a 4 h *E. coli* LPS challenge was evaluated in a weaned piglet model. Dietary glycine supplementation improved the intestinal energy status and protein synthesis associated with inhibiting AMPK signaling and activating mTOR signaling, and simultaneously reduced the intestinal inflammatory response associated with inhibiting inflammatory signaling pathways (TLR4 and NOD), and as a consequence, improved intestinal integrity.

The intestinal mucosal integrity of weanling piglets is closely related to physical health and nutrient digestion and absorption capability [24,29]. Villus height, crypt depth, and VCR were determined to indicate gross intestinal morphology [3]. In this study, the decreased jejunal crypt depth and increased jejunal VCR of the piglets fed glycine indicated that the digestive juice can be secreted easily into intestinal lumen, which showed glycine played a role in maintaining the integrity in mucosal structure in weanling piglets. Wang et al. reported supplementation with glycine increased intestinal villus height in nursery piglets [7]. Effenberger-Neidnicht et al. found that supplementation with glycine improved intestinal architecture and reduced the LPS-induced intestinal accumulation of blood in rats [11]. The above studies indicated that glycine had a positive effect on the intestinal mucosal integrity.

Protein and DNA are the basis for repair and proliferation of epithelial cells. The ratio of RNA/DNA is an indicator reflecting the cell capacity for protein synthesis [30]. The ratio of protein/DNA is a sensitive indicator of protein mass and cell size [24]. The decreased ileal RNA/DNA ratio in LPS-challenged piglets indicated that the mucosal cell capacity for repair was decreased, affected by LPS. The piglets fed diets containing glycine had increased jejunal and ileal protein mass, protein/DNA ratio, and RNA/DNA ratio in the present study. These results illustrated that glycine is beneficial to intestine mucosal repair after LPS-induced injury. Similar to our data, Lee et al. [31] reported that glycine supplementation increased intestinal protein mass in the rats under ischemia–reperfusion injury. In addition, Stoll et al. [32] demonstrated that glycine could be utilized directly to synthesize protein in the intestinal tract. Furthermore, Wang et al. [7] demonstrated that glycine was the precursor of purines, which were involved in protein synthesis and cell proliferation.

The intestinal epithelial barrier can reduce the ability of luminal pathogens and their toxins to invade the mucosa, preventing the penetration of luminal bacteria into the mucosa, to maintain gut homeostasis [33]. The disordered function of the intestinal epithelial barrier often leads to inflammatory disease in the intestine [34]. The integrity of the intestinal epithelial barrier is maintained by cohesive interactions between epithelial cells to form tight junctions [34]. Claudin-1, which is an important protein in formation of tight junctions, determines permeability characteristics in many tissues, especially the intestine [35]. Therefore, the greater abundance of claudin-1 often reflects improved function of the epithelial barrier [36]. In the present study, LPS injection decreased the ileal protein abundance of claudin-1, which showed that LPS impaired the function of intestinal epithelial barrier. Similarly, some previous reports demonstrated that LPS challenge decreased the abundance of claudin-1 [37]. However, supplementation with glycine in the diets did not affect the abundance of claudin-1 in the piglets in our study. Few reports studied the effect of glycine on the intestinal tight junctions, so the pathway to repair injured tight junction in the intestine should be further studied in the future.

Citrate synthase, ICD, and α-KGDHC are key enzymes involved in the TCA cycle, which is a central route for energy production [38]. CS catalyzes the first step of the TCA cycle by attaching molecules of acetate and attaching them to oxaloacetate [39]. ICD, which exists in mitochondria and cytoplasm, is responsible for catalyzing the oxidative decarboxylation of isocitrate into α-ketoglutarate and CO_2 [40]. α-KGDHC is a multi-enzymatic complex which converts α-ketoglutarate into succinyl-CoA [41]. In our experiment, LPS challenge reduced the activities of ICD in jejunum and CS, and α-KGDHC in ileum, which indicated that the energy production efficiency decreased. This is consistent with a previous report which showed that LPS challenge decreased the activities of jejunal CS, ICD, and α-KGDHC, and ileal ICD in weanling piglets [4]. After supplementation with glycine in the diets, the activities of α-KGDHC in jejunum and CS in ileum were increased. Therefore, it is possible that glycine could improve the energy production efficiency in intestinal mucosa by enhancing the key enzyme activities of the TCA cycle.

We hypothesized that glycine had a beneficial effect on intestinal integrity through enhancing energy production and protein synthesis in epithelial cells. AMPK, as an energy regulator, maintains the intracellular energy balance in eukaryons [42]. When the intracellular AMP/ATP ratio increases, AMPK is activated by phosphorylation [43]. The activated AMPK can switch on ATP-producing processes while synchronously switching off ATP-consuming processes to restore the cellular energy status [44]. Supplementation with glycine decreased jejunal and ileal p-AMPK/t-AMPK ratio, which illustrated dietary glycine could potentially enhance ATP-consumption to synthesize protein. mTOR is a serine–threonine kinase which controls many important aspects in mammalian cell functions, such as protein synthesis [45]. Its activity is modulated by various intracellular and extracellular factors (especially AAs and energy), meanwhile, mTOR adjusts rates of translation, transcription, protein degradation, cell signaling, and metabolism [44]. Specially, mTOR signaling plays a critical role in maintaining intestinal health [45,46]. Similar to our results, a previous report showed that LPS injection decreased jejunal mRNA expression of mTOR [47]. It is known that LPS could inhibit intestinal protein synthesis through suppressing activation of mTOR. The piglets fed diets supplemented with glycine had increased ileal p-mTOR/t-mTOR ratio, which demonstrated glycine could relieve the reduction of mucosal protein synthesis caused by LPS injury in the ileum. These results are in agreement with the decreased p-AMPK/t-AMPK ratio in piglets fed the diets supplemented with glycine. Therefore, we propose that the mechanism may be that glycine improves the intestinal mucosal energy status, and further activates mTOR signaling pathway to enhance protein synthesis and mucosal repair.

We hypothesized that supplementation with glycine improved intestinal integrity by inhibiting intestinal TLR4 and NODs, and their respective downstream signals, further reducing the inflammatory response. In the current study, LPS injection increased the mRNA abundance of TLR4 (TLR4, MyD88, and NF-κB in jejunum) and NOD signaling-related genes (NOD2 and RIPK2 in jejunum and ileum) which reflected LPS-induced intestinal inflammation by activation of TLR4 and NOD pathways. Supplementation with glycine reduced mRNA expressions of TLR4 (TLR4, LBP, MyD88, TRAF6, and NF-κB in jejunum) and NOD signaling-related genes (NOD2 and RIPK2 in jejunum and ileum) in the LPS-challenged piglets, which illustrated that glycine could relieve intestinal inflammation by inhibiting activation of inflammatory signaling pathways. In general, the inflammatory cytokines, such as TNF-α, IL-1β, and IL-6 can be overproduced when the inflammatory signaling pathways (TLR4 and NODs) are activated. Similar to our study, Tsune et al. [10] reported that glycine relieved colitis in rats by reducing mRNA expression of TNF-α and IL-1β. Furthermore, Stoffels et al. [48] demonstrated that injection of glycine before intestinal surgery could alleviate inflammation by decreasing mRNA expression of IL-6 and TNF-α.

Activation of TLR4 and NOD signaling could prevent against pathogens invading by triggering the production of pro-inflammatory cytokines and inflammatory response. However, over activation of inflammatory signaling pathways also lead to collateral host tissue injury [49]. To prevent excessive and harmful inflammatory responses, these inflammatory signaling pathways are negatively controlled by multiple mechanisms. So far, many negative regulators of TLR4 signaling (Tollip and SOCS1) and NOD signaling (ERBB2IP and centaurin β1) have been identified and characterized [50,51]. The current results showed that the piglets challenged with LPS had decreased mRNA expression of Tollip both in jejunum and ileum, as well as ERBB2IP and centaurin β1 in ileum, which is in agreement with the report of Wang et al. [52]. These results reflect that LPS challenge reduced the mRNA expression of TLR4 and NOD negative regulators, which is in agreement with increased mRNA expression of TLR4 and NOD signaling-related genes. Fujimoto et al. reported that excessive inflammatory cytokines enforced the expression of SOCS1, which resulted in decreased response of cells to TLR ligands [53]. In accordance with this, our present study showed that LPS challenge increased mRNA expressions of jejunal SOCS1, which indicates that SOCS1 might play a key role in intestinal self-protection. Supplementation with glycine increased mRNA expression of ileal ERBB2IP. Previous studies showed that Tollip bound to IL-1 receptor-associated kinase (IRAK) and inhibited IRAK phosphorylation to downregulate TLR4 signaling [54]. This indicates that supplementation with glycine could suppress the activity of IRAK by increasing the mRNA expression of Tollip, resulting in impairing the signaling from TLR4 to downstream pathways, and reducing the synthesis of pro-inflammatory cytokines.

In the present study, the effects of LPS challenge or glycine supplementation on some parameters were inconsistent in different sites of the intestine. This may be due to the difference in the anatomy and physiology among the different sites of the intestine [37]. In addition, LPS caused dynamic changes in the physiological variables, and gene and protein expression of inflammatory signaling pathways [55,56]. After 28 days feeding of Gly, the efficacy of Gly has risen. However, having only one time point (4 h) selected, to measure the effect of LPS on various physiological variables and gene and protein expression, was not perfect. Therefore, in future studies, sample collections at more time points are needed to better understand the dynamic effect of LPS on intestinal injury. Furthermore, the maintenance of normal blood flow through microcirculation plays a fundamental role in the protection and healing of intestinal mucosa [57]. Many previous studies reported that the intestinal injury induced by various factors can be alleviated through improving blood flow in microcirculation of duodenum and colon [58–61]. In the future, the blood flow of jejunal and ileal mucosa needs to be measured to better explore the mechanism of effects of LPS or glycine on the intestine.

4. Materials and Methods

4.1. Animal Care and Diets

The experiments and animal care were approved by Animal Care and Use Committee of Wuhan Polytechnic University, Hubei Province, China (EM628, 18 June 2016). Twenty-four weanling crossbred barrows (Duroc × (Landrace × Large White), 28 ± 1 days of age), with an average body weight (BW) of 7.17 ± 0.41 kg were used in this study. Piglets were housed individually in stainless steel metabolic cages (1.80 × 1.10 m^2). All piglets had free access to feed and water in an environmentally controlled house, with ambient temperature maintained at 22–25 °C. The piglets were getting routine immunization, and in good health condition without fever or diarrhea. The living environment was in accordance with animal welfare guidelines in the whole experimental period. The basal diet (Table 7) met or exceeded the National Research Council requirements for all nutrients [62].

Table 7. Ingredient composition of experimental diets (%, as-fed basis).

Ingredients		Nutrient Level [4]	
Corn	57.02	Digestible energy (MJ/Kg)	13.5
Soybean meal (44% CP)	21.40	Crude protein	18.7
Wheat middling	5.00	Crude fat	4.75
Fish meal	3.60	Ca	0.88
Soy protein concentrate	1.40	Total P	0.67
Fat powder	2.00	Lysine	1.02
Defatted milk-replacer powder	3.00	Methionine + Cystine	0.72
Limestone	0.94	Threonine	0.74
Dicalcium phosphate	1.22	Glycine	0.70
Salt	0.34		
Alanine [1]	2.38		
L-Lysine HCl (78.8%)	0.27		
DL-Methionine (99%)	0.10		
L-Threonine (98%)	0.08		
Acidifier [2]	0.20		
Butylated hydroquinone	0.05		
Vitamin and mineral premix [3]	1.00		

[1] In the 1.0% glycine diet, we used 1.0% glycine, 1.19% alanine and 0.19% cornstarch to replace 2.38% alanine. In the 2.0% glycine diet, we used 2.0% glycine and 0.38% cornstarch to replace 2.38% alanine. We made all diets isonitrogenous. [2] A compound acidifier (lactic acid and phosphoric acid), was purchased from Wuhan Fanhua Biotechnology Company, Wuhan, China. [3] Premix (defatted rice bran as carrier) supplied per kg diet: retinol acetate, 2700 μg; cholecalciferol, 62.5 μg; DL-α-tocopheryl acetate, 20 mg; menadione sodium bisulfite complex, 3 mg; riboflavin, 4 mg; D-calcium-pantothenate, 15 mg; niacin, 40 mg; choline chloride, 400 mg; folic acid, 700 μg; thiamin, 1.5 mg; pyridoxine, 3 mg; biotin, 100 μg; Mn ($MnSO_4 \cdot 5H_2O$), 20 mg; Fe ($FeSO_4 \cdot H_2O$), 83 mg; Zn ($ZnSO_4 \cdot 7H_2O$), 80 mg; Cu ($CuSO_4 \cdot 5H_2O$), 25 mg; I (KI), 0.48 mg; Se ($Na_2SeO_3 \cdot 5H_2O$), 0.36 mg. [4] The nutrients level was analyzed value except digestible energy which is calculated value.

4.2. Experimental Design

All piglets were randomly divided into 4 treatments: (1) non-challenged control (CONTR; piglets fed a basal diet and injected with 0.9% NaCl solution); (2) LPS-challenged control (LPS; piglets fed the same basal diet and injected with *E. coli* LPS (*E. coli* serotype 055:B5; potency ≥ 5,000,000 EU/mg; Sigma Chemical, St. Louis, MO, USA)); (3) LPS + 1.0% glycine group (piglets fed a 1.0% glycine-supplemented diet and injected with LPS); and (4) LPS + 2.0% glycine group (piglets fed a 2.0% glycine-supplemented diet and injected with LPS). We selected a supplementary dose of glycine in accordance with a previous study by Wang et al. [7]. In order to get isonitrogenous diets, 2.38%, 1.19%, and 0% alanine (purity > 99%; Amino Acid Bio-Chemical Co., Wuhan, China) were supplemented to the control, 1.0% glycine, and 2.0% glycine diets, respectively. After a 28 days feeding period, the piglets in the control, 1.0% glycine, and 2.0% glycine groups were given an intraperitoneal injection of LPS at 100 μg/kg BW or the same volume of 0.9% NaCl solution. After administration with LPS or NaCl solution, all piglets were deprived of feed for 4 h until slaughter, so as to avoid the potential influence on the intestinal mucosa caused by feed intake change [63].

4.3. Intestinal Sample Collections

Four hours after injection with saline or LPS, all piglets were euthanized via injection of sodium pentobarbital. A 3 cm and 10 cm segments were collected from each mid-jejunum and mid-ileum, referred to in our previous study [64]. Many previous studies demonstrated that LPS resulted in intestinal morphologic impairment and dysfunction within 3–6 h postinjection, and it is caused by increased production of pro-inflammatory cytokines [3,64,65]. Therefore, we choose the time point of 4 h following saline or LPS administration for sample collection. The 3 cm intestinal segments were gently flushed and stored in 4% paraformaldehyde/PBS until histological analysis [3,64]. The 10 cm intestinal samples were opened longitudinally, and gently flushed to remove digesta. The mucosa samples were collected by scraping using sterile glass slides, then rapidly frozen in liquid nitrogen and stored at $-80\ ^\circ$C until measuring of DNA, RNA, protein content, tricarboxylic acid (TCA) cycle key enzyme activities, mRNA, and protein expression levels. The intestinal samples were collected within 15 min after slaughter.

4.4. Intestinal Morphology Analysis

The 3 cm intestinal segment samples were cut into small pieces, not exceeding 2 mm, and enclosed into plastic tissue cassettes then processed over a 19 h period in an automatic tissue processor. Fixed intestinal samples were prepared according to the conventional paraffin-embedding techniques [66]. Samples were cut to 5 μm thickness and then stained with hematoxylin and eosin. Villus height and crypt depth were determined at $40\times$ magnification using a microscope (Olympus CX31, Tokyo, Japan). Ten well-oriented and intact villi were selected at least. Villus height was defined from the tip of the villus to the villus–crypt junction, and crypt depth was obtained as the depth of the invagination between adjacent villi [3].

4.5. Intestinal Mucosal Protein, DNA, and RNA Contents

Frozen mucosal samples were firstly ground using a pestle with supplementation of liquid nitrogen, and then homogenized in ice-cold saline water at a 1:10 (w/v) ratio, then centrifuged at 2500 rpm for 10 min (4 $^\circ$C) to collect the supernatant. The supernatant was used to measure protein, RNA, and DNA contents. The content of intestinal mucosal protein was measured referred to the method of Zhu et al. [67]. The content of DNA was measured through a fluorometric assay [68]. The content of RNA was determined through spectrophotometry described by Schmidt-Tannhauser [69].

4.6. Intestinal Mucosal TCA Cycle Key Enzyme Activities

The key enzymes involved in TCA cycle include citrate synthase (CS), isocitrate dehydrogenase (ICD), and α-ketoglutarate dehydrogenase complex (α-KGDHC). The activities of these enzymes in the intestinal mucosal supernatant were determined according to the previous report by Wang et al. [4], and assayed via commercial enzyme assay kits (#45126 for CS, #45234 for ICD and #45157 for α-KGDHC; Shanghai Yuanye Biotechnology Company, Shanghai, China).

4.7. mRNA Abundance Analysis

All the procedures, such as RNA extraction, quantification, reverse transcription, as well as real-time PCR, were carried out according to a previous report [65]. Primer pairs used in the present study are shown in Table 8. The relative expression of target genes to housekeeping gene (GAPDH) was analyzed via the $2^{-\Delta\Delta Ct}$ method [70]. The results showed that the gene expression of GAPDH was similar among 4 treatments. Relative mRNA expression of the target gene was normalized to the control group (piglets fed a control diet and injected with 0.9% NaCl solution). In detail, the $2^{-\Delta\Delta Ct}$

method is also named the comparative cycle threshold (C_t) method, where C_T is the number of cycles required to reach an arbitrary threshold. The C_t for target gene of each sample was corrected by subtracting the C_t for GAPDH (ΔC_t). The jejunal and ileal segments of the CONTR group were chosen as reference samples, and the ΔC_t for all experimental samples was subtracted by the average ΔC_T for the reference samples ($\Delta\Delta C_t$). Finally, experimental mRNA abundance relative to control mRNA abundance was calculated with use of the formula $2^{-\Delta\Delta Ct}$. We tested several housekeeping genes by analyzing gene stability as described by Vandesompele et al. [71]. We found the expression of GAPDH to be more stable than that of other housekeeping genes, and there was no variation in the expression of GAPDH among intestinal segments and treatments.

Table 8. Primer sequences used for real-time PCR.

Gene	Forward (5′-3′)	Reverse (5′-3′)	Product Length (bp)	Accession Numbers
TLR4	TCAGTTCTCACCTTCCTCCTG	GTTCATTCCTCACCCAGTCTTC	166	GQ503242.1
LBP	GAACACAGCCGAATGGTCTAC	GGAAGGAGTTGGTGGTCAGT	151	NM 001128435.1
MyD88	GATGGTAGCGGTTGTCTCTGAT	GATGCTGGGGAACTCTTTCTTC	148	AB292176.1
IRAK1	CAAGGCAGGTCAGGTTTCGT	TTCGTGGGGCGTGTAGTGT	115	XM_003135490.1
TRAF6	CAAGAGAATACCCAGTCGCACA	ATCCGAGACAAAGGGGAAGAA	122	NM_001105286.1
RP105	CGAGGCTTCTGACTGTTGTG	GGTGCTGATTGCTGGTGTC	245	AB190767.1
SOCS1	GCGTGTAGGATGGTAGCA	GAGGAGGAGGAGGAGGAAT	101	NM_001204768.1
Tollip	GCAGCAGCAACAGCAGAT	GGTCACGCCGTAGTTCTTC	133	AB490123.1
SIGIRR	ACCTTCACCTGCTCCATCCA	TTCCGTCATTCATCTCCACCTC	205	AB490122.1
NOD1	CTGTCGTCAACACCGATCCA	CCAGTTGGTGACGCAGCTT	57	AB187219.1
NOD2	GAGCGCATCCTCTTAACTTTCG	ACGCTCGTGATCCGTGAAC	66	AB195466.1
RIPK2	CAGTGTCCAGTAAATCGCAGTTG	CAGGCTTCCGTCATCTGGTT	206	XM_003355027.1
NF-κB	AGTACCCTGAGGCTATAACTCGC	TCCGCAATGGAGGAGAAGTC	133	EU399817.1
ERBB2IP	ACAATTCAGCGACAGAGTAGTG	TGACATCATTGGAGGAGTTCTTC	147	GU990777.1
centaurin β1	GAAGCCGAAGTGTCCGAATT	AGGTCACAGATGCCAAGAATG	125	XM_003358258.2
GAPDH	CGTCCCTGAGACACGATGGT	GCCTTGACTGTGCCGTGGAAT	194	AF017079.1

ERBB2IP: Erbb2 interacting protein, GAPDH: glyceraldehyde-3-phosphate dehydrogenase, IRAK1: IL-1 receptor-associated kinase, LBP: lipopolysaccharide-binding protein, MD2: myeloid differentiation protein 2, MyD88: myeloid differentiation factor 88, NF-κB: nuclear factor-κB, NOD1: nucleotide-binding oligomerization domain protein 1, NOD2: nucleotide-binding oligomerization domain protein 2, RIPK2: receptor-interacting protein kinase 2, RP105: radioprotective 105, SIGIRR: single immunoglobulin IL-1 related receptor, SOCS1: suppressor of cytokine signaling 1, TLR4: toll-like receptor 4, Tollip: toll-interacting protein, TRAF6: TNF receptor-associated factor 6.

4.8. Protein Abundance Analysis

Protein abundance of claudin-1, AMPK, mTOR, and 4EBP1 in intestinal mucosa was measured in accordance with a previous study [66]. In brief, the intestinal mucosa samples (150–200 mg) were homogenized in 600 μL lysis buffer containing phenylmethanesulfonylfluoride, protease, and phosphatase inhibitors, and centrifuged at $12,000 \times g$ for 15 min at 4 °C to collect supernatants. Equal amounts of intestinal mucosa protein (65 μg) were transferred onto 10–15% polyacrylamide gel and separated via SDS-PAGE, and then transferred to polyvinylidene difluoride membranes for immunoblotting. Immunoblots were blocked with 5% nonfat milk in Tris-buffered saline/Tween-20 for 3 h at room temperature (21–25 °C). The membranes were incubated overnight at 4°C with primary antibodies, and then with the secondary antibodies for 2 h at room temperature. Specific primary antibodies were used, including rabbit anti-claudin-1 (1:1000) (No. 519000, Invitrogen Technology Inc., Danvers, MA, USA), rabbit anti-total AMPKα (t-AMPKα, 1:1000) (No. 2532, Cell Signalling Technology Inc., Danvers, MA, USA), rabbit anti-phosphorylated AMPKα (p-AMPKα, 1:1000) (No. 2535, Cell Signalling Technology Inc.), rabbit anti-total mTOR (t-mTOR, 1:1000) (No. 2972, Cell Signalling Technology Inc.), rabbit anti-phosphorylated mTOR (p-mTOR, 1:1000) (No. 2971, Cell Signalling Technology Inc.), rabbit anti-total eukaryotic initiation factor 4E binding protein 1 (t-4EBP1, 1:1000) (No. 9452, Cell Signalling Technology Inc.), rabbit anti-phosphorylated 4EBP1 (p-4EBP1, 1:1000) (No. 9455, Cell Signalling Technology Inc. and mouse anti-β-actin (1:10,000) (No. A2228, Sigma Aldrich Inc., St. Louis, MO, USA). The secondary antibodies included goat anti-rabbit IgG-HRP (1:5000) (No. ANT019, Antgene Biotech Inc., Wuhan, China), and rabbit anti-goat IgG-HRP (1:5000)

(No. ANT020, Antgene Biotech Inc.). Blots were developed using an Enhanced Chemiluminescence Western blotting kit (Amersham Biosciences, Solna, Sweden), and visualized using a Gene Genome bioimaging system. Bands were analyzed by densitometry using Gene Tool software (Syngene, Frederick, MD, USA). β-Actin was used as a loading control in Western blotting analysis. The relative protein abundance of claudin-1 was expressed as claudin-1/β-actin. Phosphorylated form of AMPKα, mTOR and 4EBP1 were normalized with the total protein content.

4.9. Statistical Analysis

All data were analyzed as a randomized complete block design using the mixed procedure (SAS Inst. Inc., Cary, NC, USA). Individual piglets were used as the experimental unit for all statistic procedures. The data were firstly adjusted by homogeneity of variance. The model included treatment as main effect, and replicates as random effects. The treatment effects were tested using the following contrasts: (1) CONTR vs. LPS was used to test the effect of LPS challenge; (2) The different dose–response effects of glycine were tested using linear and quadratic trends for the three glycine levels (0, 1.0, and 2.0% glycine) among piglets challenged with LPS. Results were shown as mean and pooled SEM. Significant differences were declared at $p < 0.05$.

5. Conclusions

In summary, supplementation with glycine improves intestinal integrity in piglets after LPS challenge. The beneficial effects of glycine on the intestine may be related to (1) enhancing mucosal energy status and protein synthesis via maintaining mTOR and inhibiting AMPK signaling; (2) decreasing intestinal inflammation through inhibiting TLR4 and NOD signaling pathways.

Author Contributions: Y.L. designed research; Y.L., X.W., H.W., C.L., B.D. and X.L. conducted research; Y.L., H.Z. and X.X. analyzed data; Y.L. and X.X. wrote the paper; Y.L., X.X. and Y.H. edited and revised the manuscript; Y.L. had primary responsibility for final content. All authors read and approved the final manuscript.

Funding: This research received no external funding.

Acknowledgments: This study was financially supported by the State's Key Project of Research and Development Plan (2016YFD0501210), the National Natural Science Foundation of China (31772615), the Project of the Hubei Provincial Department of Education (T201508), and the Project of Hangzhou Qianjiang Distinguished Expert (3122).

Conflicts of Interest: The authors declare no conflict of interest.

References

1. Ellen, L.Z. Inflammatory disease caused by intestinal pathobionts. *Curr. Opin. Microbiol.* **2017**, *35*, 64–69.
2. Liu, Y.L. Fatty acids, inflammation and intestinal health in pigs: A review. *J. Anim. Sci. Biotechnol.* **2015**, *6*, 41. [CrossRef] [PubMed]
3. Liu, Y.L.; Huang, J.J.; Hou, Y.Q.; Zhu, H.L.; Zhao, S.J.; Ding, B.Y.; Yin, Y.L.; Yi, G.; Shi, J.; Fan, W. Dietary arginine supplementation alleviates intestinal mucosal disruption induced by *Escherichia coli* lipopolysaccharide in weaned pigs. *Br. J. Nutr.* **2008**, *100*, 552–560. [CrossRef] [PubMed]
4. Wang, X.Y.; Liu, Y.L.; Li, S.; Pi, D.; Zhu, H.L.; Hou, Y.Q.; Shi, H.; Leng, W. Asparagine attenuates intestinal injury, improves energy status and inhibits AMP-activated protein kinase signalling pathways in weaned piglets challenged with *Escherichia coli* lipopolysaccharide. *Br. J. Nutr.* **2015**, *114*, 553–565. [CrossRef] [PubMed]
5. Wang, Y.J.; Liu, W.; Chen, C.; Yan, L.M.; Song, J.; Guo, K.Y.; Wang, G.; Wu, Q.H.; Gu, W.W. Irradiation induced injury reduces energy metabolism in small intestine of Tibet minipigs. *PLoS ONE* **2013**, *8*, e58970. [CrossRef] [PubMed]
6. Wang, W.W.; Wu, Z.L.; Dai, Z.L.; Yang, Y.; Wang, J.J.; Wu, G.Y. Glycine metabolism in animal and humans: Implications for nutrition and health. *Amino Acids* **2013**, *45*, 463–477. [CrossRef] [PubMed]

7. Wang, W.W.; Dai, Z.L.; Wu, Z.L.; Lin, G.; Jia, S.C.; Hu, S.D.; Dahannayaka, S.; Wu, G.Y. Glycine is a nutritionally essential amino acid for maximal growth of milk-fed young pigs. *Amino Acids* **2014**, *46*, 2037–2045. [CrossRef] [PubMed]
8. Wu, G.Y. *Amino Acids: Biochemistry and Nutrition*; CRC Press: Boca Raton, FL, USA, 2013.
9. Wu, G.Y. Dietary requirements of synthesizable amino acids by animals: A paradigm shift in protein nutrition. *J. Anim. Sci. Biotechnol.* **2014**, *5*, 34. [CrossRef] [PubMed]
10. Tsune, I.; Ikejima, K.; Hirose, M.; Yoshikawa, M.; Enomoto, N.; Takei, Y.; Sato, N. Dietary glycine prevents chemical-induced experimental colitis in the rat. *Gastroenterology* **2003**, *125*, 775–785. [CrossRef]
11. Effenberger-Neidnicht, K.; Jägers, J.; Verhaegh, R.; de Groot, H. Glycine selectively reduces intestinal injury during endotoxemia. *J. Surg. Res.* **2014**, *192*, 592–598. [CrossRef] [PubMed]
12. Li, X.; Bradford, B.U.; Wheeler, M.D.; Stimpson, S.A.; Pink, H.M.; Brodie, T.A.; Schwab, J.H.; Thurman, R.G. Dietary glycine prevents peptidoglycan polysaccharide-induced reactive arthritis in the rat: Role for glycine-gated chloride channel. *Infect. Immun.* **2001**, *69*, 5883–5891. [CrossRef] [PubMed]
13. Kazuyuki, N.; Masaaki, H.; Chie, K.; Takeshi, T.; Koji, M.; Yuichi, Y.; Hirokazu, S.; Yoshikiyo, O.; Chikako, W.; Shunshuke, K.; et al. Toll-like receptor (TLR) 2 agonists ameliorate indomethacin-induced murine ileitis by suppressing the TLR4 signaling. *J. Gastroenterol. Hepatol.* **2015**, *30*, 1610–1617.
14. Takeuchi, O.; Akira, S. Pattern recognition receptors and inflammation. *Cell* **2010**, *140*, 805–820. [CrossRef] [PubMed]
15. Kim, H.; Zhao, Q.; Zheng, H.; Li, X.; Zhang, T.; Ma, X. A novel crosstalk between TLR4- and NOD2-mediated signaling in the regulation of intestinal inflammation. *Sci. Rep.* **2015**, *5*, 12018. [CrossRef] [PubMed]
16. Alvarez, B.; Revilla, C.; Domenech, N.; Perez, C.; Riva, L.; Paloma, M.; Alonso, F.; Ezquerra, A.; Javier, D. Expression of toll-like receptor 2 (TLR2) in porcine leukocyte subsets and tissues. *Vet. Res.* **2008**, *39*, 13. [CrossRef] [PubMed]
17. Tohno, M.; Ueda, W.; Azuma, Y.; Shimazu, T.; Katoh, S.; Wang, J.M.; Aso, H.; Takada, H.; Kawai, Y.; Saito, T.; et al. Molecular cloning and functional characterization of porcine nucleotide-binding oligomerization domain-2 (NOD2). *Mol. Immunol.* **2008**, *45*, 194–203. [CrossRef] [PubMed]
18. Parlato, M.; Yeretssian, G. NOD-like receptors in intestinal homeostasis and epithelial tissue repair. *Int. J. Mol. Sci.* **2014**, *15*, 9594–9627. [CrossRef] [PubMed]
19. Curtiss, L.K.; Tobias, P.S. Emerging role of Toll-like receptors in atherosclerosis. *J. Lipid Res.* **2009**, *50*, S340–S345. [CrossRef] [PubMed]
20. Doyle, A.; Zhang, G.H.; Fattah, E.A.A.; Eissa, T.N.; Li, Y.P. Toll-like receptor 4 mediates lipopolysaccharide-induced muscle catabolism via coordinate activation of ubiquitin-proteasome and autophagy-lysosome pathway. *FASEB J.* **2011**, *25*, 99–110. [CrossRef] [PubMed]
21. Lee, M.S.; Kim, I.H.; Kim, C.T.; Kim, Y. Reduction of body weight by dietary garlic is associated with an increase in uncoupling protein mRNA expression and activation of AMP-activated protein kinase in diet-induced obese mice. *J. Nutr.* **2011**, *141*, 1947–1953. [CrossRef] [PubMed]
22. Wu, N.; Zheng, B.; Shaywitz, A.; Dagon, Y.; Tower, C.; Bellinger, G.; Shen, C.H.; Wen, J.; Asara, J.; McGraw, T.E.; et al. AMPK-dependent degradation of TXNIP upon energy stress leads to enhanced glucose uptake via GLUT1. *Mol. Cell* **2013**, *49*, 1167–1175. [CrossRef] [PubMed]
23. Jung, T.W.; Park, H.S.; Choi, G.H.; Kim, D.; Lee, T. β-aminoisobutyric acid attenuates LPS-induced inflammation and insulin resistance in adipocytes through AMPK-mediated pathway. *J. Biomed. Sci.* **2018**, *25*, 27. [CrossRef] [PubMed]
24. Scharl, M.; Paul, G.; Barrett, K.E.; McCole, D.F. AMP-activated protein kinase mediates the interferon-gamma-induced decrease in intestinal epithelial barrier function. *J. Biol. Chem.* **2009**, *284*, 27952–27963. [CrossRef] [PubMed]
25. Jaworek, J.; Tudek, B.; Kowalczyk, P.; Kot, M.; Szklarczyk, J.; Leja-Szpak, A.; Pierzchalski, P.; Bonior, J.; Dembinski, A.; Ceranowicz, P.; et al. Effect of endotoxemia in suckling rats on pancreatic integrity and exocrine function in adults: A review report. *Gastroenterol. Res. Pract.* **2018**. [CrossRef] [PubMed]
26. Crossland, H.; Constantin-Teodosiu, D.; Gardiner, S.M.; Constantin, D.; Greenhaff, P.L. A potential role for Akt/FOXO signalling in both protein loss and the impairment of muscle carbohydrate oxidation during sepsis in rodent skeletal muscle. *J. Physiol.* **2008**, *586*, 5589–5600. [CrossRef] [PubMed]
27. Dunshea, F.R.; Cox, M.L. Effect of dietary protein on body composition and insulin resistance using a pig model of the child and adolescent. *Nutr. Diet.* **2008**, *65*, S60–S65. [CrossRef]

28. Spurlock, M.E.; Gabler, N.K. The development of porcine models of obesity and the metabolic syndrome. *J. Nutr.* **2008**, *138*, 397–402. [CrossRef] [PubMed]

29. Zhu, H.L.; Liu, Y.L.; Xie, X.L.; Huang, J.J.; Hou, Y.Q. Effect of L-arginine on intestinal mucosal immune barrier function in weaned pigs after *Escherichia coli* LPS challenge. *Innate Immun.* **2013**, *19*, 242–252. [CrossRef] [PubMed]

30. Smith, G.I.; Atherton, P.; Reeds, D.N.; Mohammed, B.S.; Rankin, D.; Rennie, M.J.; Mittendorfer, B. Omega-3 polyunsaturated fatty acids augment the muscle protein anabolic response to hyperinsulinaemia-hyperaminoacidaemia in healthy young and middle-aged men and women. *Clin. Sci.* **2011**, *121*, 267–278. [CrossRef] [PubMed]

31. Lee, M.A.; McCauley, R.D.; Kong, S.E.; Hall, J.C. Influence of glycine on intestinal ischemia-reperfusion injury. *J. Parenter. Enter. Nutr.* **2002**, *26*, 130–135. [CrossRef] [PubMed]

32. Stoll, B.; Henry, J.; Reeds, P.J.; Yu, H.; Jahoor, F.; Burrin, D.G. Catabolism dominates the first-pass intestinal metabolism of dietary essential amino acids in milk protein-fed piglets. *J. Nutr.* **1998**, *128*, 606–614. [CrossRef] [PubMed]

33. Blikslager, A.T.; Moeser, A.J.; Gookin, J.L.; Jones, S.L.; Odle, J. Restoration of barrier function in injured intestinal mucosa. *Physiol. Rev.* **2016**, *87*, 545–564. [CrossRef] [PubMed]

34. Hwieh, C.Y.; Osaka, T.; Moriyama, E.; Date, Y.; Kikuchi, J.; Tsuneda, S. Strengthening of the intestinal epithelial tight junction by *Bifidobacterium bifidum*. *Physiol. Rep.* **2015**, *3*, e12327.

35. Travis, S.; Menzies, I. Intestinal permeability: Functional assessment and significance. *Clin. Sci.* **1992**, *82*, 471–488. [CrossRef] [PubMed]

36. Pinheiro, D.F.; Pacheco, P.D.G.; Alvarenga, P.V.; Buratini, J., Jr.; Castilho, A.C.S.; Lima, P.F.; Sartori, D.R.S.; Vicentini-Paulino, M.L.M. Maternal protein restriction affects gene expression and enzyme activity of intestinal disaccharidases in adult rat offspring. *Braz. J. Med. Biol. Res.* **2013**, *46*, 287–292. [CrossRef] [PubMed]

37. Chen, S.K.; Liu, Y.L.; Wang, X.Y.; Wang, H.B.; Li, S.; Shi, H.F.; Zhu, H.L.; Zhang, J.; Pi, D.A.; Hu, C.A.; et al. Asparagine improves intestinal integrity, inhibits TLR4 and NOD signaling, and differently regulates p38 and ERK1/2 signaling in weanling piglets after LPS challenge. *Innate Immun.* **2016**, *22*, 577–587. [CrossRef] [PubMed]

38. Browne, J.L.; Sanford, P.A.; Smyth, S.H. Transport and metabolic processes in the small intestine. *Proc. R. Soc. Lond. B Biol. Sci.* **1977**, *195*, 307–321. [CrossRef] [PubMed]

39. Wiegand, G.; Remington, S.J. Citrate synthase: Structure, control, and mechanism. *Annu. Rev. Biophys. Chem.* **1986**, *15*, 97–117. [CrossRef] [PubMed]

40. Corpas, F.J.; Barroso, J.B.; Sandalio, L.M.; Palma, J.M.; Lupianez, J.A.; del Rio, L.A. Peroxisomal NADP-dependent isocitrate dehydrogenase. Characterization and activity regulation during natural senescence. *Plant. Physiol.* **1999**, *121*, 921–928. [CrossRef] [PubMed]

41. Bunik, V.I.; Strumilo, S. Regulation of catalysis within cellular network: Metabolic and signaling implications of the 2-oxoglutarate oxidative decarboxylation. *Curr. Chem. Biol.* **2009**, *3*, 279–290.

42. Canto, C.; Auwerx, J. PGC-1α, SIRT1 and AMPK, an energy sensing network that controls energy expenditure. *Curr. Opin. Lipidol.* **2009**, *20*, 98–105. [CrossRef] [PubMed]

43. Hardie, D.G. Minireview: The AMP-activated protein kinase cascade: The key sensor of cellular energy status. *Endocrinology* **2003**, *144*, 5179–5183. [CrossRef] [PubMed]

44. Switon, K.; Kotulska, K.; Janusz-kaminska, A.; Zmorzynska, J.; Jaworski, J. Molecular neurobiology of mTOR. *Neuroscience* **2017**, *341*, 112–153. [CrossRef] [PubMed]

45. Shao, Y.X.; Wolf, P.G.; Guo, S.S.; Guo, Y.M.; Gaskins, H.R.; Zhang, B.K. Zinc enhances intestinal epithelial barrier function through the PI3K/AKT/mTOR signaling pathway in Caco-2 cells. *J. Nutr. Biochem.* **2017**, *43*, 18–26. [CrossRef] [PubMed]

46. Yang, H.S.; Xia, X.; Li, T.J.; Yin, Y.L. Ethanolamine enhances the proliferation of intestinal epithelial cells via the mTOR signaling pathway and mitochondrial function. *In Vitro Cell. Dev. Biol. Anim.* **2016**, *52*, 562–567. [CrossRef] [PubMed]

47. Yi, D.; Hou, Y.Q.; Xiao, H.; Wang, L.; Zhang, Y.; Chen, H.; Wu, T.; Ding, B.; Hu, C.A.; Wu, G. N-Acetylcysteine improves intestinal function in lipopolysaccharides-challenged piglets through multiple signaling pathways. *Amino Acids* **2017**, *49*, 1915–1929. [CrossRef] [PubMed]

48. Stoffels, B.; Turler, A.; Schmidt, J.; Nazir, A.; Tsukamoto, T.; Moore, B.A.; Schnurr, C.; Kalff, J.C.; Bauer, A.J. Anti-inflammatory role of glycine in reducing rodent postoperative inflammatory ileus. *Neurogastroenterol. Motil.* **2011**, *23*, 76–87. [CrossRef] [PubMed]

49. Coll, R.C.; O'Neill, L.A. New insights into the regulation of signalling by toll-like receptors and nod-like receptors. *J. Innate Immun.* **2010**, *2*, 406–421. [CrossRef] [PubMed]

50. Piñeros, A.A.R.; Glosson-Byers, N.; Brandt, S.; Wang, S.; Wong, H.; Sturgeon, S.; McCarthy, B.P.; Territo, P.R.; Alves-Filho, J.C.; Serezani, C.H. SOCS1 is a negative regulator of metabolic reprogramming during sepsis. *JCI Insight* **2017**, *2*, 92530. [CrossRef] [PubMed]

51. Hsu, V.W. *Downregulation of ErbB2 by Perturbing Its Endocytic Recycling*; Brigham and Women's Hospital: Boston, MA, USA, 2007. Available online: https://www.researchgate.net/publication/235119665 (accessed on 1 March 2017).

52. Wang, H.B.; Liu, Y.L.; Shi, H.F.; Wang, X.Y.; Zhu, H.L.; Pi, D.A.; Leng, W.B.; Li, S. Aspartate attenuates intestinal injury and inhibits TLR4 and NODs/NF-κB and p38 signaling in weaned pigs after LPS challenge. *Eur. J. Nutr.* **2017**, *56*, 1433–1443. [CrossRef] [PubMed]

53. Fujimoto, M.; Naka, T. SOCS1, a negative regulator of cytokine signals and TLR responses, in human liver diseases. *Gastroenterol. Res. Pract.* **2010**, *210*. [CrossRef] [PubMed]

54. Wu, W.; Wang, Y.; Zou, J.; Long, F.; Yan, H.; Zeng, L.; Chen, Y. Bifidobacterium adolescentis protects against necrotizing enterocolitis and upregulates TOLLIP and SIGIRR in premature neonatal rats. *BMC Pediatr.* **2017**, *17*, 1. [CrossRef] [PubMed]

55. Leng, W.B.; Liu, Y.L.; Shi, H.F.; Li, S.; Zhu, H.L.; Pi, D.A.; Hou, Y.Q.; Gong, J. Aspartate alleviates liver injury and regulates mRNA expressions of TLR4 and NOD signaling-related genes in weaned pigs after *lipopolysaccharide* challenge. *J. Nutr. Biochem.* **2014**, *25*, 592–599. [CrossRef] [PubMed]

56. Wu, H.T.; Liu, Y.L.; Pi, D.A.; Leng, W.B.; Zhu, H.L.; Hou, Y.Q.; Li, S.; Shi, H.F.; Wang, X.Y. Asparagine attenuates hepatic injury caused by lipopolysaccharide in weaned piglets associated with modulation of Toll-like receptor 4 and nucleotide-binding oligomerisation domain protein signalling and their negative regulators. *Br. J. Nutr.* **2015**, *114*, 189–201. [CrossRef] [PubMed]

57. Leung, F.W.; Su, K.C.; Pique, J.M.; Thiefin, G.; Passoro, E.; Guth, P.H. Superior mesenteric artery is more important than inferior mesenteric artery in maintaining colonic mucosal perfusion and integrity in rats. *Dig. Dis. Sci.* **1992**, *37*, 1329–1335. [CrossRef] [PubMed]

58. Warzecha, Z.; Ceranowicz, D.; Dembiński, A.; Ceranowicz, P.; Cieszkowski, J.; Kuwahara, A.; Kato, I.; Dembiński, M.; Konturek, P.C. Ghrelin accelerates the healing of cysteamine-induced duodenal ulcers in rats. *Med. Sci. Monit.* **2012**, *18*, 181–187. [CrossRef]

59. Ceranowicz, P.; Warzecha, Z.; Cieszkowski, J.; Ceranowski, D.; Kuśnierz-Cabala, B.; Bonior, J.; Jaworek, J.; Ambroży, T.; Gil, K.; Olszanecki, R.; et al. Essential role of growth hormone and IGF-1 in therapeutic effect of ghrelin in the course of acetic acid-induced colitis. *Int. J. Mol. Sci.* **2017**, *18*, 1118. [CrossRef] [PubMed]

60. Matuszyk, A.; Ceranowicz, P.; Warzecha, Z.; Cieszkowski, J.; Bonior, J.; Jaworek, J.; Kuśnierz-Cabala, B.; Konturek, P.; Ambroży, T.; Dembiński, A. Obestatin accelerates the healing of acetic acid-induced colitis in rats. *Oxid. Med. Cell. Longev.* **2016**, *2016*, 2834386. [CrossRef] [PubMed]

61. Matuszyk, A.; Ceranowicz, P.; Warzecha, Z.; Cieszkowski, J.; Ceranowicz, D.; Gałazka, K.; Bonior, J.; Jaworek, J.; Bartuś, K.; Gil, K.; et al. Exogenous ghrelin accelerates the healing of acetic acid-induced colitis in rats. *Int. J. Mol. Sci.* **2016**, *17*, 1455. [CrossRef] [PubMed]

62. NRC. *Nutrient Requirements of Swine*, 10th ed.; National Academic Press: Washington, DC, USA, 1998.

63. Xu, X.; Chen, S.K.; Wang, H.B.; Tu, Z.X.; Wang, S.H.; Wang, X.Y.; Zhu, H.L.; Wang, C.W.; Zhu, J.D.; Liu, Y.L. Medium-chain TAG improve intestinal integrity by suppressing toll-like receptor 4, nucleotide-binding oligomerisation domain proteins and necroptosis signalling in weanling piglets challenged with lipopolysaccharide. *Br. J. Nutr.* **2018**, *119*, 1019–1028. [CrossRef] [PubMed]

64. Pi, D.A.; Liu, Y.L.; Shi, H.F.; Li, S.; Odle, J.; Lin, X.; Zhu, H.L.; Chen, F.; Hou, Y.Q.; Leng, W.B. Dietary supplementation of aspartate enhances intestinal integrity and energy status in weanling piglets after lipopolysaccharide challenge. *J. Nutr. Biochem.* **2014**, *25*, 456–462. [CrossRef] [PubMed]

65. Mercer, D.W.; Smith, G.S.; Cross, J.M. Effect of lipopolysaccharide on intestinal injury: Potential role of nitricoxide and lipid peroxidation. *J. Surg. Res.* **1996**, *63*, 185–192. [CrossRef] [PubMed]

66. Luna, L.G. *Manual of Histologic Staining Methods of the Armed Forces Institute of Pathology*, 3rd ed.; McGraw-Hill Book Company: New York, NY, USA, 1968; p. 258.

Int. J. Mol. Sci. **2018**, *19*, 1980

67. Zhu, H.L.; Liu, Y.L.; Chen, S.K.; Wang, X.Y.; Pi, D.A.; Leng, W.B.; Chen, F.; Zhang, J.; Kang, P. Fish oil enhances intestinal barrier function and inhibits corticotropin-releasing hormone/corticotropin-releasing hormone receptor 1 signalling pathway in weaned pigs after lipopolysaccharide challenge. *Br. J. Nutr.* **2016**, *115*, 1947–1957. [CrossRef] [PubMed]

68. Labarca, C.; Paigen, K. A simple, rapid, and sensitive DNA assay procedure. *Anal. Biochem.* **1980**, *102*, 344–352. [CrossRef]

69. Munro, H.N.; Fleck, A. Analysis of tissues and body fluids for nitrogenous constituents. In *Mammalian Protein Metabolism*; Academic Press: New York, NY, USA, 1969; pp. 465–483.

70. Livak, K.J.; Schmittgen, T.D. Analysis of relative gene expression data using real-time quantitative PCR and the 2(-Delta Delta C(T)) Method. *Methods* **2001**, *25*, 402–408. [CrossRef] [PubMed]

71. Vandesompele, J.; De Preter, K.; Pattyn, F.; Poppe, B.; Van Roy, N.; De Paepe, A.; Speleman, F. Accurate normalization of real-time quantitative RT-PCR data by geometric averaging of multiple internal control genes. *Genome Biol.* **2002**, *3*. [CrossRef]

International Journal of
Molecular Sciences

Review

Lipopolysaccharide-Induced Neuroinflammation as a Bridge to Understand Neurodegeneration

Carla Ribeiro Alvares Batista [1,†], **Giovanni Freitas Gomes** [1,†], **Eduardo Candelario-Jalil** [2], **Bernd L. Fiebich** [3,*,†] and **Antonio Carlos Pinheiro de Oliveira** [1,*,†]

1   Department of Pharmacology, Universidade Federal de Minas Gerais, Av. Antonio Carlos 6627, Belo Horizonte 31270-901, Brazil; cacaribeiro@gmail.com (C.R.A.B.); gvnngomes@gmail.com (G.F.G.)
2   Department of Neuroscience, University of Florida, Gainesville, FL 32610, USA; ecandelario@ufl.edu
3   Neuroimmunology and Neurochemistry Research Group, Department of Psychiatry and Psychotherapy, Medical Center–University of Freiburg, Faculty of Medicine, University of Freiburg, D-79104 Freiburg, Germany
*   Correspondence: bernd.fiebich@uniklinik-freiburg.de (B.L.F.); antoniooliveira@icb.ufmg.br or acpoliveira@gmail.com (A.C.P.d.O.); Tel.: +49-761-270-68980 (B.L.F.); +55-31-3409-2727 (A.C.P.d.O.); Fax: +49-761-270-69170 (B.L.F.); +55-31-3409-2695 (A.C.P.d.O.)
†   These authors contributed equally to this work.

Received: 4 April 2019; Accepted: 5 May 2019; Published: 9 May 2019

Abstract: A large body of experimental evidence suggests that neuroinflammation is a key pathological event triggering and perpetuating the neurodegenerative process associated with many neurological diseases. Therefore, different stimuli, such as lipopolysaccharide (LPS), are used to model neuroinflammation associated with neurodegeneration. By acting at its receptors, LPS activates various intracellular molecules, which alter the expression of a plethora of inflammatory mediators. These factors, in turn, initiate or contribute to the development of neurodegenerative processes. Therefore, LPS is an important tool for the study of neuroinflammation associated with neurodegenerative diseases. However, the serotype, route of administration, and number of injections of this toxin induce varied pathological responses. Thus, here, we review the use of LPS in various models of neurodegeneration as well as discuss the neuroinflammatory mechanisms induced by this toxin that could underpin the pathological events linked to the neurodegenerative process.

Keywords: lipopolysaccharide; inflammation; neurodegeneration; Alzheimer's disease; Parkinson's disease; amyotrophic lateral sclerosis; Huntington's disease

---

1. Introduction

Neurodegenerative diseases are devastating conditions for which there is no cure so far. In general, the mechanisms involved in disease onset and development are still poorly understood. Therefore, increasing efforts are being made to better comprehend their pathogenesis. Among the different factors involved in these conditions, inflammation is considered a key contributor. Several lines of experimental evidence have demonstrated that neuronal cell death may induce an inflammatory process, and inflammation by itself may lead to cell death [1]. Thus, it is necessary to induce inflammation in models of neurodegeneration in order to evaluate its intricate consequences.

Induction of inflammation may be achieved in different manners, and lipopolysaccharide (LPS) is an important tool for this purpose. LPS is a molecule present in the outer membrane of Gram-negative bacteria. Its main target is the toll-like receptor (TLR) 4, although it is known to act on other receptors [2–4]. The activation of TLR4 by LPS recruits a series of downstream adaptors, such as myeloid differentiation primary response protein 88 (MyD88), TIR-domain-containing adaptor-inducing interferon-β (TRIF) and TRIF-related adaptor molecule (TRAM), which are crucial for the signaling of the receptor [5,6]. The recruitment of these adaptors can further activate downstream pathways which culminate in the activation of transcription factors, which, in turn, induce a plethora of pro-inflammatory genes [6–8]. The TLR4 signaling pathway has been fully reviewed elsewhere [9].

Although most of the work in this field uses LPS in order to stimulate glial cells, mainly microglia, it is known that neurons also express TLR4. Indeed, activation of this receptor leads to the neuronal production of different inflammatory mediators [10–13].

LPS is used in a variety of in vivo and in vitro protocols. This compound not only is used to stimulate cell cultures, but also is injected either in the central nervous system (CNS) or in the periphery by single or multiple injections. Thus, its effects may vary according to the experimental protocol. Therefore, here, we review the various protocols that use LPS in order to provide an overview of the current state of the art. We also discuss the advantages and limitations of the LPS models used to understand the complex molecular and cellular mechanisms underlying the neuroinflammatory process associated with neurodegeneration.

2. LPS-Induced Inflammation in Models of Alzheimer's Disease

Alzheimer's disease (AD) is the most common neurodegenerative disorder worldwide, and its main clinical manifestation is progressive dementia [14]. It is characterized by the inability to form new memories, reflecting the dysfunction of the episodic memory system [15,16]. AD is associated with neuropathological changes such as the formation of tau aggregates seen as intraneuronal neurofibrillary tangles and the presence of extracellular amyloid-beta (Aβ) plaques [17,18]. It was demonstrated that activated microglia are present in regions of the brain where there are Aβ deposition and neuronal loss, which culminates in memory impairment. Published data showed that chronic LPS administration produced impaired spatial memory in Sprague Dawley [19] and Fisher rats [20].

Neuroinflammation frequently precedes the development of neurodegenerative pathologies such as AD [21] and is one of the pathogenic factors for neurodegeneration [22]. Significant studies from basic cellular neuroscience and human genetics support the important role of inflammation in the pathogenesis of AD [23–25]. The myeloid cells of the CNS, microglia, can be beneficial and detrimental to AD pathogenesis, since they can degrade amyloid plaques and promote neurotoxicity due to excessive inflammatory cytokine release [23,26]. LPS-induced inflammation is used in experimental in vitro and in vivo models of neuroinflammation and has been shown to also promote amyloid deposition in vivo [27,28].

Some studies have associated AD neuropathology with LPS levels in the brain. The presence of LPS and Aβ1–40/42 in amyloid plaques in gray and white matter of AD brains has been demonstrated [29]. Another study showed that LPS is abundant in the neocortex and hippocampus of AD-affected brains and that there is a strong adherence of LPS to the nuclear periphery in AD brain cell nuclei [30]. Finally, LPS was also found in lysates from the hippocampus and superior temporal lobe neocortex of AD brains [31]. The role of LPS in the development of AD is reviewed by [32,33].

In this context, experimental models using LPS could serve as a link between neuroinflammation and AD and are useful to understand the disease process and some events that occur in human AD.

2.1. Contribution of Central LPS Injection Models to Our Understanding AD Pathology

Animals can respond to LPS stimuli differently depending on age and species. In addition, the source of the stimulus, the dose, the route, and the duration of the administration used in each study may also influence the outcome [34]. LPS injection in different regions of the CNS leads to a variety of responses in animals. In this section, we will discuss the data obtained from LPS-induced inflammation in the CNS associated with AD.

Single intracerebroventricular (i.c.v.) injections of LPS resulted in increased levels of interleukin-1β (IL-1β) in the brainstem and diencephalon of rats 2 h after injection, and in all the brain regions, except cerebellum, 6 h after injection [35]. Besides, the induction of IL-1β mRNA in the nucleus basalis magno-cellularis and hippocampus was observed, as well as the presence of mRNA for tumor necrosis factor-α (TNF-α) in the nucleus basalis magno-cellularis [19].

Microglia play an important role in immune defense and inflammatory responses in the CNS [36]. When microglia are exposed to stimulatory molecules such as LPS, their receptors such as TLRs recognize LPS, inducing a series of intracellular signaling pathways [37,38]. Activation of microglia and astrocytes was observed after both single i.c.v. LPS injection and chronic LPS injection in the 4th ventricle with osmotic pumps [19,39]. In addition, single intrahippocampal LPS injections produced elevations of glial fibrillary acidic protein (GFAP) after 24 h [40]. On the other hand, 28 days after a single intrahippocampal LPS injection, chronic microglial activation was observed, marked by the increase of CR3 and CD45 in the mouse hippocampus [40]. These are important findings, since glial activation after intrahippocampal LPS injection has been related to AD-like amyloidogenic axonal pathology and dendritic degeneration [41]. Chronic i.c.v. administration of LPS induced β-amyloid precursor protein (β-APP) mRNA in the nucleus basalis magno-cellularis of rats [19]. In a marmoset monkey model, LPS co-injected with Aβ fibrils in frontal, sensorimotor, and parietal cortices accelerated the amyloidosis process, with all monkeys showing an early AD immune blood cell expression profile of the apoptosis receptor CD95 [42], suggesting a potential synergic action.

In many neuroinflammatory conditions, including in mouse models of AD, microglia activation and infiltration of peripheral immune cells are found in the brain parenchyma [43]. Microglia activation is also associated with hyperphosphorylation and aggregation of the protein tau, another important AD marker. Single intrahippocampal injection of LPS enhanced tau phosphorylation by about 2.5-fold via microglial activation in rTg4510 mice, which carry a mutant tau [44].

Microglia response after stimulation with LPS may differ between transgenic and non-transgenic mice. Although microglia in 12-month-old non-transgenic mice showed a stronger response to LPS than in 2-month-old mice of the same strain, microglia in transgenic APP/PS1 mice exhibited diminished immune response to LPS during aging. Microglial TLR4 signaling was altered in transgenic mice, suggesting that changes in TLR4 signaling may have impaired the Aβ clearance capacity of microglia [45]. In Tg2576 mice, which express a mutant form of APP, a single LPS intrahippocampal injection reduced hippocampal Aβ levels in a time- and glial activation-dependent manner [46,47]. Another study showed that intrahippocampal LPS injection increased by about sixfold the bone marrow cells recruitment from the periphery and reduced Aβ clearance in bone marrow-transplanted AD transgenic mice [48].

2.2. Systemic LPS Challenge Models Utilized to Understand AD Pathology

Systemic inflammation may affect the brain. Cytokines, such as IL-1β, IL-6, and TNF-α, produced by a systemic inflammatory response, can reach the CNS through the blood circulation [49]. The intraperitoneal (i.p.) injection of LPS, for example, leads to the detection of IL-1 in the plasma and brain regions [35]. The levels of TNF-α, IL-1α, IL-1β, and IL-6 mRNAs were increased in the hippocampus and cerebral cortex of mutated presenilin (PS) 1 transgenic mice compared to wild-type mice after i.p. injection of LPS [50]. The increase in mRNAs levels of IL-1β and IL-6 due to a single LPS i.p. injection was associated with changes in APP expression in the cerebellum of Staggerer mutant mice, which show a severe Purkinje cell deficiency in the cerebellum, whereas the cerebral cortex is not affected [51]. Similarly, a single LPS injection increased IL-1β and TNF-α by about twofold in cortices and hippocampi of aged Tg2576 mice 1, 2, 4, and 6 h after stimulus [52] and increased the blood and brain levels of IL-1β, IL-6, and TNF-α in Sprague Dawley rats [53]. In addition, in a model of LPS-induced cognitive impairment in rats, TNF-α levels were increased by about 1.6-fold in the hippocampus and frontal cortex after 7 days of a single LPS injection. Interestingly, TNF-α and IL-18 were increased in the same areas after 10 months of a single LPS injection [54]. TNF-α plays an important role in the induction of inflammatory processes, being recruited after the LPS stimulus and inducing the production of pro-inflammatory cytokines, which are involved in the pathophysiology of neurodegeneration. IL-18 might act later, when the disease is already established, participating in the progression of neurodegeneration and cognitive dysfunction.

Systemic administration of LPS also induces microglial activation. A single LPS injection increased microglial density in Sprague Dawley rats [53]. The brain metabolic response to LPS-inducing microglial activation was studied using magnetic resonance spectroscopy. Intraperitoneal injection of LPS also increased the number of Iba-1(+) microglia and induced Aβ(1–16)(+) neurons in the hippocampus in C57/CJ mice [55].

LPS has been used in different studies to stimulate the production of β-APP. Peripheral stimulation with LPS induced an increase in IL-1β and IL-6 mRNAs, followed by changes in the expression of APP isoforms in the cerebellum [51]. LPS administration for 7 days increased Aβ 1–42 cerebral expression and triggered AD-like neuronal degeneration [56]. On the other hand, chronic LPS administration increased by about twofold the number of Aβ and APP immunoreactive neurons in the neocortex of APPswe mice [28]. A similar increase in Aβ was seen in the hippocampus of EFAD mice (a model that expresses human APOE3 or APOE4 and overproduces human Aβ42) [57] and in the hippocampus, cortex, and amygdala of APPswe mice receiving chronic LPS administration [28]. In all these transgenic models, increased Aβ neuronal immunoreactivity was associated with an elevated number of F4/80-immunoreactive microglia [28] and an increase in the 6E10-immunoreactive protein, which contains Aβ fragments [58]. Repeated LPS systemic injections (three or seven times) promoted Aβ 1–42 accumulation in the hippocampus and cerebral cortex of ICR albino mice, as a result of an increase in beta- and gamma-secretase activities as well as in the activation of astrocytes in parallel to cognitive impairment [59]. A reduction of Aβ accumulation in hippocampus, cortex, and amygdala was demonstrated by chronic LPS injection in 3xTgAD mice, which exhibit both Aβ and tau pathologies, in combination with an inhibitor of soluble TNF-α signaling [58]. In addition, young and old transgenic mice showed an increase in Aβ 1–40 in the cortices between 4 and 6 h after LPS administration, which returned to baseline 18 h after a single injection [52]. However, LPS once a week for 13 weeks ameliorated amyloid pathology in the neocortex of $APP_{SWE}/PS11_{\Delta E9}$ mice [60], which was associated with increased aggregation of activated microglia around the Aβ deposits and by CNS myeloid cells inducing Aβ clearance pathways and elevated levels of the lysosomal protease cathepsin Z as well as clusterin [60]. Contradictory data suggest that there are differences in the amyloid production and that the accumulation depends on the degree of severity of inflammatory stimuli and the animal model used to evaluate the consequences of LPS injection. Indeed, it has been demonstrated that LPS-induced inflammation can contribute to the progression of a series of neurodegenerative processes [61,62]. On the other hand, immune system stimulation with low doses of LPS can induce the activation of cells that act on the resolution of the pathology in neurodegeneration [63–65].

A deficiency in Aβ clearance due to an impairment of the blood–brain barrier (BBB) has been associated with AD development [66]. In this way, the integrity of the BBB is important, since Aβ clearance ameliorates AD neuropathology [67]. Besides, an association between AD and lipoprotein receptor-related protein-1 (LRP-1)—a member of the low-density lipoprotein receptor family—has been demonstrated to participate in Aβ metabolism [68]. In this sense, some studies demonstrated that LPS induced an Aβ transport dysfunction at the BBB dependent on LRP-1 [67,69]. Repeated i.p. injection of LPS altered the BBB transport of Aβ by increasing the brain influx and decreasing the efflux of the peptide. In addition, LPS also increased the expression of neuronal LRP-1, which can be responsible for the increased production and accumulation of Aβ in the brain [69]. Similarly, another study showed a decrease in Aβ efflux by LPS-induced dysfunction of LRP-1 at the BBB [70]. A disruption of the BBB by LPS was observed in aging 5XFAD mice, which overexpress both mutant human APP and presenilin 1. On the other hand, inflammation induced by LPS may also be an interesting tool for the crossing of drugs through the BBB. Indeed, Barton et al. (2018) demonstrated that LPS may disrupt the BBB in 5XFAD mice, which improved the delivery of small molecules, such as thioflavin S, to the brain [71]. Therefore, the neuroinflammatory process could also play an important role in the pathophysiology of

AD by disrupting the BBB and impairing the removal of Aβ from the brain, as well as in facilitating a pharmacological treatment.

Increased levels of Aβ induced by LPS can promote tangle formation [53]. In fact, single LPS injection increased the levels of soluble Aβ and phosphorylated tau in the brain of rats [53] and mice [72]. Acute systemic LPS administration enhanced tau phosphorylation in wild-type and corticotropin-releasing-factor-receptors (CRFR)-deficient mice, which was associated with the activation of glycogen synthase kinase-3 (GSK-3) and cyclin-dependent kinase-5 (CDK5) [73]. Similarly, tau hyperphosphorylation in 3xTgAD mice was also mediated by the activation of CDK5 after chronic LPS administration [74].

Cognitive deficits were shown by studies using single LPS i.p. administration in rats [56] and by repeated LPS injection in EFAD mice [57]. Besides the cognitive impairment and the increase by more than tenfold in the levels of Aβ with a single i.p. administration of LPS, the elevation of nitric oxide (NO) concentrations and the overexpression of *N*-methyl-D-aspartate receptor subunit 2B (NMDAR2B) in the brain were described [75].

Finally, neuroinflammation is regulated through the cholinergic anti-inflammatory pathway by the α7 nicotinic acetylcholine receptor (α7 nAChR), involved in regulating cognitive functions and inflammatory reactions. It was demonstrated that systemic LPS injection in mice decreased α7 nAChR in the brain [76,77]. Thus, this may be another mechanism by which LPS induces neuroinflammation and cognitive impairment in models of AD.

The data presented in Section 2 demonstrate the large number of studies using LPS to induce neuroinflammation in models associated with AD. There is enough evidence to support the singular role of neuroinflammation in neurodegeneration in addition to the importance of animal models to study Aβ accumulation and tau hyperphosphorylation. In summary, it can be assumed that LPS injection models mimic memory loss and the neuropathology observed in AD. All these studies help understand the role of neuroinflammation in the progression of AD.

3. LPS-Induced Models of Parkinson's Disease

Parkinson's disease (PD) is the second most prevalent neurodegenerative disorder [78], and its neuropathology is characterized by the degeneration of dopaminergic neurons in the substantia nigra (SN), followed by the loss of axonal projections to the striatum, resulting in malfunction of the dopaminergic system [79,80]. Dopaminergic dysfunction manifests in the characteristic motor disabilities found in the disease, such as tremor, rigidity, bradykinesia, postural, and gait abnormalities [81,82]. Cytoplasmic inclusions, known as Lewy bodies, which are essentially constituted by protein deposits of α-synuclein [83], are the main hallmark feature of PD. Although the etiology of PD is not well known, it has been described that inflammation contributes to PD progression and is an important factor related to neuronal loss [84–86].

To focus on the potential role of inflammation in PD, several LPS-induced Parkinson models have been validated and used. Different routes of injection, doses, species models, and sources of endotoxin are described. In the following two sections, we will present the main contributions of LPS-induced models to providing more insights into the pathophysiology of PD.

3.1. Contribution of Central LPS Injection Models to the Elucidation of PD Pathology

Part of the knowledge about the involvement of neuroinflammation in PD was obtained from models of central injection of LPS into the SN or striatum (ST). Both models of injection can induce the dopaminergic neurodegeneration and motor symptoms characteristic of the disease. A first intranigral LPS injection was established in 1998, inducing microglial activation after 2 days, followed by a reduction in dopamine levels in the SN and ST and a decrease in tyrosine hydroxylase (TH) activity up to at least 21 days [87]. Later studies tested the impact of LPS injection on dopaminergic neurodegeneration and microglial activation. A permanent dopaminergic neuron loss after a single LPS injection into the SN was observed up to one year after the injection. Neuronal loss was associated with a strong macrophage/microglial reaction in the SN [88–91]. The inflammatory involvement was supported by the use of drugs that reduce the effects mediated by microglia. Intranigral or systemic administration of naloxone, an opioid receptor antagonist, prevented neuronal loss induced by local LPS injection in the SN [92]. Eight or 15 days of systemic dexamethasone administration prevented the reduction of TH activity and TH immunostaining induced by intranigral LPS injection, suggesting a reduction of dopamine dysfunction in addition to the reduction of microglia activation [93]. This first set of investigations supported the idea that microglia-mediated neuroinflammation plays an important role in the neurodegenerative process of PD.

LPS injection into the CNS clearly increased the expression of inflammatory mediators in the brain. Elevated levels of TNF-α and IL-6 in the SN and elevated IL-6 in the ST 90 minutes after intranigral LPS injection were found in C57BL/6 mice. Interestingly, the authors observed a 29-fold and 36-fold increase in peripheral circulating levels of TNF-α and IL-6, respectively. The peripheral levels of IL-2 and IFN-γ were also increased at day 7 post-injection, whereas no changes in these inflammatory mediators were detected in the SN. These effects were accompanied by increased CD11b immunostaining in the SN [94], which suggests an ongoing microglial activation and neuroinflammation. A comparable cytokine profile was observed after intranigral LPS injection [95,96]. In a chronic LPS injection approach that mimics the early stages of many chronic neurodegenerative diseases, the injection of LPS into the 4th ventricle for 21 or 56 days induced different responses that depended on the animal's age and the stimulus duration [97]. Gene expression and protein levels of both pro- and anti-inflammatory parameters were upregulated in the brainstem, with IL-1β, TNF-α, TGF-β, and CX3CR1 being the most important ones. Importantly, these changes in cytokine expression and loss of TH-positive neurons were more pronounced in middle-aged and aged rats compared to young rats.

A recent study evaluated the time-dependent expression of pro- and anti-inflammatory cytokines after intranigral LPS injection in adult Wistar rats. The levels of TNF-α and IL-1β mRNA were significantly increased at early time points, with a maximum after 5 h (~threefold and ~fourfold increase, respectively), while IL-6 mRNA levels were maximal after 8 h (about fivefold increase). Interestingly, IL-1β mRNA levels remained significantly increased up to 168 h after LPS injection [98]. On the other hand, anti-inflammatory mRNA expression was altered only at late time points (after 24 h and 168 h for IL-10 and IL-4, respectively). These effects were followed by microglial and astrocytic activation and dopaminergic neurodegeneration in the SN [98]. Moreover, the changes in the inflammatory mediator profile were in line with the increased expression of nuclear factor kappa B (NF-κB) after intracerebral LPS injection, which can lead to a significant increase in the transcription of pro-inflammatory cytokines (e.g., TNF-α and IL-1β) [99–102]. Furthermore, LPS intrastriatal injection caused an oxidative stress response and apoptosis, which are strongly associated with the activation of TLR/NF-κB signaling and the inhibition of the anti-oxidant Nrf/HO-1 pathway [103]. These data suggest that the LPS injection models induce an acute initial pro-inflammatory profile and that the neuronal degeneration process in the SN and ST are mediated by these inflammatory mediators, which are therefore crucial for the progression of the pathology.

Mitochondrial dysfunction is also associated with neuronal cell death in the pathogenesis of PD [104,105]. In this way, LPS injection models can contribute to evaluate possible impairments in mitochondrial activity to elucidate their impact in the pathophysiology of PD. Intrastriatal LPS injection induced changes in the mitochondrial respiratory chain, evidenced by increased levels of oxidative stress markers including protein carbonyls, 4-hydroxynonenal (4-HNE), and 3-nitrotyrosine (3-NT), and caused structural modifications in the mitochondrial cristae, leading to energy dysfunction and neuronal loss in the striatum [106]. Mitochondrial dysfunction was also supported by increased PPAR-γ, UCP2, and mitoNEET expression—three proteins involved in energy metabolism—in the SN [107]. Moreover, intrastriatal injection of LPS induced extensive S-nitrosylation/nitration of the mitochondrial complex prior to dopaminergic neuronal loss [108]. Related to this previous finding, inhibition of inducible nitric oxide synthase (iNOS) by L-N6-(l-iminoethyl)-lysine reduced mitochondrial injury and dopaminergic degeneration induced by LPS injection into the SN, indicating that iNOS-derived NO is associated with mitochondrial dysfunction. iNOS activation is mediated by p38 MAP kinase, and cell death was reduced by the inhibition of p38 [109].

Intranigral LPS injection upregulated iNOS expression (~twofold) and elevated total reactive oxygen species (ROS) production (~twofold) and NADPH oxidase activity (~fivefold) [99]. Supranigral administration of LPS induced an intense expression of NADPH-diaphorase and iNOS-immunoreactivity in macrophage-like cells, followed by an important decrease of tyrosine hydroxylase-positive neurons [110]. Pre-treatment of animals with the iNOS inhibitors *S*-methylisothiourea or L-NIL prevented dopaminergic neuronal loss, suggesting that NO mediates the neurodegeneration observed in the LPS-induced PD model [110,111]. Moreover, a single intrastriatal LPS injection was found to be associated with increased striatal cyclooxygenase-2 (COX-2) and iNOS expression three days post-injection and, in the SN, dopaminergic neuronal loss and an increase in microglia activation were observed seven days post-injection [102,112]. Furthermore, a two-week intracerebral infusion of LPS (5 ng/h, delivered using osmotic minipumps) induced a rapid activation of microglia that reached a plateau at the end of the treatment, followed by a delayed and gradual loss of nigral dopaminergic neurons starting between four and six weeks after treatment [90], suggesting that the initial activation of the immune response preceded neuronal loss.

In line with the studies described above, LPS injection was shown to alter iron and ferritin levels in glial cells of the SN of rats, which was associated with 1.5-fold and 2.5-fold decreases in TH expression in the globus pallidus [113] and in the striatum [107], respectively. It was also demonstrated that iron chelation with desferrioxamine attenuated behavior deficits, neuronal loss of dopaminergic neurons, and striatal dopamine (DA) reduction induced by intrastriatal LPS injection in C57BL/6 mice [114]. The data from studies involving mitochondrial activity and the NO cascade suggest that oxidative stress and mitochondrial dysfunction are important in PD progression, including dopaminergic dysfunction and α-syn accumulation, which can promote neurodegeneration in SN and deficits in locomotor activity.

Familial PD cases account for 10% of total cases of the disease [115,116], but the molecular mechanisms involved in the onset of familial forms still need to be elucidated. Neuroinflammation can also contribute to the progression of the genetic forms of PD. Mutations in the gene encoding for leucine-rich repeat kinase 2 (LRRK2) are associated with familial PD [117], with an increased lifetime risk for developing sporadic PD [118]. In an intranigral LPS-injection model of neuroinflammation, a robust induction of LRRK2 in microglial cells was observed [119]. In addition, injection of LPS into the SNpc of LRRK2 KO rats resulted in less pronounced TH-positive neuron loss, microglial activation, and elevated level of iNOS compared to wild-type rats [120]. A morphological evaluation revealed that the fractal dimension—a quantitative computer-based analysis for cell complexity evaluation—of Lrrk2$^{-/-}$ microglia was significantly lower than that of Lrrk2$^{+/+}$ cells in the striatum injected with LPS [121]. The expression of the protein deglycase DJ-1 (PARK7)—whose gene is related to autosomal recessive forms of PD [122,123]—can be also impacted by inflammatory challenges. It is known that mutations in the *PARK7* gene are associated with loss of dopaminergic neurons due to the upregulation of inflammatory mediators within the SN, which was demonstrated by LPS intranigral injection in *PARK7* DJ-1$^{-/-}$ KO mice [124]. These data suggest that inflammatory events that occur throughout life can contribute to the progression of diseases related to autosomal dominant or autosomal recessive mutations, as shown by results from several experimental investigations.

Experimental data obtained from local injections of LPS into the CNS have contributed to the elucidation of the pathophysiology of PD, including the familial form of the disease. In the next section, data from models that used systemic LPS challenges will be presented. Inflammatory processes in the periphery can induce both acute and adaptive responses and contribute to deleterious effects on the CNS because of the action of inflammatory mediators from the periphery that are released into the brain [35,125,126]. Thus, peripheral inflammatory challenges can contribute to a better understanding of the crosstalk between inflammation, neuroinflammation, and basic aspects involved in neurodegenerative conditions.

3.2. Contribution of Systemic LPS Challenge Models to the Elucidation of PD Pathology

Systemic LPS challenge is another model to elucidate neuroinflammation in PD. Single or multiple LPS injections were used to provide valuable insights into the potential pathogenesis of PD. Molecular and cellular alterations were found after LPS i.p. injection in C57BL/6 mice. Brain TNF-α was elevated for up to 10 months after LPS injection, suggesting a sustained brain TNF-α overproduction that was parallel to microglial activation and delayed and progressive loss of nigral TH-positive neurons [127]. Extensive neuronal loss, decline in dopamine levels, glial activation, altered cytokine profile on SN, and deficits in locomotor behavior were also observed after four consecutive days of peripheral LPS injections [128]. Additionally, authors described a time-course shift of cytokine profiles from pro- to anti-inflammatory. Five to 19 days after exposure, pro-inflammatory mediators were predominant, in parallel with neuronal loss, while anti-inflammatory molecules were predominant between days 19 and 38 post-injection. Interestingly, a single dose of LPS failed to elicit neuroinflammatory responses in female mice [129]. On the other hand, i.p. injections of LPS for five weeks (one injection per week) or for five months (one injection per month) could cause loss of TH-positive neurons in the SN 9 and 20 months after injection, respectively. In addition, motor impairment as well as a more intense immuno-staining for α-syn and inflammatory markers were observed [129]. The augmentation of protein aggregation and nigral inflammatory process was also observed in a study that compared the effect of LPS i.p. injections in wild-type mice and in transgenic mice that overexpressed α-syn. It was demonstrated that transgenic mice, but not wild-type mice, developed a delayed chronic and progressive degeneration of nigral TH-positive DA neurons, with a more prominent effect five months after LPS injection. In addition, transgenic mice treated with LPS accumulated ~1.3-fold more α-syn aggregation than non-treated or wild-type mice [61]. The synergic impact of α-syn and inflammation on the BBB was also evaluated. Knockout mice for α-syn (Snca$^{-/-}$) were subjected to LPS exposure, and it was noticed that α-syn did not alter BBB permeability in the absence of an LPS challenge. However, LPS injection induced significant augmentation in BBB permeability in normal wild-type, but not in knockout, mice [130].

α-Syn overproduction and its accumulation appear to be associated with an impaired autophagy process. Alterations in autophagic protein levels were noticed after LPS injection. Early-period evaluations (starting at day 1) revealed increased levels of microtubule-associated protein 1 light chain 3-II (LC3-II) and histone deacetylase (HDAC) 6. On the other hand, p62 level remained increased until late stages (from one day to seven months after LPS injection). A significant increase in α-syn protein in the midbrain was also found in this study, suggesting that LPS might cause an impairment of α-syn clearance [131]. Therefore, peripheral inflammatory stimuli may be an important synergic factor for α-syn-induced pathology in PD, and autophagy activity failure might be involved in the increased protein aggregation induced by the LPS challenge.

The participation of NO, oxidative stress, and mitochondrial impairment was also investigated after peripheral LPS injection. Wide ultrastructural changes were observed in SN neuronal cells, including axons alterations, the swelling of mitochondria and the Golgi complex, and the presence of autophagolysosomes, lysosomes, and dense bodies in the cytoplasm. In addition, the presence of apoptotic cells and glial activation was also observed [132]. iNOS induction was observed at the initial phase of response to the peripheral LPS injection [128]. NOS activity in the midbrain and in SN was increased 6 h after LPS treatment [132]. Furthermore, exposure of C57BL/6 mice to LPS resulted in a large increase in NOX2 mRNA expression in the midbrain 24 h after exposure, associated with a rapidly increased ROS production at 1 and 24 h [133]. Treatment of NOX2$^{-/-}$ mice with LPS demonstrated the contribution of this mediator to the pathology-associated neuroinflammation, since knockout mice presented less dopaminergic neuronal loss and reduction of microglial activation in the midbrain after LPS i.p. injection [133]. It was also observed that, despite a lack of changes in caspase-3 activity, LPS injection induced apoptosis-inducing factor (AIF) translocation from the mitochondria to the nucleus. Moreover, iNOS and nNOS (the neuronal constitutive form of NOS) inhibition prevented LPS-evoked release of AIF from the mitochondria, indicating that the increased synthesis of NO occurring in the brain during systemic inflammation might be responsible for the activation of apoptotic pathways [132]. Lastly, iNOS and NADPH oxidase inhibition was also associated with the reduction of chronic neuroinflammation and prevented α-syn pathology and dopaminergic neuronal loss in transgenic mice that overexpressed human A53T mutant α-syn submitted to LPS i.p. injection [61].

The role of oxidative stress in PD seems to be age-dependent. The upregulation of pro-oxidant and inflammatory factors was shown in the midbrain of aged C57BL/6 mice submitted to acute i.p. injection of LPS, compared with young mice injected with LPS [134]. In addition, LPS induced a more severe loss of DA neurons in aged female C57BL/6 mice. The upregulation of TLR2, p-NF-κB-p65, IL-1β, TNF-α, iNOS, and gp91phox was also associated with aging [135]. These data indicate an important aspect of aging in the neuroinflammatory process found in PD and evidence the overexpression and overproduction of factors associated with oxidative stress in aged rodents injected with LPS.

In summary, SN and ST are highly sensitive and strongly affected by systemic LPS administration. Findings from studies using peripheral LPS injection can contribute to the understanding of the progression of PD, in particular, to the comprehension of its neuroinflammatory aspect.

4. LPS Models to Understanding Inflammatory and Neuroinflammatory Aspects in Amyotrophic Lateral Sclerosis and Huntington's Disease

4.1. Amyotrophic Lateral Sclerosis

Amyotrophic lateral sclerosis (ALS) is a neuromuscular disorder associated with the voluntary motor system, characterized by the progressive degeneration of anterior-lateral horn spinal cord motor neurons leading to weakness and eventual death of the affected individuals by paralysis in a few years [136,137]. The degenerating neurons present an abnormal accumulation of cytoplasmic inclusions containing ubiquitinated proteins [138]. A role of inflammation in the pathogenesis of ALS has been suggested [139,140]. In this sense, LPS-induced inflammation may contribute to the knowledge of the involvement of neuroinflammation in the pathophysiology of ALS.

The overexpression of mutant copper, zinc superoxide dismutase (SOD) in mice is utilized as a model of ALS, inducing severe hind limb motor deficits in animals [141]. These G93A-SOD1 mice were challenged with LPS to evaluate the possible impact of systemic inflammation in this model. LPS injection increased the nuclear expression of the transcription factor CCAAT/enhancer binding protein δ (C/EBPδ), whose gene is associated with familial ALS, in the spinal cord of G93A-SOD1 mice [142]. Moreover, astroglial and microglial activation were also associated with LPS-induced inflammation in an ALS experimental model [142,143].

About 5% of ALS cases are familial forms of the disease [144]. TAR DNA-binding protein (TDP-43), a major component of cytoplasmic inclusions in sporadic and most familial ALS cases, appeared accumulated and aggregated in the cytoplasm of spinal motor neurons of TDP-43^{A315T} transgenic mice after chronic LPS administration [145].

However, there are only a few studies using LPS to induce inflammation in animal models of ALS, despite the knowledge about the role of immune and inflammatory components in this neurodegenerative disease [146,147]. More studies are necessary to clarify the gaps associated with this disorder.

4.2. Huntington's Disease

Huntington's disease (HD) is a neurodegenerative disease characterized by motor, cognitive, and behavioral dysfunctions [148,149]. HD is originated by an autosomal mutation that is characterized by an increase in the number of CAG repeats in the huntingtin (*HTT*) gene [150], resulting in the expansion of a polyglutamine tract in the resulting mutated HTT (mHTT) protein that is neurotoxic. mHTT aggregates are abundant in the nuclei and processes of neuronal cells and lead to several damages, including protein malformation, transcriptional dysfunction, irregular protein and vesicle transport, altered secretion of neurotrophic factors, and others [151–154]. The immune and inflammatory component has also been linked to the progression of HD. Changes in the cytokine profile were reported in the post-mortem brain [155] and in the plasma and serum of patients [156], and several lines of evidence of inflammation involvement have been provided by animal models [62,157–159].

Studies on the impact of inflammatory challenges in this neurodegenerative disease are rare. Peripheral injection of LPS enhanced some aspects of HD, such as microglial alterations and vascular dysfunction, as shown in 12-month-old YAC128 transgenic mice—a model that expresses human mutant huntingtin protein—challenged chronically (four months) with LPS. Changes were characterized by an increased number and morphological changes of microglia in the ST. Furthermore, an increased vessel diameter and wall thickness in the same region and disruption of the BBB permeability were observed [159]. These data indicate that LPS enhances the inflammatory response in this model of HD. Levels of proinflammatory cytokines after a single LPS i.p. injection were higher in the cortex and ST of brains obtained from Hdh^{150Q} mice (which carry 150 CAG repeats in the first exon of the endogenous gene) and R6/2 mice (which express exon 1 of the human *HD* gene with 150 CAG repeats) compared with wild-type animals [62]. The authors observed that LPS exposure caused an increased nuclear localization of p65—a NF-κB subunit—in both astrocytes and microglia in the cortex of R6/2 mice compared with wild-type mice, contributing to neuroinflammation. In addition, the levels of TNF-α remained elevated in brain, serum, and liver of the two HD mouse models after systemic LPS injection [62]. Thus, a peripheral inflammatory process contributed to the progression of HD and to a more prolonged neuroinflammation mediated by glial cells.

Interestingly, a sex-dependent response of HD R6/1 mice to an LPS single injection was demonstrated. Authors noticed that LPS-induced TNF-α expression was ~1.5-fold higher in the hypothalamus of female HD mice as compared with female wild-type mice. In contrast, LPS treatment induced an opposite effect in male HD subjects, with largely diminished TNF-α gene expression, compared with wild-type mice [160]. More lines of evidence are necessary for a better exploration of these sex-dependent aspects, but these observations might suggest differences in HD patients, depending on their gender.

Nevertheless, chronically low-dose LPS injections activating the immune system showed a significantly prolonged survival of HD R6/2 animals, less pronounced body weight loss, and an attenuated clinical score of the clasping phenotype compared with wild-type animals treated with the endotoxin [65]. Therefore, the role of inflammatory processes in HD needs to be further elucidated, and the link between neuroinflammation and HD progression may be dependent on age, gender, and severity of the inflammatory challenge.

As it can be observed by reading the reports mentioned above, different factors may be important for the outcome of the studies, which include the source of LPS, dose, route and scheme of administration. Therefore, we built tables (Tables 1–6) that further detail all these differences that must be considered for the planning of an experimental protocol design. In the tables, only papers that provide full information about the type of LPS used were included.

5. LPS in Cell Culture Models

The basic aspects of the neurodegenerative process were elucidated by numerous in vitro studies. Inflammation triggered by microglia plays an important role in promoting neurodegeneration by inducing the expression of pro-inflammatory factors [102,161–163]. In this way, LPS-induced inflammatory neurotoxicity depends on the excessive production of pro-inflammatory factors by microglia [164]. Activation of TLR4 on the cell membrane by LPS activates various signal cascades, including NF-κB via the MyD88–IRAK–TRAF6–TAK1 signaling complex [38,165–167]. Upon LPS stimulation, the transcription factor NF-κB plays an important role in the expression of pro-inflammatory genes via its translocation to the nucleus [168] which can trigger a series of inflammatory pathways.

LPS stimulation of BV-2 microglial cells [169], co-cultures of neurons, astrocytes, and microglia [170], or hippocampal neurons cultures [171] resulted in increased synthesis and release of IL-1β and TNF-α. Besides its pro-inflammatory activity, LPS affected the viability of neurons, leading to highly condensed nuclei and the absence/retraction of neurites [170]. Treatment with LPS activated microglia also in rat basal forebrain mixed neuron–glial cultures. Additionally, the number of choline acetyltransferase-immunopositive neurons were decreased in these cultures treated with LPS [172]. Recently, a study also showed the activation of microglia by LPS, which induced corpus callosum nerve fiber malfunction and fast axonal transport [173]. Microglial response induced by LPS was also associated with the activation of COX-2 and the NOS pathway, resulting in a dramatic increase in prostaglandin E_2 (PGE_2) and nitric oxide production [172,174–183], which contributed to neurotoxicity and cellular dysfunction in neuron-glia cultures.

In mesencephalic mixed neuron–glia cultures, LPS exposure induced the reduction of TH-positive neurons in the presence of glia. However, LPS treatment did not affect dopaminergic cells when neurons were cultured in the absence of glia [89,184], suggesting that the glial-mediated neuronal damage was induced by LPS. Moreover, the increased release of inflammatory mediators IL-1β and TNF-α induced by LPS was associated with decreased TH-positive cells in primary mesencephalic cultures, which was prevented by using neutralizing antibodies against IL-1β or TNF-α [185]. In contrast, pretreatment with the anti-inflammatory cytokine IL-10 prevented dopaminergic neuron loss induced by LPS in primary ventral mesencephalic cultures due to a reduced production of proinflammatory cytokines and protection against a reduction of neurotrophic factors [186].

Finally, LPS treatment reduced the DA reuptake capacity of dopaminergic neurons in the neuron–glia cultures [187], exposing other aspects that might contribute to PD pathology.

Many protein kinases, such as p38 mitogen-activated protein kinases (p38 MAPK) and protein kinase C-δ (PKCδ) have been implicated in the release of inflammatory mediators from glia, resulting in neuronal death [188–190]. p38 MAPK mediates LPS-induced neurodegeneration in mesencephalic neuron–glia cultures through the induction of nitric oxide synthase resulting in increased NO production [179]. Another study using U373 cells showed an increased IL-6 production by stimulation with LPS, mediated by the p38/Src kinase inhibitors-dependent pathway [77]. Treatment of primary and BV-2 microglial cultures with LPS resulted in increased activation of phospho-p38 MAPK [178,181–183,191–193]. In addition, PKCδ was highly upregulated during chronic microglial activation, and a significant increase in PKCδ kinase activity was observed [190], followed by ROS generation, NO production, and proinflammatory cytokine and chemokine release. Proteolytic activation of PKCδ occurred during dopaminergic degeneration and was mediated by caspase-3 [194–196]. Silencing of caspase-3 or AIF by small interfering RNAs, exclusively in DA MN9D cells, protected DA cells from LPS-induced death, demonstrating the key role of these molecules in LPS-induced neurotoxicity [96].

Finally, LPS increased the expression levels of β-site APP cleaving enzyme 1 (BACE-1), PS-1, β-APP, and Aβ1-42 in neuron cultures treated with LPS [171]. LPS exposure also contributed synergistically to the negative effects of α-synuclein on progressive dopaminergic degeneration, associated with increased microglial superoxide production [197]. In addition, LPS could also induce conformational changes in α-synuclein protein, which might accelerate the progression of PD [198].

It is noteworthy that in vitro investigations are widely used for the evaluation of mechanisms associated with cell homeostasis or dysfunction. Data from cell cultures therefore also contribute to the better understanding of gaps in intracellular signaling, molecular aspects, gene transcription, mRNA translation, and protein synthesis involved in cell physiology. In this context, LPS-induced in vitro models are very relevant to support the elucidation of the pathophysiology of neurodegenerative diseases.

Table 1. Lipopolysaccharide (LPS) source, species used, dose and route of administration, duration, evaluated parameters of models of central LPS challenges for the elucidation of Alzheimer's disease (AD).

LPS	Species Used	Dose and Route of Administration	LPS Injection (Duration)	Evaluated Parameters	References
E. coli O127:B8 (Sigma-Aldrich)	Charles River CD-VAF rats	10 ng/animal (intracerebroventricular)	Acute	IL-1 in brain regions: cerebellum, cortex, brainstem, diencephalon, or hippocampus	[35]
E. coli O55:B5 (Sigma-Aldrich)	Sprague Dawley	1.0 µg/mL (4th ventricle)	Chronic (four weeks)	Spatial working memory Activation of astrocytes and microglia	[19]
E. coli O55:B5 (Sigma-Aldrich)	Fisher-344 rats	0.25 µL/h (4th ventricle)	Chronic (28 days)	Long-term depression (LTD) Underlying mechanism of LTD impairment by neuroinflammation	[20]
S. abortus equi (Sigma-Aldrich)	Tg2576 APP mice	4 µg/µL or 10 µg/µL (intrahippocampal)	Acute	Amyloid-beta (Aβ) load Microglial and astrocytes activation over time	[46]
S. abortus equi (Sigma-Aldrich)	Nontransgenic mice obtained during breeding of our amyloid precursor protein (APP)1 + presenilin (PS)1 transgenic mouse colony	1 µL of 4 µg/µL (intrahippocampal; bilateral)	Acute	Time course of microgliosis Time course of astrogliosis Time course of TLR4 levels Quantification of glial markers (GFAP, CD45) TNF-α and IL-1β levels	[40]
S. abortus equi (Sigma-Aldrich)	Tg2576 APP mice	10 µg/µL (intrahippocampal; unilateral)	Acute	Brain amyloid burden Markers of microglial activation (CD45, CR3 or CD11b, CD68, Fcg receptor, and scavenger receptor A)	[47]
S. typhimurium (Sigma-Aldrich)	APP1PS1 transgenic mice were transplanted with eGFP-over-expressing bone marrow	4 µg of LPS (4 µg/µL in saline); (intrahippocampal; unilateral)	Acute	Proliferation, expression of markers for activated microglia Aβ removal	[48]
S. abortus equii (Sigma-Aldrich)	rTg4510 mice and non-transgenic mice	5 µg/µL (frontal cortex and hippocampus)	Acute	Activation of CD45 and arginase 1 Expression of Ser199/202 and phospho-tau Ser396	[44]
E. coli O55:B5 (Sigma-Aldrich)	Sprague Dawley rats	2.5 µg/µL (intrahippocampal; unilateral)	Acute	β-secretase-1 (BACE1) and GFAP levels Amyloidogenic protein expression Golgi preparations of cortical layer III pyramidal neurons	[41]
S. abortus equi (Sigma-Aldrich)	TgAPP/PS1 and C57BL/6	4 µg/µL (2-month-old mice) or 2 µg/µL (12-month-old mice) (intrahippocampal)	Acute	Aβ deposits in the hippocampus and cortex Activation of microglia	[45]

Table 2. LPS source, species used, dose and route of administration, duration, evaluated parameters of models of systemic LPS challenges for the elucidation of AD.

LPS	Species Used	Dose and Route of Administration	LPS Injection (Duration)	Evaluated Parameters	References
E. coli O127:B8 (Sigma-Aldrich)	Charles River CD-VAF rats	1 mg/kg (intraperitoneal)	Acute	Detection of IL-1 by thymocyte stimulation	[35]
E. coli O111:B4 (Sigma-Aldrich)	TgN(APP-Sw) 2576	0.25 µg/µL (intravenously)	Acute	Aβ levels in cortex and hippocampus IL-1β levels in cortex and hippocampus	[52]
E. coli O55:B5 (Sigma-Aldrich)	3xTg-AD or nontransgenic mice	0.1 mg/mL; 0.5 mg/kg body weight (intraperitoneal)	Chronic (twice a week for six weeks)	Characterization of time course of microglia activation in the brain Microglial activation and tangle pathology	[74]
E. coli O55:B5 (Sigma-Aldrich)	ICR mice	250 µg/kg (intraperitoneal)	Acute (daily for three or seven days)	Memory impairment Aβ accumulation in the cortex and hippocampus Expression of amyloidogenic proteins Astrocytes activation	[59]
E. coli O111:B4 (Sigma-Aldrich)	3xTgAD mice	0.25 mg/kg (intraperitoneal)	Chronic (twice weekly for four weeks)	Effect of inhibition of soluble TNF signaling on accumulation of 6E10-immunoreactive protein in hippocampus, cortex, and amygdala and amyloid-associated pathology	[58]
S. typhimurium (Sigma-Aldrich)	CD-1 mice	3, 30, 300, or 3000 µg/kg (intraperitoneal)	Acute	Transport of Aβ across the blood–brain barrier	[69]
E. coli O55:B5 (Sigma-Aldrich)	Wistar	5 mg/kg (intraperitoneal)	Acute	Cognitive functions (amnesic, discriminative, and attentional functions) Anxiety TNF and IL-18 protein levels in frontal cortex, hippocampus, striatum, cerebellum, and hypothalamus	[54]
S. typhimurium (Sigma-Aldrich)	CD-1 mice	3 mg/kg (intraperitoneal)	Acute	Aβ transporter across the blood-brain barrier Oxidative stress markers in brain and serum Brain influx of I-albumin IL-1α, IL-1β, IL-6, IL-12, IL-13, MIP-1α, MIP-1β, G-CSF, KC, MCP-1, RANTES, and TNF-α levels in cortex and hippocampus	[67]
S. typhimurium (Sigma-Aldrich)	CD-1 mice	3 mg/kg (intraperitoneal)	Acute	Quantification of LRP-1 LRP-1-dependent partitioning between the brain vasculature and parenchyma and peripheral clearance	[70]
E. coli O55:B5 (Sigma-Aldrich)	Wistar rats	500 µg/kg/day (intraperitoneal)	For seven consecutive days.	Nitric oxide (NO) production NO synthase (NOS2) Aβ 1-42 cerebral expression Memory	[56]

Table 2. *Cont.*

LPS	Species Used	Dose and Route of Administration	LPS Injection (Duration)	Evaluated Parameters	References
E. coli O8:K27 (Innaxon)	EFAD mice (express human APOE3 or APOE4 and overproduce human Aβ	0.5 mg/kg/week (intraperitoneal)	Chronic (from 4 to 6 months of age)	Cognitive dysfunction Cerebrovascular leakiness Aβ42 levels Cerebral amyloid angiopathy-like deposition IL-10, G-CSF, RANTES, IL-12, IL-17, KC levels	[57]
E. coli O111:B4 (Sigma-Aldrich)	APP$_{SWE}$/PS11$_{\Delta E9}$ Tg and wild-type	0.5 mg/kg (intraperitoneal)	Chronic (Once a week for 13 weeks)	TNF and IL-1β mRNA levels Amyloid pathology in the neocortex CD11b+ cells clustering around Aβ plaques APP, APOE, Clu, and Hexb protein expression in neocortex	[60]
E. coli O111:B4 (Sigma-Aldrich)	5XFAD and C57BL/6 mice	0.01 mg/kg, 0.1 mg/kg, 1 mg/kg, 3 mg/kg (intravenously)	Acute	Disruption of blood–brain barrier Delivery of large molecules through the blood–brain barrier Weight loss	[71]

Table 3. LPS source, species used, dose and route of administration, duration, evaluated parameters of models of central LPS challenges for the elucidation of Parkinson's disease (PD).

LPS	Species Used	Dose and Route of Administration	LPS Injection (Duration)	Evaluated Parameters	References
E. coli O26:B6 (Sigma-Aldrich)	Wistar	1 mg/mL (2 μL intranigral)	Acute	Dopamine (DA) and DA metabolites Loss of tyrosine hydroxylase (TH)-positive cells TH activity Microglial activation NOS inhibition	[87,88]
E. coli O111:B4 (Life Technologies)	Fischer 344	5 or 10 μg in 2 μL (intrastriatal, intrahippocampal or intracortical)	Acute	Loss of TH-positive cells MAP-2-positive cell loss Microglial activation	[89]
E. coli O111:B4 (Sigma-Aldrich)	Sprague–Dawley rats	5 μg in 2 μL (intranigral)	Acute	Loss of TH-positive cells Microglial activation Naloxone effects on LPS consequences	[92]

Table 3. *Cont.*

LPS	Species Used	Dose and Route of Administration	LPS Injection (Duration)	Evaluated Parameters	References
E. coli O26:B6 (Sigma-Aldrich)	Wistar	5 µg (intranigral)	Acute	Dopamine and DA metabolites Serotonin and DA metabolites TH activity Loss of TH-positive cells Glial reaction Effects of dexamethasone on LPS consequences	[93]
E. coli O111:B4 (Sigma-Aldrich)	Fischer 344	5 ng/h (intranigral)	Chronic (2 weeks)	Loss of TH-positive cells Loss of NeuN-positive cells Microglial activation	[90]
E. coli O26:B6 (Sigma-Aldrich)	Wistar	10 µg (intranigral)	Acute	Loss of TH-positive cells Loss of FG-labelled neurons NADPH-d expression iNOS expression	[110]
E. coli O55:B5 (Calbiochem)	Wistar	10 µg (supranigral)	Acute	Loss of TH-positive cells Motor evaluation Astrocyte reaction Microglial activation iNOS expression Neurotophin-3 expression	[91]
E. coli O111:B4 (Sigma-Aldrich)	Fischer 344	10 µg (intrapallidal)	Acute	Loss of TH-positive cells Microglial activation Ferritin expression Iron levels A-synuclein expression Ubiquitin expression Effect of aging on LPS consequences	[113]
E. coli O26:B6 (Sigma-Aldrich)	Wistar	2 mg/mL (intranigral)	Acute	Loss of TH-positive cells Microglial activation TH expression Cytokine mRNA expression iNOS expression Caspase-11 expression Effects of p38 MAPK inhibition in LPS consequences Effects of iNOS blockage on LPS consequences	[109]

Table 3. *Cont.*

LPS	Species Used	Dose and Route of Administration	LPS Injection (Duration)	Evaluated Parameters	References
S. minnesota (Sigma-Aldrich)	Sprague-Dawley	16, 32 or 60 µg (intrastriatal)	Acute	DA and DA metabolites Loss of TH-positive cells Microglial activation Pro-inflammatory cytokine expression Insulin receptor expression Mitochondrial activity Effects of cyclooxygenase-2 (COX-2) inhibition and PPAR-c agonist on LPS consequences	[112]
S. minnesota (Sigma-Aldrich)	Sprague-Dawley	16 µg (intrastriatal)	Acute	UCP2 expression mitoNEET expression Effects of PPAR-c agonist on LPS consequences	[107]
S. minnesota (Sigma-Aldrich)	C57BL/6	5, 7.5, or 10 µg (intrastriatal)	Acute	Loss of TH-positive cells Motor evaluation NOS expression Effects of NOS inhibition in LPS consequences Effects of iNOS knockout on LPS consequences	[111]
S. minnesota (Sigma-Aldrich)	Wistar	2.5 µg/µL (intrastriatal)	Acute	DA Nigrostriatal system evaluation a-synuclein expression Ubiquitin expression Motor evaluation Microglial activation iNOS expression Mitochondrial activity	[108]
E. coli O111:B4 (Calbiochem)	Fischer 344	5 µg (intranigral)	Acute	Loss of TH-positive cells Microglial activation Effects of IκB Kinase-β inhibition on LPS consequences	[187]
E. coli O26:B6 (Sigma-Aldrich)	ABH-Biozzi	0.5 mg/kg	Acute	NFκB mRNA expression Cell death evaluation	[95]
E. coli (Sigma-Aldrich)	C57BL/6	10 µg (intrastriatal)	Acute	Motor evaluation DA neuron loss DA and DA metabolites Microglial activation Iron concentration Effects of desferrioxamine on the LPS consequences	[114]

Table 3. *Cont.*

LPS	Species Used	Dose and Route of Administration	LPS Injection (Duration)	Evaluated Parameters	References
E. coli O55:B5 (Sigma-Aldrich)	Fischer 344	0.25 µg/h (intracerebroventricular)	Chronically (21 or 56 days)	Cytokine protein levels Cytokine mRNA expression Loss of TH-positive cells MHC II-IR microglial density Effects of aging on LPS consequences	[97]
E. coli O111:B4 (Sigma-Aldrich)	Sprague-Dawley	5 µg/5 µL (intranigral)	Acute	Astrocyte reaction Microglial activation NFκB transcription Cytokine transcription NOX2 activation NADPH-Oxidase Activity Reactive oxygen species (ROS) production Lipid peroxidation iNOS and NO expression. DA and DA metabolites Effects of NADPH-oxidase inhibition on LPS consequences	[99]
E. coli (Sigma-Aldrich)	SD rats	5 mg/mL (intrastriatal)	Acute	Motor evaluation Glial activation Oxidative stress Apoptosis	[103]
S. minnesota (Sigma-Aldrich)	Sprague-Dawley	32 µg (intrastriatal)	Acute	Mitochondrial activity and structure Oxidative stress Loss of TH-positive cells	[106]
E. coli O55:B5 (Sigma-Aldrich)	Wistar	5 µg/2 µL (intranigral)	Acute	Fever and Sickness Microglial Activation and phagocytic activity Astrocyte Activation Oxidative Stress Cytokine levels Leukocyte brain Infiltration	[98]
E. coli O111:B4 (Enzo Life Science)	*LRRK2* KO C57BL/6 and wild-type	5 mg/mL (intrastriatal)	Acute	Microglial activation Role of LRRK2 on LPS consequences	[121]
E. coli (Sigma-Aldrich)	DJ-1 KO C57BL/6 and wild-type	1 µg/µL (intranigral)	Acute	Dopaninergic normal loss sICAM-1, IFN-γ, IL-1β, IL-1Ra, IL-16, IL-17, and I-TAC expression Role of DJ-1 on LPS consequences	[124]

Table 4. LPS source, species used, dose and route of administration, duration, evaluated parameters of models of systemic LPS challenges for the elucidation of PD.

LPS	Species Used	Dose and Route of Administration	LPS Injection (Duration)	Evaluated Parameters	References
E. coli O55:B5 (Sigma-Aldrich)	C57BL/6	1 mg/kg (intraperitoneal)	Acute (single dose)	Ultrastructural Alterations in SN NOS Activity NOS and TNF expression Apoptotic Pathways	[132]
E. coli O111:B4 (Calbiochem)	C57BL/6, TNFR1/R2$^{-/-}$ KO, TNFR1/R2$^{+/+}$ WT	5 mg/kg (intraperitoneal)	Acute (single dose)	TNFα level Loss of TH-positive cells Effects of TNFR knock-out on LPS consequences	[127]
E. coli O111:B4 (Sigma-Aldrich)	C57BL/6	5 mg/kg (intraperitoneal)	Weekly injected with five doses of LPS Monthly injected with two to five doses of LPS	Motor evaluation Loss of TH-positive cells α-synuclein accumulation Microglial activation Sex differences in LPS consequences	[129]
E. coli O111:B4 (Sigma-Aldrich)	B6C3F1 WT and transgenic mice for mutant α-synuclein	3×10^6 EU/kg (intraperitoneal)	Acute (single injection)	Nigral TH-positive cells evaluation α-synuclein aggregation Cytokine levels Microglial activation Differences in acute and chronic neuroinflammation Effects iNOS inhibition of iNOS inhibition and NADPH oxidase blockage on LPS consequences	[61]
E. coli O111:B4	C57BL/6	0.2 mg/kg (intraperitoneal)	Acute (single injection)	Cytokine expression. TH-positive cells evaluation Microglial activation iNOS mRNA expression NF-κB mRNA expression. gp91phox level Oxidative stress Effects of HCT1026 on LPS consequences	[134]
E. coli O55:B5 (Sigma-Aldrich)	129/SvEv and α-syn gene-ablated mice	1 mg/kg (intraperitoneal)	Acute (single dose)	Blood–brain barrier integrity	[130]
E. coli O111:B4 (Calbiochem)	B6.129S6-Cybbtm1Din (NOX2−/−) and C57BL/6 000664 (NOX2+/+)	5 mg/kg (intraperitoneal)	Acute (single injection)	NOX2 expression ROS production Microglial activation Effects of oxidases inhibition on LPS consequences	[133]

Table 4. *Cont.*

LPS	Species Used	Dose and Route of Administration	LPS Injection (Duration)	Evaluated Parameters	References
E. coli (Sigma-Aldrich)	C57BL/6	5 mg/kg (intraperitoneal)	Acute (single injection)	TH-positive cells evaluation α-syn aggregation and levels Microglial activation Autophagic activity	[131]
E. coli O111:B4 (Sigma-Aldrich)	C57BL/6 and PKCδ KO mice	5 mg/kg (intraperitoneal)	Acute (single injection)	Motor evaluation Cytokine release and expression. Effects of PKCδ KO on LPS consequences	[190]
S. abortus equi (Enzo Life Sciences)	C57BL/6	1 μg/g (intraperitoneal)		Motor evaluation TH-positive cells evaluation DA and DA metabolites Microglial and astrocytic activation Cytokine levels and expression	[128]

Table 5. LPS source, species used, dose and route of administration, duration, evaluated parameters of models of systemic LPS challenges for the elucidation of amyotrophic lateral sclerosis (ALS).

LPS	Species Used	Dose and Route of Administration	LPS Injection (Duration)	Evaluated Parameters	References
E. coli O55:B5 (Calbiochem)	C57BL/6 EP4 floxed mice	5 mg/kg (intraperitoneal)	Acute	Quantification of COX-2, iNOS, TNF-α, IL-6, and IL-1β mRNA levels in hippocampus	[175]
E. coli O55:B5 (Sigma-Aldrich)	G93A-SOD1 C/EBPδ$^{(-/-)}$ mice	200 μg/animal (intraperitoneal) 1 μg/μL (intraperitoneal)	Acute 2, 8, 16, 24, and 48 h	C/EBPδ expression in mouse brain Quantification of NOS-2, COX-2, TNF-α, IL-1β, and IL-6 mRNA TNF-α, IL-1β and IL-6 serum levels	[142]
E. coli O55:B5 (Sigma-Aldrich)	TDP-43^{A315T} and C57BL/6 mice	1 mg/kg of body weight (intraperitoneal)	Chronic (Once a week for two months)	TDP-43 accumulation in the cytoplasm of spinal motor neurons TDP-43 aggregation	[145]

Table 6. LPS source, species used, dose and route of administration, duration, evaluated parameters of models of systemic LPS challenges for the elucidation of Huntington's disease (HD).

LPS	Species Used	Dose and Route of Administration	LPS Injection (Duration)	Evaluated Parameters	References
E. coli (Sigma-Aldrich)	Transgenic YAC128 and wild type	1 mg/kg (intraperitoneal)	Chronic (Once a week for four months)	Microglial activation Neurovascular integrity Blood brain barrier integrity	[159]
E. coli O111:B4 (Sigma-Aldrich)	Transgenic R6/2 and wild type	2 mg/kg (intraperitoneal)	Acute	NF-κB activation Inflammatory evaluation Motor evaluation	[62]
E. coli O127:B8 (Sigma-Aldrich)	Transgenic R6/2 and wild type	0.3 mg/kg (intraperitoneal)	Acute	TNF gene expression. IL-6 gene expression Sex-dependent effects of LPS injection	[160]
E. coli O111:B4 (Sigma-Aldrich)	Transgenic R6/2 and wild type	2 μg/animal (intraperitoneal)	Chronic (Once a week for seven weeks)	Splenic immune cells evaluation T-cell activity Motor evaluation	[65]

6. Final Considerations and Conclusions

In the context of AD, models that use LPS contribute to the understanding of the intricate relationship between neuroinflammation and the progression of the disease, mainly in regard to Aβ processing and deposition. Besides, activation of TLR4 and of the inflammatory pathways leads to glial reaction and neuronal loss, which contributes to memory impairment and behavioral changes. Importantly, both acute and chronic inflammation seem to play a role in this neurodegenerative disease.

On the other hand, injection of LPS per se may be used as an animal model of PD, mainly because of the high susceptibility of mesencephalic neurons to this toxin [89,131,199]. In this sense, injection of LPS can contribute to the elucidation of the inflammatory pathways that induce glial activation and the additional causes of neuronal death, dopamine signaling disbalance, α-syn aggregation, and behavioral symptoms. Finally, in regard to ALS and HD, the role of inflammatory processes in these two neurodegenerative diseases needs to be better studied and elucidated. The studies may consider to include the use of the already established models to evaluate the impact of inflammatory challenges in the development of these pathological conditions.

Importantly, the variety of protocols and serotypes of LPS used in the studies may induce a plethora of results. This wide range of outcomes may contribute to the better understanding of the intricate link between neurodegenerative diseases and peripheral and central inflammation.

In conclusion, LPS is an important tool for the evaluation of different parameters associated with inflammatory processes and may be used in studies that aim to investigate the pathophysiological mechanisms of neurodegenerative diseases. However, the serotype, route of administration, doses, and other parameters should be considered when planning experimental protocols because of the varied responses induced by the endotoxin.

Funding: We thank Fundação de Amparo à Pesquisa do Estado de Minas Gerais (FAPEMIG—process number: APQ-02044-15), Conselho Nacional de Desenvolvimento Científico e Tecnológico (CNPq—process number 424588/2016-1), and Coordenação de Aperfeiçoamento de Pessoal de Nível Superior (CAPES) for the financial support. A.C.P.d.O. acknowledge CNPq for the productivity fellowship (310347/2018-1). The article processing charge was funded by the University of Freiburg Library in the funding program « Open Access Publishing ».

Conflicts of Interest: The authors declare no conflict of interest.

References

1. Ransohoff, R.M. How neuroinflammation contributes to neurodegeneration. *Science* **2016**, *353*, 777–783. [CrossRef] [PubMed]
2. Boonen, B.; Alpizar, Y.A.; Sanchez, A.; Lopez-Requena, A.; Voets, T.; Talavera, K. Differential effects of lipopolysaccharide on mouse sensory TRP channels. *Cell Calcium* **2018**, *73*, 72–81. [CrossRef]
3. Alpizar, Y.A.; Boonen, B.; Sanchez, A.; Jung, C.; Lopez-Requena, A.; Naert, R.; Steelant, B.; Luyts, K.; Plata, C.; De Vooght, V.; et al. TRPV4 activation triggers protective responses to bacterial lipopolysaccharides in airway epithelial cells. *Nat. Commun.* **2017**, *8*, 1059. [CrossRef] [PubMed]
4. Meseguer, V.; Alpizar, Y.A.; Luis, E.; Tajada, S.; Denlinger, B.; Fajardo, O.; Manenschijn, J.A.; Fernandez-Pena, C.; Talavera, A.; Kichko, T.; et al. TRPA1 channels mediate acute neurogenic inflammation and pain produced by bacterial endotoxins. *Nat. Commun.* **2014**, *5*, 3125. [CrossRef] [PubMed]
5. Fitzgerald, K.A.; McWhirter, S.M.; Faia, K.L.; Rowe, D.C.; Latz, E.; Golenbock, D.T.; Coyle, A.J.; Liao, S.M.; Maniatis, T. IKKepsilon and TBK1 are essential components of the IRF3 signaling pathway. *Nat. Immunol.* **2003**, *4*, 491–496. [CrossRef] [PubMed]
6. Ruckdeschel, K.; Pfaffinger, G.; Haase, R.; Sing, A.; Weighardt, H.; Hacker, G.; Holzmann, B.; Heesemann, J. Signaling of apoptosis through TLRs critically involves toll/IL-1 receptor domain-containing adapter inducing IFN-beta, but not MyD88, in bacteria-infected murine macrophages. *J. Immunol.* **2004**, *173*, 3320–3328. [PubMed]
7. Zughaier, S.M.; Zimmer, S.M.; Datta, A.; Carlson, R.W.; Stephens, D.S. Differential induction of the toll-like receptor 4-MyD88-dependent and -independent signaling pathways by endotoxins. *Infect. Immun.* **2005**, *73*, 2940–2950. [PubMed]

8. Gray, P.; Dagvadorj, J.; Michelsen, K.S.; Brikos, C.; Rentsendorj, A.; Town, T.; Crother, T.R.; Arditi, M. Myeloid differentiation factor-2 interacts with Lyn kinase and is tyrosine phosphorylated following lipopolysaccharide-induced activation of the TLR4 signaling pathway. *J. Immunol.* **2011**, *187*, 4331–4337. [CrossRef]

9. Park, B.S.; Lee, J.O. Recognition of lipopolysaccharide pattern by TLR4 complexes. *Exp. Mol. Med.* **2013**, *45*, e66. [CrossRef]

10. Acosta, C.; Davies, A. Bacterial lipopolysaccharide regulates nociceptin expression in sensory neurons. *J. Neurosci. Res.* **2008**, *86*, 1077–1086. [CrossRef] [PubMed]

11. Leow-Dyke, S.; Allen, C.; Denes, A.; Nilsson, O.; Maysami, S.; Bowie, A.G.; Rothwell, N.J.; Pinteaux, E. Neuronal Toll-like receptor 4 signaling induces brain endothelial activation and neutrophil transmigration in vitro. *J. Neuroinflamm.* **2012**, *9*, 230. [CrossRef] [PubMed]

12. Chistyakov, D.V.; Azbukina, N.V.; Lopachev, A.V.; Kulichenkova, K.N.; Astakhova, A.A.; Sergeeva, M.G. Rosiglitazone as a Modulator of TLR4 and TLR3 Signaling Pathways in Rat Primary Neurons and Astrocytes. *Int. J. Mol. Sci.* **2018**, *19*, 113. [CrossRef] [PubMed]

13. Rolls, A.; Shechter, R.; London, A.; Ziv, Y.; Ronen, A.; Levy, R.; Schwartz, M. Toll-like receptors modulate adult hippocampal neurogenesis. *Nat. Cell Biol.* **2007**, *9*, 1081–1088. [CrossRef] [PubMed]

14. Jahn, H. Memory loss in Alzheimer's disease. *Dialogues Clin. Neurosci.* **2013**, *15*, 445–454. [PubMed]

15. Nelson, P.T.; Alafuzoff, I.; Bigio, E.H.; Bouras, C.; Braak, H.; Cairns, N.J.; Castellani, R.J.; Crain, B.J.; Davies, P.; Del Tredici, K.; et al. Correlation of Alzheimer disease neuropathologic changes with cognitive status: A review of the literature. *J. Neuropathol. Exp. Neurol.* **2012**, *71*, 362–381. [CrossRef] [PubMed]

16. Small, S.A.; Schobel, S.A.; Buxton, R.B.; Witter, M.P.; Barnes, C.A. A pathophysiological framework of hippocampal dysfunction in ageing and disease. *Nat. Rev. Neurosci.* **2011**, *12*, 585–601. [CrossRef]

17. Braak, H.; Braak, E. Neuropathological stageing of Alzheimer-related changes. *Acta Neuropathol.* **1991**, *82*, 239–259. [CrossRef]

18. Ittner, L.M.; Gotz, J. Amyloid-beta and tau–a toxic pas de deux in Alzheimer's disease. *Nat. Rev. Neurosci.* **2011**, *12*, 65–72. [CrossRef]

19. Hauss-Wegrzyniak, B.; Dobrzanski, P.; Stoehr, J.D.; Wenk, G.L. Chronic neuroinflammation in rats reproduces components of the neurobiology of Alzheimer's disease. *Brain Res.* **1998**, *780*, 294–303. [CrossRef]

20. Min, S.S.; Quan, H.Y.; Ma, J.; Lee, K.H.; Back, S.K.; Na, H.S.; Han, S.H.; Yee, J.Y.; Kim, C.; Han, J.S.; et al. Impairment of long-term depression induced by chronic brain inflammation in rats. *Biochem. Biophys. Res. Commun.* **2009**, *383*, 93–97. [CrossRef]

21. Wee Yong, V. Inflammation in neurological disorders: A help or a hindrance? *Neurosci. Rev. J. Bringing Neurobiol. Neurol. Psychiatry* **2010**, *16*, 408–420. [CrossRef]

22. Heppner, F.L.; Ransohoff, R.M.; Becher, B. Immune attack: The role of inflammation in Alzheimer disease. *Nat. Rev. Neurosci.* **2015**, *16*, 358–372. [CrossRef]

23. Salter, M.W.; Stevens, B. Microglia emerge as central players in brain disease. *Nat. Med.* **2017**, *23*, 1018–1027. [CrossRef]

24. Bauer, J.; Strauss, S.; Schreiter-Gasser, U.; Ganter, U.; Schlegel, P.; Witt, I.; Yolk, B.; Berger, M. Interleukin-6 and alpha-2-macroglobulin indicate an acute-phase state in Alzheimer's disease cortices. *FEBS Lett.* **1991**, *285*, 111–114. [CrossRef]

25. Mrak, R.E.; Sheng, J.G.; Griffin, W.S. Glial cytokines in Alzheimer's disease: Review and pathogenic implications. *Hum. Pathol.* **1995**, *26*, 816–823. [CrossRef]

26. Wyss-Coray, T.; Rogers, J. Inflammation in Alzheimer disease—A brief review of the basic science and clinical literature. *Cold Spring Harb. Perspect. Med.* **2012**, *2*, a006346. [CrossRef]

27. Miklossy, J. Chronic inflammation and amyloidogenesis in Alzheimer's disease—Role of Spirochetes. *J. Alzheimer's Dis. JAD* **2008**, *13*, 381–391. [CrossRef]

28. Sheng, J.G.; Bora, S.H.; Xu, G.; Borchelt, D.R.; Price, D.L.; Koliatsos, V.E. Lipopolysaccharide-induced-neuroinflammation increases intracellular accumulation of amyloid precursor protein and amyloid beta peptide in APPswe transgenic mice. *Neurobiol. Dis.* **2003**, *14*, 133–145. [CrossRef]

29. Zhan, X.; Stamova, B.; Jin, L.W.; DeCarli, C.; Phinney, B.; Sharp, F.R. Gram-negative bacterial molecules associate with Alzheimer disease pathology. *Neurology* **2016**, *87*, 2324–2332. [CrossRef]

30. Zhao, Y.; Cong, L.; Jaber, V.; Lukiw, W.J. Microbiome-Derived Lipopolysaccharide Enriched in the Perinuclear Region of Alzheimer's Disease Brain. *Front. Immunol.* **2017**, *8*, 1064. [CrossRef]

31. Zhao, Y.; Jaber, V.; Lukiw, W.J. Secretory Products of the Human GI Tract Microbiome and Their Potential Impact on Alzheimer's Disease (AD): Detection of Lipopolysaccharide (LPS) in AD Hippocampus. *Front. Cell. Infect. Microbiol.* **2017**, *7*, 318. [CrossRef]

32. Zhao, Y.; Lukiw, W.J. Bacteroidetes Neurotoxins and Inflammatory Neurodegeneration. *Mol. Neurobiol.* **2018**, *55*, 9100–9107. [CrossRef]

33. Zhan, X.; Stamova, B.; Sharp, F.R. Lipopolysaccharide Associates with Amyloid Plaques, Neurons and Oligodendrocytes in Alzheimer's Disease Brain: A Review. *Front. Aging Neurosci.* **2018**, *10*, 42. [CrossRef]

34. Zakaria, R.; Wan Yaacob, W.M.; Othman, Z.; Long, I.; Ahmad, A.H.; Al-Rahbi, B. Lipopolysaccharide-induced memory impairment in rats: A model of Alzheimer's disease. *Physiol. Res.* **2017**, *66*, 553–565.

35. Quan, N.; Sundar, S.K.; Weiss, J.M. Induction of interleukin-1 in various brain regions after peripheral and central injections of lipopolysaccharide. *J. Neuroimmunol.* **1994**, *49*, 125–134. [CrossRef]

36. Minghetti, L.; Ajmone-Cat, M.A.; De Berardinis, M.A.; De Simone, R. Microglial activation in chronic neurodegenerative diseases: Roles of apoptotic neurons and chronic stimulation. *Brain Res. Brain Res. Rev.* **2005**, *48*, 251–256. [CrossRef]

37. Lucin, K.M.; Wyss-Coray, T. Immune activation in brain aging and neurodegeneration: Too much or too little? *Neuron* **2009**, *64*, 110–122. [CrossRef]

38. Fiebich, B.L.; Batista, C.R.A.; Saliba, S.W.; Yousif, N.M.; de Oliveira, A.C.P. Role of Microglia TLRs in Neurodegeneration. *Front. Cell. Neurosci.* **2018**, *12*, 329. [CrossRef]

39. Ophir, G.; Meilin, S.; Efrati, M.; Chapman, J.; Karussis, D.; Roses, A.; Michaelson, D.M. Human apoE3 but not apoE4 rescues impaired astrocyte activation in apoE null mice. *Neurobiol. Dis.* **2003**, *12*, 56–64. [CrossRef]

40. Herber, D.L.; Maloney, J.L.; Roth, L.M.; Freeman, M.J.; Morgan, D.; Gordon, M.N. Diverse microglial responses after intrahippocampal administration of lipopolysaccharide. *Glia* **2006**, *53*, 382–391. [CrossRef]

41. Deng, X.; Li, M.; Ai, W.; He, L.; Lu, D.; Patrylo, P.R.; Cai, H.; Luo, X.; Li, Z.; Yan, X. Lipolysaccharide-Induced Neuroinflammation Is Associated with Alzheimer-Like Amyloidogenic Axonal Pathology and Dendritic Degeneration in Rats. *Adv. Alzheimer's Dis.* **2014**, *3*, 78–93. [CrossRef]

42. Philippens, I.H.; Ormel, P.R.; Baarends, G.; Johansson, M.; Remarque, E.J.; Doverskog, M. Acceleration of Amyloidosis by Inflammation in the Amyloid-Beta Marmoset Monkey Model of Alzheimer's Disease. *J. Alzheimer's Dis. JAD* **2017**, *55*, 101–113. [CrossRef]

43. Mrdjen, D.; Pavlovic, A.; Hartmann, F.J.; Schreiner, B.; Utz, S.G.; Leung, B.P.; Lelios, I.; Heppner, F.L.; Kipnis, J.; Merkler, D.; et al. High-Dimensional Single-Cell Mapping of Central Nervous System Immune Cells Reveals Distinct Myeloid Subsets in Health, Aging, and Disease. *Immunity* **2018**, *48*, 380–395.e6. [CrossRef]

44. Lee, D.C.; Rizer, J.; Selenica, M.L.; Reid, P.; Kraft, C.; Johnson, A.; Blair, L.; Gordon, M.N.; Dickey, C.A.; Morgan, D. LPS-induced inflammation exacerbates phospho-tau pathology in rTg4510 mice. *J. Neuroinflamm.* **2010**, *7*, 56. [CrossRef]

45. Go, M.; Kou, J.; Lim, J.E.; Yang, J.; Fukuchi, K.I. Microglial response to LPS increases in wild-type mice during aging but diminishes in an Alzheimer's mouse model: Implication of TLR4 signaling in disease progression. *Biochem. Biophys. Res. Commun.* **2016**, *479*, 331–337. [CrossRef]

46. Herber, D.L.; Roth, L.M.; Wilson, D.; Wilson, N.; Mason, J.E.; Morgan, D.; Gordon, M.N. Time-dependent reduction in Abeta levels after intracranial LPS administration in APP transgenic mice. *Exp. Neurol.* **2004**, *190*, 245–253. [CrossRef]

47. Herber, D.L.; Mercer, M.; Roth, L.M.; Symmonds, K.; Maloney, J.; Wilson, N.; Freeman, M.J.; Morgan, D.; Gordon, M.N. Microglial activation is required for Abeta clearance after intracranial injection of lipopolysaccharide in APP transgenic mice. *J. Neuroimmune Pharmacol. Off. J. Soc. NeuroImmune Pharmacol.* **2007**, *2*, 222–231. [CrossRef]

48. Malm, T.M.; Magga, J.; Kuh, G.F.; Vatanen, T.; Koistinaho, M.; Koistinaho, J. Minocycline reduces engraftment and activation of bone marrow-derived cells but sustains their phagocytic activity in a mouse model of Alzheimer's disease. *Glia* **2008**, *56*, 1767–1779. [CrossRef]

49. Perry, V.H. The influence of systemic inflammation on inflammation in the brain: Implications for chronic neurodegenerative disease. *Brain Behav. Immun.* **2004**, *18*, 407–413. [CrossRef]

50. Lee, J.; Chan, S.L.; Mattson, M.P. Adverse effect of a presenilin-1 mutation in microglia results in enhanced nitric oxide and inflammatory cytokine responses to immune challenge in the brain. *Neuromol. Med.* **2002**, *2*, 29–45.

51. Brugg, B.; Dubreuil, Y.L.; Huber, G.; Wollman, E.E.; Delhaye-Bouchaud, N.; Mariani, J. Inflammatory processes induce beta-amyloid precursor protein changes in mouse brain. *Proc. Natl. Acad. Sci. USA* **1995**, *92*, 3032–3035. [CrossRef]

52. Sly, L.M.; Krzesicki, R.F.; Brashler, J.R.; Buhl, A.E.; McKinley, D.D.; Carter, D.B.; Chin, J.E. Endogenous brain cytokine mRNA and inflammatory responses to lipopolysaccharide are elevated in the Tg2576 transgenic mouse model of Alzheimer's disease. *Brain Res. Bull.* **2001**, *56*, 581–588. [CrossRef]

53. Wang, L.M.; Wu, Q.; Kirk, R.A.; Horn, K.P.; Ebada Salem, A.H.; Hoffman, J.M.; Yap, J.T.; Sonnen, J.A.; Towner, R.A.; Bozza, F.A.; et al. Lipopolysaccharide endotoxemia induces amyloid-beta and p-tau formation in the rat brain. *Am. J. Nucl. Med. Mol. Imaging* **2018**, *8*, 86–99. [PubMed]

54. Bossu, P.; Cutuli, D.; Palladino, I.; Caporali, P.; Angelucci, F.; Laricchiuta, D.; Gelfo, F.; De Bartolo, P.; Caltagirone, C.; Petrosini, L. A single intraperitoneal injection of endotoxin in rats induces long-lasting modifications in behavior and brain protein levels of TNF-alpha and IL-18. *J. Neuroinflamm.* **2012**, *9*, 101. [CrossRef]

55. Katafuchi, T.; Ifuku, M.; Mawatari, S.; Noda, M.; Miake, K.; Sugiyama, M.; Fujino, T. Effects of plasmalogens on systemic lipopolysaccharide-induced glial activation and beta-amyloid accumulation in adult mice. *Ann. N. Y. Acad. Sci.* **2012**, *1262*, 85–92. [CrossRef] [PubMed]

56. Behairi, N.; Belkhelfa, M.; Rafa, H.; Labsi, M.; Deghbar, N.; Bouzid, N.; Mesbah-Amroun, H.; Touil-Boukoffa, C. All-trans retinoic acid (ATRA) prevents lipopolysaccharide-induced neuroinflammation, amyloidogenesis and memory impairment in aged rats. *J. Neuroimmunol.* **2016**, *300*, 21–29. [CrossRef]

57. Marottoli, F.M.; Katsumata, Y.; Koster, K.P.; Thomas, R.; Fardo, D.W.; Tai, L.M. Peripheral Inflammation, Apolipoprotein E4, and Amyloid-beta Interact to Induce Cognitive and Cerebrovascular Dysfunction. *ASN Neuro* **2017**, *9*, 1759091417719201. [CrossRef]

58. McAlpine, F.E.; Lee, J.K.; Harms, A.S.; Ruhn, K.A.; Blurton-Jones, M.; Hong, J.; Das, P.; Golde, T.E.; LaFerla, F.M.; Oddo, S.; et al. Inhibition of soluble TNF signaling in a mouse model of Alzheimer's disease prevents pre-plaque amyloid-associated neuropathology. *Neurobiol. Dis.* **2009**, *34*, 163–177. [CrossRef]

59. Lee, J.W.; Lee, Y.K.; Yuk, D.Y.; Choi, D.Y.; Ban, S.B.; Oh, K.W.; Hong, J.T. Neuro-inflammation induced by lipopolysaccharide causes cognitive impairment through enhancement of beta-amyloid generation. *J. Neuroinflamm.* **2008**, *5*, 37. [CrossRef]

60. Thygesen, C.; Ilkjaer, L.; Kempf, S.J.; Hemdrup, A.L.; von Linstow, C.U.; Babcock, A.A.; Darvesh, S.; Larsen, M.R.; Finsen, B. Diverse Protein Profiles in CNS Myeloid Cells and CNS Tissue From Lipopolysaccharide- and Vehicle-Injected APPSWE/PS1DeltaE9 Transgenic Mice Implicate Cathepsin Z in Alzheimer's Disease. *Front. Cell. Neurosci.* **2018**, *12*, 397. [CrossRef]

61. Gao, H.M.; Zhang, F.; Zhou, H.; Kam, W.; Wilson, B.; Hong, J.S. Neuroinflammation and alpha-synuclein dysfunction potentiate each other, driving chronic progression of neurodegeneration in a mouse model of Parkinson's disease. *Environ. Health Perspect.* **2011**, *119*, 807–814. [CrossRef]

62. Hsiao, H.Y.; Chen, Y.C.; Chen, H.M.; Tu, P.H.; Chern, Y. A critical role of astrocyte-mediated nuclear factor-kappaB-dependent inflammation in Huntington's disease. *Hum. Mol. Genet.* **2013**, *22*, 1826–1842. [CrossRef]

63. Turner, R.C.; Naser, Z.J.; Lucke-Wold, B.P.; Logsdon, A.F.; Vangilder, R.L.; Matsumoto, R.R.; Huber, J.D.; Rosen, C.L. Single low-dose lipopolysaccharide preconditioning: Neuroprotective against axonal injury and modulates glial cells. *Neuroimmunol. Neuroinflamm.* **2017**, *4*, 6–15. [CrossRef]

64. Wang, D.; Liu, Y.; Zhao, Y.R.; Zhou, J.L. Low dose of lipopolysaccharide pretreatment can alleviate the inflammatory response in wound infection mouse model. *Chin. J. Traumatol. = Zhonghua Chuang Shang Za Zhi* **2016**, *19*, 193–198. [CrossRef]

65. Lee, S.W.; Park, H.J.; Im, W.; Kim, M.; Hong, S. Repeated immune activation with low-dose lipopolysaccharide attenuates the severity of Huntington's disease in R6/2 transgenic mice. *Anim. Cells Syst.* **2018**, *22*, 219–226. [CrossRef]

66. Cockerill, I.; Oliver, J.A.; Xu, H.; Fu, B.M.; Zhu, D. Blood-Brain Barrier Integrity and Clearance of Amyloid-beta from the BBB. *Adv. Exp. Med. Biol.* **2018**, *1097*, 261–278.

67. Erickson, M.A.; Hansen, K.; Banks, W.A. Inflammation-induced dysfunction of the low-density lipoprotein receptor-related protein-1 at the blood-brain barrier: Protection by the antioxidant N-acetylcysteine. *Brain Behav. Immun.* **2012**, *26*, 1085–1094. [CrossRef]

68. Kanekiyo, T.; Bu, G. The low-density lipoprotein receptor-related protein 1 and amyloid-beta clearance in Alzheimer's disease. *Front. Aging Neurosci.* **2014**, *6*, 93. [CrossRef]

69. Jaeger, L.B.; Dohgu, S.; Sultana, R.; Lynch, J.L.; Owen, J.B.; Erickson, M.A.; Shah, G.N.; Price, T.O.; Fleegal-Demotta, M.A.; Butterfield, D.A.; et al. Lipopolysaccharide alters the blood-brain barrier transport of amyloid beta protein: A mechanism for inflammation in the progression of Alzheimer's disease. *Brain Behav. Immun.* **2009**, *23*, 507–517. [CrossRef]

70. Erickson, M.A.; Hartvigson, P.E.; Morofuji, Y.; Owen, J.B.; Butterfield, D.A.; Banks, W.A. Lipopolysaccharide impairs amyloid beta efflux from brain: Altered vascular sequestration, cerebrospinal fluid reabsorption, peripheral clearance and transporter function at the blood-brain barrier. *J. Neuroinflamm.* **2012**, *9*, 150. [CrossRef]

71. Barton, S.M.; Janve, V.A.; McClure, R.; Anderson, A.; Matsubara, J.A.; Gore, J.C.; Pham, W. Lipopolysaccharide Induced Opening of the Blood Brain Barrier on Aging 5XFAD Mouse Model. *J. Alzheimer's Dis. JAD* **2018**. [CrossRef]

72. Ma, L.; Zhang, H.; Liu, N.; Wang, P.Q.; Guo, W.Z.; Fu, Q.; Jiao, L.B.; Ma, Y.Q.; Mi, W.D. TSPO ligand PK11195 alleviates neuroinflammation and beta-amyloid generation induced by systemic LPS administration. *Brain Res. Bull.* **2016**, *121*, 192–200. [CrossRef]

73. Roe, A.D.; Staup, M.A.; Serrats, J.; Sawchenko, P.E.; Rissman, R.A. Lipopolysaccharide-induced tau phosphorylation and kinase activity–modulation, but not mediation, by corticotropin-releasing factor receptors. *Eur. J. Neurosci.* **2011**, *34*, 448–456. [CrossRef]

74. Kitazawa, M.; Oddo, S.; Yamasaki, T.R.; Green, K.N.; LaFerla, F.M. Lipopolysaccharide-induced inflammation exacerbates tau pathology by a cyclin-dependent kinase 5-mediated pathway in a transgenic model of Alzheimer's disease. *J. Neurosci. Off. J. Soc. Neurosci.* **2005**, *25*, 8843–8853. [CrossRef]

75. Maher, A.; El-Sayed, N.S.; Breitinger, H.G.; Gad, M.Z. Overexpression of NMDAR2B in an inflammatory model of Alzheimer's disease: Modulation by NOS inhibitors. *Brain Res. Bull.* **2014**, *109*, 109–116. [CrossRef]

76. Lykhmus, O.; Voytenko, L.; Koval, L.; Mykhalskiy, S.; Kholin, V.; Peschana, K.; Zouridakis, M.; Tzartos, S.; Komisarenko, S.; Skok, M. alpha7 Nicotinic acetylcholine receptor-specific antibody induces inflammation and amyloid beta42 accumulation in the mouse brain to impair memory. *PLoS ONE* **2015**, *10*, e0122706. [CrossRef]

77. Lykhmus, O.; Mishra, N.; Koval, L.; Kalashnyk, O.; Gergalova, G.; Uspenska, K.; Komisarenko, S.; Soreq, H.; Skok, M. Molecular Mechanisms Regulating LPS-Induced Inflammation in the Brain. *Front. Mol. Neurosci.* **2016**, *9*, 19. [CrossRef]

78. Poewe, W.; Seppi, K.; Tanner, C.M.; Halliday, G.M.; Brundin, P.; Volkmann, J.; Schrag, A.E.; Lang, A.E. Parkinson disease. *Nat. Rev. Dis. Prim.* **2017**, *3*, 17013. [CrossRef]

79. Javoy-Agid, F.; Agid, Y. Is the mesocortical dopaminergic system involved in Parkinson disease? *Neurology* **1980**, *30*, 1326–1330. [CrossRef]

80. Bugiani, O.; Perdelli, F.; Salvarani, S.; Leonardi, A.; Mancardi, G.L. Loss of striatal neurons in Parkinson's disease: A cytometric study. *Eur. Neurol.* **1980**, *19*, 339–344. [CrossRef]

81. Parkinson, J. An essay on the shaking palsy. *J. Neuropsychiatry Clin. Neurosci.* **2002**, *14*, 223–236. [CrossRef] [PubMed]

82. Dauer, W.; Przedborski, S. Parkinson's disease: Mechanisms and models. *Neuron* **2003**, *39*, 889–909. [CrossRef]

83. Cookson, M.R. alpha-Synuclein and neuronal cell death. *Mol. Neurodegener.* **2009**, *4*, 9. [CrossRef] [PubMed]

84. Zhang, W.; Wang, T.; Pei, Z.; Miller, D.S.; Wu, X.; Block, M.L.; Wilson, B.; Zhang, W.; Zhou, Y.; Hong, J.S.; et al. Aggregated alpha-synuclein activates microglia: A process leading to disease progression in Parkinson's disease. *FASEB J. Off. Publ. Fed. Am. Soc. Exp. Biol.* **2005**, *19*, 533–542.

85. McGeer, P.L.; Itagaki, S.; Boyes, B.E.; McGeer, E.G. Reactive microglia are positive for HLA-DR in the substantia nigra of Parkinson's and Alzheimer's disease brains. *Neurology* **1988**, *38*, 1285–1291. [CrossRef]

86. Yang, W.; Yu, S. Synucleinopathies: Common features and hippocampal manifestations. *Cell. Mol. Life Sci.* **2017**, *74*, 1485–1501. [CrossRef]

87. Castano, A.; Herrera, A.J.; Cano, J.; Machado, A. Lipopolysaccharide intranigral injection induces inflammatory reaction and damage in nigrostriatal dopaminergic system. *J. Neurochem.* **1998**, *70*, 1584–1592. [CrossRef]

88. Herrera, A.J.; Castano, A.; Venero, J.L.; Cano, J.; Machado, A. The single intranigral injection of LPS as a new model for studying the selective effects of inflammatory reactions on dopaminergic system. *Neurobiol. Dis.* **2000**, *7*, 429–447. [CrossRef]

89. Kim, W.G.; Mohney, R.P.; Wilson, B.; Jeohn, G.H.; Liu, B.; Hong, J.S. Regional difference in susceptibility to lipopolysaccharide-induced neurotoxicity in the rat brain: Role of microglia. *J. Neurosci. Off. J. Soc. Neurosci.* **2000**, *20*, 6309–6316. [CrossRef]

90. Gao, H.M.; Jiang, J.; Wilson, B.; Zhang, W.; Hong, J.S.; Liu, B. Microglial activation-mediated delayed and progressive degeneration of rat nigral dopaminergic neurons: Relevance to Parkinson's disease. *J. Neurochem.* **2002**, *81*, 1285–1297. [CrossRef]

91. Iravani, M.M.; Leung, C.C.; Sadeghian, M.; Haddon, C.O.; Rose, S.; Jenner, P. The acute and the long-term effects of nigral lipopolysaccharide administration on dopaminergic dysfunction and glial cell activation. *Eur. J. Neurosci.* **2005**, *22*, 317–330. [CrossRef]

92. Lu, X.; Bing, G.; Hagg, T. Naloxone prevents microglia-induced degeneration of dopaminergic substantia nigra neurons in adult rats. *Neuroscience* **2000**, *97*, 285–291. [CrossRef]

93. Castano, A.; Herrera, A.J.; Cano, J.; Machado, A. The degenerative effect of a single intranigral injection of LPS on the dopaminergic system is prevented by dexamethasone, and not mimicked by rh-TNF-alpha, IL-1beta and IFN-gamma. *J. Neurochem.* **2002**, *81*, 150–157. [CrossRef]

94. Mangano, E.N.; Hayley, S. Inflammatory priming of the substantia nigra influences the impact of later paraquat exposure: Neuroimmune sensitization of neurodegeneration. *Neurobiol. Aging* **2009**, *30*, 1361–1378. [CrossRef]

95. Couch, Y.; Alvarez-Erviti, L.; Sibson, N.R.; Wood, M.J.; Anthony, D.C. The acute inflammatory response to intranigral alpha-synuclein differs significantly from intranigral lipopolysaccharide and is exacerbated by peripheral inflammation. *J. Neuroinflamm.* **2011**, *8*, 166. [CrossRef]

96. Burguillos, M.A.; Hajji, N.; Englund, E.; Persson, A.; Cenci, A.M.; Machado, A.; Cano, J.; Joseph, B.; Venero, J.L. Apoptosis-inducing factor mediates dopaminergic cell death in response to LPS-induced inflammatory stimulus: Evidence in Parkinson's disease patients. *Neurobiol. Dis.* **2011**, *41*, 177–188. [CrossRef]

97. Bardou, I.; Kaercher, R.M.; Brothers, H.M.; Hopp, S.C.; Royer, S.; Wenk, G.L. Age and duration of inflammatory environment differentially affect the neuroimmune response and catecholaminergic neurons in the midbrain and brainstem. *Neurobiol. Aging* **2014**, *35*, 1065–1073. [CrossRef]

98. Flores-Martinez, Y.M.; Fernandez-Parrilla, M.A.; Ayala-Davila, J.; Reyes-Corona, D.; Blanco-Alvarez, V.M.; Soto-Rojas, L.O.; Luna-Herrera, C.; Gonzalez-Barrios, J.A.; Leon-Chavez, B.A.; Gutierrez-Castillo, M.E.; et al. Acute Neuroinflammatory Response in the Substantia Nigra Pars Compacta of Rats after a Local Injection of Lipopolysaccharide. *J. Immunol. Res.* **2018**, *2018*, 1838921. [CrossRef]

99. Sharma, N.; Kapoor, M.; Nehru, B. Apocyanin, NADPH oxidase inhibitor prevents lipopolysaccharide induced alpha-synuclein aggregation and ameliorates motor function deficits in rats: Possible role of biochemical and inflammatory alterations. *Behav. Brain Res.* **2016**, *296*, 177–190. [CrossRef]

100. Sharma, N.; Nehru, B. Apocyanin, a Microglial NADPH Oxidase Inhibitor Prevents Dopaminergic Neuronal Degeneration in Lipopolysaccharide-Induced Parkinson's Disease Model. *Mol. Neurobiol.* **2016**, *53*, 3326–3337. [CrossRef]

101. Sharma, N.; Sharma, S.; Nehru, B. Curcumin protects dopaminergic neurons against inflammation-mediated damage and improves motor dysfunction induced by single intranigral lipopolysaccharide injection. *Inflammopharmacology* **2017**, *25*, 351–368. [CrossRef]

102. Gu, C.; Hu, Q.; Wu, J.; Mu, C.; Ren, H.; Liu, C.F.; Wang, G. P7C3 Inhibits LPS-Induced Microglial Activation to Protect Dopaminergic Neurons Against Inflammatory Factor-Induced Cell Death in vitro and in vivo. *Front. Cell. Neurosci.* **2018**, *12*, 400. [CrossRef]

103. Xu, W.; Zheng, D.; Liu, Y.; Li, J.; Yang, L.; Shang, X. Glaucocalyxin B Alleviates Lipopolysaccharide-Induced Parkinson's Disease by Inhibiting TLR/NF-kappaB and Activating Nrf2/HO-1 Pathway. *Cell. Physiol. Biochem. Int. J. Exp. Cell. Physiol. Biochem. Pharmacol.* **2017**, *44*, 2091–2104. [CrossRef]

104. Moon, H.E.; Paek, S.H. Mitochondrial Dysfunction in Parkinson's Disease. *Exp. Neurobiol.* **2015**, *24*, 103–116. [CrossRef]

105. Park, J.S.; Davis, R.L.; Sue, C.M. Mitochondrial Dysfunction in Parkinson's Disease: New Mechanistic Insights and Therapeutic Perspectives. *Curr. Neurol. Neurosci. Rep.* **2018**, *18*, 21. [CrossRef]

106. Hunter, R.; Ojha, U.; Bhurtel, S.; Bing, G.; Choi, D.Y. Lipopolysaccharide-induced functional and structural injury of the mitochondria in the nigrostriatal pathway. *Neurosci. Res.* **2017**, *114*, 62–69. [CrossRef]

107. Hunter, R.L.; Choi, D.Y.; Ross, S.A.; Bing, G. Protective properties afforded by pioglitazone against intrastriatal LPS in Sprague-Dawley rats. *Neurosci. Lett.* **2008**, *432*, 198–201. [CrossRef]

108. Choi, D.Y.; Liu, M.; Hunter, R.L.; Cass, W.A.; Pandya, J.D.; Sullivan, P.G.; Shin, E.J.; Kim, H.C.; Gash, D.M.; Bing, G. Striatal neuroinflammation promotes Parkinsonism in rats. *PLoS ONE* **2009**, *4*, e5482. [CrossRef]

109. Ruano, D.; Revilla, E.; Gavilan, M.P.; Vizuete, M.L.; Pintado, C.; Vitorica, J.; Castano, A. Role of p38 and inducible nitric oxide synthase in the in vivo dopaminergic cells' degeneration induced by inflammatory processes after lipopolysaccharide injection. *Neuroscience* **2006**, *140*, 1157–1168. [CrossRef]

110. Iravani, M.M.; Kashefi, K.; Mander, P.; Rose, S.; Jenner, P. Involvement of inducible nitric oxide synthase in inflammation-induced dopaminergic neurodegeneration. *Neuroscience* **2002**, *110*, 49–58. [CrossRef]

111. Hunter, R.L.; Cheng, B.; Choi, D.Y.; Liu, M.; Liu, S.; Cass, W.A.; Bing, G. Intrastriatal lipopolysaccharide injection induces parkinsonism in C57/B6 mice. *J. Neurosci. Res.* **2009**, *87*, 1913–1921. [CrossRef]

112. Hunter, R.L.; Dragicevic, N.; Seifert, K.; Choi, D.Y.; Liu, M.; Kim, H.C.; Cass, W.A.; Sullivan, P.G.; Bing, G. Inflammation induces mitochondrial dysfunction and dopaminergic neurodegeneration in the nigrostriatal system. *J. Neurochem.* **2007**, *100*, 1375–1386. [CrossRef]

113. Zhang, J.; Stanton, D.M.; Nguyen, X.V.; Liu, M.; Zhang, Z.; Gash, D.; Bing, G. Intrapallidal lipopolysaccharide injection increases iron and ferritin levels in glia of the rat substantia nigra and induces locomotor deficits. *Neuroscience* **2005**, *135*, 829–838. [CrossRef]

114. Zhang, Z.; Zhang, K.; Du, X.; Li, Y. Neuroprotection of desferrioxamine in lipopolysaccharide-induced nigrostriatal dopamine neuron degeneration. *Mol. Med. Rep.* **2012**, *5*, 562–566. [CrossRef]

115. Kim, C.Y.; Alcalay, R.N. Genetic Forms of Parkinson's Disease. *Semin. Neurol.* **2017**, *37*, 135–146. [CrossRef]

116. Sai, Y.; Zou, Z.; Peng, K.; Dong, Z. The Parkinson's disease-related genes act in mitochondrial homeostasis. *Neurosci. Biobehav. Rev.* **2012**, *36*, 2034–2043. [CrossRef]

117. International Parkinson Disease Genomics Consortium; Nalls, M.A.; Plagnol, V.; Hernandez, D.G.; Sharma, M.; Sheerin, U.M.; Saad, M.; Simon-Sanchez, J.; Schulte, C.; Lesage, S.; et al. Imputation of sequence variants for identification of genetic risks for Parkinson's disease: A meta-analysis of genome-wide association studies. *Lancet* **2011**, *377*, 641–649.

118. Mata, I.F.; Checkoway, H.; Hutter, C.M.; Samii, A.; Roberts, J.W.; Kim, H.M.; Agarwal, P.; Alvarez, V.; Ribacoba, R.; Pastor, P.; et al. Common variation in the LRRK2 gene is a risk factor for Parkinson's disease. *Mov. Disord. Off. J. Mov. Disord. Soc.* **2012**, *27*, 1822–1825. [CrossRef]

119. Moehle, M.S.; Webber, P.J.; Tse, T.; Sukar, N.; Standaert, D.G.; DeSilva, T.M.; Cowell, R.M.; West, A.B. LRRK2 inhibition attenuates microglial inflammatory responses. *J. Neurosci. Off. J. Soc. Neurosci.* **2012**, *32*, 1602–1611. [CrossRef]

120. Daher, J.P.; Volpicelli-Daley, L.A.; Blackburn, J.P.; Moehle, M.S.; West, A.B. Abrogation of alpha-synuclein-mediated dopaminergic neurodegeneration in LRRK2-deficient rats. *Proc. Natl. Acad. Sci. USA* **2014**, *111*, 9289–9294. [CrossRef]

121. Ma, B.; Xu, L.; Pan, X.; Sun, L.; Ding, J.; Xie, C.; Koliatsos, V.E.; Cai, H. LRRK2 modulates microglial activity through regulation of chemokine (C-X3-C) receptor 1 -mediated signalling pathways. *Hum. Mol. Genet.* **2016**, *25*, 3515–3523. [CrossRef]

122. Van Duijn, C.M.; Dekker, M.C.; Bonifati, V.; Galjaard, R.J.; Houwing-Duistermaat, J.J.; Snijders, P.J.; Testers, L.; Breedveld, G.J.; Horstink, M.; Sandkuijl, L.A.; et al. Park7, a novel locus for autosomal recessive early-onset parkinsonism, on chromosome 1p36. *Am. J. Hum. Genet.* **2001**, *69*, 629–634. [CrossRef]

123. Bonifati, V.; Rizzu, P.; van Baren, M.J.; Schaap, O.; Breedveld, G.J.; Krieger, E.; Dekker, M.C.; Squitieri, F.; Ibanez, P.; Joosse, M.; et al. Mutations in the DJ-1 gene associated with autosomal recessive early-onset parkinsonism. *Science* **2003**, *299*, 256–259. [CrossRef] [PubMed]

124. Chien, C.H.; Lee, M.J.; Liou, H.C.; Liou, H.H.; Fu, W.M. Microglia-Derived Cytokines/Chemokines Are Involved in the Enhancement of LPS-Induced Loss of Nigrostriatal Dopaminergic Neurons in DJ-1 Knockout Mice. *PLoS ONE* **2016**, *11*, e0151569. [CrossRef] [PubMed]

125. Skelly, D.T.; Hennessy, E.; Dansereau, M.A.; Cunningham, C. A systematic analysis of the peripheral and CNS effects of systemic LPS, IL-1beta, [corrected] TNF-alpha and IL-6 challenges in C57BL/6 mice. *PLoS ONE* **2013**, *8*, e69123. [CrossRef]

126. Cazareth, J.; Guyon, A.; Heurteaux, C.; Chabry, J.; Petit-Paitel, A. Molecular and cellular neuroinflammatory status of mouse brain after systemic lipopolysaccharide challenge: Importance of CCR2/CCL2 signaling. *J. Neuroinflamm.* **2014**, *11*, 132. [CrossRef]

127. Qin, L.; Wu, X.; Block, M.L.; Liu, Y.; Breese, G.R.; Hong, J.S.; Knapp, D.J.; Crews, F.T. Systemic LPS causes chronic neuroinflammation and progressive neurodegeneration. *Glia* **2007**, *55*, 453–462. [CrossRef] [PubMed]

128. Beier, E.E.; Neal, M.; Alam, G.; Edler, M.; Wu, L.J.; Richardson, J.R. Alternative microglial activation is associated with cessation of progressive dopamine neuron loss in mice systemically administered lipopolysaccharide. *Neurobiol. Dis.* **2017**, *108*, 115–127. [CrossRef] [PubMed]

129. Liu, Y.; Qin, L.; Wilson, B.; Wu, X.; Qian, L.; Granholm, A.C.; Crews, F.T.; Hong, J.S. Endotoxin induces a delayed loss of TH-IR neurons in substantia nigra and motor behavioral deficits. *Neurotoxicology* **2008**, *29*, 864–870. [CrossRef] [PubMed]

130. Jangula, A.; Murphy, E.J. Lipopolysaccharide-induced blood brain barrier permeability is enhanced by alpha-synuclein expression. *Neurosci. Lett.* **2013**, *551*, 23–27. [CrossRef]

131. Zheng, H.F.; Yang, Y.P.; Hu, L.F.; Wang, M.X.; Wang, F.; Cao, L.D.; Li, D.; Mao, C.J.; Xiong, K.P.; Wang, J.D.; et al. Autophagic impairment contributes to systemic inflammation-induced dopaminergic neuron loss in the midbrain. *PLoS ONE* **2013**, *8*, e70472. [CrossRef] [PubMed]

132. Czapski, G.A.; Cakala, M.; Chalimoniuk, M.; Gajkowska, B.; Strosznajder, J.B. Role of nitric oxide in the brain during lipopolysaccharide-evoked systemic inflammation. *J. Neurosci. Res.* **2007**, *85*, 1694–1703. [CrossRef]

133. Qin, L.; Liu, Y.; Hong, J.S.; Crews, F.T. NADPH oxidase and aging drive microglial activation, oxidative stress, and dopaminergic neurodegeneration following systemic LPS administration. *Glia* **2013**, *61*, 855–868. [CrossRef] [PubMed]

134. L'Episcopo, F.; Tirolo, C.; Testa, N.; Caniglia, S.; Morale, M.C.; Impagnatiello, F.; Marchetti, B. Switching the microglial harmful phenotype promotes lifelong restoration of subtantia nigra dopaminergic neurons from inflammatory neurodegeneration in aged mice. *Rejuvenation Res.* **2011**, *14*, 411–424. [CrossRef]

135. Zhao, Y.F.; Qiong, Z.; Zhang, J.F.; Lou, Z.Y.; Zu, H.B.; Wang, Z.G.; Zeng, W.C.; Kai, Y.; Xiao, B.G. The Synergy of Aging and LPS Exposure in a Mouse Model of Parkinson's Disease. *Aging Dis.* **2018**, *9*, 785–797. [CrossRef]

136. Hudson, A.J. Amyotrophic lateral sclerosis and its association with dementia, parkinsonism and other neurological disorders: A review. *Brain J. Neurol.* **1981**, *104*, 217–247. [CrossRef]

137. Charles, T.; Swash, M. Amyotrophic lateral sclerosis: Current understanding. *J. Neurosci. Nurs. J. Am. Assoc. Neurosci. Nurses* **2001**, *33*, 245–253. [CrossRef]

138. Leigh, P.N.; Whitwell, H.; Garofalo, O.; Buller, J.; Swash, M.; Martin, J.E.; Gallo, J.M.; Weller, R.O.; Anderton, B.H. Ubiquitin-immunoreactive intraneuronal inclusions in amyotrophic lateral sclerosis. Morphology, distribution, and specificity. *Brain J. Neurol.* **1991**, *114 Pt 2*, 775–788. [CrossRef]

139. Alexianu, M.E.; Kozovska, M.; Appel, S.H. Immune reactivity in a mouse model of familial ALS correlates with disease progression. *Neurology* **2001**, *57*, 1282–1289. [CrossRef] [PubMed]

140. Graves, M.C.; Fiala, M.; Dinglasan, L.A.; Liu, N.Q.; Sayre, J.; Chiappelli, F.; van Kooten, C.; Vinters, H.V. Inflammation in amyotrophic lateral sclerosis spinal cord and brain is mediated by activated macrophages, mast cells and T cells. *Amyotroph. Lateral Scler. Other Mot. Neuron Disord.* **2004**, *5*, 213–219. [CrossRef]

141. Gurney, M.E.; Pu, H.; Chiu, A.Y.; Dal Canto, M.C.; Polchow, C.Y.; Alexander, D.D.; Caliendo, J.; Hentati, A.; Kwon, Y.W.; Deng, H.X.; et al. Motor neuron degeneration in mice that express a human Cu,Zn superoxide dismutase mutation. *Science* **1994**, *264*, 1772–1775. [CrossRef] [PubMed]

142. Valente, T.; Straccia, M.; Gresa-Arribas, N.; Dentesano, G.; Tusell, J.M.; Serratosa, J.; Mancera, P.; Sola, C.; Saura, J. CCAAT/enhancer binding protein delta regulates glial proinflammatory gene expression. *Neurobiol. Aging* **2013**, *34*, 2110–2124. [CrossRef] [PubMed]

143. Ohgomori, T.; Yamada, J.; Takeuchi, H.; Kadomatsu, K.; Jinno, S. Comparative morphometric analysis of microglia in the spinal cord of SOD1(G93A) transgenic mouse model of amyotrophic lateral sclerosis. *Eur. J. Neurosci.* **2016**, *43*, 1340–1351. [CrossRef]

144. Byrne, S.; Walsh, C.; Lynch, C.; Bede, P.; Elamin, M.; Kenna, K.; McLaughlin, R.; Hardiman, O. Rate of familial amyotrophic lateral sclerosis: A systematic review and meta-analysis. *J. Neurol. Neurosurg. Psychiatry* **2011**, *82*, 623–627. [CrossRef]

145. Correia, A.S.; Patel, P.; Dutta, K.; Julien, J.P. Inflammation Induces TDP-43 Mislocalization and Aggregation. *PLoS ONE* **2015**, *10*, e0140248. [CrossRef] [PubMed]

146. Lyon, M.S.; Wosiski-Kuhn, M.; Gillespie, R.; Caress, J.; Milligan, C. Inflammation, Immunity, and amyotrophic lateral sclerosis: I. Etiology and pathology. *Muscle Nerve* **2019**, *59*, 10–22. [CrossRef] [PubMed]

147. Liu, J.; Wang, F. Role of Neuroinflammation in Amyotrophic Lateral Sclerosis: Cellular Mechanisms and Therapeutic Implications. *Front. Immunol.* **2017**, *8*, 1005. [CrossRef]

148. Huang, W.J.; Chen, W.W.; Zhang, X. Huntington's disease: Molecular basis of pathology and status of current therapeutic approaches. *Exp. Ther. Med.* **2016**, *12*, 1951–1956. [CrossRef]

149. Folstein, S.E. The psychopathology of Huntington's disease. *Res. Publ. Assoc. Res. Nerv. Ment. Dis.* **1991**, *69*, 181–191. [CrossRef]

150. MacDonald, M.E.; Ambrose, C.M.; Duyao, M.P.; Myers, R.H.; Lin, C.; Srinidhi, L.; Barnes, G.; Taylor, S.A.; James, M.; Groot, N.; et al. A novel gene containing a trinucleotide repeat that is expanded and unstable on Huntington's disease chromosomes. *Cell* **1993**, *72*, 971–983. [CrossRef]

151. Labbadia, J.; Morimoto, R.I. Huntington's disease: Underlying molecular mechanisms and emerging concepts. *Trends Biochem. Sci.* **2013**, *38*, 378–385. [CrossRef]

152. Gunawardena, S.; Goldstein, L.S. Polyglutamine diseases and transport problems: Deadly traffic jams on neuronal highways. *Arch. Neurol.* **2005**, *62*, 46–51. [CrossRef]

153. Li, S.H.; Li, X.J. Huntingtin-protein interactions and the pathogenesis of Huntington's disease. *Trends Genet. TIG* **2004**, *20*, 146–154. [CrossRef]

154. Sugars, K.L.; Rubinsztein, D.C. Transcriptional abnormalities in Huntington disease. *Trends Genet. TIG* **2003**, *19*, 233–238. [CrossRef]

155. Silvestroni, A.; Faull, R.L.; Strand, A.D.; Moller, T. Distinct neuroinflammatory profile in post-mortem human Huntington's disease. *Neuroreport* **2009**, *20*, 1098–1103. [CrossRef]

156. Bjorkqvist, M.; Wild, E.J.; Thiele, J.; Silvestroni, A.; Andre, R.; Lahiri, N.; Raibon, E.; Lee, R.V.; Benn, C.L.; Soulet, D.; et al. A novel pathogenic pathway of immune activation detectable before clinical onset in Huntington's disease. *J. Exp. Med.* **2008**, *205*, 1869–1877. [CrossRef]

157. Dalrymple, A.; Wild, E.J.; Joubert, R.; Sathasivam, K.; Bjorkqvist, M.; Petersen, A.; Jackson, G.S.; Isaacs, J.D.; Kristiansen, M.; Bates, G.P.; et al. Proteomic profiling of plasma in Huntington's disease reveals neuroinflammatory activation and biomarker candidates. *J. Proteome Res.* **2007**, *6*, 2833–2840. [CrossRef]

158. Bouchard, J.; Truong, J.; Bouchard, K.; Dunkelberger, D.; Desrayaud, S.; Moussaoui, S.; Tabrizi, S.J.; Stella, N.; Muchowski, P.J. Cannabinoid receptor 2 signaling in peripheral immune cells modulates disease onset and severity in mouse models of Huntington's disease. *J. Neurosci. Off. J. Soc. Neurosci.* **2012**, *32*, 18259–18268. [CrossRef]

159. Franciosi, S.; Ryu, J.K.; Shim, Y.; Hill, A.; Connolly, C.; Hayden, M.R.; McLarnon, J.G.; Leavitt, B.R. Age-dependent neurovascular abnormalities and altered microglial morphology in the YAC128 mouse model of Huntington disease. *Neurobiol. Dis.* **2012**, *45*, 438–449. [CrossRef]

160. Renoir, T.; Pang, T.Y.; Shikano, Y.; Li, S.; Hannan, A.J. Loss of the Sexually Dimorphic Neuro-Inflammatory Response in a Transgenic Mouse Model of Huntington's Disease. *J. Huntington's Dis.* **2015**, *4*, 297–303. [CrossRef]

161. Kempuraj, D.; Thangavel, R.; Natteru, P.A.; Selvakumar, G.P.; Saeed, D.; Zahoor, H.; Zaheer, S.; Iyer, S.S.; Zaheer, A. Neuroinflammation Induces Neurodegeneration. *J. Neurol. Neurosurg. Spine* **2016**, *1*, 1003.

162. Glass, C.K.; Saijo, K.; Winner, B.; Marchetto, M.C.; Gage, F.H. Mechanisms underlying inflammation in neurodegeneration. *Cell* **2010**, *140*, 918–934. [CrossRef]

163. Liu, M.; Bing, G. Lipopolysaccharide animal models for Parkinson's disease. *Parkinson's Dis.* **2011**, *2011*, 327089. [CrossRef]

164. Okun, E.; Griffioen, K.J.; Lathia, J.D.; Tang, S.C.; Mattson, M.P.; Arumugam, T.V. Toll-like receptors in neurodegeneration. *Brain Res. Rev.* **2009**, *59*, 278–292. [CrossRef]

165. Schmalz, G.; Krifka, S.; Schweikl, H. Toll-like receptors, LPS, and dental monomers. *Adv. Dent. Res.* **2011**, *23*, 302–306. [CrossRef]

166. Madera-Salcedo, I.K.; Cruz, S.L.; Gonzalez-Espinosa, C. Morphine prevents lipopolysaccharide-induced TNF secretion in mast cells blocking IkappaB kinase activation and SNAP-23 phosphorylation: Correlation with the formation of a beta-arrestin/TRAF6 complex. *J. Immunol.* **2013**, *191*, 3400–3409. [CrossRef]

167. Lee, S.J.; Lee, S. Toll-like receptors and inflammation in the CNS. *Curr. Drug Targets Inflamm. Allergy* **2002**, *1*, 181–191.

168. Tak, P.P.; Firestein, G.S. NF-kappaB: A key role in inflammatory diseases. *J. Clin. Investig.* **2001**, *107*, 7–11. [CrossRef]

169. Bachstetter, A.D.; Xing, B.; de Almeida, L.; Dimayuga, E.R.; Watterson, D.M.; Van Eldik, L.J. Microglial p38alpha MAPK is a key regulator of proinflammatory cytokine up-regulation induced by toll-like receptor (TLR) ligands or beta-amyloid (Abeta). *J. Neuroinflamm.* **2011**, *8*, 79. [CrossRef]

170. Francois, A.; Terro, F.; Janet, T.; Rioux Bilan, A.; Paccalin, M.; Page, G. Involvement of interleukin-1beta in the autophagic process of microglia: Relevance to Alzheimer's disease. *J. Neuroinflamm.* **2013**, *10*, 151. [CrossRef]

171. Wu, D.; Zhang, X.; Zhao, M.; Zhou, A.L. The role of the TLR4/NF-kappaB signaling pathway in Abeta accumulation in primary hippocampal neurons. *Sheng Li Xue Bao Acta Physiol. Sin.* **2015**, *67*, 319–328.

172. McMillian, M.; Kong, L.Y.; Sawin, S.M.; Wilson, B.; Das, K.; Hudson, P.; Hong, J.S.; Bing, G. Selective killing of cholinergic neurons by microglial activation in basal forebrain mixed neuronal/glial cultures. *Biochem. Biophys. Res. Commun.* **1995**, *215*, 572–577. [CrossRef]

173. Yang, X.; Zhang, J.D.; Duan, L.; Xiong, H.G.; Jiang, Y.P.; Liang, H.C. Microglia activation mediated by toll-like receptor-4 impairs brain white matter tracts in rats. *J. Biomed. Res.* **2018**, *32*, 136–144.

174. Hoozemans, J.J.; Veerhuis, R.; Janssen, I.; van Elk, E.J.; Rozemuller, A.J.; Eikelenboom, P. The role of cyclo-oxygenase 1 and 2 activity in prostaglandin E(2) secretion by cultured human adult microglia: Implications for Alzheimer's disease. *Brain Res.* **2002**, *951*, 218–226. [CrossRef]

175. Shi, J.; Johansson, J.; Woodling, N.S.; Wang, Q.; Montine, T.J.; Andreasson, K. The prostaglandin E2 E-prostanoid 4 receptor exerts anti-inflammatory effects in brain innate immunity. *J. Immunol.* **2010**, *184*, 7207–7218. [CrossRef]

176. Huang, Y.Y.; Zhang, Q.; Zhang, J.N.; Zhang, Y.N.; Gu, L.; Yang, H.M.; Xia, N.; Wang, X.M.; Zhang, H. Triptolide up-regulates metabotropic glutamate receptor 5 to inhibit microglia activation in the lipopolysaccharide-induced model of Parkinson's disease. *Brain Behav. Immun.* **2018**, *71*, 93–107. [CrossRef]

177. Barger, S.W.; Chavis, J.A.; Drew, P.D. Dehydroepiandrosterone inhibits microglial nitric oxide production in a stimulus-specific manner. *J. Neurosci. Res.* **2000**, *62*, 503–509. [CrossRef]

178. Dewil, M.; dela Cruz, V.F.; Van Den Bosch, L.; Robberecht, W. Inhibition of p38 mitogen activated protein kinase activation and mutant SOD1(G93A)-induced motor neuron death. *Neurobiol. Dis.* **2007**, *26*, 332–341. [CrossRef]

179. Jeohn, G.H.; Cooper, C.L.; Wilson, B.; Chang, R.C.; Jang, K.J.; Kim, H.C.; Liu, B.; Hong, J.S. p38 MAP kinase is involved in lipopolysaccharide-induced dopaminergic neuronal cell death in rat mesencephalic neuron-glia cultures. *Ann. N. Y. Acad. Sci.* **2002**, *962*, 332–346. [CrossRef]

180. Fiebich, B.L.; Butcher, R.D.; Gebicke-Haerter, P.J. Protein kinase C-mediated regulation of inducible nitric oxide synthase expression in cultured microglial cells. *J. Neuroimmunol.* **1998**, *92*, 170–178. [CrossRef]

181. Akundi, R.S.; Candelario-Jalil, E.; Hess, S.; Hull, M.; Lieb, K.; Gebicke-Haerter, P.J.; Fiebich, B.L. Signal transduction pathways regulating cyclooxygenase-2 in lipopolysaccharide-activated primary rat microglia. *Glia* **2005**, *51*, 199–208. [CrossRef]

182. Bauer, M.K.; Lieb, K.; Schulze-Osthoff, K.; Berger, M.; Gebicke-Haerter, P.J.; Bauer, J.; Fiebich, B.L. Expression and regulation of cyclooxygenase-2 in rat microglia. *Eur. J. Biochem.* **1997**, *243*, 726–731. [CrossRef]

183. Yousif, N.M.; de Oliveira, A.C.P.; Brioschi, S.; Huell, M.; Biber, K.; Fiebich, B.L. Activation of EP2 receptor suppresses poly(I: C) and LPS-mediated inflammation in primary microglia and organotypic hippocampal slice cultures: Contributing role for MAPKs. *Glia* **2018**, *66*, 708–724. [CrossRef]

184. Bronstein, D.M.; Perez-Otano, I.; Sun, V.; Mullis Sawin, S.B.; Chan, J.; Wu, G.C.; Hudson, P.M.; Kong, L.Y.; Hong, J.S.; McMillian, M.K. Glia-dependent neurotoxicity and neuroprotection in mesencephalic cultures. *Brain Res.* **1995**, *704*, 112–116. [CrossRef]

185. Gayle, D.A.; Ling, Z.; Tong, C.; Landers, T.; Lipton, J.W.; Carvey, P.M. Lipopolysaccharide (LPS)-induced dopamine cell loss in culture: Roles of tumor necrosis factor-alpha, interleukin-1beta, and nitric oxide. *Brain Res. Dev. Brain Res.* **2002**, *133*, 27–35. [CrossRef]

186. Zhu, Y.; Chen, X.; Liu, Z.; Peng, Y.P.; Qiu, Y.H. Interleukin-10 Protection against Lipopolysaccharide-Induced Neuro-Inflammation and Neurotoxicity in Ventral Mesencephalic Cultures. *Int. J. Mol. Sci.* **2016**, *17*, 25. [CrossRef]

187. Zhang, F.; Qian, L.; Flood, P.M.; Shi, J.S.; Hong, J.S.; Gao, H.M. Inhibition of IkappaB kinase-beta protects dopamine neurons against lipopolysaccharide-induced neurotoxicity. *J. Pharmacol. Exp. Ther.* **2010**, *333*, 822–833. [CrossRef]

188. Bhat, N.R.; Zhang, P.; Lee, J.C.; Hogan, E.L. Extracellular signal-regulated kinase and p38 subgroups of mitogen-activated protein kinases regulate inducible nitric oxide synthase and tumor necrosis factor-alpha gene expression in endotoxin-stimulated primary glial cultures. *J. Neurosci. Off. J. Soc. Neurosci.* **1998**, *18*, 1633–1641. [CrossRef]

189. Cuenda, A.; Rousseau, S. p38 MAP-kinases pathway regulation, function and role in human diseases. *Biochim. Biophys. Acta* **2007**, *1773*, 1358–1375. [CrossRef]

190. Gordon, R.; Singh, N.; Lawana, V.; Ghosh, A.; Harischandra, D.S.; Jin, H.; Hogan, C.; Sarkar, S.; Rokad, D.; Panicker, N.; et al. Protein kinase Cdelta upregulation in microglia drives neuroinflammatory responses and dopaminergic neurodegeneration in experimental models of Parkinson's disease. *Neurobiol. Dis.* **2016**, *93*, 96–114. [CrossRef]

191. Bhatia, H.S.; Roelofs, N.; Munoz, E.; Fiebich, B.L. Alleviation of Microglial Activation Induced by p38 MAPK/MK2/PGE2 Axis by Capsaicin: Potential Involvement of other than TRPV1 Mechanism/s. *Sci. Rep.* **2017**, *7*, 116. [CrossRef]

192. Fiebich, B.L.; Lieb, K.; Engels, S.; Heinrich, M. Inhibition of LPS-induced p42/44 MAP kinase activation and iNOS/NO synthesis by parthenolide in rat primary microglial cells. *J. Neuroimmunol.* **2002**, *132*, 18–24. [CrossRef]

193. Fiebich, B.L.; Schleicher, S.; Butcher, R.D.; Craig, A.; Lieb, K. The neuropeptide substance P activates p38 mitogen-activated protein kinase resulting in IL-6 expression independently from NF-kappa B. *J. Immunol.* **2000**, *165*, 5606–5611. [CrossRef]

194. Anantharam, V.; Kitazawa, M.; Wagner, J.; Kaul, S.; Kanthasamy, A.G. Caspase-3-dependent proteolytic cleavage of protein kinase Cdelta is essential for oxidative stress-mediated dopaminergic cell death after exposure to methylcyclopentadienyl manganese tricarbonyl. *J. Neurosci. Off. J. Soc. Neurosci.* **2002**, *22*, 1738–1751. [CrossRef]

195. Kaul, S.; Kanthasamy, A.; Kitazawa, M.; Anantharam, V.; Kanthasamy, A.G. Caspase-3 dependent proteolytic activation of protein kinase C delta mediates and regulates 1-methyl-4-phenylpyridinium (MPP+)-induced apoptotic cell death in dopaminergic cells: Relevance to oxidative stress in dopaminergic degeneration. *Eur. J. Neurosci.* **2003**, *18*, 1387–1401. [CrossRef]

196. Zhang, D.; Anantharam, V.; Kanthasamy, A.; Kanthasamy, A.G. Neuroprotective effect of protein kinase C delta inhibitor rottlerin in cell culture and animal models of Parkinson's disease. *J. Pharmacol. Exp. Ther.* **2007**, *322*, 913–922. [CrossRef]

197. Zhang, W.; Gao, J.H.; Yan, Z.F.; Huang, X.Y.; Guo, P.; Sun, L.; Liu, Z.; Hu, Y.; Zuo, L.J.; Yu, S.Y.; et al. Minimally Toxic Dose of Lipopolysaccharide and alpha-Synuclein Oligomer Elicit Synergistic Dopaminergic Neurodegeneration: Role and Mechanism of Microglial NOX2 Activation. *Mol. Neurobiol.* **2018**, *55*, 619–632. [CrossRef] [PubMed]

198. Kim, C.; Lv, G.; Lee, J.S.; Jung, B.C.; Masuda-Suzukake, M.; Hong, C.S.; Valera, E.; Lee, H.J.; Paik, S.R.; Hasegawa, M.; et al. Exposure to bacterial endotoxin generates a distinct strain of alpha-synuclein fibril. *Sci. Rep.* **2016**, *6*, 30891. [CrossRef]

199. Arai, H.; Furuya, T.; Yasuda, T.; Miura, M.; Mizuno, Y.; Mochizuki, H. Neurotoxic effects of lipopolysaccharide on nigral dopaminergic neurons are mediated by microglial activation, interleukin-1beta, and expression of caspase-11 in mice. *J. Biol. Chem.* **2004**, *279*, 51647–51653. [CrossRef] [PubMed]

International Journal of
Molecular Sciences

Article

Comparative Genomics of the *Aeromonadaceae* Core Oligosaccharide Biosynthetic Regions

Gabriel Forn-Cuní, Susana Merino and Juan M. Tomás *

Department of Genética, Microbiología y Estadística, Universidad de Barcelona, Diagonal 643, 08071 Barcelona, Spain; gabrifc@gmail.com (G.-F.C.); smerino@ub.edu (S.M.)
* Correspondence: jtomas@ub.edu; Tel.: +34-93-4021486

Academic Editor: William Chi-shing Cho
Received: 7 February 2017; Accepted: 26 February 2017; Published: 28 February 2017

Abstract: Lipopolysaccharides (LPSs) are an integral part of the Gram-negative outer membrane, playing important organizational and structural roles and taking part in the bacterial infection process. In *Aeromonas hydrophila*, *piscicola*, and *salmonicida*, three different genomic regions taking part in the LPS core oligosaccharide (Core-OS) assembly have been identified, although the characterization of these clusters in most aeromonad species is still lacking. Here, we analyse the conservation of these LPS biosynthesis gene clusters in the all the 170 currently public *Aeromonas* genomes, including 30 different species, and characterise the structure of a putative common inner Core-OS in the *Aeromonadaceae* family. We describe three new genomic organizations for the inner Core-OS genomic regions, which were more evolutionary conserved than the outer Core-OS regions, which presented remarkable variability. We report how the degree of conservation of the genes from the inner and outer Core-OS may be indicative of the taxonomic relationship between *Aeromonas* species.

Keywords: *Aeromonas*; genomics; inner core oligosaccharide; outer core oligosaccharide; lipopolysaccharide

---

1. Introduction

Aeromonads are an heterogeneous group of Gram-negative bacteria emerging as important pathogens of both gastrointestinal and extraintestinal diseases in a great evolutionary range of animals: from fish to mammals, including humans [1]. In recent years, as our knowledge about *Aeromonas* taxonomy, biology and pathogenicity has increased, and the number of reported infections caused by these microorganisms in healthy and immunocompromised patients has also spiked [2]. Despite the fact that, in humans, the most common complications derived from these pathogens are mild and easily tractable, they can present a serious risk in immunocompromised patients, causing severe septicaemia and even death [3]. Therefore, increasing our knowledge on the virulence factors governing aeromonad pathogenicity is of crucial importance to prevent the increasing complications caused by these bacteria.

The virulence and pathogenicity of *Aeromonas* is multifactorial, varies between species and strains, and has been linked to, among others, toxins, flagella, secretion systems, outer-membrane proteins and capsules, and surface polysaccharides, such as lipopolysaccharides (LPSs) [4]. LPSs, also known as endotoxins, are an integral part of the outer membrane for the great majority of Gram-negative bacteria, covering approximately the 75% of its surface and playing a crucial role on its organization and structure [5]. Although the exposed sections of LPSs are highly variable between species—and sometimes even between strains—LPSs molecules follow the same structural architecture depicted in Figure 1: a hydrophobic lipid component, the lipid A, bound to a hydrophilic polysaccharide [5]. The polysaccharide is composed by the core oligosaccharide (Core-OS), and the more variable O-specific chain (O-antigen), which may be present (Smooth LPS) or not (Rough LPS). The Core-OS can be further subdivided into the inner core and outer core. On the one side, the inner core, containing a high proportion of unusual sugars, predominantly one to three 3-deoxy-D-manno-oct-2-ulosonic acid (Kdo) residues and two or three heptoses (Hep), tends to be evolutionarily conserved within and a taxonomic family or genus, and is the part bound to the lipid A [6]. On the other side, the outer core is usually formed by common sugars—as hexoses and hexosamines—present more variability, and this is the region bound to the O-antigen, if present [6].

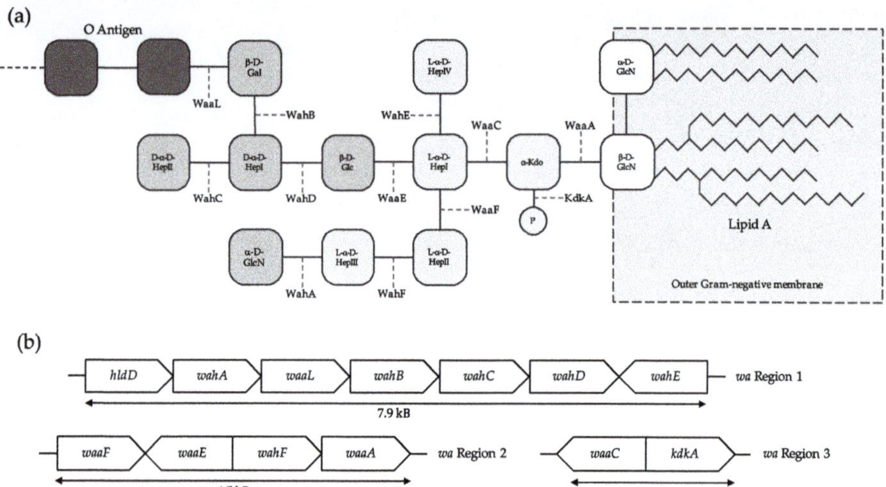

Figure 1. Structure and genomic organization of the *Aeromonas piscicola* AH-3 lipopolysaccharide (LPS). (**a**) Chemical structure of the *A. piscicola* AH-3 LPS. From clearer to darker, the carbohydrates of the inner core oligosaccharide, outer core oligosaccharide, and O-antigen (only the first two molecules) are shown. The name of the proteins catalysing each reaction is shown in each link between components; and (**b**) Genomic organization of the *A. piscicola* AH-3 genes coding for the proteins taking part in the LPS biosynthesis.

While the *Aeromonas* LPS follows the same architectural pattern as the rest of Gram-negative bacteria, this molecule displays remarkable diversity across species of this genus. We previously reported the molecular structure of the LPS from wild-type *A. hydrophila* AH-1 [7], *A. piscicola* AH-3 (formerly *A. hydrophila* AH-3) [8] and *A. salmonicida subsp. salmonicida* [9]. Furthermore, we characterized the genes coding for the proteins taking part in the assembly of both the inner and outer LPS core of these species, which are located in three different genomic regions: the *wa* Regions 1 to 3 [8,9]. Roughly, *wa* Region 1 contains genes related to the outer core, whereas *wa* Regions 2 and 3 code for enzymes related to the inner core biosynthesis (Figure 1). To date, the conservation and characterization of these LPS biosynthetic regions, which may be a key factor governing pathogenicity, remains to be studied in the rest of *Aeromonadaceae*. In this article, we analyse the synteny of the three Core-OS biosynthesis gene clusters across all the publicly available *Aeromonas* genomes and predict, for the first time, a conserved inner LPS oligosaccharide core substructure in all the species of *Aeromonas* genus.

2. Results

A total of 170 genomes comprising a variable number of strains and biovars from the Aeromonas species *A. allosaccharophila*, *A. aquatica*, *A. australiensis*, *A. bestiarum*, *A. bivalvium*, *A. caviae*, *A. dhakensis*, *A. diversa*, *A. encheleia*, *A. enteropelogenes*, *A. eucreophila*, *A. finlandiensis*, *A. fluvialis*, *A. hydrophila*, *A. jandaei*, *A. lacus*, *A. media*, *A. molluscorum*, *A. piscicola*, *A. popoffii*, *A. rivuli*, *A. salmonicida*, *A. sanarellii*, *A. schubertii*, *A. simiae*, *A. sobria*, *A. taiwanensis*, *A. tecta*, *A. veronii*, and other uncharacterized strains

(*Aeromonas sp.*) were retrieved from the National Center for Biotechnology Information genome database (Available online: https://www.ncbi.nlm.nih.gov/genome/). The complete list of genomes and strains used in this study is available at the Table S1. Using the sequence of the *A. hydrophila* ATCC7966 TetR, WaaA, and WaaC proteins, we located the *wa* Regions 1, 2, and 3, respectively, in each of the genome assemblies, thus confirming the conservation of these LPS biosynthetic regions inside *Aeromonas*.

2.1. Synteny of the wa Regions 2 and 3

The genomic *wa* Regions 2 and 3 were found in all the *Aeromonadaceae* genomes studied, highlighting the importance of a conserved inner Core-OS structure in this family. While no differences were found for *wa* Region 3—the genes *waaC* and *kdkA* were found highly conserved next to each other across all *Aeromonas* genomes studied—we identified four different genomic organizations regarding *wa* Region 2.

Overall, the synteny of the *wa* Region 2 gene cluster was predominantly conserved across most of the *Aeromonas* species, with *waaF*, *waaE*, *wahF*, and *waaA* positioned sequentially, as we previously described for *A. piscicola* AH-3 and *A. salmonicida* (Figure 2).

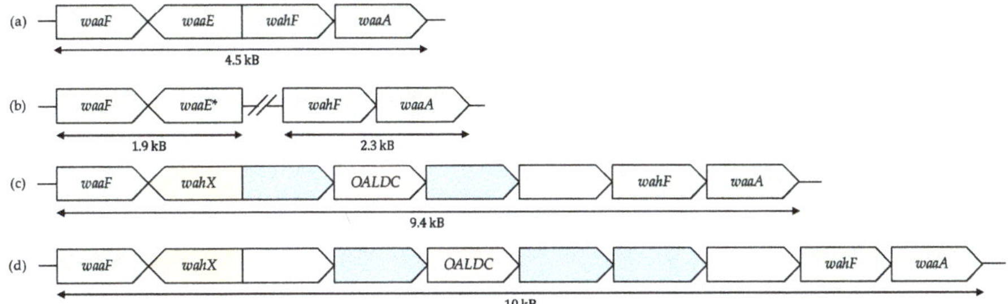

Figure 2. Comparison of the *wa* Region 2 in *Aeromonas*. (**a**) The most common genomic organization for this region found in the majority of *Aeromonas* species, with *waaF*, *waaE*, *wahF*, and *waaA* positioned sequentially in a gene cluster; (**b**) In the species *A. diversa*, *A. schubertii*, and *A. simiae*, this cluster is divided in two different genomic regions. Moreover, *A. simiae* presents *wahX* instead of *waaE*, marked with an asterisk; (**c**) A gene insertion of six genes containing *wahX*, a gene coding for an O-antigen ligase domain-containing protein (OALDC, in orange), and two glycosyltransferases (in blue) between *waaF* and *wahF* is found in the species *A. bivalvium*, *A. molluscorum*, and *A. rivuli*. Hypothetical proteins with unknown function are shown in white; and (**d**) Similarly, in the *A. fluvialis* genome, there is a gene insertion of seven genes containing *wahX*, a gene coding for an O-antigen ligase domain-containing protein, and three glycosyltransferases between *waaF* and *wahF*.

In the sequenced genomes of *A. schubertii*, *A. diversa*, and *A. simiae* strains, these genes were found split in two regions: *waaF* and *waaE* on one genomic site (*wa* Region 2.1); and *wahF* and *waaA* in another (*wa* Region 2.2) (Figure 2). It is worth mentioning that a fully closed genome from a species with this alternative genomic organization is not publicly available and, given the position of these regions in the ending part of unassembled contigs, we cannot rule out that this genomic distribution may be caused by an artefact in the genomic assembly. Although we did not detect major differences in the gene sequences of *A. schubertii* and *A. diversa*, the highly-conserved glycosyltransferase *waaE* is substituted in *A. simiae* by a gene which translated protein shows the same β-1-4-glucosyltransferase domain but less than 30% aminoacidic homology with WaaE. We named this gene, found in *Aeromonas* for the first time, *wahX*. The number of appearances and degree of conservation of *wahX* in *Aeromonadaceae*—together with the rest of genes discussed in this article—can be found in the Table 1.

Table 1. Common genes of the *wa* Regions 1, 2, and 3, and their conservation in the analysed genomes.

wa Genomic Region	Gene	Number of Genomes in Which Appears	Conservation in Studied *Aeromonas*
1	*hldD*	166/166	100%
	O-antigen ligase	166/166	100%
	wahA	154/166	93%
	wahY	13/166	7%
	wahE	101/166	61%
	wahZ	52/166	31%
2	*waaA*	170/170	100%
	wahF	170/170	100%
	waaE	165/170	97%
	wahX	5/170	3%
	waaF	170/170	100%
3	*waaC*	170/170	100%
	kdkA	170/170	100%

The gene *waaE* codes for the L-glycero-D-manno-heptose β-1,4-D-glucosyltransferase, which catalyses the union of a Glc residue to the highly conserved L-α-D-HepI of the inner Core-OS not only in *Aeromonas* species (as *A. piscicola* [8] and *A. salmonicida* [9]), but also in other *Enterobacteriaceae* as *Klebsiella pneumoniae* and *Serratia marcescens* [10]. Thus, the substitution of this gene inside the *Aeromonadaceae* family is striking. Instead of with the rest of *Aeromonas* species, the protein sequence of *wahX* displays high homology with glycosyltransferases of other *Proteobacteria*, as *Methylomarinum* and *Desulfovibrio*. The chemical composition of the *Desulfovibrio desulfuricans* LPS polysaccharide chain was recently revealed [11], containing residues of Kdo, rhamnose, methylopentose-fucose, 3 hexoses-mannoses, glucose, and galactose, but the structure of this bacterium Core-OS chain, and thus the exact function of *wahX*, remains unknown for the time being.

We also detected major changes in the *wa* Region 2 of *A. fluvialis*, as well as in the evolutionary-related group of *A. bivalvium*, *A. molluscorum*, and *A. rivuli*. As in *A. simiae*, these species presented *wahX* instead of *waaE*. Furthermore, an insertion of six genes in *A. fluvialis* and four genes in *A. bivalvium*, *A. molluscorum*, and *A. rivuli*, between *wahX* and *wahF* was found in these genomes (Figure 2).

The four different subgroups of *Aeromonas* species as categorized by their *wa* Region 2 genomic organization correlate well with the asumed evolution of this genus. Colston et al. recently compared the genomes from 56 different strains to revise and improve the phylogenetic and taxonomic relationships between *Aeromonas* species [12]. Despite that the evolutionary reconstruction of the genus is still not completely resolved and varies depending on the gene set used for the phylogeny, their results clearly defined of eight major monophyletic groups inside *Aeromonas*. Supporting our findings, *A. schubertii*, *A. diversa*, and *A. simiae* on one side and *A. bivalvium*, *A. molluscorum*, and *A. rivuli* on the other were precisely classified as two of these eight major *Aeromonas* monophyletic groups.

However, the resemblance between alternative *wa* Region 2 Types 3 and 4 despite their differential evolutionary position is intriguing. While *A. bivalvium*, *A. molluscorum*, and *A. rivuli* are well-defined evolutionary-related species [13], the *A. fluvialis* genomic sequence is more related to species of the *A. veronii* monophyletic group (see [12]) both by DNA-DNA hybridization [14] and by in silico whole genome analyses [12]. Therefore, an in-depth study of horizontal gene transfer between these species may be warranted.

2.2. Synteny of the wa Region 1

Regarding *wa* Region 1, at least two glycosyltransferases were located next to *tetR* and *hldD* in all genomes. The *A. diversa* 2478-85, *A. lacus* AE122, *A. salmonicida subsp. achromogenes* AS03, and *A. veronii* VBF557 genomes were excluded from the analysis as the assembly of this region in these species was broken between different contigs. Thus, 166 genomes were analysed. Overall, we found considerable variability in the outer core genes of the *wa* Region 1 cluster across the available *Aeromonas* genomes, with at least 29 different genomic organizations containing different genes, based in their sequence homology, in the 166 analysed genomes. The complete assignment of *wa* Region 1 types to each strain can be found at the Table S1.

While the description and characterization of each of the outer LPS core biosynthetic clusters expands far beyond the scope of this article, we can provide a rough set of patterns describing the *Aeromonadaceae wa* Region 1 derived from our genome analysis, which are summarized in Figure 3. First, the *wa* Region 1, containing *hldD*, at least two glycosyltransferases and one gene coding for a protein with an O-antigen ligase domain (i.e., *waaL* in *A. piscicola* and *A. salmonicida*), was found upstream of *tetR*. We found two exceptions to this pattern: the *A. media* strains CECT 4232 and WS, which showed only one glycosyltransferase (*wahY*, characterized below) instead of more; and the species *A. fluvialis*, *A. bivalvium*, *A. molluscorum*, and *A. rivuli*, in which this genomic region was composed only by *hldD* and one glycosyltransferase (*wahY*), while the O-antigen ligase-domain containing gene was located in the previously mentioned genomic insertion of the *wa* Region 2 (Figure 2).

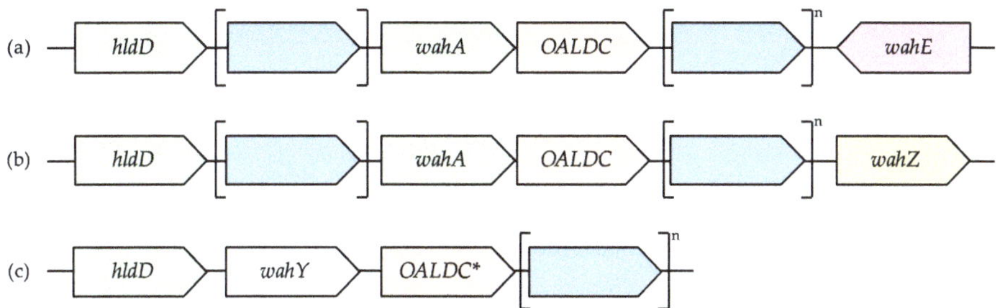

Figure 3. Schematic description of the different *Aeromonas wa* Region 1 types reported. (**a**) The gene *hldD* was found at the start of the *wa* Region 1 in all *Aeromonas* species. In *A. salmonicida*, a glycosyltransferase-coding gene (in blue) is present between *hldD* and *wahA*. A gene coding for an O-antigen ligase domain-containing was always found next to *wahA*, followed by a variable number of hypothetical genes and glycosyltransferases, and, finally, *wahE*; (**b**) Most of the species of the *A. veronii* monophyletic group present the same genomic architecture but with *wahZ* instead of *wahE*; and (**c**) In the species in which genome *wahA* is not present, *wahY* is found next to *hldD*, followed by a gene coding for an O-antigen ligase domain-containing protein (except in the species described above where this gene is found in the *wa* Region 2, marked with an asterisk) and a variable number of glycosyltransferases and hypothetical proteins. OALDC = O-antigen ligase domain containing gene.

Despite presenting high sequence variability between species, all the O-antigen ligases from the different *wa* Region 1 genomic organizations shared 10 aminoacid positions that appear to be critical for its function: four glycines, three arginines, one serine, one tryptophan and one histidine residues (positions 123, 196, 245, 264, 268, 272, 286, 288, 318, and 329 for *A. piscicola* AH-3 WaaL) (Figure S1a). Furthermore, all putative O-antigen ligases presented between 8 and 11 predicted transmembrane domains, with 10 being the most common prediction (Figure S1b), as is characteristic for lipid-A-core, O-antigen ligases [8,15].

The second pattern that we found is the high conservation of WahA, the protein that links GlcNAc to the LPS outer Core-OS using UDP-GlcNAc as a substrate. A gene coding with a protein with high homology to *wahA* was found in 153 of the 166 *Aeromonas* genomes. It was absent in the genomes of *A. bivalvium* CECT 7113, *A. caviae* AE398, CECT 4221, and FDAARGOS_76, *A. fluvialis* LMG 24681, *A. media* CECT 4232 and WS, *A. molluscorum* 848, *A. rivuli* DSM 22539, *A. salmonicida* Y47, *A. simiae* CIP 107798, and *A. veronii* AVNIH1 and AVNIH2. Interestingly though, all these genomes shared the presence of a common glycosyltransferase next to *hldD* with high homology to others from

Vibrio coralliilyticus. We named this gene *wahY*, and it was not present in any of *Aeromonas* genomes with *wahA*. Interestingly, *wahY* was presend in all the species with a *wa* Region 2 of the newly-described Types 3 and 4.

The importance of WahA remains in its bifunctional nature: it has two domains, one glycosyltransferase domain that catalyses the incorporation of GlcNAc to the LPS outer Core-OS; and one carbohydrate esterase domain deacetylates the GlcNAc residue to GlcN [16]. The gene *wahA*, is not only conserved in *Aeromonas*, but also in *Vibrio cholerae* and *shilonii* species. Thus, its substitution for *wahY*, which is annotated as type I glycosyltransferase in the online databases but shows no protein domains, technically negates the characteristic presence of GlcN in the outer Core-OS of these species.

Thirdly, the presence in the *wa* Region 1 of all genomes with *wahA* of a glycosyltransferase belonging to the inner LPS Core-OS biosynthesis. More importantly, when present in the genome, this gene was always found delimiting the end of the *wa* Region 1 cluster. We found two different gene sequences based on their homology: those that displayed high homology with *wahE* and were positioned contrary to the *hldD* direction; and those that displayed high homology to an uncharacterized glycosyltransferase, which we named *wahZ*, and were positioned in the same direction as *hldD*. Given the exclusivity of each of these genes in the genomes, we hypothesize that both genes code for a protein that catalyse the union of a carbohydrate (HepIV in the case of *wahE*) in the same position of the L-α-D-HepI.

An homolog of *wahE* was found on 101 of the 153 genomes with *wahA*, comprising the species *A. aquatica, A. bestiarum, A. caviae, A. dhakensis, A. encheleia, A. enteropelogenes, A. eucreophila, A. hydrophila, A. jandaei, A. media, A. piscicola, A. popoffii, A. salmonicida, A. sanarellii, A. schubertii, A. taiwanensis, A. tecta,* and some undefined *sp.* strains; while *wahZ* was found in 52 of the 153 genomes, comprising the species *A. allosaccharophila, A. australiensis, A. diversa, A. enteropelogenes, A. finlandiensis, A. jandaei, A. sobria, A. veronii,* and a few undefined *Aeromonas sp.* strains. In consequence, the presence of *wahZ* seems to be a defining characteristic of the *A. veronii* monophyletic group described by Colston et al. [12].

Most Common *wa* Region 1 Genomic Organizations and Notable Inconsistencies

A schematic representation of the six most common gene organizations for *wa* Region 1, depicting 122 of the 170 genomes, can be found at Figure 4. The most common *wa* Region 1 genomic structure, or Type 1, was the one described for *A. piscicola* AH-3 [8]. We found this organization in the genomes of the species *A. bestiarum, A. dhakensis, A. hydrophila, A. piscicola,* and *A. schubertii.* The second type of *wa* Region 1 (Type 2) that we found was that of *A. salmonicida,* including the subspecies *masoucida, salmonicida,* and *smithia,* but not *pectinolytica* or *achromogenes.* A group of *A. caviae* and *A. media* species, together with *A. aquatica* and *A. sanarelli* also showed common genes for this genomic region, which we assigned Type 3. The second most common genomic organization for this cluster, Type 4, comprised 23 different *A. veronii* strains, *A. australiensis,* and *A. jandaei.* The Type 5 *wa* Region 1—assigned to the genomic cluster found in *A. enteropelogenes* and *A. finlandensis* species—and Type 6—in *A. popoffi*—highlight the high recombination capacity of this region in the genus: both types present the same genomic architecture and high homology percentage between the genes, but *wahZ* is found in the Type 5 while *wahE* is found in the Type 6 *wa* Region 1.

Our results at the genomic level largely agree with the presumed evolution of the *Aeromonas* genus, and are corroborated by the published reports regarding LPS structure in the aforementioned species. For example, *A. bestiarum* presents the same *wa* Region 1 Type 1 than *A. hydrophila* and *A. piscicola,* and the mass spectrometry of the Core-OS from *A. bestiarum* Strain K296 (serotype O18) shows the identical $Hep_6Hex_1HexN_1Kdo_1P_1$ structure that we characterized for *A. piscicola* AH-3 and *A. hydrophila* AH-1 [7,17]. Similarly, although the genome of the strain *A. veronii* strain Bs19 (serotype O16), is not available, the LPS chemical structure of this strain is $Hep_5Hex_3HexN_1Kdo_1P_1$, that is, one more carbohydrate than the ones described for the Type 1 *wa* Region 1 [18]. Coinciding, the *wa* Region 1 of the majority of *A. veronii* (Type 4) is composed by one more gene than the ones described for Type 1. Moreover, given the chemical structure, we hypothesize that *wahZ* catalyses the union of a hexose to HepI.

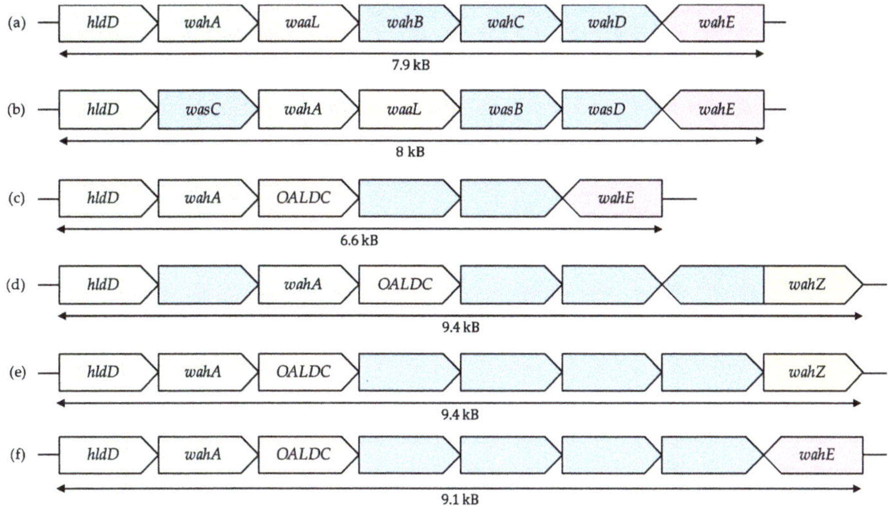

Figure 4. Comparison of the six most common *wa* Region 1 genomic organization in *Aeromonas*. Common genes (*hldD* and *wahA*) in these genomic organizations are shown in green, O-antigen ligase domain containing proteins in orange, *wahE* in purple, *wahZ* in yellow, and both characterized and uncharacterized glycosyltransferases are shown in blue. (**a**) Type 1 *wa* Region 1, found in *A. bestiarum*, *A. dhakensis*, *A. hydrophila*, *A. piscicola*, and *A. schubertii*; (**b**) Type 2 *wa* Region 1, found in *A. salmonicida* species; (**c**) Type 3 *wa* Region 1, as in *A. aquatica*, *A. caviae*, *A. media* and *A. sanarelli*; (**d**) Type 4 *wa* Region 1, found in *A. australiensis*, *A. jandaei*, and *A. veronii*; (**e**) Type 5 *wa* Region 1, in *A. enteropelogenes* and *A. finlandensis*; and (**f**) Type 6 *wa* Region 1, as found in *A. popoffi*.

However, the previous description of these *wa* Region 1 organizations also arise notable inconsistencies between taxonomic classifications in some species. Beyond the species mentioned above, the *wa* Region 1 Type 1 was also found the strains *A. enteropelogenes* LK14 (that according to the rest of *A. enteropelogenes* strains should be of Type 5), *A. jandaei* L14h (Type 4), and *A. salmonicida* strains CBA100 and *subsp. pectonolytica* 34meI (both Type 2). Of note, we previously pointed out that *A. salmonicida subsp. pectinolytica* presented a different outer LPS Core-OS than the rest of *A. salmonicida* species and similar to *A. hydrophila*, supporting this exception [19].

Similarly, *A. hydrophila* 4AK4 and *A. hydrophila* BWH65 present a Type 3 *wa* Region 1 instead of the Type 1 common for the rest of *A. hydrophila* species. *A. hydrophila* 4AK4 shows extreme resemblance to *A. caviae* (also Type 3 for some strains) in many areas, such as the—almost exclusive in aeromonads—production of poly(3-hydroxybutyrate-co-3-hydroxyhexanoate) [20] and the higher sequence homology of virulence factors to *A. caviae* than *A. hydrophila* [21]. In fact, *A. hydrophila* 4AK4 shows low average nucleotide identity (ANI) values with *A. hydrophila* and has been proposed to be a novel species relative to *A. media* [22], also represented in the Type 3 *wa* Region 1 organization. Likewise, *A. hydrophila* BWH65 shows high genome homology with *A. hydrophila* 4AK4, and both are found in the same branch as *Aeromonas* SSU (recently reclassified as *A. dhakensis* SSU) in the genomic blast-based dendogram of the *A. hydrophila* webpage in the NCBI genomes (Available online: https://www.ncbi.nlm.nih.gov/genome/1422).

Of note, notable variability between these regions was found in the species *A. veronii*, *A. caviae*, and *A. media*, which may indicate a higher recombination capacity or taxonomic nomenclature inconsistencies in this species.

2.3. Reconstruction of the Common Aeromonas Inner LPS Core-OS

Using the above information, we could reconstruct a common LPS inner core in the entire *Aeromonas* genus composed by three L-α-D-Heptoses and one α-Kdo-P, based on the conserved genes *kdkA*, *waaA*, *waaC*, *waaF*, and *wahF* (Figure 5). Moreover, with the exception of the species *A. bivalvium*, *A. molluscorum*, *A. rivuli*, *A. fluvialis*, and *A. simiae*, *waaE* was also found in all genomes. Similarly, only in eight strains (*A. caviae* strains AE398, CECT4221, and FDAARGOS76; *A. media* strains CECT 4232 and WS; *A. salmonicida* Y47; and *A. veronii* strains AVNIH1 and AVNIH2) more than in the aforementioned five species *wahA* was not found in the genome—all of them presented *wahY*. Thus, we consider *waaE* and *wahA* as highly conserved genes in aeromonads, and in consequence a high conservation of the hexose β-1,4-Glc linked to HepI, and of the α-1,7-GlcN linked to HepIII in the aeromonad LPS core (Figure 5). Furthermore, we found *wahE* to be well conserved across most of *Aeromonas* species except in the *A. veronii* monophyletic group, which presented the gene *wahZ* instead, suggesting that they may act in the carbohydrate same position.

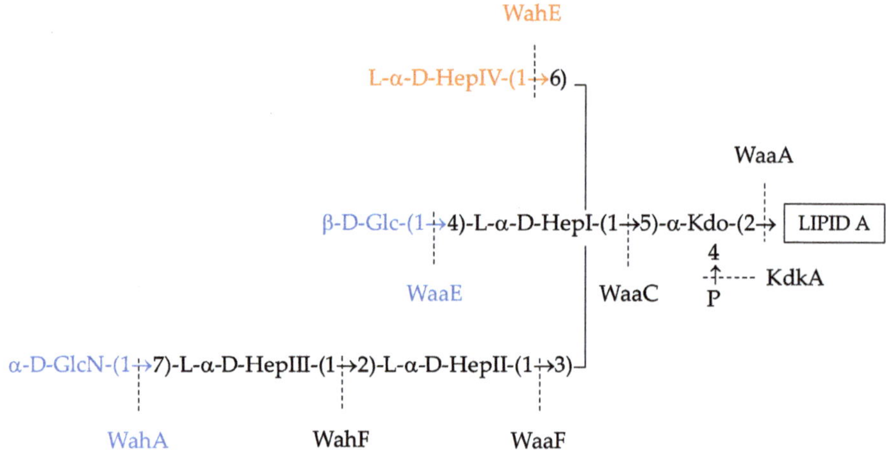

Figure 5. Reconstruction of the *Aeromonadaceae* common Core-OS. The enzymes found in all genomes are shown in black. WaaE and WahA (in blue), were also present in more than the 90% genomes analysed. Finally, WahE (in orange) is substituted by WahZ (although the carbohydrate linked is still unknown) in the species of the *A. veronii* monophyletic group.

Finally, it is worth mentioning that the absence of typical outer core homologs (as *wahB/wasB*, *wahD/wasD*) and *waaE* in *A. bivalvium*, *A. fluvialis*, *A. molluscorum*, *A. rivuli*, and *A. simiae*, suggests that either these species do not present O-antigen or their whole outer Core-OS may be entirely different than in the rest of *Aeromonas*. The presence O-antigen ligase domain-containing protein with similar homology in the *wa* Region 1 on *A. simiae* and *wa* Region 2 gene insertions of *A. bivalvium*, *A. fluvialis*, *A. molluscorum*, and *A. rivuli* supports the latter. To prove the presence of O-antigen in these species, we analysed the LPS profile gel of the species *A. hydrophila* AH-1, *A. piscicola* AH-3, and *A. bivalvium* 868E[T] by SDS-PAGE (Figure 6). The LPS profile gel confirmed the presence of O-antigen in *A. bivalvium* (and therefore, probably in the other species questioned) as well as a considerable difference in the LPS Core-OS size.

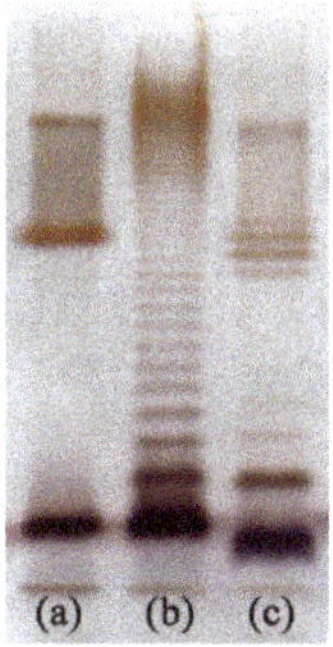

Figure 6. LPS profiles of (**a**) *A. hydrophila* AH-1; (**b**) *A. piscicola* AH-3; and (**c**) *A. bivalvium* 868E^T by SDS-PAGE.

3. Discussion

Aeromonas are a ubiquitous, rod-shaped and flagellated genus of Gram-negative bacteria emerging as important animal pathogens, especially in mammals and fish. Despite once being subdivided into solely two subgroups, mesophilic (*A. hydrophila*) and psychrophilic (*A. salmonicida*) aeromonads [1], there are currently dozens of characterized *Aeromonas* species in the literature, not including subspecies or biovars (Available online: http://www.bacterio.net/aeromonas.html for a complete list).

However, the taxonomic and phylogenetic relationships of many of these species are still today a matter of debate: practical evidence for some of these species is still lacking [1]; different names have been found to be synonyms for the same species (e.g., *A. enteropelogenes* and *A. trota* [23,24]; and *A. culicicola*, *A. ichthiosmia* and *A. veronii* [23]); characterized species have been reclassified outside the *Aeromonas* genus (as *A. sharmana* [25]); and, historically, misclassifications in this family have been usual (e.g., *A. hydrophila* AH-3 was recently reclassified into *A. piscicola* AH-3 [26]). This taxonomic classification problem is further exacerbated by a considerably high genetic homologous recombination rate and horizontal gene transfer capacity, which in fact has been demonstrated to be an important driving force in the evolution of aeromonads [27]. When taking all of this together, it is not difficult to understand why the evolutionary reconstruction of this family is still not resolved. The importance of a correct classification of *Aeromonas* species resides not only in correctly knowledge of taxonomic evolution, but also in studying the pathogenic source and potential of *Aeromonas* species and strains.

As the virulence of *Aeromonas* species is multifactorial and depends on the specific virulence factors present on each strain as well as on environmental conditions [4], the description of common processes governing aeromonad pathogenicity is of crucial importance. The LPS, an important component of the Gram-negative membrane present in all aeromonads, also plays a key role in the adhesion and infectivity process, thus directly affecting their pathogenicity. Therefore, the study of the specific LPS biosynthesis routes and structure may be further exploited in, for example, the search for common treatments against *Aeromonas* species or specific inhibitors for characteristic strains. To this end, in this study, we compared the genomic structure of the three described locus for the LPS Core-OS biosynthesis—*wa* Regions 1, 2, and 3—in all publicly available *Aeromonas* genomes, covering 170 strains spawning 30 different *Aeromonas* species to study their degree of conservation.

As expected, the genes and genomic organizations of the inner core regions were more evolutionary conserved than the outer core region, but adding to our previously characterized *wa* Region 2 in *A. piscicola* and *A. salmonicida*, we describe three new genomic organizations for these regions affecting the species *A. bivalvium*, *A. diversa*, *A. fluvialis*, *A. molluscorum*, *A. rivuli*, *A. schubertii*, and *A. simiae*.

In contrast, the outer Core-OS biosynthetic region presented remarkable variability within *Aeromonas*. In fact, *hldD* (formerly *rfaB*, coding for ADP-L-glycero-D-manno-heptose 6-epimerase) was the only common gene of the *wa* Region 1 that we could locate in all analysed genomes. Therefore, we consider *hldD* as a good marker for *wa* Region 1 position. Despite the differences between species, we also found a gene coding for a O-antigen ligase domain-containing protein and at least two glycosyltransferases in the *wa* Region 1 of all *Aeromonas* species except *A. bivalvium*, *A. fluvialis*, *A. molluscorum*, and *A. rivuli*. We also detected high degree of conservation of *wahA* (in 154 of the 166 genomes analysed for *wa* Region 1), and the common presence of *wahY* in the 12 species lacking this gene (Table 1). Similar exclusivity patterns were found between *wahE* and *wahZ*, which appears to be an exclusive marker of the *A. veronii* monophyletic group. Furthermore, our results are supported by both genome-wide phylogenetic and chemical mass spectrometry studies in the literature.

We report that, while horizontal gene transfer—not involving only *Aeromonas* species but also other proteobacteria—seems to be a potential key process for the evolution of this regions, the genes of the LPS Core-OS are considerably well conserved throughout *Aeromonas* evolution. Therefore, these genomic regions can be useful when studying the taxonomic relationships between *Aeromonas* species.

Our study found notable inconsistencies between the conservation of the genomic organization and the taxonomy of specific strains, such as *A. enteropelogenes* LK14, *A. hydrophila* strains 4AK4 and BWH65, *A. jandaei* L14h, and *A. salmonicida* strains CBA100 and *subsp. pectonolytica* 34meI. Given that extensive scrutiny of at least two of these strains arise important criticism and suspicion about their taxonomic classification, an in-depth study of this species taxonomic classification may be necessary.

Finally, the recent reclassifications of several *Aeromonas hydrophila* species, as *A. hydrophila* AH-3 (now *A. piscicola* AH-3), led recently to a serious questioning if previous research involving pathogenic factors on strains of now different *Aeromonas* species was still applicable to *A. hydrophila* [28]. The results from this study support that the LPS Core-OS from *A. hydrophila* is identical to that of *A. piscicola*, and thus previous research regarding this virulence factor should be applicable to *A. hydrophila*.

4. Materials and Methods

We retrieved the 170 different genomes for the *Aeromonas* genus available on NCBI Genome website as of September 2016 (Table S1), regardless of their assembly completeness. To roughly locate the position of the LPS core biosynthetic regions in the genomes, we performed a local tblastn of *A. hydrophila* ATCC7966 *tetR* (*wa* Region 1), *waaA* (*wa* Region 2), and *waaC* (*wa* Region 3) in each genome assembly. Due to the incomplete annotation of some of the analysed genomes, the 16,000 bp region upstream of *tetR*, 10,000 bp region upstream of *waaA*, and 5,000 bp region downstream of *waaC* were reannotated in each genome. Gene prediction was performed with Glimmer v.3.0.2 [29] and hand-curated. Comparison of the genetic regions was performed using the CloVR Comparative Pipeline [30] and explored using Sybil [31].

The predicted protein sequence of one arbitrarily selected O-antigen ligase from each *wa* Region 1 genomic organization was retrieved and aligned with MUSCLE [32]. Prediction of the transmembrane domains was performed with TMHMM Server v2.0 [33].

Cultures of *A. hydrophila* AH-1, *A. piscicola* AH-3 and *A. bivalvium* 868E^T where grown overnight in Tryptic Soy Agar at 30 °C, except for *A. piscicola* AH-3 which was grown at 25°C. LPS was obtained after proteinase K digestion of whole cells, separated by SDS-PAGE and visualized by silver staining as previously published [16].

5. Conclusions

We compared the LPS biosynthetic regions from 170 *Aeromonas* genomes and analysed the conservation of the genes taking part in the assembly of this cell-wall component. We describe high conservation of the genes related to the inner Core-OS biosynthesis, with only one organization for *wa* Region 3 and 4 different genomic organizations regarding the *wa* Region 2; and remarkable variability in the *wa* Region 1, composed roughly by genes affecting the outer Core-OS. Besides describing a common LPS Core-OS structure of all the *Aeromonas* sequenced to date, we also report how these regions can be useful for establishing evolutionary relationships between species.

Supplementary Materials: Supplementary materials can be found at www.mdpi.com/1422-0067/18/3/519/s1.

Acknowledgments: This work was partially funded by BIO2016-80329-P from the Spanish Ministerio de Economía y Competitividad, and from the Generalitat de Catalunya (Centre de Referència en Biotecnologia). We thank Maite Polo for her technical assistance and the Servicios Científico-Técnicos from the University of Barcelona.

Author Contributions: Gabriel Forn-Cuní, Susana Merino, and Juan M. Tomas conceived, designed, and performed the experiments; analysed the data; and wrote the paper.

Conflicts of Interest: The authors declare no conflict of interest.

Abbreviations

LPS	Lipopolysaccharide
Core-OS	Core-OligoSaccharide
Kdo	3-Deoxy-D-manno-oct-2-ulosonic acid
OALDC	O-antigen Ligase Domain Containing gene
ANI	Average Nucleotide Identity

1. Janda, J.M.; Abbott, S.L. The genus *Aeromonas*: Taxonomy, pathogenicity, and infection. *Clin. Microbiol. Rev.* **2010**, *23*, 35–73. [CrossRef] [PubMed]

2. Igbinosa, I.H.; Igumbor, E.U.; Aghdasi, F.; Tom, M.; Okoh, A.I. Emerging *Aeromonas* species infections and their significance in public health. *Sci. World J.* **2012**. [CrossRef] [PubMed]

3. Parker, J.L.; Shaw, J.G. *Aeromonas* spp. clinical microbiology and disease. *J. Infect.* **2011**, *62*, 109–118. [CrossRef] [PubMed]

4. Tomás, J.M. The main *Aeromonas* pathogenic factors. *ISRN Microbiol.* **2012**, *2012*, 256261. [CrossRef] [PubMed]

5. Aquilini, E.; Tomás, J.M. Lipopolysaccharides (Endotoxins). In *Reference Module in Biomedical Sciences*; Caplan, M.J., Ed.; Elsevier: Amsterdam, The Netherlands, 2015.

6. Holst, O. The structures of core regions from enterobacterial lipopolysaccharides—An update. *FEMS Microbiol. Lett.* **2007**, *271*, 3–11. [CrossRef] [PubMed]

7. Merino, S.; Canals, R.; Knirel, Y.A.; Tomás, J.M. Molecular and chemical analysis of the lipopolysaccharide from *Aeromonas hydrophila* strain AH-1 (Serotype O11). *Mar. Drugs* **2015**, *13*, 2233–2249. [CrossRef] [PubMed]

8. Jimenez, N.; Canals, R.; Lacasta, A.; Kondakova, A.N.; Lindner, B.; Knirel, Y.A.; Merino, S.; Regué, M.; Tomás, J.M. Molecular analysis of three *Aeromonas hydrophila* AH-3 (serotype O34) lipopolysaccharide core biosynthesis gene clusters. *J. Bacteriol.* **2008**, *190*, 3176–3184. [CrossRef] [PubMed]

9. Jimenez, N.; Lacasta, A.; Vilches, S.; Reyes, M.; Vazquez, J.; Aquillini, E.; Merino, S.; Regué, M.; Tomás, J.M. Genetics and proteomics of *Aeromonas salmonicida* lipopolysaccharide core biosynthesis. *J. Bacteriol.* **2009**, *191*, 2228–2236. [CrossRef] [PubMed]

10. Izquierdo, L.; Abitiu, N.; Coderch, N.; Hita, B.; Merino, S.; Gavin, R.; Tomás, J.M.; Regué, M. The inner-core lipopolysaccharide biosynthetic *waaE* gene: Function and genetic distribution among some *Enterobacteriaceae*. *Microbiology* **2002**, *148*, 3485–3496. [CrossRef] [PubMed]

11. Lodowska, J.; Wolny, D.; Jaworska-Kik, M.; Kurkiewicz, S.; Dzierżewicz, Z.; Węglarz, L. The chemical composition of endotoxin isolated from intestinal strain of *Desulfovibrio desulfuricans*. *Sci. World J.* **2012**. [CrossRef] [PubMed]

12. Colston, S.M.; Fullmer, M.S.; Beka, L.; Lamy, B.; Gogarten, J.P.; Graf, J. Bioinformatic genome comparisons for taxonomic and phylogenetic assignments using *Aeromonas* as a test case. *MBio* **2014**, *5*, e02136. [CrossRef] [PubMed]

13. Figueras, M.J.; Alperi, A.; Beaz-Hidalgo, R.; Stackebrandt, E.; Brambilla, E.; Monera, A.; Martinez-Murcia, A.J. *Aeromonas rivuli* sp. nov., isolated from the upstream region of a karst water rivulet. *Int. J. Syst. Evol. Microbiol.* **2011**, *61*, 242–248. [CrossRef] [PubMed]

14. Alperi, A.; Martínez-Murcia, A.J.; Monera, A.; Saavedra, M.J.; Figueras, M.J. *Aeromonas fluvialis* sp. nov., isolated from a Spanish river. *Int. J. Syst. Evol. Microbiol.* **2010**, *60*, 72–77. [CrossRef] [PubMed]

15. Abeyrathne, P.D.; Daniels, C.; Poon, K.K.H.; Matewish, M.J.; Lam, J.S. Functional characterization of WaaL, a ligase associated with linking O-antigen polysaccharide to the core of *Pseudomonas aeruginosa* lipopolysaccharide. *J. Bacteriol.* **2005**, *187*, 3002–3012. [CrossRef] [PubMed]

16. Jimenez, N.; Vilches, S.; Lacasta, A.; Regué, M.; Merino, S.; Tomás, J.M. A bifunctional enzyme in a single gene catalyzes the incorporation of GlcN into the *Aeromonas* core lipopolysaccharide. *J. Biol. Chem.* **2009**, *284*, 32995–33005. [CrossRef] [PubMed]

17. Turska-Szewczuk, A.; Lindner, B.; Komaniecka, I.; Kozinska, A.; Pekala, A.; Choma, A.; Holst, O. Structural and immunochemical studies of the lipopolysaccharide from the fish pathogen, *Aeromonas bestiarum* strain K296, serotype O18. *Mar. Drugs* **2013**, *11*, 1235–1255. [CrossRef] [PubMed]

18. Turska-Szewczuk, A.; Duda, K.A.; Schwudke, D.; Pekala, A.; Kozinska, A.; Holst, O. Structural studies of the lipopolysaccharide from the fish pathogen *Aeromonas veronii* strain Bs19, serotype O16. *Mar. Drugs* **2014**, *12*, 1298–1316. [CrossRef] [PubMed]

19. Merino, S.; Tomás, J.M. The *Aeromonas salmonicida* Lipopolysaccharide Core from Different Subspecies: The Unusual *subsp. pectinolytica*. *Front. Microbiol.* **2016**, *7*, 125. [CrossRef] [PubMed]

20. Han, J.; Qiu, Y.-Z.; Liu, D.-C.; Chen, G.-Q. Engineered *Aeromonas hydrophila* for enhanced production of poly(3-hydroxybutyrate-co-3-hydroxyhexanoate) with alterable monomers composition. *FEMS Microbiol. Lett.* **2004**, *239*, 195–201. [CrossRef] [PubMed]

21. Gao, X.; Jian, J.; Li, W.-J.; Yang, Y.-C.; Shen, X.-W.; Sun, Z.-R.; Wu, Q.; Chen, G.-Q. Genomic study of polyhydroxyalkanoates producing *Aeromonas hydrophila* 4AK4. *Appl. Microbiol. Biotechnol.* **2013**, *97*, 9099–9109. [CrossRef] [PubMed]

22. Beaz-Hidalgo, R.; Hossain, M.J.; Liles, M.R.; Figueras, M.-J. Strategies to avoid wrongly labelled genomes using as example the detected wrong taxonomic affiliation for aeromonas genomes in the GenBank database. *PLoS ONE* **2015**, *10*, e0115813.

23. Collins, M.D.; Martinez-Murcia, A.J.; Cai, J. *Aeromonas enteropelogenes* and *Aeromonas ichthiosmia* are identical to *Aeromonas trota* and *Aeromonas veronii*, respectively, as revealed by small-subunit rRNA sequence analysis. *Int. J. Syst. Bacteriol.* **1993**, *43*, 855–856. [CrossRef] [PubMed]

24. Huys, G.; Denys, R.; Swings, J. DNA-DNA reassociation and phenotypic data indicate synonymy between *Aeromonas enteropelogenes* Schubert et al. 1990 and *Aeromonas trota* Carnahan et al. 1991. *Int. J. Syst. Evol. Microbiol.* **2002**, *52*, 1969–1972. [CrossRef] [PubMed]

25. Martínez-Murcia, A.J.; Figueras, M.J.; Saavedra, M.J.; Stackebrandt, E. The recently proposed species *Aeromonas sharmana* sp. nov., isolate GPTSA-6T, is not a member of the genus *Aeromonas*. *Int. Microbiol.* **2007**, *10*, 61–64. [PubMed]

26. Beaz-Hidalgo, R.; Alperi, A.; Figueras, M.J.; Romalde, J.L. *Aeromonas piscicola* sp. nov., isolated from diseased fish. *Syst. Appl. Microbiol.* **2009**, *32*, 471–479. [CrossRef] [PubMed]

27. Ghatak, S.; Blom, J.; Das, S.; Sanjukta, R.; Puro, K.; Mawlong, M.; Shakuntala, I.; Sen, A.; Goesmann, A.; Kumar, A.; et al. Pan-genome analysis of *Aeromonas hydrophila*, *Aeromonas veronii* and *Aeromonas caviae* indicates phylogenomic diversity and greater pathogenic potential for *Aeromonas hydrophila*. *Antonie Van Leeuwenhoek* **2016**, *109*, 945–956. [CrossRef] [PubMed]

28. Rasmussen-Ivey, C.R.; Figueras, M.J.; McGarey, D.; Liles, M.R. Virulence Factors of *Aeromonas hydrophila*: In the Wake of Reclassification. *Front. Microbiol.* **2016**, *7*, 1337. [CrossRef] [PubMed]

29. Delcher, A.L.; Bratke, K.A.; Powers, E.C.; Salzberg, S.L. Identifying bacterial genes and endosymbiont DNA with Glimmer. *Bioinformatics* **2007**, *23*, 673–679. [CrossRef] [PubMed]

30. Angiuoli, S.V.; Matalka, M.; Gussman, A.; Galens, K.; Vangala, M.; Riley, D.R.; Arze, C.; White, J.R.; White, O.; Fricke, W.F. CloVR: A virtual machine for automated and portable sequence analysis from the desktop using cloud computing. *BMC Bioinform.* **2011**, *12*, 356. [CrossRef] [PubMed]

31. Riley, D.R.; Angiuoli, S.V.; Crabtree, J.; Dunning Hotopp, J.C.; Tettelin, H. Using Sybil for interactive comparative genomics of microbes on the web. *Bioinformatics* **2012**, *28*, 160–166. [CrossRef] [PubMed]

32. Edgar, R.C. MUSCLE: Multiple sequence alignment with high accuracy and high throughput. *Nucleic Acids Res.* **2004**, *32*, 1792–1797. [CrossRef] [PubMed]

33. Krogh, A.; Larsson, B.; von Heijne, G.; Sonnhammer, E.L. Predicting transmembrane protein topology with a hidden Markov model: Application to complete genomes. *J. Mol. Biol.* **2001**, *305*, 567–580. [CrossRef] [PubMed]

MDPI
St. Alban-Anlage 66
4052 Basel
Switzerland
Tel. +41 61 683 77 34
Fax +41 61 302 89 18
www.mdpi.com

MDPI Books Editorial Office
E-mail: books@mdpi.com
www.mdpi.com/ijms